Neurophotonics
and
Brain Mapping

Endorsements

Society for Brain Mapping & Therapeutics (SBMT)
Brain Mapping Foundation (BMF)
National Center for NanoBioElectronics (NCNBE)
Western University of Health Science College of Optometry
California Neurosurgical Institute (CNI)
California Institute of Neuroscience (CIN)

MISSION STATEMENTS

SOCIETY FOR BRAIN MAPPING & THERAPEUTICS MISSION STATEMENT

The Society for Brain Mapping & Therapeutics is a nonprofit society organized for the purpose of encouraging basic and clinical scientists who are interested in the areas of brain mapping, engineering, stem cell, nanotechnology, imaging, and medical device to improve the diagnosis, treatment, and rehabilitation of patients afflicted with neurological disorders.

This society promotes the public welfare and improves patient care through the translation of new technologies/therapies into life-saving diagnostic and therapeutic procedures. The society is committed to excellence in education and scientific discovery. The society achieves its mission through multidisciplinary collaborations with government agencies, patient advocacy groups, educational institutes, and industries, as well as philanthropic organizations.

BRAIN MAPPING FOUNDATION MISSION STATEMENT

Brain Mapping Foundation (BMF) is a nonprofit 501c3 charitable organization, which is established for the purpose of facilitating multidisciplinary brain and spinal cord research and expediting integration and translation of cutting-edge technologies into the field of neuroscience. BMF is focused on translating state-of-the-art technologies from space and defense industries into neuroscience in order to bring the most advanced medicine to wounded warriors as well as civilians.

NATIONAL CENTER FOR NANOBIOELECTRONICS MISSION STATEMENT

The National Center for NanoBioElectronics's (NCNBE) mission is to establish collaborating research laboratories and network throughout California and beyond in order to rapidly develop solutions for neurological disorders employing advances in nanotechnology, stem cell research, and medical devices (nanobioelectronics) while fostering biotech spin-offs for the purpose of job creation.

The NCNBE promotes public welfare and improves patient care through the translation of new technologies into life-saving diagnostic and therapeutic procedures. The center is committed to excellence in education and scientific discovery. The NCNBE achieves its mission through multidisciplinary collaborations/consortium with government agencies, patient advocacy groups, educational institutions, private sector, industries, and philanthropic organizations.

WESTERN UNIVERSITY OF HEALTH SCIENCE COLLEGE OF OPTOMETRY MISSION STATEMENT

The Western University of Health Science College of Optometry's mission is to graduate caring and compassionate healthcare professionals who serve the need of diverse global society. The college emphasizes the rehabilitation of visual system, neuro-optometry, and interprofessional education. The college advances the profession of optometry through innovation in healthcare education, research, and patient care.

CALIFORNIA NEUROSURGICAL INSTITUTE MISSION STATEMENT

California Neurosurgical Institute is the preeminent neurosurgical practice in Los Angeles, serving the greater part of the Los Angeles, Ventura, and Kern Counties, providing the quality neurosurgical care to four million individuals.

Its mission is to provide the highest quality neurosurgical care to patients in their own communities by working closely with physicians in their communities and their local hospitals. The institute strives to accomplish this by translating and integrating advances in science and technology into people's daily clinical practice. As an "academic department without walls," the institute's neurosurgeons have subspecialty training and interests that allow people to manage the full spectrum of neurosurgical diseases and treatments while orienting themselves toward elevating the field through basic science and clinical research efforts and collaborations.

CALIFORNIA INSTITUTE OF NEUROSCIENCE MISSION STATEMENT

California Institute of Neuroscience's mission is to provide state-of-the-art diagnostics and treatment of neurological and neurosurgical diseases and conditions involving the brain, spine, and peripheral nerves as well as perform research and developments in the aforementioned areas.

Neurophotonics
and
Brain Mapping

Edited by

Yu Chen

Associate Professor and Associate Chair, Director of Graduate Studies, Fischell
Department of Bioengineering, Fellow of American Society for Laser Medicine
and Surgery, University of Maryland School of Engineering, College Park,
Maryland

Babak Kateb

Scientific Director and Chief Strategy Officer, California Neurosurgical Institute;
Chairman of the Board and CEO of Society for Brain Mapping and Therapeutics
(SBMT); President of Brain Mapping Foundation; Director of National Center
for NanobioElectronics (NCNBE); Director of National Brain Innovation
and Technology Park (NBITP); 8159 Santa Monica Blvd, Suite # 200, West
Hollywood, California 90046

CRC Press
Taylor & Francis Group
Boca Raton London New York

CRC Press is an imprint of the
Taylor & Francis Group, an **informa** business

CRC Press
Taylor & Francis Group
6000 Broken Sound Parkway NW, Suite 300
Boca Raton, FL 33487-2742

First issued in paperback 2020

ISBN-13: 978-1-4822-3685-9 (hbk)
ISBN-13: 978-0-367-65793-2 (pbk)

Library of Congress Cataloging-in-Publication Data

Names: Chen, Yu, 1973 December 28- editor. | Kateb, Babak, editor.
Title: Neurophotonics and brain mapping / [edited by] Yu Chen & Babak Kateb.
Description: Boca Raton : Taylor & Francis, 2017. | Includes bibliographical
references and index.
Identifiers: LCCN 2016040650 | ISBN 9781482236859 (hardback : alk. paper)
Subjects: | MESH: Brain Mapping--methods | Optical Phenomena |
Neurosurgery--methods | Spectroscopy, Near-Infrared
Classification: LCC RC386.6.B7 | NLM WL 335 | DDC 612.82--dc23
LC record available at https://lccn.loc.gov/2016040650

Visit the Taylor & Francis Web site at
http://www.taylorandfrancis.com

and the CRC Press Web site at
http://www.crcpress.com

Dedication

*This work is dedicated to our families, friends, spouse,
students, colleagues, and mentors.*

*This book is dedicated to Dr. Babak Kateb's mentor, Ferenc A. Jolesz,
MD. Dr. Jolesz was B. Leonard Holman professor of radiology in Brigham
and Women's Hospital at Harvard Medical School, recipient of 2005 Pioneer
in Medicine Award by the Society for Brain Mapping & Therapeutics.
Dr. Jolesz served as the second president of the society from 2004 to 2005.*

Dr. Anna Jolesz, Dr. Ferenc Jolesz, and Dr. Babak Kateb in 2004
during the first annual meeting of Society for Brain Mapping
& Therapeutics, Westin Hotel, Pasadena, California.

This book is also dedicated to Dr. Yu Chen's mentor, Professor Britton Chance (1913–2010), who was a pioneer in neurophotonics and brain mapping, for his innovation and discovery in research and his mentoring and training of generations of scientists in this exciting field. Dr. Chance was Eldridge Reeves Johnson University professor emeritus of biochemistry and biophysics as well as professor emeritus of physical chemistry and radiological physics at the University of Pennsylvania School of Medicine.

Dr. Yu Chen and Dr. Britton Chance

Contents

SECTION I Optical Spectroscopy and Spectral Imaging of Brain

SECTION II   Brain Imaging and Mapping
from Micro to Macroscales

SECTION III   Intraoperative Neuroimaging

SECTION IV   Image-Guided Intervention and Phototherapy

Preface

Using light to determine the function of brain and malfunction of neurological diseases has been a dream for scientists, but recent advances in bio-optics and neuroscience, imaging, and medical devices have made such dream a reality. We have also covered advanced brain mapping techniques and applications of such technologies in this book because neurophotonics itself is defined as the application of optical physics for studying neuronal structures and functions for the purpose of better understanding of the nervous system, which is inherently part of the larger and more advanced field of brain mapping and therapeutics.

The capability to noninvasively image cells, proteins, genes, molecules, and nanoparticles for their function and therapeutic applications is no longer a science fiction. Today, we use such technologies in the diagnosis and treatment of neurological disorders such as traumatic brain injury, epilepsy, Alzheimer's disease, and brain cancer. However, rapid integration, translation, and commercialization of such technologies from physical sciences into clinical neuroscience/neurology/neurosurgery and psychiatry meet many challenges from proper grant funding to proper regulatory reviews by the Food and Drug Administration (FDA). This gap between deep understanding of scientific concepts to new generation of devices and optogenetic drugs could be addressed by closer interaction between academic scientists, clinicians, industries, and government agencies, which could bridge the gap of knowledge between disciplines in biology, physical sciences, and medicine.

Neurophotonic is still a young and developing field with great promises for helping scientists to track cells/genes/molecules, study their functions, and modify cellular behavior. In recent years, we have witnessed an amazing level of scientific innovation and discovery emerging from the integration of neurophotonics, nanotechnology, imaging devices, artificial intelligence, supercomputing, and cellular and stem cell therapy into "nanobioelectronics." In this book, we have attempted to highlight at least a few of such remarkable scientific progress in the field of neurophotonics. We strongly believe that neurophotonics, nanoneuroscience, nanoneurosurgery, and nanobioelectronics will revolutionize medicine and remarkably improve the prevention, diagnosis, and treatment of neurological disorders over the next 10–20 years. In our lifetime, we will see "Internet of things" include cellular, genomic, and molecular tracking.

Careful regulation of neurophotonic techniques, application, and ethics will ensure a healthy development of the field, which in turn could help scientists validate the accuracy of diagnostic tests and verify the safety and efficacy of therapies that utilize neurophotonics, including drugs, devices, cellular therapy, and their combinations. In this book, we have tried our best to make sure such topics are covered in the context of each neurophotonic technology and its application.

Thus, this book is a fine review of the latest technologies, application, and regulatory guidelines as well as policies in the young and emerging field of neurophotonics and its correlative application in brain mapping and therapeutics. This book is a collective teamwork among more than 100 world-renowned scientists who have contributed 30 chapters in neurophotonic and brain mapping with more than 2000 references. This broad coverage of neurophotonic and brain mapping technologies and applications in diverse areas of neuroscience, neurosurgery, neuroradiology, and neurology and the role of FDA regulation and healthcare policy provides a foundation to support and impact the future of personalized targeted therapy of neurological disorders.

This textbook is devoted to implementing new approaches to perform less radical procedures (surgical and nonsurgical), eventually leading to approaches with minimal or no invasiveness to treat nervous system disorders, including brain tumors, brain and spinal cord injuries, vascular malformations, Alzheimer's disease, and Parkinson's disease. We hope that this book provides scientists and physicians with helpful references as they devise successful treatments for these neurological disorders.

We thank all the contributors for their scientific contributions to this remarkable textbook. We appreciate the visionary support of CRC/Taylor & Francis Group. Without their support, this book would not have been possible. We thank the Brain Mapping Foundation, the National Center for NanoBioElectronics, and the Society for Brain Mapping & Therapeutics, California Neurosurgical Institute, the Western University School of Optometry, and California Institute for Neuroscience for their generous support and endorsement of this truly unique book.

Finally, we thank our parents, Jabbar Kateb and Maleheh Samsame-Shirzad, and Li Chen and Yaoshen Wu, for their years of support and personal guidance. We thank Lance Wobus (senior editor), Jo Koster (publisher), Randy Brehm (senior editor), Laurie Oknowsky (production coordinator), Robert Sims (project editor), S. Valantina Jessie (project manager), and Amanda Parida (editorial assistant) for their tremendous support.

We trust that readers will take the ideas and methods described in this textbook to formulate novel and more effective treatments of presently incurable diseases.

Respectfully yours,
Yu Chen
Babak Kateb

MATLAB® is a registered trademark of The MathWorks, Inc. For product information, please contact:

The MathWorks, Inc.
3 Apple Hill Drive
Natick, MA 01760-2098 USA
Tel: 508-647-7000
Fax: 508-647-7001
E-mail: info@mathworks.com
Web: www.mathworks.com

Contributors

Nimer Adeeb
Department of Neurosurgery
California Institute of Neuroscience
Thousand Oaks, California

Ashfaq Ahmed
Department of Biomedical Engineering
Florida International University
Miami, Florida

Osama Ahmed
Department of Neurosurgery
Louisiana State University of Health Sciences
Shreveport, Louisiana

Taner Akkin
Department of Biomedical Engineering
University of Minnesota
Minneapolis, Minnesota

Michael J. Alexander
Department of Neurosurgery
Cedars-Sinai Medical Center
Los Angeles, California

Sameer Allahabadi
Rice University
Division of Neurology
City of Hope National Medical Center
Houston, Texas

Sonal Ambwani
St. Jude Medical
Westford, Massachusetts

Yasaman Ardeshirpour
Section on Analytical and Functional
    Biophotonics
National Institutes of Child Health and Human
    Development
National Institutes of Health
Bethesda, Maryland

Afrouz Azari-Anderson
Section on Analytical and Functional
    Biophotonics
National Institutes of Child Health and Human
    Development
National Institutes of Health
Bethesda, Maryland

Derek Backer
California Neurosurgical Institute
Valencia, California

Yuqiang Bai
Department of Biomedical Engineering
Florida International University
Miami, Florida

Eric M. Bailey
NeuroLogica Corporation
Danvers, Massachusetts

Paola Ballesteros-Zebadua
Medical Physics Laboratory
National Institute of Neurology and
    Neurosurgery
México City, Mexico

Zeinab Barati
Institute of Arctic Biology
University of Alaska
Fairbanks, Alaska

Ibrahim Bechwati
Samsung USA
Danvers, Massachusetts

Keith Black
Department of Neurosurgery
Maxin Dunitz Neurosurgical Institute
Cedars-Sinai Medical Center
Beverly Hills, California

Frank Boehm
Society for Brain Mapping & Therapeutics
and
Brain Mapping Foundation
West Hollywood, California

Pramod Butte
Department of Neurosurgery
Cedars-Sinai Medical Center
Los Angeles, California

Merve Çebi
Department of Psychology
Uskudar University
İstanbul, Turkey

Miguel Ángel Celis-López
Clinical Neurology and Neurosurgery
    Department
National Institute of Neurology
    and Neurosurgery
México City, Mexico

Sam Chang
NASA/Jet Propulsion Laboratory
California Institute of Technology
Pasadena, California

Yu Chen
Fiscehll Department of Bioengineering
University of Maryland
College Park, Maryland
and
Society for Brain Mapping & Therapeutics
and
Brain Mapping Foundation
West Hollywood, California

Zhongping Chen
Beckman Laser Institute
Department of Biomedical Engineering
University of California
Irvine, California

Viktor Chernomordik
Section on Analytical and Functional
    Biophotonics
National Institutes of Child Health and Human
    Development
National Institutes of Health
Bethesda, Maryland

Fatima A. Chowdhry
Section on Analytical and Functional
    Biophotonics
National Institutes of Child Health and Human
    Development
National Institutes of Health
Bethesda, Maryland

Ray Chu
Department of Neurosurgery
Cedars-Sinai Medical Center
Beverly Hills, California

Francisco F. De-Miguel
Cellular Physiology Institute
Universidad Nacional Autónoma de
    México
Mexican Autonomous National
    University
México City, Mexico

Matthew Dickman
Oculus VR
Menlo Park, California

Congwu Du
Department of Biomedical Engineering
Stony Brook University
Stony Brook, New York

Paula Eboli
Department of Neurosurgery
Cedars-Sinai Medical Center
Los Angeles, California

Şehadet Ekmen
Department of Psychology
Uskudar University
İstanbul, Turkey

Türker Ergüzel
Department of Computer Engineering
Uskudar University
İstanbul, Turkey

Reha S. Erzurumlu
Department of Anatomy
    and Neurobiology
University of Maryland School
    of Medicine
Baltimore, Maryland

Salman Abbasi Fard
Department of Neurosurgery
California Institute of Neuroscience
National Institute of Neurology
   and Neurosurgery
Thousand Oaks, California

Joseph Fonte
NeuroLogica Corporation
Danvers, Massachusetts

Javier Franco-Perez
Reticular Formation Laboratory
National Institute of Neurology and Neurosurgery
México City, Mexico

Vance L. Fredrickson
Department of Neurosurgery
University of Southern California
Los Angeles, California

Amir H. Gandjbakhche
Section on Analytical and Functional
   Biophotonics
Eunice Kennedy Shriver National Institutes
   of Child Health and Human Development
National Institutes of Health
Bethesda, Maryland

Warren Grundfest
Department of Bioengineering
University of California, Los Angeles
Los Angeles, California

Rao P. Gullapalli
Department of Diagnostic Radiology
   and Nuclear Medicine
University of Maryland School of Medicine
Baltimore, Maryland

Daniel X. Hammer
Division of Biomedical Physics
Office of Science and Engineering Laboratories
Center for Devices and Radiological Health
Food and Drug Administration
Silver Spring, Maryland

Guillermo Hernández-Mendoza
Cellular Physiology Institute
Universidad Nacional Autónoma de México
Mexican Autonomous National University
México City, Mexico

Javier Herrera-Vega
Department of Computer Science
Instituto Nacional de Astrofísica, Óptica y
   Electrónica
National Institute for Astrophysics, Optics
   and Electronics
Puebla, Mexico

Alexandra Jalali
Society for Brain Mapping & Therapeutics
and
Brain Mapping Foundation
West Hollywood, California
and
Brentwood School
Los Angeles, California

Bahram Jalali
School of Engineering
University of California, Los Angeles
Los Angeles, California
and
Society for Brain Mapping & Therapeutics
and
Brain Mapping Foundation
West Hollywood, California

Caroline Jia
Division of Neurology
University of California, Los Angeles
Los Angeles, California

Ferenc A. Jolesz
Department of Radiology
Brigham and Women's Hospital
Harvard Medical School
Boston, Massachusetts

Ranu Jung
Department of Biomedical Engineering
Florida International University
Miami, Florida

Nader Shahni Karamzadeh
Section on Analytical and Functional
   Biophotonics
National Institutes of Child Health and Human
   Development
National Institutes of Health
Bethesda, Maryland

Babak Kateb
California Neurosurgical Institute
Valencia, California

and

Society for Brain Mapping & Therapeutics
and
Brain Mapping Foundation
West Hollywood, California

and

Department of Neurosurgery
Cedars-Sinai Medical Center
Beverly Hills, California

David S. Kittle
Department of Neurosurgery
Cedars-Sinai Medical Center
Los Angeles, California

Fatma Zehra Keskin Krzan
Department of Psychology
Uskudar University
İstanbul, Turkey

Carmen Kut
Department of Biomedical Engineering
Johns Hopkins University School of Medicine
Baltimore, Maryland

Asimina Lazaridou
NMR Surgical Laboratory
Department of Surgery
Massachusetts General Hospital and Shriners
    Burn Institute
Harvard Medical School
and
Athinoula A. Martinos Center of Biomedical
    Imaging
Department of Radiology
Massachusetts General Hospital
Boston, Massachusetts

Conor Leahy
Neurophotonics Laboratory
Department of Biomedical Engineering
University of California, Davis
Davis, California

Wonhye Lee
Department of Radiology
Harvard Medical School
Brigham and Women's Hospital
Boston, Massachusetts

Xingde Li
Department of Biomedical Engineering
Johns Hopkins University School of Medicine
Baltimore, Maryland

Chia-Pin Liang
Wellman Center for Photomedicine
Massachusetts General Hospital
Harvard Medical School
Boston, Massachusetts

and

Fiscehll Department of Bioengineering
University of Maryland
College Park, Maryland

Linda Liau
Department of Neurosurgery
University of California, Los Angeles
Los Angeles, California

Mark A. Liker
Department of Neurosurgery
University of Southern California
Los Angeles, California

Adam N. Mamelak
Department of Neurosurgery
Cedars-Sinai Medical Center
Los Angeles, California

Richard P. Menger
Department of Neurosurgery
Louisiana State University of Health Sciences
Shreveport, Louisiana

Yukifumi Monden
Department of Pediatrics
Jichi Medical University
Shimotsuke, Japan

and

Department of Pediatrics
Jichi Medical University
Tochigi, Japan

and

Department of Pediatrics
International University of Health and Walfare
Tochigi, Japan

Martin M. Mortazavi
Department of Neurosurgery
California Institute of Neuroscience
Thousand Oaks, California

Zhaojun Nie
Department of Neurosurgery
Cedars-Sinai Medical Center
Los Angeles, California

Shouleh Nikzad
NASA/Jet Propulsion Laboratory
California Institute of Technology
Pasadena, California

Ifije E. Ohiorhenuan
Department of Neurosurgery
University of Southern California
Los Angeles, California

Felipe Orihuela-Espina
Department of Computer Science
Instituto Nacional de Astrofísica, Óptica y
    Electrónica
National Institute for Astrophysics, Optics
    and Electronics
Puebla, Mexico

Serhat Özekes
Department of Computer Engineering
Uskudar University
İstanbul, Turkey

Yingtian Pan
Department of Biomedical Engineering
Stony Brook University
Stony Brook, New York

Asael Papour
Wellman Center for Photomedicine
Massachusetts General Hospital
Harvard Medical School
Boston, Massachusetts

Chirag G. Patil
Department of Neurosurgery
Cedars-Sinai Medical Center
Los Angeles, California

Brian Pikul
Department of Neurosurgery
Kaiser Permanent
Los Angeles, California

Kambiz Pourrezaei
School of Biomedical Engineering, Science
    and Health Systems
Drexel University
Philadelphia, Pennsylvania

Ahmad Pourshoghi
School of Biomedical Engineering, Science
    and Health Systems
Drexel University
Philadelphia, Pennsylvania

Neal Prakash
Department of Medicine
Division of Neurology
City of Hope National Medical Center
Duarte, California

Alfredo Quinones-Hinojosa
Department of Neurological Surgery
Mayo Clinic Hospital
Jacksonville, Florida

Jessica C. Ramella-Roman
Department of Biomedical Engineering
Florida International University
and
Herbert Wertheim College of Medicine
Florida International University
Miami, Florida

Jordina Rincon-Torroella
Department of Neurological Surgery
Johns Hopkins University School of Medicine
Baltimore, Maryland

Ueli Rutishauser
Department of Neurosurgery
Cedars-Sinai Medical Center
Los Angeles, California
and
Division of Biology and Biological Engineering
California Institute of Technology
Pasadena, California

Karla J. Sanchez-Perez
Optics Department
Instituto Nacional de Astrofísica, Óptica y
    Electrónica
National Institute for Astrophysics, Optics
    and Electronics
Puebla, Mexico

Shahdad Sherkat
Society for Brain Mapping & Therapeutics
and
Brain Mapping Foundation
Los Angeles, California

Elizabeth G. Smith
National Institutes of Mental Health
National Institutes of Health
Bethesda, Maryland

Chandler Sours
Department of Diagnostic Radiology
    and Nuclear Medicine
University of Maryland School of Medicine
Baltimore, Maryland

Vivek J. Srinivasan
Neurophotonics Laboratory
Department of Biomedical Engineering
University of California, Davis
Davis, California

Oscar Stafsudd
Department of Electrical Engineering
University of California, Los Angeles
Los Angeles, California

Luis Enrique Sucar
Department of Computer Science
Instituto Nacional de Astrofísica, Óptica y
    Electrónica
National Institute for Astrophysics, Optics
    and Electronics
Puebla, Mexico

Cha-Min Tang
Department of Neurology
University of Maryland School of Medicine
and
Research Service, VA Medical Center
Baltimore, Maryland

Nevzat Tarhan
Department of Psychology
Uskudar University
İstanbul, Turkey

Cumhur Taş
Department of Psychology
Uskudar University
İstanbul, Turkey
and
Division of Cognitive Neuropsychiatry
    and Psychiatric Preventative Medicine
Ruhr University
Bochum, Germany

Zach Taylor
Department of Bioengineering
University of California, Los Angeles
Los Angeles, California

Carlos G. Treviño-Palacios
Optics Department
Instituto Nacional de Astrofísica, Óptica y
    Electrónica
National Institute for Astrophysics, Optics and
    Electronics
Puebla, Mexico

Vassiliy Tsytsarev
Department of Anatomy and Neurobiology
University of Maryland School of Medicine
Baltimore, Maryland
and
Society for Brain Mapping & Therapeutics
and
Brain Mapping Foundation
West Hollywood, California

A. Aria Tzika
The NMR Surgical Laboratory
Department of Surgery
Massachusetts General Hospital and Shriners
    Burns Institute
Harvard Medical School
and
Department of Radiology
Athinoula A. Martinos Center of Biomedical
Imaging
Massachusetts General Hospital
Boston, Massachusetts

Baris Ünsalver
Department of Computer Engineering
Uskudar University
İstanbul, Turkey

Fartash Vasefi
Department of Neurosurgery
Cedars-Sinai Medical Center
Los Angeles, California

and

Society for Brain Mapping &
    Therapeutics
and
Brain Mapping Foundation
West Hollywood, California

Hui Wang
Athinoula A. Martinos
Center for Biomedical Imaging
Massachusetts General Hospital/Harvard
    Medical School
Charlestown, Massachusetts

Lihong V. Wang
Optical Imaging Laboratory
Department of Biomedical
    Engineering
Washington University in St. Louis
St. Louis, Missouri

Geethika Weliwitigoda
Samsung USA
Danvers, Massachusetts

Cristin G. Welle
Division of Biomedical Physics
Office of Science and Engineering Laboratories
Center for Devices and Radiological Health,
    Food and Drug Administration
Silver Spring, Maryland

and

Departments of Neurosurgery
    and Bioengineering
University of Colorado Denver
Aurora, Colorado

Michael E. Wolf
NeuroCite LLC
Los Angeles, California

Jun Xia
Optical Imaging Laboratory
Department of Biomedical Engineering
The State University of New York
Buffalo, New York

Vicky Yamamoto
Department of Otolaryngology—Head and
    Neck Surgery
USC-Norris Comprehensive Cancer Center
    Keck School of Medicine of USC
Society for Brain Mapping & Therapeutics
and
Brain Mapping Foundation
Los Angeles, California

Junjie Yao
Optical Imaging Laboratory
Department of Biomedical Engineering
Washington University in St. Louis
St. Louis, Missouri

Parham Yashar
California Neurosurgical Institute
Valencia, California

William H. Yong
Brain Tumor Translational Resource
University of California, Los Angeles
Los Angeles, California

Seung-Schik Yoo
Department of Radiology, Brigham and
    Women's Hospital
Harvard Medical School
Boston, Massachusetts

Issa Zakeri
School of Public Health
Drexel University
Philadelphia, Pennsylvania

Oscar Javier Zapata-Nava
Optics Department
Instituto Nacional de Astrofísica, Óptica y
    Electrónica
National Institute for Astrophysics, Optics and
    Electronics
Puebla, Mexico

and

Mechatronics Departments
Instituto Superior Tecnológico de
    Zacapoaxtla
Zacapoaxtla Tecnological Institute
Zacapoaxtla, Pue, Mexico

Deborah Zelinsky
Mind-Eye Connection
Northbrook, Illinois

Jiang Zhu
Beckman Laser Institute
Department of Biomedical Engineering
University of California, Irvine
Irvine, California

Jiachen Zhuo
Department of Diagnostic Radiology
    and Nuclear Medicine
University of Maryland School of Medicine
Baltimore, Maryland

1 History of Brain Mapping and Neurophotonics

From Technological Discoveries to Brain Initiatives

Babak Kateb, Frank Boehm, Alexandra Jalali,
Vassiliy Tsytsarev, Vicky Yamamoto, Bahram Jalali,
Derek Backer, Brian Pikul, Parham Yashar, and Yu Chen

CONTENTS

INTRODUCTION

According to the Society for Brain Mapping and Therapeutics (SBMT), brain mapping is defined as the study and elucidation of the anatomy and functionality of the brain and spinal cord through the use of imaging (including intraoperative, microscopic, endoscopic, molecular, and multimodality imaging), immunohistochemistry, optogenetics, stem cell and cellular biology, engineering (material, electrical, and biomedical), advanced technologies (ultrasound, photonics, etc.), neurophysiology, and nanotechnology (Kateb and Heiss, 2013). Diverse types of brain mapping exist, encompassing anatomical, structural, cellular, functional, and metabolic. Cellular maps, for instance, may comprise nanoscale imaging and may be alternately defined via genomics. This volume will endeavor to survey and articulate the latest advances in neurophotonics. These exciting technological advances will lead to the further enhancement of the diagnoses and treatments of myriad brain conditions toward improved patient outcomes globally.

HISTORICAL MILESTONES IN BRAIN MAPPING

The first known written instance of the word "brain," in addition to articulations of the meninges, the cerebrospinal fluid, and various brain injuries, dates back to ~3000 BC in what is referred to as *The Edwin Smith Surgical Papyrus*, which is an ancient medical text that was authored by the prominent Egyptian physician Imhotep. At the time, however, the brain was thought to be an inconsequential organ in contrast to the heart, which was deemed to be its life force (Minagar et al., 2003). The Greek medical theorist and dissectionist Alcmaeon of Croton (500 BC) is thought to have been the earliest author to propose that the brain was the center of cognition and sensation. His dissection studies of the eye led him to posit that the optic nerves "came together behind the forehead" and comprised brain terminating "light bearing paths" (Lloyd, 1975). Democritus (460–370 BC) put forward that all matter in the universe was made of various constituent atoms and that the human mind was composed of the most rapidly motile atoms (Gross, 1995). Plato (~428–347 BC) expanded on these musings in his *Timaeus*, where he states that the brain "…is the divinest part of us and lord over all the rest" (Plato, 1959), whereas Hippocrates (460–370 BC) affirmed that mental operations were "completely undertaken by the brain." He also conveyed his firm belief that epilepsy originated within the brain in his well-known essay "On the Sacred Disease" (Hippocrates, 1950; Gross, 1995).

Galen (AD 121–201) was among the first well-documented ancient scientists who provided insights into the detailed gross anatomy of the human brain (Toga and Mazziotta, 2002). The Persian polymath and physician Rhazes (AD 865–925) first conveyed the existence of seven cranial nerves and 31 spinal nerves in *Kitab al-Hawi fi al-tibb* (*The Comprehensive Book on Medicine*) (Tubbs et al., 2007). Avicenna (Abu Ali Sina AD 980–1037) conducted thorough anatomical studies of the spinal cord, described its functionality, and identified various spinal traumas and their treatments (Ghaffari et al., 2015). Leonardo da Vinci (AD 1452–1519) was the first individual to convey a precise depiction of the brain through the fusion of art perspective with dissection (Pevsner, 2002). In 1536, the Venetian physician Niccolo Massa initially discovered cerebrospinal fluid during the course of an autopsy (Herbowski, 2013); the Italian neuroanatomist Costanzo Varolio, in 1573, was the first to dissect the brain from its base (Zago and Meraviglia, 2009); and in 1586, Archangelo Piccolomini initially distinguished the neocortex, white matter, and subcortical nuclei (Compston, 2013).

From the sixteenth till the eighteenth centuries, increasing advances in the elucidation of brain structure and functionality were made, with some of the most prominent being the following*:

1641—Fissures on brain surfaces are revealed by Franciscus de la Boe Sylvius.
1664—Thomas Willis discovers the eleventh cranial nerve.
1695—Publication of *The Anatomy of the Brain* by Humphrey Ridley.
1717—Cross sections of nerve fibers are initially described by Antony van Leeuwenhoek.
1778—Classification of the twelve cranial nerves and optic chiasm (1786) are articulated by Samuel Thomas von Soemmerring.
1809—Luigi Rolando stimulates the cortex via galvanic current.
1836—Gabriel Gustav Valentin recognizes the nucleus and nucleolus of neurons.
1836—Myelinated and unmyelinated axons are discovered by Robert Remak, who later (1844) illustrates the six layers of the neocortex.

---

* Adapted from Chudler (2016).

Modern cognitive neuroscience, which couples investigative cognitive psychology with the elucidation of how brain function drives mental activity, has garnered intense interest and development over the last three decades. This research is facilitated to a great extent through functional brain imaging/mapping that employs techniques such as positron emission tomography (PET), functional magnetic resonance imaging (fMRI), electroencephalography (EEG), electrocorticography, and magnetoencephalography, along with the most recent addition of optical near-infrared spectroscopy (NIRS) imaging (Raichle, 2009).

X-Ray

German physicist Röntgen's accidental discovery of x-rays in 1895 earned him the Nobel Prize in Physics in 1901. While studying emission of light by cathode-ray tubes, he observed that the impact of cathode rays on the glass vacuum tube was generating a new type of radiation that could penetrate human tissue, wood, copper, and aluminum and could be recorded on photographic plates. Röntgen named it the x-ray to denote the fact that it was an unknown type of radiation. Röntgen did not patent his discovery as he was convinced that it had important medical applications and he wanted the world to benefit from it. As a result, x-rays were quickly adopted throughout the world. In 1896, the British army began to use x-ray machines to locate bullets and diagnose bone fractures. Also, in 1896, John Hall-Edwards in Birmingham, England, used it to locate a needle, which was stuck in the hand of an associate (Meggitt, 2008). While x-ray is still used for such applications, today it is primarily employed in the context of digital radiology and computed tomography (CT) scanners, as discussed in the following text.

Computed Tomography

X-ray computed tomography (CT), otherwise known as computed axial tomography (CAT), is a scanner that provides 3D images of the brain by taking x-ray images from different angles. It operates by passing highly focused x-ray beams through the body and recording their attenuation. These projections from multiple directions are fed to a tomographic reconstruction software algorithm. CT scanning is generally used to distinguish benign and malignant tumors, as well as injuries to the brain. The information is typically presented as a series of cross-sectional images. Referred to as planography, an earlier version of CT imaging was reported in 1953 by B. Pollak (1953). The mathematical modeling of transverse axial scanning was a key in the development of CT, which was due to the work of South African-born Allan McLeod Cormack and UK scientist Godfrey Hounsfield. The modern and first commercial CT scanner was introduced by Godfrey Hounsfield in 1971 at Atkinson Morley Hospital in London (Littleton, 2014). This revolutionary tool replaced highly invasive techniques, such as pneumoencephalography, whereby a lumbar puncture through the skull facilitated the drainage of the cerebrospinal fluid that surrounded the brain to be exchanged with air to enable clearer x-ray imagery. In the 1980s, the speed of CT scanners was significantly enhanced through the addition of spiral scanning, developed by Kalender at the University of Wisconsin. This allowed the creation of CT angiograms and the addition of time-dependent measurements of blood perfusion through the use of a dye to enhance contrast; CT angiography depicts vascular structures and ruptures of the blood vessels for diagnosis of brain injuries. Hence, CT scanning has had a tremendous impact in the field of brain mapping (Castillo, 2014).

Magnetic Resonance Imaging

Magnetic resonance imaging, or MRI, is an important tool used in brain mapping that uses a magnetic field and radio-frequency (RF) pulses to create images of the internal structure of the body. The predecessor to MRI (nuclear magnetic resonance or NMR) was invented by Felix Bloch from Harvard University and Edward Purcell from Stanford University in 1946. The next major development, in 1973, was attributed to Paul Lauterbur who invented a technique by which NMR could create images in the same way as a CT scanner, albeit with much more detail. It was then that the system was renamed as MRI (Raichle, 2009). The operation of MRI relies on the magnetic moment of protons in tissues and their interaction with a magnetic field and a RF wave. In the presence of a magnetic field, protons align in parallel with the field, which is the lowest proton energy state. When the RF signal is turned on, its energy excites the protons to a higher energy state, which is associated with the magnetic moment being misaligned with the magnetic field. Turning off the field allows the protons to relax back to their original state. During this relaxation period, energy is released via the emission of electromagnetic energy, which is detected by a coil. Since the rate of relaxation is different for various tissues, MRI can distinguish circulatory and metabolic changes in brain activity. Functional MRI, or fMRI, refers to the use of MRI to visualize the functionality of the brain while the subject is performing specific tasks. The technique was introduced by Seiji Ogawa and his colleagues at AT&T Bell Laboratories, where they studied the concentration of deoxygenated blood in the brains of living rodents (Miyamoto et al., 2014). fMRI has emerged as the most important tool for brain mapping. Chapters 13 and 14 review advanced MRI technologies for brain mapping.

Positron Emission Tomography

Other strategies for functional imaging comprise nuclear medicines such as single-photon emission computerized tomography (SPECT) or PET. PET selectively highlights areas of the body with relatively higher metabolic activity, characterized by high sugar or oxygen uptake.

A type of glucose (fluorodeoxyglucose, or FDG) is typically used to deliver a tracer molecule that emits gamma rays (through positron emission) to areas with higher metabolic activity. Alternatively, a radioactive isotope of oxygen is employed to image the volume of blood flow in different regions of the brain. Subsequently, areas of the brain with higher than normal (cancer tissue) or lower than normal (e.g., Alzheimer's disease) oxygen consumption may be identified.

A modern PET imaging system was pioneered in 1975 by researchers at Washington University in St. Louis (Phelps et al., 1975; Ter-Pogossian et al., 1975). This development benefited from the earlier work that was done in the 1950s, which involved the use of positron emitters to localize brain tumors at Massachusetts General Hospital (Sweet and Brownell, 1953), and in the 1960s at Brookhaven National Laboratory (DOE report, 2010).

HISTORICAL MILESTONES IN NEUROPHOTONICS

Neurophotonics is generally defined as the use of photonic methods for the imaging and manipulation of brain structure and function, spanning from the visualization of intracellular organelles and protein assemblies to the noninvasive macroscopic investigation of cortical activity in human subjects (Boas, 2014). Light has several advantages, such that it is nonionizing, has low cost, is portable, and possesses high molecular and biochemical specificity. At the onset of the neurophotonics era, the strategy for imaging brain function was based on slow optical changes that occurred in the neural tissue, which allowed for visualization with good spatial and temporal resolution. The existence of these optical changes associated with metabolic activity has been observed since the first part of the twentieth century, when optical alterations were demonstrated in cytochromes (Kelin, 1925) and hemoglobin (Millikan, 1937, 1942). Decades later, scientists demonstrated optical changes that were associated with neural activity (Hill and Keynes, 1949). It has only been within the last decades of the twentieth century that it became possible to utilize the detection of subtle optical changes for the imaging of neural activity (Kalchenko et al., 2014). Computerization coupled with light-sensitive semiconductor devices makes neurophotonics a very powerful facet of neuroscience (Grinvald et al., 2015). Imaging methods were dramatically improved with approaches that employed voltage- and calcium-sensitive dyes, which served to enhance spatial and temporal resolution. Following further decades, genetically encoded fluorescent probes were introduced into the armamentarium of experimental neuroscience. Currently, neurophotonics is experiencing rapid growth toward facilitating our understanding of brain activity under normal and pathological conditions.

NEAR-INFRARED SPECTROSCOPY AND IMAGING

The spectroscopy system was pioneered at Cambridge University (Hartridge and Roughton, 1923; Roughton and Millikan, 1936). Since tissue is typically opaque, light absorption is disrupted by scattering. In 1950, Britton Chance invented the "double beam spectrometer" using two wavelengths in the visible region with a small spectral interval to eliminate scattering effects (Chance, 1951). The first demonstration of near-infrared (NIR) light with human tissue in vivo was reported by Franz Jobsis in 1977 (Jobsis, 1977). Since then, many papers have been published on the use of NIR spectroscopy (NIRS) for brain mapping. Many researchers use continuous-wave (CW) technology, as the system is simple, has low cost, and is robust (Chance et al., 2005). Time-resolved spectroscopy provided a solution for the absolute quantification of chemical concentrations in tissues (Chance et al., 1988; Patterson et al., 1989). Alternatively, frequency-domain NIRS may be also used for quantization (Fishkin and Gratton, 1993; Madsen et al., 1994; Chen et al., 2003).

Since Seiji Ogawa's discovery in the early 1990s that signal changes associated with deoxyhemoglobin in NMR can detect brain cognition (Ogawa et al., 1990), researchers have been interested in using NIR light to detect brain function (Chance et al., 1993; Villringer and Chance, 1997; Hoshi et al., 2003). NIRS may quantify the concentrations of both Hb and HbO_2, thereby revealing the blood volume and oxygenation saturation changes associated with various brain functions. NIRS may be extended to imaging mode by employing multiple source-detector channels. Images can be formed using back-projection and interpolation algorithms, sometimes referred to as diffuse optical topography, which may provide an estimate of the 2D spatial distribution of the optical properties of tissues (Hielscher et al., 2002). Another approach is to perform 3D tomographic reconstruction, which is typically referred to as diffuse optical tomography (DOT). DOT is similar to x-ray CT and involves image reconstruction by solving the inverse problem (Arridge, 1999).

NIR imaging has found extensive applications in clinical settings (Hillman, 2007; Wolf et al., 2007). One major research area involves the elucidation of how the brain functions. Optical imaging offers the unique capability to noninvasively monitor functional activations in vivo without disturbing the organ. Various applications such as visual responses (Takahashi et al., 2000; Zeff et al., 2007), somatosensory responses (Becerra et al., 2009), auditory responses (Sato et al., 1999), language stimuli (Watanabe et al., 1998; Pena et al., 2003), and problem solving (Chance et al., 2007) have been explored. Another important area for optical brain imaging is to diagnose and monitor conditions such as stroke (Kirkpatrick et al., 1998; Durduran et al., 2009), epilepsy (Watanabe et al., 2002), brain injury (Kim et al., 2010), and post-traumatic stress disorder (Matsuo et al., 2003). Optical techniques are well suited for the early detection of hemorrhage (Stankovic et al., 2000) and to discriminate between ischemic and hemorrhagic stroke, which leads to improved management of therapeutics for patients (Intes and Chance, 2005). Chapters 2 through 5 review the applications of fNIRS for pain management and neurological disease monitoring. Chapters 15 through 17 review the application of optical imaging that is based on ultraviolet (UV), visible, and infrared (IR) light for neurosurgical guidance.

OPTICAL COHERENCE TOMOGRAPHY

Concurrently with other photonic imaging techniques, microscopic imaging using light has been actively pursued over the last several decades. One representative technology is optical coherence tomography (OCT), which enables the cross-sectional imaging of tissue microstructures in situ and in real time (Huang et al., 1991). OCT may achieve resolution at from 1 to 10 μm and a penetration depth of 1 to 2 mm, approaching those of standard excisional biopsy and histopathology, however, without the requirement for the removal and processing of tissue specimens (Fujimoto et al., 1995). OCT is analogous to ultrasound imaging, except that imaging is performed by measuring the echo delay time and intensity of back-reflected or backscattered light, rather than sound. OCT imaging may be performed fiber-optically using delivery devices such as handheld probes, endoscopes, catheters, laparoscopes, and needles that enable noninvasive or minimally invasive internal body imaging (Yaqoob et al., 2006).

OCT is performed using a Michelson interferometer with a low coherence length (broadband) light source. One arm of the interferometer illuminates the tissue and collects the backscattered light (referred to as the "sample arm"). Another arm has a reference path delay, which is scanned as a function of time (referred to as the "reference arm"). Optical interference between the light from the sample and reference arms occurs only when the optical delays match to the coherence length of the light source. This technique is generally referred to as "time-domain OCT." Alternatively, OCT interference signals may be detected in the frequency or Fourier domain. In Fourier-domain OCT, the reference mirror position is fixed and the light scattering profile is obtained by a Fourier transform of the interference spectrum. These techniques are somewhat analogous to Fourier transform spectroscopy and have significant sensitivity and speed advantages in contrast to time-domain OCT in that they measure the optical signals from different depths along the entire axial scan simultaneously, rather than sequentially (Choma et al., 2003; de Boer et al., 2003; Leitgeb et al., 2003). These advances not only greatly improve the performance of OCT but also enable three-dimensional OCT (3D-OCT) imaging in vivo. Time-stretch OCT is a Fourier-domain technique that delivers the highest imaging speed of OCT (Goda and Jalali, 2013). This novel system uses a photonic time-stretch technology (Jalali and Coppinger, 2001) to capture images with nanosecond time resolution and subsequently optically slows them in time so they can be digitized and analyzed by the computer in real time. This method is very attractive in neuroimaging, as the action potentials (the electrical impulses used by the brain to send signals) and physiological changes they produce occur on millisecond time scales, which is too fast to be detected by conventional OCT techniques.

Chapters 7 through 11 review the application of OCT for imaging brain function with animal models. Neurosurgical guidance using OCT represents another exciting application area (Jafri et al., 2005). Chapters 20 and 21 review the development and translation of OCT imaging to intraoperative guidance for neurosurgery.

FLUORESCENCE SPECTROSCOPY, IMAGING, AND MICROSCOPY

Fluorescence-based brain imaging uses the emission of visible or NIR light via fluorescent dyes or proteins. It visualizes processes and structures at cellular and molecular scales, that is, with much better resolution than other techniques, however, with a lower penetration depth due to the high rate of light scattering and absorption within tissues. Fluorescence imaging is key for the reconstruction of dense neural circuits known as connectomes (Helmstaedter, 2013) and the observation of gene and protein expressions through genetically encoded fluorescent protein tracers. High speed is necessary in brain mapping to capture the dynamics of action potentials. Fluorescence imaging via radio-frequency-tagged excitation (FIRE) is the world's fastest fluorescence imaging modality (Diebold et al., 2013). The microscope uses methods for data multiplexing borrowed from wireless communications to acquire fluorescent images at microsecond time intervals (see Figure 1.1). As is the case with all optical techniques, due to the low penetration depth of photons in tissue, the creation of 3D maps of the brain via fluorescence microscopy is challenging. There are currently two approaches to this: tissue sectioning and light-sheet/multiphoton fluorescence microscopy. Chapter 6 reviews the use of fluorescence-based voltage-sensitive dyes for brain imaging, whereas Chapter 11 reviews the application of multiphoton microscopy for monitoring implanted neural interfaces. Chapters 18 and 19 review the application of autofluorescence-based technologies for intraoperative guidance during neurosurgery. In addition, Chapter 12 describes another very exciting multimodal optics method referred to as photoacoustic tomography which was pioneered by Professor Lihong Wang.

OPTOGENETICS

Optogenetics has revolutionized neuroscience over the last few decades by allowing scientists to activate or inhibit specific neurons and monitor neural activity using light, without the use of

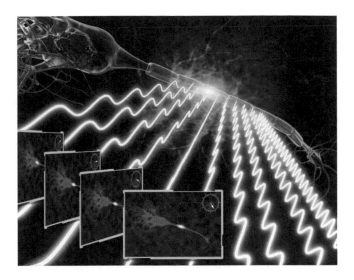

FIGURE 1.1 Conventional cameras form images of fluorescent samples pixel by pixel. FIRE creates an image by reading an entire row of pixels at once. It does this by encoding the fluorescence from each pixel on a different RF and forms images at a rate approximately 10 times faster than other technologies. (Courtesy of Jalali-Lab, Los Angeles, CA, www.photonics.ucla.edu.)

electrodes (Cho and Li, 2016). The cornerstone of optogenetics was the discovery of light-gated ion channels and the development of the technology that uses photons in the visible and IR wavelengths to control neural activities. Since then, optogenetics has had a fundamental and dramatic impact on neuroscience.

The earliest successful experiments for the control of neurons by light were conducted at the beginning of the twenty-first century (Zemelman et al., 2001; Szobota et al., 2007; Deisseroth, 2015). A distinct single-component approach involving microbial opsin genes, introduced in 2005, turned out to be widely applied, as described in the following text. Optogenetics allows not only the active influence of neural tissues but also the monitoring of neural activity using genetically encoded calcium-sensitive or voltage-sensitive fluorescent proteins (O'Connor et al., 2013; Akemann et al., 2015; Yang et al., 2016).

From a technical point of view, one of the most important steps in optogenetics is the delivery of voltage- or calcium-sensitive, or opsin-encoding, plasmids into targeted neurons. A rapid and relatively simple approach for the use of these plasmids may be provided by viral gene delivery. Typically, lentivirus or adeno-associated viruses are employed to package the opsin-construct, which are subsequently delivered in close proximity to the neurons in vitro or in vivo. Viral vectors are integrated into the cell genome of the targeted neurons and induce them to express voltage- or calcium-sensitive fluorescent proteins or photosensitive proteins. Vectors may be rendered cell type-specific using various molecular genetic methods. This allows neuroscientists to conduct experiments that were deemed impossible just a few years ago and enable strong potential for fundamental research and new therapeutic approaches for different neural disorders.

PHOTODYNAMIC THERAPY

Phototherapy pertains to the treatment of diseases and disorders using light-based technologies, which has a very long history. The initial use of UV light in the treatment of various diseases was documented in the nineteenth century. The emergence of phototherapy was indeed bright: In 1903, Niels Finsen received the Nobel Prize for his work in the field of phototherapy of different disorders, particularly lupus vulgaris (Grzybowski and Pietrzak, 2012). UV irradiation therapy may evolve as an effective anticancer therapy for several types of skin cancers. Phototherapy may also be successfully utilized in the treatment of different types of depression, including major depressive and bipolar disorders.

In the arsenal of anticancer methods, photodynamic therapy (PDT) holds an important position (Huang, 2005). PDT is based on special chemical agents called photosensitizers, in combination with a particular source of monochromatic light (Robertson et al., 2009). When photosensitizers are exposed to photons of a particular wavelength, they generate molecules that can destroy nearby tissues, including tumor cells (Allison and Sibata, 2010). Chapters 22 through 28 review the use of photonics, ultrasound, and EEG-based methods for intervention and therapy.

INTRAOPERATIVE BRAIN MAPPING

The modern era of intraoperative brain mapping was ushered in by neurosurgical luminaries such as Canadian, Wilder Penfield (1891–1976), who contributed significantly to the early mapping of the cerebral cortex through direct electrical brain stimulation and the assignment of functionality to specific brain structures. In 1937, Penfield developed what is referred to as the "homunculus," which graphically portrayed and compared specific functional domains of the brain, presumably to facilitate the more widespread propagation of the knowledge of brain science (Snyder and Whitaker, 2013). Earlier, John Hughlings Jackson (1835–1911), through his expertise in epilepsy, put forward a hierarchical representation of brain functionality that encompassed various nervous centers, levels of consciousness, and sensory–motor associations. Jackson's work inspired Penfield's own, which involved the electrocortical stimulation of epileptic patients, and led to his observation of a "record of the steams of consciousness." Further to this work, Penfield proposed that beyond the elucidation of reflexive activities via higher and lower centers, a distinct, yet at that time unknown, capacity was responsible for the direction of voluntary movements. He referred to these dual functional realms as the "automatic sensory–motor mechanism" or "computer" and the "highest brain mechanism" or "mind's mechanism" (Hogan and English, 2012).

More recently, the research of Ferenc Jolesz (1946–2014) with image-guided therapy enabled the enhancement of targeting and the resolution of monitoring toward the improved management of surgical procedures employing the data acquired from diagnostic imaging. Along with Peter Black (professor of neurosurgery at Harvard Medical School), he established the area of intraoperative MRI (iMRI)-guided interventions and developed various novel medical procedures, including for the treatment of brain tumors, derived from unique synergies between imaging and the delivery of therapies.

Skrap et al. assessed the application of novel, complex real-time neuropsychological testing (RTNT) in order to perform the continuous evaluation of cognitive function to enhance the extent of tumor tissue resection. A total of 92 patients were involved in this study, which undertook the appraisal of intraoperative neuropsychological dysfunctions when the patients were tasked with specific cognitive activities. It was found that RTNT facilitated a clearer neuropsychological representation of the effects of surgery at particular brain sites; hence, it might be employed to provide improved intraoperative feedback to facilitate more extensive and precise resection (Skrap et al., 2016).

An assessment of the utility of a subcortico-cortical evoked potential (SCEP) strategy to intraoperatively identify subcortical fibers was undertaken by Enatsu et al. In one application of this technique, a multielectrode plate was positioned on the central cortical sites, subsequent to resection, in order to demarcate the temporoparietal cortical and pyramidal tract regions of the arcuate fasciculus. The stimulation point spacing and targeted subcortical fibers under diffusion tensor imaging were quantified using 3D MRI concurrent with intraoperative neuronavigation. Since subcortical fibers are typically challenging to detect, it was demonstrated that SCEP has strong potential to enhance the efficacy and safety of lesion resections, particularly for those tumors that are in intimate proximity to critical functional sites within the brain (Enatsu et al., 2016).

GE Double Donut MRI

During the late 1980s, Ferenc Jolesz and Peter Black developed an open configuration "double donut" MRI scanner design that enabled surgical procedures to be performed in conjunction with guidance from intraoperative imagery. In this configuration, the most intense magnetic field resides in the area between each of the superconductive magnet coils (Mislow et al., 2010). This design configuration later emerged as the GE Signa SP system, which (General Electric Medical Systems, Milwaukee, WI), which was dedicated to the provision of imagery during surgical interventions (Schenck et al., 1995; Silverman et al., 1995, 1997). In contrast to a closed-bore configuration, the double donut strategy removes the central portion between two separate (albeit communicating) cryostats, to provide an open area between coils of 54 cm at the imaging isocenter, thereby allowing surgical access (laterally, from both sides, and from above) to the patient. This has enabled image guidance and the monitoring of myriad biopsies, aspirations, and surgeries (Silverman et al., 1995; Alexander et al., 1997; Black et al., 1997). All surgical instrumentation must, of course, be MR compatible (Lewin, 1999).

The integration of a flashpoint (LED-based) navigational strategy (Flashpoint, integrated technologies, Boulder, CO) enables the capacity for 3D tracking at the central axis of the magnet, which facilitates a spherical imaging area of Ø30 cm (Moriarty et al., 1996; Lipson et al., 2001; Albayrak et al., 2004). Increased patient access was made possible via the enhanced cooling of the magnetic coils through the use of niobium tin, and thermal shielding that incorporated a dual-stage cryocooler, which negated the necessity for liquid helium baths. A customized vertically positioned microscope and a range of MR-safe surgical instrumentation and integrated anesthesia ventilators and monitors were developed to be accommodated by this system. Further, a number of LCD screens were arranged in close proximity to the surgeons to allow easy viewing of the acquired imagery (single-plane imagery is typically acquired within 60–120 s) (Albayrak et al., 2004).

What is advantageous and unique to this double donut configuration is that it enables craniotomy in conjunction with direct real-time intraoperative navigation to successfully merge imagery and surgery into a single volumetric domain. Additionally, iMRI guidance allows for the more precise resection of brain tumors, and the dimensions of both incision sites and flaps may potentially be reduced. Imagery acquired prior to and following the traversal of the dura assists with the determination of the accurate location of lesions in relation to the cortex. It was found that in 30%+ of cases, where the surgical intent was to have no remaining tumor elements, concurrent MR scanning revealed residual tumor tissue, which necessitated further resection (Black et al., 1999). Typically, a final image is acquired subsequent to craniotomy closure to ensure the absence of postoperative hemorrhage. A number of disadvantages include lower image quality, increased cost (e.g., necessity for nonferromagnetic surgical instruments), relatively constrained operating space, and no possibility for functional, diffusion–perfusion, or spectroscopic investigations due to the lower intensity of the magnetic field. However, work is proceeding to address this last issue through the development of a new iMRI suite iteration, which incorporates a high magnetic field (Jolesz, 2001, 2003).

Intraoperative Magnetic Resonance Imaging Integrated Neurosurgical Suite

The Intraoperative Magnetic Resonance Imaging Integrated Neurosurgical Suite (IMRIS) produced by IMRIS, Deerfield Imaging (Minnetonka, MN, USA), provides a seamlessly integrated intraoperative imaging platform, wherein MRI and CT scanners are mounted on ceiling rails, which negates the necessity of moving or repositioning the patient. Hence, on-demand real-time imaging is available at any point, prior to surgical procedures, or at any time during or following surgeries. In addition, the integrity of the sterilized surgical field remains intact, as do the capacities for anesthesia and patient monitoring. On the basis of increasing validity revealed in clinical publications, there is strong support for IMRIS iMRI in terms of its safety and utility toward the optimization of intraoperative brain tumor removal. This system enables the capacity for minimally invasive neurosurgery, inclusive of epilepsy-associated thermal ablation, lead device positioning to facilitate deep brain stimulation, and the mediation of "tumor resection with fiber tracking for arteriovenous malformation and aneurysms" (IMRIS, 2016).

Zhuang et al. reported on the use of iMRI for the resection of 30 gliomas that involved the dominant insular lobe. They described 20 craniotomies that were performed with awake patients using cortical electrical stimulation mapping to demarcate language domains and 10 patients under general anesthesia, for whom functional navigation was conducted. For all patients, motor pathways were scrutinized by employing continuous motor-evoked potential monitoring, subcortical electrical stimulation mapping, and diffusion tensor imaging tractography–based navigation, whereas iMRI was utilized to evaluate the degree of resection. It was found through intraoperative imaging that residual tumor tissue remained in 26 patients, which led to nine additional resections. The mean precision of resection via iMRI was subsequently increased from 90% to 93% (P = 0.008) across all patients and enhanced the proportion of bulk and fine total resection from 53% to 77% (P = 0.016). Further, the levels of permanent language and motor function losses subsequent to tumor removal were 11% and 7.1%, respectively (Zhuang et al., 2016).

ADVANCED MULTIMODALITY IMAGE GUIDED OPERATING

The Advanced Multimodality Image Guided Operating (AMIGO) suite comprises a leading edge medical and surgical research platform, which integrates an extensive complement of state-of-the-art imaging and interventional surgical instrumentation between three rooms, to facilitate precisely guided interventions. A patient transfer board enables the safe transfer of the patient between the MRI, PET/CT, and operating rooms (Brigham and Women's Hospital).

In the realm of neurology, Zaidi et al. investigated 20 cases of pituitary macroadenomas, which were addressed via an endoscopic transsphenoidal strategy using a 3-T AMIGO suite. Though gross total resections were achieved in 12 cases prior to iMRI, it was found that iMRI had the capacity to improve this number, from 12 (60%) to 16 (80%). The conclusion was that in spite of the improved visual acumen of advanced endoscopes, iMRI has positive utility for the detection of unanticipated residual tumor tissue (Zaidi et al., 2016).

BRAIN INITIATIVE, G20 WORLD BRAIN MAPPING AND THERAPEUTICS INITIATIVE, AND NATIONAL PHOTONIC INITIATIVE

In order to compensate for the shortfalls of President Obama's BRAIN Initiative, and the EU Brain Project, the SBMT initiated its G20+ World Brain Mapping and Therapeutics, in 2014, through which the ultimate goals are to promote a sustainable global economy by supporting and facilitating (a) the development of improved diagnoses and treatments for neurological disorders; (b) job creation and the commercialization of innovative medical devices, methods, and technologies for neurological disorders; (c) the establishment of extensive, genuinely open, and collaborative global partnerships.

The G20+ (group of 20 developing nations plus others) or Neuroscience-20+ Initiative is aimed at engaging national and international groups on a global scale toward the innovation, integration, translation, and commercialization of neurotechnologies, advanced diagnostics, and therapeutics. The first (2014) G20+ Brain Mapping Summit was held in Brisbane, Australia, and was jointly cosponsored by the SBMT, BMF, Compumedics, Amen Clinics, and the National Center for NanoBioElectronics (NCNBE). This summit resulted in the establishment of the USA-Australia Brain Mapping Initiative and the first Brain Mapping Day at the Australian Parliament (Figure 1.2).

The second G20+ World Brain Mapping Summit commenced in Istanbul and Antalya, Turkey, in 2015, in partnership with Uskudar University, the SBMT, the BMF, and the NCNBE (see Figure 1.3). This well-attended and successful summit culminated in the establishment of

FIGURE 1.2 The US and Australian delegates (USA: Drs. Babak Kateb, Mark Liker, Aaron Filler, Uttam Sinha, Brian Hemling/Katarina Novakova; Australia: Dr. Kuldip Sidhu, Kiran Sidhu, Dimity Dornan, Matthew Kiernan, Jeff Rosenfeld, and Former member of the Parliament of QLD, Freya Ostapovitch) at the joint sessions of the Senate and Parliament of Australia. The aim of this Brain Mapping Day at the joint session of Senate and Australian parliament was to formulate, advocate, and establish a ground-breaking Brain initiative in order to encourage innovation, translation, and commercialization of innovative neurotechnology. Prior to this event Australia had no Brain Initiative. The event was organized by Society for Brain Mapping and Therapeutics (USA and Australian chapters), National Center for NanoBioElectronic, Brain Mapping Foundation (USA and Australian chapter), and Blue Horizon Charitable Foundation (October 12, 2015).

the Turkish Brain Mapping Initiative, and the following six strategic points were endorsed in a memorandum of understanding (MOU).

1. Adoption of a consortium approach for the study of the human brain
2. Global harmonization of related policies and the standardization of data
3. Economic assessment of future impacts related to the prevention and diagnoses of neurological disorders
4. Facilitating the translation, integration, and commercialization of neurotechnologies

FIGURE 1.3   2015 G20+ World Brain Mapping and Therapeutics delegation in Istanbul, Turkey.

5. Unification of global regulations and guidelines related to clinical trials and combined drug/device discovery and development
6. Global partnership and new funding for Brain Mapping Initiatives (basic and clinical science) encompassing academic, educational, governmental, industry, as well as for profit and nonprofit organizations

The third summit was held in Chongqing, China, in 2016, with the support of Daping Hospital and the Chongqing International Neurology Forum. Leading brain mapping experts from G20+ countries included those from China, the United States., Australia, Japan, Turkey, and the Middle East, all of whom participated in the event, which resulted in a corporative agreement similar to that which was established in Turkey.

The N20+ is driven by an international alliance of dedicated and passionate scientists for the purpose of significantly expanding scientific discovery as relates to the brain and spinal cord, unlike the BRAIN Initiative and European Human Brain Project. This initiative is aimed at stimulating the rapid discovery and elucidation of new aspects of brain function and neurological disorders, in conjunction with the introduction of paradigm shifting technologies and tools toward reducing the cost of healthcare globally, while increasing healthcare efficiencies and improving patient outcomes.

SBMT-IEEE BRAIN MAPPING INITIATIVE

Executives from both the SBMT and IEEE met in June 2015 to formulate the IEEE Brain Initiative (Figure 1.4). Drs. Babak Kateb, Aaron Filler (Institute for Nerve Medicine), and Dipen Sinha (Los Almos National Lab) from the SBMT met with Drs. Howard Michel and Jim Prendergast and the other IEEE executives at the IEEE headquarters. As a result of this strategic meeting, the following 10 potential areas of joint partnership were proposed and recommended by the SBMT toward the establishment of a joint SBMT-IEEE Brain Mapping Initiative:

1. Participate in the 2016 World Congress for Brain Mapping and Therapeutics.
2. Establish a potential special journal issue: IEEE-SBMT/Brain Mapping and Therapeutics.
3. Create a Kids Corner.
4. Establish significant social media *presence*.
5. Establish the standardization of neurotechnologies.

FIGURE 1.4 Executives from Society for Brain Mapping and Therapeutics (Drs. Babak Kateb, Dipen Sinha and Aaron Filler—not in the picture) and the Institute of Electrical and Electronics Engineers (IEEE) President and executive committee; the meeting was held at the IEEE headquarter office to formulate the IEEE brain mapping initiative (June 2015), which was later on introduced as IEEE Brain.

6. Partner with the SBMT and the G20 World Brain Mapping Initiative.
7. Establish fellowships and scholarships in partnership with the IEEE Foundation.
8. Establish an awards program.
9. Partner with SBMT University on Brain Mapping TV.
10. Engage IEEE-USA in Brain Policy/Brain Mapping.

NATIONAL PHOTONIC INITIATIVE

According to the official National Photonics Initiative (NPI) website (1998), the National Research Council released a report: *Harnessing Light: Optical Science and Engineering for the 21st Century.* This report presented comprehensive aspects of the emerging field of optics and photonics, and its profound impact on industries, including healthcare. Countries such as Germany, China, and the European Union have been at the forefront of the field of optics and photonics sectors. Unfortunately, the United States did not develop a unified strategy until recently (NPI, 2015).

In 2012, The Optical Society (OSA), the SPIE, the IEEE Photonics Society (IPS), the Laser Institute of America, and the American Physical Society Division of Laser Science formed the NPI for the United States, which is "an umbrella organization to identify and advance areas of photonics critical to maintaining competitiveness and national security" (NPI 2015).

In 2015, the initiative found a new alliance with President Obama's BRAIN Initiative, since the NPI began to convene the Photonics Industry Neuroscience Group (PING). The NPI-PING initially aimed to engage U.S. industry leaders in biomedical imaging, microscopy, lasers and advanced light sources, optical devices and components, and image analysis software; however, international industry leaders soon also became engaged with this partnership.

The NPI comprises an alliance of top scientific societies, uniting industry and academia, to raise awareness of photonics and the impacts of photonics on our everyday lives; increase cooperation and coordination among U.S. industries, government, and academia to advance photonics-driven fields; and drive U.S. funding and investment in areas of photonics that are critical toward maintaining and enhancing U.S. economic competitiveness and national security.

This initiative has been increasingly focused on the development and implementation of advanced optical tools and strategies for the further elucidation of brain activities and structural maps, while providing potential neurophotonic diagnostics and treatments for serious neurological disorders such as Alzheimer's and Parkinson's disease.

CONCLUSION

The notion of the mapping of the human brain is certainly not new, and indeed, spans back thousands of years. Increasingly precise human brain maps have evolved in step with the development of more advanced tools with broader capabilities. Core discoveries and scientific progress have always come from more open and collaborative teams who allow opposing ideas and competing opinions to push the boundaries of science, technology, art, medicine, and healthcare policy. New and exciting ideas, in conjunction with extensive international collaboration, have the capacity to attract significant funding and resources, which will lead to medical capabilities that are beyond our imagination. It is therefore critical for us to open our minds and release ourselves from the past practices of insular thinking, as the last fine instruments to facilitate the study of the human brain goes back to MRI and CT scans, which were not developed with government funds. Remarkable future discoveries will blossom through the funding of very high-risk/high-gain, or so-called "game changing" science, technology, and medicine, nourishing younger scientists in this area and the establishment of extensive global consortia.

This volume aims to highlight some of the exciting emerging sophisticated technologies that are being advanced and employed for space exploration by NASA. This book comprises the cumulative efforts of respected scientific works that have been achieved by pioneering scientists, physicians, and surgeons, who unwaveringly redefine the boundaries of the possible in brain mapping science and technology.

REFERENCES

Akemann W, Song C, Mutoh H, Knopfel T. (April–June 2015). Route to genetically targeted optical electrophysiology: Development and applications of voltage-sensitive fluorescent proteins. *Neurophotonics.* 2(2): pii: 021008.

Albayrak B, Samdani AF, Black PM. (2004). Intra-operative magnetic resonance imaging in neurosurgery. *Acta Neurochir (Wien).* 146: 543–557.

Alexander E, Moriarty TM, Kikinis R, Black P, Jolesz FM. (1997). The present and future role of intraoperative MRI in neurosurgical procedures. *Stereotact Funct Neurosurg.* 68: 10–17.

Allison RR, Sibata CH. (June 2010). Oncologic photodynamic therapy photosensitizers: A clinical review. *Photodiagnosis Photodyn Ther.* 7(2): 61–75.

Annese J, Toga AW. (2002). Chapter 20: Postmortem anatomy. In Toga AW, Mazziotta JC, eds., *Brain Mapping: The Methods.* San Diego, CA: Academic Press, pp. 537–539.

Arridge SR. (1999). Optical tomography in medical imaging. *Inverse Problems.* 15(2): R41–R93.

Becerra L, Harris W et al. (2009). Diffuse optical tomography activation in the somatosensory cortex: Specific activation by painful vs. non-painful thermal stimuli. *PLoS One.* 4(11): e8016.

Black PM, Alexander E 3rd, Martin C, Moriarty T, Nabavi A, Wong TZ, Schwartz RB, Jolesz F. (1999). Craniotomy for tumor treatment in an intraoperative magnetic resonance imaging unit. *Neurosurgery.* 45(3): 423–431; discussion 431–433.

Black PM, Moriarty T et al. (1997). Development and implementation of intraoperative magnetic resonance imaging and its neurosurgical applications. *Neurosurgery.* 41: 831–842.

Boas DA. (2014). Welcome to neurophotonics. *Neurophoton.* 1(1): 010101.

Brigham and Women's Hospital. (2015). Inside the advanced multimodality image guided operating suite. http://www.brighamandwomens.org/Research/amigo/inside_suite.aspx (accessed April 24, 2016).

Castillo M. (2014). History and evolution of brain tumor imaging: Insights through radiology. *Neuroradiology* 273(2): 111–125.

Chance B, Leigh JS et al. (1988). Comparison of time-resolved and -unresolved measurements of deoxyhemoglobin in brain. *Proc Natl Acad Sci USA.* 85(14): 4971–4975.

Chance B, Nioka S et al. (2005). Breast cancer detection based on incremental biochemical and physiological properties of breast cancers: A six-year, two-site study. *Acad Radiol.* 12(8): 925–933.

Chance B, Nioka S et al. (2007). A wearable brain imager. *IEEE Eng Med Biol Mag.* 26(4): 30–37.

Chance B, Zhuang Z et al. (1993). Cognition-activated low-frequency modulation of light absorption in human brain. *Proc Natl Acad Sci USA.* 90(8): 3770–3774.

Chance B. (1951). Rapid and sensitive spectrophotometry III. A double beam apparatus. *Rev Sci Instrum.* 22: 634–638.

Chen Y, Intes X et al. (2003). Probing rat brain oxygenation with near-infrared spectroscopy (NIRS) and magnetic resonance imaging (MRI). *Adv Exp Med Biol.* 510: 199–204.

Cho YK, Li D. (2016). Optogenetics: Basic concepts and their development. *Methods Mol Biol.* 1408: 1–17.

Choma MA, Sarunic MV et al. (2003). Sensitivity advantage of swept source and Fourier domain optical coherence tomography. *Opt Express.* 11(18): 2183–2189.

Chudler EH. (2016). Milestones in neuroscience research. https://faculty.washington.edu/chudler/hist.html (accessed April 19, 2016).

Compston A. (2013). Editorial. *Brain.* 136: 2645–2648.

de Boer JF, Cense B et al. (2003). Improved signal-to-noise ratio in spectral-domain compared with time-domain optical coherence tomography. *Opt Lett.* 28(21): 2067–2069.

Deisseroth K. (2015). Optogenetics: 10 years of microbial opsins in neuroscience. *Nat Neurosci.* 18(9): 1213–1225.

Diebold ED, Buckley BW, Gossett DR, Jalali B. (2013). Digitally synthesized beat frequency multiplexing for sub-millisecond fluorescence microscopy. *Nat Photon.* 7(10): 806–810.

Durduran T, Zhou C et al. (2009). Transcranial optical monitoring of cerebrovascular hemodynamics in acute stroke patients. *Opt Express.* 17(5): 3884–3902.

Enatsu R, Kanno A, Ohtaki S, Akiyama Y, Ochi S, Mikuni N. (February 2016). Intraoperative subcortical fiber mapping with subcortico-cortical evoked potentials. *World Neurosurg.* 86: 478–483.

Fishkin JB, Gratton E. (1993). Propagation of photon-density waves in strongly scattering media containing an absorbing semi-infinite plane bounded by a straight edge. *J Opt Soc Am A.* 10(1): 127–140.

Fujimoto JG, Brezinski ME et al. (1995). Optical biopsy and imaging using optical coherence tomography. *Nat Med.* 1(9): 970–972.

Ghaffari F, Naseri M, Movahhed M, Zargaran A. (July 2015). Spinal traumas and their treatments according to avicenna's canon of medicine. *World Neurosurg.* 84(1): 173–177.

Goda K, Jalali B. (2013). Dispersive Fourier transformation for fast continuous single-shot measurements. *Nat Photon.* 7(2): 102–112.

Grinvald A, Omer D, Naaman S, Sharon D. (2015). Imaging the dynamics of mammalian neocortical population activity in-vivo. *Adv Exp Med Biol.* 859: 243–271.

Gross CG. (1995). Aristotle on the brain. *The Neuroscientist.* 1: 245–250.

Grzybowski A, Pietrzak K. (July–August 2012). From patient to discoverer—Niels Ryberg Finsen (1860–1904)—The founder of phototherapy in dermatology. *Clin Dermatol.* 30(4): 451–455

Hartridge H, Roughton FJW. (1923). A method of measuring the velocity of very rapid chemical reactions. *Proc R Soc A.* 104: 376–394.

Helmstaedter M. (2013). Cellular-resolution connectomics: Challenges of dense neural circuit reconstruction. *Nat Methods.* 10: 501–507.

Herbowski L. (2013). The maze of the cerebrospinal fluid discovery. *Anat Res Int.* 2013: 596027.

Hielscher AH, Bluestone AY et al. (2002). Near-infrared diffuse optical tomography. *Dis Markers.* 18(5–6): 313–337.

Hill DK, Keynes RD. (1949). Opacity changes in stimulated nerve. *J Physiol.* 108: 278–281.

Hillman EMC. (2007). Optical brain imaging in vivo: Techniques and applications from animal to man. *J Biomed Opt* 12(5): 051402.

Hippocrates. (1950). On the sacred disease. In Chadwick J, Mann WN, trans. *The Medical Works of Hippocrates.* Oxford, U.K.: Blackwells, pp. 179–189.

Hogan RE, English EA. (July 2012). Epilepsy and brain function: Common ideas of Hughlings-Jackson and Wilder Penfield. *Epilepsy Behav.* 24(3): 311–313.

Hoshi Y, Tsou BH et al. (2003). Spatiotemporal characteristics of hemodynamic changes in the human lateral prefrontal cortex during working memory tasks. *Neuroimage.* 20(3); 1493–1504.

Huang D, Swanson EA et al. (1991). Optical coherence tomography. *Science* 254(5035): 1178–1181.

Huang Z. (June 2005). A review of progress in clinical photodynamic therapy. *Technol Cancer Res Treat.* 4(3): 283–293.

IMRIS. Deerfield imaging. http://www.imris.com/intraoperative-solutions/visius-imri/neurosurgery (accessed April 24, 2016).

Intes X, Chance B. (2005). Non-PET functional imaging techniques: Optical. *Radiol Clin North Am.* 43(1): 221–234, xii.

Jafri MS, Farhang S et al. (2005). Optical coherence tomography in the diagnosis and treatment of neurological disorders. *J Biomed Opt.* 10(5): 051603.

Jalali B, Coppinger FMA. (2001). Data conversion using time manipulation. U.S. Patent No. 6,288,659. Jalali-Lab. www.photonics.ucla.edu.

Jobsis FF. (1977). Noninvasive, infrared monitoring of cerebral and myocardial oxygen sufficiency and circulatory parameters. *Science.* 198(4323): 1264–1267.

Jolesz FA. (2001). Neurosurgical suite of the future II. *Neuroimaging Clin North Am.* 11(4): 581–592.

Jolesz FA. (2003). Future perspectives in intraoperative imaging. *Acta Neurochir (Wien).* 85: 7–13.

Kalchenko V, Israeli D, Kuznetsov Y, Harmelin A. (July 25, 2014). Transcranial optical vascular imaging (TOVI) of cortical hemodynamics in mouse brain. *Sci Rep.* 4: 5839.

Kateb B, Heiss J. (2013). *The Textbook of Nanoneuroscience and Nanoneurosurgery.* Boca Raton, FL: Taylor & Francis.

Kelin D. (1925). On cytochrome, a respiratory pigment, common to animals, yeast, and higher plants. *Proc R Soc B.* 98: 312–313.

Kim MN, Durduran T et al. (2010). Noninvasive measurement of cerebral blood flow and blood oxygenation using near-infrared and diffusecorrelation spectroscopies in critically brain-injured adults. *Neurocrit Care.* 12(2): 173–180.

Kirkpatrick PJ, Lam J et al. (1998). Defining thresholds for critical ischemia by using near-infrared spectroscopy in the adult brain. *J Neurosurg.* 89(3): 389–394.

Leitgeb R, Hitzenberger CK et al. (2003). Performance of Fourier domain vs. time domain optical coherence tomography. *Opt Express.* 11(8): 889–894.

Levy KM. (May 2013). Brain mapping project: Clinical aspects and role of neuroradiology. *Am J Neuroradiol.* 34(5): 942–943.

Lewin JS. (May 1999). Interventional MR imaging: Concepts systems, and applications in neuroradiology. *Am J Neuroradiol.* 20: 735–748.

Lipson AC, Gargollo PC, Black PM. (2001). Intraoperative magnetic resonance imaging: Considerations for the operating room of the future. *J Clin Neurosci.* 8(4): 305–310.

Littleton JT. (2014). Conventional tomography. In Gagliardi RA, ed., *A History of the Radiological Sciences (PDF).* Reston, VA: American Roentgen Ray Society. Retrieved November 29, 2014.

Lloyd GER. (1975). Alcmaeon and the early history of dissection. *Sudhoff's Archiv fur die Geschichte der Medizin.* 59: 113–147.

Madsen SJ, Anderson ER et al. (1994). Portable, high-bandwidth frequency-domain photon migration instrument for tissue spectroscopy. *Opt Lett.* 19(23): 1934–1936.

Matsuo K, Kato T et al. (2003). Activation of the prefrontal cortex to trauma-related stimuli measured by near-infrared spectroscopy in posttraumatic stress disorder due to terrorism. *Psychophysiology.* 40(4): 492–500.

Meggitt G. (2008). Taming the rays: A history of radiation and protection. S.l.:Lulu.com.

Millikan GA. (1937). Experiments on muscle hemoglobin in vivo; the instantaneous measurement of muscle metabolism. *Proc R Soc B.* 123: 218–241.

Millikan GA. (1942). The oximeter, an instrument for measuring continuously the oxygen saturation of arterial blood in man. *Rev Sci Instrum.* 13: 434–444.

Minagar A, Ragheb J, Kelley RE. (May 2003). The Edwin Smith surgical papyrus: Description and analysis of the earliest case of aphasia. *J Med Biogr.* 11(2):114–117.

Mislow JM, Golby AJ, Black PM. (February 2010). Origins of intraoperative MRI. *Magn Reson Imaging Clin North Am.* 18(1): 1–10.

Miyamoto K, Osada T, Adachi Y. (2014). Remapping of memory encoding and retrieval networks: Insights from neuroimaging in primates. *Rev Behav Brain Res.* 275: 53–61.

Moriarty TM, Kikinis R, Jolesz FA, Black PM, Alexander E 3rd. (1996). Magnetic resonance imaging therapy. Intraoperative MR imaging. *Neurosurg Clin North Am.* 7(2): 323–331.

National Photonic Initiative (NPI). (2015). What is the NPI? http://www.lightourfuture.org/home/about-npi/history/ (accessed April 24, 2016).

O'Connor DH, Hires SA, Guo ZV, Li N, Yu J, Sun QQ, Huber D, Svoboda K. (July 2013). Neural coding during active somatosensation revealed using illusory touch. *Nat Neurosci.* 16(7): 958–965.

Ogawa S, Lee TM et al. (1990). Brain magnetic resonance imaging with contrast dependent on blood oxygenation. *Proc Natl Acad Sci USA*. 87(24): 9868–9872.

Patterson MS, Chance B et al. (1989). Time resolved reflectance and transmittance for the non-invasive measurement of tissue optical properties. *Appl Opt*. 28(12): 2331–2336.

Pena M, Maki A et al. (2003). Sounds and silence: An optical topography study of language recognition at birth. *Proc Natl Acad Sci USA*. 100(20): 11702–11705.

Pevsner J. (April 2002). Leonardo da Vinci's contributions to neuroscience. *Trends Neurosci*. 25(4): 217–220.

Phelps ME, Hoffman EJ, Mullani NA, Ter-Pogossian MM. (March 1, 1975). Application of annihilation coincidence detection to transaxial reconstruction tomography. *J Nucl Med*. 16(3): 210–224.

Plato. (1959). *Timaeus*. Cornford FM, trans. New York: Bobbs-Merrill.

Pollak B. (December 1953). Experiences with planography. *Chest*. 24(6): 663–669.

Raichle ME. (February 2009). A brief history of human brain mapping. *Trends Neurosci*. 32(2): 118–126.

Robertson CA, Evans DH, Abrahamse H. (July 17, 2009). Photodynamic therapy (PDT): A short review on cellular mechanisms and cancer research applications for PDT. *J Photochem Photobiol B*. 96(1): 1–8.

Roughton FJW, Millikan GA. (1936). Photoelectric methods of measuring the velocity of rapid reactions. I. General principles and controls. *Proc R Soc Lond*. A 155: 258–269.

Sato H, Takeuchi T et al. (1999). Temporal cortex activation during speech recognition: An optical topography study. *Cognition*. 73(3): B55–B66.

Schenck JF, Jolesz FA, Roemer PB et al. (1995). Superconducting open configuration MR imaging system for image-guided therapy. *Radiology*. 195: 805–814.

Sidhu K, Kateb B. (2015). World brain mapping and terapeutic initiative: A proposed G20 priority due to major impact of the cost of neurological disorders on the world economy. *J Neurol Disord*. 2: 6.

Silverman SG, Collick BD et al. (1995). Interactive MR guided biopsy in an open-configuration MR imaging system. *Radiology*. 197: 175–181.

Silverman SG, Jolesz FA et al. (1997). Design and implementation of an interventional MR imaging suite. *Am J Roentgenol*. 168: 1465–1471.

Skrap M, Marin D, Ius T, Fabbro F, Tomasino B. (February 2016). Brain mapping: A novel intraoperative neuropsychological approach. *J Neurosurg*. 5: 1–11.

Snyder PJ, Whitaker HA. (2013). Neurologic heuristics and artistic whimsy: The cerebral cartography of Wilder Penfield. *J Hist Neurosci*. 22(3): 277–291.

Stankovic MR, Maulik D et al. (2000). Role of frequency domain optical spectroscopy in the detection of neonatal brain hemorrhage—A newborn piglet study. *J Matern Fetal Med*. 9(2): 142–149.

Sweet WH, Brownell GL. (1953). Localization of brain tumors with positron emitters. *Nucleonics*. 11: 40–45.

Szobota S, Gorostiza P et al. (May 24, 2007). Remote control of neuronal activity with a light-gated glutamate receptor. *Neuron*. 54(4): 535–545.

Takahashi K, Ogata S et al. (2000). Activation of the visual cortex imaged by 24-channel near-infrared spectroscopy. *J Biomed Opt*. 5(1): 93–96.

Ter-Pogossian MM, Phelps ME, Hoffman EJ, Mullani NA. (1975). A positron-emission transaxial tomograph for nuclear imaging (PET). *Radiology*. 114(1): 89–98.

Tubbs RS, Shoja MM, Loukas M, Oakes WJ. (November 2007). Abubakr Muhammad Ibn Zakaria Razi, Rhazes (865–925 AD). *Childs Nerv Syst*. 23(11): 1225–1226.

U.S. Department of Energy. (September 2010). *A Vital Legacy: Biological and Environmental Research in the Atomic Age*. The Office of Biological and Environmental Research, Berkeley, CA, pp. 25–26.

Villringer A, Chance B. (1997). Non-invasive optical spectroscopy and imaging of human brain function. *Trends Neurosci*. 20(10): 435–442.

Watanabe E, Maki A et al. (1998). Non-invasive assessment of language dominance with near-infrared spectroscopic mapping. *Neurosci Lett*. 256(1): 49–52.

Watanabe E, Nagahori Y et al. (2002). Focus diagnosis of epilepsy using near-infrared spectroscopy. *Epilepsia*. 43(Suppl. 9): 50–55.

Wolf M, Ferrari M et al. (2007). Progress of near-infrared spectroscopy and topography for brain and muscle clinical applications. *J Biomed Opt*. 12(6): 062104.

Yang H, Kwon SE, Severson KS, O'Connor DH. (January 2016). Origins of choice-related activity in mouse somatosensory cortex. *Nat Neurosci*. 19(1): 127–134.

Yaqoob Z, Wu J et al. (2006). Methods and application areas of endoscopic optical coherence tomography. *J Biomed Opt*. 11(6): 063001.

Zago S, Meraviglia MV. (July 2009). Costanzo Varolio (1543–1575). *J Neurol*. 256(7): 1195–1196.

Zaidi HA, De Los Reyes K et al. (March 2016). The utility of high-resolution intraoperative MRI in endoscopic transsphenoidal surgery for pituitary macroadenomas: Early experience in the Advanced Multimodality Image Guided Operating suite. *Neurosurg Focus.* 40(3): E18.

Zeff BW, White BR et al. (2007). Retinotopic mapping of adult human visual cortex with high-density diffuse optical tomography. *Proc Natl Acad Sci USA.* 104(29): 12169–12174.

Zemelman BV, Miesenböck G. (August 2001). Genetic schemes and schemata in neurophysiology. *Curr Opin Neurobiol.* 11(4): 409–414.

Zhuang DX, Wu JS, Yao CJ, Qiu TM, Lu JF, Zhu FP, Xu G, Zhu W, Zhou LF. (February 3, 2016). Intraoperative multi-information-guided resection of dominant-sided insular gliomas in a 3T intraoperative MRI integrated neurosurgical suite. *World Neurosurg* 89: 84–92.

Section I

Optical Spectroscopy and
Spectral Imaging of Brain

2 Pain Assessment Using Near-Infrared Spectroscopy

Kambiz Pourrezaei, Ahmad Pourshoghi,
Zeinab Barati, and Issa Zakeri

CONTENTS

INTRODUCTION

Pain serves as an alarm to prevent possible damage to the organism and is often a symptom of injury or disease. Pain is the most frequently encountered symptom in daily medical practice; however, its perception and expression are highly subjective. The gold standard for evaluating the presence, intensity, quality, and location of pain is self-reporting questionnaires—if the patient is capable of reliable communication. These questionnaires, which are widely used in pain clinics, are highly subjective on the part of the patient as well as on the part of the healthcare providers, especially when there is more than one provider. This shortcoming of pain questionnaires may result in under/ overprescription of analgesic drugs, with their associated complications.

In conventional pain practice, physiological parameters, such as heart rate, blood pressure, respiratory rate, galvanic skin response, cutaneous blood flow, and behavioral measures such as facial grimacing and guarding of the painful area have been used to monitor the response to a noxious stimulus (Champion et al., 1998; Sweet and McGrath, 1998; Blount and Loiselle, 2009). However, physiological parameters can be unstable and nonspecific, and behavioral measures change due to many factors other than pain, such as distress (Srouji et al., 2010). Thus, clinical experience has proved that these are not reliable for pain assessment and should be interpreted cautiously (Ranger and Gélinas, 2012).

During the past two decades, neurophysiological techniques that measure cerebral metabolism and circulation changes have been widely employed to open a window into the human cerebral response to pain, with a long-term goal of obtaining a more direct measurement of pain perception. Neuroimaging studies using advanced imaging modalities such as positron emission tomography (PET) and functional magnetic resonance imaging (fMRI) have revealed brain regions that become activated during a physical or psychological experience of pain (Casey et al., 1996; Disbrow et al., 1998; Porro et al., 1998; Tracey et al., 2000; Seifert and Maihofner, 2007). Some research has found a relationship between subjects' report of an ongoing pain and the BOLD (blood oxygen level dependent) signal acquired by fMRI (Porro et al., 1998). These modalities have advanced our understanding of the underlying mechanisms in nociception and have made a great impact on basic science. However, clinical practice has not accepted them for routine examinations for several reasons, including cost, availability, and enclosed environment.

Functional near-infrared spectroscopy (fNIRS) is a novel optical imaging modality for noninvasive, continuous monitoring of tissue oxygenation and regional blood flow (Rolfe, 2000). fNIRS can measure changes in hemodynamics simultaneously across different sites of sympathetic innervations, such as hand, forearm, and face, as well as the cortex. Recently, several studies have suggested the use of fNIRS for monitoring cortical activation in response to noxious stimuli in newborn infants (Bartocci et al., 2006; Slater et al., 2006, 2008; Ozawa et al., 2011; Ranger et al., 2013) and adults (Becerra et al., 2008, 2009; Viola et al., 2010; Gelinas et al., 2010; Watanabe et al., 2011; Barati et al., 2013; Lee et al., 2013; Re et al., 2013; Yennu et al., 2013). fNIRS can be portable and has low equipment and maintenance costs. It is relatively robust to motion artifacts and therefore, no immobilization is required during measurement, unlike movement constraints imposed by other functional imaging techniques.

Pain is a complex and multifaceted process and can be best examined through the assessment and integration of multiple physiological, cortical, and behavioral measures that most closely describe the pain experience. The purpose of this chapter is to summarize the studies that have employed fNIRS for the assessment of some aspects of human response to pain. It also includes a brief overview of the most accepted neuroimaging modalities for the study of brain response to noxious procedures. Assessment of the cerebral and physiological activities along with behavioral and subjective measures may help identify the best clinically practical measures for humans.

FUNCTIONAL NEUROIMAGING

Functional neuroimaging is a term used for the study of human brain function to understand the physiology, functional architecture, and dynamics of the brain. Neuroimaging uses various techniques to directly or indirectly image the structure or function of the brain or to monitor the activities in certain brain areas. Some methods of neuroimaging record intracellular effects of neural activities, including changes in membrane potential or electrical and magnetic fields that are induced by ion fluxes (mainly Na^+, K^+, Cl^-, and Ca^{2+}) across a neuron's membrane. Electroencephalography (EEG) and magnetoencephalography (MEG) directly measure electric current and magnetic field changes secondary to neural activities on the scalp. On the other hand, vascular-based neuroimaging techniques such as fMRI and fNIRS record hemodynamic changes that are indirectly correlated to neural activities.

fNIRS is an optical neuroimaging technique that allows for a noninvasive study of neurovascular coupling, that is, the relationship between local neural activity and subsequent changes in cerebral blood flow (CBF). This occurs through a complex sequence of coordinated events involving neurons, glia, and vascular cells. In a simple model, neural activation demands adenosine triphosphate (ATP), which increases oxygen and glucose consumption, followed by an increase in CBF. The increase in oxygen consumption is relatively lower than the increase in CBF, which leads to a net increase in blood oxygen level. This imbalance between oxygen consumption and CBF explains the principle behind the BOLD signal of fMRI. Different studies have validated this relationship

(Logothetis et al., 2001; Arthurs and Boniface, 2003), suggesting that hemodynamic changes could provide a marker for assessing neural activity (León-Domínguez, 2012). fNIRS signal has shown to be strongly correlated with PET measures of changes in regional cerebral blood flow (rCBF) and the fMRI BOLD signal (Kleinschmidt et al., 1996; Hock et al., 1997; Villringer and Chance, 1997; Toronov et al., 2001, 2003; MacIntosh et al., 2003; Huppert et al., 2006).

NEUROIMAGING OF PAIN

The early pain imaging studies used PET and reported on cerebral responses to noxious heat (Borsook et al., 2010). Since then, different functional modalities and brain imaging techniques have been used to study brain reactivity to pain in both normal subjects and patients with clinical pain conditions. fMRI, PET, and scalp EEG are commonly used to study the neural basis of pain. Researchers are also increasingly using other magnetic resonance-based measures, such as diffusion tensor imaging, spectroscopy, and volumetric imaging, to assess pain-related changes in the brain's wiring, chemistry, and structure in order to gain further insights into the neurobiology of pain, particularly chronic pain (Lee and Tracey, 2013).

POSITRON EMISSION TOMOGRAPHY

Positron emission tomography (PET) is an imaging modality that provides information about the regional cerebral blood flow (rCBF) and tissue metabolism by detecting emissions from active chemicals that have been injected into the blood stream. In the first imaging study of pain (Talbot et al., 1991), Talbot and colleagues used PET and reported on activation in several brain regions in response to noxious heat, including the contralateral anterior cingulate cortex and primary and secondary somatosensory cortices. Since then, several different studies have confirmed activation of the same brain regions among others (often bilateral), including the primary and secondary somatosensory cortices (Bushnell et al., 1999; Coghill et al., 1999; Peyron et al., 1999; Chen et al., 2002), the posterior, mid, and anterior insula (Rainville et al., 1997; Tolle et al., 1999; Fulbright et al., 2001), the anterior cingulate (Rainville et al. 1997; Willoch et al., 2004), the prefrontal cortices (Lorenz et al., 2003; Porro et al., 2003; Ochsner et al., 2006), and thalamus (Wager et al., 2004). Neuroimaging has challenged the classical theory of pain mechanisms by uncovering new brain regions involved in pain processing such as nucleus accumbens (Scott et al., 2006), insula (Wiech et al., 2006), dorsolateral prefrontal cortex (Lorenz et al., 2003), basal ganglia, and cerebellum that reflect sensory, cognitive, and affective dimensions of pain.

FUNCTIONAL MAGNETIC RESONANCE IMAGING

Functional magnetic resonance imaging (fMRI) is another functional imaging technique widely used in pain studies. The fMRI BOLD signal is correlated with changes in the local concentration of deoxyhemoglobin, which is the hemoglobin molecule that is desaturated with oxygen. So, it can objectively measure pain-related neural activities and provide a valuable tool for studying the mechanisms of pain (Sava et al., 2009; Wager et al., 2013). Most fMRI studies of pain have utilized thermal stimuli (contact Peltier thermodes or laser) to activate pain circuits. Other types of stimuli, including electrical and mechanical (pressure), have not been as extensively used (Borsook and Becerra, 2006).

fMRI data has been used to decode whether a stimulus was perceived as painful (Brodersen et al., 2012). The results show that during pain anticipation, activity in the periaqueductal gray (PAG) and orbitofrontal cortex (OFC) afforded the most accurate trial-by-trial discrimination between painful and nonpainful experiences; whereas during the actual stimulation, primary and secondary somatosensory cortex, anterior insula, dorsolateral and ventrolateral prefrontal cortex, and OFC were most discriminative. The most accurate prediction of pain perception during the stimulation period, however, was made by a combined activity in the pain regions, commonly referred to as the "pain matrix," a name given to an extensive network of brain regions activated during pain perception, including somatosensory, insular, and cingulate areas, as well as frontal and parietal areas.

fMRI has been also used to study the influence of both placebo and necebo (positive and negative expectation) on the human pain process (Tracey, 2010). It has been shown that an expectation of decreased pain reduces both the subjective report as well as the activation of sensory, insular, and cingulate ("pain matrix") cortices (Koyama et al., 2005). Another fMRI study (Kong et al., 2008) has shown that nocebo effects are mediated by the hippocampus and regions involved with anticipatory anxiety and, as such, are distinct from placebo effects at a neural level.

In another group of studies, machine learning methods, such as support vector machine (SVM), have been used to analyze and classify fMRI signals during painful stimulations. Whole-brain patterns of activity were used to train an SVM to distinguish painful from nonpainful thermal stimulation (Brown et al., 2011). An accuracy of 81% was reported for distinguishing painful from nonpainful stimuli. Pain-processing regions of the brain, including the primary somatosensory cortex, secondary somatosensory cortex, insular cortex, primary motor cortex, and cingulate cortex had major contribution to the SVM performance; however, region of interest (ROI) analyses revealed that whole-brain patterns of activity led to a more accurate classification than localized activity from individual brain regions (Brown et al., 2011). In another study (Wager et al., 2013), machine-learning analyses was used to identify a pattern of fMRI activity across brain regions—a neurologic signature—that was associated with heat-induced pain. The pattern included the thalamus, the posterior and anterior insulae, the secondary somatosensory cortex, the anterior cingulate cortex, the periaqueductal gray matter, and other regions. This group of studies shows that the BOLD signal is sufficiently consistent between individuals to potentially train a physiology-based pain classifier that performs accurately when trained on one group of subjects and tested on another (Brown et al., 2011).

More recently, real-time fMRI (rtfMRI) has been proposed for neuro-feedback studies in which subjects are trained to regulate activations in defined brain regions using feedback information extracted by real-time processing of the ongoing fMRI data (Chapin et al., 2012). Anterior cingulate has been suggested as an initial target for the use of rtfMRI feedback to alter pain experience in both healthy controls and patients with chronic pain (deCharms et al., 2005). Subjects successfully learned to control the activation of the anterior cingulate cortex and this process led to significant reductions in the magnitude of their chronic pain.

Electroencephalography

EEG is a noninvasive technique that detects electrical impulses in the brain due to neuronal activity using electrodes placed on the patient's scalp. It has been used for the study of cortical activation during external painful stimulation.

Earlier EEG studies on healthy subjects using induced tonic pain have shown the following: (1) increase in low-frequency delta power, (2) rare change in theta power, (3) decrease in alpha power, and (4) increase in high-frequency beta power (Chen, 2001). Also, a general conclusion from event-related potentials (ERPs) and the phasic pain-related signal is that the early components (<50 ms) are more related to physical stimulus parameters, while the late components (>150 ms) are related to pain perception. Further, the late components (150–400 ms) are largely due to the activation of thin myelinated A-delta fibers, while the very-late components (700–900 ms) and ultra-late components (1100–1500 ms) are related to thin nonmyelinated C-fiber activation (Chen, 2001).

In more recent studies, EEG recordings have been used to study pain network. Interactions between cortical modules, such as S1 (primary somatosensory cortex), PS (parasylvian cortex including secondary somatosensory and insular cortex), and MF (medial frontal cortex including cingulate and supplementary motor cortex), due to a painful cutaneous laser stimuli, have been demonstrated by measuring functional interactions between local field potentials from the cortex using implanted electrodes (Ohara et al., 2004, 2006, 2008; Liu et al., 2011a,b). Also, it has been shown that these interactions change dynamically with tasks, such as attention to or distraction from a cutaneous laser stimulus (Ohara et al., 2006; Liu et al., 2011b). In another study, scalp EEG has been used to demonstrate that EEG channels with post stimulus event-related causality (ERC) interactions were consistently different during the painful laser stimulus versus the nonpainful electric stimulus (Markman et al., 2013).

OTHER IMAGING MODALITIES

fMRI, PET scan, and EEG have been dominant techniques in pain researches, but other methods such as magnetoencephalography (MEG) (Kringelbach et al., 2007), magnetic resonance spectroscopy (MRS) (Pattany et al., 2002; Dichgans et al., 2005; Siddall et al., 2006; Prescot et al., 2009) (Borsook et al., 2007, 2010), functional transcranial Doppler sonography (fTCD) (Duschek et al., 2012), and Voxel-based morphometry (VBM) (Jovicich et al., 2009; Borsook et al., 2010; Mao, 2013) have been also used in pain studies. Another widely used modality is diffusion tensor imaging (DTI), which is a magnetic resonance technique that enables measuring microstructural changes in water diffusion in the brain. It has been used to study the white matter architecture and integrity of normal and diseased brains in a number of pain disorders such as migraine (Whitcher et al., 2007), central post-stroke pain (Seghier et al., 2005), and fibromyalgia (Lutz et al., 2008).

This review of imaging studies shows that fMRI is the most dominant and probably powerful imaging technique in pain research that can provide superior spatial information and significant insight into the understanding of brain response to painful stimuli. PET scan is another powerful technique; however, its application in research is particularly limited in the United States due to exposure to radioactive materials and the need for high-level clinical expertise. Nonetheless, both fMRI and PET have limited routine clinical use because of high equipment and maintenance costs. On the other hand, while EEG is potentially a powerful and well-understood technique that provides excellent temporal information, its usage in pain studies has not been very popular. Possible reasons for this may include sensitivity to noise and body movement, lack of specificity to pain, and the requirement of many electrodes to provide localized information. However, modern EEG systems that benefit from optimized hardware and improved signal processing fused with other techniques such as fNIRS can play an important role in future researches.

FUNCTIONAL NEAR-INFRARED SPECTROSCOPY

fNIRS works based on the fact that brain activity is associated with changes in the optical properties of the brain tissue in the near-infrared (NIR) range. Propagated light in any substance undergoes several absorptive and scattering phenomena. In living tissues, scattering predominates over absorption in the NIR range (700–1000 nm). The highly scattered light reflects back to the tissue surface. By putting a photo detector on the surface of the skin, one can measure the amount of absorption changes within the sampled volume (mostly within a banana-shaped optical pathway) (Figure 2.1).

The major absorbing components—chromophores—in soft tissues in the NIR range are water, oxyhemoglobin (HbO_2), and deoxyhemoglobin (Hb). There are also other minor absorbents such as melanin and lipids. The concentration of water and other absorbents can be assumed constant during the test period (static absorbers), and they have a negligible contribution to the overall attenuation. On the other hand, concentration of dynamic absorbers—oxygenated and deoxygenated hemoglobin—changes during the experiment according to the function and metabolism of the tissue.

When NIR light strikes a blood vessel, it is absorbed by oxyhemoglobin and deoxyhemoglobin. This absorption changes the intensity of the light which scatters back to the surface. There is a direct relationship between the concentration of oxyhemoglobin and deoxyhemoglobin in the blood and the changes in the intensity of light measured on the surface. The equation that governs this relationship is known as Modified Beer-Lambert law (MBLL) (Cope et al., 1988).

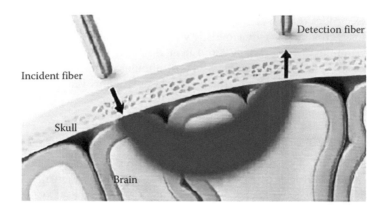

FIGURE 2.1 Volume of tissue sampled by an fNIRS measurement. (From Barati, Z. et al., *Ann. Biomed. Eng.*, 41(2):223–237, 2013. With permission.)

Modified Beer–Lambert Law

The modified Beer–Lambert law states that changes in the concentration of light-absorbing components are proportional to changes in light attenuation, divided by the mean optical pathlength and extinction coefficients of the chromophores in the tissue. The mean optical pathlength is a measure of the average distance that light travels between the source and detector after several episodes of scattering and absorption.

Scattering increases the probability of photon absorption in a medium because deviated photons travel longer distances. The extended optical pathlength (DP) is defined as

$$DP = DPF * d \qquad (2.1)$$

where d is the geometrical distance between the source and detector, and DPF is the differential pathlength factor which estimates how many times farther than d the detected light has traveled (Scholkmann and Wolf, 2013). DPF can be considered approximately constant for a given tissue. MBLL models light transportation through tissue by

$$I(\lambda) = I_0(\lambda) e^{-\mu_a(\lambda)DP(\lambda)+G(\lambda)} \qquad (2.2)$$

where
   $I(\lambda)$ is the measured wavelength-dependent light intensity
   $I_0(\lambda)$ is the incident light intensity
   $\mu_a(\lambda)$ is the absorption coefficient of the tissue
   DP is the optical pathlength given in Equation 2.1
   $G(\lambda)$ is a wavelength, medium, and geometry-dependent constant reflecting scattering losses
      (Scholkmann and Wolf, 2013)

Absorbance (A) represents the loss in light intensity and is usually measured in units of optical density (OD).

$$A(\lambda) = \log_{10}\left(\frac{I_0(\lambda)}{I(\lambda)}\right) = \mu_a(\lambda)DP(\lambda) - G(\lambda) \qquad (2.3)$$

Continuous wave fNIRS measures changes in absorption rather than absolute values. Assuming that $G(\lambda)$ and $DP(\lambda)$ are constant during the measurement, changes in absorbance between time 1 and 2 at a given λ can be written as

$$\Delta A(\lambda) = A_2(\lambda) - A_1(\lambda) = \Delta\mu_a(\lambda)DP(\lambda) \tag{2.4}$$

Beer determined that absorption coefficient for an absorbing compound, dissolved in a nonabsorbing medium, is linearly proportional to its concentration in the solution [C].

$$\mu_a = \alpha[C] \tag{2.5}$$

Rewriting Equation 2.4 using Equation 2.5 gives

$$A_2(\lambda) - A_1(\lambda) = \alpha(\lambda) * \Delta[C] * DP(\lambda) \tag{2.6}$$

Choosing two wavelengths in the NIR spectrum and measuring the absorbance change at two different time points result in

$$\Delta A(\lambda_1) = \left(\alpha_{HbO_2}(\lambda_1)\Delta[HbO_2] + \alpha_{Hb}(\lambda_1)\Delta[Hb]\right) * DP$$
$$= \left(\alpha_{HbO_2}(\lambda_1)\Delta[HbO_2] + \alpha_{Hb}(\lambda_1)\Delta[Hb]\right) * DP(\lambda_1) \tag{2.7}$$

$$\Delta A(\lambda_2) = \left(\alpha_{HbO_2}(\lambda_2)\Delta[HbO_2] + \alpha_{Hb}(\lambda_2)\Delta[Hb]\right) * DP(\lambda_2) \tag{2.8}$$

The relative change in the concentration of Hb and HbO_2 molecules can be calculated by

$$\Delta[HbO_2] = \frac{\alpha_{HbO_2}(\lambda_2) \cdot \dfrac{\Delta A(\lambda_1)}{DP(\lambda_1)} - \alpha_{HbO_2}(\lambda_1) \cdot \dfrac{\Delta A(\lambda_2)}{DP(\lambda_2)}}{\alpha_{Hb}(\lambda_1) \cdot \alpha_{HbO_2}(\lambda_2) - \alpha_{Hb}(\lambda_2) \cdot \alpha_{HbO_2}(\lambda_1)}$$

$$\Delta[Hb] = \frac{\alpha_{Hb}(\lambda_1) \cdot \dfrac{\Delta A(\lambda_2)}{DP(\lambda_2)} - \alpha_{Hb}(\lambda_2) \cdot \dfrac{\Delta A(\lambda_1)}{DP(\lambda_1)}}{\alpha_{Hb}(\lambda_1) \cdot \alpha_{HbO_2}(\lambda_2) - \alpha_{Hb}(\lambda_2) \cdot \alpha_{HbO_2}(\lambda_1)}$$

fNIRS Applications

Methods such as MEG and fMRI need bulky and expensive instruments and a dedicated building to eliminate effects of external magnetic fields. Yet, both systems require that the subject remains motionless during measurement (Branco, 2007). fNIRS is not only noninvasive and safe but also portable, affordable, and has good temporal resolution. Even wireless fNIRS system is now available that enables monitoring brain activity in moving subjects, for example during walking or running. Moreover, it is particularly suited for subjects who are hard to remain motionless in an fMRI magnet or a PET scanner, such as newborns, infants, young children, and patients with attention deficits, dementia, and bedridden. Furthermore, fNIRS is ideal for long-term bedside monitoring of cerebral activities in operating rooms and intensive care units.

fNIRS does not involve ionizing radiation (like x-ray or CT) and contrast agents (like PET scan); it can provide continuous monitoring of Hb, HbO_2 and cerebral blood flow. A disadvantage, however, is the spatial resolution which decreases with increasing depth below the surface. fNIRS can be used as a partial replacement for fMRI, but it cannot fully replace it because of its lower penetration depth (0.5–2 cm), which makes the measurements limited to the cortex only, as compared to fMRI, which has access to the subcortical tissues too.

Nevertheless, fNIRS is an effective tool to study normal brain physiology and its alternation in diseases, which covers a vast area of studies (Boas et al., 2014), including behavioral and cognitive studies, motor control, anesthesia, psychiatric conditions and disorders (Ehlis et al., 2014) such as bipolar disorder and depression (Matsubara et al., 2014; Takizawa et al., 2014), schizophrenia (Marumo et al., 2014), Alzheimer (Hock et al., 1996, 1997), Parkinson (Murata et al., 2000), and clinical neurology and epilepsy (Obrig, 2014). Studies suggest that fNIRS is a valid addition to the arsenal of neuroscientific methods available for the assessment of the neural mechanism and can be a useful tool in clinical settings (Boas et al., 2014).

fNIRS FOR PAIN ASSESSMENT

Literature reporting on employing fNIRS for pain assessment is very recent but fast-growing. In a comprehensive literature search, we accessed 21 research articles using fNIRS for the assessment of pain-related cerebral hemodynamic changes. We classified these articles, plus an unpublished manuscript of our group, into three categories: experimental, clinical, and infants' populations (Table 2.1).

Possibly due to the ease of measurements on the frontal region, 8 studies measured hemodynamic changes in the frontal areas only, whereas 13 studies included the sensory cortex. Only one study in preterm neonates included the occipital cortex, which is not typically activated in response to pain, to test the specificity of the response and showed that pain-related cortical activity is specific to somatosensory cortex (Bartocci et al., 2006). Moreover, since autonomic nervous system arousal during nociception may cause significant systemic changes in the hemodynamic activity contributing to the fNIRS signal changes, it is essential to include physiological measures, such as heart rate and blood pressure, and examine their relation with fNIRS parameters. Only 2 out of 14 studies in adults reported systemic physiological parameters (Gelinas et al., 2010; Re et al., 2013). Six out of eight infant studies reported heart rate (HR) and arterial oxygen saturation (SaO_2) (Bartocci et al., 2006; Slater et al., 2006, 2008; Ranger et al., 2013; Hwang and Seol, 2015; Olsson et al., 2015), with one study reporting the mean arterial pressure as well (Ranger et al., 2013). None of the studies assessed the association between the physiological parameters and the fNIRS measurements.

Studies with experimental models of pain in healthy adults employed a variety of noxious stimuli, including hot plates, pressure, electrical stimulus, and cold water. The cold water stimulus is quite different as it evokes a significant autonomic response in addition to any cortical activity due to nociception. All of the eight studies using experimental models of pain reported bilateral HbO_2 increase in frontal and/or somatosensory areas (Barati et al., 2013; Becerra et al., 2008, 2009; Lee et al., 2013; Re et al., 2013; Yennu et al., 2013; Yucel et al., 2015) from which five studies reported bilateral decrease in Hb (Barati et al., 2013; Lee et al., 2013; Re et al., 2013). Seven studies included innocuous sensory stimuli as negative control experiments to test the specificity of the noxious stimulus. Studies were consistent in recruiting right-handed subjects and mostly delivered the stimulus to the right hand (Becerra et al., 2008, 2009; Lee et al., 2013; Re et al., 2013; Yennu et al., 2013). However, it is still impractical to quantitatively compare the results as they used a variety of commercial and homemade fNIRS devices as well as different pain modalities.

Six studies reported the hemodynamic response to pain induced as a result of a clinical intervention or in a clinical population in human adults. Three studies were conducted during migraine attacks (Viola et al., 2010; Watanabe et al., 2011; Pourshoghi et al., 2015) and 1 study during cardiac surgeries (Gelinas et al., 2010). One of the migraine studies (Watanabe et al., 2011) reported that pain relief after a migraine attack and secondary to injection of sumatriptan was consistent with a decrease in HbO_2 as a measure of intracerebral blood flow, whereas injection of saline in control subjects did not cause any change. The other migraine study (Viola et al., 2010) reported that during prolonged migraine with aura, cerebral tissue oxygen saturation ($SctO_2$) increased

TABLE 2.1

Summary of Pain Studies Using NIRS

Authors	Population	Noxious Stimulus/Site Applied	NIRS Parameter Change/ Activated Site(s)	Device	Negative Control Experiment	Systemic Physiological Measure
Experimental						
Becerra et al. (2008)	Healthy volunteers (n = 9)/ right-handed	Thermal (46°C)/right hand	Bilateral somatosensory and frontal/HbO$_2$ ↑	TechEN CW5	Yes (brush stimulation)	Not reported
Becerra et al. (2009)	Healthy volunteers (n = 6)/ right-handed	Thermal (mean: 45.3°C)/face	Bilateral somatosensory/ HbO$_2$ ↑	TechEN CW5	Yes (innocuous heat 41.4°C)	No
Lee et al. (2013)	Healthy volunteers (n = 7)/ right handed	Pressure/Right index finger	Prefrontal/HbO$_2$ ↑ and Hb ↓	ETG-4000	Yes (itch stimulation)	No
Yennu et al. (2013)	Healthy volunteers (n = 9)/ right-handed	Thermal (mean: 46.1°C)/right temporo-mandibular joint	Bilateral prefrontal/HbO$_2$ ↑	Cephalogics	Yes (innocuous heat 43°C)	No
Re et al. (2013)	Healthy volunteers (n = 9)	Electrical/right forearm	Bilateral sensorimotor and prefrontal/HbO$_2$ ↑ and Hb ↓	Homemade	Yes (innocuous electrical stimulation)	Yes (HR, respiration rate, skin conductance)
Barati et al. (2013)	Healthy volunteers (n = 20)/right-handed	Cold water (0°C)/right hand	Bilateral frontal/HbO$_2$ ↑ and Hb ↓	Homemade	No	No
Barati (2013)	Healthy volunteers (n = 21)/right-handed	Cold water (1°C, 5°C, 10°C, 15°C)/right hand	Bilateral frontal/HbO$_2$ ↑ and Hb ↓	Homemade	Yes (innocuous cold water 15°C)	Not reported
Yucel et a. (2015)	Healthy volunteers (n = 11)/right-handed	Electrical/left thumb	Bilateral somatosensory (HbO$_2$ ↑, Hb ↓) and frontal gyrus (HbO$_2$ ↓, Hb ↓)	TechEN CW6	Yes (innocuous electrical stimulation)	No

(Continued)

TABLE 2.1 (*Continued*)
Summary of Pain Studies Using NIRS

Authors	Population	Noxious Stimulus/Site Applied	NIRS Parameter Change/ Activated Site(s)	Device	Negative Control Experiment	Systemic Physiological Measure
Clinical						
Gelinas et al. (2010)	Adults undergoing cardiac surgery (n = 40)	Intravenous/arterial line insertions (awake)/Sternal bone incision/thorax opening (anesthetized)	Bilateral frontal/rSO$_2$ ↑	INVOS-4100	Yes (skin disinfection)	Yes (MAP, HR, expired oxygen fraction, end-tidal CO$_2$)
Viola et al. (2010)	Migraineurs (n = 8)	Migraine attack	Ipsilateral frontal, parietal, temporal, and occipital/ SctO$_2$ ↑	Homemade	No	No
Watanabe et al. (2011)	Migraineurs (n = 4)	Migraine attack	Bilateral temporo-parietal/ HbO$_2$ ↓ (post migraine attack)	ETG-100	Yes (saline injection)	No
Pourshoghi et al. (2015)	Migraineurs (n = 41)	Migraine attack	Frontal/HbO$_2$ ↓ (after pain relief with medication)	Homemade	No	No
Uceyler et al. (2015)	Fibromyalgia patients (n = 25)/right-handed	Pressure/finger extensor muscle of the right hand	Bilateral somatosensory and frontal/HbO$_2$ ↑	ETG-4000	Yes (subthreshold pressure stimulation)	No
Racek et al. (2015)	Hypersensitive teeth (n = 21)	Cold plate/a hypersensitive tooth	Contralateral somatosensory HbO$_2$↑ and bilateral prefrontal HbO$_2$ ↓	TechEN-CW6	Yes (percussion on the same tooth)	No

(*Continued*)

TABLE 2.1 (Continued)
Summary of Pain Studies Using NIRS

Authors	Population	Noxious Stimulus/Site Applied	NIRS Parameter Change/Activated Site(s)	Device	Negative Control Experiment	Systemic Physiological Measure
Infants						
Bartocci et al. (2006)	Preterm neonates (n = 40)	Venipuncture/left or right hand	Bilateral somatosensory/HbO$_2$ ↑	NIRO-300	Yes (occipital recording, skin disinfection)	Yes (HR, SaO$_2$)
Slater et al. (2006)	Term and preterm neonates (n = 18)	Heel lance	Contralateral somatosensory/HbT ↑	NIRO-200	Yes (von Frey hairs test)	No
Slater et al. (2008)	Term and preterm infants (n = 12)	Heel lance	Contralateral somatosensory/HbT ↑	NIRO-200	No	Yes (HR, SaO$_2$)
Ozawa et al. (2011)	Term and preterm infants (n = 80)	Venipuncture/left or right hand	Bilateral prefrontal/HbO$_2$ ↑	NIRO-200	No	Yes (HR, SaO$_2$)
Ranger et al. (2013)	Critically ill infants (less than 12 months) (n = 20)	Chest drain removal	Right primary somatosensory, fronto-temporal, temporo-parietal/Hb ↑	NIRO-300	No	Yes (HR, SaO$_2$, MAP)
Hwang and Seol (2015)	Preterm infants with respiratory distress syndrome (n = 24)	Heel lance	Frontal/rScO$_2$ ↓ (regional cerebral oxygen saturation)	INVOS 5100	No	Yes (HR and SaO$_2$)
Bembich et al. (2015)	Healthy full-term infants (n = 25)	Heel lance	Bilateral somatosensory and motor/HbO$_2$ ↑	ETG-100	No	No
Olsson et al. (2015)	Preterm infants (n = 10)	Venipuncture	Contralateral somatosensory/HbO$_2$ ↑	NIRO 200NX	Yes (sham stimulus)	Yes (HR, SaO$_2$)

ipsilateral to the headache side. Gelinas et al. studied the adults' response to painful procedures performed for a cardiac surgery in two different periods: (1) awake period during intravenous and arterial line insertions and (2) anesthetized period during the sternal bone incision and thorax opening (Gelinas et al., 2010). This study that benefited from a relatively large sample size (n = 40) found that during painful procedures, regional cerebral oxygenation (rSO_2) significantly increased in bilateral frontal cortex, while no significant activity was seen during tactile stimulus (skin disinfection).

Infants' studies were the first to propose the use of fNIRS for pain assessment in humans (Bartocci et al., 2006; Slater et al., 2006). Six out of the eight infant studies to date utilized the same commercial NIRS system (NIRO by Hamamatsu). Two studies with term and preterm infants during heel lance consistently showed contralateral activation in the somatosensory cortex (Slater et al., 2006, 2008) and one study found bilateral somatosensory and motor activation (Bembich et al., 2015). Two studies during venipuncture found bilateral activation in the somatosensory (Bartocci et al., 2006) or prefrontal cortices (Ozawa et al., 2011) and one found contralateral somatosensory activation (Olsson et al., 2015). One study with critically ill babies during chest-drain removal after cardiac surgery found a significant increase in Hb in the right primary somatosensory or fronto-temporal/temporo-parietal areas (Ranger et al., 2013).

Study of correlation between cerebral activation in response to pain and behavioral measures or self-reports has recently received special attention. Two studies using experimental models of pain in healthy adults examined the correlation between subjects' self-reports and fNIRS parameters. Lee et al. reported that as the intensity of the noxious pressure stimuli increases, the HbO_2 in the frontal cortex increases as well, consistent with an increase in the perceived pain (Lee et al., 2013). They also observed that in response to repeated constant stimuli, subjects report decaying perceived pain consistent with a decrease in HbO_2. Gelinas et al. did not find any association between activation in the frontal cortex, pain behaviors, and subjective pain scores during painful procedures in awake patients. This may be partly explained by low variability of the measures due to premedication in the majority of patients with morphine (Gelinas et al., 2010).

Four studies with infants also assessed the linear relationship between pain behaviors and fNIRS measures. Slater et al. studies the association between the premature infant pain profile (PIPP) scores and cortical activity in the somatosensory cortex during heel lance in 12 infants (aged 25–43 weeks postmenstrual) (Slater et al., 2008). They found a moderate correlation between the PIPP score and the level of cortical activity in the contralateral somatosensory cortex, with the facial expression component of PIPP having a larger correlation with cortical activation and the physiological component (heart rate and oxygen saturation) having a weaker correlation. In 13/33 test occasions (8 infants) no change in facial expression was observed. Despite this observation, a cortical response was observed in 10/13 occasions. Ozawa et al. studied the effect of previous exposure to a painful procedure (venipuncture) on the correlation between prefrontal cortical pain response and PIPP scores in 80 newborns (aged 37–42 weeks of gestational age; 50 full-term, 30 premature) (Ozawa et al., 2011). For full-term infants with no experience of painful procedure, bilateral change in HbO_2 in the prefrontal cortex was significantly correlated to facial expression score on the PIPP and the total PIPP score. Full-term infants with prior experience of painful procedure showed no correlation between HbO_2 change and physiological, facial expression, or total PIPP scores. For preterm infants with experience of painful procedure, they found moderate correlation between HbO_2 change in both sides of prefrontal cortex and physiological score of PIPP and between HbO_2 change in the left prefrontal area and total PIPP score, but not with the facial expression score. Ranger et al. did not find any association between cerebral Hb changes, physiological measures, and behavioral pain scores (Face Leg Activity Cry Consolability; FLACC) during chest drain removal in sick infants (Ranger et al., 2013). Finally, Bembich et al. found no correlation between the Prechtl's behavior classification or the Neonatal Infant Pain Scale (NIPS) score with an observed increase in HbO_2 in the contralateral somatosensory cortex (Bembich et al., 2015).

HEMODYNAMIC RESPONSE TO COLD WATER STIMULUS

EXPERIMENT 1: REPEATED COLD PRESSOR TESTS

In the first set of experiments, 20 healthy, right-handed individuals (10 females, 10 males) performed three consecutive trials of the cold pressor test (CPT) at 0°C; each lasted for 45 s, followed by a 2 min hand immersion in tepid water (~23°C). Hemodynamic parameters Hb and HbO$_2$ were simultaneously recorded on the forehead by two fNIRS sensors; each consisted of three optodes: two optodes distances of 2.8 cm (far channels) and one optode distance of 1 cm (near channel). The averaged hemodynamic response on the right forehead across 20 subjects is shown in Figure 2.2a.

A consistent bilateral activation was observed in far and near channels. As expected, the trend of change in far and near channels were very similar since CPT-induced hemodynamic response is massively regulated by the sympathetic nervous system. We found laterality of response in far channels (larger changes on right side) but not in near channels. A pattern of adaptation to repeated cold stimuli was observed in all channels as well as in the reported pain scores on the numeric rating scale from 0 to 10 (NIRS-11), where 0 means no pain and 10 means the worst pain imaginable (Figure 2.2b).

EXPERIMENT 2: TOLERANCE/THRESHOLD TEST WITH COLD PRESSOR TESTS
AT VARYING TEMPERATURES

In the second set of experiments, 21 healthy, right-handed subjects (11 females, 10 males) were recruited to perform four CPTs at 1°C, 5°C, 10°C, and 15°C. Each experiment started with a 2 min immersion of the right hand in tepid water, followed by immersion in cold water (1°C, 5°C, 10°C, or 15°C) for as long as the subject could tolerate the pain/unpleasantness of the stimulus, but not longer than 5 min, and ended with a 2 min immersion in tepid water (~23°C) for recovery. The same fNIRS sensors and placement as in experiment 1 were employed here. Eight out of 21 subjects could complete the CPTs for 5 min for all the four temperatures. The trajectories of the hemodynamic response (Hb and HbO$_2$) averaged for two far channels on the right forehead and the pain scores reported every 15 s, which averaged across these eight subjects, are shown in Figure 2.3.

As expected, while the water temperature decreased, the reported pain score increased, and the pain threshold and tolerance decreased. The reported pain scores slightly decreased through the CPT at higher temperatures of water, whereas they mainly remained high at lower temperatures (Figure 2.3).

A consistent bilateral hemodynamic activation was observed in far and near channels. A pattern of adaptation of the hemodynamic response to the cold water stimulus was seen in all channels for all temperatures, that is, Hb/HbO$_2$ sharply decreased/increased upon hand immersion in cold water and after peaking at around 1 min, they slowly increased/decreased to return to their pre-stimulus values. Statistical analyses for 21 subjects revealed asymmetrical activity in far channels (larger changes on right side) but not in near channels. While no gender differences were found in subjective pain measures (perhaps due to small sample size), sex specific response was observed in far channels (but not in near channels). Interestingly, males and females were found to be significantly different in HbO$_2$ response in the far channel on right forehead.

ASSOCIATION BETWEEN THE fNIRS SIGNAL AND PAIN SELF-REPORTS

In experiment 1, we found a pattern of adaptation to repeated cold water (0°C) stimuli in fNIRS parameters as well as the subjects' self-reports (Barati et al., 2013). A similar observation was made by Lee et al. (2013). We also found a significant correlation between the reported pain scores and the maximum change in total hemoglobin (HbT). In the experiment 2, we observed that as the temperature of cold water decreased from 15°C to 1°C, the reported pain increased and so did Hb and HbO$_2$. We also found a significant correlation between the pain tolerance (the maximum duration that the subject could hold his/her hand in cold water) and Hb at 1°C (but not at higher temperatures).

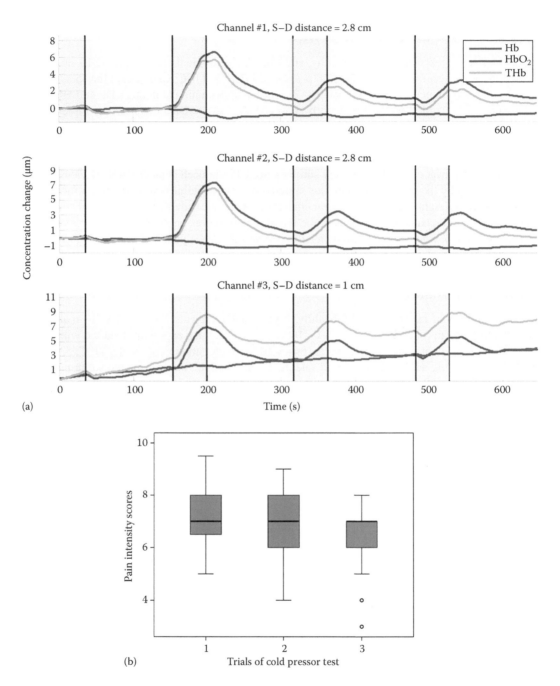

FIGURE 2.2   (a) The hemodynamic response to three trials of the cold pressor test (CPT) averaged across 20 subjects on the right sides of the forehead. The black vertical lines represent time points at which subjects switched their hands from the tepid water to the ice water and vice versa. The pink shaded field represents the baseline period, the orange shaded fields represent the immersion in tepid water condition, and the blue shaded fields represent the CPT conditions. Hb, HbO_2, and THb are abbreviations for deoxyhemoglobin, oxy-hemoglobin, and total hemoglobin, respectively. "S–D" symbol stands for "source–detector." (b) Boxplots of post-stimulus pain rating scores of 20 subjects across three trials of cold pressor test (CPT). Two outliers were identified for pain scores reported after the third trial of CPT and are represented as circles.

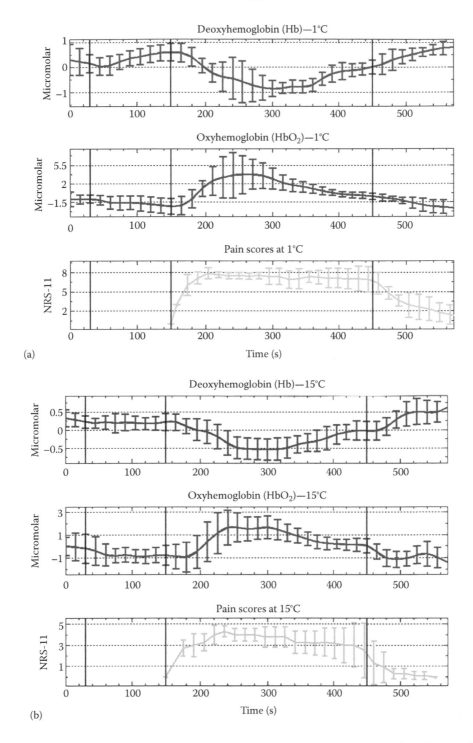

FIGURE 2.3 The hemodynamic response (Hb and HbO$_2$) averaged for two far channels on the right forehead and the numeric rating scales (NRS-11) reported every 15 s for the cold pressor test at 1°C (a) and 15°C (b). The trajectories are averaged across 8 subjects that could complete the cold pressor tests for the entire 5 min for all four temperatures (1°C, 5°C, 10°C, and 15°C). The top panels (blue) represent deoxyhemoglobin (Hb), the middle panels (red) represent oxyhemoglobin (HbO$_2$), and the bottom panels (green) represent pain scores on the numeric rating scale (NRS-11). The black vertical lines from left to right represent hand immersion into tepid water, hand immersion into cold water, and hand immersion back into tepid water.

CONCLUSION

This chapter provided a brief overview of the most accepted neuroimaging modalities for the study of brain response to noxious procedures, and in more details summarized the studies that have employed fNIRS—an optical imaging modality for noninvasive, continuous monitoring of tissue oxygenation and regional blood flow—for the assessment of some aspects of human response to pain. Advanced imaging modalities such as PET and fMRI have revealed brain regions that become activated during a physical or psychological experience of pain and have advanced our understanding of the underlying mechanisms in nociception. However, clinical practice has not accepted them for routine examinations for several reasons, including cost, availability, and enclosed environment. fNIRS application for the assessment of pain is recent, but literature shows a fast growing interest in such a novel solution. We believe that in near future fNIRS can serve as an affordable and portable alternative to advanced neuroimaging modalities for bedside assessment of pain.

Neuroimaging techniques have the potential to introduce new approaches to pain management in the clinic. For example, fMRI signal has been successfully used as a biofeedback to manage the activation of brain areas involved in pain. A simpler and more clinically friendly brain monitoring technique such as fNIRS, EEG, or their combination, can be used to provide biofeedback in a home/clinical setting. Another potential application of fNIRS could be the assessment of preoperative sensitivity to acute pain such as cold water stimulus. Studies have shown that pain sensitivity to cold stimuli is a risk factor for developing post-operative pain (Bisgaard et al., 2001). fNIRS may help in effective and objective examination of pre- and post-surgery sensitivity to pain for improved management of post-operative pain.

REFERENCES

Arthurs, O. J. and Boniface, S. J. 2003. What aspect of the fMRI BOLD signal best reflects the underlying electrophysiology in human somatosensory cortex? *Clin Neurophysiol* 114(7):1203–1209.

Barati, Z. 2013. Monitoring and analysis of hemodynamic response to cold noxious stimuli using functional near infrared spectroscopy. Ph.D. Dissertation, Drexel University, pp.141. Publication Number 3591014.

Barati, Z., Shewokis, P. A., Izzetoglu, M., Polikar, R., Mychaskiw, G., and Pourrezaei, K. 2013. Hemodynamic response to repeated noxious cold pressor tests measured by functional near infrared spectroscopy on forehead. *Ann Biomed Eng* 41(2):223–237. doi: 10.1007/s10439-012-0642-0.

Bartocci, M., Bergqvist, L. L., Lagercrantz, H., and Anand, K. J. 2006. Pain activates cortical areas in the preterm newborn brain. *Pain* 122(1–2):109–117. doi: 10.1016/j.pain.2006.01.015.

Becerra, L., Harris, W., Grant, M., George, E., Boas, D., and Borsook, D. 2009. Diffuse optical tomography activation in the somatosensory cortex: Specific activation by painful vs. non-painful thermal stimuli. *PLoS One* 4(11):e8016. doi: 10.1371/journal.pone.0008016.

Becerra, L., Harris, W., Joseph, D., Huppert, T., Boas, D. A., and Borsook, D. 2008. Diffuse optical tomography of pain and tactile stimulation: Activation in cortical sensory and emotional systems. *Neuroimage* 41(2):252–259. doi: 10.1016/j.neuroimage.2008.01.047.

Bembich, S., Brovedani, P., Cont, G., Travan, L., Grassi, V., and Demarini, S. 2015. Pain activates a defined area of the somatosensory and motor cortex in newborn infants. *Acta Paediatr* 104(11):e530–e533. doi: 10.1111/apa.13122.

Bisgaard, T., Klarskov, B., Rosenberg, J., and Kehlet, H. 2001. Characteristics and prediction of early pain after laparoscopic cholecystectomy. *Pain* 90(3):261–269.

Blount, R. L. and Loiselle, K. A. 2009. Behavioural assessment of pediatric pain. *Pain Res Manage* 14(1):47–52.

Boas, D. A., Elwell, C. E., Ferrari, M., and Taga, G. 2014. Twenty years of functional near-infrared spectroscopy: Introduction for the special issue. *Neuroimage* 85(Pt 1):1–5. doi: 10.1016/j.neuroimage.2013.11.033.

Borsook, D. and Becerra, L. R. 2006. Breaking down the barriers: fMRI applications in pain, analgesia and analgesics. *Mol Pain* 2:30. doi: 10.1186/1744-8069-2-30.

Borsook, D., Moulton, E. A., Schmidt, K. F., and Becerra, L. R. 2007. Neuroimaging revolutionizes therapeutic approaches to chronic pain. *Mol Pain* 3:25. doi: 10.1186/1744-8069-3-25.

Borsook, D., Sava, S., and Becerra, L. 2010. The pain imaging revolution: Advancing pain into the 21st century. *Neuroscientist* 16(2):171–185. doi: 10.1177/1073858409349902.

Branco, G. 2007. *The Development and Evaluation of Head Probes for Optical Imaging of the Infant Head.* Department of Medical Physics & Bioengineering, University College London (UCL), London, U.K.

Brodersen, K. H., Wiech, K., Lomakina, E. I., Lin, C. S., Buhmann, J. M., Bingel, U., Ploner, M., Stephan, K. E., and Tracey, I. 2012. Decoding the perception of pain from fMRI using multivariate pattern analysis. *Neuroimage* 63(3):1162–1170. doi: 10.1016/j.neuroimage.2012.08.035.

Brown, J. E., Chatterjee, N., Younger, J., and Mackey, S. 2011. Towards a physiology-based measure of pain: Patterns of human brain activity distinguish painful from non-painful thermal stimulation. *PLoS One* 6(9):e24124. doi: 10.1371/journal.pone.0024124.

Bushnell, M. C., Duncan, G. H., Hofbauer, R. K., Ha, B., Chen, J. I., and Carrier, B. 1999. Pain perception: Is there a role for primary somatosensory cortex? *Proc Natl Acad Sci USA* 96(14):7705–7709.

Casey, K. L., Minoshima, S., Morrow, T. J., and Koeppe, R. A. 1996. Comparison of human cerebral activation pattern during cutaneous warmth, heat pain, and deep cold pain. *J Neurophysiol* 76(1):571–581.

Champion, G. D., Goodenough, B., Von Baeyer, C. L., and Thomas, W. 1998. Measurement of pain by self-report. *Prog Pain Res Manage* 10:123–160.

Chapin, H., Bagarinao, E., and Mackey, S. 2012. Real-time fMRI applied to pain management. *Neurosci Lett* 520(2):174–181. doi: 10.1016/j.neulet.2012.02.076.

Chen, A. C. 2001. New perspectives in EEG/MEG brain mapping and PET/fMRI neuroimaging of human pain. *Int J Psychophysiol* 42(2):147–159.

Chen, J. I., Ha, B., Bushnell, M. C., Pike, B., and Duncan, G. H. 2002. Differentiating noxious- and innocuous-related activation of human somatosensory cortices using temporal analysis of fMRI. *J Neurophysiol* 88(1):464–474.

Coghill, R. C., Sang, C. N., Maisog, J. M., and Iadarola, M. J. 1999. Pain intensity processing within the human brain: A bilateral, distributed mechanism. *J Neurophysiol* 82(4):1934–1943.

Cope, M., Delpy, D. T., Reynolds, E. O., Wray, S., Wyatt, J., and van der Zee, P. 1988. Methods of quantitating cerebral near infrared spectroscopy data. *Adv Exp Med Biol* 222:183–189.

deCharms, R. C., Maeda, F., Glover, G. H., Ludlow, D., Pauly, J. M., Soneji, D., Gabrieli, J. D., and Mackey, S. C. 2005. Control over brain activation and pain learned by using real-time functional MRI. *Proc Natl Acad Sci USA* 102(51):18626–18631. doi: 10.1073/pnas.0505210102.

Dichgans, M., Herzog, J., Freilinger, T., Wilke, M., and Auer, D. P. 2005. 1H-MRS alterations in the cerebellum of patients with familial hemiplegic migraine type 1. *Neurology* 64(4):608–613. doi: 10.1212/01.wnl.0000151855.98318.50.

Disbrow, E., Buonocore, M., Antognini, J., Carstens, E., and Rowley, H. A. 1998. Somatosensory cortex: A comparison of the response to noxious thermal, mechanical, and electrical stimuli using functional magnetic resonance imaging. *Hum Brain Mapp* 6(3):150–159.

Duschek, S., Hellmann, N., Merzoug, K., Reyes del Paso, G. A., and Werner, N. S. 2012. Cerebral blood flow dynamics during pain processing investigated by functional transcranial Doppler sonography. *Pain Med* 13(3):419–426. doi: 10.1111/j.1526-4637.2012.01329.x.

Ehlis, A. C., Schneider, S., Dresler, T., and Fallgatter, A. J. 2014. Application of functional near-infrared spectroscopy in psychiatry. *Neuroimage* 85(Pt 1):478–488. doi: 10.1016/j.neuroimage.2013.03.067.

Fulbright, R. K., Troche, C. J., Skudlarski, P., Gore, J. C., and Wexler, B. E. 2001. Functional MR imaging of regional brain activation associated with the affective experience of pain. *AJR Am J Roentgenol* 177(5):1205–1210. doi: 10.2214/ajr.177.5.1771205.

Gelinas, C., Choiniere, M., Ranger, M., Denault, A., Deschamps, A., and Johnston, C. 2010. Toward a new approach for the detection of pain in adult patients undergoing cardiac surgery: Near-infrared spectroscopy—A pilot study. *Heart and Lung* 39(6):485–493. doi: 10.1016/j.hrtlng.2009.10.018.

Hock, C., Villringer, K., Muller-Spahn, F., Hofmann, M., Schuh-Hofer, S., Heekeren, H., Wenzel, R., Dirnagl, U., and Villringer, A. 1996. Near infrared spectroscopy in the diagnosis of Alzheimer's disease. *Ann N Y Acad Sci* 777:22–29.

Hock, C., Villringer, K., Muller-Spahn, F., Wenzel, R., Heekeren, H., Schuh-Hofer, S., Hofmann, M. et al. 1997. Decrease in parietal cerebral hemoglobin oxygenation during performance of a verbal fluency task in patients with Alzheimer's disease monitored by means of near-infrared spectroscopy (NIRS)—Correlation with simultaneous rCBF-PET measurements. *Brain Res* 755(2):293–303.

Huppert, T. J., Hoge, R. D., Diamond, S. G., Franceschini, M. A., and Boas, D. A. 2006. A temporal comparison of BOLD, ASL, and NIRS hemodynamic responses to motor stimuli in adult humans. *Neuroimage* 29(2):368–382. doi: 10.1016/j.neuroimage.2005.08.065.

Hwang, M. J. and Seol, G. H. 2015. Cerebral oxygenation and pain of heel blood sampling using manual and automatic lancets in premature infants. *J Perinat Neonatal Nurs* 29(4):356–362. doi: 10.1097/jpn.0000000000000138.

Jovicich, J., Czanner, S., Han, X., Salat, D., van der Kouwe, A., Quinn, B., Pacheco, J. et al. 2009. MRI-derived measurements of human subcortical, ventricular and intracranial brain volumes: Reliability effects of scan sessions, acquisition sequences, data analyses, scanner upgrade, scanner vendors and field strengths. *Neuroimage* 46(1):177–192. doi: 10.1016/j.neuroimage.2009.02.010.

Kleinschmidt, A., Obrig, H., Requardt, M., Merboldt, K. D., Dirnagl, U., Villringer, A., and Frahm, J. 1996. Simultaneous recording of cerebral blood oxygenation changes during human brain activation by magnetic resonance imaging and near-infrared spectroscopy. *J Cereb Blood Flow Metab* 16(5):817–826. doi: 10.1097/00004647-199609000-00006.

Kong, J., Gollub, R. L., Polich, G., Kirsch, I., Laviolette, P., Vangel, M., Rosen, B., and Kaptchuk, T. J. 2008. A functional magnetic resonance imaging study on the neural mechanisms of hyperalgesic nocebo effect. *J Neurosci* 28(49):13354–13362. doi: 10.1523/jneurosci.2944-08.2008.

Koyama, T., McHaffie, J. G., Laurienti, P. J., and Coghill, R. C. 2005. The subjective experience of pain: Where expectations become reality. *Proc Natl Acad Sci USA* 102(36):12950–12955. doi: 10.1073/pnas.0408576102.

Kringelbach, M. L., Jenkinson, N., Green, A. L., Owen, S. L., Hansen, P. C., Cornelissen, P. L., Holliday, I. E., Stein, J., and Aziz, T. Z. 2007. Deep brain stimulation for chronic pain investigated with magnetoencephalography. *Neuroreport* 18(3):223–228. doi: 10.1097/WNR.0b013e328010dc3d.

Lee, C. H., Sugiyama, T., Kataoka, A., Kudo, A., Fujino, F., Chen, Y. W., Mitsuyama, Y., Nomura, S., and Yoshioka, T. 2013. Analysis for distinctive activation patterns of pain and itchy in the human brain cortex measured using near infrared spectroscopy (NIRS). *PLoS One* 8(10):e75360. doi: 10.1371/journal.pone.0075360.

Lee, M. C. and Tracey, I. 2013. Imaging pain: A potent means for investigating pain mechanisms in patients. *Br J Anaesth* 111(1):64–72. doi: 10.1093/bja/aet174.

León-Carrión, J. and León-Domínguez, U. 2012. Functional near-infrared spectroscopy (fNIRS): Principles and neuroscientific applications. In: Bright, P., ed. *Neuroimaging—Methods*. InTech, Rijeka, Croatia. doi: 10.5772/23146.

Liu, C. C., Ohara, S., Franaszczuk, P. J., and Lenz, F. A. 2011a. Attention to painful cutaneous laser stimuli evokes directed functional connectivity between activity recorded directly from human pain-related cortical structures. *Pain* 152(3):664–675. doi: 10.1016/j.pain.2010.12.016.

Liu, C. C., Ohara, S., Franaszczuk, P. J., Crone, N. E., and Lenz, F. A. 2011b. Attention to painful cutaneous laser stimuli evokes directed functional interactions between human sensory and modulatory pain-related cortical areas. *Pain* 152(12):2781–2791. doi: 10.1016/j.pain.2011.09.002.

Logothetis, N. K., Pauls, J., Augath, M., Trinath, T., and Oeltermann, A. 2001. Neurophysiological investigation of the basis of the fMRI signal. *Nature* 412(6843):150–157. doi: 10.1038/35084005.

Lorenz, J., Minoshima, S., and Casey, K. L. 2003. Keeping pain out of mind: The role of the dorsolateral prefrontal cortex in pain modulation. *Brain* 126(Pt 5):1079–1091.

Lutz, J., Jager, L., de Quervain, D., Krauseneck, T., Padberg, F., Wichnalek, M., Beyer, A. et al. 2008. White and gray matter abnormalities in the brain of patients with fibromyalgia: A diffusion-tensor and volumetric imaging study. *Arthritis Rheum* 58(12):3960–3969. doi: 10.1002/art.24070.

MacIntosh, B. J., Klassen, L. M., and Menon, R. S. 2003. Transient hemodynamics during a breath hold challenge in a two part functional imaging study with simultaneous near-infrared spectroscopy in adult humans. *Neuroimage* 20(2):1246–1252. doi: 10.1016/s1053-8119(03)00417-8.

Mao, C., Wei, L., Zhang, Q., Liao, X., Yang, X., and Zhang, M. 2013. Differences in brain structure in patients with distinct sites of chronic pain. *Neural Regen Res* 8(32):2981–2990.

Markman, T., Liu, C. C., Chien, J. H., Crone, N. E., Zhang, J., and Lenz, F. A. 2013. EEG analysis reveals widespread directed functional interactions related to a painful cutaneous laser stimulus. *J Neurophysiol* 110(10):2440–2449. doi: 10.1152/jn.00246.2013.

Marumo, K., Takizawa, R., Kinou, M., Kawasaki, S., Kawakubo, Y., Fukuda, M., and Kasai, K. 2014. Functional abnormalities in the left ventrolateral prefrontal cortex during a semantic fluency task, and their association with thought disorder in patients with schizophrenia. *Neuroimage* 85 Pt 1:518–526. doi: 10.1016/j.neuroimage.2013.04.050.

Matsubara, T., Matsuo, K., Nakashima, M., Nakano, M., Harada, K., Watanuki, T., Egashira, K., and Watanabe, Y. 2014. Prefrontal activation in response to emotional words in patients with bipolar disorder and major depressive disorder. *Neuroimage* 85(Pt 1):489–497. doi: 10.1016/j.neuroimage.2013.04.098.

Murata, Y., Katayama, Y., Oshima, H., Kawamata, T., Yamamoto, T., Sakatani, K., and Suzuki, S. 2000. Changes in cerebral blood oxygenation induced by deep brain stimulation: Study by near-infrared spectroscopy (NIRS). *Keio J Med* 49(Suppl 1):A61–A63.

Obrig, H. 2014. NIRS in clinical neurology—A 'promising' tool? *Neuroimage* 85 Pt 1:535–546. doi: 10.1016/j.neuroimage.2013.03.045.

Ochsner, K. N., Ludlow, D. H., Knierim, K., Hanelin, J., Ramachandran, T., Glover, G. C., and Mackey, S. C. 2006. Neural correlates of individual differences in pain-related fear and anxiety. *Pain* 120(1–2):69–77. doi: 10.1016/j.pain.2005.10.014.

Ohara, S., Crone, N. E., Weiss, N., Kim, J. H., and Lenz, F. A. 2008. Analysis of synchrony demonstrates that the presence of pain networks prior to a noxious stimulus can enable the perception of pain in response to that stimulus. *Exp Brain Res* 185(2):353–358. doi: 10.1007/s00221-008-1284-1.

Ohara, S., Crone, N. E., Weiss, N., and Lenz, F. A. 2006. Analysis of synchrony demonstrates 'pain networks' defined by rapidly switching, task-specific, functional connectivity between pain-related cortical structures. *Pain* 123(3):244–253. doi: 10.1016/j.pain.2006.02.012.

Ohara, S., Crone, N. E., Weiss, N., Vogel, H., Treede, R. D., and Lenz, F. A. 2004. Attention to pain is processed at multiple cortical sites in man. *Exp Brain Res* 156(4):513–517. doi: 10.1007/s00221-004-1885-2.

Olsson, E., Ahlsen, G., and Eriksson, M. 2015. Skin-to-skin contact reduces near-infrared spectroscopy pain responses in premature infants during blood sampling. *Acta Paediatr* 105:376–380. doi: 10.1111/apa.13180.

Ozawa, M., Kanda, K., Hirata, M., Kusakawa, I., and Suzuki, C. 2011. Influence of repeated painful procedures on prefrontal cortical pain responses in newborns. *Acta Paediatrica* 100(2):198–203. doi: 10.1111/j.1651-2227.2010.02022.x.

Pattany, P. M., Yezierski, R. P., Widerstrom-Noga, E. G., Bowen, B. C., Martinez-Arizala, A., Garcia, B. R., and Quencer, R. M. 2002. Proton magnetic resonance spectroscopy of the thalamus in patients with chronic neuropathic pain after spinal cord injury. *AJNR Am J Neuroradiol* 23(6):901–905.

Peyron, R., Garcia-Larrea, L., Gregoire, M. C., Costes, N., Convers, P., Lavenne, F., Mauguiere, F., Michel, D., and Laurent, B. 1999. Haemodynamic brain responses to acute pain in humans: Sensory and attentional networks. *Brain* 122(Pt 9):1765–1780.

Porro, C. A., Cettolo, V., Francescato, M. P., and Baraldi, P. 1998. Temporal and intensity coding of pain in human cortex. *J Neurophysiol* 80(6):3312–3320.

Porro, C. A., Cettolo, V., Francescato, M. P., and Baraldi, P. 2003. Functional activity mapping of the mesial hemispheric wall during anticipation of pain. *Neuroimage* 19(4):1738–1747.

Pourshoghi, A., Danesh, A., Tabby, D. S., Grothusen, J., and Pourrezaei, K. 2015. Cerebral reactivity in migraine patients measured with functional near-infrared spectroscopy. *Eur J Med Res* 20:96. doi: 10.1186/s40001-015-0190-9.

Prescot, A., Becerra, L., Pendse, G., Tully, S., Jensen, E., Hargreaves, R., Renshaw, P., Burstein, R., and Borsook, D. 2009. Excitatory neurotransmitters in brain regions in interictal migraine patients. *Mol Pain* 5:34. doi: 10.1186/1744-8069-5-34.

Racek, A. J., Hu, X., Nascimento, T. D., Bender, M. C., Khatib, L., Chiego, Jr., D., Holland, G. R. et al. 2015. Different brain responses to pain and its expectation in the dental chair. *J Dent Res* 94(7):998–1003. doi: 10.1177/0022034515581642.

Rainville, P., Duncan, G. H., Price, D. D., Carrier, B., and Bushnell, M. C. 1997. Pain affect encoded in human anterior cingulate but not somatosensory cortex. *Science* 277(5328):968–971.

Ranger, M., Celeste Johnston, C., Rennick, J. E., Limperopoulos, C., Heldt, T., and du Plessis, A. J. 2013. A multidimensional approach to pain assessment in critically ill infants during a painful procedure. *Clin J Pain* 29(7):613–620. doi: 10.1097/AJP.0b013e31826dfb13.

Ranger, M. and Gélinas, C. 2012. Innovating in pain assessment of the critically ill: Exploring cerebral near-infrared spectroscopy as a bedside approach. *Pain Manage Nursing* 15(2):519–529. doi: http://dx.doi.org/10.1016/j.pmn.2012.03.005.

Re, R., Muthalib, M., Zucchelli, L., Perrey, S., Contini, D., Caffini, M., Spinelli, L., Kerr, G., and Torricelli, A. 2013. Multichannel time domain fNIRS mapping of cortical activation and superficial systemic responses during neuromuscular electrical stimulation. Paper read at In: Pavone, F., Hillman, E., Daria, V., and Charpak, S., eds. *SPIE Proceedings (Optical Society of America, 2013)*, Neurophotonics, Munich, Germany, vol. 8804, paper 880404.

Rolfe, P. 2000. In vivo near-infrared spectroscopy. *Annu Rev Biomed Eng* 2:715–754.

Sava, S., Lebel, A. A., Leslie, D. S., Drosos, A., Berde, C., Becerra, L., and Borsook, D. 2009. Challenges of functional imaging research of pain in children. *Mol Pain* 5:30. doi: 10.1186/1744-8069-5-30.

Scholkmann, F. and Wolf, M. 2013. General equation for the differential pathlength factor of the frontal human head depending on wavelength and age. *J Biomed Opt* 18(10):105004. doi: 10.1117/1.jbo.18.10.105004.

Scott, D. J., Heitzeg, M. M., Koeppe, R. A., Stohler, C. S., and Zubieta, J. K. 2006. Variations in the human pain stress experience mediated by ventral and dorsal basal ganglia dopamine activity. *J Neurosci* 26(42):10789–10795. doi: 10.1523/jneurosci.2577-06.2006.

Seghier, M. L., Lazeyras, F., Vuilleumier, P., Schnider, A., and Carota, A. 2005. Functional magnetic resonance imaging and diffusion tensor imaging in a case of central poststroke pain. *J Pain* 6(3):208–212. doi: 10.1016/j.jpain.2004.11.004.

Seifert, F. and Maihofner, C. 2007. Representation of cold allodynia in the human brain—A functional MRI study. *Neuroimage* 35(3):1168–1180. doi: 10.1016/j.neuroimage.2007.01.021.

Siddall, P. J., Stanwell, P., Woodhouse, A., Somorjai, R. L., Dolenko, B., Nikulin, A., Bourne, R. et al. 2006. Magnetic resonance spectroscopy detects biochemical changes in the brain associated with chronic low back pain: A preliminary report. *Anesth Analg* 102(4):1164–1168. doi: 10.1213/01.ane.0000198333.22687.a6.

Slater, R., Boyd, S., Meek, J., and Fitzgerald, M. 2006. Cortical pain responses in the infant brain. *Pain* 123(3):332; author reply 332-4. doi: 10.1016/j.pain.2006.05.009.

Slater, R., Cantarella, A., Franck, L., Meek, J., and Fitzgerald, M. 2008. How well do clinical pain assessment tools reflect pain in infants? *PLoS Med* 5(6):e129. doi: 10.1371/journal.pmed.0050129.

Srouji, R., Ratnapalan, S., and Schneeweiss, S. 2010. Pain in children: Assessment and nonpharmacological management. *Int J Pediatr* 2010:11. doi: 10.1155/2010/474838.

Sweet, S. D. and McGrath, P. J. 1998. Physiological measures of pain. *Prog Pain Res Manage* 10:59–82.

Takizawa, R., Fukuda, M., Kawasaki, S., Kasai, K., Mimura, M., Pu, S., Noda, T., Niwa, S., and Okazaki, Y. 2014. Neuroimaging-aided differential diagnosis of the depressive state. *Neuroimage* 85(Pt 1):498–507. doi: 10.1016/j.neuroimage.2013.05.126.

Talbot, J. D., Marrett, S., Evans, A. C., Meyer, E., Bushnell, M. C., and Duncan, G. H. 1991. Multiple representations of pain in human cerebral cortex. *Science* 251(4999):1355–1358.

Tolle, T. R., Kaufmann, T., Siessmeier, T., Lautenbacher, S., Berthele, A., Munz, F., Zieglgansberger, W. et al. 1999. Region-specific encoding of sensory and affective components of pain in the human brain: A positron emission tomography correlation analysis. *Ann Neurol* 45(1):40–47.

Toronov, V., Walker, S., Gupta, R., Choi, J. H., Gratton, E., Hueber, D., and Webb, A. 2003. The roles of changes in deoxyhemoglobin concentration and regional cerebral blood volume in the fMRI BOLD signal. *Neuroimage* 19(4):1521–1531.

Toronov, V., Webb, A., Choi, J. H., Wolf, M., Michalos, A., Gratton, E., and Hueber, D. 2001. Investigation of human brain hemodynamics by simultaneous near-infrared spectroscopy and functional magnetic resonance imaging. *Med Phys* 28(4):521–527.

Tracey, I. 2010. Getting the pain you expect: Mechanisms of placebo, nocebo and reappraisal effects in humans. *Nat Med* 16(11):1277–1283. doi: 10.1038/nm.2229.

Tracey, I., Becerra, L., Chang, I., Breiter, H., Jenkins, L., Borsook, D., and Gonzalez, R. G. 2000. Noxious hot and cold stimulation produce common patterns of brain activation in humans: A functional magnetic resonance imaging study. *Neurosci Lett* 288(2):159–162.

Uceyler, N., Zeller, J., Kewenig, S., Kittel-Schneider, S., Fallgatter, A. J., and Sommer, C. 2015. Increased cortical activation upon painful stimulation in fibromyalgia syndrome. *BMC Neurol* 15:210. doi: 10.1186/s12883-015-0472-4.

Villringer, A. and Chance, B. 1997. Non-invasive optical spectroscopy and imaging of human brain function. *Trends Neurosci* 20(10):435–442.

Viola, S., Viola, P., Litterio, P., Buongarzone, M. P., and Fiorelli, L. 2010. Pathophysiology of migraine attack with prolonged aura revealed by transcranial Doppler and near infrared spectroscopy. *Neurol Sci* 31(Suppl 1):S165–S166. doi: 10.1007/s10072-010-0318-1.

Wager, T. D., Atlas, L. Y., Lindquist, M. A., Roy, M., Woo, C. W., and Kross, E. 2013. An fMRI-based neurologic signature of physical pain. *N Engl J Med* 368(15):1388–1397. doi: 10.1056/NEJMoa1204471.

Wager, T. D., Rilling, J. K., Smith, E. E., Sokolik, A., Casey, K. L., Davidson, R. J., Kosslyn, S. M., Rose, R. M., and Cohen, J. D. 2004. Placebo-induced changes in fMRI in the anticipation and experience of pain. *Science* 303(5661):1162–1167. doi: 10.1126/science.1093065.

Watanabe, Y., Tanaka, H., Dan, I., Sakurai, K., Kimoto, K., Takashima, R., and Hirata, K. 2011. Monitoring cortical hemodynamic changes after sumatriptan injection during migraine attack by near-infrared spectroscopy. *Neurosci Res* 69(1):60–66. doi: 10.1016/j.neures.2010.09.003.

Whitcher, B., Wisco, J. J., Hadjikhani, N., and Tuch, D. S. 2007. Statistical group comparison of diffusion tensors via multivariate hypothesis testing. *Magn Reson Med* 57(6):1065–1074. doi: 10.1002/mrm.21229.

Wiech, K., Kalisch, R., Weiskopf, N., Pleger, B., Stephan, K. E., and Dolan, R. J. 2006. Anterolateral prefrontal cortex mediates the analgesic effect of expected and perceived control over pain. *J Neurosci* 26(44):11501–11509. doi: 10.1523/jneurosci.2568-06.2006.

Willoch, F., Schindler, F., Wester, H. J., Empl, M., Straube, A., Schwaiger, M., Conrad, B., and Tolle, T. R. 2004. Central poststroke pain and reduced opioid receptor binding within pain processing circuitries: A [11C]diprenorphine PET study. *Pain* 108(3):213–220. doi: 10.1016/j.pain.2003.08.014.

Yennu, A., Tian, F., Liu, H., Rawat, R., Manry, M. T., and Gatchel, R. 2013. A preliminary investigation of human frontal cortex under noxious thermal stimulation over the temporomandibular joint using functional near infrared spectroscopy. *J Appl Biobehav Res* 18:134–155.

Yucel, M. A., Aasted, C. M., Petkov, M. P., Borsook, D., Boas, D. A., and Becerra, L. 2015. Specificity of hemodynamic brain responses to painful stimuli: A functional near-infrared spectroscopy study. *Sci Rep* 5:9469. doi: 10.1038/srep09469.

3 Application of Functional Near-Infrared Spectroscopy in Brain Mapping

Afrouz Azari-Anderson, Fatima A. Chowdhry,
Yasaman Ardeshirpour, Nader Shahni Karamzadeh,
Elizabeth G. Smith, Viktor Chernomordik,
and Amir H. Gandjbakhche

CONTENTS

INTRODUCTION

Recent advancements in neurophotonics and brain mapping techniques have opened new insights into brain function and structure. The constant challenge has been to develop imaging systems that are noninvasive and subject-friendly and can provide better spatial and temporal resolutions toward understanding cognitive functions of the brain. Functional magnetic resonance imaging (fMRI), positron emission tomography (PET), and electroencephalography (EEG) are currently the three leading imaging modalities in the field of neuroimaging research.

This chapter is devoted to the review of functional near-infrared spectroscopy (fNIRS), an emerging noninvasive imaging technique that uses the near-infrared light to assess cerebral hemodynamics in the cortical regions. This assessment is based on measuring the changes of concentrations of oxyhemoglobin (HbO) and deoxyhemoglobin (HbR) (Benavides-Varela et al., 2011; Boas et al., 2004; Gratton et al., 2005; Huppert et al., 2009; Yodh and Boas, 2003). The technique takes advantage of the fact that biological tissues are relatively transparent to light in the near-infrared (700–1000 nm) range, while HbO and HbR are the main absorbers in this spectral region. Since the increase in regional cerebral blood flow (CBF) is correlated with regional neural activation, measurements of temporal variations of HbO and HbR through fNIRS can capture functionally evoked changes in the outermost cortex. By revealing functional changes occurring in the cortex, this method can further elucidate the likely correlation between cognitive deficiencies and changes in cerebral hemodynamics. Thus, fNIRS can be used as a simple tool to follow hemodynamic variations and quantify brain function. Additionally, patterns of brain activation can be evaluated using fNIRS to map variations of hemoglobin concentration in response to functional tasks.

Compared to other well-established brain mapping modalities, such as fMRI and PET, fNIRS offers unique features, including higher temporal resolution of several milliseconds, and spectroscopic information about temporal variations of both the components of hemoglobin, HbO and HbR, while fMRI can assess only HbR changes. When used for functional imaging of the brain, the captured changes in HbR in response to brain stimulations are similar to the blood-oxygen-level-dependent (BOLD) signals in fMRI. NIRS instruments are much smaller and less restraining compared to fMRI or PET and can tolerate subject motion to a larger extent than fMRI. These features make this patient-friendly technique well suited to study children and populations with neurodevelopmental disorders when keeping the subjects still for long periods of time becomes extremely challenging.

This chapter provides an overview of fNIRS and its applications in brain mapping studies. It then focuses on two of our recent studies involving fNIRS. The first one used a novel algorithm to coregister functional NIRS data with structural images, in order to assess the accuracy of fNIRS system compared with that of fMRI. The second study aimed at the application of fNIRS in studying brain development and cerebral autoregulation (CA) in children.

PRINCIPLES OF fNIRS

NEUROVASCULAR COUPLING

Neurovascular coupling is the process by which activation of groups of neurons results in changes in regional CBF. Such mechanisms activate vasculature to maintain blood flow ensuring the proper CBF at the site of the activation. Based on this theory, active cortical regions will have higher values of CBF. Increase in CBF at the regions of activation is due to escalating demand of oxygen consumption by neurons (Zhang et al., 1998). Therefore, active regions are characterized by increased levels of HbO and decreased levels of HbR. As a result, neural activation and local brain function, induced by a functional task, can be indirectly measured by evaluating hemodynamic response function (HRF) (Figure 3.1) derived from changes in HbO and HbR. To capture the hemodynamic response, functional stimuli/task must be maintained long enough to evoke increases in the blood volume above the detection threshold of fNIRS. At the end of the stimulation/task, a decrease of regional CBF will result in changes in concentration of HbO and HbR to the pretask level.

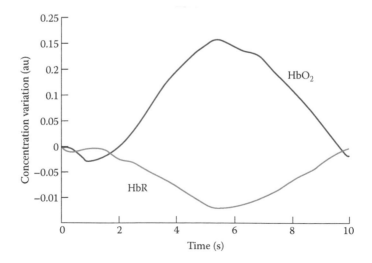

FIGURE 3.1 Example of HRF during activation.

Oxyhemoglobin and Deoxyhemoglobin Measurement

Scattering and absorption in biological tissues determine the penetration depth of light. In the near-infrared spectral band, the main absorbers of light are hemoglobin species (HbO and HbR) and water. Figure 3.2 illustrates the absorption spectra of these three components in the Near-Infrared (NIR) region. For wavelengths above 900 nm, water becomes the dominant absorber. However, in the so-called therapeutic window (690–900 nm), HbO and HbR have considerably higher absorption coefficients than water and therefore can be used as functional biomarkers. NIR absorption spectrum indicates that absorption of HbR is higher than that of HbO at wavelengths below 800 nm, whereas HbO has higher absorption coefficient at wavelengths above 800 nm. Therefore, with data on light attenuation between source and detector at two wavelengths (below and above 800 nm), analyzed in the framework of modified Beer–Lambert law (refer to Chapter 2 for details) (Huppert, 2013; Obrig et al., 2000b), it is possible to evaluate temporal changes in the concentrations of these two hemoglobin species.

The information about the changes in total hemoglobin (HbT) can be calculated by adding the respective values of HbO and HbR concentrations. Comparatively, additional wavelength at isosbestic point (800 nm) (Figure 3.1), where the spectrum of HbO and HbR cross, can be used as a third wavelength in Beer–Lambert law to calculate the changes in HbT.

fNIRS System

Overall, the fNIRS system employs sets of light sources and detectors. During the measurements, light penetrates through the scalp and reaches the outermost cortical areas. Photons, migrating through tissue, undergo several absorption and scattering events. The probabilistic trajectory (path) of photons from source to detector, based on diffusion theory, is well described by a "banana shape"

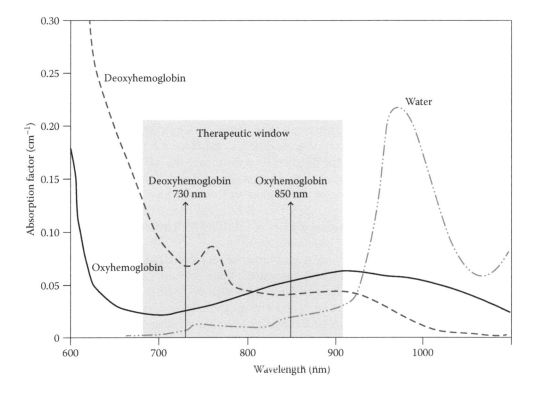

FIGURE 3.2 Absorption spectra of HbO and HbR in near-infrared region.

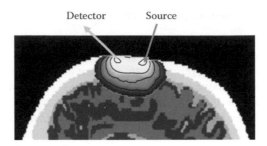

Detector   Source

FIGURE 3.3 Photon diffusion path from location of source to detector, described as "banana shape."

(Figure 3.3) (Bassan et al., 2005). The backscattered light is then collected by detectors. Source–detector separation is an important parameter that defines effective light penetration depth. For instance, too short a distance between source and detector results in shorter penetration depth that is not sufficient to probe the cortex, whereas too large source–detector separation does not allow in collecting enough backscattered photons. The optimal distance between source and detector for brain study is between 2 and 3 cm and provides a spatial resolution of 0.5–1 cm and a penetration depth of ~0.5–0.8 cm in brain tissue (Ayaz et al., 2013).

There are three types of fNIRS systems that are currently used in research studies: (1) continuous wave (CW), (2) frequency domain (FD), and (3) time domain (TD). In the CW system, light of a given intensity enters the tissue, and changes in intensity due to scattering and absorption are then measured. This system can provide an adequate approximation of changes in hemoglobin species concentrations, using the modified Beer–Lambert law for data analysis (Huppert, 2013; Obrig et al., 2000b). The CW system is simple, inexpensive, and less vulnerable to subject movements, and it becomes suitable for studying animal and human subjects. In FD instrument, the frequency-modulated light illuminates the tissue, and both intensity and phase shift of the propagating light can be measured. In the case of TD, a very short pulse of light (pulse duration in order of picoseconds) penetrates through the tissue, and information about the intensity of the detected light and its time delay (taken into account by temporal point spread function of system) are measured. TD and FD systems are very sensitive to movements and are more expensive compared to CW systems; therefore, they are usually used to provide the information about absorption and scattering properties of the tissue.

There are two different types of instrumentation within the CW system; fiber based system that allows the user to change the source–detector distance and place the probes over different cortical regions. Although, the issue of hair pigmentation disturbing the NIR light should be taken into account when placing the sensor over regions that are covered with hair. The second type is based on source–detector pairs that have been placed together forming a single sensor. This type is more suitable for studying prefrontal cortex, where there is no effect of hair on the signal. Although there are limitations of choosing what cortical regions to measure, the molded sensor will provide more source–detector stability in terms of head movement.

fNIRS system can consist of only one source–detector set (optodes) or multiple sets of optodes to cover larger cortical areas for purposes of mapping the brain activation and functional connectivity. The fNIRS Devices LLC research described in *Studies* section (fNIR Devices LLC, Potomac, MD, USA). This instrument consists of an array of 4 sources and 10 detectors, with a total of 16 source–detector pairs (Figure 3.4). It collects data at two wavelengths—730 and 850 nm—with an acquisition frequency of 2 Hz. Light is delivered to the head at the source position, and backscattered light is measured at the position of the detector, with a source–detector distance of 2.5 cm.

The 2-D activation map reflecting changes in hemodynamic response function, by means of analogous color intensity map, can be used to visualize the brain activation in response to functional tasks.

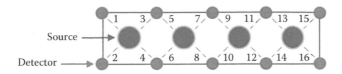

FIGURE 3.4 Schematic of CW fNIRS device using 4 sources and 10 detectors, comprising 16 source–detectors (optodes).

APPLICATIONS

fNIRS has been a valuable tool in monitoring the brain activation at cortical regions during the execution of functional tasks. Several studies have compared the accuracy of fNIRS with that of fMRI, signifying the ability of fNIRS to probe the cognitive function of the brain (Maggio, 2014; Müller and Osterreich, 2014; Panerai et al., 2005; Wong et al., 2008). fNIRS measurements, performed during several widely used function tasks, such as N-Back and Go/No-Go, are used to investigate the neural basis of working memory processes and response inhibition in prefrontal cortex (PFC) region, respectively. In short, in a classical N-Back paradigm, 0-back, 1-back, and 2-back conditions can be investigated. For a 0-back task, subjects will press the button when the target signal (e.g., "X") appears. For 1-back and 2-back tasks, the subject is instructed to press the button if the presented cue matches one and two letters/images back, respectively. In Go/No-Go task, the Go and No-Go cues are displayed on the screen and subjects are asked to respond when they observe the Go signal and vice versa when No-Go cue is shown. Visual, auditory, and motor tasks are other examples of functional tasks that are often used in fNIRS studies investigating brain function and activation at the occipital lobe, Broca's area, and motor cortex regions. Table 3.1 provides examples of studies applying fNIRS at different cortical domains and using various functional tasks.

The importance of genetic biomarkers in the studies of certain brain disorders is inevitable. On the other hand, functional biomarkers, such as HbO and HbR, can be used to pinpoint the underlying complications regarding brain function in certain disorders. Combined studies using genetic and functional biomarkers can feasibly help to identify the specific genetic changes that can control

TABLE 3.1
Summary of fNIRS Studies Based on Region of Interest and Specific Task

Author	Study of Interest	Region of Interest	Task
León-Carrión et al. (2010)	Learning process	Dorsolateral prefrontal cortex	Classical and modified N-Back
Vermeij et al. (2012)	Effects of aging on cerebral oxygenation	Prefrontal cortex	N-Back
Sato et al. (2007)	Brain function during motor task execution	Motor cortex	Finger tapping
Schroeter et al. (2002)	Executive function	Lateral prefrontal cortex	Color–word matching Stroop task
Taga et al. (2003)	Development of the brain in early infancy	Occipital cortex	Presentation of a checkerboard pattern
Inoue et al. (2012)	Hemodynamic response in ADHD population	Prefrontal cortex	Go/No-Go
Rodriguez Merzagora et al. (2013)	TBI	Dorsolateral prefrontal cortex	Verbal working memory
Nishimura et al. (2011)	Schizophrenia	Prefrontal cortex	Go/No-Go
Khan et al. (2010)	Cerebral palsy	Motor cortex	Finger tapping
Ishii-Takahashi et al. (2013)	ASD and ADHD	Prefrontal cortex	Stop signal

the relevant brain function. Specific use of fNIRS includes assessment of brain activity for behavioral domains deficient in specific disorders such as autism spectrum disorder (ASD) and attention deficit hyperactivity disorder (ADHD) (Liu et al., 2010; Naseer and Hong, 2013; Pierro et al., 2014). The system has been proposed to be a suitable medical tool to classify and distinguish brain disorders. Ishii-Takahashi et al. (2013) compared the prefrontal activation in control ASD and ADHD subjects based on changes in the concentration of HbO during the performance of stop signal task. Verbal fluency task was used as a control index of PFC to examine if the findings are task specific. Based on the grand-averaged HbO signal from each group during the stop signal task, the ASD group showed significantly smaller changes in HbO compared to ADHD groups in the area of left ventrolateral prefrontal cortex (VLPFC). Changes in HbO in the ASD group were significantly smaller in the bilateral dorsolateral prefrontal cortex (DLPFC), left VLPFC, left premotor area (PMA), left presupplementary motor area (pre-SMA), and frontal pole when compared to the healthy group. Moreover, the ADHD group showed significantly smaller changes than healthy groups in the frontal pole, bilateral DLPFC, right PMA, and right pre-SMA. Based on this conclusion, the time course of HbO signal in the left VLPFC may be used as a diagnostic marker to distinguish between ADHD and ASD groups. Monden et al. (2012) used fNIRS as a diagnostic tool to monitor the effect of methylphenidate (MPH) in children with ADHD. The activation was measured based on changes in HbO signal while subjects performed Go/No-Go tasks. Their results indicated that on average, there is a reduction in activation in the right inferior frontal gyrus and middle frontal gyrus in nonmedicated ADHD children when compared with controls. However, this reduction was extremely normalized after the administration of MPH but not after placebo administration. Such activation difference, assessed by fNIRS, could serve as a biomarker for early clinical diagnosis and for monitoring the acute effects of MPH in ADHD children. The study by Inoue et al. (2012) also revealed the reduced activation in the PFC (based on the HbO signal) during the No-Go condition for ADHD group compared to that of controls. However, such differences were not observed during the performance of the Go condition.

Likewise, fNIRS can be used to evaluate cognitive disabilities such as traumatic brain injury (TBI), where the PFC is particularly vulnerable and can lead to impaired executive function resulting in disability and social displacement (Jin et al., 2014; Rodriguez Merzagora et al., 2013). Rodriguez Merzagora et al. (2013) used fNIRS and working memory N-Back task to compare hemodynamic responses in DLPFC region in both TBI and control group. Maximum values of HbO (HbO_{max}), HbR (HbR_{max}), and HbT (HbT_{max}) were extracted from hemodynamic response signals during each condition of N-Back task. Moreover, the maximum ratio (MR) between HbO and HbR signal was calculated to account for the metabolic ratio, since HbR is a by-product of the consumption of the HbO at the activation site. The signals from left channels (left DLPFC) and right channels (right DLPFC) were averaged separately to reduce data dimensionality. Analysis of left and right DLPFC revealed significantly higher activation in TBI group at the left DLPFC based on HbO_{max}, HbR_{max}, and $HbTOT_{max}$ but not MR. Such increase in the TBI group can be due to higher regional CBF or increase in hemodynamic and metabolic activity during working memory load in the DLPFC in TBI participants as suggested by the author. Behavioral data based on correct and incorrect response during N-Back task however did not show a difference between the TBI and control group. Further, there was a nonparametric significant positive correlation between response time and MR in left and right DLPFC. The TBI group on the other hand showed significant negative correlation between HbR_{max} (by-product of metabolic consumption) and percentage of correct answer indicating impairments in the information manipulation of working memory.

Study of motor cortex abnormalities in children with cerebral palsy (CP) has also been a subject of investigation by fNIRS (Khan et al., 2010). In a study by Khan et al. (2010), a CW fNIRS system was used to image the area of the motor cortex during the execution of finger tapping task. In the finger tapping task, subjects were asked to simply tap their fingers for a certain period of time. Functional image was then reconstructed from the averaged temporal changes of HbO. To distinguish the spatial location of the activation site, clustering algorithms were applied to images followed by a similarity algorithm. Similarity algorithms are used to compare an area with lower changes in HbO with those with primary changes (maximum activation areas), which would otherwise be considered a baseline. From here, two metrics based on spatial distance from images and temporal distance (duration over time to peak) from temporal HbO plot were quantified. Using these metrics, it was possible to distinguish the area of activation between control subjects and children with CP. The accuracy of fNIRS sensor positioning in repeated measurement was also evaluated and confirmed by testing the subjects twice.

Most importantly, fNIRS allows testing in children and adults who are limited by their age or developmental status from being able to withstand fMRI or other procedures that require no movement. Unlike fMRI that can provide coregistered functional and anatomical images, fNIRS imaging requires registration with individual data or a standard anatomical system for interpretation and analysis. Several techniques have been proposed for coregistering the NIRS data with fMRI (Amyot et al., 2012; Ayaz et al., 2006; Cui et al., 2011; Okamoto et al., 2009). Amyot et al. (2012) used the novel angular method to compare the location of activation during the cognitive activation paradigm, obtained by fNIRS, with corresponding areas previously found in fMRI study. *fNIRS Normative Database to Judge a Complexity Task* section presents the overview of this study.

Brain computer interface (BCI) is another potential area that can use fNIRS as a real-time feedback system (Girouard, 2013; Hirshfield et al., 2009). The fNIRS–BCI system will use the brain's hemodynamic response (captured by fNIRS) as a measure of the brain's activity level. BCI systems are designed to create a direct pathway of communication between the brain and an external device such as a computer. Used as neural feedback, these systems can be employed to help the brain regulate its states of activity more adaptively, enhancing the cognitive function of patients with disabilities, or to be used for rehabilitation purposes. Several clinical studies have reported the positive effectiveness of the technique in improving the brain's cognitive status (Daly, 2006; Haller et al., 2010; Shih et al., 2012; Subramanian et al., 2011), although there are very few research studies employing fNIRS–BCI modality (Ayaz et al., 2011; Hirshfield et al., 2009; Jawad Khan et al., 2014; Nagaoka et al., 2010; Solovey et al., 2009). In the study by Nagaoka et al. (2010), fNIRS signal based on HbT was collected from bilateral hemispheres above the motor cortex and was transferred to the NIRS–BCI software at the workstation via Ethernet online connection. Subjects performed two types of tasks: mainly hand-grasping task (as a motor task) and motor-imagery task based on the imagination of hand-grasping task. Data collected from channels with proper hemodynamic response were chosen for analysis.

Using fNIRS technique in combination with methods of frequency analysis, it is also possible to identify and distinguish between physiological mechanisms underlying temporal variations in CBF. In this sense, the captured HbO and HbR signals from fNIRS provide information on different physiological processes that are associated with various frequency bands of the signal. For example, low-frequency oscillations (LFOs) at 0.07–0.1 Hz are attributed to CA (Obrig et al., 2000a; Sassaroli et al., 2012; Van Beek et al., 2008), while higher-frequency oscillations at 0.2–0.3 Hz are due to respiration (i.e., they result from corresponding vascular variations and blood pooling). Thus, fNIRS can be applied as a noninvasive method to study the underlying physiological mechanisms such as CA that in turn can be associated with cognitive function of the brain. *Prefrontal Cortex Hemodynamics and Age: A Pilot Study Using fNIRS in Children* section describes a recent fNIRS study of CA and its correlation with age in children (Anderson et al., 2014).

STUDIES

fNIRS NORMATIVE DATABASE TO JUDGE A COMPLEXITY TASK

The bulk of this study was to quantitatively compare the accuracy of fNIRS in imaging the location of activation with that of fMRI and to estimate the individual topographical variation in the fNIRS measurements. Coregistering physical points (optodes coordinates) with 2- or 3-D anatomical images can be done with 3-D digitization and stereotaxic systems. The challenge remains in finding the appropriate registration algorithm. Here, the novel angular algorithm was applied to a combined fNIRS–stereotaxic system to coregister the points with an fMRI study (Amyot et al., 2012).

The judgments of complexity task were replicated from an fMRI study (Krueger et al., 2009), where it produced robust anterior frontal activation. There were two standard conditions used within the task: an experimental condition (Complexity task) and a control condition (Font task). At the beginning of each trial, an instruction indicating the type of task (Complexity or Font) and the name of a daily life activity, for example, "stirring a cup of coffee," were displayed on a computer monitor for 4 s. In the Complexity task, participants were asked to make a dichotomous decision about the complexity of the activity: simple, for example, "stirring a cup of coffee," versus complex, for example, "planning a wedding," using a two-button response pad. In the Font task, participants were asked to decide whether the instruction and the activity were presented in the same or different fonts.

Changes in HbO and HbR concentrations were recorded for the entire experiment. HbO and HbR changes for the 16 source–detector pairs were filtered with a low-pass filter to exclude noise. For each event (33 Font trials and 33 Complexity trials), changes in HbO and HbR for 10 s after stimulus presentation were extracted and averaged for each condition (Font or Complexity task). Finally, for each source–detector pair, the Complexity condition was contrasted with the Font condition and normalized to the subject's maximum intensity value from the 16 source–detector pairs for that condition. Figure 3.5 shows the average hemodynamic response during the task performance for 20 individuals.

Sixteen sets of Cartesian coordinates (based on source–detector pairs) were registered in the Montreal Neurological Institute (MNI) atlas and transformed into spherical coordinates. A spherical system is appropriate for fNIRS since measurement is located on the 2-D head surface. The locations were described using two parameters: θ (azimuth angle) and φ (elevation angle), where the origins of this coordinate system are based on an MNI atlas. The activation map for each subject was created from the normalized integrated HRF and interpolated over the surface. The activation map (Figure 3.5a) was created based on the average of 20 individual activation maps. Each pixel of the activation map corresponds to one spherical coordinate in the MNI atlas. The centers of mass of the group activation maps were at the angle $\theta = 95.0° \pm 11.1°$ and $\varphi = 8.5° \pm 6.7°$ for HbO and at the angle $\theta = 95.2° \pm 9.1°$ and $\varphi = 8.5° \pm 7.0°$ for HbR (Figure 3.5a). The average of these two angles gives the Cartesian coordinates x = −7, y = 85, and z = 13 in the MNI atlas. The corresponding maxima in the Krueger et al.'s study were at −3, 66, and 19, corresponding to Brodmann area 10 (Krueger et al., 2009). Projection of this location on the surface of the MNI atlas head yields a spherical coordinate location of $\theta = 93.4°$ and $\varphi = 11.0°$. The group analysis gives a mean activation at the angle $\theta = 95.0°$ and $\varphi = 8.5°$ for HbO and at the angle $\theta = 95.2$ and $\varphi = 8.5°$ for HbR. The difference between the fMRI activation projected on the surface and our measured activation was 1.3 cm. Finally, the 2-D images were wrapped on the surface of the cortex for better visualization (Figure 3.5b).

Moreover, to evaluate the spatial variability of the peak Complexity–Font HRF difference across subjects, 2-D probability maps for HbO and HbR (Figure 3.6) were created. Angular variation in the φ plane appeared normally distributed, whereas the variability in θ was irregular.

In summary, a simple task was used to produce increased anterior frontal lobe blood flow, and by using MRI coregistration, group analysis, and an angular mapping technique, it showed that the scalp activity detected with a simple NIRS apparatus corresponded to the blood flow change detected with fMRI during the same task. The combination of a somewhat easy activation task, simple and relatively inexpensive NIRS apparatus, application of the emitter/detector array over the hairless forehead, and single-subject analysis, which could be easily automated, makes this technique an attractive candidate for further validation in clinical populations. The ability to assess frontal lobe function in a rapid, objective, and standardized way, without the need for expertise in cognitive test administration, might be particularly helpful in mild TBI, where objective measures are needed (Ruff, 2011).

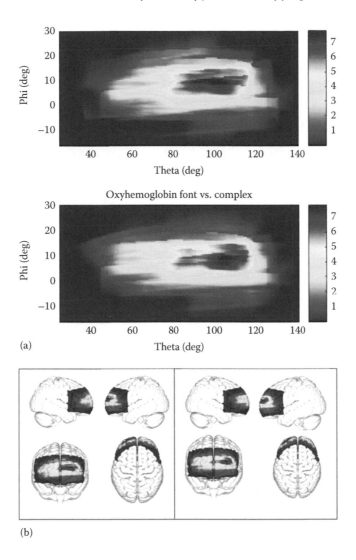

(a)

Oxyhemoglobin font vs. complex

(b)

FIGURE 3.5 (a) HbO (above) and HbR (below) activation map for all subjects in the spherical coordinate. φ represents the elevation and θ the azimuth. The colors represent the value of the integrate HRF measured for each source–detector pair over 10 s interpolated. (b) HbR (left) and HbO (right) concentration map on the surface cortex of the MNI atlas. (Data from Amyot F. et al., *Neuroimage*, 60(2), 879, 2012.)

PREFRONTAL CORTEX HEMODYNAMICS AND AGE: A PILOT STUDY USING fNIRS IN CHILDREN

Cerebral hemodynamics results from dynamic cognitive processes and underlying physiological processes, both of which can be captured by fNIRS. The purpose of this study was to find a metric to quantify the hemodynamic variations due to CA. Furthermore, such metrics have typically been applied in developing children to better understand the early changes and development of CA.

Due to its high metabolic demand and dynamic nature, the brain requires a rapid yet precise mechanism of oxygen delivery by means of CBF. CA is the physiological process that keeps CBF constant regardless of changes in atrial blood pressure and is a vital mechanism for brain function (Cipolla, 2009; Clarke and Sokoloff, 1999; Van Beek et al., 2008).

Further information about CA can be extracted from hemodynamic oscillations by measuring changes in the concentration of HbO and HbR using fNIRS in response to functional tasks. Different physiological processes result in hemodynamic oscillations at specified characteristic frequencies

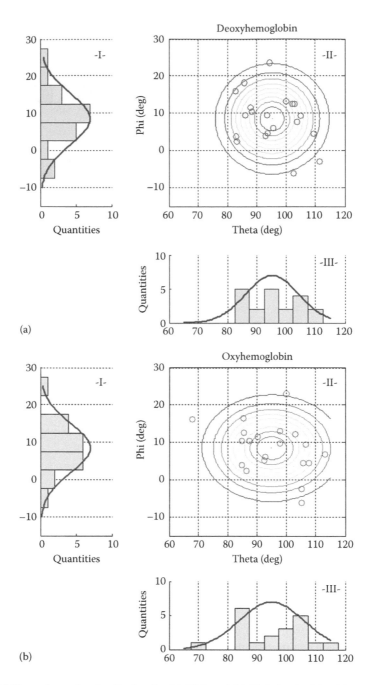

FIGURE 3.6 (a) Normal angular distribution for (a) HbR and (b) HbO for the 20 subjects. The 2-D angular distribution (II-a and II-b) is given by the normal distribution of elevation (red curve in I-a and I-b) and the normal distribution of the azimuth (red curve in I-a and I-b) and the normal distribution of the azimuth (red curve in III-a and III-b). The lines in Figure II-a and II-b are the isoprobability lines calculated; the blue circles represent the subject measurement. (Data from Amyot F. et al., *Neuroimage*, 60(2), 879, 2012.)

(Sassaroli et al., 2012). For example, LFOs at 0.07–0.1 Hz are attributed to CA (Obrig et al., 2000a; Sassaroli et al., 2012; Van Beek et al., 2008), while higher-frequency oscillations at 0.2–0.3 Hz are due to respiration (i.e., they result from corresponding vascular variations and blood pooling). However, the frequency range associated with autoregulation in children and infants likely ranges from 0 to 0.1 Hz (Bassan et al., 2005; Wong et al., 2008). Moreover, this broader frequency range also captures the influence of cerebral vasomotion and oscillation of CBF. Spontaneous oscillations, induced by CA in tone of blood vessels (due to vasomotion), are not related to heart pulsation (1–1.5 Hz) or respiration (0.2–0.3 Hz) (Naseer and Hong, 2013; Pierro et al., 2014), and they can potentially become a predictor of the degree of autoregulation and its impairment.

In this study, two narrow frequency bands, <0.1 and 0.2–0.3 Hz, related to autoregulation and respiration, respectively, have been applied to the temporal HbO and HbR signals. Instantaneous amplitudes $\Lambda(t)$ of HbO and HbR variations were obtained from the filtered NIRS using an algorithm, based on an analytic signal continuation approach.

The new oxygenation variability (OV) index was introduced to characterize variations in oxygen saturation ($SO_2 = [HbO]/([HbO] + [HbR])$), where HbO and HbR are amplitudes of analytic signal in the given frequency band, corresponding to both hemoglobin species. The OV index is defined as the dimensionless ratio of standard deviation and mean of oxygen saturation data. This index therefore assesses the variability of the oxygen saturation, SO_2, during the functional task at a given frequency band.

We hypothesize that the OV index, measured with the noninvasive and patient-friendly fNIRS modality, will reveal a developmental pattern similar to those found in the previous studies of CBF (Biagi et al., 2007; Chiron, 1992; Kilroy et al., 2011; Schöning and Hartig, 1996; Takahashi et al., 1999). While PET studies use injections of radioactive contrast to measure changes in blood flow, the present study of children uses the more suitable and feasible noninvasive fNIRS method to measure changes in hemodynamics at frequencies related to cerebral autoregulation.

fNIRS data were collected from 17 children (ages 4–8 years), while they performed a standard Go/No-Go task. Using a simple ("classical") Go/No-go task, subjects were instructed to push a button quickly when a green "spaceship" would appear on the monitor screen and to refrain from pushing when they saw a red "spaceship." The entire task lasted for approximately 3 min, with 8 task blocks alternating between 8 rest periods. Each task contained 4 Go and 4 No-Go trials in a random order.

Next, the relationship between the OV index and children's age was evaluated to examine the possible developmental changes in CA. Based on the results, variations in OV index in children indicated a dynamic relationship between age and hemodynamic CA oscillations in a frequency band below 0.1 Hz. This relationship between age and OV index indicated a significant quadratic relationship for frequencies <0.1 Hz, across all conditions (Go, No-Go, and rest) ($F(14) = 10.6$, $p = 0.006$). To determine the quadratic effect of age and condition, we then used repeated measures Analysis of Variance (ANOVA) with age as the between-subjects factor and condition (Go, No-Go, and rest) as the within-subjects factor. There was no condition by age interaction ($F(14) = 0.73$, $p = 0.5$). Thus, we removed the nonsignificant interaction from the model and found a significant main effect for condition after accounting for the quadratic effect of age ($F(16) = 5.0$, $p = 0.02$). Post hoc comparisons showed higher OV index, for frequencies <0.1 Hz, for both No-Go and Go conditions when compared to the rest ($F(16) = 5.3$, $p = 0.04$, $F(16) = 8.3$, $p = 0.01$, respectively), but no difference in OV index between No-Go and Go conditions ($F(16) = 0.25$, $p = 0.62$).

After establishing statistically similar OV index values for the Go and No-Go conditions, we combined the two into one condition, "task," and reanalyzed the relationship with age. Across all conditions (task, rest), the quadratic effect of age remained significant ($F(14) = 9.94$, $p = 0.007$). The earlier quadratic fit reveals a maximum at 5.94 years; therefore, we examined linear trends for children above and below 6 years. OV index increased significantly with age for children between the ages of 4–6 years ($r = .68$, $p = 0.039$) and showed a trend toward decreasing with age for children ages 6–8 years ($r = -0.62$, $p = 0.1$; see Figure 3.7). The OV index did not show a significant

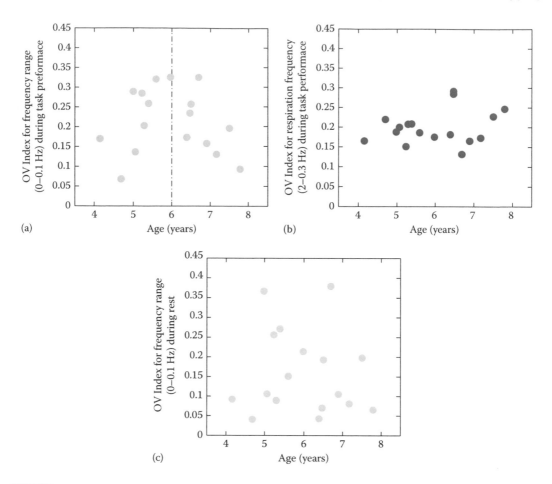

FIGURE 3.7 OV index vs. age for 17 subjects (a) for frequencies related to autoregulation (<0.1 Hz) during the task (collapsed across No-Go and Go conditions), (b) for frequencies related to respiration (0.2–0.3 Hz) during the task, and (c) for frequencies related to autoregulation (<0.1 Hz) during rest.

correlation with age for either age groups during rest (below 6, r = 0.34, p = 0.35; ages 6–8, r = −0.09, p = 0.81) (Figure 3.7c). We also investigated both linear and nonlinear trends for the respiration frequencies and found no significant effect of age (linear: F(15) = 0.98, p = 0.34; quadratic: F(14) = 0.01, p = 0.91) (Figure 3.7b).

The significant relationship between age and OV index is in line with studies using other modalities to show a similar relationship between CBF and age. Specifically, those studies illustrate an increase in CBF until about 6–7 years, a decrease afterward, and with an eventual plateau (Biagi et al., 2007; Chiron, 1992; Kilroy et al., 2011; Schöning and Hartig, 1996; Takahashi et al., 1999). Here, we characterize that pattern further by showing a quadratic trajectory of change in OV index in frequencies related to CA. Additional data, particularly longitudinal data, are needed to improve the power to model this effect. Expanding the sample to children over age 6 will also be important, in order to characterize this trajectory across development.

The task and rest were used in conjugation to see whether CA would show any differences related to the performance of functional task conditions. Though the OV index, for frequencies <0.1 Hz, was similar during Go and No-Go conditions, OV index was lower during rest. This finding is in line with those showing increases in efficiency of CA during the completion of cognitive tasks (Panerai et al., 2005). While the current study focuses on the prefrontal cortex, future studies should include other functional tasks known to elicit activity in other brain regions in order to further investigate the link between OV index, age, and task. Such analyses might reveal a significant role for frequencies <0.1 Hz across a broad range of tasks/stimulus types and brain regions.

This pilot study indicates that the variability in OV index is related to developmental changes of CA in children. The OV index can ultimately be used to elucidate the relationship between CA development and brain function. Further, we show that the choice of frequency plays a crucial role in understanding physiological processes at different stages of physiological development. While analysis of hemodynamic oscillations originating from CA does not provide a full picture of this complicated phenomenon, the noninvasive and patient-friendly fNIRS method can be used to monitor brain development and possibly detect impairments of CA in typical and atypical groups.

CONCLUSION

In summary, fNIRS can be used as a fast, noninvasive, portable, and user-friendly imaging platform to assess the cognitive function of the brain. Such a system can be utilized in real-time situations and over a wide range of developmental stages and subjects to follow changes in cerebral hemodynamics and, potentially, link such characteristics to the cognitive status. The establishment of biomarkers, such as HbR and HbO, by means of fNIRS, will open new possibilities for diagnostics pertaining to neurodevelopmental disorders and, ultimately, for therapeutic intervention. fNIRS has been shown to be an effective technique in detecting cerebral brain activation in comparison with fMRI. Moreover, this technique allows real-time imaging of subjects, particularly those with disabilities that have difficulties in performing under fMRI settings.

Although in recent years the research related to fNIRS has revealed rapid growth, there is a need to standardize the fNIRS data across studies. Such standardization perhaps would allow fNIRS to be established as a reliable device in the field of neuroimaging, for both research and bedside medical use. Combining fNIRS system with other imaging modalities, such as EEG and fMRI, can likewise help to integrate functional brain data for enhanced comparison and evaluation of cognitive function.

REFERENCES

Amyot F., Zimmermann T., Riley J., Kainerstorfer J.M., Chernomordik V., Mooshagian E., Najafizadeh L., Krueger F., Gandjbakhche A.H., and Wassermann E.M. (2012). Normative database of judgment of complexity task with functional near infrared spectroscopy—Application for TBI. *Neuroimage*, *60*(2), 879–883.

Anderson A.A., Smith E., Chernomordik V., Ardeshirpour Y., Chowdhry F., Thurm A., Black D., Matthews D., Rennert O., and Gandjbakhche A.H. (2014). Prefrontal cortex hemodynamics and age: A pilot study using functional near infrared spectroscopy in children. *Front. Neurosci.*, *8*, 393. doi: 10.3389/fnins.2014.00393.

Ayaz H., Izzetoglu M., Platek S.M., Bunce S., Izzetoglu K., Pourrezaei K., and Onaral B. (2006). Registering fNIR data to brain surface image using MRI templates. *Conf. Proc. IEEE Eng. Med. Biol. Soc.*, *2006*, 2671–2674.

Ayaz H., Onaral B., Izzetoglu K., Shewokis P.A., McKendrick R., and Parasuraman R. (2013). Continuous monitoring of brain dynamics with functional near infrared spectroscopy as a tool for neuroergonomic research: Empirical examples and a technological development. *Front. Hum. Neurosci.*, *7*, 871.

Ayaz H., Shewokis P., Bunce S., and Onaral B. (2011). An optical brain computer interface for environmental control. *Conf. Proc. IEEE Eng. Med. Biol. Soc.*, *2011*, 6327–6330.

Bassan H., Gauvreau K., Newburger J.W., Tsuji M., Limperopoulos C., Soul J.S., Walter G., Laussen P.C., Jonas R.A., and du Plessis A.J. (2005). Identification of pressure passive cerebral perfusion and its mediators after infant cardiac surgery. *Pediatr. Res.*, *57*(1), 35–41.

Benavides-Varela S., Gómez D.M., Macagno F., Bion R.A., Peretz I., and Mehler J. (2011). Memory in the neonate brain. *PloS One*, *6*(11), e27497.

Blagi L., Abbruzzese A., Bianchi M.C., Alsop D.C., Del Guerra A., and Tosetti M. (2007). Age dependence of cerebral perfusion assessed by magnetic resonance continuous arterial spin labeling. *Magn. Reson. Imag.*, *25*(4), 696–702.

Boas D.A., Dale A.M., and Franceschini M.A. (2004). Diffuse optical imaging of brain activation: Approaches to optimizing image sensitivity, resolution, and accuracy. *Neuroimage*, *23*(Suppl. 1), S275–S288.

Chiron C., Raynaud C., Maziere B., Zilbovicius M., Laflamme L., Masure M.C., Dulac O., Bourguignon M., and Syrota A. (1992). Changes in regional cerebral blood flow during brain maturation in children and adolescents. *J. Nucl. Med.*, *33*, 696–703.

Cipolla M.J. (2009). *The Cerebral Circulation*. San Rafael, CA: Morgan & Claypool Life Sciences.

Clarke D.D. and Sokoloff L. (1999). Circulation and energy metabolism of the brain. In Siegel G.J., Albers R.W., Fisher S.K., and Uhler M. (Eds.), *Basic Neurochemistry: Molecular, Cellular and Medical Aspects*. Philadelphia, PA: Lippincott-Raven.

Cui X., Bray S., Bryant D.M., Glover G.H., and Reiss A.L. (2011). A quantitative comparison of NIRS and fMRI across multiple cognitive tasks. *Neuroimage*, *54*(4), 2808–2821.

Daly J.J. and Wolpaw J.R. (2006). Brain-computer interfaces in neurological rehabilitation. *Lancet Neurol.*, *7*(11), 1032–1043.

Girouard A., Solovey E.T., and Jacob R.J.K. (2013). Designing a passive brain computer interface using real time classification of functional near-infrared spectroscopy. *Int. J. Autonom. Adapt. Comm. Syst.*, *6*(1), 26–44.

Gratton E., Toronov V., Wolf U., Wolf M., and Webb A. (2005). Measurement of brain activity by near-infrared light. *J. Biomed. Opt.*, *10*(1), 11008.

Haller S., Birbaumer N., and Veit R. (2010). Real-time fMRI feedback training may improve chronic tinnitus. *Eur. Radiol.*, *20*(3), 696–703.

Hirshfield L.M., Chauncey K., Gulotta R., Girouard A., Solovey E.T., Jacob R.J.K., Sassaroli A., and Fantini S. (2009). Combining electroencephalograph and functional near infrared spectroscopy to explore users' mental workload. Paper presented at *the Foundations of Augmented Cognition Neuroergonomics and Operational Neuroscience HCII*, San Diego, CA.

Huppert T.J. (2013). History of diffuse optical spectroscopy of human tissue. In Madsen S.J. (Ed.), *Optical Methods and Instrumentation in Brain Imaging and Therapy*. New York: Springer.

Huppert T.J., Diamond S.G., Franceschini M.A., and Boas D.A. (2009). HomER: A review of time-series analysis methods for near-infrared spectroscopy of the brain. *Appl. Opt.*, *48*(10), D280–D298.

Inoue Y., Sakihara K., Gunji A., Ozawa H., Kimiya S., Shinoda H., Kaga M., and Inagaki M. (2012). Reduced prefrontal hemodynamic response in children with ADHD during the Go/NoGo task: A NIRS study. *Neuroreport*, *23*(2), 55–60.

Ishii-Takahashi A., Takizawa R., Nishimura Y., Kawakubo Y., Kuwabara H., Matsubayashi J., Hamada K. et al. (2013). Prefrontal activation during inhibitory control measured by near-infrared spectroscopy for differentiating between autism spectrum disorders and attention deficit hyperactivity disorder in adults. *Neuroimage Clin.*, *4*, 53–63.

Jawad Khan M., Hong M.J., and Hong K.S. (2014). Decoding of four movement directions using hybrid NIRS-EEG brain-computer interface. *Front. Hum. Neurosci.*, *8*, 244.

Jin S.H., Lee S.H., Jang G.H., and Lee Y.J. (2014). Cortical activation by fNIRS and fMRI during grasping in patient with traumatic brain injury: A case study. Paper presented at *the 2014 International Winter Workshop on Brain-Computer Interface (BCI)*, Jeongsun-kun, South Korea.

Khan B., Tian F., Behbehani K., Romero M.I., Delgado M.R., Clegg N.J., Smith L., Reid D., Liu H., and Alexandrakis G. (2010). Identification of abnormal motor cortex activation patterns in children with cerebral palsy by functional near-infrared spectroscopy. *J. Biomed. Opt.*, *15*(3), 036008.

Kilroy E., Liu C.Y., Yan L., Kim Y.C., Dapretto M., Mendez M.F., and Wang D.J.J. (2011). Relationships between cerebral blood flow and IQ in typically developing children and adolescents. *J. Cognit. Sci.*, *12*(2), 151–170.

Krueger F., Spampinato M.V., Barbey A.K., Huey E.D., Morland T., and Grafman J. (2009). The frontopolar cortex mediates event knowledge complexity: A parametric functional MRI study. *Neuroreport*, *20*, 1093–1097.

León-Carrión J., Izzetoglu M., Izzetoglu K., Martín-Rodríguez J.F., Damas-López J., Barroso Martin J.M., and Domínguez-Morales M.R. (2010). Efficient learning produces spontaneous neural repetition suppression in prefrontal cortex. *Behav. Brain Res.*, *208*(2), 502–508.

Liu J., Simpson D.M., and Allen R. (2010). Tracking the dynamics of the cerebral autoregulation response to sudden changes of $PaCO_2$. Paper presented at *the World Congress on Medical Physics and Biomedical Engineering*, Munich, Germany.

Maggio P., Salinet A.S., Robinson T.G., and Panerai R.B. (2014). Influence of CO_2 on neurovascular coupling: Interaction with dynamic cerebral autoregulation and cerebrovascular reactivity. *Physiol. Rep.*, *2*(3), e00280.

Monden Y., Dan H., Nagashima M., Dan I., Kyutoku Y., Okamoto M., Yamagata T., Momoi M.Y., and Watanabe E. (2012). Clinically-oriented monitoring of acute effects of methylphenidate on cerebral hemodynamics in ADHD children using fNIRS. *Clin. Neurophysiol.*, *123*(6), 1147–1157.

Müller M.W. and Osterreich M. (2014). A comparison of dynamic cerebral autoregulation across changes in cerebral blood flow velocity for 200 s. *Front. Physiol.*, *5*, 327.

Nagaoka T., Sakatani K., Awano T., Yokose N., Hoshino T., Murata Y., Katayama Y., Ishikawa A., and Eda H. (2010). Development of a new rehabilitation system based on a brain-computer interface using near-infrared spectroscopy. *Adv. Exp. Med. Biol.*, *662*, 497–503.

Naseer N. and Hong K. (2013). Classification of functional near-infrared spectroscopy signals corresponding to the right- and left-wrist motor imagery for development of a brain-computer interface. *Neurosci. Lett.*, *553*, 84–89.

Nishimura Y., Takizawa R., Muroi M., Marumo K., Kinou M., and Kasai K. (2011). Prefrontal cortex activity during response inhibition associated with excitement symptoms in schizophrenia. *Brain Res.*, *1370*, 194–203.

Obrig H., Neufang M., Wenzel R., Kohl M., Steinbrink J., Einhäupl K., and Villringer A. (2000a). Spontaneous low frequency oscillations of cerebral hemodynamic and metabolism in human adults. *Neuroimage*, *12*(6), 623–639.

Obrig H., Wenzel R., Kohl M., Horst S., Wobst P., Steinbrink J., Thomas F., and Villringer A. (2000b). Near-infrared spectroscopy: Does it function in functional activation studies of the adult brain? *Int. J. Psychophysiol.*, *35*(2–3), 125–142.

Okamoto M., Tsuzuki D., Clowney L., Dan H., Singh A.K., and Dan I. (2009). Structural atlas-based spatial registration for functional near-infrared spectroscopy enabling inter-study data integration. *Clin. Neurophysiol.*, *120*, 1320–1328.

Panerai R.B., Moody M., Eames P.J., and Potter J.F. (2005). Dynamic cerebral autoregulation during brain activation paradigms. *Am. J. Physiol. Heart Circ. Physiol.*, *289*(3), H1202–H1208.

Pierro M.L., Hallacoglu B., Sassaroli A., Kainerstorfer J.M., and Fantini S. (2014). Validation of a novel hemodynamic model for coherent hemodynamics spectroscopy (CHS) and functional brain studies with fNIRS and fMRI. *Neuroimage*, *85*(1), 222–233.

Rodriguez Merzagora A.C., Izzetoglu M., Onaral B., and Schultheis M.T. (2013). Verbal working memory impairments following traumatic brain injury: An fNIRS investigation. *Brain Imaging Behav.*, *8*(3), 446–459.

Ruff R.M. (2011). Mild traumatic brain injury and neural recovery: Rethinking the debate. *NeuroRehabilitation*, *28*(3), 167–180.

Sassaroli A., Pierro M., Bergethon P.R., and Fantini S. (2012). Low-frequency spontaneous oscillations of cerebral hemodynamics investigated with near-infrared spectroscopy: A review. *IEEE J. Sel. Topics Quant. Electron.*, *18*(4), 1478–1492.

Sato T., Ito M., Suto T., Kameyama M., Suda M., Yamagishi Y., Ohshima A., Uehara T., Fukuda M., and Mikuni M. (2007). Time courses of brain activation and their implications for function: A multichannel near-infrared spectroscopy study during finger tapping. *Neurosci. Res.*, *58*(3), 297–304.

Schöning M. and Hartig B. (1996). Age dependence of total cerebral blood flow volume from childhood to adulthood. *J. Cerebr. Blood Flow Metabol.*, *16*, 827–833.

Schroeter M.L., Zysset S., Kupka T., Kruggel F., and Yves von Cramon D. (2002). Near-infrared spectroscopy can detect brain activity during a color–word matching stroop task in an event-related design. *Hum. Brain Mapp.*, *17*(1), 61–71.

Shih J.J., Krusienski D., and Wolpaw J.R. (2012). Brain-computer interfaces in medicine. *Mayo Clin. Proc.*, *87*(3), 268–279.

Solovey E.T., Girouard A., Chauncey K., Hirshfield L.M., Sassaroli A., Zheng F., Fantini S., and Jacob R.J.K. (2009). Using fNIRS brain sensing in realistic HCI settings: Experiments and guidelines. Paper presented at *the UIST'09 Proceedings of the 22nd Annual ACM Symposium on User Interface Software and Technology*, Victoria, British Columbia, Canada.

Subramanian L., Hindle J., Johnston S., Roberts M.V., Husain M., Goebel R., and Linden D. (2011). Real-time functional magnetic resonance imaging neurofeedback for treatment of Parkinson's disease. *J. Neurosci.*, *31*(45), 16309–16317.

Taga G., Asakawa K., Maki A., Konishi Y., and Koizumi H. (2003). Brain imaging in awake infants by near-infrared optical topography. *Proc. Natl. Acad. Sci. USA*, *100*(19), 10722–10727.

Takahashi T., Shirane R., Sato S., and Yoshimoto T. (1999). Developmental changes of cerebral blood flow and oxygen metabolism in children. *AJNR Am. J. Neuroradiol.*, *20*(5), 917–922.

Van Beck A.H., Claassen J.H., Olde Rikkert M.J., and Jansen R.W. (2008). Cerebral autoregulation: An overview of current concepts and methodology with special focus on the elderly. *J. Cerebr. Blood Flow Metabol.*, *28*, 1071–1085.

Vermeij A., van Beek A., Olde Rikkert M.G., Claassen J.A., and Kessels R.P. (2012). Effects of aging on cerebral oxygenation during working-memory performance: A functional near-infrared spectroscopy study. *PLoS One*, *7*(9), e46210.

Wong F.Y., Leung T., Austin T., Wilkinson M., Meek J.H., Wyatt J.S., and Walker A.M. (2008). Impaired autoregulation in preterm infants identified by using spatially resolved spectroscopy. *Pediatrics*, *121*(3), e604–e611.

Yodh A. and Boas D. (2003). Functional imaging with diffusing light. In Vo-Dinh T. (Ed.), *Biomedical Photonics Handbook*, Vol. *21*(1), pp. 21–45. Boca Raton, FL: CRC Press

Zhang R., Zuckerman J.H., Giller C.A., and Levine B.D. (1998). Transfer function analysis of dynamic cerebral autoregulation in humans. *Am. J. Physiol.*, *274*(1), H233–H241.

4 Therapeutic Monitoring of Children with Attention Deficit Hyperactivity Disorder Using fNIRS Assessment

Yukifumi Monden

CONTENTS

INTRODUCTION

Attention deficit hyperactivity disorder (ADHD) is a pediatric neuropsychiatric disorder, and is the most prevalent psychiatric disorder of childhood, affecting 3%–12% of school-age children in the United States (Dittmann et al. 2009, Thomas et al. 2015). ADHD is characterized with a behavioral phenotype of inattention, hyperactivity, and impulsivity. Symptoms of ADHD are usually identified during the early elementary school years (Drechsler et al. 2005). ADHD children tend to suffer from emotional problems, which can lead to academic difficulties and antisocial behaviors (Classi et al. 2012). Moreover, such ADHD symptoms become chronic in 75%–85% of patients (Lahey et al. 2004). Consequently, ADHD often persists into adolescence and adulthood, presenting patients with difficulty in educational and vocational performance and increased risk of depression and suicide (Chronis-Tuscano et al. 2010). Indeed, 4%–5% of adults have recently been reported as having ADHD (Safren et al. 2010).

Current ADHD diagnosis mainly depends on an interview-based evaluation of the degrees of the phenotypes listed in the diagnostic criteria of the DSM-IV and DSM-5 as observed by a patient's parents or teachers (Wehmeier et al. 2010, Zhu et al. 2008) and thus often entails subjective evaluation.

Treatment with medication and behavioral therapy are recommended in all ADHD clinical guidelines for ADHD children (Pliszka 2007, Taylor et al. 2004, Wolraich et al. 2011). In addition, the Multimodal Treatment of Attention Deficit Hyperactivity Disorder study funded by the National Institute of Health (1999) and the American Academy of Pediatrics reported that treatment with medication was superior to behavioral therapy for school-age children (Hodgkins et al. 2012). Medication therapy is recommended in all ADHD clinical guidelines for ADHD children (Pliszka 2007, Thomas et al. 2015, Wolraich et al. 2011). Based on considerable evidence for medication treatment over several decades, the nonstimulant drug atomoxetine (ATX; Strattera, Eli Lilly and Co., Indianapolis, IN, USA) and the stimulant drug methylphenidate (MPH; OROS-methylphenidate commercially available as Concerta) have been recommended as primary medications for the improvement of executive function in ADHD patients (Cubillo et al. 2014, Faraone et al. 2007, Faraone and Buitelaar 2010, Newcorn et al. 2008, Safren et al. 2010).

In order to confirm the effectiveness of the medication treatment, a pharmacological biomarker would be of great use. However, the clinical therapeutic effects of these medications in ADHD children are not yet clearly understood, and there is no evidence-based method with objective markers for selecting effective medications. Thus, there is a compelling need for more objective approaches, preferably based on biomarkers, for the early diagnosis of ADHD as well as for the monitoring of the effectiveness of pharmacological treatment (Wehmeier et al. 2011, Zhu et al. 2008).

One promising approach is noninvasive functional neuroimaging in combination with neuropsychological testing. Functional near-infrared spectroscopy (fNIRS) is an increasingly popular neuroimaging technique for the noninvasive monitoring of human brain activity. It utilizes the tight coupling between neural activity and regional cerebral hemodynamic changes with a high affinity for studying developing brains (Ferrari and Quaresima 2012, Lloyd-Fox et al. 2010, Minagawa-Kawai et al. 2008, Obrig and Villringer 2003, Strangman et al. 2002). fNIRS offers distinct advantages, such as its compactness, which makes it useful in confined experimental settings, affordable price, tolerance to body motion, and accessibility (Herrmann et al. 2005, Hock et al. 1997, Moriai-Izawa et al. 2012, Okamoto et al. 2004a,b, 2006, Strangman et al. 2002, Suto et al. 2004). These merits allow fNIRS to be contrasted with conventional imaging modalities such as single-photon emission computed tomography, positron emission tomography, magnetoencephalography, and functional magnetic resonance imaging (fMRI), which require large apparatuses and are susceptible to motion artifacts. Hence, we expect fNIRS to occupy a unique position among neuroimaging modalities to provide complementary usage most exemplified in clinical situations, such as a bedside setting, for diagnostic and treatment purposes (Suto et al. 2004).

Indeed, fNIRS has been implemented in various clinical domains, including pathological gate monitoring in neurologic rehabilitation (Miyai et al. 2001), monitoring of ischemia (Murata et al. 2002), assessment of language dominance before neurosurgery (Watanabe et al. 2000), detection of epileptic focus (Nguyen et al. 2012, Watanabe et al. 2002), and diagnosis of various psychiatric diseases (Hahn et al. 2012, Suto et al. 2004), among others. Prognosticating the possible clinical spread of fNIRS in the near future, in Japan, the first clinical applications of fNIRS in neurosurgery, to offer assessment of hemispheric dominance for language function (Watanabe et al. 2000) and detection of epileptic focus (Watanabe et al. 2002), have been included in the National Health Insurance coverage since 2002. Subsequently, the use of fNIRS in psychiatry was approved as an advanced medical technology in 2009 (Ferrari and Quaresima 2012) as an aid for the differential diagnosis of depressive symptoms. Further, there are great expectations for the application of fNIRS in various clinical situations, such as the exploration of objective diagnoses for psychiatric disorders, pediatric developmental disorders, and dementia as well as assessment of medication and rehabilitation.

Among these, one of the most promising clinical applications of fNIRS, in which its convenience and robustness merits would be highly appreciated, is the functional monitoring of children with ADHD, who have difficulty performing active cognitive tasks in the enclosed environments of other imaging modalities. A growing body of fNIRS studies has started to investigate the cortical hemodynamics of ADHD patients (Ehlis et al. 2008, Inoue et al. 2011, Negoro et al. 2010, Schecklmann et al. 2008, 2010, Weber et al. 2005).

This has led us to postulate that fNIRS would be effective in monitoring the effect of MPH and ATX in ADHD children, especially in younger children who are difficult to assess in an fMRI environment. The lack of evidence associating a neuropharmacological mechanism to therapeutic improvement is tantamount to a missed opportunity for appreciating how MPH and ATX work, and such understanding is a vital step toward developing an objective, evidence-based neuropharmacological treatment for ADHD children. Thus, we performed the current fNIRS study in order to assess the acute neuropharmacological effects of MPH and ATX on the inhibitory and attentional functions of ADHD children.

In addition, to elucidate the neural basis of the effects of MPH and ATX on ADHD children, experimental design should be optimized for a neuropharmacological context, including a randomized, double-blind design with a comparison to healthy control subjects. Thus we enrolled 69 ADHD children and age- and sex-matched control subjects and examined the neuropharmacological effects of MPH and ATX on the inhibition and attention control, utilizing a within-subject, double-blind, placebo-controlled design.

MONITORING NEUROPHARMACOLOGICAL EFFECTS OF ATX AND MPH USING fNIRS MEASUREMENT DURING A RESPONSE INHIBITION TASK (GO/NO-GO TASK)

INTRODUCTION

A go/no-go task was selected as the experimental task for the following reasons. First, response inhibition as measured by go/no-go tasks has emerged as one of the principal paradigms for studying ADHD (Aron and Poldrack 2005). Former fMRI studies successfully elucidated neural substrates for ADHD using motor response inhibition tasks, including go/no-go, stop signal, and Stroop tasks (Bush et al. 1999, Dillo et al. 2010, Durston et al. 2003, Rubia et al. 1999, Vaidya et al. 1998). Second, among these tasks, Stroop task performance matures latest at around 17–19 years of age (Comalli et al. 1962), followed by stop signal tasks at 13–17 years (Williams et al. 1999) and go/no-go tasks at approximately 12 years (Levin et al. 1991). Therefore, a go/no-go task is the primary choice for a study of school-age children. Third, fMRI studies have elucidated the neural substrate for go/no-go tasks, including the bilateral dorsolateral prefrontal cortex (DLPFC), ventrolateral prefrontal cortex (VLPFC), premotor cortex, inferior parietal lobe, lingual gyrus, caudate, and right anterior cingulate (Mcnon et al. 2001). Among these, the right VLPFC was found most responsible for response inhibition (Rubia et al. 2003), while another similar response inhibition task recruited the right DLPFC (Garavan et al. 1999). Fourth, VLPFC activation during a go/no-go task was replicated in an fNIRS study (Herrmann et al. 2005). An extensive review of functional neuroimaging in healthy adults indicates that widespread regions of the frontal cortex, especially the right inferior frontal gyrus (IFG), are associated with response inhibition (Aron and Poldrack 2005). Structural neuroimaging in ADHD has fairly consistently indicated gray matter density reductions in the striatum and right IFG (Durston et al. 2004). A former fMRI study on ADHD children with an MPH history reported that MPH increased the activation of the frontal cortices and striatum during go/no-go tasks (Vaidya et al. 1998). The specificity of the implicated brain regions in healthy subjects, as well as functional and structural changes to those regions in ADHD patients, suggests that response inhibition is a good neurofunctional biomarker candidate for ADHD (Aron and Poldrack 2005).

From a genetic perspective, genetic factors have also been reported to play an important role in the development and course of ADHD. In recent years, several studies have identified different candidate genes for ADHD, such as the catechol-O-methyltransferase (COMT) gene (Tomlinson et al. 2015), the dopamine (DA) active transporter 1 gene (DAT1, also known as SLC6A3), and the DA receptor D4 (DRD4) gene (Faraone et al. 2014). These genes are thought to be involved in the monoamine system, and their dysfunction in the prefrontal cortex is considered to be the core pathomechanism of ADHD.

Therefore, using the decreased IFG activation during a response inhibition task as a potential neurofunctional biomarker for ADHD, we aimed to establish a robust procedure for detecting its recovery with MPH administration. Our initial effort (Monden et al. 2011) was to test whether fNIRS-based diagnosis could be introduced in actual clinical situations. We demonstrated that fNIRS could monitor the cortical hemodynamics of ADHD children (6–14 years old) performing a go/no-go task before and 1.5 h after MPH administration, allowing the observation of the acute effect of MPH as a significant increase in the oxygenated hemoglobin (oxy-Hb) signal in the right lateral prefrontal cortex. As the monitoring takes only a few minutes, we further showed that the entire process can be implemented within a 1-day hospital visit.

However, since that study was optimized for assessing the feasibility of introducing fNIRS as an actual clinical tool that allows the pre- and post-medication comparison to be performed in a 1-day hospital visit, a neuropharmacological examination of the effects of MPH on ADHD children had yet to be performed. Thus, subsequently, enrolling a total of 64 ADHD children and age- and sex-matched healthy control children, we examined the pharmacological effects of MPH and ATX on the cortical hemodynamics of ADHD during a go/no-go task. The subjects received either MPH, ATX, or a placebo in a randomized, double-blind, placebo-controlled, crossover design. Moreover, we desired to validate the feasibility of introducing fNIRS-based diagnosis of the effects of ATX and MPH administration to ADHD children as young as 6 years old (mean age of 8.8; SD 2.2, range 6–14 years), the earliest age at which the FDA recommends starting ATX or MPH administration.

EXPERIMENTAL DESIGN

Figure 4.1 summarizes the experimental procedure. We examined the effects of ATX and MPH in a randomized, double-blind, placebo-controlled, crossover study while the subjects performed a go/no-go task. All patients were pre-medicated with ATX ($n = 16$) and MPH ($n = 16$) as part of their regular medication regimen. We examined ADHD subjects twice (the times of day for both measurements were scheduled to be as close as possible), at least 2 days apart, but within 30 days. Control subjects only underwent a single, nonmedicated session. On each examination day, ADHD subjects underwent two sessions: one before drug (MPH/ATX or placebo) administration and the other 1.5 h after drug administration. Before each preadministration session, all ADHD subjects underwent a washout period of 2 days. Each session consisted of 6 block sets, each containing alternating go (baseline) and go/no-go (target) blocks. Each block lasted 24 s and was preceded by instructions displayed for 3 s, giving an overall block-set time of 54 s and a total session time of 6 min. In the go block, we presented subjects with a random sequence of two pictures and asked them to press a button for both pictures. In the go/no-go block, we presented subjects with a no-go picture 50% of the time, thus requiring subjects to respond to half the trials (go trials) and inhibit their response to the other half (no-go trials). After ADHD subjects performed the first session, ATX, MPH, or a placebo was administered orally. We generated stimuli and collected responses using E-Prime 2.0 (Psychology Software Tools). Stimuli were presented to the subject on a 17″ desktop computer screen. The distance between the subject's eyes and the screen was about 50 cm.

fNIRS MEASUREMENT

We used the multichannel fNIRS system ETG-4000 (Hitachi Medical Corporation, Kashiwa, Japan), using two wavelengths of near-infrared light (695 and 830 nm, respectively). We analyzed

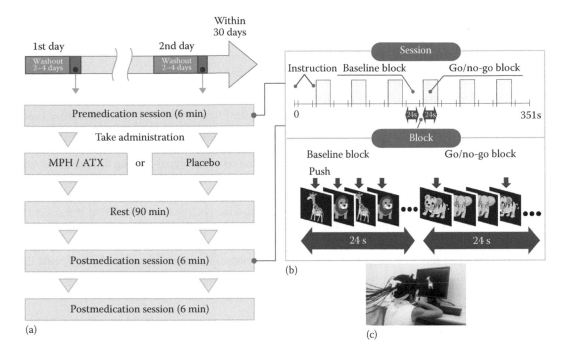

FIGURE 4.1 Experimental design. (a) A schematic showing the flow of pre- and post-medication administration sessions for ADHD subjects. (b, c) fNIRS measurements. Brain activity was measured while ADHD and control subjects performed the go/no-go task.

the optical data based on the modified Beer–Lambert law (Cope et al. 1988), as previously described (Maki et al. 1995). This method allowed us to calculate signals reflecting the oxy-Hb, deoxygenated hemoglobin (deoxy-Hb), and total hemoglobin signal changes, calculated in units of millimolar-millimeter (mM-mm) (Maki et al. 1995).

We set the fNIRS probes to cover the lateral prefrontal cortices and inferior parietal lobe in reference to previous studies (Garavan et al. 1999, Herrmann et al. 2004, 2005, Liddle et al. 2001, Rubia et al. 2003). Specifically, we used two sets of 3 × 5 multichannel probe holders, consisting of eight illuminating and seven detecting probes arranged alternately at an interprobe distance of 3 cm, resulting in 22 channels (CH) per set. The midpoint of a pair of illuminating and detecting probes was defined as a channel location. The bilateral probe holders were attached in the following manner: (1) their upper anterior corners, where the left and right probe holders were connected by a belt, were symmetrically placed across the sagittal midline; (2) the lower anterior corners of the probe holder were placed over the supraorbital prominence; (3) the lower edges of the probe holders were attached to the upper part of the auricles. For spatial profiling of fNIRS data, we employed virtual registration (Tsuzuki and Dan 2014, Tsuzuki et al. 2007) to register fNIRS data to the Montreal Neurological Institute (MNI) standard brain space (Brett et al. 2002). Briefly, this method allows us to place a virtual probe holder on the scalp by simulating the holder's deformation and by registering probes and channels onto reference brains in an MRI database (Okamoto et al. 2004a,b, Okamoto and Dan 2005). Specifically, the positions for channels and reference points consisting of the Nz (nasion), Cz, and left and right preauricular points are measured using a 3D digitizer in real-world (RW) space. The reference points in the RW are affine-transformed to the corresponding reference points in each entry in reference to the MRI database in MNI space. Adopting the same transformation parameters, we obtained the MNI coordinate values for the fNIRS channels to obtain the most likely estimate of the location of given channels for the group of subjects and the spatial variability associated with the estimation (Singh and Dan 2006). Finally, we anatomically labeled the estimated locations using a MATLAB® function that reads anatomical labeling information coded in macro-anatomical brain atlases (Rorden and Brett 2000, Shattuck et al. 2008) (LBPA40 and Brodmann).

Analysis of fNIRS Data

We preprocessed individual timeline data for the oxy-Hb and deoxy-Hb signals of each channel with a first-degree polynomial fitting and high-pass filter using cutoff frequencies of 0.01 Hz to remove baseline drift and a 0.8 Hz low-pass filter to remove heartbeat pulsations. Note that Hb signals analyzed in the current study do not directly represent cortical Hb concentration changes, but contain an unknown optical path length that cannot be measured. Direct comparison of Hb signals among different channels and regions should be avoided as optical path length is known to vary among cortical regions (Katagiri et al. 2010). Hence, we performed statistical analyses in a channel-wise manner. From the preprocessed time series data, we computed channel-wise and subject-wise contrasts by calculating the intertrial mean of differences between the peak Hb signals (4–24 s after go/no-go block onset) and baseline (14–24 s after go block onset) periods. For the six go/no-go blocks, we visually inspected the motion of each subject and removed the blocks with sudden, obvious, discontinuous noise. We subjected the resulting contrasts to second-level, random-effects group analyses.

Statistical Analyses

We statistically analyzed oxy-Hb signals in a channel-wise manner. Specifically, for control subjects, who were examined only once, we generated a target versus baseline contrast for the session. For ADHD subjects, we generated the following contrasts: (1) pre-medication contrasts, target versus baseline contrasts for pre-medication conditions for the first day exclusively; (2) post-medication contrasts, respective target versus baseline contrasts for post-placebo and post-ATX/MPH conditions; and (3) intermedication contrasts, differences between ATX/MPH[post–pre] and placebo[post–pre] contrasts.

Results

To screen the channels involved in go/no-go tasks in normal control subjects, we performed paired t-tests (two-tails) on target versus baseline contrasts. Significant increase was found in four channels on the right hemisphere (CH 5, 10, 15, and 20). Among these channels, the right CH 10 exhibited the most significant activation and was the only channel remaining after family-wise error correction ($t = 4.61$, $p = 0.0003$, Cohen's $d = 1.152$). This channel was located in the border region between the right middle frontal gyrus (MFG) and IFG (MNI coordinates x, y, z [SD]: 46, 43, 30 [14], MFG 78%, IFG 22% with reference to macroanatomical brain atlases) (Rorden and Brett 2000, Shattuck et al. 2008). Thus, we set the right CH 10 as a region of interest (ROI) for the rest of the study.

In ADHD conditions, the right CH 10 exhibited significant oxy-Hb increase in the post-ATX ($t = 4.66$, $p = 0.0003$, Cohen's $d = 1.166$) and post-MPH ($t = 4.41$, $p = 0.0006$, Cohen's $d = 1.138$) conditions. Conversely, we found that no channels were activated in the pre-medication and post-placebo conditions. Finally, the effects of medications were investigated in the intermedication contrast: we found the right CH 10 to be significantly different in the comparison between the placebo condition and the ATX (paired t-test, $p < 0.05$, Cohen's $d = 0.663$) and MPH (one-sample t-test, $p < 0.05$, Cohen's $d = 0.952$) conditions. These results demonstrate that ATX and MPH, but not the placebo, induced an oxy-Hb signal increase during the go/no-go task (Figure 4.2).

Previous neuroimaging studies have elucidated the neural correlates of go/no-go tasks (Simmonds et al. 2008), including the bilateral IFG, MFG, and superior frontal gyrus (SFG), the supplementary motor area, the anterior cingulate gyrus, the inferior parietal and temporal lobes, the caudate nucleus, and the cerebellum (Rubia et al. 2003). The IFG may be specifically related to motor response inhibition, while the MFG, SFG, and inferior parietal cortices possibly mediate more

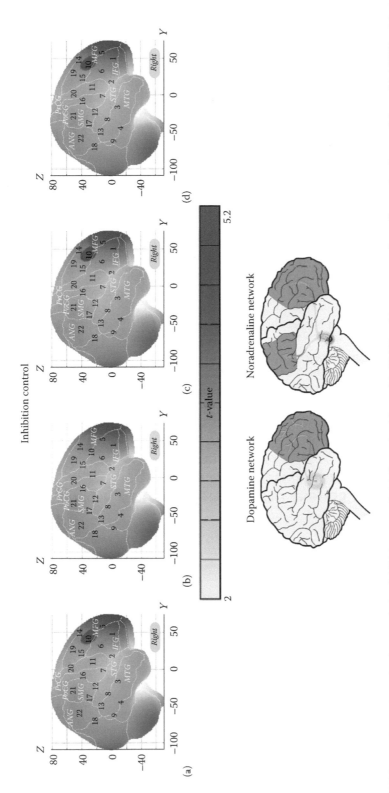

FIGURE 4.2 Hemodynamic changes during inhibition control. (a) Control subjects, (b) ADHD subjects; pre-medicated/post-placebo conditions, (c) ADHD subjects; post-MPH condition, (d) ADHD subjects; post-ATX condition, shown as t-maps of oxy-Hb signal, with significant t-values (one-sample t-test, $p < 0.05$ corrected) being shown according to the color bar.

general meta-motor executive control functions, such as motor attention, conflict monitoring, and response selection, necessary for inhibition task performance (Rubia et al. 2001).

Among these regions, our fNIRS measurements covered the right and left IFG (BA 44/45), MFG (BA 46/9), and SMG (supramarginal gyrus, BA40), and we found activation in the right MFG and IFG (BA9, 46, 45) during the go/no-go task period in the control subjects, but not in the first-day, pre-medicated ADHD subjects. These results suggest that the right prefrontal function associated with go/no-go task performance was impaired in ADHD children. The administration of a placebo did not result in right prefrontal activation, while administration of ATX and MPH led to a degree of right prefrontal activation in the ADHD children, comparable to that of the normal control subjects.

Discussion

These results suggest that normalized right IFG/MFG activation during a go/no-go task, as observed using fNIRS, may serve as a robust neurobiological marker for evaluating ATX and MPH effects on ADHD children as with evaluating MPH effects.

In a different vein of studies using animals, both ATX and MPH led to increased noradrenalin (NA) and DA in the prefrontal cortex of mice (Koda et al. 2010) and rats (Ago et al. 2014). Taken together, it would be natural to conclude that administration of either ATX or MPH increases NA and DA concentration in the prefrontal cortex, leading to normalization of inhibitory control in ADHD children. However, this does not necessarily suggest that both medications affect prefrontal functions via the same neuropharmacological mechanism. We must note here that ATX and MPH have an almost opposite affinity to DA and NA transporters. While MPH has a 10-fold higher affinity to DA than to NA transporters, ATX has a 300-fold higher affinity to NA than to DA transporters (Bymaster et al. 2002). Based on this evidence, we speculate that MPH has by far larger effects on the DA system between prefrontal and striatal regions, while ATX has far larger effects on the locus coeruleus NA system between the prefrontal and coeruleus areas (Singh-Curry and Husain 2009). Thus, what appears to be similar activation patterns induced by ATX and MPH in the prefrontal cortex may reflect different neural substrates.

MONITORING NEUROPHARMACOLOGICAL EFFECTS OF ATX AND MPH USING fNIRS MEASUREMENT DURING AN ATTENTION TASK (ODDBALL TASK)

Introduction

While our former studies focused on the effects of inhibitory functions reflected in the go/no-go task performance, inhibition alone is insufficient for explaining the overall mechanism of ADHD for the following reasons. First, recently, several groups have reported that children with ADHD may not always develop hyperactive and frenetic behavior; rather, some are more accurately characterized with hypoactivity, sluggishness, and slow response (Maedgen and Carlson 2000). Second, inattention is considered to represent a distinct neurofunctional impairment in ADHD. Several neuroimaging studies have suggested that ADHD patients, who mainly suffer from inattention, tend to exhibit dysfunction of the frontoparietal network, which has been implicated as one of the main neuronetworks for attention (e.g., Chochon et al. 1999, Peers et al. 2005, Rivera et al. 2005). However, ADHD patients with both inhibition and attention deficit would exhibit dysfunction of the DA system in the prefrontal cortex innervating from the ventral tegmentum (mesocortical pathway) (e.g., Casey et al. 1997, Castellanos 1997, Hale et al. 2000). Thus, more neurocognitive and pharmacological data for attention function are necessary in order to clarify the pathophysiology of ADHD as well as to set neuropharmacological biomarkers.

To date, there are only two fMRI studies that have performed MPH-related neuropharmacological assessments of ADHD children during attention tasks, both utilizing a double-blind, placebo-controlled design (Rubia et al. 2009, Shafritz et al. 2004). They investigated the neural correlates for the effects of MPH associated with selective and divided attention for 15 adolescents with ADHD (ages 14–17) and 14 healthy comparison subjects (ages 12–20) without MPH administration. The divided attention task evoked significantly less activation in the left ventral basal ganglia and the middle temporal gyrus in unmedicated ADHD subjects than in the healthy comparison subjects. Administration of MPH to ADHD subjects normalized activation of the left ventral basal ganglia, while no effect was observed for the middle temporal gyrus. Rubia examined the effects of MPH on medication-naïve children with ADHD during a continuous performance task, enrolling 13 right-handed male adolescent boys (10–15 years) (Rubia et al. 2009). Under the placebo condition, ADHD children exhibited reduced activation and functional interconnectivity in the bilateral fronto-striato-parieto-cerebellar networks, which were normalized upon medication with MPH. However, the participants of these fMRI studies have been limited to adolescents over 10 years old. To establish early diagnosis for elementary school-age children, introduction of fNIRS diagnosis focusing on attentional function would be beneficial.

Thus, to explore the neural substrate for the effects of ATX and MPH on attentional control in school-age ADHD children, we conducted a randomized, double-blind, placebo-controlled study employing an oddball task to evaluate ATX- and MPH-related specific neuro-activation in ADHD children using fNIRS analysis.

We adopted a visual-based oddball task that represents measures of attention. The task consists of detection of and response to infrequent (oddball) target events included in a series of repetitive events. An oddball task is often referred to as a response selection task. A wealth of studies using different modalities, including fMRI, electroencephalography, and event-related potential (ERP), have explored the neural correlates of attentional control using oddball tasks (Bledowski et al. 2004).

We enrolled a total of 74 ADHD children and age- and sex-matched healthy control children as young as 6 years old (mean age of 9.5; SD 2.2, range 6–14 years), the earliest age at which the FDA recommends starting ATX and MPH administration. Then we examined the pharmacological effects of MPH and ATX on the cortical hemodynamics of ADHD subjects during an oddball task. Subjects received either MPH, ATX, or a placebo in a randomized, double-blind, placebo-controlled, crossover design. All patients were pre-medicated with ATX ($n = 22$) and MPH ($n = 15$) as part of their regular medication regimen.

EXPERIMENTAL DESIGN

The experimental procedure for the oddball task was similar to the go/no-go paradigm depicted in Figure 4.1 (see Figure 4.3).

During the session, the subjects viewed a series of pictures once every second and responded by pressing a key for every picture. Each session consisted of 6 block sets, each containing alternating baseline and oddball blocks. In the baseline block, we presented the subjects with one picture and asked them to press a red button on the response box for that picture. Following the baseline block, we presented tiger (standard stimulus, 80% of trials) or elephant (target stimulus, 20% of trials) pictures sequentially for 200 ms, with an interstimulus interval of 800 ms. A target versus a standard ratio of 2:8 was selected so as to maintain consistency with former neuroimaging studies (Hahn et al. 2012, Katagiri et al. 2010). The total number of trials, presented in a single run, was 325. Participants were instructed to respond to the standard stimuli (tiger) by pressing a red button and to target stimuli (elephant) by pressing a blue button on the response box. The blue button was located next to the red button on the response box. Specific instructions (in Japanese) were as follows: "For this task, you will be shown tigers and elephants on the computer screen. You should press the red button for tigers and press the blue button for elephants, as quickly as possible. Don't forget that you want to be prompt but also accurate, so do not go too fast." Participants responded using the forefinger of their right hand.

Each session consisted of 6 block sets, each containing alternating baseline and oddball blocks. Each block lasted 25 s and was preceded by instructions displayed for 3 s to inform a subject of the

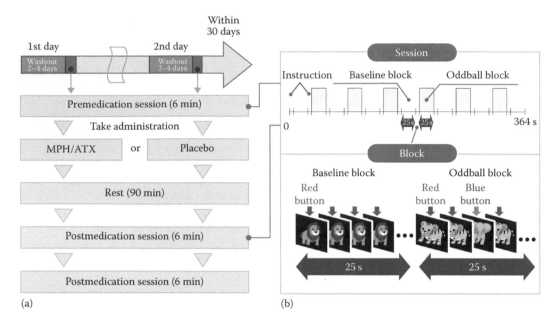

FIGURE 4.3 Experimental design. (a) A schematic showing the flow of pre- and post-medication administration sessions for ADHD subjects. (b) fNIRS measurements. Brain activity was measured while ADHD and control subjects performed the oddball task.

new block, giving an overall block-set time of 56 s and a total session time of 6.0 min. Each subject performed a practice block before any measurements to ensure their understanding of the instructions. Analysis of fNIRS data and statistical analysis were the same as of the go/no-go paradigm.

STATISTICAL ANALYSES

We statistically analyzed oxy-Hb signals in a channel-wise manner. Specifically, for control subjects, who were examined only once, we generated a target versus a baseline contrast for the session. For ADHD subjects, we generated the following contrasts: (1) pre-medication contrasts, the target versus baseline contrasts for pre-medication conditions (either placebo or ATX/MPH administration) for the first day exclusively; (2) post-medication contrasts, the respective target versus baseline contrasts for post-placebo and post-ATX/MPH conditions; and (3) intermedication contrasts, differences between ATX/MPH$^{\text{post–pre}}$ and placebo$^{\text{post–pre}}$ contrasts.

RESULTS

We screened for any fNIRS channels involved in the oddball task in control subjects. Significant oxy-Hb increase was found in the right CH 10 ($t = 4.59$, $p = 0.0003$, Cohen's $d = 0.987$) and the right CH 22 ($t = 4.79$, $p = 0.0001$, Cohen's $d = 1.013$) in control subjects. The right CH 10 was located in the border region between the MFG and IFG (MNI coordinates x, y, z [SD]: 45, 44, 31 [15], MFG 78%, IFG 22%), and the right CH 22 was located in the border region between the right angular gyrus and the right supramarginal gyrus (MNI coordinates x, y, z [SD]: 57, −58, 46 [19], angular gyrus 96%, supramarginal gyrus 4%) with reference to macroanatomical brain atlases (Rorden and Brett 2000, Schecklmann et al. 2008) . Thus, we set the right CH 10 and 22 as an ROI for the rest of the study (Figure 4.4).

In ADHD conditions, there was a significant oxy-Hb increase in the right CH 10 ($t = 3.55$, $p = 0.0032$, Cohen's $d = 0.915$) and the right CH 22 ($t = 2.55$, $p = 0.0231$, Cohen's $d = 0.658$) in the

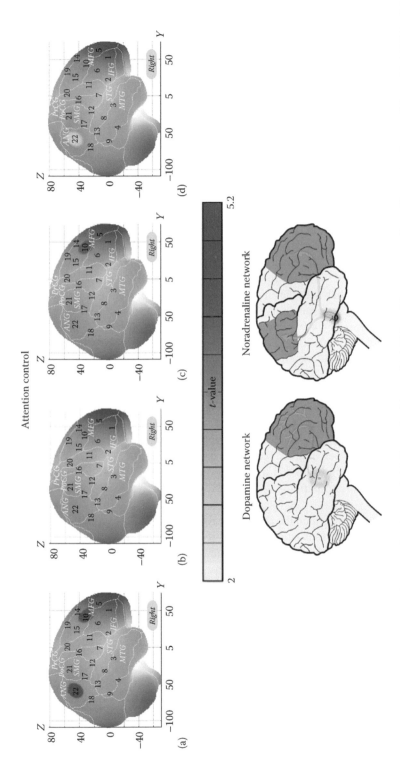

FIGURE 4.4 Hemodynamic changes during attentional control. (a) Control subjects, (b) ADHD subjects; pre-medicated/post-placebo conditions, (c) ADHD subjects; post-MPH condition, (d) ADHD subjects; post-ATX condition, shown as t-maps of oxy-Hb signal, with significant t-values (one-sample t-test, p < 0.05 corrected) being shown according to the color bar.

post-ATX condition. On the other hand, there was significant oxy-Hb increase in the right CH 10 ($t = 4.16$, $p = 0.0004$, Cohen's $d = 0.887$), but not in the right CH 22 in the post-MPH condition. We found that no channels were activated in pre-medication or post-placebo conditions.

The effects of medications were investigated in the intermedication contrast: we found the right CH 10 to be significantly different in the comparison between the placebo condition and the ATX (paired t-test, $p < 0.05$, Cohen's $d = 1.032$) and MPH (paired t-test, $p < 0.05$, Cohen's $d = 1.019$) conditions. These results demonstrate that ATX and MPH, but not the placebo, induced an oxy-Hb signal increase during the oddball task. Conversely, we found that the right CH 22 exhibited a significant increase only in the ATX (paired t-test, $p < 0.05$, Cohen's $d = 0.633$) condition, but not in the MPH condition.

Many studies have explored the neural correlates of attentional control using oddball tasks. It is known that an oddball task recruits several brain regions, including the bilateral superior, inferior and dorsolateral prefrontal cortices, the supplementary motor area, the anterior cingulate gyrus, the parietal and temporal lobes, the caudate nucleus, and the amygdala (Adler et al. 2001, Ardekani et al. 2002, Bledowski et al. 2004, Clark et al. 2000, Gur et al. 2007, Kiehl et al. 2001a,b, 2005, Mccarthy et al. 1997, Stevens et al. 2000).

Among these regions, the current study covered the prefrontal and parietal cortices. Indeed, in the control subjects, we observed cortical activations during oddball tasks on the border of the right inferior (BA 45), middle (BA 9/44/46) frontal, and angular gyri (BA 39).

The MFG and IFG, together with the angular gyrus, are the components of the attentional system and have extensive reciprocal connections (Petrides and Pandya 1984). These networks are thought to play an important role in the executive control needed to guide goal-directed and stimulus-driven attention (Corbetta and Shulman 2002). Moreover, recent fMRI and ERP studies on healthy adults have provided experimental evidence for involvement of the prefrontal and parietal networks using oddball tasks (Adler et al. 2001, Ardekani et al. 2002, Bledowski et al. 2004, Clark et al. 2000, Gur et al. 2007, Kiehl et al. 2001a,b, 2005, Mccarthy et al. 1997, Stevens et al. 2000). Thus, it is relevant that our current fNIRS-based study successfully detected concurrent activations in the attentional network between the prefrontal and inferior parietal cortices in control subjects. On the other hand, an fMRI study of ADHD children, in comparison with the control children, has indicated reduced functional connectivity between the IFC, basal ganglia, parietal cortices, and cerebellum during sustained attention, which may indicate that dysfunctions are not limited to specific brain regions, but involve the whole fronto-striato-parieto-cerebellar network underlying sustained attention processes (Rubia et al. 2009).

In the current study, we confirmed the absence of right frontoparietal activations in pre-medicated ADHD children, as in our previous study. These observations coincide with the results of an fMRI study by indicating reduced activation and functional interconnectivity in the bilateral fronto-striato-parieto-cerebellar networks during a continuous performance task under a placebo condition for ADHD children. Thus, our results add to the experimental evidence for the dysfunction of the attention-associated regions in ADHD children.

The impaired right prefrontal and inferior parietal activations were significantly normalized by ATX administration in ADHD children. Although a placebo effect was found in parietal normalization, we confirmed that the effects of ATX significantly surpassed those of the placebo as revealed by intermedication contrasts, both in the right IFG/MFG and IPL channels. Thus, the current results suggest that ATX administration leads to functional normalization of right prefrontal and inferior parietal activations of the attention network in ADHD children. Conversely, MPH administration significantly normalized reduced activation in the right MFG/IFG but not in the angular gyri. Activation in the right MFG/IFG after MPH administration was also confirmed in comparison with placebo administration. The failure of MPH to affect the inferior parietal regions is not limited to our former study. Shafritz reported that the reduced middle temporal activation in ADHD adolescents compared to control subjects was not normalized by MPH administration (Shafritz et al. 2004).

DISCUSSION

The differences between ATX- and MPH-induced normalization appear relevant when considering the pharmacological effects of both substances. ATX, an approved nonstimulant medication used to treat ADHD, is a selective NA reuptake inhibitor (Bolden-Watson and Richelson 1993) with a predominantly higher affinity to DA than to NA transporters (Bymaster et al. 2002). Given the prevalence of NA neurons in the right IFG/MFG and IPL innervating from the locus coeruleus (Nagashima et al. 2014b, Singh-Curry and Husain 2009), it appears relevant that normalization of cortical activation in ADHD children occurs in the right IPL with ATX administration.

On the other hand, MPH is known to affect both DA and NA systems (Arnsten and Li 2005, Volkow et al. 2005), but has a 10-fold higher affinity to DA than to NA receptors (Bymaster et al. 2002), resulting in by far larger effects on the DA system. Given the predominant distribution of DA neurons in the right IFG/MFG reflecting the frontostriatal DA system but not in the right IPL (Faraone and Biederman 1998, Singh-Curry and Husain 2009), it is reasonable that MPH-induced normalization of the cortical activation in ADHD children occurs solely in the right IFG/MFG.

GENERAL DISCUSSION AND CONCLUSIONS

fNIRS-BASED DIFFERENTIATION BETWEEN DA AND NA SYSTEMS IN ADHD CHILDREN

We could conclude that, when taken together, the current fNIRS results illustrate the specific neuropharmacological mechanism of ATX and MPH underlying the functional normalization of the attention network components. Furthermore, these results extend the conclusions of our combined fNIRS studies revealing ATX- and MPH-induced normalization in the right MFG/IFG (Monden et al. 2011, 2012, Nagashima et al. 2014a): fNIRS-based measurement can distinguish between neural substrates of the DA and NA systems differentially involved in inhibitory and attentional controls (Figure 4.5).

CLINICAL IMPLICATIONS

It is clinically important to note the surprisingly low post-scan data exclusion rate: In the current series of studies, which enrolled right-handed ADHD children with IQs over 70, no subjects were rejected. The data exclusion rate for an fMRI study with a similar patient population (all less than 11 years old), where compliance and motion were issues, was 50% (Durston et al. 2003), as is often the case given the young sample of children with ADHD. Lengthy test administration time, often observed with vigilance tasks, may also inflate performance variability by increasing fatigue and/or decreasing participant compliance (e.g., task duration greater than 20 min) (van der Meere et al. 1999). The go/no-go task of the current experiment took 6.0 min, and total measurement time was less than 15 min (including probe setup and position digitizing). The fNIRS-based examination of the effects of ATX and MPH was applicable to ADHD children as young as 6 years old and thus would contribute to the early clinical diagnosis and treatment of ADHD children.

CONCLUSIONS

fNIRS-based measurement is sufficiently sensitive to dissociate the neuropharmacological functional differences of ATX and MPH during different cognitive operations, at least at a group level. Furthermore, activation in the right IFG, MFG, and IPL could serve as an objective neurofunctional biomarker to indicate the effects of ATX, as well as MPH, in the case of IFG/MFG activation, on attentional and inhibitory controls in ADHD children. Further exploration with individual-level analysis will strengthen the clinical utility of fNIRS-based measurement for the functional, neuropharmacological monitoring of ADHD children.

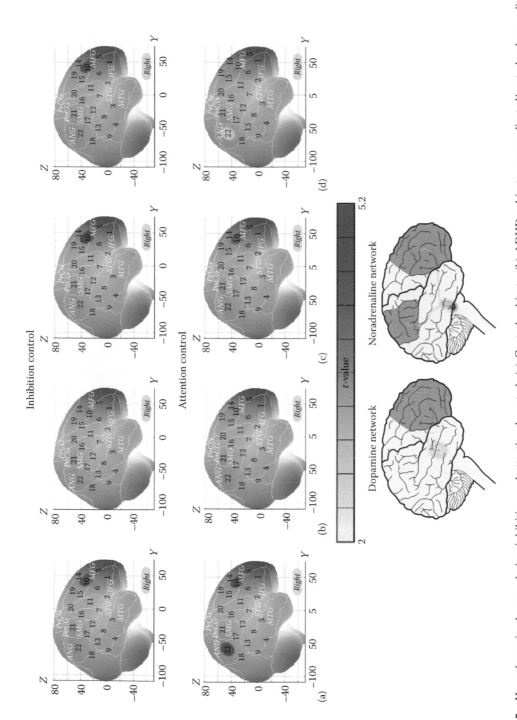

FIGURE 4.5 Hemodynamic changes during inhibition and attentional control. (a) Control subjects, (b) ADHD subjects; pre-medicated/post-placebo condition, (c) ADHD subjects; post-MPH condition, (d) ADHD subjects; post-ATX condition, shown as *t*-maps of oxy-Hb signal, with significant *t*-values (one-sample *t*-test, $p < 0.05$ corrected) being shown according to the color bar.

REFERENCES

Adler, C. M., K. W. Sax, S. K. Holland et al. Changes in neuronal activation with increasing attention demand in healthy volunteers: An fMRI study. *Synapse* 42, 4 (2001): 266–272.

Ago, Y., M. Umehara, K. Higashino et al. Atomoxetine-induced increases in monoamine release in the prefrontal cortex are similar in spontaneously hypertensive rats and Wistar-Kyoto rats. *Neurochemical Research* 39, 5 (2014): 825–832.

Ardekani, B. A., S. J. Choi, G.-A. Hossein-Zadeh et al. Functional magnetic resonance imaging of brain activity in the visual oddball task. *Cognitive Brain Research* 14, 3 (2002): 347–356.

Arnsten, A. F. T. and B.-M. Li. Neurobiology of executive functions: Catecholamine influences on prefrontal cortical functions. *Biological Psychiatry* 57, 11 (2005): 1377–1384.

Aron, A. R. and R. A. Poldrack. The cognitive neuroscience of response inhibition: Relevance for genetic research in attention-deficit/hyperactivity disorder. *Biological Psychiatry* 57, 11 (2005): 1285–1292.

Bledowski, C., D. Prvulovic, R. Goebel et al. Attentional systems in target and distractor processing: A combined ERP and fMRI study. *NeuroImage* 22, 2 (2004): 530–540.

Bolden-Watson, C. and E. Richelson. Blockade by newly-developed antidepressants of biogenic amine uptake into rat brain synaptosomes. *Life Sciences* 52, 12 (1993): 1023–1029.

Brett, M., I. S. Johnsrude, and A. M. Owen. The problem of functional localization in the human brain. *Nature Reviews. Neuroscience* 3, 3 (March 2002): 243–249.

Bush, G., J. A. Frazier, S. L. Rauch et al. Anterior cingulate cortex dysfunction in attention-deficit/hyperactivity disorder revealed by fmri and the counting stroop. [In eng]. *Biological Psychiatry* 45, 12 (1999): 1542–1552.

Bymaster, F. P., J. S. Katner, D. L. Nelson et al. Atomoxetine increases extracellular levels of norepinephrine and dopamine in prefrontal cortex of rat: A potential mechanism for efficacy in attention deficit/hyperactivity disorder. *Neuropsychopharmacology* 27, 5 (2002): 699–711.

Casey, B. J., F. X. Castellanos, J. N. Giedd et al. Implication of right frontostriatal circuitry in response inhibition and attention-deficit/hyperactivity disorder. *Journal of the American Academy of Child and Adolescent Psychiatry* 36, 3 (1997): 374–383.

Castellanos, F. X. Toward a pathophysiology of attention-deficit/hyperactivity disorder. *Clinical Pediatrics (Phila)* 36, 7 (1997): 381–393.

Chochon, F., L. Cohen, P. F. van de Moortele, and S. Dehaene. Differential contributions of the left and right inferior parietal lobules to number processing. *Journal of Cognitive Neuroscience* 11, 6 (November 1999): 617–630.

Chronis-Tuscano, A., B. S. Molina, W. E. Pelham et al. Very early predictors of adolescent depression and suicide attempts in children with attention-deficit/hyperactivity disorder. *Archives of General Psychiatry* 67, 10 (2010): 1044–1051.

Clark, V. P., S. Fannon, S. Lai et al. Responses to rare visual target and distractor stimuli using event-related fMRI. *Journal of Neurophysiology* 83, 5 (2000): 3133–3139.

Classi, P., D. Milton, S. Ward et al. Social and emotional difficulties in children with ADHD and the impact on school attendance and healthcare utilization. *Child and Adolescent Psychiatry and Mental Health* 6, 1 (2012): 33.

Comalli, P. E., Jr., S. Wapner, and H. Werner. Interference effects of Stroop color-word test in childhood, adulthood, and aging. *The Journal of Genetic Psychology* 100 (1962): 47–53.

Cope, M., D. T. Delpy, E. O. Reynolds et al. Methods of quantitating cerebral near infrared spectroscopy data. *Advances in Experimental Medicine and Biology* 222 (1988): 183–189.

Corbetta, M. and G. L. Shulman. Control of goal-directed and stimulus-driven attention in the brain. *Nature Reviews Neuroscience* 3, 3 (2002): 201–215.

Cubillo, A., A. B. Smith, N. Barrett et al. Shared and drug-specific effects of atomoxetine and methylphenidate on inhibitory brain dysfunction in medication-naive ADHD boys. *Cerebral Cortex* 24, 1 (2014): 174–185.

Dillo, W., A. Goke, V. Prox-Vagedes et al. Neuronal correlates of ADHD in adults with evidence for compensation strategies—A functional MRI study with a Go/No-Go paradigm. *German Medical Science* 8 (2010): Doc09.

Dittmann, R. W., P. M. Wehmeier, A. Schacht et al. Atomoxetine treatment and ADHD-related difficulties as assessed by adolescent patients, their parents and physicians. *Child and Adolescent Psychiatry and Mental Health* 3, 1 (2009): 21.

Drechsler, R., D. Brandeis, M. Foldenyi et al. The course of neuropsychological functions in children with attention deficit hyperactivity disorder from late childhood to early adolescence. *Journal of Child Psychology and Psychiatry* 46, 8 (2005): 824–936.

Durston, S., H. E. Hulshoff Pol, H. G. Schnack et al. Magnetic resonance imaging of boys with attention-deficit/hyperactivity disorder and their unaffected siblings. *Journal of the American Academy of Child and Adolescent Psychiatry* 43, 3 (2004): 332–340.

Durston, S., N. T. Tottenham, K. M. Thomas et al. Differential patterns of striatal activation in young children with and without ADHD. *Biological Psychiatry* 53, 10 (2003): 871–878.

Ehlis, A. C., C. G. Bahne, C. P. Jacob et al. Reduced lateral prefrontal activation in adult patients with attention-deficit/hyperactivity disorder (ADHD) during a working memory task: A functional near-infrared spectroscopy (fNIRS) study. *Journal of Psychiatric Research* 42, 13 (2008): 1060–1067.

Faraone, S. V., C. Bonvicini, and C. Scassellati. Biomarkers in the diagnosis of ADHD—Promising directions. *Current Psychiatry Reports* 16, 11 (2014): 497.

Faraone, S. V. and J. Biederman. Neurobiology of executive functions: Catecholamine influences on prefrontal cortical functions. *Biological Psychiatry* 44, 10 (1998): 951–958.

Faraone, S. V. and J. Buitelaar. Comparing the efficacy of stimulants for ADHD in children and adolescents using meta-analysis. *European Child & Adolescent Psychiatry* 19, 4 (2010): 353–364.

Faraone, S. V., S. B. Wigal, and P. Hodgkins. Forecasting three-month outcomes in a laboratory school comparison of mixed amphetamine salts extended release (Adderall XR) and atomoxetine (Strattera) in school-aged children with ADHD. *Journal of Attention Disorders* 11, 1 (2007): 74–82.

Ferrari, M. and V. Quaresima. A brief review on the history of human functional near-infrared spectroscopy (fNIRS) development and fields of application. *NeuroImage* 63, 2 (2012): 921–935.

Garavan, H., T. J. Ross, and E. A. Stein. Right hemispheric dominance of inhibitory control: An event-related functional MRI study. *Proceedings of the National Academy of Sciences of the United States of America* 96, 14 (1999): 8301–8306.

Gur, R. C., B. I. Turetsky, J. Loughead et al. Hemodynamic responses in neural circuitries for detection of visual target and novelty: An event-related fMRI study. *Human Brain Mapping* 28, 4 (2007): 263–274.

Hahn, T., A. F. Marquand, M. M. Plichta et al. A novel approach to probabilistic biomarker-based classification using functional near-infrared spectroscopy. *Human Brain Mapping* 34, 5 (2012): 1102–1114.

Hale, T. S., A. R. Hariri, and J. T. McCracken. Attention-deficit/hyperactivity disorder: perspectives from neuroimaging. *Mental Retardation and Developmental Disabilities Research Reviews* 6, 3 (2000): 214–219.

Herrmann, M. J., A. C. Ehlis, and A. J. Fallgatter. Bilaterally reduced frontal activation during a verbal fluency task in depressed patients as measured by near-infrared spectroscopy. *The Journal of Neuropsychiatry and Clinical Neurosciences* 16, 2 (2004): 170–175.

Herrmann, M. J., M. M. Plichta, A. C. Ehlis et al. Optical topography during a Go-NoGo task assessed with multi-channel near-infrared spectroscopy. *Behavioural Brain Research* 160, 1 (2005): 135–140.

Hock, C., K. Villringer, F. Müller-Spahn et al. Decrease in parietal cerebral hemoglobin oxygenation during performance of a verbal fluency task in patients with Alzheimer's disease monitored by means of near-infrared spectroscopy (NIRS)—Correlation with simultaneous rCBF-PET measurements. *Brain Research* 755, 2 (1997): 293–303.

Hodgkins, P., M. Shaw, D. Coghill, and L. Hechtman. Amfetamine and methylphenidate medications for attention-deficit/hyperactivity disorder: Complementary treatment options. *European Child and Adolescent Psychiatry* 21, 9 (2012): 477–492.

Inoue, Y., K. Sakihara, A. Gunji et al. Reduced prefrontal hemodynamic response in children with ADHD during the Go/NoGo task: A NIRS study. *Neuroreport* 23, 25 (2011): 55–60.

Katagiri, A., I. Dan, D. Tuzuki et al. Mapping of optical pathlength of human adult head at multi-wavelengths in near infrared spectroscopy. In E. Takahashi, D. F. Bruley, eds. *Oxygen Transport to Tissue XXXI*, pp. 205–212. Springer, New York, 2010.

Kiehl, K. A., K. R. Laurens, T. L. Duty et al. Event-related fMRI of auditory and visual oddball tasks. *Journal of Psychophysiology* 15, 4 (2001a): 221–240.

Kiehl, K. A., K. R. Laurens, T. L. Duty et al. Neural sources involved in auditory target detection and novelty processing: An event-related fMRI study. *Psychophysiology* 38, 1 (2001b): 133–142.

Kiehl, K. A., M. C. Stevens, K. R. Laurens et al. An adaptive reflexive processing model of neurocognitive function: Supporting evidence from a large scale (n = 100) fMRI study of an auditory oddball task. *NeuroImage* 25, 3 (2005): 899–915.

Koda, K., Y. Ago, Y. Cong et al. Effects of acute and chronic administration of atomoxetine and methylphenidate on extracellular levels of noradrenaline, dopamine and serotonin in the prefrontal cortex and striatum of mice. *Journal of Neurochemistry* 114, 1 (2010): 259–270.

Lahey, B. B., B. Applegate, I. D. Waldman et al. The structure of child and adolescent psychopathology: Generating new hypotheses. *Journal of Abnormal Psychology* 113, 3 (2004): 358–385.

Levin, H. S., K. A. Culhane, J. Hartmann et al. Developmental changes in performance on tests of purported frontal lobe functioning. *Developmental Neuropsychology* 7, 3 (1991): 377–395.

Liddle, P. F., K. A. Kiehl, and A. M. Smith. Event-related fMRI study of response inhibition. *Human Brain Mapping* 12, 2 (2001): 100–109.

Lloyd-Fox, S., A. Blasi, and C. E. Elwell. Illuminating the developing brain: The past, present and future of functional near infrared spectroscopy. *Neuroscience and Biobehavioral Reviews* 34, 3 (2010): 269–284.

Maedgen, J. W. and C. L. Carlson. Social functioning and emotional regulation in the attention deficit hyperactivity disorder subtypes. *Journal of Clinical Child Psychology* 29, 1 (2000): 30–42.

Maki, A., Y. Yamashita, Y. Ito et al. Spatial and temporal analysis of human motor activity using noninvasive NIR topography. *Medical Physics* 22, 12 (1995): 1997–2005.

Mccarthy, G., M. Luby, J. Gore et al. Infrequent events transiently activate human prefrontal and parietal cortex as measured by functional MRI. *Journal of Neurophysiology* 77, 3 (1997): 1630–1634.

Menon, V., N. E. Adleman, C. D. White et al. Error-related brain activation during a Go/NoGo response inhibition task. *Human Brain Mapping* 12, 3 (2001): 131–143.

Minagawa-Kawai, Y., K. Mori, J. C. Hebden et al. Optical imaging of infants' neurocognitive development: Recent advances and perspectives. *Developmental Neurobiology* 68, 6 (2008): 712–728.

Miyai, I., H. C. Tanabe, I. Sase et al. Cortical mapping of gait in humans: A near-infrared spectroscopic topography study. *NeuroImage* 14, 5 (2001): 1186–1192.

Monden, Y., H. Dan, M. Nagashima et al. Clinically-oriented monitoring of acute effects of methylphenidate on cerebral hemodynamics in ADHD children using fNIRS. *Clinical Neurophysiology* 123, 6 (2011): 1147–1157.

Monden, Y., H. Dan, M. Nagashima et al. Right prefrontal activation as a neuro-functional biomarker for monitoring acute effects of methylphenidate in ADHD children: An fNIRS study. *NeuroImage: Clinical* 1, 1 (2012): 131–140.

Moriai-Izawa, A., H. Dan, I. Dan et al. Multichannel fNIRS assessment of overt and covert confrontation naming. *Brain and Language* 121, 3 (2012): 185–193.

Murata, Y., K. Sakatani, Y. Katayama, and C. Fukaya. Increase in focal concentration of deoxyhaemoglobin during neuronal activity in cerebral ischaemic patients. *Journal of Neurology, Neurosurgery, and Psychiatry* 73, 2 (2002): 182–184.

Nagashima, M., Y. Monden, I. Dan et al. Acute neuropharmacological effects of atomoxetine on inhibitory control in ADHD children: A fNIRS study. *NeuroImage: Clinical* 10, 6 (2014a): 192–201.

Nagashima, M., Y. Monden, I. Dan et al. Neuropharmacological effect of methylphenidate on attention network in children with attention deficit hyperactivity disorder during oddball paradigms as assessed using functional near-infrared spectroscopy. *Neurophotonics* 1, 1 (2014b): 015001-01.

Negoro, H., M. Sawada, J. Iida et al. Prefrontal dysfunction in attention-deficit/hyperactivity disorder as measured by near-infrared spectroscopy. *Child Psychiatry and Human Development* 41, 2 (2010): 193–203.

Newcorn, J. H., C. J. Kratochvil, A. J. Allen et al. Atomoxetine and osmotically released methylphenidate for the treatment of attention deficit hyperactivity disorder: Acute comparison and differential response. *The American Journal of Psychiatry* 165, 6 (2008): 721–730.

Nguyen, D. K., J. Tremblay, P. Pouliot et al. Noninvasive continuous functional near-infrared spectroscopy combined with electroencephalography recording of frontal lobe seizures. *Epilepsia* 54, 2 (2012): 331–340.

Obrig, H. and A. Villringer. Beyond the visible—Imaging the human brain with light. *Journal of Cerebral Blood Flow and Metabolism* 23, 1 (2003): 1–18.

Okamoto, M. and I. Dan. Automated cortical projection of head-surface locations for transcranial functional brain mapping. *NeuroImage* 26, 1 (2005): 18–28.

Okamoto, M., H. Dan, K. Sakamoto et al. Three-dimensional probabilistic anatomical cranio-cerebral correlation via the international 10-20 system oriented for transcranial functional brain mapping. *NeuroImage* 21, 1 (2004a): 99–111.

Okamoto, M., H. Dan, K. Shimizu et al. Multimodal assessment of cortical activation during apple peeling by NIRS and fMRI. *NeuroImage* 21, 4 (2004b): 1275–1288.

Okamoto, M., M. Matsunami, H. Dan et al. Prefrontal activity during taste encoding: An fNIRS study. *NeuroImage* 31, 2 (2006): 796–806.

Peers, P. V., C. J. Ludwig, C. Rorden et al. Attentional functions of parietal and frontal cortex. *Cerebral Cortex* 15, 10 (2005): 1469–1484.

Petrides, M. and D. N. Pandya. Projections to the frontal cortex from the posterior parietal region in the rhesus monkey. *Journal of Comparative Neurology* 228, 1 (1984): 105–116.

Pliszka, S. Practice parameter for the assessment and treatment of children and adolescents with attention-deficit/hyperactivity disorder. *Journal of the American Academy of Child and Adolescent Psychiatry* 46, 7 (2007): 894–921.

Rivera, S. M., A. L. Reiss, M. A. Eckert et al. Developmental changes in mental arithmetic: Evidence for increased functional specialization in the left inferior parietal cortex. *Cerebral Cortex* 15, 11 (2005): 1779–1790.

Rorden, C. and M. Brett. Stereotaxic display of brain lesions. *Behavioural Neurology* 12, 4 (2000): 191–200.

Rubia, K., R. Halari, A. Cubillo et al. Methylphenidate normalises activation and functional connectivity deficits in attention and motivation networks in medication-Naive children with ADHD during a rewarded continuous performance task. *Neuropharmacology* 57, 7–8 (2009): 640–652.

Rubia, K., S. Overmeyer, E. Taylor et al. Hypofrontality in attention deficit hyperactivity disorder during higher-order motor control: A study with functional MRI. *The American Journal of Psychiatry* 156, 6 (1999): 891–896.

Rubia, K., T. Russell, S. Overmeyer et al. Mapping motor inhibition: Conjunctive brain activations across different versions of Go/No-Go and Stop tasks. *NeuroImage* 13, 2 (2001): 250–261.

Rubia, K., A. B. Smith, M. J. Brammer et al. Right inferior prefrontal cortex mediates response inhibition while mesial prefrontal cortex is responsible for error detection. *NeuroImage* 20, 1 (2003): 351–358.

Safren, S. A., S. Sprich, M. J. Mimiaga et al. Cognitive behavioral therapy vs relaxation with educational support for medication-treated adults with ADHD and persistent symptoms: A randomized controlled trial. *JAMA* 304, 8 (2010): 875–880.

Schecklmann, M., A. C. Ehlis, M. M. Plichta et al. Diminished prefrontal oxygenation with normal and above-average verbal fluency performance in adult ADHD. *Journal of Psychiatric Research* 43, 2 (2008): 98–106.

Schecklmann, M., M. Romanos, F. Bretscher et al. Prefrontal oxygenation during working memory in adhd. *Journal of Psychiatric Research* 44, 10 (2010): 621–628.

Shafritz, K. M., K. E. Marchione, J. C. Gore et al. The effects of methylphenidate on neural systems of attention in attention deficit hyperactivity disorder. *The American Journal of Psychiatry* 161, 11 (2004): 1990–1997.

Shattuck, D. W., M. Mirza, V. Adisetiyo et al. Construction of a 3D probabilistic atlas of human cortical structures. *NeuroImage* 39, 3 (2008): 1064–1080.

Simmonds, D. J., J. J. Pekar, and S. H. Mostofsky. Meta-analysis of Go/No-Go tasks demonstrating that fMRI activation associated with response inhibition is task-dependent. *Neuropsychologia* 46, 1 (2008): 224–232.

Singh, A. K. and I. Dan. Exploring the false discovery rate in multichannel NIRS. *NeuroImage* 33, 2 (2006): 542–549.

Singh-Curry, V. and M. Husain. The functional role of the inferior parietal lobe in the dorsal and ventral stream dichotomy. *Neuropsychologia* 47, 6 (2009): 1434–1438.

Stevens, A. A., P. Skudlarski, J. C. Gatenby et al. Event-related fMRI of auditory and visual oddball tasks. *Magnetic Resonance Imaging* 18, 5 (2000): 495–502.

Strangman, G., D. A. Boas, and J. P. Sutton. Non-invasive neuroimaging using near-infrared light. *Biological Psychiatry* 52, 7 (2002): 679–693.

Suto, T., M. Fukuda, M. Ito et al. Multichannel near-infrared spectroscopy in depression and schizophrenia: Cognitive brain activation study. *Biological Psychiatry* 55, 5 (2004): 501–511.

Taylor, E., M. Dopfner, J. Sergeant et al. European clinical guidelines for hyperkinetic disorder—First upgrade. *European Child and Adolescent Psychiatry* 13, Suppl. 1 (2004): 17–30.

Thomas, R., S. Sanders, J. Doust et al. Prevalence of attention-deficit/hyperactivity disorder: A systematic review and meta-analysis. *Pediatrics* 135, 4 (2015): e994–e1001.

Tomlinson, A., B. Grayson, S. Marsh et al. Putative therapeutic targets for symptom subtypes of adult ADHD: D4 receptor agonism and comt inhibition improve attention and response inhibition in a novel translational animal model. *European Neuropsychopharmacology* 25, 4 (2015): 454–467.

Tsuzuki, D. and I. Dan. Spatial registration for functional near-infrared spectroscopy: From channel position on the scalp to cortical location in individual and group analyses. *NeuroImage* 85, Pt 1 (2014): 92–103.

Tsuzuki, D., V. Jurcak, A. K. Singh et al. Virtual spatial registration of stand-alone fNIRS data to mni space. *NeuroImage* 34, 4 (2007): 1506–1518.

Vaidya, C. J., G. Austin, G. Kirkorian et al. Selective effects of methylphenidate in attention deficit hyperactivity disorder: A functional magnetic resonance study. *Proceedings of the National Academy of Sciences of the United States of America* 95, 24 (1998): 14494–14499.

van der Meere, J., B. Gunning, and N. Stemerdink. The effect of methylphenidate and clonidine on response inhibition and state regulation in children with ADHD. *Journal of Child Psychology and Psychiatry* 40, 2 (1999): 291–298.

Volkow, N. D., G.-J. Wang, Y. Ma et al. Activation of orbital and medial prefrontal cortex by methylphenidate in cocaine-addicted subjects but not in controls: Relevance to addiction. *The Journal of Neuroscience* 25, 15 (2005): 3932–3939.

Watanabe, E., A. Maki, F. Kawaguchi et al. Noninvasive cerebral blood volume measurement during seizures using multichannel near infrared spectroscopic topography. *Journal of Biomedical Optics* 5, 3 (2000): 287–290.

Watanabe, E., Y. Nagahori, and Y. Mayanagi. Focus diagnosis of epilepsy using near-infrared spectroscopy. *Epilepsia* 43, Suppl. 9 (2002): 50–55.

Weber, P., J. Lutschg, and H. Fahnenstich. Cerebral hemodynamic changes in response to an executive function task in children with attention-deficit hyperactivity disorder measured by near-infrared spectroscopy. *Journal of Developmental and Behavioral Pediatrics* 26, 2 (2005): 105–111.

Wehmeier, P. M., A. Schacht, and R. A. Barkley. Social and emotional impairment in children and adolescents with ADHD and the impact on quality of life. *The Journal of Adolescent Health* 46, 3 (2010): 209–217.

Wehmeier, P. M., A. Schacht, C. Wolff et al. Neuropsychological outcomes across the day in children with attention-deficit/hyperactivity disorder treated with atomoxetine: Results from a placebo-controlled study using a computer-based continuous performance test combined with an infra-red motion-tracking device. *Journal of Child and Adolescent Psychopharmacology* 21, 5 (2011): 433–444.

Williams, B. R., J. S. Ponesse, R. J. Schachar et al. Development of inhibitory control across the life span. *Developmental Psychology* 35, 1 (1999): 205–213.

Wolraich, M., L. Brown, R. T. Brown et al. ADHD: Clinical practice guideline for the diagnosis, evaluation, and treatment of attention-deficit/hyperactivity disorder in children and adolescents. *Pediatrics* 128, 5 (2011): 1007–1022.

Zhu, C. Z., Y. F. Zang, Q. J. Cao et al. Fisher discriminative analysis of resting-state brain function for attention-deficit/hyperactivity disorder. *NeuroImage* 40, 1 (2008): 110–120.

5 Functional Near-Infrared Spectroscopy

In Search of the Fast Optical Signal

Carlos G. Treviño-Palacios, Karla J. Sanchez-Perez,
Javier Herrera-Vega, Felipe Orihuela-Espina,
Luis Enrique Sucar, Oscar Javier Zapata-Nava,
Francisco F. De-Miguel, Guillermo Hernández-Mendoza,
Paola Ballesteros-Zebadua, Javier Franco-Perez,
and Miguel Ángel Celis-López

CONTENTS

INTRODUCTION

Diffuse optical imaging (DOI) is an emerging modality of medical imaging based on the use of near-infrared light. This technique permits to study living tissue noninvasively (Villringer and Chance 1997; Strangman et al. 2002). Its working principle capitalizes on the characteristic spectroscopic signatures of the molecules of interest. Its practical manifestation involves irradiating a narrow collimated beam over the biological tissue, where the light is scattered, and in response, part of it abandons the tissue to be collected by a detector. The sensed light encodes information about the physiological changes in the tissue in the form of changes in absorption and scattering. In the adult head, DOI becomes a practical form of functional neuroimaging known as functional near-infrared spectroscopy (fNIRS).

Optical imaging is currently the only neuroimaging modality with the potential to measure both the direct neural activity and the indirect hemodynamic responses manifested through the neurovascular coupling. The physiological changes resulting from brain activity can be split in three stages (Villringer and Chance 1997; Franceschini and Boas 2004) with different latencies. First, the neuronal membrane potential is altered, which in turn changes the index of refraction at this boundary. This occurs just a few milliseconds (50—150 ms) after the initiation of the activation episode and is commonly referred to as fast optical signal (FOS). Second, the change in membrane potential is followed by a flux of ions and water associated with a glial swelling, which further affects the scattering of the tissue. The latency of this signal is between 0.5 and 1 s. Finally, the neural activity induces an arteriolar vasodilation and subsequently an increase in the regional cerebral blood flow. This late hemodynamic response is referred to as neurovascular coupling, and it is manifested a few seconds (2–5 s) after the onset of brain activation. The typical hemodynamic response starts with a small increase in deoxyhemoglobin (HHb), followed by a much larger increase of the oxyhemoglobin (HbO_2) (Orihuela-Espina et al. 2010). These changes in chromophore concentration alter the absorption properties of the tissue. The affectation of the tissue's optical properties in all stages means that fNIRS is a suitable candidate for observing the brain activity directly, or more commonly the hemodynamic response.

This chapter will discuss the underlying principles that govern optical imaging using infrared radiation in instrumental and in numerical reconstruction efforts to pave the way for an effective direct observation of neural activity through the FOS.

fNIRS

fNIRS is a noninvasive, nonionizing method for functional monitoring and imaging of brain hemodynamics that is used to study human brain function. This technique is based on changes in blood oxygenation and in the optical properties of neural tissue caused by physiological activity (Zepeda et al. 2004).

The discovery of this technique could be traced back to 1862, when Hoppe-Seyler described the spectrum of HbO_2 (Perutz 1995), and in 1864, Stokes from the United Kingdom added the spectrum of HHb and consequently discovered the importance of hemoglobin for oxygen transport (Perutz 1995). In 1876, von Vierordt analyzed living tissue by measuring the spectral changes of light penetrating the tissue when blood circulation was occluded (Severinghaus, 2007), and in 1894, Hüfner spectroscopically determined the absolute and relative amounts of HbO_2 and HHb in vitro (Hüfner 1894). It was not until 1977 that Jobsis first described the in vivo application of near-infrared spectroscopy (NIRS) to monitor changes in the oxygenation of the exposed suprasylvian gyrus of a cat (Jobsis 1977). In that paper, Jobsis showed that changes in the spectrum measured across the head could be related to changes in the chromophore concentration using the known spectra of hemoglobin and cytochrome oxidase (Wray et al. 1988).

WAVELENGTH SELECTION

In vivo NIRS utilizes light in the wavelength range 700–1000 nm to illuminate large sections of tissue. In this spectral region, the light transmission is sufficient to provide a large penetration depth (several millimeters) in addition to an acceptable detected intensity (Figure 5.1). Consequently, spectroscopic methods can, be used to measure the concentration of typical tissue constituents, such as hemoglobin, which present absorption bands in that spectral region (Fantini et al. 1995).

In the particular case of slow hemodynamic response based on absorption, the choice of observation wavelengths follows from the absorption spectra of the major tissue chromophores (Delpy 1997). The extinction coefficient of hemoglobin and melanin becomes higher at shorter wavelengths ($\lambda < 700$ nm), thus limiting light penetration (Horecker 1943). Water, the dominant tissue chromophore, absorbs strongly below 300 nm and above 900 nm (Boulnois 1986; Delpy and Cope 1997), leaving only the NIR region where the overall absorption is sufficiently low for light to be detected across several millimeters of tissue. In the NIR region, many chromophores still absorb light, but three are of major importance for monitoring neural activity: HbO_2 and HHb in red blood cells and the oxidized form of cytochrome oxidase (CytOx), an enzyme involved in oxidative metabolism in the mitochondrial membrane (Delpy and Cope 1997).

To maximize the sensitivity of the optical measurement to changes in the tissue oxygenation, the wavelength must meet $\lambda_1 < \lambda_{iso} < \lambda_2$, where $\lambda_{iso} \approx 800$ nm is the isosbestic wavelength at which the extinction coefficients of HbO_2 and HHb have the same value. Several studies have explored the optimal combination of wavelengths to reduce crosstalk and enhance spectra separability (Uludag et al. 2004).

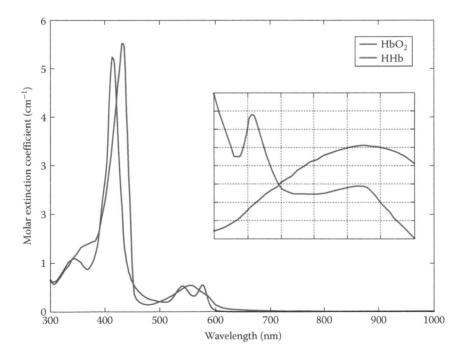

FIGURE 5.1 Molar extinction coefficients for the two species of hemoglobin: HbO_2 and HHb. The different spectroscopic signature permits fNIRS to monitor blood oxygenation in the brain. This is the slow hemodynamic response.

LIGHT SOURCES

NIRS utilizes (1) laser diodes and/or light-emitting diode (LED) sources, spanning between 650 and 1000 nm, and (2) flexible fiber optics to carry the NIR light from the source to the tissue and from the tissue to the detector. Fiber optics are suitable to carry information regardless of the head position and posture (Ferrari and Quaresima 2012).

(1) Continuous-wave (CW) NIRS makes use of a steady-state light source (e.g., an LED or a laser with constant intensity) whose amplitude may be modulated at a low (a few kHz) frequency to exploit the significant improvements in sensitivity obtained from phase-locked detection techniques. It also makes use of a detector sensitive to light attenuation changes (e.g., photodiode); (2) frequency-domain (FD) NIRS is based on amplitude-modulated light sources (at frequencies from 100 MHz to ~1 GHz) and on the detection of light amplitude demodulation and phase shift; and (3) time-domain (TD) NIRS employs a pulsed light source, typically a laser that provides light pulses lasting a few tens of picoseconds, and a detector with temporal resolution in the sub-nanosecond scale (for review, see Wolf et al. 2007; Torricelli et al. 2014).

MODALITIES

In a separate manner from the measurement geometry, three different imaging approaches can be implemented based on a specific irradiation strategy: (1) the CW modality, which is based on constant tissue illumination, that measures light attenuation through the head; (2) the FD method, which measures the attenuation and phase delay of emerging light by illuminating the head with intensity-modulated light; and (3) the TD technique, which detects the shape of the pulse after propagation through tissues by illuminating the head with short pulses of light (Ferrari and Quaresima 2012) (Figure 5.2).

Continuous Wave

CW fNIRS uses light sources with constant frequency and amplitude emissions. Changes in the light intensity reflect changes in the relative concentration of hemoglobin through the modified Beer–Lambert law (MBLL) (Cope et al. 1988). fNIRS measurements of hemoglobin concentration changes upon brain activation are almost exclusively performed with CW devices, as reviewed by Scholkmann et al. (2014).

CW devices are the simplest, usually utilizing inexpensive laser diodes or even LEDs. These devices measure intensity changes at different wavelengths, permitting the estimation of cerebral hemoglobin concentration changes. The disadvantage of CW systems is that they cannot separate light scattering and absorption; therefore, HbO_2 and HHb cannot be determined absolutely (Scholkmann et al. 2014).

Time Domain

In TD NIRS or time-resolved spectroscopy, a very short NIR pulse is irradiated on the tissue, with a pulse length usually in the order of picoseconds (Torricelli et al. 2014). Information about hemodynamic changes arises from the attenuation, the decay, and the time profile of the backscattered signal.

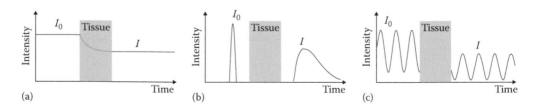

FIGURE 5.2 Schematic illustrations of the three modalities of DOI. (a) CW irradiation permits detection of light extinction, (b) TD irradiation monitors photon's time of flight, and (c) FD reconstructions are made from attenuation and phase shifts. (Data from Ferrari, M. and Quaresima, V., *NeuroImage*, 63(2), 921, 2012.)

TD NIRS relies on the ability to measure the photon distribution of time-of-flight (DTOF) in a diffusive medium (in the literature the DTOF is also called temporal point spread function). Following the injection of a light pulse within a turbid medium, with a typical emission of less than 10 s or picoseconds, the DTOF measured at a fixed distance from the injection point (typically in the range of 10–40 mm) is delayed, broadened, and attenuated. The delay is a consequence of the finite time that light takes to travel the distance between the source and the detector; broadening is mainly due to the different paths that photons undergo because of multiple scattering; attenuation appears because absorption reduces the probability of detecting a photon, and diffusion into other directions within the medium decreases the number of detected photons in the considered direction. Increasing the source–detector distance yields an increased delay and broadening of the DTOF and decreases the number of detected photons. Similar behavior is observed when the scattering increases. Finally, absorption affects both the signal intensity and the trailing edge (i.e., slope of the tail) of the DTOF, while leaving the temporal position of the DTOF substantially unchanged (Torricelli et al. 2014). Because of the need for high-speed detection and high-speed emitters, time-resolved methods are the most expensive and technically demanding. TD devices may soon be commercialized for brain activation studies (Boas et al. 2014).

TD NIRS has the following advantage compared to CW and FD fNIRS; that is, it yields the highest amount of information about the migration of photons through tissue. Therefore, it also promises to provide images with the highest spatial resolution in 3D (Scholkmann et al. 2014).

Frequency Domain

FD NIRS technology modulates the emitted light intensity and then measures the intensity of the detected light as well as the phase shift, which corresponds to the time of flight (Scholkmann et al. 2014). The advantage of FD NIRS is that it can easily be miniaturized.

Systems working in this modality have traditionally used derivations of the diffusion equation or direct derivations of the Boltzmann radiative transfer equation model of how light travels through highly scattering tissue (Arridge et al. 1992; Klose and Larsen 2006). This model allow not only to calculate the optical properties of bulk tissue but also to reconstruct chromophore concentrations in a 3D space (Scholkmann et al. 2014). There is more detailed information on the instrumentation for optical studies of tissue in the FD by Chance et al. (1998) and Fantini and Franceschini (2002).

In FD, the light intensity is usually sinusoidally modulated and subsequently the phase shift and intensity of the detected light is measured. The phase shift corresponds to the time of flight (Scholkmann et al. 2014). Although the photons traveling through biological tissue suffer absorption, scattering, and fluorescence, we are only interested in the absorption and scattering effects.

Our Approach

To interrogate the brain activity includes the use of an adjustable helmet to mount a set of oximeters (Figure 5.3a). The reconfigurable attachment for fNIRS analysis (RAFA) mount is presented by our group position source or detector channels around EEG 10/20 distribution for comparison. Different head mounts have been proposed (Ferrari and Quaresima 2012). Each oximeter pair, or channel, operates with a combination of CW and FD modalities to illuminate the tissue simultaneously (Figure 5.3b). Due to the low level reflection signal levels, lock-in detection is used.

The sources we use are LEDs at $\lambda_1 = 940$ nm and $\lambda_2 = 632$ nm. The source at λ_1 operates with a pulse train at 10 kHz frequency, while simultaneously, the λ_2 source operates in CW mode. The light leaving the tissue is collected by silicon photodiodes (Thorlabs FDS100). Using the absorption spectrum of light, the detected light level is interpreted as changes in the concentration of HbO_2 and HHb (Kleinschmidt et al. 1996; Villringer and Chance 1997). The light illumination and attenuation through living tissue at two selected wavelengths allow the calculation of HbO_2 and HHb concentrations using the MBLL.

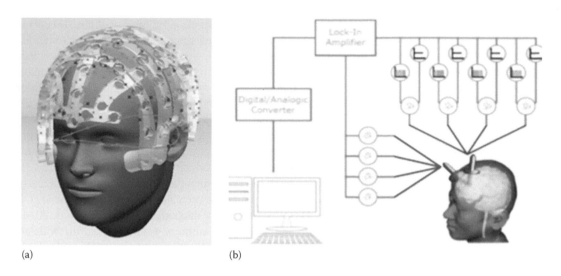

FIGURE 5.3 (a) RAFA use for fNIRS signal detection on a modular support for an oximeter array. (b) Experimental configuration for fNIRS data acquisition presenting four channels as example.

Using the selected dual CW-pulsed operation as described, the light detected at time t_1 (both λ_1 and λ_2 present) and t_2 (only λ_2 present) provide different information at different times. Later in this chapter, we will discuss in more detail the BLL. In this case, the set of equation is given as

$$I(t_1) = \left[I_0(\lambda_1) + I_0(\lambda_2) \right] e^{-\left[A_1(\lambda_1) + A_1(\lambda_2) \right]}$$
$$I(t_2) = I_0(\lambda_1) e^{-A_2(\lambda_2)}$$

$$(5.1)$$

where

$I(t_i)$ is the detected light intensity at time t_i

$I_0(\lambda_i)$ is the intensity of the sources

$A(\lambda_i)$ is the wavelength-dependent tissue attenuation expressed as the product of the absorption coefficients ($\varepsilon_i(\lambda_i)$ shown in Figure 5.1), the unknown concentration for each chromophore ($c_i(\lambda_i)$), and the total mean path length of detected photons (L): $A(\lambda_i) = \varepsilon_i(\lambda_i) * c_i(\lambda_i) * L$.

From these two measurements, we can determine the relative concentration of chromophores as the ratio between $c_i(\lambda_i)$, eliminating the total mean path length dependence, as follows:

$$\Delta HbO_2 = \frac{\log\left(\dfrac{I(t_1)}{I_0(\lambda_1) + I_0(\lambda_2)}\right)\varepsilon_{HHb}(\lambda_1) - \log\left(\dfrac{I(t_2)}{I_0(\lambda_2)}\right)\left[\varepsilon_{HHb}(\lambda_1) + \varepsilon_{HHb}(\lambda_2)\right]}{\varepsilon_{HbO_2}(\lambda_1)\left[\varepsilon_{HHb}(\lambda_1) + \varepsilon_{HHb}(\lambda_2)\right] - \varepsilon_{HbO_2}(\lambda_1)\left[\varepsilon_{HHb}(\lambda_1) + \varepsilon_{HbO_2}(\lambda_2)\right]}$$

$$\Delta HHb = \frac{\log\left(\dfrac{I(t_2)}{I_0(\lambda_2)}\right)\left[\varepsilon_{HbO_2}(\lambda_1) + \varepsilon_{HbO_2}(\lambda_2)\right] - \log\left(\dfrac{I(t_1)}{I_0(\lambda_1) + I_0(\lambda_2)}\right)\varepsilon_{HbO_2}(\lambda_1)}{\varepsilon_{HbO_2}(\lambda_1)\left[\varepsilon_{HHb}(\lambda_1) + \varepsilon_{HHb}(\lambda_2)\right] - \varepsilon_{HbO_2}(\lambda_1)\left[\varepsilon_{HHb}(\lambda_1) + \varepsilon_{HbO_2}(\lambda_2)\right]}$$

$$(5.2)$$

These relations are used in modern pulse oximeters to determine blood oxygen saturation. Depending on the light source modulation (CW, TD, or FD) and the number of wavelengths used, particular relations for oxygen saturation are obtained. The earlier equations are the result of our particular λ_1–CW, λ_2–FD combination, which is convenient because the CW signal can also directly provide arterial pulse rhythm.

Using a series of oximeters mounted on top of the head, we can proceed to produce a brain functional image reconstruction.

IMAGE RECONSTRUCTION IN DIFFUSE OPTICAL IMAGING

The optical imaging process can be characterized in terms of an inverse problem with two major mathematical transformations projecting in opposite directions. The first transformation, known as *image formation*, refers to the physical process of light–tissue interaction, having as a result a remitted light spectrum. In DOI, the interest is not in the raw optical measurement but in the recovery of the optical properties of the tissue and from there the histophysiological information encoded in sensed light. This decoding transformation is known as *image reconstruction*.

In general, image reconstruction refers to the inverse problem (ill-posed) of reconstructing optical properties or indirectly the histophysiological information of the interrogated tissue from only boundary measurements (Arridge and Schotland 2009).

In order to solve the inverse problem, it is first convenient to model the image formation forward process (light propagation in tissue). Several approximations to the physics of radiation transport in matter have been developed for this purpose, from Maxwell's equations to the diffusion theory (Arridge 2011). This model of physical process in mathematical terms, as we will detail later in this chapter, is a function projecting points from a space representing all the possible optical configurations of the tissue, which in turn corresponds to histophysiological configurations to paired points in a certain imaging space. Without loss of generality, the function can incorporate the details of the illuminant and/or the details of the image acquisition device.

The image reconstruction transformation is not a model of a physical process but a mathematical function capable of recovering sufficiently good approximations to the original locations in the optical configurations space from the observed locations in the imaging space, in other words from pixel vector values. Some approaches to solve the inverse problem rise from linear algorithms (Delpy et al. 1988; Colak et al. 1997) representing iterative nonlinear methods (Jiang et al. 2000; Dehghani et al. 2009a; Jiang 2011) and machine learning (Baillet and Garnero 1997; Eppstein et al. 2001; Claridge and Hidovic-Rowe 2014). This image reconstruction projection is, however, a nonlinear and ill-posed problem where uniqueness of the solution is not guaranteed. The inherent ill-posedness of the inversion problem requires regularization to achieve stability of the solution (Wang et al. 2011).

FORWARD PROBLEM

The image formation corresponds to the mathematical projection expressing the relation between the space of matter properties Π to the space of remitted spectrum Λ. Matter properties vary continuously in every location of the tissue, thus making Π a space of infinite dimensionality. However, in nature, these variations are not random, thus facilitating spatial discretization to different degrees of faithfulness. The practical modeling of the image formation relation thus demands a reduction of this continuous space from its true infinite dimensional expression into the finite expression:

$$G : \Pi \to \Lambda \quad \Pi \in \mathbb{R}^n \tag{5.3}$$

Similarly, the wavelength is a continuous variable, that is, Λ has infinite dimensions, but admits discretization. Access to a remitted spectrum is not continuous but is only observable through the action of a finite number of sensors. Although sensors continuously sense infinite wavelengths in a certain range, no matter how narrow, nominal representation at discrete wavelengths to the required level of precision projects the Λ to a computable image space (sensor space) I:

$$F : \Lambda \rightarrow I \tag{5.4}$$

Equation 5.4 represents the *image acquisition* subsequent to the formation of the image representation. The complete forward model may then be stated as a double mapping (Orihuela-Espina 2005; Herrera-Vega and Orihuela-Espina 2015), first from the space of matter properties Π to the space Λ of the remitted spectra and then from the remitted spectra Λ to an image space I:

$$F \circ G : \Pi \rightarrow \Lambda \rightarrow I \quad \Pi \in \mathbb{R}^n \quad I \in \mathbb{R}^m \tag{5.5}$$

Equation 5.5 represents the generic transformation and the specific form of the forward model given by the physical approximation of the radiation transport phenomenon. It is convenient here to recall the fundamental events occurring during light–matter interaction.

Physics of Electromagnetic Radiation Transport

Several phenomena occur during light propagation through matter (van de Hulst 1957). We describe them here briefly, but the reader may consult more specialized literature about light transport (van de Hulst 1957; Newton 1982; Prahl 1988; Branco 2007; Bass et al. 2009).

As light propagates inside the medium, photons hit particles, thus transferring part of the energy in a phenomenon known as absorption.

The photon, with its remaining energy, continues its journey but may suffer a scattering, which is a change in its travelling direction as a consequence of the collision. The angular distribution of the scattered light is expressed by the anisotropy function (Hall et al. 2012). Energy reemission may occur neither immediately (time delay–phosphorescence) nor at the same wavelength (wavelength shift–fluorescence) (Richards-Kortum and Sevick-Muraca 1996). The relative size between the photon wavelength and the scatter dictates the scattering regime (van de Hulst 1957), which in biological tissue is known to follow Mie scattering.

Refraction is the change of light direction when it crosses from one medium to another. It is ruled by Snell's law (Boreman 1998). Reflection is the bouncing of light when hitting the boundary between two media with different refraction index. Refraction and reflection occur at the boundaries between two media, and thus, the geometry of the frontier between the two media determines whether reflection is specular or diffuse. These are specific cases of the more general phenomenon of scattering (Jacob 1999).

Diffraction, which is similar to scattering, is also a deviation from the incoming direction of the photon. However, diffraction does not require a collision between the photon and the molecule. The bending in the trajectory occurs as the photon passes sufficiently close to the molecule, transforming an incident wavefront into a diffracted wavefront (Nolte 2012). Traditionally diffraction has imposed a limit on image resolution characterized by Abbe's limit, but this has been recently overcome by stimulated emission depletion nanoscopy technologies (Willig et al. 2006).

Each biological tissue exhibits distinctive optical properties. Optical and geometrical properties of matter dictate how light propagates within the medium. The combined level of absorption and/or scattering referred to as the extinction of a tissue depends on its macroscopic absorption coefficient μ_a and its scattering coefficient μ_s representing the respective probability of an extinction event occurring by length unit. Figure 5.4 shows an example of the effect of these coefficients.

We can model the propagation of light through the biological medium by using different models. Each of these models has assumptions that make tractable the complex process of light propagation and permit the image formation process.

Common models include the radiative transport equation (Branco 2007), diffusion theory (Arridge and Hebden 1997; Arridge 2009; Jacques and Pogue 2008; Jiang 2011), stochastic methods

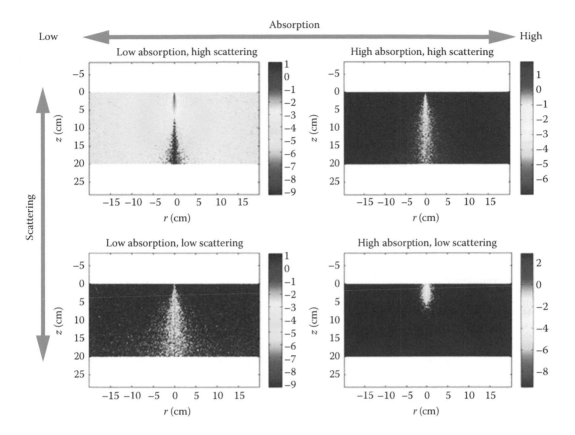

FIGURE 5.4 Effect of varying the absorption and scattering coefficients on the photon trajectory.

based on Monte Carlo techniques (Wang et al. 1995; Fang and Boas 2009; Zhu and Liu 2013), and the popular MBLL (Delpy et al. 1988).

Among these models, the BLL is a valuable for brain optical imaging. This model describes the loss of light intensity (I), mainly due to absorption, as a function of the concentration of a substance in a nonscattering medium:

$$I = I_0 e^{-\varepsilon(\lambda)cd} \tag{5.6}$$

where

I_0 is the incident light
(λ) is the specific extinction coefficient of the medium (wavelength dependent)
c is the concentration of the chromophore (responsible for the extinction)
d is the thickness of the medium

Since biological tissues are highly scattering, to use this law, an extension known as the MBLL was developed by Delpy et al. (1988). This new model adds two elements: the differential pathlength factor that accounts for the distance traveled by light due to scattering and a factor G accounting for the tissue geometry and the light attenuation due to scattering (Boas et al. 2011; Scholkmann et al. 2014). The optical density (OD) following the MBLL in general form is

$$OD(t, \lambda) = -\log_{10}\left(\frac{I(t, \lambda)}{I_0(t, \lambda)}\right) = \sum_i \varepsilon_i(\lambda)c_i d\, DPF(\lambda) + G(\lambda) \tag{5.7}$$

This law is regularly used in optical imaging under the assumption that the change in scattering is small when compared to changes in absorption over time. The factor $G(\lambda)$ is assumed to be time invariant. In this way, when differentiating the changes in OD for two consecutive measures, t_0 and t_1, the term $G(\lambda)$ nullifies and the relative changes ΔOD can be estimated from

$$\Delta OD\left(\Delta t, \lambda\right) = OD\left(t_1, \lambda\right) - OD\left(t_0, \lambda\right) \tag{5.8}$$

and

$$\Delta OD\left(t, \lambda\right) = -\log_{10}\left(\frac{I(t, \lambda)}{I_0(t, \lambda)}\right) = \sum_i \varepsilon_i\left(\lambda\right) \Delta c_i d \, \mathrm{DPF}\left(\lambda\right) \tag{5.9}$$

where the chromophore concentration $\Delta c_i = c_i(t_1) - c_i(t_0)$ is the relative change in concentration with time. The MBLL is perhaps the most common forward model currently used in fNIRS (Scholkmann et al. 2014). However, there are two main strong assumptions (discussed later) that limit its confidence.

Tissue Compounds in the Optical Path

Biological tissue is a highly scattering medium since its compounds affect the light propagation in particular ways. The molecules within the tissue may absorb or scatter part of the incident light. For instance, when a beam of light passes through or is reflected by a colored substance, a characteristic portion of the mixed wavelengths is absorbed. The remaining light will then assume the complementary color to the wavelength(s) absorbed. The part (atoms and electrons) of a molecule responsible for its color is called a chromophore and determines the spectral properties of the molecule. Each molecule thus presents specific extinction properties, which in turn characterizes the macroscopic extinction properties of the tissue.

The macroscopic extinction coefficients are the net expression of the individual specific molecular extinction coefficients of its components. Often a linear summary of them weighted by their concentration is assumed to explain the macroscopic term:

$$\mu_e = \sum_{k \in K} c_k \mu_e^k \tag{5.10}$$

where k identifies individual molecules contributing to the extinction from all those present in the tissue K. This formulation assumes single scattering in which different events are independent of each other (Bonner et al. 1987) and often suffices since high concentrations of materials are unusual in biological media. However, a more elaborated multiple scattering formulation is also possible.

The different specific extinction profiles are not whimsical. They reflect the molecule's geometry and size. A naive, but effective, model assumes that the molecules are spherical and that absorption by a molecule is proportional to its size (cross-sectional area) and the shadow it casts, namely, its absorption efficiency.

INVERSE PROBLEM

Inverse problems arise under situations where it is not possible to measure directly the parameters of a system, and instead they need to be inferred from measurements in the boundary (Kabanikhin 2008; Wang et al. 2011). This is the case in optical imaging; the aim is to achieve optical properties of the tissue from light measurements in the boundary, known as image reconstruction. Figure 5.5 illustrates the data transformation from a fictitious discretized space of histological parameters Π to the image space I through the projection of the forward model F, and finally, the nosy sensor

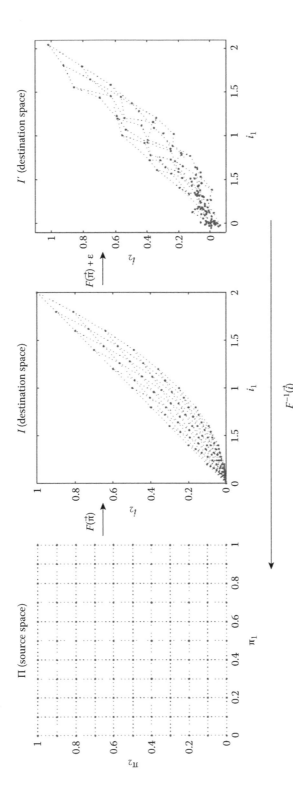

FIGURE 5.5 Fictitious nonlinear mapping from parameter space Π (left), to image space I (center) and noisy space I' (right).

space I'. Thus, the inverse problem is a mapping from image values to histological properties F^{-1}: $I' \rightarrow \Pi$ (Claridge and Hidovic-Rowe 2014).

Inverse problems are often ill-posed, meaning that they are not fulfilled with conditions of well-posed problems: the solution exists, is unique, and is continuous given the data (Hadamard 1902).

To deal with ill-posed problems, it is important to have a proper design of the inversion strategy, the regularization (to achieve stability), and the optimization algorithm (Vogel 2002; Wang and Yuan 2005).

Image Reconstruction Methods

Sensed light at tissue surface is a function of the changes in the optical properties inside a tissue. This relation is inherently nonlinear (Dehghani et al. 2009a). However, the methods to solve the inverse problem range from linear to nonlinear (Branco 2007; Arridge and Schotland 2009; Dehghani et al. 2009b). In the following lines, we provide a brief overview of the reconstruction methods, but the reader is invited to check the study of Arridge and Hebden (1997), Dehghani et al. (2009b), Arridge (2011), and Jiang (2011) for a more profound treatment of the topic.

Linear Methods

Linear methods are the simplest to address the reconstruction problem. These methods are able to produce images of temporal changes in the optical properties of the tissue (Dehghani et al. 2009a) and require a differential measure of sensed light in time $\Delta y = (y_1 - y_0)$ as the difference in changes of the optical properties $\Delta \pi = (\pi_1 - \pi_0)$ occur.

This is the case for the MBLL, which is used to recover directly the information of interest in optical neuroimaging (Elwell et al. 1994; Hueber et al. 2001; Ferrari et al. 2004; Boas et al. 2011). The reconstruction algorithm resolves the system of equations derived from the MBLL (Equation 5.7) itself for the number of chromophores of interest, that is, HbO_2 and HHb for fNIRS.

As already explained earlier, to apply this method, measurements in times t_0 and t_1 are required to get a relative change of light intensity ($\Delta OD(\Delta t, \lambda)$). To obtain the concentration changes of "n" chromophores of interest, that equation should be evaluated at "n" different wavelengths. Then, the resulting system to solve is

$$\Delta C_i = \left[\frac{\Delta OD\left(\Delta t, \lambda_j\right)}{DPF\left(\lambda_j\right)} \right] \frac{1}{d\epsilon_i\left(\lambda_j\right)} \tag{5.11}$$

However, this approach's main drawbacks are as follows: (1) the assumption on geometry and scattering (the G term in Equation 5.7) does not hold true for measurements from the brain and (2) the changes in light attenuation are attributable directly to changes in hemoglobin in the cerebral cortex, ignoring systemic interference and other losses (Orihuela-Espina et al. 2010). Despite these strong assumptions, its use in imaging neural activity has been widely accepted in neuroscience research because of the high correlation of observable/measurable responses to stimuli.

Nonlinear Methods

Nonlinear reconstruction methods are iterative methods that are based in optimization looking to minimize the difference between the values calculated by the forward model Φ^C and the experimental data Φ^M (Jiang et al. 1996). Two main approaches for such optimization have been used (Dehghani et al. 2009a). These are classical methods to solve nonlinear systems equations, like the gradient-based reconstruction (Zhu et al. 1997) and Newton-like methods (Jiang et al. 2000; Jiang 2011). The latter requires the calculation of the Jacobian matrix and their inversion in each iteration.

In general, the procedure followed in these methods proceeds as follows:

Given an initial guess of the optical parameter under study (μ_0), a measure $\hat{y} = G\left(\mu_0\right)$ is computed with the forward model given the initial parameter. From these, the Jacobian matrix is calculated (J). Then the distance between $\hat{y}$ and an experimental data y is measured.

If the distance is greater than a given threshold, the Jacobian matrix has to be inverted and regularized in order to estimate a new value for μ. The method iterates until the μ value has been closely approximated.

An important point of success of these methods relies on how closely the initial estimate approaches the correct solution (Arridge and Hebden 1997).

Alternative Approaches

Alternative methods to reconstruct image problems have been proposed. One of them is the so-called coloration map. This is a lookup table in which new observations are approximated looking for the estimated by parameter $\hat{\pi}$ (Claridge et al. 2002; Orihuela-Espina 2005; Claridge and Hidovic-Rowe 2014). In this method, the space of the parameters to be recovered is discretized and each vector is projected to the spectral space through the forward model. Each remitted spectrum is convolved with a response function simulating image acquisition process by the device and projecting each spectrum to the image space. Then the reconstruction is given by

$$\hat{\pi} = \arg \min_{\hat{\pi}} \left\| F\left(\hat{\pi}\right) - \vec{i} \right\| \tag{5.12}$$

However, this approach is severely affected by the coarseness of the parameter space discretization (Orihuela-Espina 2005). We have used this approach as a preliminary step to show the distortive effect of scalp as seen in the next section (Herrera-Vega and Orihuela-Espina 2015).

Another model used by Claridge and Hidovic-Rowe (2014) uses a discrete Markov random field (MRF) optimization by using an iterated conditional model algorithm that maximizes the probability of obtaining each variable in the MRF to get the parameter $\hat{\pi}$ with the highest probability. A range of machine learning approaches have also been attempted (Fesller 2000).

AN EXAMPLE OF IMAGE RECONSTRUCTION IN fNIRS

Here we present a simple example of image formation and reconstruction applied to neuroimaging using fNIRS and illustrate the noise introduced by the scalp even when scalp blood flow is neglected. As a forward model, we have used the Monte Carlo technique to estimate light propagation in tissue and the remitted spectra as a function of the changes in optical properties due to the physiological changes in the tissue. A homogeneous flat model of the adult human head (Figure 5.6) was defined,

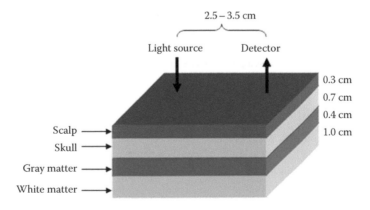

FIGURE 5.6 A naïve flat homogeneous model of the adult human head used to compute the remitted spectrum at the tissue surface. The model is inspired by the work of Okada and Delpy (2003), but was modified to introduce the effect of hemoglobin species in the gray matter.

inspired by the work of Okada and Delpy (2003). The model consists of four main layers, namely, the scalp, skull, gray matter, and white matter, with each layer being defined by its optical properties (n, μ_a, μ_s, g, respectively) in the wavelength range from 700 to 1000 nm.

fNIRS brain imaging is carried out through measuring the hemoglobin concentration in the cerebral cortex (Strangman et al. 2002). To simulate such activity in our forward model (Figure 5.6), we consider that tissue absorption depends linearly on the concentrations of chromophores in the tissue (Durduran et al. 2010). The absorption coefficient of the gray matter layer was altered in function of the changes in hemoglobin species (HbO$_2$, HHb) concentrations as follows:

$$\mu^{GM}(\lambda) = \mu_a^{GM}(\lambda) + SAC^{HbO_2}(\lambda) * C^{HbO_2} + SAC^{HHb}(\lambda) * C^{HHb} \qquad (5.13)$$

where

$\mu_a^{GM}(\lambda)$ is the wavelength-dependent absorption coefficient for the gray matter

$SAC^{HbO_2}(\lambda)$ and $SAC^{HHb}(\lambda)$ are the specific absorption coefficients for HbO$_2$ and HHb, as described in Equation 5.1, respectively

C^{HbO_2} and C^{HHb} are the mentioned hemoglobin concentrations.

Figure 5.7 shows the remitted spectrum obtained by considering the variations of this model with different number of layers, thus permitting an appreciation of the effect produced by each layer on the final spectrum. In the figure, the simulated model considers only two layers (the gray and white matters), thus corresponding to a fictitiously exposed brain. Without other layers on the top absorbing or scattering light, this model predicts a monotonic behavior of the remitted spectrum as a function of the change in the hemoglobin concentration. In the middle plot, the skull layer was included and the spectrum becomes noisy, but still the monotonic behavior depending on the cortical hemoglobin concentration can be appreciated. Figure 5.7 shows how the spectrum with the scalp layer included (note, however, that the scalp blood flow was not introduced) increases the noise in the signal, hindering reconstruction (Figure 5.8).

For a more complete version of the model including all four layers, Figure 5.9 shows a reconstruction made by considering a histologically plausible but arbitrary range of hemoglobin concentrations in the gray matter. The forward remitted spectrum was convolved with idealized Gaussian responses of optical filters that project the spectrum to a new point in the coloration map. A naive reconstruction was carried out by the nearest neighbor approach by using Euclidean distance to get the most similar point in the coloration map. The approximated concentration pair of HbO$_2$–HHb can be obtained from the transfer of the parameter space to the coloration map. Highlighting the poor signal-to-noise ratio, in this example, the physiological parameters could not be retrieved correctly. Equivalent simulations (Herrera-Vega and Orihuela-Espina 2015) agree with the literature that this effect is mainly attributable to the noisy spectrum produced by the scalp layer. A correct reconstruction was obtained by using the three-layer model (i.e., without the scalp layer) as shown in Figure 5.9.

FAST OPTICAL SIGNAL

In 1945, it was discovered that neuronal activity causes changes in the optical properties of nerve cells (Hill and Keynes 1949). Several studies have then reported transient optical changes in isolated large nerves or neurons during excitation (Hill 1950; Cohen et al. 1968; Stepnoski et al. 1991; Carter et al. 2004; Yao et al. 2005).

Attempting to do noninvasive brain imaging is difficult because the detection of optical changes require well-defined geometries for the measurements of collective neural activity in the bulk tissue that contains a considerable number of neurons and other cells. Compared with the slow hemodynamic signal, the fast signal in NIRS occurs within milliseconds after the stimulation.

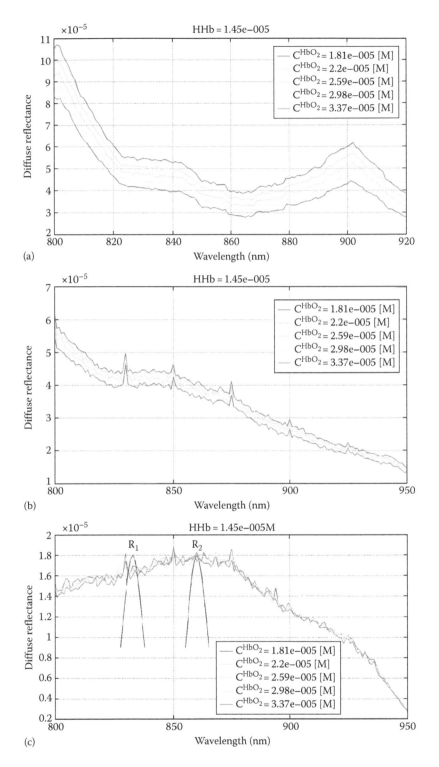

FIGURE 5.7 Remitted spectrum measured between 2.5 and 3.5 cm from the light source position for a model with (a) two layers (wm, gm), (b) three layers (wm, gm, and skull), and (c) four layers (wm, gm, skull, and scalp), respectively. Each case is modeled with five concentrations of HbO₂ and HHb. The Gaussians labeled as R_1 and R_2 represent the idealized responses of two virtual optical filters used to produce the coloration map in Figure 5.8.

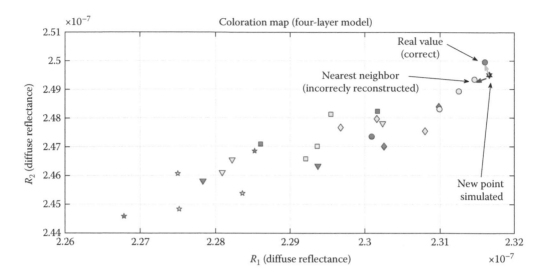

FIGURE 5.8 Coloration map for the four-layer model. The marker shapes represent levels of HbO₂ concentration, and the marker color indicates HHb concentrations.

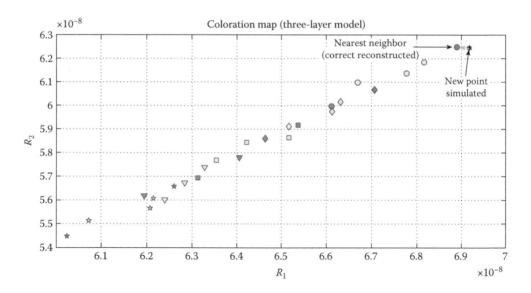

FIGURE 5.9 Coloration map for the three-layer model. Symbols indicate HbO₂ and their color indicates the HHb concentrations.

Although Stepnoski et al. (1991) showed that the neural activity is linked to a change in the light scattering, there is still a controversy of whether the in vivo detection of the fast signal is ultimately due to changes in absorption or scattering (Franceschini and Boas 2004). At least three hypotheses have been proposed to explain the biophysical phenomena underlying the FOS (Gratton and Fabiani 2001):

- The FOS is produced by changes in the scattering properties of the tissues upon neuronal activity. As indicated, these changes are the result of changes in the cell membrane and flux of ions and water.

- The FOS recorded in vivo with the intact scalp is the manifestation of rapid deoxygenation of the hemoglobin contained in brain tissue due to rapid oxygen consumption in a localized cortical region.
- The FOS obtained with intact scalp reflects rapid changes in brain blood flow or blood volume.

The work by Gratton and Fabiani (2001) tested the first two hypotheses and concludes in favor of the scattering hypothesis. Franceschini and Boas also indicate this as the most likely explanation (Franceschini and Boas 2004). However, failure to consistently detect the FOS using the phase of the light means that an absorption origin of the FOS cannot be excluded (Morren et al. 2004).

Other research suggests that FOS originates from nerve impulses, also known as action potentials, from the swelling of the nerve or companion cells, or from a brief period of anaerobic metabolism that alters the tissue transparency (Cohen 1973; Rector et al. 1995; Malonek and Grinvald 1996; Cannestra et al. 2001).

Evidence suggests that diffuse optical methods can detect cell swelling that occurs in the 50–200 ms following neuronal firing (Steinbrink et al. 2000; Gratton and Fabiani 2001). This type of "fast" optical signal appears to be significantly smaller than the hemodynamic signals (in the order of a 0.01% signal change) (Strangman et al. 2002). Therefore, the controversy over the possibility of detecting FOS in vivo has seen some light from promising results (Steinbrink et al. 2005; Radhakrishnan et al. 2009; Chiarelli et al. 2014).

Single Neuron Observation

To investigate the source of optical changes in neurons, we have performed single neuron measurements. Retzius neurons from the leech (*Hirudo medicinalis*) were isolated by using a glass pipet and cultured at 18°C–20°C. Recordings were made between 1 and 7 days later.

Observations were carried out on a homemade microscope using a titanium sapphire–pulsed laser Chameleon Ultra II from Coherent Inc. as an illumination source. The laser produces 120 fs pulses at 80 MHz repetition rate. We used 20 mW average power to prevent neuron damage. The image in Figure 5.10 was obtained with a CCD camera using a 40× microscope objective (Nikon CFI Fluor 4× W).

The FOS was recorded by focusing the light on the neuron and recording the transmitted light using a PIN silicon detector. The optical signal converted into an output current in the detector was then converted into a voltage, which was digitized with a PCI 6110 (National Instruments) converter.

The −60 mV resting potential of the neuron was depolarized by increasing the external potassium concentration from 4 to 53 mM upon addition of solution to the close proximity of the neuron. This manipulation produced an average 2 ms fast transient decrease in the transmitted light lasting 1–3 ms (Figure 5.11). This change is extremely fast with respect to any detectable morphological change (De-Miguel et al. 2015). Therefore, we suspect that these changes are due to the fast depolarization of the neuron membrane prior to the morphological changes (Steinbrink et al. 2000; Gratton and Fabiani 2001).

FUTURE WORK

To study optical changes in the nervous system, we will perform two different approaches.

Neuron Model

Following the observations using potassium burst, we looked into other fields of science and found that optical gating, widely used in nonlinear optical characterization, has the potential of detecting extremely small optical response.

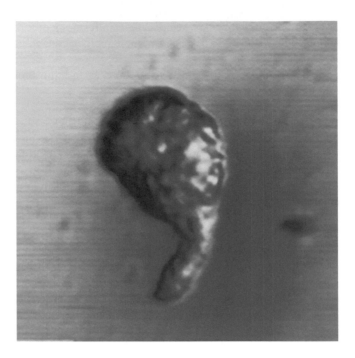

FIGURE 5.10   Isolated Retzius neuron observed on a confocal microscope using 40× objective. The diameter of the neuron is about 70 µm, and from it the remaining portion of a primary axon remains pointing downward.

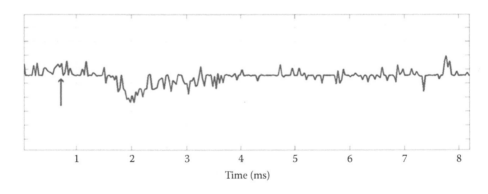

Time (ms)

FIGURE 5.11   Transmitted light recorded from a Retzius neuron upon potassium depolarization burst, applied at the time pointed by the arrow.

In order to detect FOS, we use a second harmonic ultrafast fiber laser in combination with our modified multiphoton confocal microscope. Analysis will be carried out using second harmonic generation frequency-resolved optical gating (Trebino et al. 1997). This technique, widely used in nonlinear optical characterization, has a potential of detecting small optical response. The signal is correlated with a periodic stimuli applied on individual neurons placed on the modified microscope.

EXPERIMENTS IN ISOLATED NEURONS

Leech Retzius neurons in culture will be electrically stimulated by intracellular injection of square current pulses or pseudo-random noise to change in the resting membrane potential. Injection of positive current pulses will also be used to a produce action potentials.

Voltage responses recorded by bridge-balanced microelectrodes will be correlated with the optical signals evoked by electrical stimulation.

Additional experiments will be made in neurons maintained inside their ganglia to explore the contribution of extracellular space and glial responses to the FOSs. The intracellular spaces mediating neurons and glia have been characterized by the use of electron microscopy; therefore, focusing the light in particular areas of the ganglion will provide evidence about the contribution of glial cells, neurons, and extracellular spaces to the FOS.

RAT MODEL

Anesthetized male Wistar rats will be implanted with a fixation device using dental cement, allowing the optic fiber to be near the occipital cortex. Additionally, stainless steel skull screws will be implanted through the skull into the visual cortex (7 mm behind the bregma and 3 mm lateral of the midline) as the positive electrodes. A reference screw electrode will be placed on the midline, 3 mm rostral to the bregma. Dental cement will also be used to fix the screws. One week after, the skin on the head is closed and the animal has recovered from the surgery. After correct signal amplification, filtration, and processing, to evaluate the correlation between the FOS and the electroencephalographic signal, we will evoke visual potentials to evaluate the function of the visual pathway. Light stimulation of the eye evokes visual electrical signals in the visual cortex. The visual potentials are analogous to auditory somatosensory evoked potential, which produce electroencephalographic cortical evoked potentials.

The cortical activity will be measured on three separate days within 2 weeks. The animals will be anesthetized, placed in a dark room, and allowed to adapt dark for 5 min. The pupils will be dilated with 1.0% tropicamide eye drops. The positive screw over the contralateral visual cortex of the stimulated eye and the reference screw will be connected to an electrical signal amplifier. A needle electrode will be inserted into the tail as the ground electrode. Visual stimuli will be generated by a flash LED delivered 100 times at a frequency of 1 Hz. The time of measurement will be 200 ms, at 5 kHz sampling. Responses will be amplified 20,000 times and filtered by bandpass filter, with low and high cutoff frequencies of 1 and 100 Hz, respectively.

PERSPECTIVES

FOS has been treated in the literature as a change in the optical properties of neuron due to a single event. The extremely fast response observed in our experiment suggests that it is possible that there are more than one processes over time producing the FOS. Based on the previous neuron and rat models, we will be in the position to further clarify the source for the FOS, which we suspect is due to a cascade of mechanism and not just a single process, or ratify that there is a single event producing the optical changes.

SUMMARY

DOI is a neuroimaging modality that can measure noninvasively functional brain mapping. DOI is noninvasive and does not require sophisticated preparation. Without being openly mentioned, it has the tremendous advantage of being mounted on a portable device, which is compatible with other techniques such as EEG.

A major drawback is the complexity in the reconstruction process, which is far from being trivial. When we observe the hemodynamic response, fNIRS measures the indirect brain activity manifested through neurovascular coupling. The same noninvasive procedure can potentially be used to detect direct neural activity through neuron optical changes manifested in the FOS.

Therefore, DOI in its both modalities (fNIRS and FOS) is an emerging technique that complements neurophotonic imaging. We have glimpsed its principles and reconstruction process.

The observations in our single neuron model open the possibility for complex optical-induced physiological process with more than one source for FOS. Furthermore, when this technique matures, it can be a feasible candidate for a successful brain–human interface.

ACKNOWLEDGMENTS

The authors acknowledge funding from CONACyT Mexico under grant CB-2011-01-169558 and the contribution of Ana Isabel Rua Graciano to the present work.

REFERENCES

Arridge S.R. Methods in diffuse optical imaging. *Philosophical Transactions of the Royal Society A* 369 (2011):4558–4576.

Arridge S.R., Cope M., Delpy D.T. The theoretical basis for the determination of optical pathlengths in tissue: Temporal and frequency analysis. *Physics in Medicine and Biology* 37, 7 (1992):1531–1560.

Arridge S.R., Hebden J.C. Optical imaging in medicine: II. Modelling and reconstruction. *Physics in Medicine and Biology* 42 (1997):841–853.

Arridge S.R., Schotland J.C. Optical tomography: Forward and inverse problems. *Inverse Problems* 25, 12 (2009):123010.

Baillet S., Garnero L. A Bayesian approach to introducing anatomo-functional priors in the eeg/meg inverse problem. *IEEE Transactions on Biomedical Engineering* 44, 5 (1997):374–385.

Bass M., DeCusatis C., Enoch J. et al. *Handbook of Optics*. McGraw-Hill Education, New York, 2009.

Boas D.A., Elwell C.E., Ferrari M. et al. Twenty years of functional near-infrared spectroscopy: Introduction for the special issue. *NeuroImage* 85 (2014):1–5.

Boas D.A., Gaudette T., Strangman G. et al. The accuracy of near infrared spectroscopy and imaging during focal changes in cerebral hemodynamics. *NeuroImage* 13 (2011):76–90.

Bonner R.F., Nossal R., Havlin S., Weiss G.H. Model for photon migration in turbid biological media. *Journal of the Optical Society of America A* 4, 3 (1987):423–432.

Boreman G.D. *Basic Electro-Optics for Electrical Engineers*, 1st edn. SPIE Press, Bellingham, WA, 1998.

Boulnois J.L. Photophysical processes in recent medical laser developments: A review. *Lasers in Medical Science* 1 (1986): 47–66.

Branco G. The development and evaluation of head probes for optical imaging of the infant head. PhD thesis, University College London, London, U.K., 2007.

Cannestra A.F., Pouratian N., Bookheimer S. et al. Temporal spatial differences observed by functional MRI and human intraoperative optical imaging. *Cerebral Cortex* 11, 8 (2001):773–782.

Carter K.M., George J.S., Rector D.M. Simultaneous birefringence and scattered light measurements reveal anatomical features in isolated crustacean nerve. *Journal of Neuroscience Methods* 135 (2004):9–16.

Chance B., Cope M., Gratton E. et al. Phase measurement of light absorption and scatter in human tissue. *The Review of Scientific Instruments* 69 (1998):3457–3481.

Chiarelli A.M., Romani G.L., Merla A. Fast optical signals in the sensorimotor cortex: General Linear Convolution Model applied to multiple source-detector distance-based data. *NeuroImage* 85 (2014):245–254.

Claridge E., Cotton S., Hall P. et al. From colour to tissue histology: Physics based interpretation of images of pigmented skin lesions. In *Medical Image Computing and Computer-Assisted Intervention*, Volume 2488 of Lecture Notes in Computer Science, pp. 730–738. Springer, Berlin, Germany, 2002.

Claridge E., Hidovic-Rowe D. Model based inversion for deriving maps of histological parameters characteristics of cancer from ex-vivo multispectral images of the colon. *IEEE Transactions on Medical Imaging* 33, 4 (2014):822–835.

Cohen, L.B. Changes in neuron structure during action potential propagation and synaptic transmission. *Physiological Reviews* 53, 2 (1973):373–417.

Cohen L.B., Keynes R.D., Hille B. Light scattering and birefringence changes during nerve activity. *Nature* 218 (1968):438–441.

Colak S.B., Papaioannou D.G., Hooft G.W.T. et al. Tomographic image reconstruction from optical projections in light-diffusing media. *Applied Optics* 36, 1 (1997):180–213.

Cope M., Delpy D.T., Reynolds E.O.R. et al. Methods of quantitating cerebral near infrared spectroscopy data. *Advances in Experimental Medicine and Biology* 215 (1988):183–189.

Dehghani H., Srinivasan S., Pogue B.W. et al. Numerical modelling and image reconstruction in diffuse optical tomography. *Philosophical Transactions. Series A, Mathematical, Physical, and Engineering Sciences* 367 (2009a):3073–3093.

Dehghani H., White B.R., Zeff Benjamin W. et al. Deep sensitivity and image reconstruction analysis of dense imaging arrays for mapping brain function with diffuse optical tomography. *Communications in Numerical Methods in Engineering* 48 (2009b):D137–D143.

Delpy D.T., Cope M. Quantification in tissue near-infrared spectroscopy. *Philosophical Transactions of the Royal Society of London. Series B, Biological Sciences* 352, 1354 (1997):649–659.

Delpy D.T., Cope M., Van der Zee P. Estimation of optical path length through tissue from direct time of flight measurement. *Physics in Medicine and Biology* 33, 12 (1988):1433–1442.

De-Miguel F.F., Leon-Pinzon C., Noguez P. et al. Serotonin release from the neuronal cell body and its long-lasting effects on the nervous system. *Philosophical Transactions of the Royal Society B* 370 (2015):20140196.

Durduran T., Choe R., Baker W.B. et al. Diffuse optics for tissue monitoring and tomography. *Reports on Progress in Physics* 73 (2010):1–43.

Elwell C.E., Cope M., Edwards A.D. et al. Quantification of adult cerebral hemodynamics by near-infrared spectroscopy. *Journal of Applied Physiology* 77, 6 (1994):2753–2760.

Eppstein M.J., Dougherty D.E., Hawrysz D.J. et al. Three-dimensional Bayesian optical image reconstruction with domain decomposition. *IEEE Transactions on Medical Imaging* 20, 3 (2001):147–163.

Fang Q., Boas D.A. Monte Carlo simulation of photon migration in 3d turbid media accelerated by graphics processing units. *Optics Express* 17, 22 (2009):20178–20190.

Fantini S., Franceschini M.A. Frequency-domain techniques for tissue spectroscopy and imaging. In Tuchhin V.V., ed. *Handbook of Optical Biomedical Diagnostics*, Vol. 7, pp. 405–453. SPIE Press, Bellingham, WA, 2002.

Fantini S., Franceschini M.A., Maier J.S. et al. Frequency-domain multichannel optical detector for noninvasive tissue spectroscopy and oximetry. *Optical Engineering* 34, 1 (1995):32–42.

Ferrari M., Mottola L., Quaresima V. Principles, techniques, and limitations of near infrared spectroscopy. *Canadian Journal of Applied Physiology* 229 (2004):463–487.

Ferrari M., Quaresima V. A brief review on the history of human functional near-infrared spectroscopy (fNIRS) development and fields of application. *NeuroImage* 63, 2 (2012):921–935.

Fesller J.A. Statistical image reconstruction methods for transmission tomography. In *Handbook of Medical Imaging*, Vol. 2: Medical Image Processing and Analysis, pp. 1–70. SPIE Press, Bellingham, WA, 2000.

Franceschini M.A., Boas D.A. Noninvasive measurement of neuronal activity with near-infrared optical imaging. *NeuroImage* 21 (2004):372–386.

Gratton G., Fabiani M. Shedding light on brain function: The event-related optical signal. *Trends in Cognitive Sciences* 5, 8 (2001):357–363.

Hadamard J. Sur les problemes aux derivées partielles et leur signification physique. *Princeton University Bulletin* 13 (1902):49–52.

Hall G., Jacques S.L., Eliceiri K.W. et al. Goniometric measurements of thick tissue using Monte Carlo simulations to obtain the single scattering anisotropy coefficient. *Biomedical Optics Express* 3, 11 (2012):2707–2719.

Herrera-Vega J., Orihuela-Espina F. Image reconstruction in functional optical neuroimage: The modelling and separation of the scalp blood flow. Technical report, Instituto Nacional de Astrofísica Óptica y Electrónica, Tonantzintla, Puebla, 2015.

Hill D.K. The volume change resulting from stimulation of a giant nerve fibre. *The Journal of Physiology* 111, 3–4 (1950):304–327.

Hill D.K., Keynes R. Opacity changes in stimulated nerve. *The Journal of Physiology* 108, 3 (1949):278–281.

Horecker B.L. The absorption spectra of hemoglobin and its derivatives in the visible and near infrared regions. *The Journal of Biological Chemistry* 148 (1943):173–183.

Hueber D.M., Franceschini M.A., Ma H.Y. et al. Non-invasive and quantitative near-infrared haemoglobin spectrometry in the piglet brain during hypoxic stress, using a frequency-domain multidistance instrument. *Physics in Medicine and Biology* 46, 1 (2001):41–62.

Hüfner G. Neue Versuche zur Bestimmung der Sauerstoffcapacität des Blutfarbstoffs. *Virchows Archiv fur pathologische Anatomie und Physiologie und fur klinische Medizin* 55 (1894): 130–176.

Jacques S.L., Pogue B.W. Tutorial on diffuse light transport. *Journal of Biomedical Optics.* 13, 4 (2008):041302.

Jiang H. *Diffuse Optical Tomography*. CRC Press, Boca Raton, FL, 2011.

Jiang H., Paulsen K.D., Osterberg U.L. et al. Optical image reconstruction using frequency domain data: Simulations and experiments. *Journal of the Optical Society of America A* 13 (1996):253–266.

Jiang H., Xu Y., Iftimia N. Experimental three dimensional optical image reconstruction of heterogeneous turbid media from continuous-wave data. *Optics Express* 7, 5 (2000):204–208.

Jobsis F.F. Noninvasive, infrared monitoring of cerebral and myocardial oxygen sufficiency and circulatory parameters. *Science* 198, 4323 (1977):1264–1267.

Kabanikhin S.I. Definitions and examples of inverse and ill-posed problems. *Journal of Inverse and Ill-posed Problem* 16 (2008):317–357.

Kleinschmidt A., Obrig H., Requardt M. et al. Simultaneous recording of cerebral blood oxygenation changes during human brain activation by magnetic resonance imaging and near-infrared spectroscopy. *Journal of Cerebral Blood Flow and Metabolism* 16, 5 (1996):817–826.

Klose A.D., Larsen E.W. Light transport in biological tissue based on the simplified spherical harmonics equations. *Journal of Computational Physics* 220, 1 (2006):441–470.

Malonek D., Grinvald A. Interactions between electrical activity and cortical microcirculation revealed by imaging spectroscopy: Implications for functional brain mapping. *Science* 272 (1996):551–554.

Morren G., Wolf U., Lemmerling P. et al. Detection of fast neuronal signals in the motor cortex from functional near infrared spectroscopy measurements using independent component analysis. *Medical & Biological Engineering & Computing* 42, 1 (2004):92–99.

Newton R.G. *Scattering Theory of Waves and Particles*. Springer-Verlag, New York, 1982.

Nolte D.D. *Optical Interferometry for Biology and Medicine*. Springer-Verlag, New York, 2012.

Okada E., Delpy D.T. Near-infrared light propagation in an adult head model. I. Modeling of low-level scattering in the cerebrospinal fluid layer. *Applied Optics* 42 (2003):2906–2914.

Orihuela-Espina F. Modelling and verification of the diffuse reflectance of the ocular fundus. PhD thesis, Department of Computer Science, University of Birmingham, Birmingham, U.K., 2005.

Orihuela-Espina F., Leff D.R., James D.R.C. et al. Quality control and assurance in functional near infrared spectroscopy (FNIRS) experimentation. *Physics in Medicine and Biology* 55 (2010):3701–3724.

Perutz, M. Stokes and haemoglobin. *Biological Chemistry Hoppe-Seyler* 376, 8 (1995): 449–450.

Prahl S.A. Light transport in tissue. PhD thesis, The University of Texas at Austin, Austin, TX, 1988.

Radhakrishnan H., Vanduffel W., Deng H.P. et al. Fast optical signal not detected in awake behaving monkeys. *NeuroImage* 45, 2 (2009):410–419.

Rector D.M., Poe G.R., Kristensen M.P. et al. Imaging the dorsal hippocampus: Light reflectance relationships to electroencephalographic patterns during sleep. *Brain Research* 696, 1–2 (1995):151–160.

Richards-Kortum R., Sevick-Muraca E. Quantitative optical spectroscopy for tissue diagnosis. *Annual Review of Physical Chemistry* 47 (1996):555–606.

Scholkmann F., Kleiser S., Metz Andreas J. et al. A review on continuous wave functional near-infrared spectroscopy and imaging instrumentation and methodology. *NeuroImage* 85 (2014):6–27.

Severinghaus J.W. Takuo Aoyagi: Discovery of pulse oximetry. *Anesthesia and Analgesia*. 105, 6 Suppl. (2007):S1–S4.

Steinbrink J., Kempf F.C., Villringer A. et al. The fast optical signal—Robust or elusive when non-invasively measured in the human adult? *NeuroImage* 26, 4 (2005):996–1008.

Steinbrink J., Kohl M., Obrig H. et al. Somatosensory evoked fast optical intensity changes detected non-invasively in the adult human head. *Neuroscience Letters* 291, 2 (2000):105–108.

Stepnoski R.A., LaPorta A., Raccuia-Behling F. et al. Noninvasive detection of changes in membrane potential in cultured neurons by light scattering. *Proceedings of the National Academy of Sciences* 88 (1991):9382–9386.

Strangman G., Boas D.A., Sutton J.P. Non-invasive neuroimaging using near-infrared light. *Biological Psychiatry* 52, 7 (2002):679–693.

Torricelli A., Contini D., Pifferi A. et al. Time domain functional NIRS imaging for human brain mapping. *NeuroImage* 85 (2014): 28–50.

Trebino R., DeLong K.W., Fittinghoff D.N. et al. Measuring ultrashort laser pulses in the time-frequency domain using frequency-resolved optical gating. *Review of Scientific Instruments* 68 (1997):3277–3295.

Uludag K., Dubowitz D.J., Yoder E.J. et al. Coupling of cerebral blood flow and oxygen consumption during physiological activation and deactivation measured with fMRI. *NeuroImage* 23, 1(2004):148–155.

Van de Hulst H.C. *Light Scattering by Small Particles*. John Wiley & Sons, Inc., New York, 1957.

Villringer A., Chance B. Non-invasive optical spectroscopy and imaging of human brain function. *Trends in Neuroscience* 20 (1997):435–442.

Vogel C.R. *Computational Methods for Inverse Problems*. SIAM, Philadelphia, PA, 2002.

Wang L., Jacques S.L., Zheng L. MCML-Monte Carlo modeling of light transport in multi-layered tissues. *Computer Methods and Programs in Biomedicine* 47 (1995):131–146.

Wang Y., Yang C., Yagola A.G. *Optimization and Regularization for Computational Inverse Problems and Applications*. Springer, Berlin, Germany, 2011.

Wang Y.F., Yuan Y.X. Convergence and regularity of trust region methods for nonlinear ill-posed inverse problems. *Inverse Problems* 21 (2005):821–838.

Willig K.I., Rizzoli S.O., Westphal V. et al. STED microscopy reveals that synaptotagmin remains clustered after synaptic vesicle exocytosis. *Nature* 440 (2006):935–939.

Wolf M., Ferrari M., Quaresima V. Progress of near-infrared spectroscopy and topography for brain and muscle clinical applications. *Journal of Biomedical Optics* 12 (2007):062104.

Wray S., Cope M., Delpy D.T. et al. Characterisation of the near infrared absorption spectra of cytochrome aa3 and haemoglobin for the non invasive monitoring of cerebral oxygenation. *Biochimica et Biophysica Acta* 933 (1988):184–192.

Yao X.C., Foust A., Rector D.M. et al. Cross-polarized reflected light measurement of fast optical responses associated with neural activation. *Biophysical Journal* 88, 6 (2005):4170–4177.

Zepeda A., Arias C., Sengpiel E. Optical imaging of intrinsic signals: Recent developments in the methodology and its applications. *Journal of Neuroscience Methods* 136, 1 (2004):1–21.

Zhu C., Liu Q. Review of Monte Carlo modeling of light transport in tissues. *Journal of Biomedical Optics* 18, 5 (2013):50902.

Zhu W., Wang Y., Yao Y. et al. Iterative total least-squares image reconstruction algorithm for optical tomography by the conjugate gradient method. *Journal of the Optical Society of America* 14, 4 (1997):799–807.

6 Voltage-Sensitive Dye and Intrinsic Signal Optical Imaging

Advantages and Disadvantages for Functional Brain Mapping

Vassiliy Tsytsarev and Reha S. Erzurumlu

CONTENTS

INTRODUCTION

Brain imaging makes it possible to associate neural activity of specific regions with a variety of physiological reactions of the organism. Functional brain mapping has improved and expanded in impressive ways within the last several years (reviewed in Frostig, 2009). The technology of detection and analysis of emitted and reflected photons in vivo provides a powerful imaging approach that permits visualization of neural activity at the systems levels. In addition, x-ray computed tomography (CT), positron emission tomography (PET), functional magnetic resonance imaging (fMRI), and diffusion tensor imaging have provided neuroscientists with tools to observe functional and connectional anatomy and to monitor brain functions in healthy and diseased humans and other species in noninvasive ways (Raichle, 2003; Zhang and Raichle, 2010; Choudhri et al., 2014; Miller et al., 2014).

In vivo brain optical imaging techniques are different from CT, PET, and fMRI because they are based on the optical features of the nervous tissue. In contrast to PET and fMRI, optical imaging methods are mainly 2D due to very low light conductivity of the brain parenchyma. Optical imaging varies at the levels of spatial (up to microns) and temporal (from tens of microseconds to seconds) resolutions. The invasiveness can also be quite different: optical imaging methods can be applied to the exposed brain after removal of the bone above the imaged area, or transcranially, through the skin, muscles, and skull.

From a physiological point of view, the brain optical imaging methods can be classified into two categories: use of external fluorescence probes (extrinsic) and use of intrinsic signals. Both are employed in vast subject areas of basic neuroscience research at the present time, and some types of intrinsic optical signals (IOS) are also suitable for use in clinical neuroscience. Here we describe the physical and chemical bases, equipment, and principles of analysis of the intrinsic and voltage-sensitive dye optical imaging (VSDi) techniques.

OPTICAL IMAGING OF INTRINSIC SIGNALS

IOS imaging is probably the oldest optical imaging technique that aims to visualize the neural activity in the cerebral cortex (Salzberg et al., 1973; Grinvald et al., 1977; Cohen et al., 1978). The brain IOS is based on different physicochemical features of the tissue metabolism, including changes in the cerebral blood flow, hemoglobin oxygenation and deoxygenation, movement of ion and water molecules through the cell membrane, and probably some other unknown elements. All these components change the level of light absorption and therefore can be monitored using different illumination wavelengths (Figure 6.1). In most cases the most effective way to visualize a functional map is based on the slow changes in the optical properties of the brain parenchyma, permitting localization of the activated areas. These optical changes are usually low in the absolute value and masked by technical and biological noise. While IOS consists of components with different biological origins, it has been demonstrated that functional maps of the neocortex generated by using different wavelengths are quite similar (Grinvald et al., 2001).

The physiological basis of the IOS is the initial increasing of deoxyhemoglobin concentration, due to rising oxygen consumption; this process is called "the initial dip" (Bahar et al., 2006; Takeshita and Bahar, 2011). In a few hundreds of milliseconds or seconds after the initial dip, the blood flow increases, bringing oxyhemoglobin to the activated cortical area. This is the second component of the IOS. The third component is the changes in blood volume, caused by the vasodilatation and the fast filling of capillaries in the activated cortical area.

All of these three components dominate the IOS at wavelengths ~400–600 nm. An additional component of the IOS, which is the only outcome not associated with the changes in light absorption, is light scattering. There is consensus that this component is based on the ion movement through the cell membranes and changes in extracellular spaces. Light scattering produces significantly high signal above 630 nm, mainly in the near-infrared part of the spectrum.

IOSs reflect metabolic activities, mainly oxygenation and deoxygenation, rather than neural spikes, but brain metabolism is directly associated with neural activity. Therefore, it is reasonable to deduce that IOS can be effectively used to visualize neural activity in the cerebral cortex. The IOS was developed as an investigative tool in functional brain mapping in laboratory animals, but now it is also used in human studies during neurosurgery or transcranial imaging.

The IOS was successfully used in combination with direct brain electrical stimulation in a study of the somatosensory cortical network in the squirrel monkey (Brock et al., 2013). This study revealed that specific microstimulation of different cortical depths and loci could evoke restricted optical responses, allowing for the tracing of functional corticocortical connections (Brock et al., 2013). Studies from the same laboratory showed that both electrical and pulsed infrared neural stimulation evoke electrophysiological and IOS responses in the neocortex (Cayce et al., 2014). In this case, infrared neural stimulation was used to enhance or diminish cortical responses in the macaque visual cortex (Cayce et al., 2014).

Conventional IOS imaging approach has been used for functional brain mapping, and in studies of hemodynamics. Cerebral blood flow imaging during neurosurgery is necessary to perform vascular reconstruction. Recently a hyperspectral camera (HSC) was used in the assessment of cerebral blood flow in rats and humans (Mori et al., 2014). Changes in the local oxygenation were derived from spectral imaging data in wide diapason (400–800 nm) by the HSC and verified by the single-photon emission computer tomography (Mori et al., 2014).

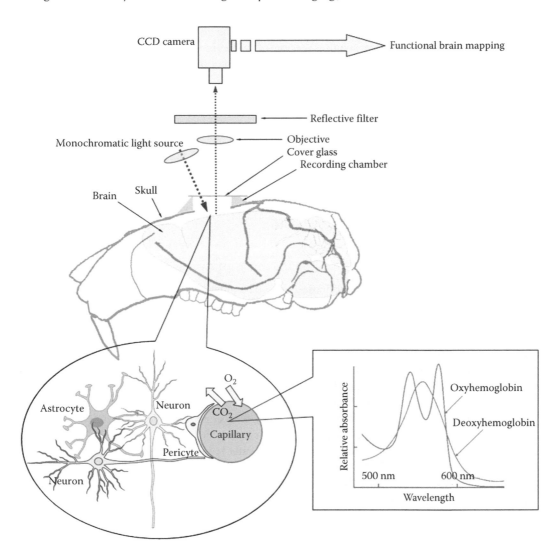

FIGURE 6.1 Schematic of the IOS setup. IOS monitors the brain activity by measuring intrinsic activity-related changes in tissue light absorbance. These signals result from changes in cerebral blood flow, hemoglobin oxy- and deoxygenation, and the light scattering changes that are coupled with neurons and astrocytes activity.

INTRAOPERATIVE HUMAN IOS

The imaging of the exposed human cerebral cortex was first realized a couple of decades ago (Haglund et al., 1992). The IOS has been informative in studying functional maps during stimulation-evoked epileptiform (Bahr et al., 2006) and cognitively evoked functional activity in the sensory cortices and in Broca's and Wernicke's areas. Ultimately this method has been improved and expanded. The IOS was employed in combination with electrical stimulation and mapping during brain surgery and different activity patterns have been recorded in the receptive (Wernicke's) and expressive (Broca's) language areas of the neocortex (Cannestra et al., 2000).

The time of an IOS is relatively limited. On the other hand, since the brain does not have pain receptors, IOS imaging sessions can be realized with an awake patient. However, it is important to note that unjustified prolonged "open brain" time within a surgical procedure presents ethical concerns. Recently, in patients with brain tumor, the abnormal tissues were surgically removed by using IOSs in combination with electrical stimulation of the contralateral median nerve (Sobottka et al., 2013a). The intraoperative imaging session of the brain surface was continuously recorded over a very short period, enough for data collection, without compromising patient safety (Sobottka et al., 2013b).

An important caveat to the IOS imaging studies is that the data are usually contaminated by different biological and technical noises that cannot be eliminated by the averaging of the recording trials. There are several data analysis methods that can improve the situation. One of them is to maximize a weighted difference between the signal and the noise variance (Yokoo et al., 2001). This method has been effective for the IOS data obtained from the visual cortex (Yokoo et al., 2001). Another mathematical filtering method is based on the polynomial subtraction of spatially modulated components of the obtained data (Ribot et al., 2006). This technique, as well as Yokoo's method, can allow visualization of iso-orientation domains in the visual cortex and it is also applicable to nonvisual sensory areas (Ribot et al., 2006).

INTRINSIC FLUORESCENCE

A special case of intrinsic signal is autofluorescence. The most-studied autofluorescence is based on nicotinamide adenine dinucleotide phosphate ($NADP^+$) and flavoproteins. The ultraviolet photons excite $NADP^+$ and $NADP^+$ emits photons in the longer part of the spectrum. This phenomenon has been shown in vitro (Lipton, 1973) as well as in vivo (Jobsis et al., 1971). Unfortunately ultraviolet kills cells, which limits its application in brain imaging.

Unlike ultraviolet, the visual and infrared parts of the spectrum are more promising in intrinsic signal imaging studies. Fluorescence of the flavoproteins in the blue light has been successfully used for functional brain mapping (Takahashi et al., 2006; Chery et al., 2011). More recently, mitochondrial flavoprotein fluorescence, excited with blue light, has been used to generate functional signals, both in vitro and in vivo (Tsytsarev et al., 2012). Flavoprotein fluorescence signals have a similar time course to hemodynamic OIS signals, but they reportedly have greater signal-to-noise ratio and are more spatially localized (Lippert et al., 2007). These methods involve simple modifications of conventional imaging apparatuses (excitation and emission filters, dichroic mirrors), but have not been implemented for intraoperative imaging.

OPTICAL COHERENCE TOMOGRAPHY

Intrinsic signal can be recorded also by optical coherence tomography (OCT), a method based on low-coherence interferometry for performing high-spatial-resolution 3D optical imaging.

In contrast to the conventional imaging system, OCT is noninvasive and label-free. It also allows to record fast IOSs, which reflects neuronal activity, rather than delayed hemodynamic responses (Lee and Boas, 2012).

In addition to sensory-evoked activities, epileptic seizures have been visualized by OCT. Epileptic seizures modulate the optical refractive index of the cortical neurons (Tsytsarev et al., 2010). OCT imaging of epileptic activity was confirmed by a simultaneously recorded electroencephalogram signal during seizure (Tsytsarev et al., 2013).

Within the last few years, OCT has been used for imaging and mapping the cortical tissue in vivo in relation to evoked neural activity or peripheral sensory stimulation (Satomura et al., 2004; Aguirre et al., 2006; Rajagopalan and Tanifuji, 2007; Chen et al., 2009; Lenkov et al., 2013). The modulation of the optical refractive index by neuronal activity during activation of the brain tissue coincides with the changes of the surface reflectivity (Tsytsarev et al., 2010). OCT promises to be a powerful direct functional brain mapping method with a depth range of a few millimeters and spatial resolution of up to few microns.

Both the visual and the infrared part of the spectrum can be used for IOS imaging. It is possible to visualize the distribution of cortical activation by using the infrared imaging technique, which reflects changes in the cerebral blood flow and metabolism (Huppert et al., 2009). Reportedly, the temperature gradient between brain arteries and veins reaches 1.5°C–2.0°C, between veins and cortical tissue 0.2°C–1.0°C (Gorbach et al., 2003). Infrared imaging can be used for relatively fast identification of diverse functional areas of the cerebral cortex, such as the somatosensory motor and language areas (Gorbach et al., 2003).

If the temporal and spatial resolution of the imaging system is sufficient, the high-speed optical imaging approach allows monitoring the cerebral blood flow response independently in individual vessels (Liao et al., 2010, 2012a,b; Chen et al. 2011). This approach allows to determine where and when the first hemodynamic response occurs in reply to the stimuli (Chen et al., 2011; Liao et al., 2012a,b). Based on these data, the authors concluded that the hemodynamic response appears to initiate in the brain tissue and then spreads rapidly to the surface arterioles, independent of the direction of blood flow within vessels (Wanger et al., 2013).

NONINVASIVE HUMAN BRAIN OPTICAL IMAGING

Transcranial optical imaging, the so-called near-infrared spectroscopy (NIRS), is based on principles similar to that of IOS (Kalchenko et al., 2015). This method uses near-infrared light to monitor the changes in oxy- and deoxyhemoglobin concentration (Liao et al., 2013) via absorption spectroscopy (Devor et al., 2012; Yang and Chen, 2013). While IOS is generally performed using an opening in the skull to access direct signal of the brain tissue, NIRS imaging is obtained through the intact skull, muscles, and skin. Like fMRI, the NIRS monitors oxygen-related brain activity; therefore, it allows localization of the activated brain areas by the level of oxygen consumption. Recently, NIRS was successfully used to study the differences in functional connectivity in frontal temporal cortices between normal children and those diagnosed with autism spectrum disorder (ASD) (Zhu et al., 2014).

ASD is a neurodevelopmental disorder with unclear etiology, which has been associated with social and behavioral problems. Some investigators propose that atypical neural synchronization is one of the bases of ASD (Zhu et al., 2014).

Using NIRS, it was shown that children at high risk of autism showed increased overall functional connectivity compared to the low-risk group at 3 months of age (Keehn et al., 2013). No significant differences were found among the high- and low-risk groups between 6 and 9 months of age, while at the age of 1 year, the high-risk group showed decreased connectivity in comparison to the low-risk group (Keehn et al., 2013). NIRS was also used for identifying functional abnormalities in the prefrontal cortex in teenagers with ASD (Devor et al., 2012; Iwanaga et al., 2013).

The details of the biological origin underlying fast IOSs remains unclear, but most investigators believe that the ion and water movement as well as small changes in the cellular volume, associated with neural activity, causes a change in the optical scattering of the brain parenchyma. A special case of the IOS is fast optical imaging and, in particular, the event-related optical signal (EROS). EROS imaging uses near-infrared or infrared light through optical fibers to monitor changes in the optical properties of the brain tissue (Baniqued et al., 2013). This method is based on the scattering properties of the neurons and therefore provides a direct measure of the neural activity within centimeters (spatial) and milliseconds (temporal) resolution. In the auditory system, a mismatch negativity (a brain response to acoustic irregularities), previously demonstrated by evoked potential recordings in humans, has been confirmed by EROS (Sable et al., 2007). Recently associative aspects of memory have been examined by EROS methods. It was found that a brain region involved in face recognition was activated not only in response to the face representation but also when viewing scenes, associated with specific faces (Walker et al., 2014).

In comparison with other brain imaging methods, EROS has its own advantages and disadvantages. One of the main disadvantages is the depth of light penetration—not more than few tens of millimeters. Signal-to-noise ratio is low, so the data require averaging of few tens of trials or more. But the important advantage of EROS is the cost and portability. It is easy to combine EROS with

EEG and electrical or magnetic brain stimulation (Gratton et al., 2006; Gratton and Fabiani, 2010). Ultimately, as a pool of methods, IOS may prove to be a useful approach to functional brain mapping in basic and clinical neuroscience.

TERAHERTZ RADIATION

T-ray (Teraherz radiation, T-waves, T-light), or submillimeter waves, is a type of electromagnetic ray (Figure 6.2) that resides in between the infrared and microwaves of the electromagnetic spectrum, or from 0.1 to 20 THz. It has potential applications in biological and medical research, including brain imaging (Wang et al., 2007; Wagh et al., 2009). T-photons do not have enough energy to ionize molecules but they interact with absorptive material. T-ray can detect the differences in the concentration of different ions, including protons; therefore, probably it can provide a new method for brain imaging. Historically, the T-ray has remained relatively untapped because both stable and inexpensive T-ray sources and detectors were not developed (Wang et al., 2003). This problem was so severe that it was also called a "the terahertz-gap" or "T-gap" by the scientific community, but within the last decade, the situation has dramatically changed. Improvements have been made to overcome technical problems, making it possible to apply T-ray in neuroscience research (Trevino-Palacios et al., 2010).

T-imaging is useful for the identification of hidden objects under the fabric and also it can spectroscopically determine chemical compounds (including explosive), so it is commonly used in security systems. Usefulness of T-ray imaging was shown by a study of historic remains. Images of the ancient Egyptian human and fish mummies were obtained by T-ray imaging (Ohrstrom et al., 2010). Conventional x-ray imaging provides higher spatial resolution, but for the study of mummified soft tissue, T-ray imaging was more valuable because the structures containing bones and the collagen could be differentiated better (Ohrstrom et al., 2010).

Metal and water molecules block T-rays. For the radio-frequency electromagnetic waves, the water molecule acts as a typical molecular dipole and absorbs the energy and its temperature increases. In contrast, at a frequency higher than ~0.3 THz, the water molecule dipole cannot react and there is no heating. However, due to high absorption, the penetration depth of T-rays is limited because the biological tissue contains more than 50% of water. The water content can act as a contrast mechanism to differentiate between activated and nonactivated tissue, or between normal and malign tissue. Numerically, the lateral resolution of T-ray imaging is limited by diffraction to about half of the wavelength, which can be a hundred microns, which is enough for many biomedical imaging purposes.

Conventional T-imaging is based on the measurement of the signal change originating from the absorption of T-photons by water molecules and heavy ions. Therefore, it poorly distinguishes

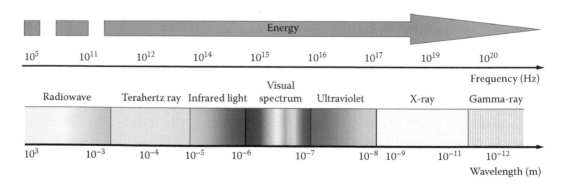

FIGURE 6.2 A diagram of the electromagnetic ray, showing frequencies and wavelengths across the range of the energy.

between the different types of living tissues, but the selectivity can be improved by adopting nanoparticle probes (Bruchez et al., 1998; Oh et al., 2011). Biologically neutral nanoparticles amplify the signal and can be designed to target specific cell types (Sukhanova et al., 2004; Oh et al., 2011). T-imaging, in combination with nanoparticle probes, is sensitive enough to distinguish different types of biological tissues in vivo. Thus, T-imaging can facilitate the study of biological phenomena at the cellular and molecular levels, which is useful in tumor differentiation and functional imaging (Oh et al., 2011).

Recently proposed T-ray imaging system (Bakopoulos et al., 2008) can be used to detect in vivo biomarkers in the brain in animal experiments and in medicine. These investigators have demonstrated a 2D THz imaging system for brain imaging applications, both at the macro- and molecular levels (Bakopoulos et al., 2008). In terms of applications in brain mapping, the large organic molecules vibrate at terahertz frequencies, and specific chemicals absorb certain characteristic T-ray frequencies. T-ray-sensitive devices can then detect this absorption and identify the molecules (Bakopoulos et al., 2008). T-ray could probably be used to obtain information about physiological processes in living tissues, with spatial resolution up to hundreds of micrometers.

VOLTAGE-SENSITIVE DYE IMAGING

VSDi is a powerful tool in understanding how neurons communicate with each other in health and disease. Elaborate neural networks carry out the processing of sensory information as well as behavioral and motor control. To study the functions and structures of neural networks, it is necessary to monitor the activity of cell ensembles with high temporal resolution, rather than that of single cells. VSDi techniques are ideally suited for this purpose (Grinvald and Hildesheim, 2004; Tsytsarev et al., 2014). VSDi offers a unique opportunity to monitor neural activity with relatively high spatial and high temporal resolution, which is comparable to the microelectrode recording technique.

The first in vivo VSDi of neural processing in the vertebrate brain was realized in the frog (Grinvald et al., 1984) and rat (Orbach et al., 1985) brain. These early studies encountered difficulties due to small signal amplitudes, large hemodynamic signals, phototoxicity, and limitations in detector and computer technology.

VSDi is based on voltage-sensitive fluorescence probes that are fluorescent chemicals that change their optical features in response to the changes of the membrane potential. The molecules of the voltage-sensitive dyes bind to the external neural surface and play the role of energy transducers that transform changes in the transmembrane voltage into the changes in light absorption or emitted photons (Fromherz et al., 2008). Photosensitive devices record these changes. The voltage across the neural membrane generates a strong electric field, which affects voltage-sensitive dye molecule fluorescence properties. Changes in the transmembrane potential are linearly related to the fluorescence of the voltage-sensitive dye molecule, which follow the transmembrane voltage. The amplitudes of the VSDi signals are linearly correlated with changes in the membrane potential; therefore, these dyes are called potentiometric dyes.

The main difference between VSDi and IOS is that VSDi signal originates directly from neural activity-dependent changes of the optical features, while IOS is based on the indirect reflection of the neural activity.

There are three main mechanisms of the voltage-sensitive dye activity: dipole rotation, electrochromism, and a potential-sensitive monomer–dimer equilibrium (Zochowski et al., 2000; Zecevic et al., 2003). The core of the voltage-sensitive dye molecule is the chromophore, which interacts with the photons in which the difference in energy between the ground state and an excited state of the chromophore matches the energy of the photon (Loew, 2010, 2011; Peterka et al., 2011).

A molecule of the voltage-sensitive dye has a paired structure and consists of hydrocarbon chains and a compact, hydrophilic group. The molecule goes through an electrochemical mechanism during change of the electric field, which changes fluorescence and light absorption.

A specific case of VSDi is the genetically encoded voltage-sensitive dye (Barnett et al., 2012; Perron et al., 2012). Additionally, the voltage-sensitive fluorescent protein (VSFP) is a promising alternative to the conventional VSDi, which is based on extrinsic fluorescence probes (Knopfel et al., 2006; Akemann et al., 2012; Baker et al., 2012). In response to the neural spike generation, VSFP produces fast fluorescence response (Mutoh et al., 2011, 2012). VSFP imaging is much less traumatic due to the absence of phototoxicity, but its application is limited by the difficulties of the VSFP expressions in experimental animals.

The experimental setup for VSDi (Figure 6.3) is similar to the IOS. A system is mounted on a vibration isolation table. Changes in fluorescence are detected in the macroscope image plane by a fast CCD camera or by another type of photosensitive device; each pixel receives photons from a small part of the recording region, depending on the optic magnification and the physical size of the sample.

VSDi of the cerebral cortex is performed by recording the fluorescence from the stained cortex with a camera axis aimed perpendicular to the brain surface. The brain tissue is stained up to a depth

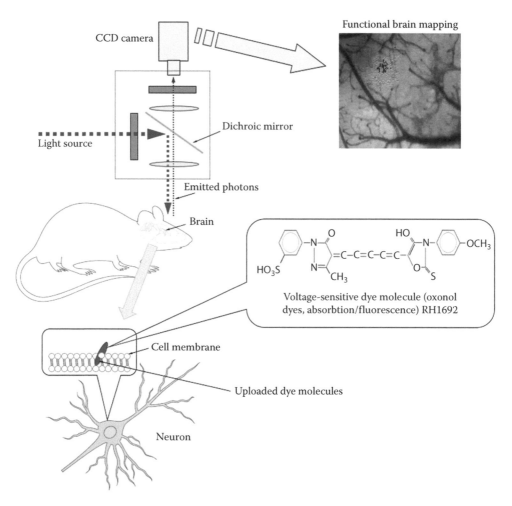

FIGURE 6.3 Schematic of the VSDi setup. VSDi is also called "extrinsic optical imaging" because it is based on the usage of voltage-sensitive dyes. After craniotomy, the dye molecules are applied on the brain surface and they bind to the external surface of the cells' membranes. Voltage-sensitive dye molecules act as energy transducers that transform changes in the transmembrane voltage into the changes in the photon emission. The intensity of the fluorescent signal is proportional to the number of membrane-connected molecules under each measuring pixel.

of 2 mm, but because of diffusion, the upper layers are stained better than the lower ones (Grinvald et al., 2001). Therefore, most of the neural activity-related fluorescence originates from the upper cortical layers, but the apical dendrites of layer V and some VI neurons do reach the upper cortical layers and also contribute to the fluorescence (Grinvald et al., 2001).

The amplitude of the VSDi signal is determined by membrane potentials and by the size of the membrane area, the number of binding molecules, and the sensitivity of the particular dye (Grinvald et al., 2001; Chemla and Chavane, 2010a,b).

The bleaching of the voltage-sensitive dye may change the structure of optical signals and limit the duration of the measurement. To minimize bleaching, the illumination time and light intensity are minimized.

HEARTBEAT ARTIFACT

The level of the activity-dependent fluorescence signals is very low: about 10^{-4} to 10^{-5} of the photons reaching the CCD-camera. Therefore, obtaining a good signal-to-noise ratio is a very important and complicated task. One of the largest difficulties for VSDi experiments is the optical signal changes caused by the heart rhythm (Grandy et al., 2012). These optical changes consist of two components: the first is the movement artifact caused by the physical pulsation of the blood vessels. The second is caused by the change of the hemoglobin light absorption determined by its oxygenation.

To minimize both these factors, it is reasonable to use voltage-sensitive dye with the excitation wavelength in the near-infrared part of the spectrum.

If these wavelengths are longer than 600 nm, oxy- and deoxyhemoglobin absorb light more or less equally; therefore, the blood oxygenation doesn't contribute to the optical signal.

Another way to decrease a cardio artifact is the synchronization of a light shutter, recording system, and the stimulator with the electrocardiogram: the light shutter opens and a photosensitive device records only during the period between two consecutive heartbeats in order to minimize the effect of the heart rhythm (Tsytsarev et al., 2010). With regard to the other sources of noise, lowering the light intensity can minimize phototoxicity and bleaching, and therefore improve the signal-to-noise ratio (Grinvald et al., 2001).

Alternatively, phototoxicity and bleaching can be decreased by reducing the dye concentration (Grinvald et al., 2001). The brain tissue is stained up to a depth of at least 2 mm; but because of the dye diffusion, the upper cortical layers are stained better than the lower ones (Grinvald et al., 2001; Tsytsarev et al., 2012).

One of the optical imaging modalities, potentially applicable not only for VSDi but also for IOS, is hyperspectral imaging, which is defined as the highly effective combination acquisition of spectral information from the different points of the sample (Studer et al., 2012). The application of this method is becoming popular with the availability of an increasing panel of fluorescent dyes with wide emission spectrum (Studer et al., 2012).

Several groups observed the propagating waves using VSDi, under anesthesia, but not in awake animals (Gao et al., 2012; Osman et al., 2012). Recently it has been hypothesized that these waves are missed in the trial-averaged data processing (Muller et al., 2005, 2014). Also, using VSDi, the spiral waves have been found in the cerebral cortex. These waves occur frequently, both during pharmacologically induced oscillations and during sleeplike conditions (Huang et al., 2010). These observations support the hypothesis that spiral waves, as specific spatiotemporal neural activity pattern, modulate neural activity and may contribute to both normal and pathological neural network processing (Huang et al., 2010). Using VSDi in combination with the newly developed method of data analysis, Muller et al. (2014) showed that the stimulus-evoked population response in the awake monkey visual cortex propagates as a travelling wave.

The whisker system of laboratory rats and mice is very suitable for neuroscience research with imaging methods. The combination of VSDi with electrophysiology and behavior testing provided evidence that the neural encoding in the thalamocortical system can be modulated through sensory adaptation (Ollerenshaw et al., 2014).

In the olfactory system, spatiotemporal responses of the olfactory bulb to various odorants have been mapped in great detail. In contrast, far less is known about the activation of the nasally projecting trigeminal ganglion cells through volatile chemical stimuli. Application of different chemical stimuli to the nasal cavity elicited stimulus-specific, spatiotemporal activation patterns (Rothermel et al., 2011). Using VSDi it was shown that stimulus-specific patterns of the neural activity occur within the trigeminal ganglia upon nasal presentation of ten different odorants (Lubbert et al., 2013); moreover, the intensity of the fluorescence and the latencies were also dependent on the odorant (Lubbert et al., 2013). These results provide the first direct insights into the spatial representation of nasal chemosensory information within the trigeminal ganglion imaged at high temporal resolution.

Even though the visual cortex is one of the most studied brain areas, the functional representation in this area is still not fully understood and is frequently the object of VSDi experiments (Eriksson et al., 2012; Preuss and Stein, 2013). Visual system, and in particular, saccades, frequently is the subject of neuroimaging research. Using VSDi neural population responses in the primary and secondary visual cortex in monkeys have been investigated with VSDi, while the animals were presented with collinear or orthogonal arrays of Gabor patches—a graphical sine wave grating seen through a Gaussian window (Gilad et al., 2013, 2014). The obtained data suggested that collinear effects are mediated by synchronization in a distributed network within both the primary and secondary visual cortex (Gilad et al., 2012).

The primary visual cortex of the awake primate was studied using VSDi (Ayzenshtat et al., 2012). In this study, the investigators used natural images for visual stimulation during the face/scramble discrimination task. They showed that the population response consisted of two distinct phases: the first was spread over most of the imaged area and the second was spatially confined. These data clarified the spatial encoding of low- and high-level features of natural images in the primary visual cortex.

VSDi has been used to understand the role of lateral interactions in the cat primary visual cortex (Chavane et al., 2011). At the same time, electrophysiological recordings showed that the loss of orientation selectivity arises from the diversity of converging synaptic input patterns originating from the area outside of the receptive field (Chavane et al., 2011). Authors conclude that the stimulus enhances the long-range orientation-selective spread (Chavane et al., 2011).

Not only stimuli encoding, but also stimuli prediction has been examined in the visual cortex by VSDi (Eriksson et al., 2012). The stimulus prediction was examined in time using both VSDi and unit recording in the cat visual cortex (Eriksson et al., 2012). It was shown that a single cortical neuron combines stimulus and error-like coding. Omer et al. (2013) designed a novel data acquisition protocol for sensory stimulation in continuous sessions of VSDi to visualize spontaneous and evoked neural activity in both the anesthetized and the awake animal. They developed a new algorithm for "temporally structured component analysis," which separates functional signal components from other signals referred to as noise. This algorithm facilitates rapid optical imaging explorations in monkeys and can be applied to other species (Omer et al., 2013).

The transfer of visual information in the cortical level has been studied using VSDi in mice (Polack and Contreras, 2012). It was found that primary visual cortex activation is rapidly followed by depolarization of three different functional groups of neural networks (Polac and Contreras, 2012). In these studies, VSDi was an essential method, which allowed the conclusion that the cortex integrates visual information simultaneously through across-area parallel and within-area serial processing (Polac and Contreras, 2012).

The auditory cortex contains a map of the sound frequencies, and has been studied using VSDi. Recent VSDi studies revealed the presence of tonotopic maps in the auditory cortex of guinea pigs (Horikawa et al., 1998; Nishimura et al., 2007; Farley and Norena, 2013) and rats (Tsytsarev et al., 2009). In the primary auditory cortex, low frequencies are represented caudally, while high frequencies are represented rostrally. In addition to tonotopicity, the representation of the interaural time difference in the auditory cortex was examined. It was found that acoustic stimuli with dissimilar interaural time difference activate different localized domains in the rat auditory cortex (Tsytsarev et al., 2009).

VSDi experiments, using a flexible fiber optic image bundle, have been carried out in freely moving mice to visualize neural activity in the barrel cortex while simultaneously filming whisker-related behavior (Ferezou et al., 2006). More recently, VSDi has been used to examine the functional map of the directional sensitivity (Tsytsarev et al., 2010) and gamma oscillations (Khazipov et al. 2013) of the rat barrel cortex.

A new implantable imaging system was developed using a complementary metal-oxide semiconductor (CMOS) technology with VSDi (Kobayashi et al., 2012). In contrast with the CCD-camera, a CMOS sensor is small enough to be implanted into the brain (Kobayashi et al., 2012). Light-emitting diodes are used as a source of the excitation light. The light microsensors are less traumatic, so such a system can combine functional VSDi in the brain, with behavioral experiments in freely moving animals (Kobayashi et al., 2012).

The combination of a microscopic view, with a macroscopic functional imaging, which keeps track of the linkage of distant cortical modules, was successfully realized using VSDi, in combination with electrophysiology and calcium-sensitive imaging (Feldmeyer et al., 2013).

Since the cellular architecture of the brain is 3D, 3D imaging of neuronal activity offers new possibilities for understanding the neural network activity (Acker and Loew, 2013). Traditional VSDi has limited spatial resolution, especially along the vertical axis (deep resolution). In combination with the two-photon technique, VSDi can resolve neural activities of different depths (Fisher et al., 2005, 2008; Homma et al., 2009; Ghouri et al., 2015). Spontaneous neural activity in the barrel field of the somatosensory cortex was monitored by this method in head-restrained awake mice (Kuhn et al., 2008; An et al., 2012).

In conclusion, various strategies of functional brain mapping are providing new glimpses into brain connectivity and specific activation patterns in health and disease. In this context voltage-sensitive dye and intrinsic signal optical imaging by themselves or coupled with conventional electrophysiological recordings and stimulations in experimental settings are paving the way for a better understanding of brain functions.

REFERENCES

Acker CD, Loew LM. Characterization of voltage-sensitive dyes in living cells using two-photon excitation. *Methods Mol Biol*, 2013, 995:147–160.

Aguirre AD, Chen Y, Fujimoto JG, Ruvinskaya L, Devor A, Boas DA. Depth-resolved imaging of functional activation in the rat cerebral cortex using optical coherence tomography. *Opt Lett*, 2006, 31:3459–3461.

Akemann W, Mutoh H, Perron A, Park YK, Iwamoto Y, Knopfel T. Imaging neural circuit dynamics with a voltage-sensitive fluorescent protein. *J Neurophysiol*, 2012, 108:2323–2337.

An S, Yang JW, Sun H, Kilb W, Luhmann HJ. Long-term potentiation in the neonatal rat barrel cortex in vivo. *J Neurosci*, 2012, 32:9511–9516.

Ayzenshtat I, Gilad A, Zurawel G, Slovin H. Population response to natural images in the primary visual cortex encodes local stimulus attributes and perceptual processing. *J Neurosci*, 2012, 32:13971–13986.

Bahar S, Suh M, Zhao M, Schwartz TH. Intrinsic optical signal imaging of neocortical seizures: The 'epileptic dip'. *Neuroreport*, 2006, 17:499–503.

Baker BJ, Jin L, Han Z, Cohen LB, Popovic M, Platisa J, Pieribone V. Genetically encoded fluorescent voltage sensors using the voltage-sensing domain of Nematostella and Danio phosphatases exhibit fast kinetics. *J Neurosci Methods*, 2012, 208:190–196.

Bakopoulos P, Karanasiou I, Zakynthinos P, Pleros N, Avramopoulos H, Uzunoglu N. Towards brain imaging using THz technology. *IEEE International Workshop on Imaging Systems and Techniques*, Chania, Greece, pp. 7–10. Imaging Systems and Techniques, 2008.

Baniqued PL, Low KA, Fabiani M, Gratton G. Frontoparietal traffic signals: A fast optical imaging study of preparatory dynamics in response mode switching. *J Cog Neurosci*, 2013, 25:887–902.

Barnett L, Platisa J, Popovic M, Pieribone VA, Hughes T. A fluorescent, genetically-encoded voltage probe capable of resolving action potentials. *PLoS One*, 2012, 7:e43454.

Brock AA, Friedman RM, Fan RH, Roe AW. Optical imaging of cortical networks via intracortical microstimulation. *J Neurophysiol*, 2013, 110:2670–2678.

Bruchez M, Moronne M, Gin P, Weiss S, Alivisatos AP. Semiconductor nanocrystals as fluorescent biological labels. *Science*, 1998, 281:2013–2016.

Cannestra AF, Bookheimer SY, Pouratian N, O'Farrell A, Sicotte N, Martin NA, Becker D, Rubino G, Toga AW. Temporal and topographical characterization of language cortices using intraoperative optical intrinsic signals. *Neuroimage*, 2000, 12:41–54.

Cayce JM, Friedman RM, Chen G, Jansen ED, Mahadevan-Jansen A, Roe AW. Infrared neural stimulation of primary visual cortex in non-human primates. *Neuroimage*, 2014, 84:181–190.

Chavane F, Sharon D, Jancke D, Marre O, Fregnac Y, Grinvald A. Lateral spread of orientation selectivity in V1 is controlled by intracortical cooperativity. *Front Syst Neurosci*, 2011, 5:4.

Chemla S, Chavane F. A biophysical cortical column model to study the multi-component origin of the VSDI signal. *NeuroImage*, 2010a, 53:420–438.

Chemla S, Chavane F. Voltage-sensitive dye imaging: Technique review and models. *J Physiol Paris*, 2010b, 104:40–50.

Chen BR, Bouchard MB, McCaslin AF, Burgess SA, Hillman EM. High-speed vascular dynamics of the hemodynamic response. *Neuroimage*, 2011, 54:1021–1030.

Chen Y, Aguirre AD, Ruvinskaya L, Devor A, Boas DA, Fujimoto JG. Optical coherence tomography (OCT) reveals depth-resolved dynamics during functional brain activation. *J Neurosci Methods*, 2009 178:162–173.

Chery R, L'Heureux B, Bendahmane M, Renaud R, Martin C, Pain F, Gurden H. Imaging odor-evoked activities in the mouse olfactory bulb using optical reflectance and autofluorescence signals. *J Vis Exp*, 2011, 56:e3336.

Choudhri AF, Chin EM, Blitz AM, Gandhi D. Diffusion tensor imaging of cerebral white matter: Technique, anatomy, and pathologic patterns. *Radiol Clin North Am*, 2014, 52:413–425.

Cohen LB, Salzberg BM, Grinvald A. Optical methods for monitoring neuron activity. *Annu Rev Neurosci*, 1978, 1:171–182.

Devor A, Sakadzic S, Srinivasan VJ, Yaseen MA, Nizar K, Saisan PA, Tian P et al. Frontiers in optical imaging of cerebral blood flow and metabolism. *J Cereb Blood Flow Metab*, 2012, 32:1259–1276.

Eriksson D, Wunderle T, Schmidt K. Visual cortex combines a stimulus and an error-like signal with a proportion that is dependent on time, space, and stimulus contrast. *Fronti Syst Neurosci*, 2012, 6:26.

Farley BJ, Norena AJ. Spatiotemporal coordination of slow-wave ongoing activity across auditory cortical areas. *J Neurosci*, 2013, 33:3299–3310.

Feldmeyer D, Brecht M, Helmchen F, Petersen CC, Poulet JF, Staiger JF, Luhmann HJ, Schwarz C. Barrel cortex function. *Prog Neurobiol*, 2013, 103:3–27.

Ferezou I, Bolea S, Petersen CCH. Visualizing the cortical representation of whisker touch: Voltage-sensitive dye imaging in freely moving mice. *Neuron*, 2006, 50:617–629.

Fisher JA, Barchi JR, Welle CG, Kim GH, Kosterin P, Obaid AL, Yodh AG, Contreras D, Salzberg BM. Two-photon excitation of potentiometric probes enables optical recording of action potentials from mammalian nerve terminals in situ. *J Neurophysiol*, 2008, 99:1545–1553.

Fisher JA, Salzberg BM, Yodh AG. Near infrared two-photon excitation cross-sections of voltage-sensitive dyes. *J Neurosci Methods*, 2005, 148:94–102.

Fromherz P, Hubener G, Kuhn B, Hinner MJ. ANNINE-6plus, a voltage-sensitive dye with good solubility, strong membrane binding and high sensitivity. *Eur Biophys J Biophy*, 2008, 37:509–514.

Frostig RD. *In Vivo Optical Imaging of Brain Function*, 2nd edn. Boca Raton, FL: CRC Press, 2009.

Gao X, Xu W, Wang Z, Takagaki K, Li B, Wu JY. Interactions between two propagating waves in rat visual cortex. *Neuroscience*, 2012, 216:57–69.

Ghouri IA, Kelly A, Burton FL, Smith GL, Kemi OJ. 2-photon excitation fluorescence microscopy enables deeper high-resolution imaging of voltage and Ca in intact mice, rat, and rabbit hearts. *J Biophotonics*, 2015, 8:112–123.

Gilad A, Meirovithz E, Leshem A, Arieli A, Slovin H. Collinear stimuli induce local and cross-areal coherence in the visual cortex of behaving monkeys. *PLoS One*, 2012, 7:e49391.

Gilad A, Meirovithz E, Slovin H. Population responses to contour integration: Early encoding of discrete elements and late perceptual grouping. *Neuron*, 2013, 78:389–402.

Gilad A, Pesoa Y, Ayzenshtat I, Slovin H. Figure-ground processing during fixational saccades in V1: Indication for higher-order stability. *J Neurosci*, 2014, 34:3247–3252.

Gorbach AM, Heiss J, Kufta C, Sato S, Fedio P, Kammerer WA, Solomon J, Oldfield EH. Intraoperative infrared functional imaging of human brain. *Ann Neurol*, 2003, 54:297–309.

Grandy TH, Greenfield SA, Devonshire IM. An evaluation of in vivo voltage-sensitive dyes: Pharmacological side effects and signal-to-noise ratios after effective removal of brain-pulsation artifacts. *J Neurophysiol*, 2012, 108:2931–2945.

Gratton G, Brumback CR, Gordon BA, Pearson MA, Low KA, Fabiani M. Effects of measurement method, wavelength, and source-detector distance on the fast optical signal. *Neuroimage*, 2006, 32:1576–1590.

Gratton G, Fabiani M. Fast optical imaging of human brain function. *Front Hum Neurosci*, 2010, 4:52.

Grinvald A, Anglister L, Freeman JA, Hildesheim R, Manker A. Real-time optical imaging of naturally evoked electrical activity in intact frog brain. *Nature*, 1984, 308:848–850.

Grinvald A, Hildesheim R. VSDI: A new era in functional imaging of cortical dynamics. *Nat Rev Neurosci*, 2004, 5:874–888.

Grinvald, A, Salzberg BM, Cohen LB. Simultaneous recording from several neurons in an invertebrate central nervous-system. *Nature*, 1977, 268:140–142.

Grinvald A, Shoham D, Shmuel A, Glaser D, Vanzetta I, Shtoyerman E, Slovin H et al. In vivo optical imaging of cortical architecture and dynamics. Technical Report GC-AG/99-6. The Grodetsky Center for Research of Higher Brain Functions, The Weizmann Institute of Science, Rehovot, Israel, 2001.

Haglund MM, Ojemann GA, Hochman DW. Optical imaging of epileptiform and functional activity in human cerebral cortex. *Nature*, 1992, 358:668–671.

Homma R, Baker BJ, Jin L, Garaschuk O, Konnerth A, Cohen LB, Bleau CX, Canepari M, Djurisic M, Zecevic D. Wide-field and two-photon imaging of brain activity with voltage- and calcium-sensitive dyes. *Methods Mol Biol*, 2009, 489:43–79.

Horikawa J, Nasu M, Taniguchi I. Optical recording of responses to frequency-modulated sounds in the auditory cortex, *Neuroreport*, 1998, 9:799–802.

Huang XY, Xu WF, Liang JM, Takagaki K, Gao X, Wu JY. Spiral wave dynamics in neocortex. *Neuron*, 2010, 68:978–990.

Huppert TJ, Diamond SG, Franceschini MA, Boas DA. HomER: A review of time-series analysis methods for near-infrared spectroscopy of the brain. *Appl Opt*, 2009, 48:D280–D298.

Iwanaga R, Tanaka G, Nakane H, Honda S, Imamura A, Ozawa H. Usefulness of near-infrared spectroscopy to detect brain dysfunction in children with autism spectrum disorder when inferring the mental state of others. *Psychiatry Clin Neurosci*, 2013, 67:203–209.

Jobsis FF, O'Connor M, Vitale A, Vreman H. Intracellular redox changes in functioning cerebral cortex. I. Metabolic effects of epileptiform activity. *J Neurophysiol*, 1971, 34:735–749.

Kalchenko V, Israeli D, Kuznetsov Y, Meglinski I, Harmelin A. A simple approach for non-invasive transcranial optical vascular imaging (nTOVI). *J Biophotonics*, 2015, 8:897–901.

Keehn B, Wagner JB, Tager-Flusberg H, Nelson CA. Functional connectivity in the first year of life in infants at-risk for autism: A preliminary near-infrared spectroscopy study. *Front Hum Neurosci*, 2013, 7:444.

Khazipov R, Minlebaev M, Valeeva G. Early gamma oscillations. *Neuroscience*, 2013, 10(250):240–252.

Knopfel T, Diez-Garcia J, Akemann W. Optical probing of neuronal circuit dynamics: Genetically encoded versus classical fluorescent sensors. *Trends Neurosci*, 2006, 29:160–166.

Kobayashi T, Motoyama M, Masuda H, Ohta Y, Haruta M, Noda T, Sasagawa K et al. Novel implantable imaging system for enabling simultaneous multiplanar and multipoint analysis for fluorescence potentiometry in the visual cortex. *Biosens Bioelectron*, 2012, 38:321–330.

Kuhn B, Denk W, Bruno RM. In vivo two-photon voltage-sensitive dye imaging reveals top-down control of cortical layers 1 and 2 during wakefulness. *Proc Natl Acad Sci USA*, 2008, 105:7588–7593.

Lee J, Boas DA. Frequency-domain measurement of neuronal activity using dynamic optical coherence tomography. *Conf Proc IEEE Eng Med Biol Soc*, 2012, 2012:2643–2646.

Lenkov DN, Volnova AB, Pope ARD, Tsytsarev V. Advantages and limitations of brain imaging methods in the research of absence epilepsy in humans and animal models. *J Neurosci Methods*, 2013, 212:195–202.

Liao LD, Li M-L, Lai H-Y, Shih Y-YI, Lo Y-C, Tsang S, Chao PC-P, Lin C-T, Jaw F-S, Chen Y-Y. Imaging brain hemodynamic changes during rat forepaw electrical stimulation using functional photoacoustic microscopy. *NeuroImage*, 2010, 52:562–570.

Liao LD, Lin CT, Shih YYI, Duong TQ, Lai HY, Wang PH, Wu R, Tsang S, Chang JY, Li ML. Transcranial imaging of functional cerebral hemodynamic changes in single blood vessels using in vivo photoacoustic microscopy. *J Cereb Blood Flow Metab*, 2012a, 32:938–951.

Liao LD, Lin CT, Shih YYI, Lai HY, Zhao WT, Duong TQ, Chang JY, Chen YY, Li ML. Investigation of the cerebral hemodynamic response function in single blood vessels by functional photoacoustic microscopy. *J Biomed Opt*, 2012b, 17:061210.

Liao LD, Tsytsarev V, Delgado-Martinez I, Li ML, Erzurumlu RS, Vipin A, Orellana J et al. Neurovascular coupling: In vivo optical techniques for functional brain imaging. *Biomed Eng Online*, 2013, 12:38.

Lippert MT, Takagaki K, Xu W, Huang X, Wu JY. Methods for voltage-sensitive dye imaging of rat cortical activity with high signal-to-noise ratio. *J Neurophysiol*, 2007, 98:502–512.

Lipton P. Effects of membrane depolarization on nicotinamide nucleotide fluorescence in brain slices. *Biochem J*, 1973, 136:999–1009.

Loew LM. Design and use of organic voltage sensitive dyes. In: M. Canepari and D. Zecevic (eds.), *Membrane Potential Imaging in the Nervous System: Methods and Applications*. New York: Springer Science, 2010.

Loew LM. Design and use of organic voltage sensitive dyes. In: M. Canepari and D. Zecevic (eds.), *Membrane Potential Imaging in the Nervous System*. New York: Springer Science, 2011, pp. 13–23.

Lubbert M, Kyereme J, Rothermel M, Wetzel CH, Hoffmann KP, Hatt H. In vivo monitoring of chemically evoked activity patterns in the rat trigeminal ganglion. *Front Syst Neurosci*, 2013 7:64.

Miller TR, Mohan S, Choudhri AF, Gandhi D, Jindal G. Advances in multiple sclerosis and its variants: Conventional and newer imaging techniques. *Radiol Clin North Am*, 2014, 52:321–336.

Mori M, Chiba T, Nakamizo A, Kumashiro R, Murata M, Akahoshi T, Tomikawa M et al. Intraoperative visualization of cerebral oxygenation using hyperspectral image data: A two-dimensional mapping method. *Int J Comput Assist Radiol Surg*, 2014, 9(6):1059–1072.

Muller J, Lupton JM, Lagoudakis PG, Schindler F, Koeppe R, Rogach AL, Feldmann J, Talapin DV, Weller H. Wave function engineering in elongated semiconductor nanocrystals with heterogeneous carrier confinement. *Nano Lett*, 2005, 5:2044–2049.

Muller L, Reynaud A, Chavane F, Destexhe A. The stimulus-evoked population response in visual cortex of awake monkey is a propagating wave. *Nat Commun*, 2014, 5:3675.

Mutoh H, Akemann W, Knopfel T. Genetically engineered fluorescent voltage reporters. *ACS Chem Neurosci*, 2012, 3:585–592.

Mutoh H, Perron A, Akemann W, Iwamoto Y, Knopfel T. Optogenetic monitoring of membrane potentials. *Exp Physiol*, 2011, 96:13–18.

Nishimura M, Shirasawa H, Kaizo H, Song WJ. New field with tonotopic organization in guinea pig auditory cortex. *J Neurophysiol*, 2007, 97:927–932.

Oh SJ, Choi J, Maeng I, Park JY, Lee K, Huh YM, Suh JS, Haam S, Son JH. Molecular imaging with terahertz waves. *Opt Express*, 2011, 19:4009–4016.

Ohrstrom L, Bitzer A, Walther RFJ. Technical note: Terahertz imaging of ancient mummies and bone. *Am J Phys Antropology*, 2010, 142:497–500.

Ollerenshaw DR, Zheng HJ, Millard DC, Wang Q, Stanley GB. The adaptive trade-off between detection and discrimination in cortical representations and behavior. *Neuron*, 2014, 81:1152–1164.

Omer DB, Hildesheim R, Grinvald A. Temporally-structured acquisition of multidimensional optical imaging data facilitates visualization of elusive cortical representations in the behaving monkey. *NeuroImage*, 2013, 82:237–251

Orbach HS, Cohen LB, Grinvald A. Optical mapping of electrical activity in rat somatosensory and visual cortex. *J Neurosci*, 1985, 5(7):1886–1895.

Osman A, Park JH, Dickensheets D, Platisa J, Culurciello E, Pieribone VA. Design constraints for mobile, high-speed fluorescence brain imaging in awake animals. *IEEE Trans Biomed Circuits Syst*, 2012, 6:446–453

Perron A, Akemann W, Mutoh H, Knopfel T. Genetically encoded probes for optical imaging of brain electrical activity. *Prog Brain Res*, 2012, 196:63–77.

Peterka DS, Takahashi H, Yuste R. Imaging voltage in neurons. *Neuron*, 2011, 69:9–21

Polack PO, Contreras D. Long-range parallel processing and local recurrent activity in the visual cortex of the mouse. *J Neurosci*, 2012, 32(32):11120–11131.

Preuss S, Stein W. Comparison of two voltage-sensitive dyes and their suitability for long-term imaging of neuronal activity. *PLoS One*, 2013, 8:e75678.

Raichle ME. Functional brain imaging and human brain function. *J Neurosci*, 2003, 23:3959–3962.

Rajagopalan UM, Tanifuji M. Functional optical coherence tomography reveals localized layer-specific activations in cat primary visual cortex in vivo. *Opt Lett*, 2007, 32(17):2614–2616.

Ribot J, Tanaka S, Tanaka H, Ajima AJ. Online analysis method for intrinsic signal optical imaging. *Neurosci Methods*, 2006, 153:8–20.

Rothermel M, Ng BS, Grabska-Barwinska A, Hatt H, Jancke D. Nasal chemosensory-stimulation evoked activity patterns in the rat trigeminal ganglion visualized by in vivo voltage-sensitive dye imaging. *PLoS One*, 2011, 6:e26158.

Sable JJ, Low KA, Whalen CJ, Maclin EL, Fabiani M, Gratton G. Optical imaging of temporal integration in human auditory cortex. *Eur J Neurosci*, 2007, 25:298–306.

Salzberg BM, Davila HV, Cohen LB. Optical recording of impulses in individual neurons of an invertebrate central nervous-system. *Nature*, 1973, 246:508–509.

Satomura Y, Seki J, Ooi Y, Yanagida T, Seiyama A. In vivo imaging of the rat cerebral microvessels with optical coherence tomography. *Clin Hemorheol Microcirc*, 2004, 31:31–40.

Sobottka SB, Meyer T, Kirsch M, Koch E, Steinmeier R, Morgenstern U, Schackert G. Evaluation of the clinical practicability of intraoperative optical imaging comparing three different camera setups. *Biomed Tech (Berl)*, 2013a, 58:237–248.

Sobottka SB, Meyer T, Kirsch M, Koch E, Steinmeier R, Morgenstern U, Schackert G. Intraoperative optical imaging of intrinsic signals: A reliable method for visualizing stimulated functional brain areas during surgery. *J Neurosurg*, 2013b, 119:853–863.

Studer V, Bobin J, Chahid M, Mousavi HS, Candes E, Dahan M. Compressive fluorescence microscopy for biological and hyperspectral imaging. *Proc Natl Acad Sci USA*, 2012, 109:E1679–E1687.

Sukhanova A, Devy M, Venteo L, Kaplan H, Artemyev M, Oleinikov V, Klinov, D, Pluot M, Cohen JHM, Nabiev I. Biocompatible fluorescent nanocrystals for immunolabeling of membrane proteins and cells. *Anal Biochem*, 2004, 324:60–67

Takahashi K, Hishida R, Kubota Y, Kudoh M, Takahashi S, Shibuki K. Transcranial fluorescence imaging of auditory cortical plasticity regulated by acoustic environments in mice. *Eur J Neurosci*, 2006, 23:1365–1376.

Takeshita D, Bahar S. Synchronization analysis of voltage-sensitive dye imaging during focal seizures in the rat neocortex. *Chaos*, 2011, 21(4):047506.

Trevino-Palacios CG, Celis-Lopez MA, Larraga-Gutierrez JM, Garcia-Garduno A, Zapata-Nava OJ, Orduna Diaz A, Torres-Jacome A, de-la-Hidalga-Wade J, Iturbe-Castillo MD. Brain imaging using T-rays instrumentation advances. *11th Mexican Symposium on Medical Physics, AIP Conference Proceedings*, Vol. 1310, Mexico City, pp. 146–149, 2010.

Tsytsarev V, Bernardelli C, Maslov KI. Living brain optical imaging: Technology, methods and applications. *J Neurosci Neuroeng*, 2012, 1:180–192.

Tsytsarev V, Fukuyama H, Pope D, Pumbo E, Kimura M. Optical imaging of interaural time difference representation in rat auditory cortex. *Front Neuroeng*, 2009, 2:2.

Tsytsarev V, Rao B, Maslov KI, Li L, Wang LV. Photoacoustic and optical coherence tomography of epilepsy with high temporal and spatial resolution and dual optical contrasts. *J Neurosci Methods*, 2013, 216(2):142–145.

Tsytsarev V, Liao L-D, Kong KV, Liu Y-H, Erzurumlu RS, Olivo M, Thakor NV. Recent progress in voltage-sensitive dye imaging for neuroscience. *J Nanosci Nanotechnol*, 2014, 14:4733–4744.

Tsytsarev V, Pope D, Pumbo E, Yablonskii A, Hofmann M. Study of the cortical representation of whisker directional deflection using voltage-sensitive dye optical imaging. *NeuroImage*, 2010, 53:233–238.

Wagh MP, Sonawane YH, Joshi OU. Terahertz technology: A boon to tablet analysis. *Indian J Pharm Sci*, 2009:71:235–241.

Walker JA, Low KA, Cohen NJ, Fabiani M, Gratton G. When memory leads the brain to take scenes at face value: Face areas are reactivated at test by scenes that were paired with faces at study. *Front Hum Neurosci*, 2014, 8:18.

Wang S, Ferguson B, Abbott D, Zhang XC. T-ray Imaging and tomography. *J Biol Phys*, 2003 29:247–256.

Wang X, Cui Y, Sun W, Zhang Y, Zhang C. Terahertz pulse reflective focal-plane tomography. *Opt Express*, 2007, 15:14369–14375.

Wanger T, Takagaki K, Lippert MT, Goldschmidt J, Ohl FW. Wave propagation of cortical population activity under urethane anesthesia is state dependent. *BMC Neurosci*, 2013, 14:78.

Watanabe H, Rajagopalan UM, Nakamichi Y, Igarashi KM, Kadono H, Tanifuji M. Functional optical coherence tomography of rat olfactory bulb with periodic odor stimulation. *Biomed Opt Express*, 2016, 7(3):841–854.

Yang Z-Y, Chen HC. Applying near-infrared spectroscopy in cognitive neuroscience. *J Neurosci Neuroeng*, 2013, 2:231–242.

Yokoo T, Knight BW, Sirovich L. An optimization approach to signal extraction from noisy multivariate data. *Neuroimage*, 2001, 14:1309–1326.

Zecevic D, Djurisic M, Cohen LB, Antic S, Wachowiak M, Falk CX, Zochowski MR. Imaging nervous system activity with voltage-sensitive dyes. *Curr Protoc Neurosci*, 2003, Chapter 6:Unit 6.17.

Zhang D, Raichle ME. Disease and the brain's dark energy. *Nat Rev Neurol*, 2010, 6:15–28.

Zhu H, Fan Y, Guo H, Huang D, He S. Reduced interhemispheric functional connectivity of children with autism spectrum disorder: Evidence from functional near infrared spectroscopy studies. *Biomed Optic Express*, 2014, 5:1263–1274.

Zochowski M, Wachowiak M, Falk CX, Cohen LB, Lam YW, Antic S, Zecevic D. Imaging membrane potential with voltage-sensitive dyes. *Biol Bull*, 2000, 198:1–21.

7 Optical Techniques to Study Drug-Induced Neurovascular and Cellular Changes *In Vivo*

Congwu Du and Yingtian Pan

CONTENTS

INTRODUCTION

Substance abuse results in profound and wide-ranging changes in brain chemistry, morphology, physiology, and function. Although researchers have made enormous strides in identifying the receptors for each major drug of abuse and in clarifying the neural circuitry involved in addictive processes, the physiological and functional changes in the brain during drug use or after chronic exposure to substances of abuse are not fully understood. For example, one of the most serious medical risks of cocaine abuse is stroke due to the drug's disruption of blood flow in the brain. About 25%–60% of cocaine-induced strokes can be attributed to cerebral vasospasm and ischemia (Johnson et al. 2001; Buttner et al. 2003; Bartzokis et al. 2004; Bolouri and Small 2004). The resultant neurologic deficits can range from mild and transient (e.g., facial paralysis) to severe with permanent disability (e.g., tetraplegia (Spivey and Euerle 1990)). Brain imaging studies have documented marked decreases in cerebral blood flow (CBF) and blood volume (CBV) in cocaine abusers (Volkow et al. 1988; Pearlson et al. 1993; Wallace et al. 1996; Gollub et al. 1998). However, the mechanisms underlying cocaine-induced CBF reduction, cerebral vasospasm, and ischemia are poorly understood. The possibilities include (1) direct vasoconstrictive effects elicited via cocaine-induced intracellular calcium ($[Ca^{2+}]_i$) increase in vascular smooth muscle cells (Zhang et al. 1996), (2) indirect vasoconstriction secondary to the release of sympathomimetic amines, (3) a local anesthetic action secondary to blockade of ion channels, or (4) an indirect effect of reduced neural activity and metabolic demand (Volkow et al. 1988). Therefore, separation of the cellular effects from the vascular effects of cocaine is crucial to understanding the mechanisms that lead to neurovascular toxicity in cocaine abusers.

The ability to distinguish the cellular effects from the vascular effects of cocaine remains a technical challenge for neurophotonics and brain mapping communities because the technological requirements necessary for this endeavor include

1. Multiparameter detection or imaging to enable simultaneous quantification of the changes in hemodynamics, tissue oxygenation, and cellular activity induced by cocaine
2. High temporal resolution to enable capturing real-time dynamic changes and brain responses during drug intoxication
3. High spatial resolution to permit separation of vascular compartments (e.g., artery, vein, and capillary) and image their vascular hemodynamic changes without fluorescence labeling
4. A relatively large field of view (FOV) to provide 2D or 3D quantitative images of the CBF velocity (CBFv) networks and cerebrovascular angiography

In the neuroimaging field, a noninvasive and high spatiotemporal-resolution imaging of cerebral hemodynamic and cellular (e.g., neuronal) response to drug challenge remains a major challenge. Although conventional neuroimaging tools such as positron emission tomography (PET) and functional MRI (fMRI) have greatly advanced our understanding of the pharmacological and physiological effects of cocaine (Volkow et al. 1988; Gollub et al. 1998), the spatial resolution of these imaging techniques (e.g., >1 mm) is insufficient to resolve individual vascular compartments or cells (London et al. 1990; Lee et al. 2003). While optical microscopy (e.g., multiphoton microscopy) has shown some promise for visualizing capillary vasculature and cellular details of the cerebral cortex of rodents *in vivo*, its FOV is too small and the image depth is limited (e.g., ~300 μm) (Helmchen and Waters 2002). Other imaging approaches using intrinsic hemoglobin contrast such as intrinsic signal imaging (IOS) (Frostig et al. 2005), laminar optical tomography (LOT) (Hillman et al. 2004), and laser speckle imaging (LSI) (Dunn et al. 2001) have been reported to map the brain's hemodynamic activity with improved spatial resolution; while IOS excels in the spatial domain with ~50 μm resolution (Frostig et al. 2005), LOT is unable to resolve individual vessels and LSI only measures "relative" changes in CBFv.

To tackle this challenge, we have developed optical-fiber-based diffusion and fluorescence (ODF) spectroscopy (Figure 7.1) and multimodality optical/fluorescence imaging (OFI) platforms to permit simultaneous assessment of cerebral hemodynamics and cellular activities in cortical brain *in vivo*.

In this chapter, we will present the state-of-the-art optical techniques developed in our labs, focusing on their technical capabilities that enable simultaneous recordings and dynamic measurements to capture the physiological changes from the cortex of a living brain. We will summarize their application to drug abuse and drug addiction—one of the common but serious brain diseases to study the complex neurovascular and cellular activity changes in the brain induced by stimulants such as cocaine.

MULTIWAVELENGTH SPECTROSCOPY FOR BRAIN FUNCTIONAL STUDIES

OPTICAL-FIBER-BASED DIFFUSION AND FLUORESCENCE SPECTROSCOPY

Figure 7.1 illustrates a multiwavelength optical-fiber-based diffusion and fluorescence (ODF) spectroscopy platform that is able to detect the changes of the local cerebral blood volume (CBV), oxygenated ([HbO]) and deoxygenated hemoglobin ([HbR]), thus hemoglobin oxygenation in tissue

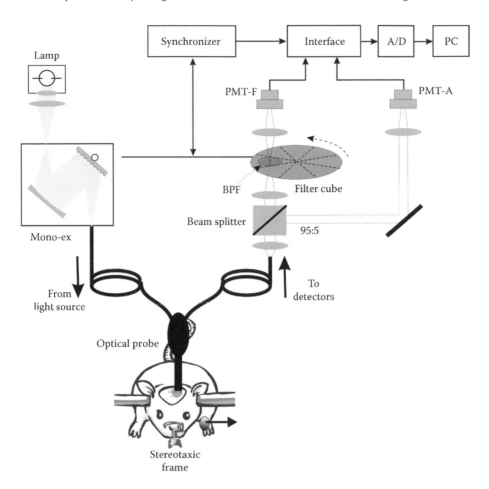

FIGURE 7.1 A schematic to illustrate optical-fiber-based diffusion and fluorescence (ODF) spectroscopy to simultaneously detect the dynamic changes in CBV, hemoglobin oxygenation (S_tO_2), and intracellular calcium from the cortical brain of an animal. (Data from Du, C. et al., *J. Innov. Opt. Health Sci.*, 2(2), 1, 2009.)

(S_tO_2), as well as intracellular calcium ($[Ca^{2+}]_i$) from the surface of the brain (Du et al. 2005). Briefly speaking, it consisted of a 150-W xenon lamp, a monochromator (Mono-Ex), and a pair of photon detectors for fluorescence (PMT-F) and diffuse reflectance (PMT-A). The lamp was connected to a computer-controlled monochromator to select the incident wavelengths of 548, 555, and 572 nm by time-sharing to sequentially deliver the light at these wavelengths onto the brain surface through one arm of a Y-shape bifurcated fiber-optic bundle. To measure intracellular calcium, fluorescence calcium indicator, Rhod2 (AM, Molecular Probes, Eugene), was infused into the brain topically (Du et al. 2005, 2009). Calcium-dependent fluorescence and diffusive reflectance from the brain tissue were collected by the fiber tip in the common leg of the Y-coupler and detected by the photon detectors of PMT-F and PMT-A accordingly. The changes in CBV and S_tO_2 could be separately quantified from the reflectance units measured by summing and subtracting the optical densities of the signals at 555 and 572 nm (i.e., symmetric wavelengths to the hemoglobin isosbestic wavelength of tissue oxygenation as described previously) (Du et al. 2005). The signals were digitized and stored in a personal computer for data processing. The ratio of Ca^{2+} fluorescence emission at 589 nm over reemitted excitation reflectance (e.g., at 548 nm) represents the calcium-dependent fluorescence with a minimal influence from the physiological interference.

Ischemic Effects versus Cocaine's Effects on Brain: Changes of CBV, S_tO_2, and $[Ca^{2+}]_i$ in Brain Cortex

Figure 7.2(A_0–C_0) shows the time course of mean blood volume (CBV), oxygenation (S_tO_2), and $[Ca^{2+}]_i$ fluorescence (labeled by the intracellular calcium indicator Rhod2 (AM, Molecular Probes, and Eugene) during a transient ischemic insult (e.g., $t = 0$–5 min marked as a gray strip in Figure 7.2(A_0–C_0) and reperfusion (e.g., $t > 5$ min in Figure 7.2(A_0–C_0)) that were obtained from the cortical brain of the rats (Du et al. 2005). The results show that the ischemic insult produced a decrease in CBV and S_tO_2 for ~57.4% ± 12.6% and ~47.3% ± 12.5%, respectively. In addition, the ischemia induced a ~8.5% ± 1.7% increase in the intracellular calcium florescence ($[Ca^{2+}]_i$). All of these signals were back to the baseline levels after the reperfusion. This study validates that the ODF technique offers the distinct advantage to separately detect the changes in free intracellular calcium alone with the hemodynamic alterations in the brain *in vivo*, which opens the opportunities to study the physiological function of a normal brain as well as the cerebral pathological processes such as drug-elicited cell injury or brain functional change.

For example, Figure 7.2(A_1–C_1) illustrates the cocaine-induced changes in CBV, S_tO_2, and $[Ca^{2+}]_i$ fluorescence of the cortex. Similar to the ischemia experiments above, the animals were anesthetized

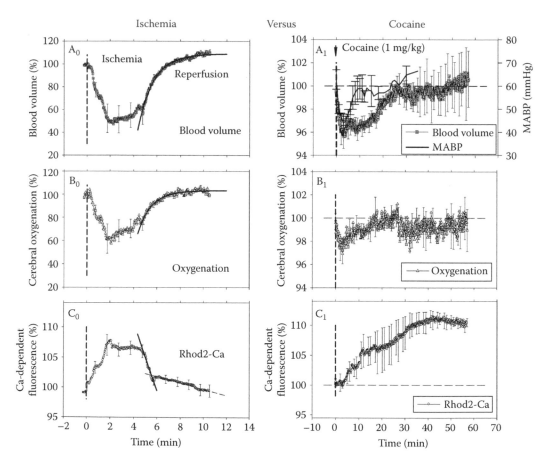

FIGURE 7.2 Changes in CBV, S_tO_2, and intracellular calcium florescence ($[Ca^{2+}]_i$) induced by a transient ischemia (left panel, A_0–C_0) (Du et al. 2005) versus those changes induced by cocaine (right panel, A_1–C_1). It indicates that cocaine-induced hemodynamic changes (CBV, S_tO_2) were ~10 folds less than those induced by ischemia. However, both ischemia and cocaine-induced $[Ca^{2+}]_i$ increase, indicating the neurotoxic effects of cocaine. (Data from Du, C. et al., *J. Neurosci.*, 26(45), 11522, 2006.)

with isoflurane and the measurements were conducted from the same cortical surface (i.e., somato-sensory cortex) of the rats. As shown in Figure 7.2(A_1 and B_1), cocaine-induced changes decrease in CBV and S_tO_2, but only ~4.2% $\pm$ 1.2% and ~3.1% $\pm$ 0.9%, respectively, which is ~10 folds lower than those changes induced by ischemia (i.e., ~57.4% $\pm$ 12.6% for CBV and ~47.3% $\pm$ 12.5% for S_tO_2, respectively). The time course of cocaine-induced changes in the mean artery blood pressure (i.e., MABP) is superposed in Figure 7.2(A_1), indicating that MABP decreased rapidly within 3–4 min after cocaine administration but recovered to the baseline at 7–8 min of the post-cocaine administration, followed by a slight overshooting afterward. A comparison of cocaine-induced temporal changes between CBV (black trace in Figure 7.2(A_1)) and MABP (blue trace in Figure 7.2(A_1)) indicates that the decrease in CBV only coincided with the decrease in MABP over the first 3–4 min. In other words, MABP quickly recovered to its normal level after 3–4 min following cocaine administration whereas CBV persisted at a low level of ~4.2% $\pm$ 1.2% till ~15 min when it started to recover. This indicates that the CBV decrease was resulted from the direct cocaine's vasoconstriction effect rather than a secondary change due to the MABP decrease induced by cocaine. Interestingly, intra-cellular calcium ($[Ca^{2+}]_i$), as measured by $Rhod_2$ fluorescence (Figure 7.2(C_1)), remained unchanged for 4–5 min after the cocaine challenge, followed by gradual increase until reaching a maximum of ~10.3% $\pm$ 1.3% over 40 min post-cocaine administration. The significant $[Ca^{2+}]_i$ increase induced by cocaine in the brain is likely to accentuate the neurotoxic effects from cocaine-induced vasoconstriction, thus facilitating the occurrence of seizures, which is one of the most frequent complications associated with cocaine overdoses (Kaye and Darke 2004; Du et al. 2006).

MULTIMODAL NEUROIMAGING TO STUDY CORTICAL BRAIN FUNCTION

MULTIMODAL OPTICAL AND FLUORESCENCE IMAGING

Although the ODF spectroscopy system in Figure 7.1 can provide simultaneous assessment of cere-bral hemodynamic (by measure of CBF), tissue oxygenation metabolism (by measure of S_tO_2) and cellular activities (by measure of $[Ca^{2+}]_i$) in cortical brain *in vivo* to access the dynamic changes in CBV, S_tO_2, and $[Ca^{2+}]_i$ fluorescence of the brain cortex induced by cocaine, it is unable to separate the contributions of cerebral blood vessels from those of the brain tissue (i.e., avascular tissue) to resolve the vasoconstriction effects of cocaine on the individual vessels (i.e., due to its poor spatial resolution). In other words, this method has appropriate temporal resolution, but it has no spatial resolution to differentiate the potentially different responses from the different types of vessels to stimulants such as cocaine.

To solve this problem, we developed a multimodal optical/fluorescence imaging (OFI) plat-form. OFI integrates the following: (1) dual-wavelength laser speckle imaging (DW-LSI; Luo et al. 2009) for concurrent detection of changes in CBFv, CBV, and tissue hemoglobin oxygenation at high spatiotemporal resolutions across a large FOV; (2) digital-frequency-ramping Doppler optical coherence tomography (DFR-OCT; Yuan et al. 2009) to permit quantitative 3D CBF imaging of the neurovascular networks; and (3) $Rhod_2$ fluorescence imaging to measure the $[Ca^{2+}]_i$ changes of the brain, which serves as an indicator of cellular activation (Du et al. 2005, 2006, 2009).

Figure 7.3 illustrates the multimodal OFI system that integrates DW-LSI and fluorescence imag-ing (upper dashed box) with 3D optical coherence Doppler tomography (ODT, lower dashed box). It is a custom imaging platform (Yuan et al. 2011) whose major subsystems and modules are sum-marized as follows.

DW-LSI and Fluorescence Imaging

A custom illumination module, which comprised two single-mode laser diodes at the wave-lengths of 785 nm (50 mW, HL7851G, Hitachi) and 830 nm (30 mW, DL5032, Sanyo) sym-metric to hemoglobin's isosbestic point of 805 nm for DW-LSI and 1 diode laser at 532 nm

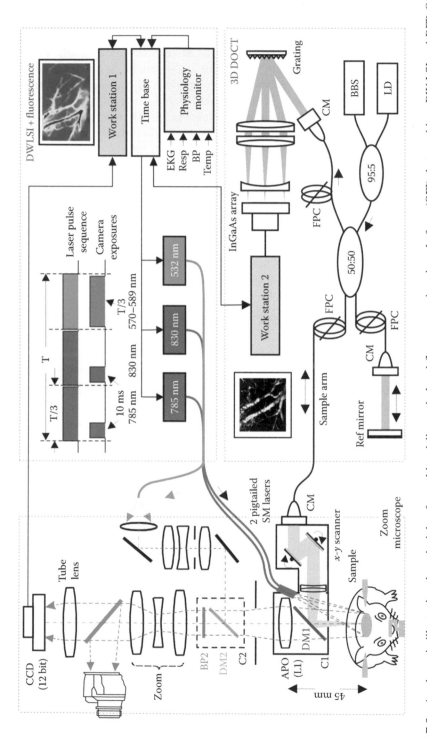

FIGURE 7.3 A schematic illustrating the principle of a multimodality optical and fluorescence imaging platform (OFI) that combines DW-LSI and DFR-OCT for fluorescence imaging used in the study (Yuan et al. 2011). Upper dashed box: DW-LSI and fluorescence imager. $LD_{1,2}$: laser diodes at $\lambda_{1,2} = 785$, 830 nm for CBF and metabolic imaging; λ_{ex}: 532 nm for excitation of Rhod$_2$-[Ca^{2+}]$_i$; fluorescence imaging (λ_{em}: 570–589 nm), SM: single mode fiber. Lower dashed box: 3D ODT. CM: collimator; BBS: broadband source ($\lambda = 1.3$ μm); LD: aiming laser ($\lambda = 670$ nm), FPC: fiber-optic polarization controller. Left dash box: modified zoom microscope. C1: epi-illumination cube 1. DM1: dichroic beam splitter ($\lambda_D = 1$ μm); L1: 2× APO ($f = 45$ mm, NA = 0.22). C2: epi-illumination cube 2. DM2: dichroic beam splitter ($\lambda_D = 550$ nm); BP2: barrier filter ($\lambda_B \geq 570$ nm). (Adapted from Yuan, Z. et al., *Neuroimage*, 54(2), 1132, 2011. With permission.)

(50 mW, G30/R100, Optlaser) for $Rhod_2$-Ca^{2+} excitation, was employed to sequentially illuminate the cortical window (>ϕ5 mm) via 3 optical fibers. A cooled 12-bit CCD camera (pixel size: 6.45 μm; Retiga Ex, QImaging) synchronized with the laser pulses acquired 2 diffuse reflectance images (exposure: $T_{1,2}$ = 10 ms) for DW-LSI and 1 $Rhod_2$-Ca^{2+} fluorescence image peaked around 570–589 nm (exposure: $T_3 \approx$ 30 ms). The pulse sequences for laser and camera exposures were generated by a time base (PCI-6221, NI) to enable sequential DW-LSI and fluorescence imaging acquisitions. The optical setup of OFI was based on a modified fluorescence zoom microscope (AZ100, Nikon) to take advantage of its long working distance (WD = 45 mm), large FOV (e.g., ~ϕ5 mm), and high NA (NA = 0.22) optics, for example, 2× Plan Apo objective (L1). A custom epi-illumination cube (C1) mounted in front of L1 integrated the 1.3 μm ODT system with a dichroic mirror (DM1, cut off at $\lambda \approx$ 1 μm). The 785 and 830 nm laser beams were delivered by 2 pigtailed SM fibers (NA $\approx$ 0.1) to obliquely (24°) illuminate the cranial window (over ϕ6 mm) to be overlapped with the 532 nm laser beam introduced by the epi-illumination cube 2 (C2). Diffuse mirrors were inserted in front of fiber endfaces to provide uniform illumination on the cranial window. The dichroic mirror (CM2, cut off at $\lambda \approx$ 550 nm) reflected the 532 nm beam to the cortical brain for fluorescence excitation and transmitted the light beams at all 3 wavelengths (i.e., 785, 830 nm, and fluorescence emission around 570–589 nm) together with a barrier filter (BP2: λ > 570 nm) to be detected by the CCD camera operated in a time-sharing mode. The captured image sequences were streamed into workstation 2 via an IEEE-1394 interface for image processing to extract 2D images of CBF, and the changes in HbO_2, HbR (thus, total hemoglobin HbT), and $[Ca^{2+}]_i$ fluorescence.

3D ODT

A spectral-domain OCT (SDOCT) system was used to perform Doppler flow imaging in which a pigtailed broadband source (BBS: 12 mW, λ = 1.3 μm, $\Delta\lambda$ = 90 nm, coherence length $L_C \approx$ 8 μm; Inphenix) was employed to illuminate a fiber-optic Michelson interferometer. The reference beam was collimated to transmit through a pair of variable-thickness wedge prisms and reflected by a stationary mirror to minimize the dispersion mismatch between the two arms and match the path length of the sample beam. The sample arm was connected to C1 in which light exiting the SM fiber was collimated (ϕ5 mm), transversely scanned by a pair of servo mirrors (x–y scanner), focused by an achromate (f/40 mm), and reflected by a dichroic mirror DM1 onto the cortical surface, while the light backscattered from the cortical brain was coupled back into the sample fiber through the same optical pathway. Then, the recombined sample and the reference beams in the detection fiber were connected to a custom spectrometer in which light was collimated (ϕ10 mm), linearly diffracted by a grating (d^{-1} = 1200/mm), and focused by a lens group (f/140 mm) to be detected by a fast InGaAs line camera to enable 2D OCT (xz cross section) at 47 fps. A 3D OCT image cube (e.g., x/1k × z/1k × y/400 voxels) could be acquired in approximately 8 s to cover a volume of 2.5 × 2 × 2.5 mm^3 on the cortical brain. Post-image processing with digital-frequency ramping (i.e., DFR-OCT) or other phase subtraction methods (PSM) was applied to retrieve Doppler flow signals and reconstruct the 3D CBF network of the brain cortex quantitatively. The transverse and axial resolutions of the SDOCT system were 12 and 8 μm, respectively, and, in more recent studies (ultrahigh-resolution ODT or μODT), were improved to around 3 μm.

Image Processing

For DW-LSI, a 2D flow index map was retrieved from the speckle patterns in an acquired raw photographic image based on the dynamic laser speckle features resulting from local CBFs (Dunn et al. 2003; Luo et al. 2008). To extract the flow contrast embedded in the speckle patterns of a raw reflectance image, 5 × 5 binning was performed to compute the speckle contrast map $K = \sigma/<I>$, where $<I>$ and σ were the mean intensity and standard deviation of each binning window (i.e., 6 × 5 $\approx$ 30 μm). The relationship between K and the dynamic features of speckles is highly complex, which can be approximated as (Bandyopadhyay et al. 2005)

$$K = \left\{ \frac{\tau_C}{T} + \frac{\tau_C^2}{2T^2} \left[\exp\left(-\frac{2T}{\tau_C}\right) - 1 \right] \right\}^{1/2} \tag{7.1}$$

where

$\tau_C = [ka <v^2>^{1/2}]^{-1}$ is the autocorrelation time of the speckle intensity fluctuation
$k = 2\pi/\lambda$ is the wave number of light
$<v^2>^{1/2}$ represents the rooted-mean-square speed of moving scattering particles (a nondirectional, statistical unit of flow rate)

As a is an unknown factor associated with v distribution and tissue scattering characteristics, the combined product $a<v^2>^{1/2}$ is defined as the speed index, which was obtained from a precalculated look-up table between K and $a<v^2>^{1/2}$ to minimize computation, which was very helpful for instantaneous image acquisition and display.

The changes in CBV and S_tO_2 can be derived from DW-LSI images under the assumption that $\Delta[HbO_2]$ and $\Delta[HbR]$ dominate the dynamic changes of light absorption and thus the measured diffuse reflectance (Dunn et al. 2003; Zhang et al. 2007), which can be given by

$$\begin{bmatrix} \Delta[HbO_2(t)] \\ \Delta[HbR(t)] \end{bmatrix} = \begin{bmatrix} \varepsilon_{HbO_2}^{\lambda_1} & \varepsilon_{HbR}^{\lambda_1} \\ \varepsilon_{HbO_2}^{\lambda_2} & \varepsilon_{HbR}^{\lambda_2} \end{bmatrix}^{-1} \cdot \begin{bmatrix} \dfrac{\ln\left(R_{\lambda_1}(0)/R_{\lambda_1}(t)\right)}{L_{\lambda_1}(t)} \\ \dfrac{\ln\left(R_{\lambda_2}(0)/R_{\lambda_2}(t)\right)}{L_{\lambda_2}(t)} \end{bmatrix} \tag{7.2}$$

where ε refers to the molar spectral absorptivity of the chromophore. $R_{\lambda_1}(t)$, $R_{\lambda_2}(t)$ are the diffuse reflectance images measured at these two wavelengths after being averaged over 5 consecutive frames for speckle noise reduction at time t, and $R_{\lambda_1}(0)$, $R_{\lambda_2}(0)$ are their baseline values prior to a stimulation (e.g., cocaine challenge). $L_{\lambda_1}(t) \approx L_{\lambda_2}(t)$, where $L_{\lambda_1}(t)$, $L_{\lambda_2}(t)$ are their path lengths. The total hemoglobin concentration change can be obtained by $\Delta[HbT] = \Delta[HbO_2] + \Delta[HbR]$, assuming that it is linearly proportional to local CBV (Dunn et al. 2003; Zhang et al. 2007). According to Equations 7.1 and 7.2, the changes in CBF, CBV, and $[HbO_2]$ of the cortical brain can be simultaneously imaged by DW-LSI at high spatiotemporal resolutions. For $[Ca^{2+}]_i$ fluorescence imaging, a cross-correlation test between baseline ($t \leq 0$) and post-cocaine injection ($t > 0$) periods was used to analyze $[Ca^{2+}]_i$ fluorescence change in response to cocaine administration. Specifically, for each pixel (x, y), cross-correlation between measured fluorescence $I_{xy}(t)$ and a step function $u(t)$ ($u(t) = 0$, $t \leq 0$; $u(t) = 1$, $t > 0$) was calculated to statistically determine a significant increase in the $[Ca^{2+}]_i$ fluorescence ($p < 0.01$), which was used to mask the active regions in the $[Ca^{2+}]_i$ fluorescence image (e.g., the overlapped $[Ca^{2+}]_i$ clouds in Figure 7.7(B_0–B_4)). The averaged $[Ca^{2+}]_i$ fluorescence change within the mask at each time point was calculated to present the time course of $[Ca^{2+}]_i$ fluorescence change in response to cocaine administration (e.g., Figure 7.7(B_5)). The influence of hemoglobin absorption on the fluorescence emission can be corrected empirically through the CBF changes measured simultaneously (Yuan et al. 2011).

For quantitative Doppler OCT imaging, the detected spectral interference can be derived as

$$I_{OCT}(k; \Delta L, x) = S(k)\left[\left(I_s + I_r\right) + 2\sqrt{I_s I_r} \cdot \cos(k\Delta L + \phi_\tau(\Delta L, x))\right] \tag{7.3}$$

where

I_s and I_r are the intensities in the sample and reference arms
$S(k)$ is the source spectrum ($k = 2\pi/\lambda$ is the wave number of light)
$\Delta L = L_s - L_r$ is the round-trip path length difference between the sample and the reference arms

The interferometric term spectrally modulating $S(k)$ at a frequency of $\Delta k = \pi/\Delta L$ (i.e., inversely proportional to ΔL) can be decoded via an inverse FFT to restore the A-scan (Dorrer et al. 2000). It is noteworthy that as the detected spectral interferogram at a transverse position x, $I_{OCT}(k; \Delta L, x)$, preserves the relative phase $\phi_\tau(\Delta L, x)$, SDOCT allows for the detection of Doppler flow velocity (i.e., phase change) along with tissue morphology (i.e., amplitude). By detecting the phase shift $\Delta\phi_\tau(\Delta L, x)$ between two consecutive A-scans with τ interval, the Doppler flow velocity can be derived by phase subtraction method (PSM) as $v_f(z, x) = [\lambda\Delta\phi_\tau(\Delta L, x)]/(4\pi n \cos \alpha)$ (Leitgeb et al. 2003; Wang et al. 2007b; Luo et al. 2008), where n is the refractive index of the tissue and α is the angle between the flow and incident light. However, this method for Doppler flow imaging is prone to phase noise induced by various effects (e.g., multiple scattering, tissue heterogeneity and micro motion, and electronic noise), which leads to severe under detection of subsurface blood flows. In addition to methods based on dynamic speckle variance, recent advances in OCT angiography (Wang et al. 2007a) dramatically improved the sensitivity for flow detection by applying Hilbert transform to the detected $I_{OCT}(k; \Delta L, x)$ along the transverse x-axis, that is, frequency offsetting to reduce background phase noise. We further developed a simple digital-frequency-ramping method, termed DFR-OCT (Yuan et al. 2009), to enhance flow detection which can be applied to conventional SDOCT with no hardware modification. Importantly, DFR-OCT enables quantitative 2D and 3D Doppler flow imaging by digitally ramping threshold frequency v_R in Hilbert transforms, that is, $\phi_\tau(\Delta L, x)$ in Equation 7.3 can be expressed as (Yuan et al. 2009)

$$\phi_\tau\left(\Delta L, x\right) = \phi_0 + \left(2\pi/\lambda\right)\cdot\left(v_f + v_b + v_R\right)\cdot\tau\cdot n_x \tag{7.4}$$

where
ϕ_τ is the initial phase
$n_x = 1, 2, \ldots, N_x$ is A-scan index (N_x is the number of scans along x-axis)
v_f and v_b are the velocities of Doppler flow and background noise flow

If the digitally imposed frequency-ramping offset v_R is chosen to ensure,

$$\left(v_f + v_b + v_R\right)\cdot\left(v_b + v_R\right) < 0 \tag{7.5}$$

Hilbert transform (sensitive to frequency thresholding, i.e., flipping sign) along x-axis can easily separate these two types of flow signals, that is, to differentiate all Doppler flows with v_f above the noise ground v_b for nonquantitative angiographic detection (Yuan et al. 2009). To further quantify Doppler flow rate v_f, which can be normalized to the range of $(-1, 1]$, v_R can be similarly ramped over N steps with a step size of $\Delta v = 1/N$ or $v_R(i) = v_R + i\cdot\Delta v$ (where $i = -N/2,\ldots, N/2$) to search for points at which the transverse modulation frequency of Hilbert transform flips sign. Here, because of phase wrapping effect, the ramping range can be further reduced to half to save computation efforts in our new algorithm. Then, based on Equations 7.4 and 7.5, the flow rate for DFR-OCT can be derived as

$$\left|v_f\right| = \Delta v\left\{\sum_{i=-N/2}^{-1}\left[\left(v_f - i\Delta v\right) < 0\right] + \sum_{i=1}^{N/2}\left[\left(v_f - i\Delta v\right) > 0\right]\right\} \tag{7.6}$$

Compared with the previous DFR algorithm (Yuan et al. 2009), the two sets of summation operations in Equation 7.6 accumulate the results from each step and further reduce the noise level because of the averaging effect. Here, $(v_f - i\cdot\Delta v) > 0$ and $(v_f - i\cdot\Delta v) < 0$ are both Boolean functions (i.e., 1 for "true," 0 for "false"), and the sign of v_f can be determined by

$$sign(v_f) = \begin{cases} (+) & if \sum_{i=1}^{N/2}\left[(v_f - i\Delta v) > 0\right] > \sum_{i=-N/2}^{-1}\left[(v_f - i\Delta v) < 0\right] \\ (-) & if \sum_{i=1}^{N/2}\left[(v_f - i\Delta v) > 0\right] < \sum_{i=-N/2}^{-1}\left[(v_f - i\Delta v) < 0\right] \end{cases} \qquad (7.7)$$

The step size was optimized and set to $\Delta v = 0.02$, which is a compromise of extensive computation and the discernible Doppler flow rate resolution. Preclinical validation studies show that although the sensitivity of DFR-OCT may not be as high as the more recent ODT methods, it is less stringent to the need for dense transverse light scans, thus yielding a higher temporal resolution for studying dynamic changes in the blood flows.

OFI TO IMAGE VASCULAR ACTION: QUANTITATIVE NEUROVASCULAR RESPONSE TO COCAINE

Figure 7.4 shows the results of simultaneous CBFv images by LSI and 3D ODT from the OFI platform before and after cocaine administration (1 mg/kg, *i.v.*, which is a clinically relevant dose). Figure 7.4a shows that LSI provides *en face* CBFv images over a large FOV of 5×3 mm^2 at a high frame rate of ~10 fps, thus permitting continuous monitoring of time-varying CBFv changes in the full field that includes large and small vessels (e.g., 1, 2) and CBF perfusion in the surrounding cortical tissue (e.g., 3, resulting from microcirculatory flows irresolvable by LSI, i.e., ϕ30 μm or less). The CBF flow indices ($v_1 \approx 7.8$k, $v_2 \approx 3.5$k) in blood vessels of different caliber can be readily differentiated by LSI despite its reliance on relative measurements. Unlike LSI, ODT enables quantitative 3D imaging of the local CBFv network at ~10 μm spatial resolution. Figure 7.4c shows a projected 3D ODT image of the vascular CBFv network ($2.5 \times 2.5 \times 2$ mm^3) whose *en face* FOV corresponds to the cortical area landmarked by the dashed rectangle in Figure 7.4a. Figure 7.4e reveals that the relative CBF indices change as a function of time in response to cocaine. The flow indices of the three flow traces (1, 2, and 3) measured by LSI (e.g., left axis) can be calibrated by DFR-OCT to convert to absolute flow rates (e.g., right axis) (Luo et al. 2008; Yuan et al. 2009), where the vertical lines at t_b and t_p represent the time points for concurrent LSI and ODT acquisitions at the baseline, for example, $t_b = -2.5$ min, as illustrated in Figure 7.4a and c, and at the elevated response to cocaine (phase 2), for example, $t_p \approx 9$ min, as illustrated in Figure 7.4b and d, respectively. Figure 7.4g summarized the mean CBF changes measured by LSI and DFR-OCT in different vascular compartments before (t_b) and after (t_p) a cocaine challenge from different animals ($n = 4$). The results indicated that the CBF rates by LSI increased 28% ± 7%, 26% ± 8%, and 32% ± 3% in arterial (A), venous (V), and tissue perfusion (T), respectively. The corresponding CBF increases (A, V) measured by DFR-OCT, that is, 22% ± 8% and 23% ± 7% correlated with the LSI measurements.

A comparison between the upper and lower panels indicated that both LSI and ODT imaging modalities were able to detect the CBFv increase in response to a cocaine challenge, as indicated by flow rates (i.e., in pseudo color) in both larger and smaller vessels. Specifically, the flow profiles for vessel 1 (~ϕ170 μm, vein) and vessel 2 (~ϕ40 μm, arteriole) before and after cocaine injection were plotted, indicating that the averaged flow rates increased from $v_1(t_b) \approx 3.1$ mm/s, $v_2(t_b) \approx 1.4$ mm/s to $v_1(t_p) \approx 4.1$ mm/s, $v_2(t_p) \approx 2.0$ mm/s. Despite the flow rate increase, no significant vasodilation (i.e., increase in vessel size) was observed in response to the cocaine challenge. The normalized CBF changes in Figure 7.5f demonstrated the high spatiotemporal resolution of LSI to distinguish the time-lapse CBF responses to cocaine (green curve) in the different sized vessels (e.g., curves 1, 2) and within the brain tissue (e.g., curve 3, perfusion flow of avascular tissue). The transient decrease in CBFv (phase 1, $t = 3.5 \pm 0.9$ min, $n = 4$) following cocaine administration was consistently observed in all experiments. This early dynamic change in CBFv (e.g., the "dipping" effect) did not appear in the vehicle animals challenged by saline, suggesting that it was a pharmacological effect from acute cocaine administration rather than measurement artifacts, for example, induced by

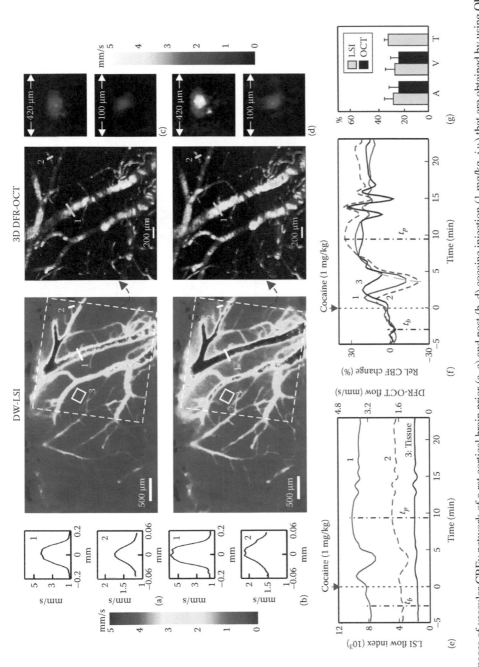

FIGURE 7.4 Images of vascular CBFv network of a rat cortical brain prior (a, c) and post (b, d) cocaine injection (1 mg/kg, *i.v.*) that are obtained by using OFI. Relative 1D and quantitative 2D flow profiles of vessel 1 (~φ170 μm, vein) and vessel 2 (~φ40 μm, arteriole) are shown in the left and right small panels. Flow depths of $z_{1,2}$ = 240, 330 μm are given by 3D OCT. The cocaine-evoked CBF changes from the selected ROIs of 1 (large), 2 (small), and 3 (tissue perfusion by irresolvable capillary flows, e.g., <φ30 μm) are shown in panels (e and f). t_b, t_p: baseline and elevated CBF periods. (g) Statistical analysis of cocaine-induced CBF increase (%) in arteriolar (A), venous (V), and perfusion (T) flows. (Adapted from Yuan, Z. et al., *Neuroimage*, 54(2), 1134, 2011. With permission.)

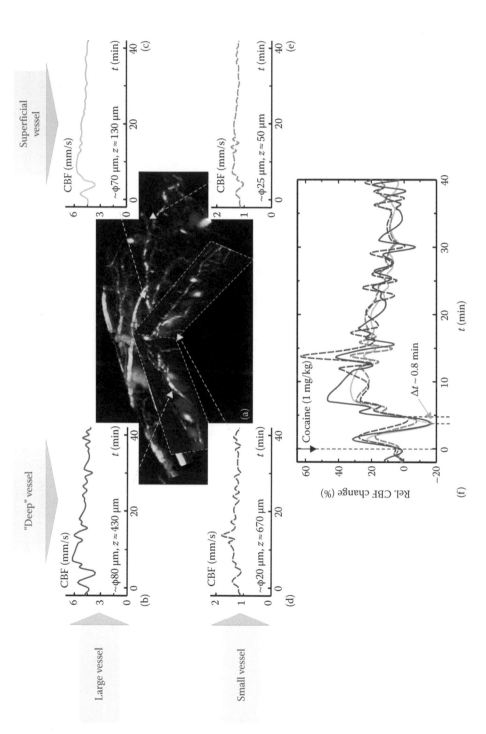

FIGURE 7.5 3D pie-cut ODT image of (a) local CBFv network in a rat cortical brain and its response to cocaine (1 mg/kg, *i.v.*) in four typical CBFs: (b) large deep (~φ80 μm, $z \approx 430$ μm, solid red curve), (c) large superficial (~φ70 μm, $z \approx 130$ μm, solid blue curve), (d) small deep (~φ20 μm, $z \approx 670$ μm, dashed pink curve), and (e) small superficial (~φ25 μm, $z \approx 50$ μm, dashed blue curve), (f) normalized time courses of cocaine-induced CBF changes in the four vessels. 3D image size: 2.5 × 2.5 × 2 mm³, spatial resolution: ~10 μm. (Adapted from Yuan, Z. et al., *Neuroimage*, 54(2), 1135, 2011. With permission.)

scattering decrease. Similar effects on CBV and BOLD were observed in previous MRI studies (Luo et al. 2003; Schwarz et al. 2003). In contrast to the "dip," which was not seen with saline administration, the immediate transient increase in CBF (peaked within 1 min) upon cocaine administration was also observed after saline administration, indicating that it was most likely caused by the bolus flush of solution.

OFI to Image Vascular Action: "Visualize" CBFv Network Response to Cocaine

OFI modality holds the unique capability to fill up the knowledge gap between conventional macroscopic and microscopic flow imaging techniques described earlier, because it combines LSI with ODT (Yuan et al. 2009) to provide *in vivo* quantitative 3D images of the cortical CBFv networks of brain at high spatiotemporal resolutions across a large FOV (Yuan et al. 2009). Figure 7.5a shows a 3D pie-cut image ($2.5 \times 2.5 \times 2$ mm^3) of the CBFv network of a rat cortical brain acquired by DFR-OCT (Yuan et al. 2009). Panels (b–e) show the time course of CBFv detected by DW-LSI and quantified by coregistering with DFR-OCT. DW-LSI provides more detailed transient features (Figure 7.5b through f) whereas DFR-OCT provides quantitative 3D images of the CBFv network (Figure 7.5a). Thus, combining these two imaging techniques allows us to acquire CBFv images quantitatively, at both high spatiotemporal resolutions (30 μm, 10 Hz) and across a large FOV (4×5 mm^2). 3D DFR-OCT detects the CBFv network at depths up to 0.8–1 mm depending on flow size, rate, and projection angle. In summary, Figure 7.5 demonstrates the unique merits of combining these two methods to quantify detailed vascular effects of cocaine on the 3D CBFv network (3D ODT), that is, both large/small and superficial/deep CBFs in real time (DW-LSI).

OFI to Image Vascular-Tissue "Interaction": Simultaneous Imaging of Cerebral Hemodynamic and Hemoglobin Oxygenation Changes Induced by Cocaine

OFI allows for simultaneous detection of the changes in CBFv (i.e., ΔCBFv), and oxygenated and deoxygenated hemoglobin concentrations (i.e., $\Delta[HbO_2]$, $\Delta[HbR]$). The changes in total hemoglobin ($\Delta[HbT]$), that is, the changes in ΔCBV, can be determined accordingly by $\Delta[HbT] = \Delta[HbO_2] + \Delta[HbR]$ (Luo et al. 2009). Figure 7.6 shows the time courses of ΔCBFv, $\Delta[HbO_2]$, $\Delta[HbR]$, and $\Delta[HbT]$ in a cortical brain (α-chloralose anesthesia). Figure 7.6a through c compares the temporal responses to a cocaine challenge (1 mg/kg, *i.v.*) at $t = 0$ s of an arteriolar flow, tissue perfusion (i.e., from microflows irresolvable by DW-LSI), and a venous flow, as indicated by three arrows in the lower middle panel. The time traces of ΔCBFv, $\Delta[HbO_2]$, $\Delta[HbR]$, and ΔCBV characterized by dotted, red, blue, and green curves show that after an initial dip (phase 1), CBFv, $[HbO_2]$, and CBV increased following cocaine injection, whereas $[HbR]$ decreased (phase 2). Here, t_b, t_p, and t_r represent three typical time points (i.e., baseline, ~20 min and ~40 min after cocaine challenge) at which the corresponding full-field CBFv images are shown in Figure 7.6d through f in which the quantitative flow rates in the segmented arteriolar, tissue perfusion, and venous flow compartments are presented by red, green, and blue colors, respectively.

OFI to Image Vascular-Tissue "Interaction": Separation of Cellular Effects from Vascular Effects of Cocaine

Figure 7.7 shows time-lapse simultaneous images of the CBF network (A_0–A_4) and the cortical brain $[Ca^{2+}]_i$ fluorescence (B_0–B_4) labeled with Rhod$_2$ in response to a cocaine challenge (1 mg/kg, *i.v.*) at $t = 0$ s in a drug-naïve rat (α-chloralose anesthesia). The detailed time course characteristics of ΔCBF and $\Delta[Ca^{2+}]_i$ are plotted in A5 and B5, respectively. The results indicate that cocaine induced a transient decrease (phase 1) in CBF within $t = 3.3 \pm 0.8$ min ($n = 4$) after cocaine injection, followed by an overshoot (phase 2) that reached the plateau around $t = 7.1 \pm 0.2$ min ($n = 4$).

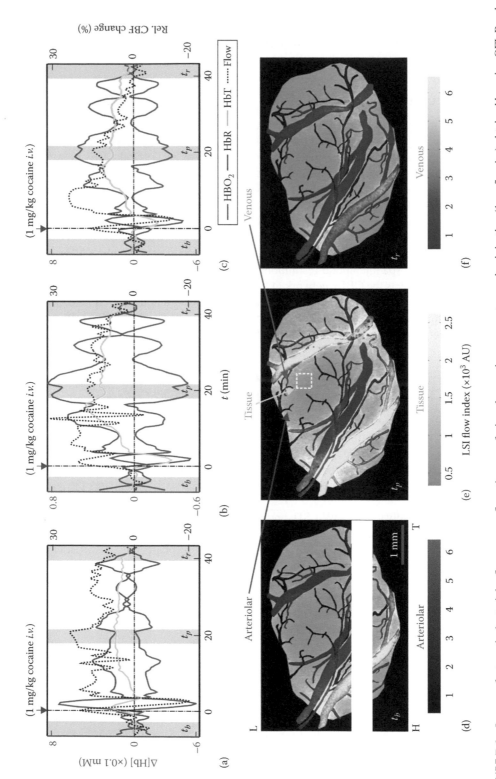

FIGURE 7.6 Images of rat cortical arteriolar flow, venous flow, tissue perfusion changes in response to cocaine injection (1 mg/kg, *i.v.*) obtained by using OFI. Panels of (a), (b), and (c) show the traces of the relative changes in CBF, [HbO₂], [HbR], and CBV (i.e., [HbT]) in selected ROIs marked by three arrows in (e). t_b, t_p, t_r; time points at baseline, 20 min and 40 min after cocaine challenge corresponding to the three bottom panels (d), (e), and (f), respectively. The rat was under α-chloralose anesthesia. H, head; T, tail; L, lateral. (Adapted from Yuan, Z. et al., *Neuroimage*, 54(2), 1136, 2011. With permission.)

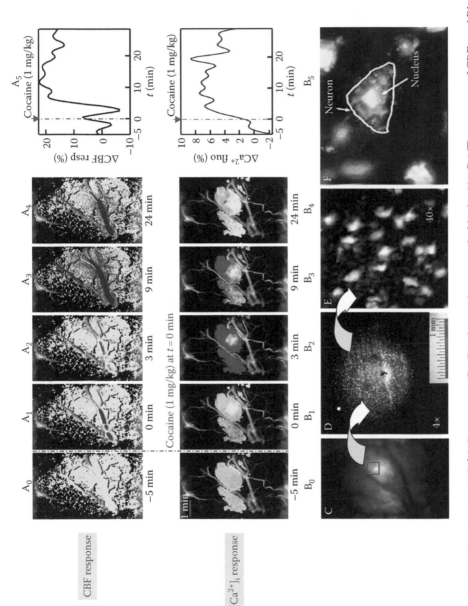

FIGURE 7.7 Cocaine-induced CBF (A_0–A_5) and $[Ca^{2+}]_i$ fluorescence (B_0–B_5) changes in a rat cortical brain. (A_5, B_5) Time courses of CBFv and Rhod$_2$-$[Ca^{2+}]_i$ changes (masked Ca^{2+} fluorescence clouds were overlapped on LSI images) in response to cocaine. (C) Raw baseline Ca^{2+} fluorescence image *in vivo* to indicate injection spot and Rhod$_2$ distribution. (D–F) *Ex vivo* cryosection fluorescence microscopic images (4×, 40×) of the brain specimen after *in vivo* imaging to indicate the distribution of Rhod$_2$ uptake of Ca^{2+} in the cortical brain. (F) An immunostained fluorescence microscopic image to indicate intracellular Rhod$_2$ localization in a neuronal cell *ex vivo*. (Adapted from Yuan, Z. et al., *Neuroimage*, 54(2), 1137, 2011. With permission.)

In contrast, $[Ca^{2+}]_i$ fluorescence increased immediately in response to the cocaine challenge (peaked at $t = 4.1 \pm 0.4$ min, $n = 4$) and persisted over 20 min, followed by a gradual decay. A comparison between panels A3 and B2 reveals a time lapse ($\Delta t = 2.9 \pm 0.5$ min, $n = 4$) between the peak cellular (e.g., $\Delta[Ca^{2+}]_i$) and phase 2 vascular (e.g., $\Delta CBFv$) responses to cocaine. Additionally, it is noteworthy that cocaine-induced blood absorption increase (e.g., $\Delta CBFv$ thus ΔCBV) might affect the measured $Rhod_2$-fluorescence emission. An empirical model was derived to correct the artifact (Yuan et al. 2011) and the result (solid curve) in Figure 7.7(B_5) indicated that although the time profile between the two curves did not differ significantly, especially in the early phase ($t < 4$ min, phase 1), the measured $\Delta[Ca^{2+}]_i$ curve (dashed curve) could be 25% underestimated in the later stage ($t > 4$ min, phase 2) due to increased CBFv and should thus be corrected. Figure 7.7(C) is the raw baseline fluorescence image to illustrate the loading spot and the $Rhod_2$ distribution on the cortical brain *in vivo* and Figure 7.7(D–F) are the corresponding cryosection fluorescence microscopic images where the bright spots in (D, E) indicate intracellular Rhod2 uptake of Ca^{2+} and the bright area in (F) shows their intracellular localizations.

IMAGING NEUROVASCULAR MORPHOLOGY AND CBFv FUNCTION: SIMULTANEOUS ANGIOGRAPHY AND CBF NETWORK IMAGING

Neuroimaging studies on the hemodynamic effects of cocaine are crucial to elucidating the mechanisms underlying its neurotoxicity including microcirculatory pathology (micro-hemorrhagic stroke) and hemodynamic dysfunction (microischemic stroke). The ultrahigh resolution and sensitivity and large FOV of ultrahigh-resolution OCA/ODT (i.e., µOCA/µODT) (Ren et al. 2012a,b) show its relevance for these studies. Figure 7.8 shows the results of mouse cortex before and after acute cocaine challenge (2.5 mg/kg, *i.v.*) and identifies the occurrence of what appears to be a cocaine-induced microischemia along with the CBF response patterns of the adjacent cerebrovascular networks. Specifically, as demonstrated in the cropped panels in Figure 7.8a and b, the CBFv in an arcade (~ϕ23 µm, pointed by blue arrows in panel µODT, Figure 7.8a) interconnecting two side branch arterioles shut down after cocaine administration (panel µODT, Figure 7.8b). Cocaine abolished the flow in this vessel, which appeared indiscernible by µODT, even though it was fully detectable

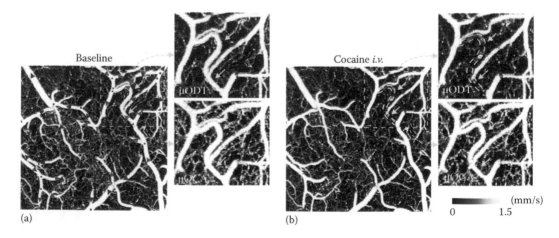

FIGURE 7.8 Cocaine-evoked microvascular disruptions on the mouse cortex: (a) baseline and (b) 30 min after cocaine injection (2.5 mg/kg, *i.v.*). The cropped images of µODT and µOCA: The quantitative CBF maps measured by µODT (presented by color images) and the cerebral angiography network measured by µOCA (presented by gray scale images) before (in a) and after (in b) cocaine injection, showing the representative CBF disruption in a ~ϕ23 µm arteriole (arcade) induced by cocaine. (Data from Ren, H. et al., *Nat. Mol. Psychiatry*, 17(10), 1022, 2012a.)

by μOCA (Ren et al. 2012a). This suggests that the cocaine-induced CBFv dysfunction probably reflected vasoconstriction of an isolated vessel rather than vessel rupture (e.g., hemorrhage that results in blurring of the local OCT imaging due to pronounced blood backscattering). Moreover, the fact that there was no CBFv drop in the surrounding microvascular networks suggests that the interruption of flow in this arcade was compensated by the microcirculatory networks, even when the shunt was long lasting (remained for >40 min). More importantly, this result suggests quantitative CBFv imaging (μODT) is more sensitive for detecting ischemic events than angiography (μOCA).

MICROISCHEMIA INDUCED BY COCAINE IN BRAIN

It is shown in Figure 7.9 that cocaine-induced progression of vasoconstrictions and local ischemia (mostly shunts of terminal vessels) after repeated acute cocaine injections (e.g., three repeated doses of 2.5 mg/kg/each, *i.v.*), although no vasculatural impairment such as vessel rupture was observed. The dashed green, blue, and yellow circles outline the deactivated branch vessels elicited by 1–3 cocaine doses, respectively (Figure 7.9a through g). A comparison between the panels (a and b) shows that terminal arterioles (~77%) were more vulnerable to ischemic shunts than terminal venules (~23%). Noticeable drops of active circulation in the immediate capillary circuits and the spreading of vasoconstrictive clouds (dashed dark circles) with repeated cocaine revealed that the microcirculation was bypassed (local cerebral microvascular network was unable to compensate). Such microischemic events were focal and probably undetectable by current imaging methods, including OCA (e.g., no obvious disruption was detected by μOCA in panel (b) even after the three repeated cocaine injections).

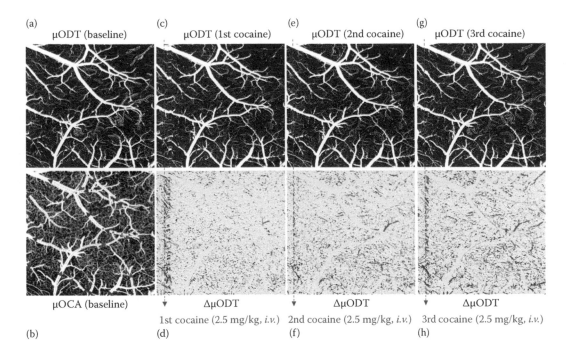

FIGURE 7.9 Cocaine-evoked microischemia in the mouse cortex. (a, b) μODT and μOCA at baseline; (c–h) μODT and ΔμODT after 1–3 cocaine doses (green/blue/yellow dashed circles: occluded vessels after 1–3 cocaine doses, dark circles: vasoconstrictive clouds). Image size: $2 \times 2 \times 1$ mm³. Red/blue arrows: flow directions of arteriolar/venial vessels; dots: vessel junctures. (Data from Ren, H. et al., *Nat. Mol. Psychiatry*, 17(10), 1022, 2012a.)

Note that the differences between the $\Delta\mu ODT$ responses of different vessels (d, f, h) are likely to reflect the heterogeneity in neurovascular responses to cocaine, for example, some areas in the bottom show increased (red) flow. This type of approach will enable to systematically evaluate the effects of acute and repeated cocaine administration on vascular architecture and CBF and to help understand the mechanisms underlying cocaine-induced ischemic and hemorrhagic strokes and provide with a tool to monitor potential therapeutic interventions.

CONCLUSIONS AND OUTLOOK

We present the multimodal optical spectroscopy (ODF) technique and imaging platform (OFI) to simultaneously detect the function of neurovascular networks, tissue oxygenation metabolism, and cellular activity of the brain *in vivo*. Additionally, the merits of the OFI technique include the following: (1) a larger FOV ($\sim 4 \times 5$ mm^2), (2) high temporal (~ 10 Hz) and spatial (~ 30 μm) resolutions, (3) *in vivo* quantitative 3D imaging of the CBF network, (4) vascular differentiation (arterial vs. venous, for $\phi > 30$ μm vessels), and (5) endogenous imaging of cerebral hemodynamic changes. Both ODF and OFI techniques (Du and Pan 2011) have been validated on anesthetized animal models, that is, to study the hemodynamic and cellular activity changes of cortex induced by ischemia or cocaine challenge. It has the potential to apply them for studying the brain functional changes in awake animals and to characterize the cellular and neurovascular changes from brain activity.

The main computational power in the brain resides in the cerebral cortex, the size and complexity of which serve to differentiate humans from other animals. As each behavior is complex and requires a number of processing stages, the workload of generating sensorimotor behavior is divided among many different areas; some handle vision, some handle touch, smell, others cognitive functions (e.g., working memory). Within each cortical area, function is parceled into fundamental compartments or modules. The organization and function of these modules remains an area of active research. Studies of these systems are prevalent throughout the functional imaging literature; however, optical imaging of the cortex is currently the only technique that can provide sufficient resolution to reveal the intricate nuances of their hemodynamic and neuronal responses.

As shown earlier, OFI enables imaging of the vascular CBF network at high spatial resolution (~ 30 μm) and over a large FOV (4×5 mm^2). By coregistering with DFR-OCT, quantitative 3D imaging of the CBF network is permitted over a cubic cortical volume of $2.5 \times 2 \times 2.5$ mm^3 at ~ 10 μm resolution. Such high spatial resolution and large FOV will bridge the gap between conventional mesoscopic (e.g., fMRI and PET) and microscopic (e.g., confocal and 2-photon) imaging modalities to study cerebral neurovascular coupling effects of brain function in behavioral animals, which requires both high resolution and large FOV.

Key technical issues for imaging studies in an awake animal concern the motion artifacts that can arise due to brain displacements during active behavior. Increasing image acquisition rate will be helpful to reduce misalignment or deformation of images induced by the motion. While the typical frame rates for *in vivo* TPLSM are 2–20 Hz, those of our current DW-LSI and 2D ODT are 10 and 47 fps and can be easily upgraded to 30 and 94 fps. Such high temporal resolution 2D and 3D CBFv imaging enables the capture of transient neurovascular events in awake animals, thus providing new insights to advance our understanding of the coupling between neuronal activity, metabolism, and hemodynamic changes and the perturbation by pharmacological agents and animal models of brain diseases. Additionally, it will likely allow us to monitor therapeutic approaches aimed at vascular as well as neuronal repair.

A remaining challenge in behavioral brain function studies is to understand the modular basis of cognition. For example, how do networks of modules within the brain or/and across cortical areas achieve different tasks such as vision or memory or emotion? Visualization of such modular activations during behavior is tantamount to "watching" the brain at work, and simultaneous multi-parameter measurement will separate vascular response, tissue metabolism, and cellular activity of functional brain, thus distinguishing behavioral differences and their effects on the brain. Advanced optical modalities like OFI that combine DW-LSI, 3D ODT, and $[Ca^{2+}]_i$ fluorescence enable concurrent imaging of the changes in hemodynamics (CBF and CBV), hemoglobin oxygenation (HbO_2), and $[Ca^{2+}]_i$ from cortical brain *in vivo*. The value of simultaneous imaging of neurovascular and cellular functions has been illustrated by our results with cocaine that clearly differentiate cocaine-induced cellular (neuronal) effects from neurovascular effects.

It should be noted that the cranial window was required to image the cortical brain using an optical imaging technique including ODF and OFI presented earlier. This is somewhat invasive as a tradeoff for high resolution and high sensitivity of imaging. Fortunately, many pioneering studies have demonstrated that optical imaging could be performed in the awake animals. For instance, Grinvald and colleagues (Grinvald et al. 1991) imaged ocular dominance maps in an awake, untrained monkey. Ocular dominance, orientation, and color maps were obtained in an awake, fixated monkey by Vnek et al. (1999). Optical imaging has also been applied to demonstrate the organization of the parietal and prefrontal cortex (Seidemann et al. 2002), and a spatial topographic map for working memory in monkeys performing a delay-match-to-sample task (Roe 2007). Recently, we have demonstrated capability of the cranial window implantation on cortex of rodents to image the neurovascular changes from the *in vivo* brain along with the chronic cocaine exposure (Park et al. 2015) and are validating a custom-developed rodent treadmill that allows high-resolution multimodality neuroimaging of an awake animal. This indicates that optical imaging will complement with other existing neuroimaging modalities such as PET and fMRI to provide new insights into brain physiology, function, and behavior.

SUMMARY

Deficits in prefrontal function play a crucial role in compulsive cocaine use, which is a hallmark of addiction. Dysfunction of the prefrontal cortex might result from the effects of cocaine on cells (e.g., neurons) as well as from disruption of cerebral blood vessels. However, the mechanisms underlying cocaine's neurotoxic effects are not fully understood, partially due to technical limitations of the current imaging techniques (e.g., PET and fMRI) to differentiate vascular from cellular effects at sufficiently high temporal and spatial resolutions. We have recently developed multimodal spectroscopic/fluorescence imaging platforms that can simultaneously characterize the changes in cerebrovascular hemodynamics, hemoglobin oxygenation, and intracellular calcium fluorescence for monitoring the effects of cocaine on the brain. In addition, our imaging platforms provide several uniquely important merits, including (1) a large FOV, (2) high spatiotemporal resolutions, (3) quantitative three dimensional (3D) imaging of CBF networks, (4) label-free imaging of hemodynamic changes, (5) separation of vascular compartments (e.g., arterial and venous vessels) and monitoring of cortical brain metabolic changes, and (6) discrimination of cellular (neuronal) responses from vascular changes. By further incorporating ultrahigh-resolution optical coherence Doppler tomography (μODT), we are able to provide 3D micro-angiography and quantitative imaging of the capillary CBF networks. These optical strategies have been used to investigate the effects of cocaine on brain physiology, to facilitate the studies of brain functional changes induced by addictive substance, and to provide new insights into neurobiological effects of the drug on the brain.

REFERENCES

Bandyopadhyay, R., Gittings, A.S., Suh, S.S., Dixon, P.K., and Durian, D.J. 2005. Speckle-visibility spectroscopy: A tool to study time-varying dynamics. *The Review of Scientific Instruments* 76:093110.

Bartzokis, G., Beckson, M., Lu, P.H., Edwards, N., Rapoport, R., Bridge, P., and Mints, J. 2004. Cortical gray matter volumes are associated with subjective responses to cocaine infusion. *American Journal on Addictions* 13(1):64–73.

Bolouri, M.R. and Small, G.A. 2004. Neuroimaging of hypoxia and cocaine-induced hippocampal stroke. *Journal of Neuroimaging* 14(3):290–291.

Buttner, A., Mall, G., Penning, R., Sache, H., and Weis, S. 2003. The neuropathology of cocaine abuse. *Legal Medicine (Tokyo)* 5(Suppl 1):S240–S242.

Dorrer, C., Belabas, N., Likforman, J.P., and Joffre, M. 2000. Spectral resolution and sampling issues in Fourier-transform spectral interferometry. *Journal of the Optical Society of America B: Optical Physics* 17:1795–1802.

Du, C., Koretsky, A.P., Izrailtyan I., and Benveniste, H. 2005. Simultaneous detection of blood volume, oxygenation, and intracellular calcium changes during cerebral ischemia and reperfusion in vivo using diffuse reflectance and fluorescence. *Journal of Cerebral Blood Flow and Metabolism* 25:1078–1092.

Du, C. and Pan, Y. 2011. Optical Imaging for Brain Functional Studies: Simultaneous imaging of vascular response, tissue metabolism and cellular activity of animal brain in vivo. *Review in Neuroscience: Behavioral Neuroimaging* 22(6):695–709.

Du, C., Yu, M., Volkow N.D., Koretsky, A.P., Fowler, J.S., and Benveniste, H. 2006. Cocaine increases the intracellular calcium concentration in brain independently of its cerebrovascular effects. *The Journal of Neuroscience* 26(45):11522–11531.

Du, C., Zhongchi, L., Yu, M., Benveniste, H., Tully, M., Pan, R., and Chance, B. 2009. Detection of Ca^{2+}-dependent neuronal activity simultaneously with dynamic changes in cerebral blood volume and tissue oxygenation from in vivo brain. *Journal of Innovative Optical Health Sciences* 2(2):1–12.

Dunn, A.K., Bolay, H., Moskowitz, M.A., and Boas, D.A. 2001. Dynamic imaging of cerebral blood flow using laser speckle. *Journal of Cerebral Blood Flow and Metabolism* 21(3):195–201.

Dunn, A.K., Devor, A., Bolay, H., Andermann, M.L., Moskowitz, M.A., Dale, A.M., and Boas, D.A. 2003. Simultaneous imaging of total cerebral hemoglobin concentration, oxygenation, and blood flow during functional activation. *Optics Letters* 28:28–30.

Frostig, R.D., Masino, S.A., Kwon, M.C., and Chen, C.H. 2005. Using light to probe the brain: Intrinsic signal optical imaging. *International Journal of Imaging System and Technology* 6:216–224.

Gollub, R.L. et al. 1998. Cocaine decreases cortical cerebral blood flow but does not obscure regional activation in functional magnetic resonance imaging in human subjects. *Journal of Cerebral Blood Flow & Metabolism* 18(7):724–734.

Grinvald, A., Frostig R.D., Siegel R.M., and Bartfeld, E. 1991. High-resolution optical imaging of functional brain architecture in the awake monkey. *Proceedings of the National Academy of Sciences* 88(24):11559–11563.

Helmchen, F. and Waters, J. 2002. Ca^{2+} imaging in the mammalian brain in vivo. *European Journal of Pharmacology* 447(2–3):119–129.

Hillman, E.M., Boas, D.A., Dale, A.M., and Dunn, A.K. 2004. Laminar optical tomography: Demonstration of millimeter-scale depth-resolved imaging in turbid media. *Optics Letters* 29(14):1650–1652.

Johnson, B.A., Devous, M.D., Ruiz, P., and Ait-Daoud, N. 2001. Treatment advances for cocaine-induced ischemic stroke: Focus on dihydropyridine-class calcium channel antagonists. *American Journal of Psychiatry* 158(8):1191–1198.

Kaye S. and Darke S. 2004. Non-fatal cocaine overdose among injecting and non-injecting cocaine users in Sydney. *Addiction* 99:1315–1322.

Lee, J.H., Telang, F.W., and Springer, C.S. 2003. Abnormal brain activation to visual stimulation in cocaine abusers. *Life Sciences* 73(15):1953–1961.

Leitgeb, R.A., Schmetterer, L., Drexler, W., Fercher, A.F., Zawadzki, R.J., and Bajraszewski, T. 2003. Real-time assessment of retinal blood flow with ultrafast acquisition by color Doppler Fourier domain optical coherence tomography. *Optics Express* 11:3116–3121.

London, E.D. et al. 1990. Cocaine-induced reduction of glucose utilization in human brain. A study using positron emission tomography and [fluorine 18]-fluorodeoxyglucose. *Archives of General Psychiatry* 47(6):567–574.

Luo, F., Wu, G.H., Li, Z., and Li, S.J. 2003. Characterization of effects of mean arterial blood pressure induced by cocaine and cocaine methiodide on BOLD signals in rat brain. *Magnetic Resonance in Medicine* 49:264–270.

Luo, Z., Wang, Z., Yuan, Z., Du, C., and Pan, Y. 2008. Optical coherence Doppler tomography quantifies laser speckle contrast imaging for blood flow imaging in the rat cerebral cortex. *Optics Letters* 33:1156–1158.

Luo, Z.C., Yuan, Z.J., Pan, Y.T., and Du, C. 2009. Simultaneous imaging of cortical hemodynamics and blood oxygenation change during cerebral ischemia using dual-wavelength laser speckle contrast imaging. *Optics Letters* 34(9):1480–1482.

Park, K., You, J., Du, C., and Pan, Y. 2015. Cranial window implantation on mouse cortex to study microvascular change induced by cocaine. *Quantitative Imaging in Medicine and Surgery* 5(1):97–107.

Pearlson, G.D., Jeffery, P.J., Harris, G.J., Ross, C.A., Fischman, M.W., and Camargo, E.E. 1993. Correlation of acute cocaine-induced changes in local cerebral blood flow with subjective effects. *The American Journal of Psychiatry* 150:495–497.

Ren, H., Du, C., and Pan, Y. 2012b. Cerebral blood flow imaged with ultrahigh-resolution optical coherence angiography and Doppler tomography. *Optics Letters* 37(8):1388–1390.

Ren, H., Du, C., Yuan, Z., Park, K., Volkow, N., and Pan, Y. 2012a. Cocaine induced cortical microischemia in the rodent brain: Clinical implications. *Nature: Molecular Psychiatry* 17(10):1017–1025.

Roe, A.W. 2007. Long-term optical imaging of intrinsic signals in anesthetized and awake monkeys. *Applied Optics* 46(10):1872–1880.

Schwarz, A.J., Reese, T., Gozzi, A., and Bifone, A. 2003. Functional MRI using intravascular contrast agents: Detrending of the relative cerebrovascular (rCBV) time course. *Magnetic Resonance Imaging* 21:1191–1200.

Seidemann, E., Arieli, A., Grinvald, A., and Slovin, H. 2002. Dynamics of depolarization and hyperpolarization in the frontal cortex and saccade goal. *Science* 295:862–865.

Spivey, W.H. and Euerle, B. 1990. Neurologic complications of cocaine abuse. *Annals of Emergency Medicine* 19:1422–1428.

Vnek, N., Ramsden, B.M., Hung, C.P., Goldman-Rakic, P.S., and Roe, A.W. 1999. Optical imaging of functional domains in the cortex of the awake and behaving monkey. *Proceedings of the National Academy of Sciences of the United States of America* 96(7):4057–4060.

Volkow, N.D., Mullani, N., Gould, K.L., Adler, S., and Krajewski, K. 1988. Cerebral blood flow in chronic cocaine users: A study with positron emission tomography. *The British Journal of Psychiatry* 152:641–664.

Wallace, E.A., Wisniewski, G., Zubal, G., vanDyck, C.H., Pfau, S.E., Smith, E.O., Rosen, M.I., Sullivan, M.C., Woods, S.W., and Kosten, T.R. 1996. Acute cocaine effects on absolute cerebral blood flow. *Psychopharmacology (Berlin)* 128:17–20.

Wang, R.K., Jacques, S.L., Ma, Z., Hurst, S., Hanson, S.R., and Gruber, A. 2007a. Three dimensional optical angiography. *Optics Express* 5:4083–4097.

Wang, Z.G., Lee, C.S.D., Waltzer, W., Liu, J., Xie, H., and Yuan, Z. 2007b. In vivo bladder imaging with MEMS-based endoscopic spectral domain optical coherence tomography. *Journal of Biomedical Optics* 12(3):034009.

Yuan, Z., Luo, Z., Volkow, N.D., Pan, Y., and Du, C. 2011. Imaging separation of neuronal from vascular effects of cocaine on rat cortical brain in vivo. *NeuroImage* 54(2):1130–1139.

Yuan, Z.J., Luo, Z.C., Ren, H.G., Du, C., and Pan, Y.T. 2009. A digital frequency ramping method for enhancing Doppler flow imaging in Fourier-domain optical coherence tomography. *Optics Express* 17(5):3951–3963.

Zhang, A., Cheng, T.P., Altura, B.T., and Altura, B.M. 1996. Acute cocaine results in rapid rises in intracellular free calcium concentration in canine cerebral vascular smooth muscle cells: Possible relation to etiology of stroke. *Neuroscience Letters* 215:57–59.

Zhang, H.F., Maslov, K., Sivaramakrishnan, M., Stoica, G., and Wang, L.H.V. 2007. Imaging of hemoglobin oxygen saturation variations in single vessels in vivo using photoacoustic microscopy. *Applied Physics Letters* 90:053901.

Section II

**Brain Imaging and Mapping
from Micro to Macroscales**

8 Three-Dimensional Optical Coherence Microscopy Angiography and Mapping of Angio-Architecture in the Central Nervous System

Conor Leahy and Vivek J. Srinivasan

CONTENTS

INTRODUCTION

Optical imaging techniques have become widespread in neuroscience and are now considered valuable tools in both neurovascular and cellular physiology research (Wilt et al., 2009). Macroscopic optical modalities include techniques such as optical intrinsic signal imaging (Grinvald et al., 1986), laser Doppler imaging (Dirnagl et al., 1989), laser speckle imaging (Dunn et al., 2001), diffuse optical imaging (Villringer and Chance, 1997), and laminar optical tomography (Hillman et al., 2004). These techniques typically achieve spatial resolutions on the order of hundreds of microns to millimeters. Two-photon microscopy is capable of resolution on the micron scale, though the imaging speed, penetration depth, and field of view are somewhat limited.

Optical coherence tomography (OCT) (Huang et al., 1991) has been a highly successful optical imaging modality in clinical applications due to the fact that it is label-free and enables high penetration depths, while still maintaining micron-scale resolution. Over the past two decades, OCT has increasingly been employed for the assessment of human tissues *in vivo*, and is now recognized as an important advancement in the diagnosis and monitoring of many pathologies. Optical coherence microscopy (OCM) is a term used to describe OCT approaches that prioritize high transverse spatial resolution, typically by employing higher numerical aperture (NA) focusing (Beaurepaire et al., 1998; Drexler and Fujimoto, 2008). Quantitative analysis based on 3-D imaging techniques such as OCM allows for the use of histological analysis tools, without the need for biopsy or the usage of contrast agents. More recently, OCT and OCM techniques have begun to expand beyond the clinic to the realm of basic research, including neuroscience.

A recent enabling advancement in OCT imaging has been the development of angiography methodologies to selectively image vasculature. Red blood cells (RBCs) within blood vessels are typically seen as scattering and dynamically changing in standard OCT intensity images. Based on this observation, angiography techniques have been developed as a means of visualizing vasculature *in vivo* (Makita et al., 2006; Wang et al., 2007; Mariampillai et al., 2008; Tao et al., 2008; Vakoc et al., 2009). By using intensity- or phase-based algorithms, OCT angiography techniques enhance the image contrast associated with moving RBCs in comparison to that of the surrounding static tissue.

This chapter describes approaches to combine OCM and angiography methods and apply them to the study of vascular architecture in the central nervous system, with particular regard to the brain cortical vasculature. Techniques for the analysis and categorization of vasculature are outlined, with emphasis on working with OCM data. Some discussion of the described methods and their applications is presented, along with suggestions for future research directions.

OPTICAL COHERENCE MICROSCOPY ANGIOGRAPHY

OCT typically employs a low-coherence light source and an interferometer to perform depth-resolved detection of backscattered light from a sample. OCT provides an improvement in imaging depth compared to confocal microscopy via the use of coherence gating, which significantly reduces unwanted scattered and out-of-focus light from within the sample (Drexler and Fujimoto, 2008). By comparison, confocal microscopy is generally limited to relatively superficial depths in highly scattering samples, due to the loss of image contrast associated with scattered and out-of-focus light. The axial resolution of an OCT system in air is determined by the coherence length of the source and can be written as

$$\Delta z = \frac{2 \ln 2}{\pi} \frac{\lambda_0^2}{\Delta \lambda} \tag{8.1}$$

where λ_0 and $\Delta \lambda$ are the center wavelength and spectral bandwidth of the low-coherence source, respectively. The axial resolution in a sample is obtained by dividing the right hand side of Equation 8.1 by the refractive index. As high axial resolution is achieved even with relatively weak focusing, cross-sectional OCT imaging with a large depth of field (DOF) is possible. OCM combines the coherence gating of OCT with tight focusing of confocal microscopy (Izatt and Choma, 2008). The increased NAs used in OCM allow for high transverse resolution (as well as high axial resolution), albeit with a reduced DOF.

OCT SIGNAL ACQUISITION

OCT is based on the principle of low-coherence interferometry, which measures the echo time delay and intensity of backscattered light from a sample by comparing it to a reference path. The goal is to reconstruct the sample reflectivity as a function of depth. Measurements are most commonly performed with a Michelson-type interferometer (Schmitt, 1999). Light from a broadband source is separated into two beams using a beamsplitter; one of the beams is incident onto the sample to be imaged, while the second beam traverses a reference path. The backscattered light from the sample undergoes interference with the reflected light from the reference arm, and is subsequently measured by a photodetector (Fujimoto et al., 2000).

OCT is implemented either in a time-domain or a Fourier-domain (frequency-domain) configuration (Tomlins and Wang, 2005). In time-domain OCT, depth-sectioning is provided by translating the reference arm over time. Fourier-domain OCT techniques eliminate the need for translation of the reference arm by measuring the interference spectrum. The delay and magnitude of the optical reflections from within the sample can be detected by Fourier analysis of the interference spectrum. Due to the simultaneous detection of many depths, Fourier-domain techniques possess an inherent sensitivity advantage over time-domain techniques (De Boer et al., 2003; Leitgeb et al., 2003). Fourier-domain approaches can be implemented with either spectral-domain OCT, which employs a spectrometer in the detection arm to capture spectral components of the interferometric signal simultaneously (Fercher et al., 1995; Haüsler and Lindner, 1998; Nassif et al., 2004), or swept-source OCT, where a tunable laser is swept across a broad wavelength range to capture spectral components of the interferometric signal sequentially (Chinn et al., 1997; Choma et al., 2003).

For spectrometer-based spectral-domain OCT systems, the measured OCT signal is the optical intensity as a function of wavelength, created by the superposition of the fields reflected back from the sample and reference arms. The reference spectrum intensity is typically subtracted digitally. In order to employ the discrete Fourier transform in reconstructing the axial scan as a function of depth, the spectrum should be evenly sampled in the conjugate variable, that is, wavenumber. Therefore, the spectrometer output must be transformed from wavelength to wavenumber (Wojtkowski et al., 2004). Fourier transformation is then used to obtain the complex OCT signal as a function of depth. A detailed description of the composition of the OCT signal is given by Izatt and Choma (2008).

ANGIOGRAM COMPUTATION AND BULK AXIAL MOTION CORRECTION

The aim of OCT angiography is to enhance the contrast of moving RBCs with respect to the surrounding tissue (Srinivasan et al., 2010a). The complex OCT signal, $S(t)$, can be considered as the superposition of a static component (S_0), a dynamic component (S_d), and an additive noise component (N) (Yousefi et al., 2011; Srinivasan et al., 2012b), that is,

$$S(t) = S_0 + S_d(t) + N(t) \tag{8.2}$$

At any given voxel, these three components are assumed to be independent, complex, random processes. The dynamic scattering component is related to RBC content; therefore, estimation of the strength of this component yields a representation of the angiogram. This can be computed by high-pass filtering of the complex OCT signal in time. If the filter kernel is chosen appropriately, the resulting angiogram will be mostly insensitive to RBC velocity changes (Srinivasan and Radhakrishnan, 2014). A protocol for rapid computation of OCT angiograms can be implemented by employing a scanning method that acquires two (or more) images at the same location but separated in time. The angiogram can then be obtained using a simple differencing of corresponding A-scans in the time-separated images (Radhakrishnan and Srinivasan, 2013).

Axial motion of the sample causes a phase shift of the complex OCT signal (White et al., 2003). Axial motion can be corrected by computing and correcting for this phase shift. The phase shift $\Delta\varphi_n$ for the nth A-line at time t, relative to time $t - T$, is given by

$$\Delta\varphi_n(t) = -arg\left\{\sum_z S_n(z,t) S_n^*(z,t-T)\right\} \tag{8.3}$$

where the superscript "*" denotes the complex conjugate, and the summation is taken over extravascular, static pixels. The corrected OCT signal for the nth A-line can then be obtained via

$$S'_n(z,t) = S_n(z,t)e^{i\Delta\varphi_n(t)}$$ (8.4)

A representation of the angiogram for a particular A-scan at time t can then be written as

$$A'_n(z,t) = \left| S'_n(z,t) - S_n(z,t-T) \right|$$ (8.5)

where T is the interframe time. By using the complex OCT signal, this method of angiogram computation incorporates both magnitude and phase information. If T is much larger than the decorrelation time of the signal, the resultant angiogram will be mostly insensitive to RBC velocity changes and will therefore relate mostly to RBC content.

Speckle Reduction

Like other coherent imaging modalities, OCT techniques are affected by speckle, that is, intensity variations due to interference produced by the coherent superposition of backscattered light waves from different points in a resolution element (Goodman, 1985). Speckle causes the intensity to fluctuate about its mean value, which can make the separation between structures with different scattering properties difficult to discern (Duncan et al., 2008). The subtraction in Equation 8.5 is coherent, and therefore the angiogram will be impacted by speckle. Image quality and segmentation performance can be improved by employing some form of speckle reduction, for example, through averaging speckle patterns. Averaging is most effective when the speckle patterns are uncorrelated; therefore, some form of diversity in the images is desirable, for example, spatial diversity, different wavelengths or polarizations, or temporal separation of dynamic images (Schmitt et al., 1999). For N averaged independent speckle patterns, the signal-to-noise ratio (SNR) is proportional to $N^{0.5}$. Where there is residual correlation between consecutive acquired images, the exponent of N takes a value of less than 0.5, and thus the benefits of averaging are reduced (Pircher et al., 2003).

Advantages of OCM Angiography

High-NA OCM imaging geometries are typically employed in order to achieve high transverse resolution. Here, we argue that an additional benefit of high-NA OCM angiography is the improved rejection of multiple-scattered light arising from vessels, yielding better localization of the vessel lumen. Thus, a narrow DOF can lead to improved data quality, which ultimately facilitates segmentation.

Confocal imaging systems have an inherent trade-off between transverse resolution and axial DOF. In most OCT and OCM implementations, light is detected through the same fiber used for illumination; hence the detection and illumination modes are identical in the focal plane. Assuming a Gaussian beam profile at the sample, the lateral resolution Δx of an OCT or OCM system, defined as the full-width at half-maximum (FWHM) of the lateral point-spread function, can be written as

$$\Delta x = \frac{\sqrt{2\ln 2}}{\pi} \frac{\lambda_0}{NA}$$ (8.6)

where *NA* is the numerical aperture. The corresponding DOF, Z, is given by

$$Z = 2\frac{\lambda_0 n}{\left(NA\right)^2} \tag{8.7}$$

where n is the sample refractive index. Using a low NA allows for a large DOF, while a higher NA leads to a reduced DOF. It also follows from Equations 8.6 and 8.7 that by selecting a higher NA to improve the lateral resolution, the achievable DOF is reduced (Tomlins and Wang, 2005). Figure 8.1a illustrates low-NA focusing for OCT imaging of an RBC. The trade-off between lateral resolution and DOF can be compared with Figure 8.1b, which illustrates higher NA focusing.

Here, we highlight that a lower DOF imaging geometry better rejects multiple-scattered photons (which have larger path delays and hence create "tail" artifacts in angiograms). Figure 8.2 illustrates the effects of NA on the localization of capillaries in OCT angiograms. The low-NA cross-sectional imaging of Figure 8.2a is degraded by light originating from multiple scattering events, as illustrated in Figure 8.1a (Leahy et al., 2016). Thus, capillaries are observed to have accompanying tails with significant axial extent, resulting mainly from RBC forward scattering followed or preceded by tissue backscattering. These tails may also be caused by changes in the optical path length, as high refractive index RBCs cross the beam path. Figure 8.2b shows the cross-sectional image (synthesized from dynamically focused datasets [Srinivasan et al., 2012a]) obtained with a higher NA imaging geometry. The increased rejection of unwanted multiple-scattered light in this case allows for improved axial localization of capillaries. Figure 8.2c illustrates mean normalized axial line profiles of signal amplitude from selected capillaries of similar size in the datasets represented in (a) and (b), showing improved localization of features in the high-NA case due to greater rejection of multiple-scattered light. The improved imaging of the vessel lumen and rejection of tail artifacts in OCM angiography facilitates the advanced image processing and quantitative analysis methods described in the forthcoming sections.

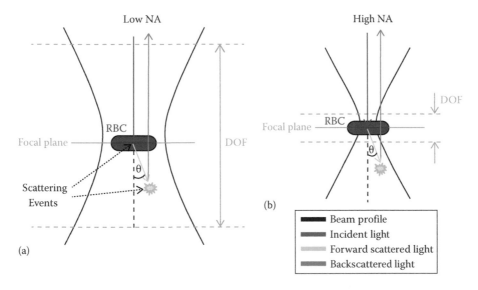

FIGURE 8.1 Low and high numerical aperture focusing on a red blood cell, showing the trade-off between transverse resolution, depth of field (DOF), and rejection of multiple-scattered light. (a) The low-NA geometry has an increased DOF, which tends to capture more multiple-scattered light from outside the focal plane. (b) The high-NA geometry provides finer transverse resolution and improved rejection of multiple-scattered light. (Data from Leahy, C. et al., *J. Biomed. Opt.*, 21(2), 020502, 2016.)

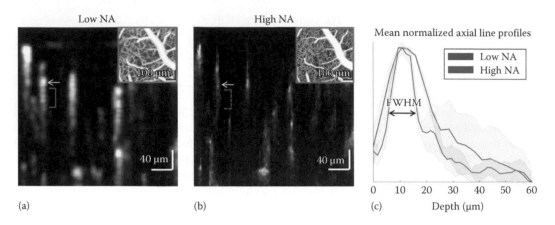

FIGURE 8.2 (a) Cross-sectional image obtained with low-NA imaging geometry. The yellow arrow shows a capillary lumen, with the approximate extent of the corresponding multiple-scattering tail marked (pink bracket). (b) Cross-sectional image obtained with high-NA imaging geometry and dynamic focusing. The yellow arrow shows the same capillary lumen, while the multiple-scattering tail (dotted pink bracket) is less prominent than in (a). Insets in (a–b) show *en face* maximum intensity projections. (c) Mean normalized axial line profiles of signal amplitude from selected capillaries of similar size, with standard deviations shown by the shaded regions. The estimated FWHM of the mean vessel axial line profile was 20.45 ± 4.64 μm using low NA and 10.65 ± 1.78 μm using high-NA and dynamic focusing. (Data from Leahy, C. et al., *J. Biomed. Opt.*, 21(2), 020502, 2016.)

IMAGE PROCESSING

A number of features of OCM angiography data present unique challenges for quantitative analysis. Figure 8.3 shows a 3-D representation of an OCM cortical angiogram volume. In general, blood vessels are diverse structures that exhibit a wide range of sizes, curvatures, and geometries. The network of blood vessels is also immersed in static scattering tissue, which may be imperfectly suppressed by the angiogram computation. Besides these physical confounds, images of vasculature are additionally corrupted by shot noise and speckle. For OCT/OCM images in highly scattering tissue, there is an inherent reduction in signal intensity with depth due to the attenuation of the sample medium. There may also be variations in signal strength across the imaged field that can be attributed to shadows caused by attenuation from large surface vessels. These confounds can be significantly detrimental to quantitative diagnostic tasks, and must be accounted for in any image processing or analysis subsequently performed on the acquired data.

This section describes image processing techniques applied to the segmentation and analysis of vascular networks obtained using OCM angiography. Figure 8.4 illustrates a data processing scheme for 3-D OCM data, starting from the complex signal and linking together the various processing steps. The methods described in this section are generally applicable to other imaging modalities; however, each application presents its own specific challenges, and thus the described procedures pertain particularly to OCM data.

IMAGE ENHANCEMENT

Image enhancement can increase the contrast of structures within the image in order to facilitate segmentation. Vessel enhancement is an important preprocessing step that can improve the representation of small or low-contrast vessels in the final segmentation. Enhancement techniques based on the computation of the Hessian matrix are particularly suited to angiography, as they provide intuitive and convenient information about the geometry of blood vessels from the image data (Yuan et al., 2011). In particular, the so-called Frangi filter (Frangi et al., 1998) is a

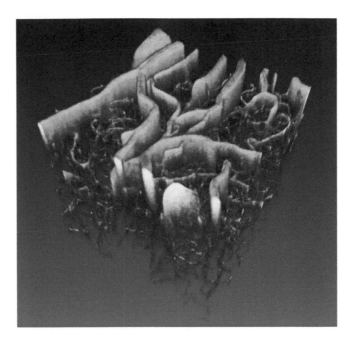

FIGURE 8.3 3-D volume rendering of a cortical OCM angiography dataset (acquired using high-NA geometry with dynamic focusing).

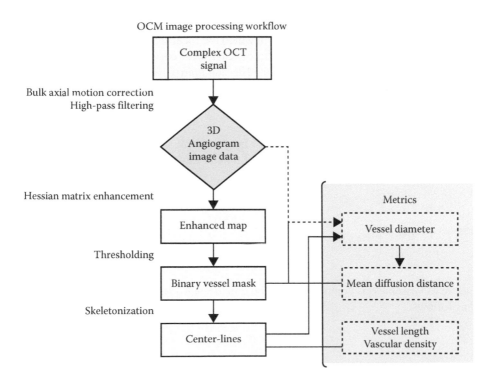

FIGURE 8.4 Flowchart depiction of steps involved in an OCM image processing scheme.

widely adopted multiscale vessel enhancement technique that utilizes a metric of tube-like structure or "vesselness," computed from the eigenvalues of the Hessian matrix.

The local structure of an image I at a particular location $\mathbf{x}$ and scale s can be described by its second-order Taylor series expansion:

$$I(\mathbf{x}+\delta\mathbf{x},s) \approx I(\mathbf{x},s) + \delta\mathbf{x}^T\nabla_s + \frac{1}{2}\delta\mathbf{x}^T H_s \delta\mathbf{x} \tag{8.8}$$

where $I(x, s)$ represents the image convolved in three dimensions with a Gaussian kernel of scale s, that is,

$$I(\mathbf{x},s) = I(\mathbf{x}) * G(\mathbf{x},s) \tag{8.9}$$

$$G(\mathbf{x},s) = \frac{1}{\left(\sqrt{2\pi s^2}\right)^3} e^{-\frac{\|\mathbf{x}\|^2}{2s^2}} \tag{8.10}$$

In Equation 8.8, ∇_s and H_s denote the gradient vector and Hessian matrix, respectively. The Hessian matrix describes the local curvature of the image structure and can be written as

$$H_s = \begin{bmatrix} I_{xx} & I_{xy} & I_{xz} \\ I_{yx} & I_{yy} & I_{yz} \\ I_{zx} & I_{zy} & I_{zz} \end{bmatrix} \tag{8.11}$$

where the subscripts denote derivatives of $I(x, s)$. Using the eigenvectors of the Hessian matrix, it is possible to determine the principal directions in which the local second-order structure of the image can be decomposed. This reveals the direction of least local curvature (i.e., the direction along a vessel). The eigenvalues and eigenvectors of the Hessian matrix can be found via

$$H_s \hat{u}_{s,k} = \lambda_{s,k} \hat{u}_{s,k} \tag{8.12}$$

where $\lambda_{s,k}$ is the eigenvalue with the kth smallest magnitude at scale s, corresponding to the eigenvector, $\hat{u}_{s,k}$. Letting $|\lambda_{s,1}| \leq |\lambda_{s,2}| \leq |\lambda_{s,3}|$, for an ideal tubular structure the following conditions hold:

$$|\lambda_{s,1}| \approx 0 \tag{8.13}$$

$$|\lambda_{s,1}| \ll |\lambda_{s,2}| \tag{8.14}$$

$$\lambda_{s,2} \approx \lambda_{s,3} \tag{8.15}$$

The measure of vesselness is computed using three metrics, $R_A(s)$, $R_B(s)$, and $R_C(s)$. The first two are defined as follows:

$$R_A(s) = \frac{|\lambda_{s,2}|}{|\lambda_{s,3}|} \tag{8.16}$$

$$R_B(s) = \frac{|\lambda_{s,1}|}{\sqrt{|\lambda_{s,2}\lambda_{s,3}|}} \tag{8.17}$$

The ratio $R_A(s)$ serves to distinguish between plate-like and line-like structures in the cross-sectional plane, while $R_B(s)$ accounts for the deviation from a blob-like structure. These two ratios are complemented by $R_C(s)$, which is computed using

$$R_C(s) = \sqrt{\lambda_{s,1}^2 + \lambda_{s,2}^2 + \lambda_{s,3}^2} \tag{8.18}$$

$R_C(s)$ is small in regions of the image where no structure is present and thus helps to distinguish image structure from background voxels. The vesselness function incorporates all three metrics and is defined as

$$v_0(s) = \begin{cases} 0 & \text{if } \lambda_{s,2} > 0 \text{ or } \lambda_{s,3} > 0 \\ \left(1 - e^{-\frac{R_A(s)^2}{2\alpha^2}}\right) e^{-\frac{R_B(s)^2}{2\beta^2}} \left(1 - e^{-\frac{R_C(s)^2}{2\theta^2}}\right) & \text{otherwise} \end{cases} \tag{8.19}$$

where α, β, and θ are parameters affecting the sensitivity of the vesselness function to each of the three metrics described here. The overall vesselness metric is computed via

$$V = \max_{s_{\min} \le s \le s_{\max}} v_0(s) \tag{8.20}$$

where $s_{\min}$ and $s_{\max}$ are the specified lower and upper bounds for the vessel scale.

The Hessian matrix-based approach is particularly advantageous in that handling of different vessel directions is implicit to the method. Thus, it is capable of performing enhancement without the requirement for multiple templates of vessels to account for orientation. However, Hessian-based approaches to vessel enhancement are subject to several drawbacks stemming from the fact that they are based upon local measures and can be degraded by local intensity inhomogeneities, such as those caused by noise. The approach also requires the *a priori* selection of a set of scales for evaluation. Finally, Hessian-based filters may be less effective than matched filters at distinguishing vessel edges (Wu et al., 2007).

THRESHOLDING

Thresholding of the image data is performed in order to obtain a binary mask of the vasculature (i.e., an equivalent-sized volume with all vessel voxels set to 1 and all other voxels set to 0). A typical thresholding scheme assigns a value of 0 to all voxels whose intensity falls below a certain prescribed limit, and a value of 1 to all values above this limit. The thresholded version $I_T(x, y, z)$ of a grayscale image $I(x, y, z)$, can be obtained via

$$I_T(x, y, z) = \begin{cases} 1 & \text{if } I(x, y, z) > T \\ 0 & \text{otherwise} \end{cases} \tag{8.21}$$

where T is some prescribed voxel intensity threshold. A drawback of the simple threshold described by Equation 8.21 is that it operates globally on the image. For angiography images, in addition to noise in the image, there may be remnants of static tissue that have not been fully suppressed. This confounds the process of segmenting the vasculature via a simple threshold operation.

Thresholding algorithms attempt to automatically determine an appropriate image threshold, while mitigating the possibility of incorrectly labeling image structure as part of the background, or vice versa. Otsu's method (Otsu, 1979) is one such approach that calculates an optimal threshold for a bimodal image by minimizing the within-class variance of both the structure and background voxels. In *adaptive thresholding*, the threshold is varied over the image, that is, $T \rightarrow T(x, y, z)$. At any given voxel $I(x, y, z)$, the applied threshold value $T(x, y, z)$ is derived from some local neighborhood of voxel intensities (Chow and Kaneko, 1972). The underlying assumption is that any nonuniformities in the image will be less dramatic on a local scale, and thus locally computed thresholds will be more effective for segmentation. A comprehensive description of the different categories of automatic thresholding methods is given by Sezgin and Sankur (2004).

MORPHOLOGICAL PROCESSING

Once the binary vessel mask has been obtained via thresholding, it is often useful to perform some morphological image processing as an additional noise removal step. Morphological processing involves manipulation of the binary pixels/voxels based on their relative positioning, probed by a predefined shape known as a structuring element. The simplest morphological operations are *erosion* and *dilation*. Erosion removes voxels at the boundaries of image objects, whereas dilation adds voxels to the boundaries. A comprehensive mathematical treatment of morphological operations is given by Serra and Vincent (1992). Two useful operations for removing morphological noise are *opening* and *closing*. These combine erosion and dilation to implement image processing operations, using a common structuring element. Morphological opening comprises an erosion followed by a dilation. It is useful for removing small objects from an image while preserving larger structures. Conversely, morphological closing is implemented as a dilation followed by an erosion, and is particularly useful for removing small "holes" in the segmented image structure that are not contiguous with the vessel map. Morphological processing can be particularly useful for OCT/OCM angiography images in mitigating the effects of noise or other artifacts.

SKELETONIZATION

The term "skeletonization" refers generally to the transformation of a binary image object into a thin subset of the original that maintains some topological equivalency. Skeletons provide a primitive representation of the object's original geometry, which can serve to emphasize particular features of interest, such as length and connectivity. This reduced representation of image objects can be very useful for subsequent image processing steps, such as feature recognition or the computation of shape-based metrics.

For blood vessel networks, skeletonization approaches typically endeavor to compute one-voxel thick center lines that are equidistant from the vessel boundaries. These center lines can be found either by a distance transformation (Tsai et al., 2009) or by employing a thinning algorithm that successively erodes the outer voxels of an image object while obeying certain topological and geometrical constraints (Lee et al., 1994). During the thinning process, voxels are systematically removed from the skeleton, until only a single voxel-thick structure remains. This thinning must be constrained in order to ensure that the center lines are equidistant from the original object boundaries. A MATLAB® implementation of 3-D medial axis thinning (Kerschnitzki et al., 2013) was used to obtain a 3-D skeleton. The desired outcome is a center-line mask that preserves the connectivity of the binary mask with minimal error. Additional processing and recentering steps can be applied to the center-line mask in order to improve its accuracy and remove any web-like loops or short free-ends that are manifested due to noise in the vicinity of the vessel boundaries (Tsai et al., 2009).

A significant challenge encountered when applying skeletonization methods to biomedical imaging data stems from the difficulty in obtaining a precise segmentation of the original data. Noise or other confounds that are manifested in an inaccurate binary vessel mask lead in turn to incorrect skeleton topology. Skeletonization errors typically take the form of small erroneous branches that appear due to inhomogeneities in the structure of the binary vessel mask and gaps in segmented branches (e.g., due to low image contrast in a portion of the vessel) that are manifested as

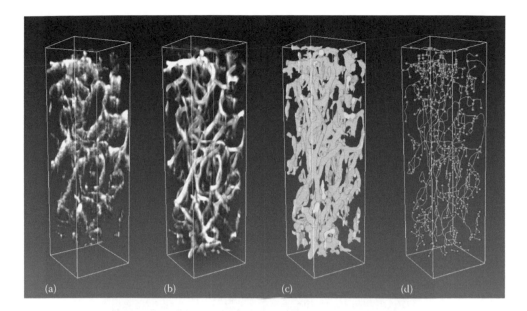

FIGURE 8.5 Overview of key steps involved in image processing of OCM cortical angiography data. (a) Columnar segment from a 3-D OCM cortical angiogram; (b) Enhanced image ("vesselness") obtained using Hessian matrix method; (c) Binary segmented vasculature map obtained via thresholding; (d) Center lines and branch points obtained using skeletonization.

unconnected end points in the skeleton. Many of the small erroneous branches can be removed by simply defining a well-considered minimum length threshold. Threshold relaxation (Kaufhold et al., 2008) and tensor voting (Risser et al., 2008) are two strategies that have been advanced for the correction of vessel-gap errors, though few results have been reported for 3-D vascular data (Kaufhold et al., 2012). Manually guided skeletonization approaches can assist the creation of skeletons with improved accuracy. For example, a graphical user interface can incorporate user inputs to identify erroneous skeleton topology (Leahy et al., 2016). For 3-D applications, however, localizing points in the skeletonized volume via user interaction with a 2-D projection on a computer screen is somewhat awkward. Various algorithms have been designed to work with input from a human user to provide improved segmentation outcomes, such as the use of shortest-path algorithms to trace vessel center-line paths based on the provision of branch end points (Abeysinghe and Ju, 2009; Lesage et al., 2009). Skeletonization of vasculature from image data can also be performed using commercially available software packages, such as *Amira* (FEITM Visualization Sciences Group).

Figure 8.5 illustrates an example of the core stages involved in obtaining a skeletonization of a cortical OCM angiography dataset: enhancement, thresholding, and skeletonization. Figure 8.6a shows an *en face* maximum intensity projection and a rendered columnar segment with the semi-transparent binary mask overlaid on the uncorrected skeleton. After skeleton correction and diameter estimation, a 3-D rendering of segmented cortical vasculature (acquired using high-NA imaging geometry and dynamic focusing) is shown in Figure 8.6b, representing both superficial (pial) vessels and the deeper capillary bed.

VECTORIZATION

A validated vascular skeleton can be converted to a vectorized graph, which can facilitate quantitative measures of the interconnectivity of the vessel network (Blinder et al., 2013). Vectorization entails explicitly labeling all branches and nodes (branch points) that constitute the corrected skeleton, as well as determining the interconnectivity between them. A vectorization can be obtained as

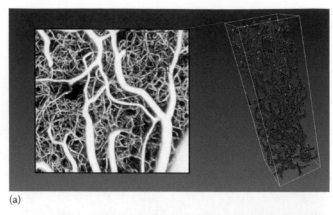

(a)

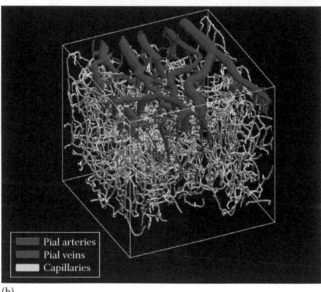

Pial arteries
Pial veins
Capillaries

(b)

FIGURE 8.6   (a, left) Maximum intensity projection of an enhanced OCM cortical vascular dataset; (a, right) 3-D representation of a binary segmented columnar segment from an OCM cortical vascular dataset (red), with overlay of center lines (black) and branch points (blue); (b) After skeleton correction and diameter estimation, 3-D representation of the segmented cortical vasculature showing pial vessels (arteries in red, veins in blue) and capillaries (gray). (Data from Leahy, C. et al., *J. Biomed. Opt.*, 21(2), 020502, 2016.)

follows. Firstly, the branch points on the skeleton are determined based on two criteria: a skeleton voxel is designated as a branch point if either (1) it has more than two neighboring skeleton voxels within its $3 \times 3 \times 3$ neighborhood or (2) if its deletion would result in the number of distinct objects in the local $3 \times 3 \times 3$ neighborhood increasing from one to three. Next, individual vessel branches can be delineated by deleting the branch points, thus breaking the skeleton into distinct branch components that can each be assigned an individual label. A sorted list of the voxels comprising each individual branch is then determined by first finding the branch end points (identified as singly connected voxels) and then successively locating the adjacent voxel in the local $3 \times 3 \times 3$ neighborhood until the opposite end is reached. The branch points are also labeled, independently from the branches. The graph connectivity information can then be found by searching the local neighborhood of each labeled branch point in turn and identifying the labels of branches that contact this neighborhood. In this manner, each labeled branch point can be associated with one or more labeled branches. Similarly, branches can then be associated with their respective connected nodes. The graph information can be stored in indexed data structures for analysis or used to generate a visual representation of interconnectivity.

VASCULAR METRICS

Blood vessels often comprise highly interconnected, irregular patterns. Forming useful descriptors of the vasculature can be aided by employing metrics of their properties. These metrics can be derived either directly from the raw image data or through the segmented or skeletonized data via statistical methods.

Vessel Length and Diameter

Several different techniques have been used to estimate vessel diameters from biomedical images (Ng et al., 2014). Often these methods operate based on a cross section defined perpendicular to the local longitudinal orientation of the vessel. The cross-sectional profile of a blood vessel may be fitted with various types of models, including Gaussian and difference of Gaussian (Lowell et al., 2004). The choice of model may be predicated upon the specific imaging modality; the point-spread function of the particular imaging protocol should be accounted for in the diameter calculation (Tsai et al., 2009).

The length of a particular segmented vessel can be estimated directly from its center lines. The distances separating the centers of adjacent voxels in a center line can be assigned a value based on the particular type of voxel geometry. Assuming isotropic voxels of unit dimension, it can be taken that face-to-face voxels have a separation of 1, edge-to-edge voxels have a separation of $\sqrt{2}$, and corner-to-corner voxels have a separation of $\sqrt{3}$. By assuming these separation values along a center-line segment, obtaining an estimate of the segment length is straightforward. The accuracy of computing the length of a center-line segment in this manner is limited by the digital sampling geometry. An alternative approach that may circumvent this problem is to fit the segments with 3-D parametric curves and evaluate their lengths numerically (Leahy et al., 2016).

Vascular Density and Fractional Vascular Volume

Two useful metrics for describing the density of vasculature as a function of depth within tissue are *fractional vascular volume* and *normalized vascular length*. The fractional vascular volume can be computed by determining the volume of all blood vessels within a region of tissue and dividing by the total volume. The density of vessels within the sample volume can also be described by the normalized vascular length, which is computed in units of inverse squared distance by dividing the total length of vessels by the total volume of the sample. Computing these metrics for a series of volume slabs at varying depths may be of particular interest in tissues with a layered structure, such as the cortex (Tsai et al., 2009).

Mean Oxygen Diffusion Distance

A useful parameter pertaining to the maximum metabolic rate of oxygen that a given network topology can support is the oxygen *diffusion distance*, which can be approximated by the distance from any tissue location to the nearest perfused blood vessel. This simplistic measure assumes that flow and oxygen distributions are homogeneous and that a given tissue location is predominantly supplied by the nearest vessel. However, it does provide a simple assay of the sparsity of oxygen-supplying capillaries (Srinivasan et al., 2015). By performing a Euclidean distance transform on the binary vessel mask in a particular region of capillaries, the mean diffusion distance can be estimated from the result by taking the mean of all nonzero distance values. The accuracy of this estimate will be degraded by any errors in the segmentation of the original binary vascular mask; alternatively, a reconstruction of the binary vessel mask based on a vectorized graph and estimated branch diameters may be used.

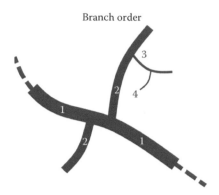

FIGURE 8.7 Diagram illustrating the notion of branch order: major supplying arteries and veins are assigned an order of 1, arterioles and venules thus have an order of 2, and subsequent branching vessels have an order of 3 or higher.

Branch Order

The vectorized vascular graph can be analyzed to describe the branching hierarchy of the blood vessel network using the notion of *branch order*, which relates to the sequence of bifurcations in the vessel network. Figure 8.7 illustrates vessel branch ordering. Assigning an order of 1 to major supplying and draining arteries and veins, the vectorized graph can be exploited to determine the relative branch order of all other branches in the network. Using this convention, arterioles and venules that branch from the major supplying vessels have an order of 2, and subsequent branching vessels are associated with orders of 3 and higher. If the major vessels are labeled as either arteries or veins, this labeling can be used in conjunction with the connectivity information provided by the graph in order to distinguish between the arterial and venous subnetworks. This information may be useful in revealing patterns of vascular interconnectivity between different layers and/or columns of cortical tissue.

CONCLUSION AND FUTURE PROSPECTS

OCT angiography has emerged as a technique specifically aimed at the visualization of vasculature *in vivo* (Makita et al., 2006; Wang et al., 2007; Srinivasan et al., 2012b). Here, OCM angiography is shown to enable high transverse resolution imaging of vasculature as well as mitigate multiple-scattering tails in images. Using these improved OCM angiography methods, a framework for segmentation, skeletonization, and vectorization of vascular networks in three dimensions is presented. The described image processing and vectorization procedures provide a means of gaining insight into the topology and interconnectivity of vasculature that would be difficult to achieve based only on qualitative observation. Further developments along these lines may yield fully automatic vascular graphing methods that can be applied across a range of tissue types.

A significant motivation for obtaining an accurate vascular graph is its potential for use with models of blood flow in vascular networks. By combining a vectorized map of blood vessels with experimentally quantified blood flow data (e.g., as computed using Doppler OCT methods [Srinivasan et al., 2010b]), it is possible to infer the flow in each segment down to the capillary level, by assuming that the vasculature behaves as a resistive network (Blinder et al., 2013). Such a framework allows for interesting hypothesis-based studies, for example, testing the capability of dilation of a single arteriole to produce localized changes in blood flow (Leahy et al., 2015). Future development of the techniques described earlier will significantly advance understanding of the relationship between flow and metabolism in the central nervous system and may constitute a baseline for characterizing changes in vascular diseases.

ACKNOWLEDGMENTS

The authors wish to acknowledge members of the Neurophotonics Laboratory at UC Davis for their contributions and useful scientific discussions. This work received support from the National Institutes of Health (R00-NS067050, R01-NS094681), the American Heart Association (11IRG5440002), and the Glaucoma Research Foundation Catalyst for a Cure 2.

REFERENCES

Abeysinghe, S. S. and Ju, T. (2009). Interactive skeletonization of intensity volumes. *Vis. Comput.*, 25(5–7):627–635.

Beaurepaire, E., Boccara, A. C., Lebec, M., Blanchot, L., and Saint-Jalmes, H. (1998). Full-field optical coherence microscopy. *Opt. Lett.*, 23(4):244–246.

Blinder, P., Tsai, P. S., Kaufhold, J. P., Knutsen, P. M., Suhl, H., and Kleinfeld, D. (2013). The cortical angiome: An interconnected vascular network with noncolumnar patterns of blood flow. *Nat. Neurosci.*, 16(7):889–897.

Chinn, S., Swanson, E., and Fujimoto, J. (1997). Optical coherence tomography using a frequency-tunable optical source. *Opt. Lett.*, 22(5):340–342.

Choma, M., Sarunic, M., Yang, C., and Izatt, J. (2003). Sensitivity advantage of swept source and Fourier domain optical coherence tomography. *Opt. Express*, 11(18):2183–2189.

Chow, C. and Kaneko, T. (1972). Automatic boundary detection of the left ventricle from cineangiograms. *Comput. Biomed. Res.*, 5(4):388–410.

De Boer, J. F., Cense, B., Park, B. H., Pierce, M. C., Tearney, G. J., and Bouma, B. E. (2003). Improved signal-to-noise ratio in spectral-domain compared with time-domain optical coherence tomography. *Opt. Lett.*, 28(21):2067–2069.

Dirnagl, U., Kaplan, B., Jacewicz, M., and Pulsinelli, W. (1989). Continuous measurement of cerebral cortical blood flow by laser-Doppler flowmetry in a rat stroke model. *J. Cereb. Blood Flow Metab.*, 9(5):589–596.

Drexler, W. and Fujimoto, J. (2008). *Optical Coherence Tomography: Technology and Applications*. Springer, Berlin, Germany.

Duncan, D. D., Kirkpatrick, S. J., and Wang, R. K. (2008). Statistics of local speckle contrast. *J. Opt. Soc. Am. A*, 25(1):9–15.

Dunn, A. K., Bolay, H., Moskowitz, M. A., and Boas, D. A. (2001). Dynamic imaging of cerebral blood flow using laser speckle. *J. Cereb. Blood Flow Metab.*, 21(3):195–201.

Fercher, A. F., Hitzenberger, C. K., Kamp, G., and El-Zaiat, S. Y. (1995). Measurement of intraocular distances by backscattering spectral interferometry. *Opt. Commun.*, 117(1):43–48.

Frangi, A. F., Niessen, W. J., Vincken, K. L., and Viergever, M. A. (1998). Multiscale vessel enhancement filtering. In C. Taylor and A. C. F. Colchester, eds., *Medical Image Computing and Computer-Assisted Intervention—MICCAI 98*, pp. 130–137. Springer, Berlin, Germany.

Fujimoto, J. G., Pitris, C., Boppart, S. A., and Brezinski, M. E. (2000). Optical coherence tomography: An emerging technology for biomedical imaging and optical biopsy. *Neoplasia*, 2(1):9–25.

Goodman, J. (1985). *Statistical Optics*. Wiley-Interscience, New York.

Grinvald, A., Lieke, E., Frostig, R. D., Gilber, C. D., and Wiesel, T. N. (1986). Functional architecture of cortex revealed by optical imaging of intrinsic signals. *Nature*, 324:361–364.

Haüsler, G. and Lindner, M. W. (1998). "Coherence radar" and "spectral radar"—New tools for dermatological diagnosis. *J. Biomed. Opt.*, 3(1):21–31.

Hillman, E. M., Boas, D. A., Dale, A. M., and Dunn, A. K. (2004). Laminar optical tomography: Demonstration of millimeter-scale depth-resolved imaging in turbid media. *Opt. Lett.*, 29(14):1650–1652.

Huang, D., Swanson, E. A., Lin, C. P. et al. (1991). Optical coherence tomography. *Science*, 254(5035):1178–1181.

Izatt, J. A. and Choma, A. (2008). Theory of optical coherence tomography. In W. Drexler and J. G. Fujimoto, eds., *Optical Coherence Tomography*, pp. 47–72. Springer, Berlin, Germany.

Kaufhold, J., Tsai, P., Blinder, P., and Kleinfeld, D. (2008). Threshold relaxation is an effective means to connect gaps in 3D images of complex microvascular networks. In *Third Workshop on Microscopic Image Analysis with Applications in Biology*, New York.

Kaufhold, J. P., Tsai, P. S., Blinder, P., and Kleinfeld, D. (2012). Vectorization of optically sectioned brain microvasculature: Learning aids completion of vascular graphs by connecting gaps and deleting open-ended segments. *Med. Image Anal.*, 16(6):1241–1258.

Kerschnitzki, M., Kollmannsberger, P., Burghammer, M. et al. (2013). Architecture of the osteocyte network correlates with bone material quality. *J. Bone Miner. Res.*, 28(8):1837–1845.

Leahy, C., Radhakrishnan, H., Bernucci, M., and Srinivasan, V. J. (2016). Imaging and graphing of cortical vasculature using dynamically focused optical coherence microscopy angiography. *J. Biomed. Opt.*, 21(2):020502.

Leahy, C., Radhakrishnan, H., Weiner, G., Goldberg, J. L., and Srinivasan, V. J. (2015). Mapping the 3D connectivity of the rat inner retinal vascular network using OCT angiography. *Invest. Ophthalmol. Vis. Sci.*, 56(10), 5785–5793.

Lee, T. C., Kashvap, R. L., and Chu, C. N. (1994). Building skeleton models via 3-D medial surface/axis thinning algorithms. *Graph. Model. Im. Process.*, 56(6):462–478.

Leitgeb, R., Hitzenberger, C., and Fercher, A. (2003). Performance of Fourier domain vs. time domain optical coherence tomography. *Opt. Express*, 11(8):889–894.

Lesage, D., Angelini, E. D., Blocj, I., and Funka-Lea, G. (2009). A review of 3D vessel lumen segmentation techniques: Models, features and extraction schemes. *Med. Image Anal.*, 13(6):819–845.

Lowell, J., Hunter, A., Steel, D., Basu, A., Ryder, R., and Kennedy, R. L. (2004). Measurement of retinal vessel widths from fundus images based on 2-D modeling. *IEEE Trans. Med. Imaging*, 23(10):1196–1204.

Makita, S., Hong, Y., Yamanari, M., Yatagai, T., and Yasuno, Y. (2006). Optical coherence angiography. *Opt. Express*, 14(17):7821–7840.

Mariampillai, A., Standish, B. A., Moriyama, E. H. et al. (2008). Speckle variance detection of microvasculature using swept-source optical coherence tomography. *Opt. Lett.*, 33(13):1530–1532.

Nassif, N., Cense, B., Park, B., Pierce, M., Yun, S., Bouma, B., Tearney, G., Chen, T., and de Boer, J. (2004). In vivo high-resolution video-rate spectral-domain optical coherence tomography of the human retina and optic nerve. *Opt. Express*, 12(3):367–376.

Ng, E., Acharya, U., Suri, J., and Campilho, A., eds. (2014). *Image Analysis and Modeling in Ophthalmology*. CRC Press, Boca Raton, FL.

Otsu, N. (1979). A threshold selection method from gray-level histograms. *IEEE Trans. Syst. Man Cybern.*, 9:62–66.

Pircher, M., Gotzinger, E., Leitgeib, R., Fercher, A. F., and Hitzenberger, C. K. (2003). Speckle reduction in optical coherence tomography by frequency compounding. *J. Biomed. Opt.*, 8(3):565–569.

Radhakrishnan, H. and Srinivasan, V. J. (2013). Compartment-resolved imaging of cortical functional hyperemia with OCT angiography. *Biomed. Opt. Express*, 4(8):1255–1268.

Risser, L., Plouraboué, F., and Descombes, X. (2008). Gap filling of 3-D microvascular networks by tensor voting. *IEEE Trans. Med. Imaging*, 27(5):674–687.

Schmitt, J. M. (1999). Optical coherence tomography (OCT): A review. *IEEE J. Sel. Top. Quant. Electron.*, 5(4):1205–1215.

Schmitt, J. M., Xiang, S. H., and Yung, K. M. (1999). Speckle in optical coherence tomography. *J. Biomed. Opt.*, 4(1):95–105.

Serra, J. and Vincent, L. (1992). An overview of morphological filtering. *Circ. Syst. Signal Process.*, 11(1):47–108.

Sezgin, M. and Sankur, B. (2004). Survey over image thresholding techniques and quantitative performance evaluation. *J. Electron. Imaging*, 13(1):146–168.

Srinivasan, V. J., Jiang, J. Y., Yaseen, M. A. et al. (2010a). Rapid volumetric angiography of cortical microvasculature with optical coherence tomography. *Opt. Lett.*, 35(1):43–45.

Srinivasan, V. J. and Radhakrishnan, H. (2014). Optical coherence tomography angiography reveals laminar microvascular hemodynamics in the rat somatosensory cortex during activation. *NeuroImage*, 102:393–406.

Srinivasan, V. J., Radhakrishnan, H., Jiang, J. Y., Barry, S., and Cable, A. (2012a). Optical coherence microscopy for deep tissue imaging of the cerebral cortex with intrinsic contrast. *Opt. Express*, 20(3):2220–2239.

Srinivasan, V. J., Radhakrishnan, H., Lo, E. H. et al. (2012b). OCT methods for capillary velocimetry. *Biomed. Opt. Express*, 3(3):612–629.

Srinivasan, V. J., Sakadzic, S., Gorczynska, I. et al. (2010b). Quantitative cerebral blood flow with optical coherence tomography. *Opt. Express*, 18(3):2477–2494.

Srinivasan, V. J., Yu, E., Radhakrishnan, H. et al. (2015). Micro-heterogeneity of flow in a mouse model of chronic cerebral hypoperfusion revealed by longitudinal Doppler optical coherence tomography and angiography. *J. Cereb. Blood Flow Metab.*, 35(10):1552–1560.

Tao, Y. K., Davis, A. M., and Izatt, J. A. (2008). Single-pass volumetric bidirectional blood flow imaging spectral domain optical coherence tomography using a modified Hilbert transform. *Opt. Express*, 16(16):12350–12361.

Tomlins, P. H. and Wang, R. (2005). Theory, developments and applications of optical coherence tomography. *J. Phys. D: Appl. Phys.*, 38(15):2519.

Tsai, P. S., Kaufhold, J. P., Blinder, P. et al. (2009). Correlations of neuronal and microvascular densities in murine cortex revealed by direct counting and colocalization of nuclei and vessels. *J. Neurosci.*, 29(46):14553–14570.

Vakoc, B. J., Lanning, R. M., Tyrrell, J. A. et al. (2009). Three-dimensional microscopy of the tumor microenvironment in vivo using optical frequency domain imaging. *Nat. Med.*, 15(10):1219–1223.

Villringer, A. and Chance, B. (1997). Non-invasive optical spectroscopy and imaging of human brain function. *Trends Neurosci.*, 20(10):435–442.

Wang, R. K., Jacques, S. L., Ma, Z., Hurst, S., Hanson, S. R., and Gruber, A. (2007). Three-dimensional optical angiography. *Opt. Express*, 15(7):4083–4097.

White, B., Pierce, M., Nassif, N. et al. (2003). In vivo dynamic human retinal blood flow imaging using ultra-high-speed spectral domain optical coherence tomography. *Opt. Express*, 11(25):3490–3497.

Wilt, B. A., Burns, L. D., Ho, E. T. W., Ghosh, K. K., Mukamel, E. A., and Schnitzer, M. J. (2009). Advances in light microscopy for neuroscience. *Annu. Rev. Neurosci.*, 32:435.

Wojtkowski, M., Srinivasan, V., Ko, T., Fujimoto, J., Kowalczyk, A., and Duker, J. (2004). Ultrahigh-resolution, high-speed, Fourier domain optical coherence tomography and methods for dispersion compensation. *Opt. Express*, 12(11):2404–2422.

Wu, C.-H., Agam, G., and Stanchev, P. (2007). A general framework for vessel segmentation in retinal images. In *International Symposium on Computational Intelligence in Robotics and Automation, 2007 (CIRA 2007)*, Jacksonville, FL. IEEE, Piscataway, NJ, pp. 37–42.

Yousefi, S., Zhi, Z., and Wang, R. K. (2011). Eigendecomposition-based clutter filtering technique for optical microangiography. *IEEE Trans. Biomed. Eng.*, 58(8):2316–2323.

Yuan, Y., Luo, Y., and Chung, A. C. S. (2011). VE-LLI-VO: Vessel enhancement using local line integrals and variational optimization. *IEEE Trans. Image Process.*, 20(7):1912–1924.

9 Doppler Optical Coherence Tomography and Its Application in Measurement of Cerebral Blood Flow

Zhongping Chen and Jiang Zhu

CONTENTS

INTRODUCTION

Optical coherence tomography (OCT) is a noninvasive medical imaging technique with micrometer resolution, millimeter penetration, and 3D information (Fujimoto et al. 1995; Fujimoto 2003; Huang et al. 1991; Low et al. 2006; Vakoc et al. 2012). It uses broadband light sources with short coherence length to perform cross-sectional imaging within the optical scattering media. Combining the Doppler principle with OCT, Doppler OCT uses the intrinsic optical signals backscattered by the moving blood particles inside blood vessels as contrast (Chen et al. 1997a; Chen et al. 1997b; Ding et al. 2002; Izatt et al. 1997; Liu and Chen 2013a; Rollins et al. 2002; Zhao et al. 2000a).

Doppler OCT is used in the monitoring of ocular diseases (Choi et al. 2012; Huang et al. 2015; Leitgeb et al. 2014; Sehi et al. 2014), determination of skin therapy (Liu et al. 2012a, 2013b; Zhao et al. 2000b), and cardiac investigation (Li et al. 2011; Peterson et al. 2012; Yazdanfar et al. 1997). Recently it has become a powerful tool for the study of cerebrovascular physiology, because it not only can provide 3D structural angiography but also measure cerebral blood flow in the cortical microvasculature (Chen and Zhang 2015; Cho et al. 2016; Lee and Boas 2015; Liu and Chen 2013b).

In this chapter, we'll introduce Doppler-based OCT methods, including phase-resolved color Doppler (PRCD), phase-resolved Doppler variance (PRDV), and intensity-based Doppler variance (IBDV). The applications, improvement, and multimodality imaging of Doppler OCT in the investigation of cerebral blood flow will also be reviewed.

METHODS OF DOPPLER OCT

Since the first 2D *in vivo* Doppler OCT imaging was reported in 1997 (Chen et al. 1997b), Doppler OCT has developed rapidly. It has become a significant method for the mapping and measurement of cerebral blood flow. Compared with other microvascular imaging techniques, Doppler OCT benefits from the Doppler principle and the OCT technique and, thus, provides simultaneous information regarding both *in vivo* blood flow dynamics and tissue structures in the biological system (Chen and Zhang 2015).

A general structure of an OCT system based on a Michelson interferometer is shown in Figure 9.1, including a low coherence broadband light source, a reference light path, a sample light path, a coupler, and a photo detector. The coupler splits the light beam from the light source into a sample arm and a reference arm. Light reflected from a reference mirror interferes in the coupler with the light scattered back from the sample. The interference light is then detected by a photo detector. For the interference of the light from the reference path and the sample path, the optical path difference between two light paths is matched within the coherence length of the light source.

A common Fourier domain OCT system can be classified into swept-source OCT and spectral-domain OCT according to the types of the light source. For swept-source OCT, swept-source generates a spectrum in single successive frequency steps, and a single element photovoltaic detector monitors the interference light at different wavelengths. In comparison, for spectral-domain OCT, a superluminescent diode simultaneously generates the broadband spectrum, and the interference light at different wavelengths is extracted via a dispersive unit and is captured by a line-array charge coupled device (CCD) simultaneously.

Intensities of interference light at different wavelengths are processed by a Fourier transformation. The acquired fringes and transformed signals are shown in Figure 9.2. F_j is the complex sequence at the jth A-line. Generally, the coherence length of a light source determines the axial resolution of an OCT system, while the beam diameter on the sample and the numerical aperture of the objective determine the lateral resolution. With the complex sequence F_j, the measurement of the flow velocity can be made based on the Doppler principle. Here, we introduce the phase-resolved Doppler OCT measurement and the Doppler variance measurement.

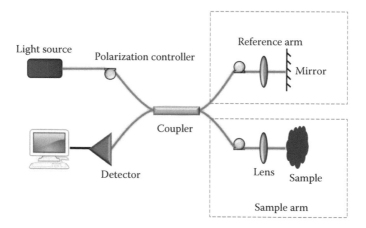

FIGURE 9.1   General structure of OCT. This OCT system includes a low coherence broadband light source, a reference light path, a sample light path, a coupler, and a photo detector.

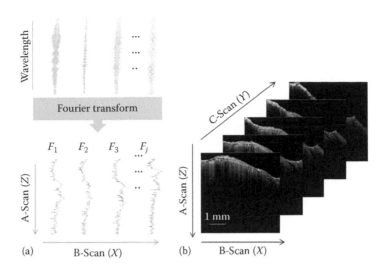

FIGURE 9.2 OCT data capture. (a) The acquired fringes and transformed signals from an OCT system. F_j is the complex sequence at the jth A-line. (b) OCT A-scan, B-scan, and C-scan.

PHASE-RESOLVED COLOR DOPPLER MEASUREMENT

Based on the Doppler principle, the frequency shift f of the scattered light from a moving particle is determined by the following equation:

$$f = \frac{1}{2\pi} \cdot \left(\overline{k_s} - \overline{k_i} \right) \cdot \vec{V} \tag{9.1}$$

where
  $\vec{V}$ is the velocity vector of the moving particle
  $\overline{k_i}$ and $\overline{k_s}$ are wave vectors of the incident light and scattered light, respectively

The simplified relationship between the Doppler shift and the flow velocity is described by the following equation:

$$f = \frac{2n \cdot V \cdot \cos(\theta)}{\lambda_0} \tag{9.2}$$

where
  λ_0 is the vacuum center wavelength of the incident light
  n is the refractive index of highly scattering media
  θ is the Doppler angle between the beam vector and the flow vector

In Equation 9.2, $V \cdot \cos(\theta)$ represents the flow velocity along the beam vector. When the Doppler angle θ is known in advance, the flow velocity can be determined according to the measurement of the Doppler frequency shift. The sign of the flow vector depends on the frequency shift f and the Doppler angle θ, and the value of the flow vector is directly proportional to the flow velocity along the beam direction.

In an OCT system, the frequency shift f can be measured with the phase change $\Delta\phi$ and the sampling time interval ΔT between adjacent A-lines, which is described by the following equation:

$$f = \frac{\Delta\phi}{2\pi \cdot \Delta T} \tag{9.3}$$

where $\Delta\phi/\Delta T$ is the rate of the phase change between adjacent A-lines.

Therefore, the flow velocity can be determined by the rate of phase change using the following equation by combining Equation 9.2 with Equation 9.3:

$$V \cdot \cos(\theta) = \frac{\lambda_0}{4\pi \cdot n} \cdot \frac{\Delta\phi}{\Delta T} \tag{9.4}$$

For the determination of the phase change $\Delta\phi$ between adjacent A-lines, the phase value $\phi_{j,i}$ at the jth A-line and the depth of the ith pixel is calculated by

$$\phi_{j,i} = \tan^{-1}\left[\frac{\mathrm{Im}\left(F_{j,i}\right)}{\mathrm{Re}\left(F_{j,i}\right)}\right] \tag{9.5}$$

where

$F_{j,i}$ is the complex signal at the jth A-line and the ith depth

$\mathrm{Im}(F_{j,i})$ and $\mathrm{Re}(F_{j,i})$ are the imaginary unit and real unit of the complex signal $F_{j,i}$, respectively

Thus, the phase change $\Delta\phi_{j,i}$ between the jth A-line and the $(j+1)$th A-line at the ith depth is calculated by the following equation:

$$\Delta\phi_{j,i} = \tan^{-1}\left[\frac{\mathrm{Im}\left(F_{j+1,i}\right)}{\mathrm{Re}\left(F_{j+1,i}\right)}\right] - \tan^{-1}\left[\frac{\mathrm{Im}\left(F_{j,i}\right)}{\mathrm{Re}\left(F_{j,i}\right)}\right] \tag{9.6}$$

In order to improve the signal-to-noise ratio and reduce the measurement error, the average phase change in the adjacent area is calculated by the following equation:

$$\Delta\phi_{j,i} = \frac{1}{M \cdot N}\sum_{n=0}^{N-1}\sum_{m=0}^{M-1}\left\{\tan^{-1}\left[\frac{\mathrm{Im}\left(F_{j+n+1,i+m}\right)}{\mathrm{Re}\left(F_{j+n+1,i+m}\right)}\right] - \tan^{-1}\left[\frac{\mathrm{Im}\left(F_{j+n,i+m}\right)}{\mathrm{Re}\left(F_{j+n,i+m}\right)}\right]\right\} \tag{9.7}$$

where

N is the number of A-lines that are averaged

M is the number of depth points that are averaged

The phase change can also be described by the following equation based on a cross-correlation algorithm (Zhao et al. 2000b):

$$\Delta\phi_{j,i} = \tan^{-1}\left[\frac{\mathrm{Im}\left(F_{j,i} \cdot F_{j+1,i}^{*}\right)}{\mathrm{Re}\left(F_{j,i} \cdot F_{j+1,i}^{*}\right)}\right] \tag{9.8}$$

Incorporating the lateral and depth averaging window, it can be presented by the following equation:

$$\Delta\phi_{j,i} = \tan^{-1}\left\{\frac{\mathrm{Im}\left[\sum_{n=0}^{N-1}\sum_{m=0}^{M-1}\left(F_{j+n,i+m} \cdot F_{j+n+1,i+m}^{*}\right)\right]}{\mathrm{Re}\left[\sum_{n=0}^{N-1}\sum_{m=0}^{M-1}\left(F_{j+n,i+m} \cdot F_{j+n+1,i+m}^{*}\right)\right]}\right\} \tag{9.9}$$

which is equivalent to the following equation:

$$\Delta\phi_{j,i} =$$
$$\tan^{-1}\left\{\frac{\sum_{n=0}^{N-1}\sum_{m=0}^{M-1}\left[\text{Re}\left(F_{j+n+1,i+m}\right)\cdot\text{Im}\left(F_{j+n,i+m}\right)-\text{Im}\left(F_{j+n+1,i+m}\right)\cdot\text{Re}\left(F_{j+n,i+m}\right)\right]}{\sum_{n=0}^{N-1}\sum_{m=0}^{M-1}\left[\text{Re}\left(F_{j+n+1,i+m}\right)\cdot\text{Re}\left(F_{j+n,i+m}\right)+\text{Im}\left(F_{j+n+1,i+m}\right)\cdot\text{Im}\left(F_{j+n,i+m}\right)\right]}\right\} \quad (9.10)$$

where the calculated phase change $\Delta\phi_{j,i}$ is in a range of $-\pi$ to $+\pi$.

Using Equations 9.4 and 9.10, the absolute flow velocity can be measured with the known Doppler angle θ between the beam vector and flow vector. From Equation 9.4, when the Doppler angle is close to $90°$, $\cos(\theta)$ is close to zero and the flow velocity cannot be calculated accurately. This method cannot be used for the velocity measurement when the flow direction is perpendicular to the beam direction.

Dynamic Range and Correction for Phase Wrapping

When the flow velocity V decreases, the phase change $\Delta\phi_{j,i}$ will also decrease. Due to the limitation of the OCT signal-to-noise ratio, the phase change $\Delta\phi_{j,i}$ may not be accurately measured if flow velocity V is below a threshold. Alternatively, with the same velocity, the increase of the A-line interval ΔT will result in an increase of the phase change $\Delta\phi_{j,i}$ to the normal measurement range of a Doppler OCT system. This operation can increase the sensitivity of the velocity measurement. However, the increased A-line interval will cause a longer time for data capture, thus decreasing the imaging speed. Inter-2D-frame or inter-3D-frame A-line correlation can overcome these limitations (Chen and Zhang 2015).

When the flow velocity continues to increase and exceeds a threshold, the phase wrap will occur because the calculated value of $\Delta\phi_{j,i}$ is limited to a range of $-\pi$ to $+\pi$ from Equation 9.10. In this case, either the decrease of the A-line interval or the correction for the phase wrapping is required for the correct measurement of the phase change. However, the minimal A-line interval is determined by the system performance, and thus, the phase unwrapping technique is necessary for quantifying high-speed flow.

Figure 9.3 shows the calculation of the phase change with and without phase wrapping. The red arrow indicates the correct value of the phase change and the blue one indicates the calculated value. When the phase change is larger than $+\pi$ and smaller than $+2\pi$, a negative value of the phase change will be calculated. Meanwhile, if the phase change exceeds $-\pi$ but doesn't exceed -2π, a positive

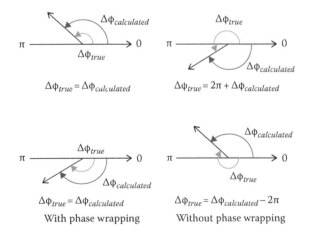

FIGURE 9.3 Calculation of phase changes with and without phase wrapping. The red arrow indicates the calculated value of the phase change and the blue one indicates the true value.

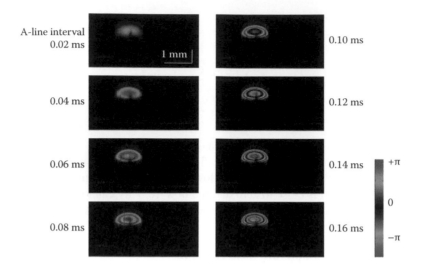

FIGURE 9.4 Phase-resolved color Doppler (PRCD) images during flow velocity measurement for scattered milk in a transparent tube. The flow velocity is controlled by a syringe pump and is measured using a Doppler OCT system based on a swept source. Phase wrapping can be observed with the increase of the A-line interval.

value of the phase change will be calculated. In order to correctly calculate the flow velocity, a correction of phase wrapping needs to be performed.

Flow velocity measurement for scattered milk in a transparent tube is shown in Figure 9.4. The flow velocity is controlled by a syringe pump and is measured using a Doppler OCT system based on a swept source. The flow velocity remains invariable, and the A-line sampling interval is increased. In Figure 9.4, the cross-sectional distribution of the flow velocity is not uniform in the tube, where the value in the central region is larger than the value in the peripheral region. When the A-line interval increases, the phase wrapping occurs first in the central region, and then appears in the peripheral region. With the continuous increase of the A-line interval, the phase change may wrap for several times. So for correction of the phase wrapping, a key problem is the determination of the wrapping times. If the phase wrapping times cannot be determined, the correction cannot be performed and the flow velocity cannot be measured. Therefore, the A-line interval should be optimized for an accurate measurement of phase changes.

Calculation of Doppler Angles

After the measurement of the phase changes, the flow velocity along the OCT beam direction can be calculated, which is represented as $V \cdot \cos(\theta)$. For the calculation of an absolute flow velocity V, the Doppler angle θ between the beam vector and the flow vector should be determined. In order to automatically calculate Doppler angles, three main steps can be performed. First, a modified region growing segmentation method is used to localize the vessels on successive B-scan PRCD images. Then the vessel skeleton, which represents the vessel geometry in 3D, can be constructed based on the coordinates of vessels in each B-scan PRCD image. Finally, an averaged gradient method is used to calculate the Doppler angle at each location along the vessel skeleton (Qi et al. 2016). Figure 9.5 shows the flow chart for an automatic calculation of flow velocities.

Correction for Bulk Motion

In a Doppler OCT system, another critical issue is that blood flow velocities are normally much lower than the typical physiological bulk motion. The noise from the bulk motion may conceal the phase change from the moving of scattered particles. In order to extract accurate phase changes, the bulk motion correction is required after the phase changes are collected. Usually, the diameters of

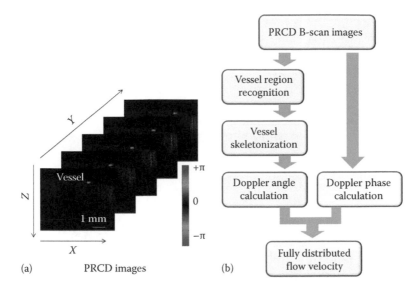

(a) PRCD images   (b)

FIGURE 9.5 Calculation of flow velocities using Doppler OCT. (a) PRCD images. (b) Flow chart for automatic calculation of flow velocities based on PRCD images.

measured vessels are much smaller compared to the axial imaging depth of the Doppler OCT system, so there is only a small range in an A-line presenting the phase change due to the blood flow. However, the bulk motion will result in a larger range of uniform phase changes in an A-line. Based on this assumption, the peak of the phase change histogram in an A-line represents the bulk motion phase. The subtraction of the bulk motion phase can correct the bulk motion distortion. Figure 9.6a and b show the cross-sectional OCT image and phase change between adjacent A-lines of the human retina and choroid. In Figure 9.6e, a typical histogram of phase changes shows a phase peak due to the bulk motion. Figures 9.6h and k show the images from the PRCD method before and after the bulk motion correction, respectively. After the correction for bulk motion, the vessel regions appear clearly from the background (Yu and Chen 2010).

DOPPLER VARIANCE MEASUREMENT

Using the PRCD method, the absolute velocity can be measured when the flow direction is not perpendicular to the OCT incident beam. If the Doppler angle is close to 90°, the Doppler shift is not sensitive to the flow velocity and, thus, the PRCD measurement cannot determine the transverse flow velocity. In order to overcome this limitation, a method that uses standard deviation (SD) of the Doppler shift is developed for the measurement of the transverse flow.

Here we consider the light scattered from the sample into the different sides of a large numeric aperture objective (shown in Figure 9.7). The Doppler shift from the left-most light beam is calculated by the following equation as the Doppler angle is $\theta + \varphi$:

$$f_1 = \frac{2V \cdot n \cdot \cos(\theta + \varphi)}{\lambda_0} \tag{9.11}$$

where as the Doppler shift from the right-most light beam is calculated by Equation 9.12 with the Doppler angle of $\theta - \varphi$:

$$f_2 = \frac{2V \cdot n \cdot \cos(\theta - \varphi)}{\lambda_0} \tag{9.12}$$

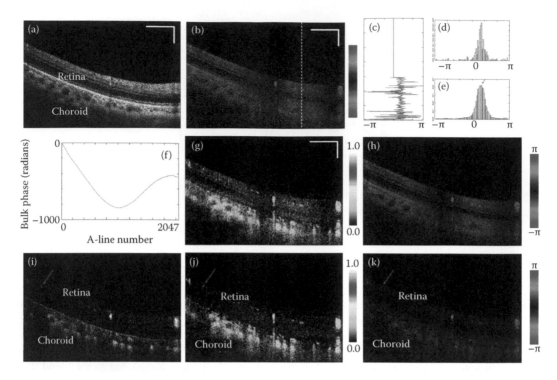

FIGURE 9.6 Correction for bulk motion. (a) Cross-sectional OCT image of the human retina and choroid. (b) Phase changes between A-lines without average; (c) Phase changes of an A-line in (b); (d) Histogram of the phase changes in an A-line; (e) Smoothed histogram; (f) Accumulated bulk motion phase of each A-line. (g, h) Images from the PRDV method and PRCD method without bulk motion correction, respectively. (i) Corresponding image from the optical microangiography method. (j, k) Images from the PRDV method and PRCD method after bulk motion correction, respectively. The scale bars represent 500 μm. (From Yu, L. and Chen, Z., *J. Biomed. Opt.*, 15(1), 016029, 2010. With permission.)

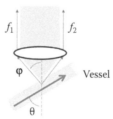

FIGURE 9.7 Schematic of Doppler variance measurement. The Doppler angles for the left-most light beam and the right-most light beam are $\theta + \varphi$ and $\theta - \varphi$, respectively.

Therefore, the bandwidth of the Doppler spectra defined by the geometrical optics is determined by the following equation:

$$f_2 - f_1 = \frac{4V \cdot n \cdot \sin(\theta) \cdot \sin(\varphi)}{\lambda_0} = \frac{4V \cdot n \cdot \sin(\theta) \cdot NA}{\lambda_0} \qquad (9.13)$$

where
    NA is the effective numerical aperture
    φ is the optical aperture angle

When a Gaussian optical beam is applied in the system, the Doppler bandwidth B (full bandwidth at $1/e$ maximum) is the inverse of the transit time for particles traveling through the focus zone and can be determined by the SD σ of the Doppler shift as well, which is described by the following equation (Ren et al. 2002):

$$B = \frac{\pi}{8} \cdot (f_2 - f_1) = 4\sigma \tag{9.14}$$

Therefore,

$$V \cdot \sin(\theta) = \frac{8\lambda_0}{n \cdot NA \cdot \pi} \cdot \sigma \tag{9.15}$$

From Equation 9.15, the SD σ is in a linear relationship to the transverse flow velocity presented by $V \cdot \sin(\theta)$; and the slope of this linear dependence is determined by the effective NA of the objective. With the known parameters of NA and θ, the flow velocity can be calculated quantitatively according to the measurement of the SD σ of the Doppler shift.

The SD of the Doppler shift can usually be determined by the OCT complex signals with the following equation (Zhao et al. 2000b):

$$\sigma_{j,i}^2 = \frac{\int_{-\infty}^{\infty} (f - \bar{f})^2 P(f) d(f)}{\int_{-\infty}^{\infty} P(f) d(f)} = \frac{1}{(2\pi \cdot \Delta T)^2} \frac{|F_{j,i} - F_{j+1,i}|^2}{|F_{j,i}|^2 + |F_{j+1,i}|^2} = \frac{1}{(2\pi \cdot \Delta T)^2} \cdot \left(1 - \frac{2 \cdot |F_{j,i} \cdot F_{j+1,i}^*|}{F_{j,i} \cdot F_{j,i}^* + F_{j+1,i} \cdot F_{j+1,i}^*}\right) \tag{9.16}$$

where
  $\bar{f}$ is the average Doppler shift
  $P(f)$ is the Doppler power spectrum
  $\sigma_{j,i}$ is the SD of the Doppler shift at the jth A-line and at the ith depth

Incorporating the lateral and depth averaging window, Equation 9.16 can be presented by the following equation:

$$\sigma_{j,i}^2 = \frac{1}{(2\pi \cdot \Delta T)^2} \cdot \left[1 - \frac{2 \cdot \left|\sum_{n=0}^{N-1} \sum_{m=0}^{M-1} \left(F_{j+n,i+m} \cdot F_{j+n+1,i+m}^*\right)\right|}{\sum_{n=0}^{N-1} \sum_{m=0}^{M-1} \left(F_{j+n,i+m} \cdot F_{j+n,i+m}^*\right) + \sum_{n=0}^{N-1} \sum_{m=0}^{M-1} \left(F_{j+n+1,i+m} \cdot F_{j+n+1,i+m}^*\right)}\right] \tag{9.17}$$

and

$$\sigma_{j,i}^2 = \frac{1}{(2\pi \cdot \Delta T)^2} \cdot \left[1 - \frac{2 \cdot \sum_{n=0}^{N-1} \sum_{m=0}^{M-1} \left|F_{j+n,i+m} \cdot F_{j+n+1,i+m}^*\right|}{\sum_{n=0}^{N-1} \sum_{m=0}^{M-1} \left(F_{j+n,i+m} \cdot F_{j+n,i+m}^*\right) + \sum_{n=0}^{N-1} \sum_{m=0}^{M-1} \left(F_{j+n+1,i+m} \cdot F_{j+n+1,i+m}^*\right)}\right] \tag{9.18}$$

where N and M are the number of A-lines and depth points in the averaging window, respectively.

Equation 9.17 involves the phase item and amplitude item. So the calculated value incorporates the phase difference and amplitude difference between adjacent A-lines in the averaging window, which is usually called phase-resolved Doppler variance. However, no phase item is considered in Equation 9.18 and only amplitude difference between adjacent A-lines will affect the SD σ, which is called intensity-based Doppler variance.

COMPARISON OF DIFFERENT DOPPLER OCT METHODS

From Equations 9.4 and 9.15, the axial velocity V_A can be determined by the phase change $\Delta\phi$ and the transversal velocity V_T can be determined by the phase variance σ, which is illustrated in Figure 9.8. If the Doppler angle θ is close to 90°, the axial velocity V_A cannot be measured accurately and the flow velocity should be calculated by the transversal velocity V_T.

For PRCD imaging, a phase wrapping will occur due to a high velocity or a long A-line interval. Meanwhile, for Doppler variance imaging, a saturated value will be calculated if the flow velocity or the A-line interval exceeds a threshold, which is shown in Figure 9.9 (Liu et al. 2012b). When the signals are unsaturated, the values from the PRDV method will be similar to the values from the IBDV method.

Because the phase item is incorporated in the PRCD method and PRDV method, these methods depend on the phase stability of the system. However, the phase term is removed from the IBDV method, so it can eliminate the distortion due to phase fluctuation. Furthermore, the IBDV method is not obviously affected by the bulk motion distortion. So the images from the IBDV method can be processed without the bulk motion correction and the requirement of the phase stability (Huang et al. 2015; Liu et al. 2011a,b; Zhu et al. 2015). For phase-resolved methods, the correction for bulk motion is necessary for high quality images.

Usually the effective numerical aperture in Equation 9.15 is not known in a Doppler OCT system, so a calibration for the IBDV or PRDV method is required for the absolute velocity measurement. The maximum intensity (MI) projection of the phase change from the PRCD method can assess the flow direction and flow velocity along the OCT beam, and the SD projection of the phase variance from the IBDV or PRDV method can be used for angiography. The SD projection of the phase variance volume can provide more details for profiling the blood flow than the MI projection of the PRCD volume.

DOPPLER OCT FOR CEREBRAL BLOOD FLOW IMAGING

Due to the ability to quantitatively measure blood flow and perform 3D mapping of vascular networks, Doppler OCT has been demonstrated as a promising method for the quantitative analysis of cerebrovascular physiology. It can quantitatively and noninvasively measure cerebral blood flow over a physiologic range with good spatiotemporal resolution (Srinivasan et al. 2011). Figure 9.10 shows microvascular imaging using Doppler OCT (Liu and Chen 2013a).

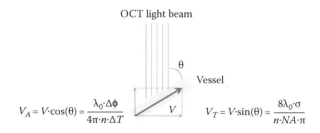

FIGURE 9.8 Illustration of flow velocity measurement with phase change and the phase variance.

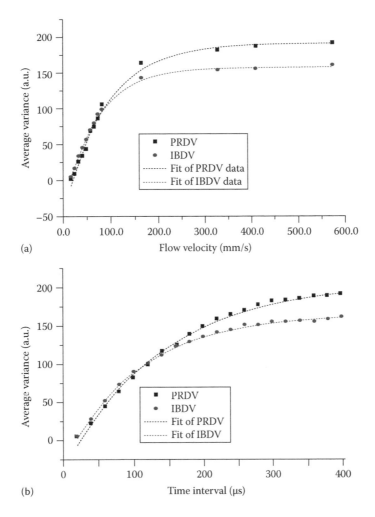

(a)

(b)

FIGURE 9.9 Comparison between the PRDV method and IBDV method. (a) The variance increases with the rise of the flow velocity for both methods. (b) The variance increases with the elongation of the time interval between adjacent A-lines for both methods. (From Liu, G. and Chen, Z., *Chin. Opt. Lett.*, 11(1), 11702, 2013a. With permission.)

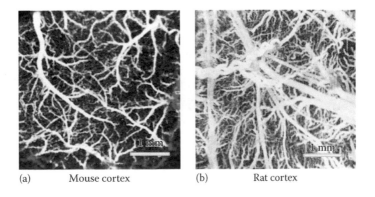

(a)   Mouse cortex   (b)   Rat cortex

FIGURE 9.10 Microvascular imaging using Doppler OCT. (a) Mouse cerebral cortex imaging. (b) Rat cerebral cortex imaging. (From Liu, G. and Chen, Z., *Chin. Opt. Lett.*, 11(1), 11702, 2013a. With permission.)

APPLICATION OF DOPPLER OCT FOR STUDY OF CEREBROVASCULAR PHYSIOLOGY

Using Doppler OCT, the responses of cerebral microvascular networks to cocaine administration revealed that cocaine induces cerebral microischemia and exacerbates these effects during repeated use in the rodent brain cortex (Ren et al. 2012a). The measurement of basal cerebral blood flow in atherosclerotic mice showed lower basal flow in the atherosclerotic group or atherosclerotic/catechin group than in the wild-type group. However, there was no significant difference between catechin treated and untreated groups (Drouin et al. 2011). The Doppler OCT also revealed different spatial configurations in the pial microvessels depending on the cerebral preparations and delayed swelling of the cortical surface in the somatosensory cortex following electrical stimulation to the hind paw in the case of dura removal (Satomura et al. 2004). For the investigation of a localized ischemic stroke, spectral Doppler OCT recorded flow dynamic information with high temporal resolution before and after the ischemia induced by the photocoagulation of Rose Bengal photodynamic therapy (PDT) in the mouse cortical microvasculature (shown in Figure 9.11) (Yu et al. 2010).

Doppler OCT has also been used for the investigation of instant and long-term cerebral microcirculatory adjustments induced by different stimulations. Optical stimulation of cortical neurons in transgenic mice resulted in a 20% increase in the diameter and a 100% increase in the blood flow inside the middle cerebral artery and neighboring vessels, and the blood flow gradually returned back to baseline after the removal of the light pulses. In comparison, for wild-type mice there are no obvious changes in the diameter and the blood flow during the optical stimulation (Atry et al. 2015). After the injection of the drug with the function of vasodilation or light stimulation in front of the rat's eye

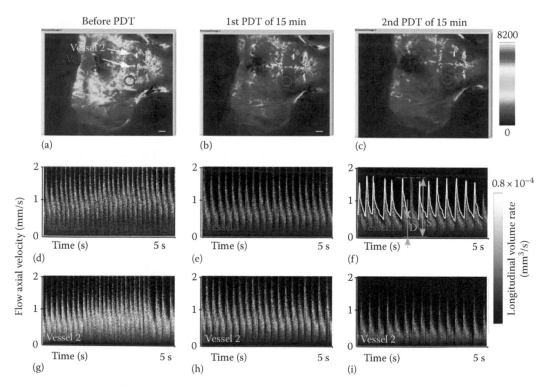

FIGURE 9.11   (a–c) Laser speckle flow index maps of the mouse skull with Rose Bengal photodynamic therapy (PDT) treatment. Flow velocity change of vessel 1 (d) before PDT, (e) after first PDT of 15 min, and (f) after second PDT of 15 min; flow velocity change of vessel 2 (g) before PDT, (h) after first PDT of 15 min, and (i) after second PDT of 15 min, respectively. (Scale bar represents 1 mm.) (From Yu, L. et al., *J. Biomed. Opt.*, 15(6), 066006, 2010. With permission.)

or electric stimulation of the sciatic nerve in the rat's paw, Doppler OCT was used to determine the flow velocity changes in the rat sensory cortex due to different stimulations (Wang et al. 2011).

In order to avoid laceration of at-risk intraparenchymal vessels and reduce the risk of cerebral hemorrhage during stereotactic neurosurgery in the brain, a forward-imaging probe with Doppler OCT can provide real-time feedback and guide the instrument to specific targets (Liang et al. 2011).

IMPROVEMENT OF DOPPLER OCT FOR THE MEASUREMENT OF CEREBRAL BLOOD FLOW

In order to improve the sensitivity and spatial resolution for cerebral blood flow imaging, a particle counting ultrahigh-resolution optical Doppler tomography is proposed based on the information of Doppler phase transients induced by red blood cell passage through a capillary (Ren et al. 2012b). Incorporating a phase summation method, absolute cerebral blood flow velocity (BFV) can be quantified in both branch vessels and capillaries (You et al. 2014). With intralipid injection, the sensitivity of Doppler OCT measurement can also be improved by obviating the errors from long latency between flowing red blood (Pan et al. 2014). In order to decrease the complexity and the cost of an OCT system, a coherence-gated Doppler method is proposed, which combines laser Doppler flowmetry and Doppler OCT. The coherence-gated Doppler system has the ability to differentiate tissue, vein, and artery and detect at-risk blood vessels in neurosurgery (Liang et al. 2013). Incorporating the 2D flow profile in the *en face* plane, blood flux in the rat cortex can be also obtained using a Doppler OCT system without the requirement of an explicit calculation of Doppler angles (Srinivasan et al. 2010).

MULTIMODALITY METHODS INCORPORATING DOPPLER OCT

The phase-resolved color Doppler method can combine PRDV method and speckle contrast method for a multiparametric imaging of murine brain. Both qualitative and quantitative information about vasculature can be obtained from the same data set (Bukowska et al. 2012). Integrated with optical/fluorescence imaging, Doppler OCT can aid in the assessment of spontaneous low-frequency blood-oxygen-level-dependent oscillations of cerebral blood flow in arteries and veins, respectively (Du et al. 2014). Combining Doppler OCT with laser speckle imaging and fluorescence imaging approaches, the multimodality platform can be used for the simultaneous assessment of cerebral hemodynamic and neuronal activities in cortical brain (Yuan et al. 2011). Using a multimodal optical imaging approach incorporating Doppler optical coherence tomography, spatial frequency domain imaging, and confocal microscopy, Alzheimer's disease–related changes are quantified in a triple transgenic mouse model (3×Tg-AD) and age-matched controls. From the Doppler OCT analysis, there is a 29% decrease of the superficial cortical vessel volume in 3×Tg-AD mice than in controls (shown in Figure 9.12) (Lin et al. 2014).

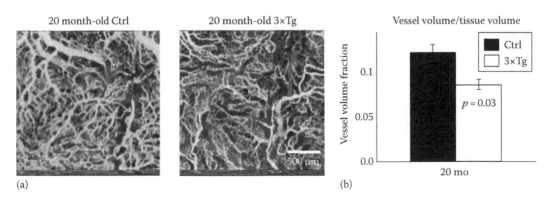

FIGURE 9.12 Doppler OCT imaging in an Alzheimer's mouse model. (a) Doppler OCT images of cortical vessels in control and 3×Tg-AD mice. Superficial vessels are in white and deeper vessels are in pink. (b) Quantification of the superficial vessel (white) area in each group. (From Lin, A.J. et al., *Neurophotonics*, 1(1), 011005, 2014. With permission.)

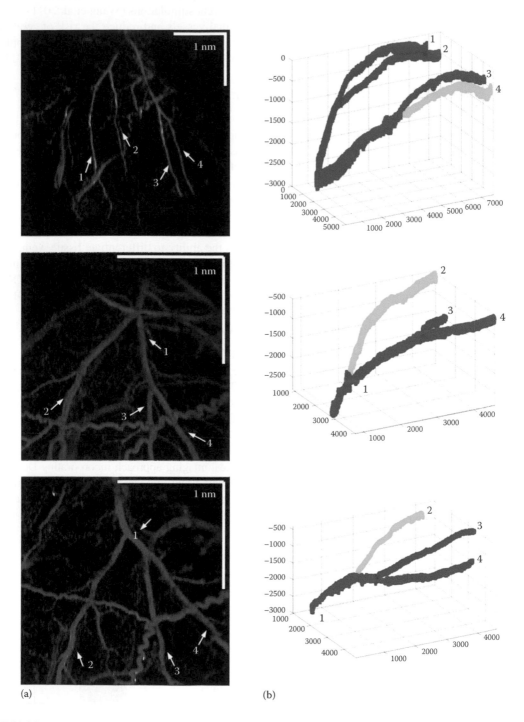

FIGURE 9.13 Fully distributed absolute blood flow velocity (BFV) measurement of the major cerebral arteries of rat brain. (a) Intensity-based Doppler variance *en face* images. (b) Reconstructed vessel skeleton.

(*Continued*)

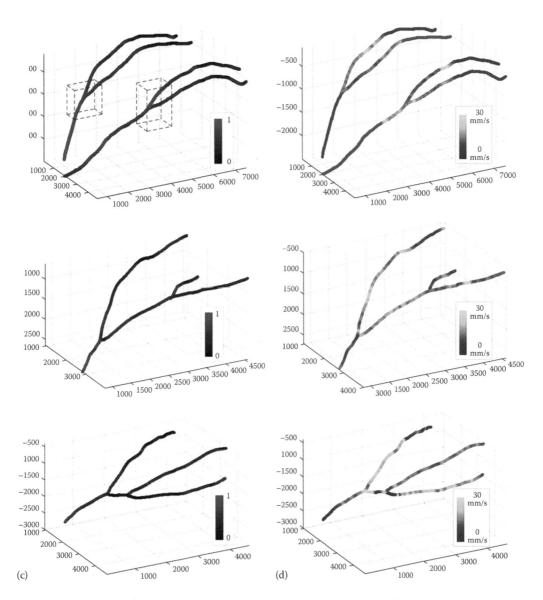

FIGURE 9.13 (Continued) Fully distributed absolute blood flow velocity (BFV) measurement of the major cerebral arteries of rat brain. (c) Doppler angle calculation results (cosine value). (d) Reconstructed absolute BFV distributed along the entire vessels. (From Qi, L. et al., *Biomed. Opt. Express*, 7(2), 601–615, 2016. With permission.)

With the ability of absolute flow velocity, Doppler OCT can also be used for the quantitative calibration of other qualitative imaging methods, such as laser speckle contrast imaging, so that qualitative imaging methods can extend their applications in the hemodynamics analysis of functional brain activations (Luo et al. 2008, 2009). Compared with speckle contrast based methods, Doppler OCT is more suitable for studying cerebral hemodynamic responses due to the ability of detection of directional flow below the Brownian motion regime (Ren et al. 2012c). Recently, we developed a method that enables the measurement of absolute BFV distributed along the entire major cerebral arteries of rat brain (Figure 9.13) (Qi et al. 2016).

CONCLUSIONS

With high-resolution 3D structural imaging from the OCT method and accurate flow velocity measurement based on the Doppler principle, Doppler OCT is the emerging technique for the investigation of *in vivo* cerebrovascular physiology. With the improvement of the OCT system, a deeper tissue may be measured with higher resolution and faster speed. Meanwhile, the combination of Doppler OCT with other optical imaging methods makes it possible to obtain and associate more and more physiological parameters in the same cerebral event.

REFERENCES

Ahn, Y. C., Jung, W., and Chen, Z. 2007. Quantification of a three-dimensional velocity vector using spectral-domain Doppler optical coherence tomography. *Opt Lett* 32(11): 1587–1589.

Atry, F., Frye, S., Richner, T. J., Brodnick, S. K., Soehartono, A., Williams, J., and Pashaie, R. 2015. Monitoring cerebral hemodynamics following optogenetic stimulation via optical coherence tomography. *IEEE Trans Biomed Eng* 62(2): 766–773.

Bukowska, D., Ruminski, D., Szlag, D., Grulkowski, I., Wlodarczyk, J., Szkulmowski, M., Wilczynski, G., Gorczynska, I., and Wojtkowski, M. 2012. Multi-parametric imaging of murine brain using spectral and time domain optical coherence tomography. *J Biomed Opt* 17(10): 101515.

Chen, Z., Milner, T. E., Dave, D., and Nelson, J. S. 1997a. Optical Doppler tomographic imaging of fluid flow velocity in highly scattering media. *Opt Lett* 22(1): 64–66.

Chen, Z., Milner, T. E., Srinivas, S., Wang, X., Malekafzali, A., van Gemert, M. J., and Nelson, J. S. 1997b. Noninvasive imaging of in vivo blood flow velocity using optical Doppler tomography. *Opt Lett* 22(14):1119–1121.

Chen, Z. and Zhang, J. 2015. Doppler optical coherence tomography. In: *Optical Coherence Tomography: Technology and Applications*, 2nd edn. Drexler, W., and Fujimoto, J. G., Eds. Cham, Switzerland: Springer International Publishing.

Choi, W., Baumann, B., Liu, J. J., Clermont, A. C., Feener, E. P., Duker, J. S., and Fujimoto, J. G. 2012. Measurement of pulsatile total blood flow in the human and rat retina with ultrahigh speed spectral/Fourier domain OCT. *Biomed Opt Express* 3(5): 1047–1061.

Cho, Y. K., Zheng, G., Augustine, G. J., Hochbaum, D., Cohen, A., Knöpfel, T., Pisanello, F., Pavone, F. S., Vellekoop, I. M., Booth, M. J. et al. (2016). Roadmap on neurophotonics. *J Opt* 18: 093007.

Ding, Z., Zhao, Y., Ren, H., Nelson, J., and Chen, Z. 2002. Real-time phase-resolved optical coherence tomography and optical Doppler tomography. *Opt Express* 10(5): 236–245.

Drouin, A., Bolduc, V., Thorin-Trescases, N., Bélanger, É., Fernandes, P., Baraghis, E., Lesage, F. et al. 2011. Catechin treatment improves cerebrovascular flow-mediated dilation and learning abilities in atherosclerotic mice. *Am J Physiol Heart Circ Physiol* 300(3): H1032–H1043.

Du, C., Volkow, N. D., Koretsky, A. P., and Pan, Y. 2014. Low-frequency calcium oscillations accompany deoxyhemoglobin oscillations in rat somatosensory cortex. *Proc Natl Acad Sci USA* 111(43): E4677–E4686.

Fujimoto, J. G. 2003. Optical coherence tomography for ultrahigh resolution in vivo imaging. *Nat Biotechnol* 21(11): 1361–1367.

Fujimoto, J. G., Brezinski, M. E., Tearney, G. J., Boppart, S. A., Bouma, B., Hee, M. R., Southern, J. F., and Swanson, E. A. 1995. Optical biopsy and imaging using optical coherence tomography. *Nat Med* 1(9): 970–972.

Huang, D., Swanson, E. A., Lin, C. P., Schuman, J. S., Stinson, W. G., Chang, W., Hee, M. R., Flotte, T., Gregory, K., and Puliafito, C. A. 1991. Optical coherence tomography. *Science* 254(5035): 1178–1181.

Huang, S., Piao, Z., Zhu, J., Lu, F., and Chen, Z. 2015. In vivo microvascular network imaging of the human retina combined with an automatic three-dimensional segmentation method. *J Biomed Opt* 20(7): 76003.

Izatt, J. A., Kulkarni, M. D., Yazdanfar, S., Barton, J. K., and Welch, A. J. 1997. In vivo bidirectional color Doppler flow imaging of picoliter blood volumes using optical coherence tomography. *Opt Lett* 22(18): 1439–1441.

Lee, J. and Boas, D. A. 2015. OCT and coherence imaging for the neurosciences. In *Optical Coherence Tomography: Technology and Applications*, 2nd edn. Drexler, W., and Fujimoto, J. G., Eds. Cham, Switzerland: Springer International Publishing.

Leitgeb, R. A., Werkmeister, R. M., Blatter, C., and Schmetterer, L. 2014. Doppler optical coherence tomography. *Prog Retin Eye Res* 41(100): 26–43.

Li, P., Liu, A., Shi, L., Yin, X., Rugonyi, S., and Wang, R. K. 2011. Assessment of strain and strain rate in embryonic chick heart in vivo using tissue Doppler optical coherence tomography. *Phys Med Biol* 56(22): 7081–7092.

Liang, C. P., Wierwille, J., Moreira, T., Schwartzbauer, G., Jafri, M. S., Tang, C. M., and Chen, Y. 2011. A forward-imaging needle-type OCT probe for image guided stereotactic procedures. *Opt Express* 19(27): 26283–26294.

Liang, C. P., Wu, Y., Schmitt, J., Bigeleisen, P. E., Slavin, J., Jafri, M. S., Tang, C. M., and Chen, Y. 2013. Coherence-gated Doppler: A fiber sensor for precise localization of blood flow. *Biomed Opt Express* 4(5): 760–771.

Lin, A. J., Liu, G., Castello, N. A., Yeh, J. J., Rahimian, R., Lee, G. et al. 2014. Optical imaging in an Alzheimer's mouse model reveals amyloid-β-dependent vascular impairment. *Neurophotonics* 1(1): 011005.

Liu, G. and Chen, Z. 2013a. Advances in Doppler OCT. *Chin Opt Lett* 11(1): 11702.

Liu, G. and Chen, Z. 2013b. Optical coherence tomography for brain imaging. In *Optical Methods and Instrumentation in Brain Imaging and Therapy*, Madsen, S. J. Ed. New York: Springer Science+Business Media.

Liu, G., Chou, L., Jia, W., Qi, W., Choi, B., and Chen, Z. 2011a. Intensity-based modified Doppler variance algorithm: Application to phase instable and phase stable optical coherence tomography systems. *Opt Express* 19(12): 11429–11440.

Liu, G., Jia, W., Sun, V., Choi, B., and Chen, Z. 2012a. High-resolution imaging of microvasculature in human skin in-vivo with optical coherence tomography. *Opt Express* 20(7): 7694–7705.

Liu, G., Lin, A. J., Tromberg, B. J., and Chen, Z. 2012b. A comparison of Doppler optical coherence tomography methods. *Biomed Opt Express* 3(10): 2669–2680.

Liu, G., Jia, W., Nelson, J. S., and Chen, Z. 2013b. In vivo, high-resolution, three-dimensional imaging of port wine stain microvasculature in human skin. *Lasers Surg Med* 45(10): 628–632.

Liu, G., Qi, W., Yu, L., and Chen, Z. 2011b. Real-time bulk-motion-correction free Doppler variance optical coherence tomography for choroidal capillary vasculature imaging. *Opt Express* 19(4): 3657–3666.

Low, A. F., Tearney, G. J., Bouma, B. E., and Jang, I. K. 2006. Technology insight: Optical coherence tomography—Current status and future development. *Nat Clin Pract Cardiovasc Med* 3(3): 154–162.

Luo, Z., Wang, Z., Yuan, Z., Du, C., and Pan, Y. 2008. Optical coherence Doppler tomography quantifies laser speckle contrast imaging for blood flow imaging in the rat cerebral cortex. *Opt Lett* 33(10): 1156–1158.

Luo, Z., Yuan, Z., Tully, M., Pan, Y., and Du, C. 2009. Quantification of cocaine-induced cortical blood flow changes using laser speckle contrast imaging and Doppler optical coherence tomography. *Appl Opt* 48(10): D247–D255.

Pan, Y., You, J., Volkow, N. D., Park, K., and Du, C. 2014. Ultrasensitive detection of 3D cerebral microvascular network dynamics in vivo. *Neuroimage* 103: 492–501.

Peterson, L. M., Jenkins, M. W., Gu, S., Barwick, L., Watanabe, M., and Rollins, A. M. 2012. 4D shear stress maps of the developing heart using Doppler optical coherence tomography. *Biomed Opt Express* 3(11): 3022–3032.

Qi, L., Zhu, J., Hancock, A. M., Dai, C., Zhang, X., Frostig, R. D., and Chen, Z. 2016. Fully distributed absolute blood flow velocity measurement for middle cerebral arteries using Doppler optical coherence tomography. *Biomed Opt Express* 7(2): 601–615.

Ren, H., Brecke, K. M., Ding, Z., Zhao, Y., Nelson, J. S., and Chen, Z. 2002. Imaging and quantifying transverse flow velocity with the Doppler bandwidth in a phase-resolved functional optical coherence tomography. *Opt Lett* 27(6): 409–411.

Ren, H., Du, C., and Pan, Y. 2012c. Cerebral blood flow imaged with ultrahigh-resolution optical coherence angiography and Doppler tomography. *Opt Lett* 37(8): 1388–1390.

Ren, H., Du, C., Park, K., Volkow, N. D., and Pan, Y. 2012b. Quantitative imaging of red blood cell velocity invivo using optical coherence Doppler tomography. *Appl Phys Lett* 100(23): 233702–2337024.

Ren, H., Du, C., Yuan, Z., Park, K., Volkow, N. D., and Pan, Y. 2012a. Cocaine-induced cortical microischemia in the rodent brain: Clinical implications. *Mol Psychiatry* 17(10): 1017–1025.

Rollins, A. M., Yazdanfar, S., Barton, J. K., and Izatt, J. A. 2002. Real-time in vivo color Doppler optical coherence tomography. *J Biomed Opt* 7(1): 123–129.

Satomura, Y., Seki, J., Ooi, Y., Yanagida, T., and Seiyama, A. 2004. In vivo imaging of the rat cerebral microvessels with optical coherence tomography. *Clin Hemorheol Microcirc* 31(1): 31–40.

Sehi, M., Goharian, I., Konduru, R., Tan, O., Srinivas, S., Sadda, S. R., Francis, B. A., Huang, D., and Greenfield, D. S. 2014. Retinal blood flow in glaucomatous eyes with single-hemifield damage. *Ophthalmology* 121(3): 750–758.

Srinivasan, V. J., Atochin, D. N., Radhakrishnan, H., Jiang, J. Y., Ruvinskaya, S., Wu, W., Barry, S. et al. 2011. Optical coherence tomography for the quantitative study of cerebrovascular physiology. *J Cereb Blood Flow Metab* 31(6): 1339–1345.

Srinivasan, V. J., Sakadzić, S., Gorczynska, I., Ruvinskaya, S., Wu, W., Fujimoto, J. G., and Boas, D. A. 2010. Quantitative cerebral blood flow with optical coherence tomography. *Opt Express* 18(3): 2477–2494.

Vakoc, B. J., Fukumura, D., Jain, R. K., and Bouma, B. E. 2012. Cancer imaging by optical coherence tomography: Preclinical progress and clinical potential. *Nat Rev Cancer* 12(5): 363–368.

Wang, C., Yang, Y., Ding, Z., Meng, J., Wang, K., Yang, W., and Xu, Y. 2011. Monitoring of drug and stimulation induced cerebral blood flow velocity changes in rat sensory cortex using spectral domain Doppler optical coherence tomography. *J Biomed Opt* 16(4): 046001.

Yazdanfar, S., Kulkarni, M., and Izatt, J. 1997. High resolution imaging of in vivo cardiac dynamics using color Doppler optical coherence tomography. *Opt Express* 1(13): 424–431.

You, J., Du, C., Volkow, N. D., and Pan, Y. 2014. Optical coherence Doppler tomography for quantitative cerebral blood flow imaging. *Biomed Opt Express* 5(9): 3217–3230.

Yu, L. and Chen, Z. 2010. Doppler variance imaging for three-dimensional retina and choroid angiography. *J Biomed Opt* 15(1): 016029.

Yu, L., Nguyen, E., Liu, G., Choi, B., and Chen, Z. 2010. Spectral Doppler optical coherence tomography imaging of localized ischemic stroke in a mouse model. *J Biomed Opt* 15(6): 066006.

Yuan, Z., Luo, Z., Volkow, N. D., Pan, Y., and Du, C. 2011. Imaging separation of neuronal from vascular effects of cocaine on rat cortical brain in vivo. *Neuroimage* 54(2): 1130–1139.

Zhao, Y., Chen, Z., Saxer, C., Shen, Q., Xiang, S., de Boer, J. F., and Nelson, J. S. 2000b. Doppler standard deviation imaging for clinical monitoring of in vivo human skin blood flow. *Opt Lett* 25(18):1358–1360.

Zhao, Y., Chen, Z., Saxer, C., Xiang, S., de Boer, J. F., and Nelson, J. S. 2000a. Phase-resolved optical coherence tomography and optical Doppler tomography for imaging blood flow in human skin with fast scanning speed and high velocity sensitivity. *Opt Lett* 25(2):114–116.

Zhu, J., Qu, Y., Ma, T., Li, R., Du, Y., Huang, S., Shung, K. K., Zhou, Q., and Chen, Z. 2015. Imaging and characterizing shear wave and shear modulus under orthogonal acoustic radiation force excitation using OCT Doppler variance method. *Opt Lett* 40(9): 2099–2102.

10 Brain Imaging and Mapping with Multi-Contrast Optical Coherence Tomography

Hui Wang and Taner Akkin

CONTENTS

INTRODUCTION

The brain is a highly heterogeneous network with billions of neurons communicating through trillions of synaptic connections. In the past decades, system-level imaging techniques such as functional MRI and PET and high-resolution microscopes have revolutionized our understanding of regional activity and single-neuron functions. However, our knowledge of how these neurons are coordinated to form complex behaviors and what go wrong under neurological disorders are yet sparsely scattered, without coherently linked apprehension. The questions need to be addressed in all aspects: the cell types and neuronal connections in the network, the dynamic circuitry and chemicals under neural activity, and the supporting systems such as blood supply and microenvironment. Systematic explorations of these subjects have imposed a pressing need for innovative neuroimaging techniques that simultaneously possess the power of high spatiotemporal resolution and considerable coverage of the brain.

The emergence of optical coherence tomography (OCT) (Huang et al., 1991) offers a viable solution for 3D reconstruction of biological tissues with high spatiotemporal resolution. OCT, analogous to ultrasound imaging, makes use of an optical interferometry to provide cross-sectional and 3D images of tissue microstructures up to a few millimeters of depth. The signal is intrinsically originated from the light backscattered from the sample, thus enabling noninvasive or minimally invasive *in vivo* studies. Since its invention, OCT has proved to be an appealing tool in a wide range of applications in neuroscience (Boppart, 2003). The technique has been extensively established in the application of ophthalmology, especially the retina, for a variety of purposes. Its ability to quantitatively examine the multiple neuronal layers in the retina and the associated finer structures has provided an easily accessible window for studying central nervous system diseases, such as multiple sclerosis (Frohman et al., 2008). Feasibility of OCT in localizing nerves and blood vessels has been investigated in rabbit (Boppart et al., 1998a). In the functional studies of neural activity, the intensity and phase changes of OCT signals without and with voltage-sensitive dyes have been correlated with changes of membrane potentials in isolated nerve (Akkin et al., 2010).

The brain consists of hundreds of neuronal types with a diversity of morphology and shapes (Bertrand et al., 2002). The interpretation of the intricate organizations and connections could be confounded on 2D microscopy images. The 3D imaging in OCT restores the compressed space to its full visibility. It is essentially important in revealing delicate features and complex architectures in brain with high resolution (2–20 μm) and remarkable sensitivity (~100 dB). The imaging depth allows for whole-brain imaging in semitransparent animal models, such as *Xenopus* and zebra fish, in the developmental stage (Boppart et al., 1996a,b). In mammal brains, the imaging depth of ~1 mm has been found in the cerebral cortex (Jeon et al., 2006), which is sufficient for studying intact cerebral laminar structures and functions in small animals such as mouse and rat. Although deep brain imaging is limited by light attenuation due to scattering and absorption events, reconstruction of the whole-brain space is possible *ex vivo* by fusing volumetric data from sequential OCT scans (Wang et al., 2014a). OCT signal comes from the inherent optical property of tissues without the need of exogenous contrast agents.

Besides the conventional OCT, there are specialized OCT techniques. These include polarization-sensitive OCT (De Boer et al., 1997), Doppler OCT (Chen et al., 1997), and optical microangiography (Makita et al., 2006; Wang et al., 2007), which are capable of probing white matter tracts, cerebral blood circulation, and microvasculature networks, respectively, in brain studies. OCT rigs that incorporate multiple specializations into one single system are called multi-contrast (MC) OCT herein. The MC-OCT enables simultaneous imaging of several contrasts in one scan. This integration has several advantages. From the spatial perspective, it allows for the interrogation of structural architectures probed by different contrasts through automatically aligned framework. From the temporal perspective, it provides the interaction of different signals revealing physiological processes through simultaneous data collection. The merits are particularly beneficial in answering important neuroscience questions such as neurovascular coupling, brain mapping, and brain-wide wiring diagram, and their topological relationship with cell types. In this chapter, we review the basics of OCT techniques and the structural brain imaging and mapping studies using multiple OCT contrasts.

OPTICAL TECHNIQUES

OPTICAL COHERENCE TOMOGRAPHY

OCT is built upon the low-temporal-coherence interferometry (LCI). Majority of the OCT systems recruit a Michelson interferometer (Figure 10.1). Light from a low coherent source is split into a reference field and a sample field. The sample field is focused on a target tissue through a lens. Light backscattered from the tissue returns to the interferometer and mixes with the reference field reflected from a mirror. The interference occurs in the beam splitter when the mismatch between the optical path lengths of the sample and reference arms is within one coherence length. The signal carrying interference is detected by a photodetector and can be written as

$$\bar{I}_d(\tau) = I_s(t) + I_r(t) + G_{sr}(\tau) \tag{10.1}$$

where I_s and I_r are halves of the mean intensities back from the sample and reference arms, respectively. The term G_{sr} represents the interference term as a function of time delay τ that depends on the optical path length difference (Δz) between the sample and reference arms and the speed of light (c), $\tau = \Delta z/c$.

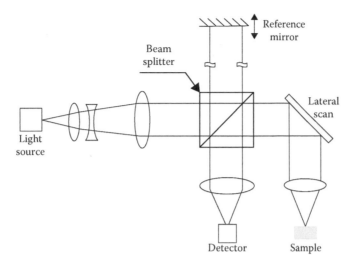

FIGURE 10.1   Basic schematic of a typical TD-OCT setup.

The mean intensities contribute to the DC signal at the detector, whereas the interference term carries information about tissue microstructure. For a single sample surface, we can write

$$G_{sr}(\tau) = \mathrm{Re}\left\{E_s^*(t+\tau)E_r(t)\right\} = \left|\Gamma(\tau)\right|\cos\left[2\pi v_0 \tau + \varnothing(\tau)\right] \tag{10.2}$$

where
  E_s and E_r are the sample and reference fields directed to detection
  $\Gamma(\tau)$ is the temporal coherence function of the light source with an argument $\varnothing(\tau)$
  $2\pi v_0 \tau$ is the phase delay between the two arms with v_0 being the center frequency of the source
     ($v_0 = c/\lambda_0$; λ_0: center wavelength)

The Wiener–Khinchin theorem states that Fourier transform of the autocorrelation function is the power spectrum. Therefore, a Fourier transform relation exists between the power spectral density $S(v)$ and the temporal coherence function $\Gamma(\tau)$ of the source:

$$\Gamma(\tau) = \int_0^\infty S(v)\exp(-j2\pi\tau)dv. \tag{10.3}$$

This equation also forms the basis of Fourier domain (FD) OCT. Clearly, the shape and width of the source spectrum are crucial as the temporal coherence function determines the axial resolution. Under the assumption of a Gaussian-shaped source spectrum, the axial resolution of OCT is determined by

$$l_c = \frac{2\ln 2}{\pi}\frac{\lambda_0^2}{\Delta\lambda} \approx 0.44\frac{\lambda_0^2}{\Delta\lambda} \tag{10.4}$$

where $\Delta\lambda$ is the bandwidth defined as the full width at half maximum (FWHM).

As in conventional microscopy, the lateral resolution Δx is related to the wavelength and the sample path optics that focuses the beam on a sample and is given by

$$\Delta x = \frac{4\lambda_0}{\pi} \frac{f}{d} \tag{10.5}$$

where
  d is the spot size on the objective lens
  f is its focal length

The estimation of lateral resolution is considered within the depth of focus, which is defined as twice the Rayleigh range:

$$DOF = \frac{2\pi\omega_0^2}{\lambda_0} \tag{10.6}$$

where ω_0 is the beam waist at focus.

The sensitivity, referred to as the weakest sample reflectivity yielding a signal power equal to the noise of the system, is an important characteristic in OCT. In shot-noise limited regime, the sensitivity of 100 dB or higher has been reported. Some of the applications rely on the phase whose noise is inversely proportional to the signal-to-noise ratio (SNR) of the interferometric signal.

The OCT systems can be constructed in time domain (TD) or FD. TD-OCT employs a depth-scan mechanism in the reference arm to localize scatters in the tissue. As the optical path length in the reference arm cycles within a range, a so-called depth profile or A-line is acquired. The reference arm may consist of a mirror on an actuator or a more complex rapid scanning delay line that may also be used for compensating dispersion mismatch between the reference and sample arms (Tearney et al., 1997). In order to create cross-sectional (B-line) or 3D images of tissues, the beam is scanned over the sample by galvanometer based scanners.

The FD-OCT has a stationary mirror in the reference arm. The depth profiles are reconstructed from the interference related oscillations on the spectrum. FD-OCT is realized either by employing a spectrometer in the detection arm, called spectral-domain (SD) OCT (Fercher et al., 1995; Häusler and Lindner, 1998), or by sweeping a wavelength-tuning light source, called swept-source (SS) OCT (Chinn et al., 1997; Golubovic et al., 1997). Comparing to TD-OCT, the advantages of FD-OCT manifest in the faster acquisition speed and superior sensitivity (Choma et al., 2003; De Boer et al., 2003; Leitgeb et al., 2003).

Light Polarization and Birefringence

Light polarization describes the direction of wave oscillation. The light wave propagating along z-direction oscillates in the orthogonal (xy) plane; therefore, its polarization state can be described by the electric field components in the x and y directions. For low-coherent light with a frequency relationship $\Delta\nu \ll \nu_o$, the electric field of the wave can be written as

$$E(t) = E_x(t)\hat{x} + E_y(t)\hat{y} \tag{10.7}$$

where

$$E_x(t) = E_{ox} \cos\left[2\pi\nu_o t\right]$$

$$E_y(t) = E_{oy} \cos\left[2\pi\nu_o t - \Delta\varphi\right]$$

$\Delta\varphi$ is the phase difference between the electric field components in the x and y directions

When the amplitude E_{ox} or E_{oy} is equal to zero, light is linearly polarized in the y or x direction, respectively. Also, when the waves are in phase or out of phase, $\Delta\varphi = n\pi$ ($n = 0, 1, 2, \ldots$), light is linearly polarized at an angle that depends on the amplitude ratio of E_{ox} and E_{oy}. Left-handed and right-handed circularly polarized light are other special cases that occur when the amplitudes are equal ($E_{ox} = E_{oy}$) and the phase difference is $\pm n\pi/2$. For all other cases, the polarization state is elliptical.

Birefringence, an optical property of many tissues, including muscle, tendon and nerve, introduces change in polarization states. Birefringence is divided into two regimes: intrinsic birefringence as a consequence of anisotropic molecules and form birefringence originated from regularly arranged structures surrounded by another isotropic medium. Birefringent samples have an optic axis that is collinear with the direction of structure anisotropy. Light waves oscillating parallel and perpendicular to the optic axis encounter different refractive indices (n_o-ordinary; n_e-extraordinary), and the difference is known to be the birefringence, $\Delta n = n_e - n_o$. Because the speed of light depends on the refractive index, light passing through birefringent tissues experiences a delay between the orthogonal components. The delay, known as retardance (δ), is equivalent to the birefringence multiplied by the distance light travels ($\delta = \Delta n \cdot z$), and it can be represented by phase as the path of one wavelength corresponds to $360°$.

POLARIZATION-SENSITIVE OCT

Polarization-sensitive (PS) OCT (de Boer et al., 1997) makes use of light polarization to produce additional contrasts in OCT imaging. It targets the birefringent tissues. Jones formalism provides a convenient mathematical tool to describe light polarization and the polarization effects (Jones, 1941). A 2×2 Jones matrix J characterizes the polarization property of any nondepolarizing optical component, which transforms the incident polarization state E to a transmitted sate E':

$$E' = JE \tag{10.8}$$

Note that the OCT technique is not sensitive to the multiply scattered diffuse light that degrades the degree of polarization, as the interference is mainly produced by single or a few times scattering directed back to the imaging system. The Jones matrix for a birefringent material with an orientation angle θ and a retardance δ is given by (Gil and Bernabeu, 1987)

$$J_b = \begin{bmatrix} e^{i\delta/2}\cos_\theta^2 + e^{-i\delta/2}\sin_\theta^2 & \left(e^{i\delta/2} - e^{-i\delta/2}\right)\cos\theta\sin\theta \\ \left(e^{i\delta/2} - e^{-i\delta/2}\right)\cos\theta\sin\theta & e^{i\delta/2}\sin_\theta^2 + e^{-i\delta/2}\cos_\theta^2 \end{bmatrix} \tag{10.9}$$

The Jones formalism has been commonly implemented in PS-OCT setups. The first PS-LCI by Hee et al. (1992) relies on the Jones formalism to measure the retardance between orthogonal polarizations of light backscattered from a birefringent sample. The basic scheme is based on a low-coherence Michelson interferometer and a dual-channel PS detection unit (Figure 10.2). Later, de Boer et al. (1997, 1998) developed a PS-OCT system for imaging biological tissues. Everett et al. (1998) and Schoenenberger et al. (1998) used PS-OCT to obtain birefringence maps of the myocardium in porcine. Hitzenberger et al. (2001) advanced the PS-OCT technique to image both phase retardance and fast axis orientation in the myocardium.

The Stokes parameters and Mueller matrices provide a more complete description of the polarization properties of light in a turbid media. The Stokes parameter S is a vector of $[I, Q, U, V]^T$, where I is the total intensity, and Q, U, and V represent the horizontal/vertical component, the $\pm 45°$ linear component, and the left/right circular component, respectively. The Mueller matrix M is a 4 by 4 matrix which linearly relates the Stokes vector of the backscattered light S' with the Stokes vector of the illuminating light S by $S' = MS$ (Huard, 1997). The PS-OCT systems capable of yielding the full 4×4 Mueller matrix images have been presented (Yao and Wang, 1999; Yasuno et al., 2002).

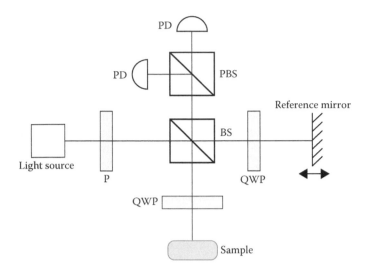

FIGURE 10.2 Basic schematic of PS-LCI setup. P, polarizer; BS, beam splitter; QWP, quarter-wave plate; PBS, polarization beam splitter; PD, photodetector.

de Boer et al. (1999) presented an approach to realize depth-resolved Stokes parameters. The effects of multiple scattering and speckle on the polarization measurement have been evaluated as well (de Boer and Milner, 2002). Multiple scattering randomly scrambles polarization and reduces the degree of polarization that can have an effect on birefringence measurement. Speckle, a main problem in OCT, is another factor as it introduces fluctuations in the interference signal, which could be uncorrelated in the orthogonal polarization channels.

The OCT and PS-OCT systems can be implemented with bulk or fiber optics. The bulk system has the advantages of simple optical component and computationally inexpensive algorithms, and the polarization states are stably kept in air. The fiber-based setups, on the other hand, have more flexibility, especially in clinics. Because environmental factors make the polarization state of light propagating in conventional fibers unpredictable, measurements with multiple-input polarization states are typically required for single-mode fiber-based PS-OCT. Very recently, single-input polarization has been reported in single-mode fiber-based PS-OCT (Trasischker et al., 2014; Ding et al., 2015), in which extensive polarization adjustment and Muller matrix–based theoretical analysis are necessary.

POLARIZATION-MAINTAINING FIBER–BASED MULTI-CONTRAST OCT

To combine the merits of fiber optics and free-space configuration, polarization-maintaining fibers (PMF) have been introduced into PS-LCI (Davé et al., 2003). The PMF has two orthogonal polarization channels on the same fiber core. For instance, when a linearly polarized light travels through a PMF channel, the linear state is preserved due to high isolation between the channels. The PMF-based PS-OCT techniques combine advantages of fiber technology with straightforward analyses of bulk setups. Such PS-OCT implementations have been reported in TD (Al-Qaisi and Akkin, 2008), SD (Götzinger et al., 2009; Wang et al., 2010), and SS technology (Al-Qaisi and Akkin, 2010). By incorporating the Doppler flow measurement (Doppler shift) into the PMF-based PS-OCT setup, a MC-OCT system has been built and applied for brain imaging (Wang et al., 2011).

Here we describe the PMF-based MC-OCT system in SD. The light source is a superluminescence diode with $\lambda_0 = 840$ nm and $\Delta\lambda = 50$ nm, yielding an axial (z-axis) resolution of 5.5 μm in tissue. Light from the source is linearly polarized and coupled into a PMF channel. A 2×2 PM coupler splits the incoming light into the reference and sample arms. In the reference arm, a quarter-wave plate (QWP) is oriented at 22.5° with respect to the incoming polarization state; therefore, light in its return couples back onto both PMF channels equally. In the sample arm, a QWP oriented at 45° ensures circular polarization on the sample whose birefringence changes the polarization into an elliptical state that is detected by PS-OCT. A scan lens (focal length: 36 mm) supports consistent imaging quality over a large area. The lateral (xy-axis) resolution is about 15 μm. Interferometric signals carrying the optical delay gate between reference light and backscattered light from the sample are detected by a specially designed spectrometer, which consists of a grating to disperse spectral components, a Wollaston prism to separate the polarization channels, achromatic collimating and focusing lenses, and a line scan camera. The camera simultaneously acquires the spectra on the two polarization channels at a rate of 25 kHz. The PMF-based PS-OCT observed ghost images that resulted from the cross talk between the orthogonal channels of the polarization components including the coupler. Such ghost images were shifted out of the imaging range with long (30 m) PMFs in the interferometer. The schematic of the optical setup is shown in Figure 10.3.

After wavelength-to-k space remapping (Dorrer et al., 2000) and dispersion compensation (Cense et al., 2004), an inverse Fourier transform was applied to generate a complex depth profile, $A_{1,2}(z)$ $\exp\{i\Phi_{1,2}(z)\}$, where A and Φ denote the amplitude and phase as a function of depth z, respectively, and 1 and 2 correspond to the cross-coupled and the main polarization channels. The imaging contrasts are derived from the amplitudes and phases of the depth profiles: reflectivity, cross-polarization, retardance, and optic axis orientation $\theta(z)$ are computed based on the Jones formalism as follows:

$$R(z) \propto A_1(z)^2 + A_2(z)^2 \tag{10.10}$$

$$C(z) \propto A_1(z)^2 \tag{10.11}$$

$$\delta(z) = \arctan\left(A_1(z)/A_2(z)\right) \tag{10.12}$$

$$\theta(z) = \left(\phi_1(z) - \phi_2(z)\right)/2 \tag{10.13}$$

It is noted that the optic axis orientation in Equation 10.14 is a relative measurement due to an arbitrary phase delay between the orthogonal PMF channels. To obtain the absolute optic axis orientation, an additional calibration step preferably at the time of measurement is needed.

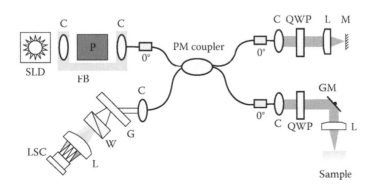

FIGURE 10.3 Schematic of MC-OCT. SLD, superluminescent diode; C, collimator; P, polarizer; QWP, quarter-wave plate; L, lens; M, mirror; GM, galvo mirror; G, grating; W, Wollaston prism; LSC, line scan camera. (Reproduced from Wang, H. et al., *Opt. Lett.*, 35, 154, 2010, Fig. 1a. With permission.)

For *in vivo* or *in vitro* studies, the Doppler flow information is also available with the same measurement. The phase difference between the successive *A*-lines is used to determine the flow velocity (Zhao et al., 2000). Bidirectional flow (ω) under the PS-OCT configuration can be calculated as

$$\omega(z) = \frac{1}{T} \frac{A_1^2(z)\Delta\Phi_1(z) + A_2^2(z)\Delta\Phi_2(z)}{A_1^2(z) + A_2^2(z)} \tag{10.14}$$

where
$$\Delta\Phi_{1,2}(z) = \Phi^{(n)}_{1,2}(z) - \Phi^{(n-1)}_{n-1}(z)$$
n and $n - 1$ denote sequential *A*-lines
T is the time interval between two *A*-lines

BRAIN IMAGING AND MAPPING

Earlier efforts in OCT applications in the central nervous system have been made to inspect the morphological change in developmental neurobiology and pathological conditions (Boppart, 2003). Bizheva et al. (2004) used OCT to image a fixed brain slice of monkey visual cortex and reported scattering coefficients of ~9 and 22 mm^{-1} in the gray matter and white matter, respectively. Using unfixed coronal sections of an adult rat brain, Jeon et al. (2006) reported an average OCT imaging depth of 0.77 mm in gray matter and 0.31 mm in white matter. Light attenuation in gray matter is less than that in white matter, but the actual attenuation coefficients depend on the species and the wavelength of light. The following subsections review the use of polarization-based brain imaging and mapping and the use of OCT techniques in cross-modality studies.

POLARIZATION-SENSITIVE OCT FOR DIFFERENTIATING GRAY MATTER AND WHITE MATTER

Birefringence of nerve structures and the subcomponents has been investigated using a polarization microscope. It was found that the peak retardance of a single microtubule was 0.07 nm and the retardance of a small amount of assembles linearly increased with the number of microtubules in the bundle (Oldenbourg et al., 1998). The radial actin bundles of the nerve growth cones have been visualized on birefringence images (Katoh et al., 1999). In the macroscopic scale, polarized light imaging has been used to image *ex vivo* mammalian and human brain slices (Axer et al., 2001; Larsen et al., 2007). Multiple parameters including the retardance and the fiber axis orientations were recovered from the systematical acquisitions of a rotating polarimeter (Axer et al., 2011).

As the birefringence property is pronounced in myelin fiber tracts, white matter in the brain can be identified by the change of light polarization state. Nakaji et al. (2008) used a bulk-optic PS-OCT setup to characterize the fiber tracts in rat brain slices. The light was horizontally polarized, and passed through a rotating half-wave plate in the sample arm. The half-wave plate alters the direction of light polarization incident on the sample. When the direction is parallel to the axis of the nerve fibers, the polarization states do not change despite the presence of birefringence. Otherwise, change of light polarization is seen on both horizontal and vertical polarization components. In such a way, the fiber tracts were identified and their orientations were matched. This method, however, requires multiple measurements for quantitative evaluation of fiber birefringence and orientation.

Wang et al. (2011) discussed the benefit of using polarization measures in MC-OCT to differentiate the white and gray matter in *ex vivo* rat brain. Figure 10.4a illustrates cross-sectional images of the reflectivity, retardance, and optic axis orientation contrasts. The gray matter of the amygdala region and the white matter of the optic tract and the internal capsule are captured on the cross section and indicated by the pink and yellow bars, respectively, above the MC-OCT images.

The reflectivity in the white matter sees a sharper decay and shallower penetration depth compared to the gray matter. In the retardance image, the signal dramatically increases along depth in the white matter, whereas it remains low in gray matter. The optic axis orientation image provides additional information to locate the fiber tracts and distinguish fiber directions in the white matter.

Identification of the white matter can be complicated due to its axis-dependent optical character-istics (Hebeda et al., 1994). The left panel of Figure 10.4b demonstrates reflectivity profiles (A-lines) for three neighboring brain regions, corresponding to the colored arrows on the cross-sectional images. The internal capsule (indicated by the blue arrow) exhibits the strongest peak signal due to its parallel alignment with respect to the illumination plane. On the other hand, the peak reflectivity of the optic tract (in green), which has a greater inclination angle through the plane, is smaller than that of the adjacent gray matter (in red). Because the optical property of fiber tracts is more close to a specular reflector, backscattering becomes considerably weaker as the direction of fiber alignment is more off the illumination plane. As a result, identification of the gray or white matter based on the back-scattering can be misleading. The retardance measure in MC-OCT, in contrast, provides a more robust differentiation. The right panel of Figure 10.4b shows the retardance curves for the same three regions. The increase of retardance showing the presence of birefringence can be the main indicator of the myelinated neuronal fiber tracts.

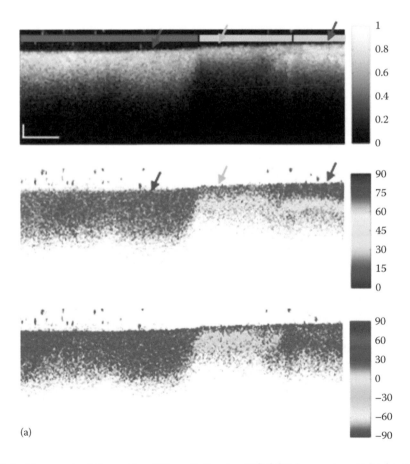

(a)

FIGURE 10.4 Gray and white matter differentiation by MC-OCT in *ex vivo* rat brain. (a) MC-OCT images showing a cross section of the reflectivity, phase retardance, and optic axis orientation contrasts (scale bars. 100 μm axial, 500 μm lateral). White and gray matter regions are indicated by the orange and pink bars above the reflectivity image. The location on the brain is indicated by the dashed line on the microscopy (left). *(Continued)*

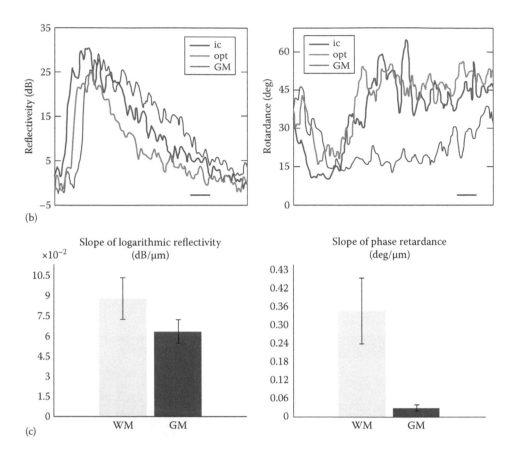

FIGURE 10.4 (*Continued*) Gray and white matter differentiation by MC-OCT in *ex vivo* rat brain. (b) Reflectivity and phase retardance profiles for three regions indicated by the color-matched arrows in (a) (scale bar: 50 μm). (c) Statistics of the attenuation (left) and the birefringence (right) for white (light gray) and gray (dark gray) matter. The error bar represents the standard deviation. (Reproduced from Wang, H. et al., *NeuroImage*, 58, 984, 2011, Fig. 2. With permission.)

The attenuation and birefringence extracted from the *A*-line slopes of reflectivity and retardance, respectively, are shown in Figure 10.4c. The attenuation was estimated to be $(8.48 \pm 1.46) \times 10^{-2}$ dB/μm in white matter and $(6.14 \pm 0.88) \times 10^{-2}$ dB/μm in gray matter. The slope of phase retardance was found $0.33° \pm 0.10°$/μm in white matter and $0.03° \pm 0.01°$/μm in gray matter, representing a birefringence of $(7.78 \pm 0.61) \times 10^{-4}$ and $(0.21 \pm 0.068) \times 10^{-4}$, respectively. Both attenuation and birefringence are significantly higher in the white matter. However, the differentiation is more pronounced in the birefringence measure. Thus, it is proved that birefringence contrast provides a more reliable means for targeting the myelinated fiber tracts in the brain.

MULTI-CONTRAST OCT FOR BRAIN ARCHITECTURE AND NEURONAL FIBER MAPS

Extracted from a single optical scan, the contrasts of MC-OCT provide different perspectives of the brain structure. These are elegantly depicted on the *en face* images. The *en face* image projects the volumetric data of an OCT scan (optical section) onto the *xy*-plane, thereby unveiling the surface and subsurface features on the 2D plane. It resembles the view of block-face microscopy and aids reading and interpretation of the images. Figure 10.5 illustrates an *en face* view of reflectivity (right) generated from the corresponding images of an optical section (left).

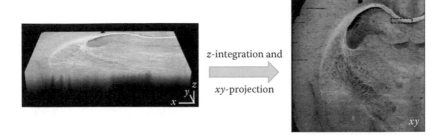

z-integration and

xy-projection

FIGURE 10.5 Production of the *en face* image (right) from a volumetric data (left) of an optical section in OCT.

The *en face* images in MC-OCT can be produced by the contrasts of reflectivity, attenuation, retardance, optic axis orientation, etc. The pixel intensity $\bar{I}(i, j)$ of the *en face* reflectivity and retardance images is computed by taking the mean value on corresponding A-lines, given by

$$\bar{I}(i, j) = \frac{1}{N_{i,j}} \sum_{k=1}^{N_{i,j}} S(i, j, k),$$ (10.15)

where
$S(i, j, k)$ represents a 3D dataset for reflectivity or retardance
i and j are the indices for lateral coordinates and k is the index for axial direction
$N_{i,j}$ is the number of elements included in the *en face* computation

The pixel intensity of *en face* attenuation map is the slope of the corresponding reflectivity profile. A linear fit is applied on each reflectivity A-line (in dB) to derive the attenuation. The pixel intensity of *en face* optic axis orientation image is determined by the peak of a histogram for the orientation values along imaging depth.

Figure 10.6 displays *en face* images of two coronal sections in a rat brain. The *en face* reflectivity distinguishes the gross structures with altered intensity and keeps track of the small features (a, d). The small fiber tracts are clearly visualized on the magnified regions of interests (ROIs) (i, ii, and iv) after applying local contrast enhancement. As the reflective intensity of the white matter relies on the orientation of fiber tracts, variation of the signal brightness exposes the complex patterns of fiber architecture in 3D space. The *en face* retardance map targets the white matter and yields an integral delineation of the fiber routes (b, e). The white matter consistently presents a positive contrast over the gray matter, which alleviates the dependence on fiber orientation in most conditions (except for fibers running perpendicular to the *en face* plane). The *en face* attenuation image supports the subdivision of local structures, thus facilitating our understanding of the brain architectures and the functional implications (c, f). The cortical layers are visible on the attenuation maps. The layers of somatosensory cortex distinguished by the MC-OCT (ROI in c) are well correlated with the cresyl violet staining image (iii). The attenuation image also provides the capability to identify the white matter, most of which highly correlates with the retardance map.

The optic axis orientation in MC-OCT offers a quantitative assessment of fiber orientations in the *xy*-plane (Wang et al., 2014a). Figure 10.7 demonstrates the *en face* orientation map (left and middle) of a sagittal section in a rat brain. The orientation values are color-coded according to the color wheel, and the brightness of the map is determined by the retardance. The anatomical image with reflectivity contrast is shown as well (right). The orientations of neuronal tracts in general agree with the geometry and features of the fibers, as revealed by the reflectivity image. Multiple colors within a fiber bundle indicate its altered direction in space, such as the big bundle of stria medullaris of thalamus (white arrow) and the small tracts at the tail of the corpus callosum (rectangular box). The orientation map also unveils the fiber groups with distinct orientations, which are hardly differentiable in the anatomical images, such as groups of fibers with distinct axes in the thalamic region (round ROI, also shown on the magnified image).

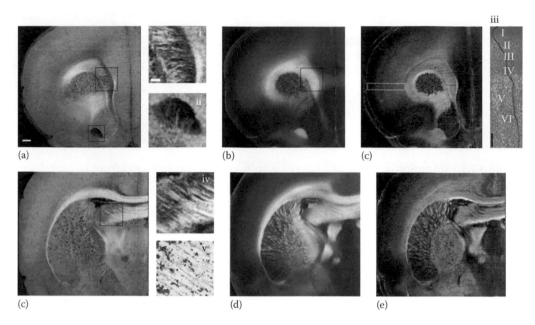

FIGURE 10.6 *En face* images of reflectivity (a, d), retardance (b, e), and attenuation (c, f) for two coronal sections. Fiber directions in iv are compared with the myelin stain in v (objective: 40×). The attenuation profile of the ROI in (c) was plotted against the cortical layers and displayed on top of the cresyl violet stain (objective: 10×). Scale bars: 500 μm for coronal sections (a–f), 200 μm for ROIs (i, ii, iii, and iv), and 30 μm for histology (v). (Reproduced from Wang, H. et al., *NeuroImage*, 84, 1007, 2014a, Fig. 2. With permission.)

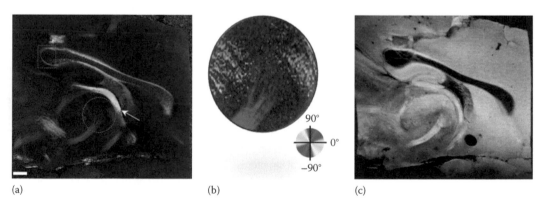

FIGURE 10.7 *En face* orientation maps of sagittal sections show the in-plane orientation of fiber tracts. (a) The value of orientation is color-coded in hue-saturation-value (HSV) space according to the color wheel, and the brightness is controlled by the retardance. The rounded ROI is magnified in (b), showing distinct fiber groups with altered orientation. The associated anatomy (reflectivity image) is shown in (c).

Depth-resolved description of neuronal fiber tracts can be achieved as well, owing to the 3D imaging capability of the MC-OCT. The birefringence contrast provides such a mechanism. As retardance is linearly cumulating as polarized light propagates in a birefringent tissue, birefringence Δn can be characterized by the derivative of phase retardance $\delta(z)$ along the light travel direction z as

$$\Delta n(z) = (\lambda/2\pi)\frac{d\delta(z)}{dz} \qquad (10.16)$$

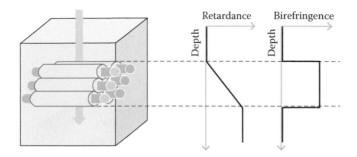

FIGURE 10.8   The relationship of phase retardance and birefringence as polarized light propagates in the nerve fibers.

It should be noted that the measured retardance is also dependent on the inclination angle between the tissue axis and the wave oscillation plane (Larsen et al., 2007). Therefore, the previous birefringence is the apparent birefringence.

The relationship of phase retardance and birefringence is illustrated in Figure 10.8. Assuming a fiber bundle is located inside an isotropic medium, the retardance increases along the depth while light meets the fiber bundle; otherwise, the retardance keeps unchanged. As a result, a birefringence map produced by the derivative of a cross-sectional retardance image indicates the localization of nerve fibers both laterally and axially. Fiber description in 3D is achieved by stacking all the cross-sectional birefringence images of an optical section (Wang et al., 2011). Figure 10.9 illustrates the 3D fiber maps for a brain volume of $6 \times 6 \times 0.45$ mm³. Two types of information are encoded in the map: the birefringence indicating fiber localization and the in-plane orientation of fiber axis. The nerve fiber tracts are identified by their spatial connections and color. One issue in creating the birefringence image is that the speckle noise inherently in coherence imaging needs to be eliminated before the local derivatives can be taken. The nonlinear anisotropic diffusion filter (Perona and Malik, 1990; Catté et al., 1992) appears to be one of the effective tools for noise reduction on the retardance images of PS-OCT (Stifter et al., 2010).

SERIAL OPTICAL COHERENCE SCANNING FOR LARGE-SCALE BRAIN MAPPING

The introduction of block-face imaging in microscopy (Odgaard et al., 1990) has leveraged an opportunity for large-scale reconstruction of biological tissues. During imaging, the entire tissue block was mounted still and thin slices were removed between consecutive scans. Thereby, the procedure yields automatically aligned image stacks of the entire specimen. The block-face imaging has been adopted in electron microscopy (Denk and Horstmann, 2004), confocal microscopy (Sands et al., 2005), and two-photon microscopy (Tsai et al., 2003; Ragan et al., 2012) with the purpose to realize comprehensive synaptic connections or axonal networks in 3D space. Wang et al. (2014a) proposed a serial optical coherence scanner (SOCS) for large-scale reconstruction of biological tissues at high resolution. SOCS integrates a vibratome slicer into a MC-OCT system, which allows scan and section of the entire brain without any sample movements. The optical system is the PMF-based MC-OCT described in the technique section, which has an axial resolution (z) of 5.5 μm and a lateral resolution (xy) of ~15 μm. A vibratome slicer is positioned under the sample path optics of the MC-OCT. The brain is mounted on the slicer and kept stationary during the whole scans. One volumetric scan (optical section) contains 300 cross-sectional frames (B-line) with 1000 A-lines in each frame, covering a field of view of $7 \times 7 \times 1.78$ mm³ (xyz), with a voxel size of $7 \times 23 \times 3.47$ μm³. After the scan of one optical section, a thin slice is cut off, allowing deeper regions to be imaged. The procedure is repeated until the whole sample block is imaged. The thickness of a slice is less than the penetration depth of light, and it is optimally determined

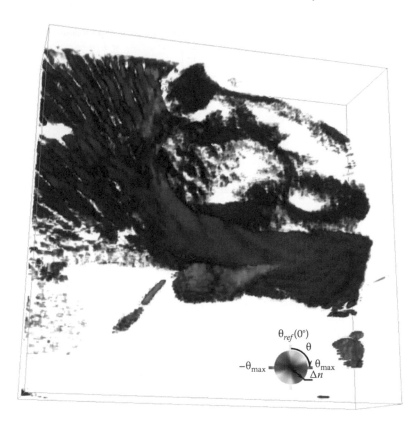

FIGURE 10.9 3D fiber maps of a rat brain section (volume: 6 × 6 × 0.45 mm³). Nerve fibers are tracked with their in-plane orientation θ indicated by the colors on the color wheel. The brightness of colors is controlled by the birefringence Δ*n*. (Reprinted from Wang, H. et al., *NeuroImage*, 58, 984, 2011, Fig. 6. With permission.)

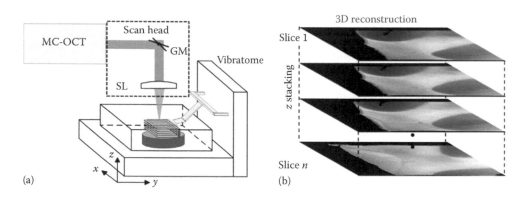

FIGURE 10.10 (a) SOCS schematic diagram. GM, galvo mirrors; SL, scan lens. (b) 3D reconstruction from *en face* images. Individual slices are stacked in z to form a 3D representation of the sample.

to ensure satisfactory SNR for high-quality 3D reconstruction of the entire sample. Schematic diagram of SOCS is demonstrated in Figure 10.10a.

Three-dimensional reconstruction of the entire sample can be realized by orderly stacking the *en face* images generated from each optical scan, without the necessity of additional registration between slices. The voxel size in z-axis is limited by the thickness of physical sections. Such constructions facilitate the quick identification and global assessment of brain structures at

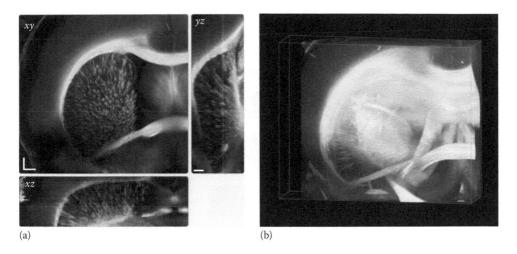

(a)   (b)

FIGURE 10.11 Global description of 3D neuronal roadmaps in a rat brain by stacking the *en face* retardance images of the serial scans. (a) Orthogonal viewing planes: *xy*, coronal, *xz*, horizontal; and *yz*, sagittal. Scale bars: *xy*, 500 µm, and *z*, 1 mm. (b) Volume rendering of the 3D dataset (volume size: $7 \times 7 \times 5.5$ mm³). (Reprinted from Wang, H. et al., *NeuroImage*, 84, 1007, 2014a, Fig. 4. With permission.)

mesoscopic resolution. For macroscopic brains that are remarkably larger than the field of view that one optical scan covers, multiple scans are necessary for one slice, and fusion of these lateral scans is a prerequisite before 3D stacking. The process of the reconstruction is illustrated in Figure 10.10b.

A stack of *en face* retardance images was shown in Figure 10.11 to depict the global fiber roadmap. Geometry and spatial organizations of neural pathways are captured on the orthogonal viewing planes (Figure 10.11a). Maximum intensity projection of volume rendering provides a perspective view and discloses the configurations of the white matter in the brain (Figure 10.11a).

The large-scale reconstruction can be accomplished at higher resolution by z-stacking the cross sections, which, due to the depth-resolved capability, unveil microstructures that are not exhibited by the *en face* images. The cross-polarization contrast has been utilized in the work of Wang et al. (2014a) to connect the serial sections at the natural resolution of SOCS ($15 \times 15 \times 5.5$ µm³). The advantage of using cross-polarized light is that it highlights the boundaries of nerve fibers and also preserves decent signal intensity in deeper regions. To achieve smooth transition between serial scans, algorithms minimizing the intensity mismatch across stitching borders are desired. One method was to multiply each A-line with the following regularization function $L(z)$:

$$L(z) = \begin{cases} \exp\left(-\dfrac{2z-N}{N}\ln\left(\dfrac{a_i + b_{i-1}}{2a_i}\right)\right), & z \le N/2 \\[3mm] \exp\left(-\dfrac{2z-N}{N}\ln\left(\dfrac{a_{i+1} + b_i}{2b_i}\right)\right), & z > Z/2 \end{cases}$$ (10.17)

where
   N is the number of points in each A-line
   i is the slice index
   a_i and b_i are local averages of intensities for the start and the end of A-lines in slice i

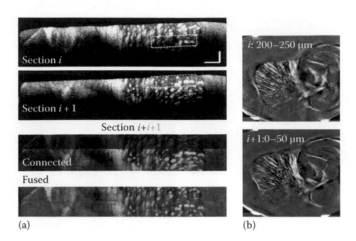

FIGURE 10.12 (a) Stitching cross sections of consecutive volume scans (i and $i + 1$). Same regions are covered in both scans, as exampled by the yellow boxes. The connected image is color-coded (i in red, and $i + 1$ in green), and fused image, after the removal of the depth-dependent trend, is in gray scale. Scale bars: horizontal, 500 μm, and vertical, 150 μm. (b) *En face* cross-polarization images from the overlap region for the two scans. Scale bar: 500 μm.

Stitching consecutive volume scans requires careful alignment on z-axis. Figure 10.12a shows two cross sections from consecutive volume scans (section i and $i + 1$). As the imaging depth is greater than the thickness of one slice, overlapping brain regions are seen on two adjacent slices (indicated by the rectangular boxes). The connective location of the lower cross section onto the upper one was determined by the thickness of the slice. The landmarks in the overlap are used to examine the effectiveness of stitching. After connected along the z-axis, the color-blended image of the two scans (in red and green) demonstrates a decent alignment of the overlapping fiber tracts (in yellow), and the fused image after removing the depth-dependent trend of intensity exhibits a smooth transition. The results suggest that mechanical distortion caused by slicing is negligible and additional registration is not in need. The continuity of fiber tracts across the stitching border demonstrates the feasibility of tracing long axonal fiber bundles in the brain. The *en face* images from the overlap region of scan i and $i + 1$ were produced (Figure 10.12b). As expected, the two images exhibit high similarity, despite a slight difference in the lateral resolution due to a limited depth of focus.

Reconstruction of the rat brain at the natural resolution of SOCS is visualized on orthogonal planes (Figure 10.13). The horizontal and coronal views compose of stitched cross sections owning a resolution of 5.5×15 μm^2, and the sagittal view denotes the *en face* plane with an isotropic resolution of 15 μm. Trajectories of small fibers are recognized on the high-resolution images. The tiny dots on the horizontal plane (arrow) represent the small fiber tracts in the midbrain region, which are estimated to have an average size of approximately 23 μm. Owing to a dependence of light intensity on the fiber orientation, it is noticed that fiber identification in certain range of angles (rectangular box) are weak or indistinctive from the surrounding tissue. This challenge needs to be overcome in the future for a full range reconstruction of high-resolution connectivity maps.

CROSS-MODALITY APPLICATIONS

The ability of OCT to image micro- and macroscale characteristics in normal and pathological brain offers a new direction for architectural brain mapping; however, its performance needs to be related and compared with other neuroanatomical methods. In this section, we review the cross-validations of OCT with histology and diffusion MRI, and discuss multimodality imaging at the end.

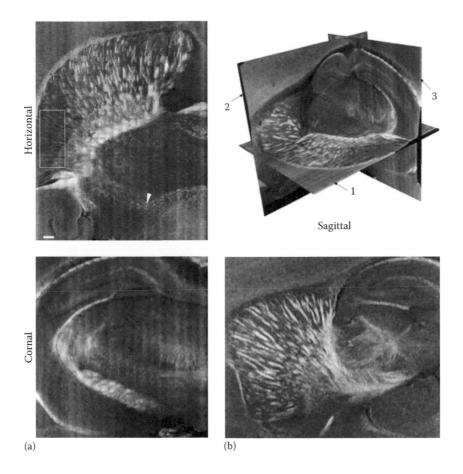

FIGURE 10.13 Reconstruction of rat brain at natural resolution of SOCS. (a) Orthogonal views with three representative planes. (b) Visualization of 3D reconstruction: 1, horizontal; 2, sagittal; and 3, coronal.

OCT and Histology

Traditional histology has provided the ground truth for the identification of structural neuroimaging. The resolution of OCT has been improved both axially and laterally to generate images at a scale that is fully comparable with histology. The resulting techniques are the so-called ultrahigh-resolution OCT (UHR-OCT) and optical coherence microscopy (OCM), respectively. UHR-OCT/OCM have proved to be beneficial in revealing the detailed information of peripheral nerve fibers and cortical organizations in living rodents (Arous et al., 2011; Srinivasan et al., 2012; Leahy et al., 2013). Srinivasan et al. (2012) used OCM with a dynamic focusing scheme to create images that unveil the cortical myelination and neuronal cell bodies and macro-vasculature networks up to a depth of 1.3 mm in *in vivo* rat cortex. Consistent with the previous findings, the OCM images illustrated considerably an increase of neuronal density from cortical layer I to layer II/III, and the presence of myelin fibers in layers I and IV–VI.

Magnain et al. (2014) utilized OCM in SD to study the laminar structure of the isocortex in *ex vivo* human brain and compared cortical cytoarchitecture with the Nissl staining. The OCM system bears an axial resolution of 3.5 μm and a lateral resolution of 3 μm with a depth of focus of about 30 μm. The field of view in one scan covers a region of 1.5×1.5 mm². Figure 10.14 shows the block-face image, the Nissl stain, and the OCT images of an isocortex human brain. The OCT and Nissl stain images showed good agreement in cortical laminar organization. The registration of SD-OCT

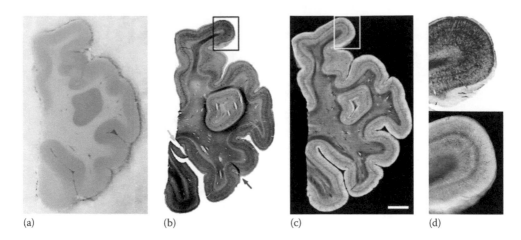

(a)   (b)   (c)   (d)

FIGURE 10.14 *Ex vivo* imaging of the isocortex in *ex vivo* human brain. (a) Block-face image, (b) Nissl stain, (c) OCT images. (d) Comparison of contrast on the boxed region (top, Nissl, and bottom, OCT). Scale bar: 5 mm. (Reprinted from Magnain, C. et al., *NeuroImage*, 84, 524, 2014, Fig. 3. With permission.)

and histology images, respectively, with block-face images indicated that SD-OCT-to-block-face registration was significantly more accurate than the other. The thickness of cortical layers was estimated. Generally, the thicknesses of cortical layers measured by OCT and Nissl stain were generally in good accordance. The mean Pearson's correlation for the two modalities resulted in 0.84 ± 0.16, with a very small p-value. As opposed to histological processing, which routinely induces distortions and damages to biological slices (red and green arrows in Figure 10.14b), one advantage of using OCM to study the cortical layers is that the optical scan preserves the integrity of the sample, hence facilitating the reconstruction of tissue blocks.

The close correspondence of OCT images and the histology stain have created intriguing opportunities for brain tumor studies. Especially, the UHR-OCT techniques overcome the limitation that only morphological changes could be observed in brain pathologies (Boppart et al., 1998b; Böhringer et al., 2009) without detailed features of tumorous cells. With an improved axial and lateral resolution, Bizheva et al. (2005) demonstrated the UHR-OCT images of a tumorous brain that were well correlated with histological staining. The identification of malignant structures was characterized by the presence of altered vessel architecture, microcalcifications, and cysts in the tumorous tissue.

The full-field (FF) OCT provides another approach to reach ultrahigh resolution in 3D imaging. Unlike the conventional OCT, FF-OCT is a wide-field imaging technique that equips a spatially incoherent light source illuminating the whole aperture of a microscope objective and an area scan camera for data acquisition (Dubois et al., 2004). The FF-OCT directly creates *en face* images analogous to the histological view, thereby facilitating pathologist's reading and interpretation. The system yields a resolution of $1.5 \times 1.5 \times 1 \ \mu m^3$ ($X \times Y \times Z$) on brain samples down to depths of approximately 200–300 μm. Using a FF-OCT, Assayag et al. (2013) imaged *ex vivo* tumorous and nontumorous tissues of the human brain. The FF-OCT images clearly identified a subpopulation of neuronal cell bodies, the myelinated axon bundles, and vasculature in the human epileptic brain and cerebellum, which were decently correlated with histological staining. Meningiomas and hemangiopericytoma, two types of meningeal tumors, showed distinctive features on the FF-OCT images. However, imposing challenges exist for the identification of gliomas, a highly aggressive and life-threatening brain cancer, due to lack of contrast for individual glial cells such as astrocytes or oligodendrocytes. The low-grade gliomas, which were identified by immunohistochemical staining, were indiscernible from normal brain tissues, whereas the high-grade gliomas disrupt the normal parenchyma architecture and the corresponding morphology changes could be observed on the FF-OCT images.

Multi-Contrast OCT and Diffusion MRI

Owing to the ability of large-scale reconstruction and the specialty in white matter imaging, SOCS can be linked to the diffusion MRI (dMRI) technique. dMRI captures the diffusion of water molecules in the brain, which run preferentially along the axonal axes, and hence indirectly describes the fiber organization and orientation. The technique provides a unique solution to noninvasively visualize white matter fiber bundles in the living human brain at a resolution of millimeter to submillimeter scale. Wang et al. (2014b) reported a cross-validation study of SOCS and diffusion tensor imaging (DTI) in human medulla oblongata. The sample was scanned by DTI, followed by SOCS, and then the 3D images generated by the two scanners were coregistered for comparisons. Figure 10.15 shows the coregistered SOCS and DTI images in two slices. The top two rows display the fractional anisotropy (FA) images of DTI and the retardance images of SOCS, both of which highlight the white matter bundles. The fiber maps exhibit remarkable similarities with a Pearson's correlation coefficient of 0.90. The microscopic resolution of SOCS enables a closer inspection of the neuroanatomical factors affecting the FA values. In general, the relationship of FA and retardance follows a linear trend with a positive slope. Greater anisotropy in DTI correlates with higher retardance observed within densely packed fiber bundles. Lower anisotropy might be explained by less density, complex geometry, and spatial interactions of the fiber tracts, which are delineated on the retardance image. However, inconsistency of high FA and low retardance are also found in certain regions. Reduction of the retardance can be attributed to a low birefringence or a large inclination angle of the fibers with respect to the xy-plane.

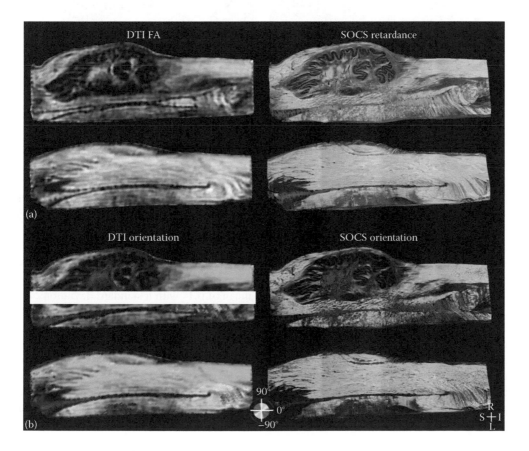

FIGURE 10.15 Coregistered DTI and SOCS images. (a) Representative FA (left) and retardance (right) images on the xy-plane are shown for two slices. (b) Their corresponding orientation maps. The color wheel indicates the orientations of fiber tracts. The brightness on SOCS and DTI maps is controlled by retardance and FA values, respectively.

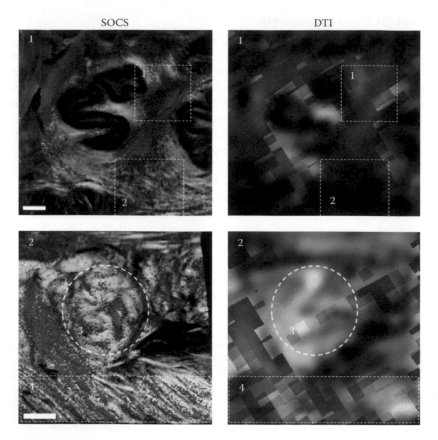

FIGURE 10.16 Crossing fibers on SOCS and DTI orientation maps. Complicated fiber patterns are revealed by SOCS, but not DIT. Scale bar: 500 μm.

The quantitative assessment of fiber orientation in SOCS enables a direct comparison with DTI orientation. The bottom two rows in Figure 10.15 demonstrate the coregistered orientation maps for the same medulla slices. The DTI orientation vectors were projected onto SOCS *en face* plane. The orientation maps share the same color-coding given in the color wheel. The brightness of colors on SOCS and DTI images was controlled by the retardance and FA values, respectively. The fiber orientations exhibit a substantial agreement between the coregistered images. Statistical analysis of pixel-wise orientation mismatch in the white matter resulted in a mean of 5.4° with a standard deviation of 32.5°. The mismatch is further reduced to a mean of 0.1° with a standard deviation of 22.6° when the thresholds of FA and retardance are elevated to only include the clearly defined fibers.

Despite the positive correlations of DTI and SOCS measures, complications stand out for the regions holding intricate fiber patterns. Figure 10.16 contains small fiber tracts less than 100 μm in diameter. The superior resolution of SOCS supports the description of individual fiber traces accompanied by altered orientations, whereas pixelated color patches are observed on DTI orientation images without interpretable structures (ROI-1, -2, -3). Although the low-resolution estimation preserves the fidelity of the dominant orientation in parallel aligned fibers (ROI-4), it loses the indication of crossing features under the complex geometries (ROI-3).

Multimodality Imaging

OCT has pronounced advantage of creating high-resolution reconstruction of tissues using intrinsic contrasts, but it does not necessarily provide cellular specificity. This deficiency can be compensated by multimodality imaging. For instance, UHR-OCT/OCM have been combined with single-photon and multi-photon microscopy for various applications (Podoleanu et al., 2004; Gaertner et al., 2012; Iftimia et al., 2013), owing to their shared merits of 3D capability and similar resolutions. Bolmont et al. (2012) have used an extended focus Fourier domain OCM (xfOCM) together with a confocal fluorescence channel to image the cerebral β-amyloidosis both *ex vivo* and in a living Alzheimeric mouse. Individual Aβ plaques in Alzheimeric brain parenchyma and vasculature were detected on xfOCM images and confirmed on biomarker-labeled confocal images. *In vivo* imaging of cerebral amyloidosis over 1 month illustrated the evolution of existing amyloid plaques and the appearance of new deposits. Another study reported a combined Doppler OCT and confocal lifetime microscopy system to investigate the cerebral oxygen delivery and metabolism in anesthetized rats (Yaseen et al., 2011). Therefore, multimodal systems have the potential to lead a more comprehensive understanding of structural alterations and metabolic dynamics in various neurovascular diseases (Yu et al., 2010; Ren et al., 2012).

OCT/OCM and multiphoton microscopy can be integrated by combining their sample arms (Jeong et al., 2011); however, the spectrum of OCT light source and the excitation wavelength used in multiphoton microscopy may be overlapping. In that case, an ultrafast pulsed laser can serve as a common light source for multimodal imaging (Joo et al., 2007; Chong et al., 2013). The integrated OCM-multiphoton system has been used in clinical dermatologic applications (Alex et al., 2013), and its application on brain studies may be possible in future.

CONCLUSION AND OUTLOOK

The use of OCT techniques to study the brain is still at the early stage. The MC-OCT delineates the brain structures and connective networks without preselecting neuronal types. This capability accelerates our understanding of the whole-brain aspects at high resolution. With the development of an automated serial scanner and routine processing engines, data acquisition of an entire mouse or rat brain can be done in several hours and reconstruction of the whole brain may take a few days. Full reconstruction of primate and human brains is more challenging due to large volume; but it is not impossible with standardized procedures in the future. Further technical developments are needed for a more accurate representation of nerve fibers, whose through-plane orientation (inclination angle) affects the optical signals. At the other side of scalability, the positive outcome of using OCT to distinguish local structures in the brain has led to plenty of inventions of intraoperative guiding probes and developments of image-guided neurosurgical procedures. Further miniaturization of the optic probe and standardized fabrication will expedite the pace toward real clinical usage.

Cross-modality investigations pivoted by MC-OCT have just risen above the horizons. This cross-scale bridge has opened up numerous opportunities to create high-dimensional data with unprecedented perspectives as well as challenges and uncertainties. As the spatial coverage of serial MC-OCT is improved with acquisition speed, comprehensive studies by dMRI and MC-OCT, which complement each other in terms of resolution and imaging depth, can be conducted on macroscopic brains, such as human and nonhuman primates. Moreover, the same sample can be reused for histology, fluorescence, immunohistochemistry, and other molecular labeling techniques to construct a higher-dimensional brain database. The sample imaged by OCT may also be used for higher resolution imaging with targeted regions, such as the synaptic junctions revealed by electron microscopy. Therefore, hierarchical architecture maps linking the macro- and microscales could be created to serve as structural constraints for the functional investigations. Potential bottlenecks include handling and managing large datasets and translating information between the imaging modalities.

OCT imaging on the brain for studying neurological diseases opens vast opportunities in future. The neuronal contents such as degraded myelination and modified axonal networks may be revealed by the MC-OCT in animal models. This could provide pathological implications for the study of disease progressions in multiple sclerosis, Parkinson's, schizophrenia, and other neurological disorders. As the MC-OCT techniques for systematical brain imaging and mapping continue to develop, we expect a greater impact of this technology to be seen in broad neuroscience and brain research communities.

ACKNOWLEDGMENTS

The authors would like to acknowledge supports from the Institute for Engineering in Medicine Seed Grant program and the Doctoral Dissertation Fellowship at the University of Minnesota, and the National Institute of Health (R01EB012538).

REFERENCES

Akkin, T., Landowne, D., and Sivaprakasam, A. (2010). Detection of neural action potentials using optical coherence tomography: Intensity and phase measurements with and without dyes. *Front. Neuroenergetics* 2, 22.

Alex, A., Weingast, J., Weinigel, M., Kellner-Höfer, M., Nemecek, R., Binder, M., Pehamberger, H., König, K., and Drexler, W. (2013). Three-dimensional multiphoton/optical coherence tomography for diagnostic applications in dermatology. *J. Biophotonics* 6, 352–362.

Al-Qaisi, M.K. and Akkin, T. (2008). Polarization-sensitive optical coherence tomography based on polarization-maintaining fibers and frequency multiplexing. *Opt. Express* 16, 13032.

Al-Qaisi, M.K. and Akkin, T. (2010). Swept-source polarization-sensitive optical coherence tomography based on polarization-maintaining fiber. *Opt. Express* 18, 3392–3403.

Arous, B.J., Binding, J., Léger, J.-F., Casado, M., Topilko, P., Gigan, S., Claude Boccara, A., and Bourdieu, L. (2011). Single myelin fiber imaging in living rodents without labeling by deep optical coherence microscopy. *J. Biomed. Opt.* 16, 116012–1160129.

Assayag, O., Grieve, K., Devaux, B., Harms, F., Pallud, J., Chretien, F., Boccara, C., and Varlet, P. (2013). Imaging of non-tumorous and tumorous human brain tissues with full-field optical coherence tomography. *NeuroImage Clin.* 2, 549–557.

Axer, H., Axer, M., Krings, T., and Keyserlingk, D.G. (2001). Quantitative estimation of 3-D fiber course in gross histological sections of the human brain using polarized light. *J. Neurosci. Methods* 105, 121–131.

Axer, M., Amunts, K., Grässel, D., Palm, C., Dammers, J., Axer, H., Pietrzyk, U., and Zilles, K. (2011). A novel approach to the human connectome: Ultra-high resolution mapping of fiber tracts in the brain. *Neuroimage* 54, 1091–1101.

Bertrand, N., Castro, D.S., and Guillemot, F. (2002). Proneural genes and the specification of neural cell types. *Nat. Rev. Neurosci.* 3, 517–530.

Bizheva, K., Unterhuber, A., Hermann, B., PovazË, B., Sattmann, H., Drexler, W., Stingl, A., Le, T., Mei, M., Holzwarth, R. et al. (2004). Imaging ex vivo and in vitro brain morphology in animal models with ultra-high resolution optical coherence tomography. *J. Biomed. Opt.* 9, 719–724.

Bizheva, K., Unterhuber, A., Hermann, B., PovazË, B., Sattmann, H., Fercher, A.F., Drexler, W., Preusser, M., Budka, H., Stingl, A., et al. (2005). Imaging ex vivo healthy and pathological human brain tissue with ultra-high-resolution optical coherence tomography. *J. Biomed. Opt.* 10, 011006–0110067.

Böhringer, H.J., Lankenau, E., Stellmacher, F., Reusche, E., Hüttmann, G., and Giese, A. (2009). Imaging of human brain tumor tissue by near-infrared laser coherence tomography. *Acta Neurochir. (Wien)* 151, 507–517.

Bolmont, T., Bouwens, A., Pache, C., Dimitrov, M., Berclaz, C., Villiger, M., Wegenast-Braun, B.M., Lasser, T., and Fraering, P.C. (2012). Label-free imaging of cerebral β-amyloidosis with extended-focus optical coherence microscopy. *J. Neurosci.* 32, 14548–14556.

Boppart, S.A. (2003). Optical coherence tomography: Technology and applications for neuroimaging. *Psychophysiology* 40, 529–541.

Boppart, S.A., Bouma, B.E., Brezinski, M.E., Tearney, G.J., and Fujimoto, J.G. (1996b). Imaging developing neural morphology using optical coherence tomography. *J. Neurosci. Methods* 70, 65–72.

Boppart, S.A., Bouma, B.E., Pitris, C., Tearney, G.J., Southern, J.F., Brezinski, M.E., and Fujimoto, J.G. (1998a). Intraoperative assessment of microsurgery with three-dimensional optical coherence tomography. *Radiology* 208, 81–86.

Boppart, S.A., Brezinski, M.E., Bouma, B.E., Tearney, G.J., and Fujimoto, J.G. (1996a). Investigation of developing embryonic morphology using optical coherence tomography. *Dev. Biol.* 177, 54–63.

Boppart, S.A., Brezinski, M.E., Pitris, C., and Fujimoto, J.G. (1998b). Optical coherence tomography for neurosurgical imaging of human intracortical melanoma. *Neurosurgery* 43, 834–841.

Catté, F., Lions, P., Morel, J., and Coll, T. (1992). Image selective smoothing and edge detection by nonlinear diffusion. *SIAM J. Numer. Anal.* 29, 182–193.

Cense, B., Nassif, N.A., Chen, T.C., Pierce, M.C., Yun, S.-H., Park, B.H., Bouma, B.E., Tearney, G.J., and de Boer, J.F. (2004). Ultrahigh-resolution high-speed retinal imaging using spectral-domain optical coherence tomography. *Opt. Express 12*, 2435.

Chen, Z., Milner, T.E., Dave, D., and Nelson, J.S. (1997). Optical Doppler tomographic imaging of fluid flow velocity in highly scattering media. *Opt. Lett. 22*, 64.

Chinn, S.R., Swanson, E.A., and Fujimoto, J.G. (1997). Optical coherence tomography using a frequency-tunable optical source. *Opt. Lett. 22*, 340.

Choma, M., Sarunic, M., Yang, C., and Izatt, J. (2003). Sensitivity advantage of swept source and Fourier domain optical coherence tomography. *Opt. Express 11*, 2183.

Chong, S.P., Lai, T., Zhou, Y., and Tang, S. (2013). Tri-modal microscopy with multiphoton and optical coherence microscopy/tomography for multi-scale and multi-contrast imaging. *Biomed. Opt. Express 4*, 1584.

Davé, D.P., Akkin, T., and Milner, T.E. (2003). Polarization-maintaining fiber-based optical low-coherence reflectometer for characterization and ranging of birefringence. *Opt. Lett. 28*, 1775–1777.

de Boer, J., Srinivas, S., Malekafzali, A., Chen, Z., and Nelson, J. (1998). Imaging thermally damaged tissue by polarization sensitive optical coherence tomography. *Opt. Express 3*, 212–218.

de Boer, J.F. and Milner, T.E. (2002). Review of polarization sensitive optical coherence tomography and Stokes vector determination. *J. Biomed. Opt. 7*, 359–371.

de Boer, J.F., Cense, B., Park, B.H., Pierce, M.C., Tearney, G.J., and Bouma, B.E. (2003). Improved signal-to-noise ratio in spectral-domain compared with time-domain optical coherence tomography. *Opt. Lett. 28*, 2067–2069.

de Boer, J.F., Milner, T.E., and Nelson, J.S. (1999). Determination of the depth-resolved Stokes parameters of light backscattered from turbid media by use of polarization-sensitive optical coherence tomography. *Opt. Lett. 24*, 300.

de Boer, J.F., Milner, T.E., van Gemert, M.J., and Nelson, J.S. (1997). Two-dimensional birefringence imaging in biological tissue by polarization-sensitive optical coherence tomography. *Opt. Lett. 22*, 934–936.

Denk, W. and Horstmann, H. (2004). Serial block-face scanning electron microscopy to reconstruct three-dimensional tissue nanostructure. *PLoS Biol. 2*, e329.

Ding, Z., Liang, C.-P., Tang, Q., and Chen, Y. (2015). Quantitative single-mode fiber based PS-OCT with single input polarization state using Mueller matrix. *Biomed. Opt. Express 6*, 1828–1843.

Dorrer, C., Belabas, N., Likforman, J.-P., and Joffre, M. (2000). Spectral resolution and sampling issues in Fourier-transform spectral interferometry. *JOSA B 17*, 1795–1802.

Dubois, A., Grieve, K., Moneron, G., Lecaque, R., Vabre, L., and Boccara, C. (2004). Ultrahigh-resolution full-field optical coherence tomography. *Appl. Opt. 43*, 2874–2883.

Everett, M.J., Schoenenberger, K., Colston, B.W., and Da Silva, L.B. (1998). Birefringence characterization of biological tissue by use of optical coherence tomography. *Opt. Lett. 23*, 228–230.

Fercher, A.F., Hitzenberger, C.K., Kamp, G., and El-Zaiat, S.Y. (1995). Measurement of intraocular distances by backscattering spectral interferometry. *Opt. Commun. 117*, 43–48.

Frohman, E.M., Fujimoto, J.G., Frohman, T.C., Calabresi, P.A., Cutter, G., and Balcer, L.J. (2008). Optical coherence tomography: A window into the mechanisms of multiple sclerosis. *Nat. Clin. Pract. Neurol. 4*, 664–675.

Gaertner, M., Cimalla, P., Meissner, S., Kuebler, W.M., and Koch, E. (2012). Three-dimensional simultaneous optical coherence tomography and confocal fluorescence microscopy for investigation of lung tissue. *J. Biomed. Opt. 17*, 071310–071310.

Gil, J.J. and Bernabeu, E. (1987). Obtainment of the polarizing and retardation parameters of a non-depolarizing optical system from the polar decomposition of its Mueller matrix. *Optik 76*, 67–71.

Golubovic, B., Bouma, B.E., Tearney, G.J., and Fujimoto, J.G. (1997). Optical frequency-domain reflectometry using rapid wavelength tuning of a Cr_4^+:forsterite laser. *Opt. Lett. 22*, 1704.

Götzinger, E., Baumann, B., Pircher, M., and Hitzenberger, C.K. (2009). Polarization maintaining fiber based ultra-high resolution spectral domain polarization sensitive optical coherence tomography. *Opt. Express 17*, 22704.

Häusler, G. and Lindner, M.W. (1998). "Coherence radar" and "spectral radar"—New tools for dermatological diagnosis. *J. Biomed. Opt. 3*, 21–31.

Hebeda, K.M., Menovsky, T., Beek, J.F., Wolbers, J.G., and van Gemert, M.J. (1994). Light propagation in the brain depends on nerve fiber orientation. *Neurosurgery 35*, 720–722; discussion 722–724.

Hee, M.R., Swanson, E.A., Fujimoto, J.G., and Huang, D. (1992). Polarization-sensitive low-coherence reflectometer for birefringence characterization and ranging. *J. Opt. Soc. Am. B 9*, 903.

Hitzenberger, C., Goetzinger, E., Sticker, M., Pircher, M., and Fercher, A. (2001). Measurement and imaging of birefringence and optic axis orientation by phase resolved polarization sensitive optical coherence tomography. *Opt. Express 9*, 780.

Huang, D., Swanson, E.A., Lin, C.P., Schuman, J.S., Stinson, W.G., Chang, W., Hee, M.R., Flotte, T., Gregory, K., Puliafito, C.A., et al. (1991). Optical coherence tomography. *Science 254*, 1178–1181.

Huard, S. (1997). *Polarization of Light*, 348 pp. Wiley-VCH, Chichester, U.K.

Iftimia, N., Ferguson, R.D., Mujat, M., Patel, A.H., Zhang, E.Z., Fox, W., and Rajadhyaksha, M. (2013). Combined reflectance confocal microscopy/optical coherence tomography imaging for skin burn assessment. *Biomed. Opt. Express 4*, 680.

Jeon, S.W., Shure, M.A., Baker, K.B., Huang, D., Rollins, A.M., Chahlavi, A., and Rezai, A.R. (2006). A feasibility study of optical coherence tomography for guiding deep brain probes. *J. Neurosci. Methods 154*, 96–101.

Jeong, B., Lee, B., Jang, M.S., Nam, H., Yoon, S.J., Wang, T., Doh, J., Yang, B.-G., Jang, M.H., and Kim, K.H. (2011). Combined two-photon microscopy and optical coherence tomography using individually optimized sources. *Opt. Express 19*, 13089.

Jones, R.C. (1941). A new calculus for the treatment of optical systems description and discussion of the calculus. *J. Opt. Soc. Am. 31*, 488.

Joo, C., Kim, K.H., and de Boer, J.F. (2007). Spectral-domain optical coherence phase and multiphoton microscopy. *Opt. Lett. 32*, 623.

Katoh, K., Hammar, K., Smith, P.J.S., and Oldenbourg, R. (1999). Arrangement of radial actin bundles in the growth cone of Aplysia bag cell neurons shows the immediate past history of filopodial behavior. *Proc. Natl. Acad. Sci. 96*, 7928–7931.

Larsen, L., Griffin, L.D., GRäßel, D., Witte, O.W., and Axer, H. (2007). Polarized light imaging of white matter architecture. *Microsc. Res. Tech. 70*, 851–863.

Leahy, C., Radhakrishnan, H., and Srinivasan, V.J. (2013). Volumetric imaging and quantification of cytoarchitecture and myeloarchitecture with intrinsic scattering contrast. *Biomed. Opt. Express 4*, 1978.

Leitgeb, R., Hitzenberger, C., and Fercher, A. (2003). Performance of fourier domain vs time domain optical coherence tomography. *Opt. Express 11*, 889.

Magnain, C., Augustinack, J.C., Reuter, M., Wachinger, C., Frosch, M.P., Ragan, T., Akkin, T., Wedeen, V.J., Boas, D.A., and Fischl, B. (2014). Blockface histology with optical coherence tomography: A comparison with Nissl staining. *NeuroImage 84*, 524–533.

Makita, S., Hong, Y., Yamanari, M., Yatagai, T., and Yasuno, Y. (2006). Optical coherence angiography. *Opt. Express 14*, 7821.

Nakaji, H., Kouyama, N., Muragaki, Y., Kawakami, Y., and Iseki, H. (2008). Localization of nerve fiber bundles by polarization-sensitive optical coherence tomography. *J. Neurosci. Methods 174*, 82–90.

Odgaard, A., Andersen, K., Melsen, F., and Gundersen, H.J.G. (1990). A direct method for fast three-dimensional serial reconstruction. *J. Microsc. 159*, 335–342.

Oldenbourg, R., Salmon, E.D., and Tran, P.T. (1998). Birefringence of single and bundled microtubules. *Biophys. J. 74*, 645–654.

Perona, P. and Malik, J. (1990). Scale-space and edge detection using anisotropic diffusion. *IEEE Trans. Pattern Anal. Mach. Intell. 12*, 629–639.

Podoleanu, A.G., Dobre, G.M., Cucu, R.G., Rosen, R., Garcia, P., Nieto, J., Will, D., Gentile, R., Muldoon, T., Walsh, J. et al. (2004). Combined multiplanar optical coherence tomography and confocal scanning ophthalmoscopy. *J. Biomed. Opt. 9*, 86–93.

Ragan, T., Kadiri, L.R., Venkataraju, K.U., Bahlmann, K., Sutin, J., Taranda, J., Arganda-Carreras, I., Kim, Y., Seung, H.S., and Osten, P. (2012). Serial two-photon tomography for automated ex vivo mouse brain imaging. *Nat. Methods 9*, 255–258.

Ren, H., Du, C., Yuan, Z., Park, K., Volkow, N.D., and Pan, Y. (2012). Cocaine-induced cortical microischemia in the rodent brain: Clinical implications. *Mol. Psychiatry 17*, 1017–1025.

Sands, G.B., Gerneke, D.A., Hooks, D.A., Green, C.R., Smaill, B.H., and Legrice, I.J. (2005). Automated imaging of extended tissue volumes using confocal microscopy. *Microsc. Res. Tech. 67*, 227–239.

Schoenenberger, K., Colston, B.W., Maitland, D.J., Da Silva, L.B., and Everett, M.J. (1998). Mapping of birefringence and thermal damage in tissue by use of polarization-sensitive optical coherence tomography. *Appl. Opt. 37*, 6026.

Srinivasan, V.J., Radhakrishnan, H., Jiang, J.Y., Barry, S., and Cable, A.E. (2012). Optical coherence microscopy for deep tissue imaging of the cerebral cortex with intrinsic contrast. *Opt. Express 20*, 2220.

Stifter, D., Leiss-Holzinger, E., Major, Z., Baumann, B., Pircher, M., Götzinger, E., Hitzenberger, C.K., and Heise, B. (2010). Dynamic optical studies in materials testing with spectral-domain polarization-sensitive optical coherence tomography. *Opt. Express 18*, 25712.

Tearney, G.J., Bouma, B.E., and Fujimoto, J.G. (1997). High-speed phase- and group-delay scanning with a grating-based phase control delay line. *Opt. Lett. 22*, 1811.

Trasischker, W., Zotter, S., Torzicky, T., Baumann, B., Haindl, R., Pircher, M., and Hitzenberger, C.K. (2014). Single input state polarization sensitive swept source optical coherence tomography based on an all single mode fiber interferometer. *Biomed. Opt. Express 5*, 2798–2809.

Tsai, P.S., Friedman, B., Ifarraguerri, A.I., Thompson, B.D., Lev-Ram, V., Schaffer, C.B., Xiong, Q., Tsien, R.Y., Squier, J.A., and Kleinfeld, D. (2003). All-optical histology using ultrashort laser pulses. *Neuron 39*, 27–41.

Wang, H., Al-Qaisi, M.K., and Akkin, T. (2010). Polarization-maintaining fiber based polarization-sensitive optical coherence tomography in spectral domain. *Opt. Lett. 35*, 154.

Wang, H., Black, A.J., Zhu, J., Stigen, T.W., Al-Qaisi, M.K., Netoff, T.I., Abosch, A., and Akkin, T. (2011). Reconstructing micrometer-scale fiber pathways in the brain: Multi-contrast optical coherence tomography based tractography. *NeuroImage 58*, 984–992.

Wang, H., Zhu, J., and Akkin, T. (2014a). Serial optical coherence scanner for large-scale brain imaging at microscopic resolution. *NeuroImage 84*, 1007–1017.

Wang, H., Zhu, J., Reuter, M., Vinke, L.N., Yendiki, A., Boas, D.A., Fischl, B., and Akkin, T. (2014b). Cross-validation of serial optical coherence scanning and diffusion tensor imaging: A study on neural fiber maps in human medulla oblongata. *NeuroImage 100*, 395–404.

Wang, R.K., Jacques, S.L., Ma, Z., Hurst, S., Hanson, S.R., and Gruber, A. (2007). Three dimensional optical angiography. *Opt. Express 15*, 4083.

Yao, G. and Wang, L.V. (1999). Two-dimensional depth-resolved Mueller matrix characterization of biological tissue by optical coherence tomography. *Opt. Lett. 24*, 537.

Yaseen, M.A., Srinivasan, V.J., Sakadžić, S., Radhakrishnan, H., Gorczynska, I., Wu, W., Fujimoto, J.G., and Boas, D.A. (2011). Microvascular oxygen tension and flow measurements in rodent cerebral cortex during baseline conditions and functional activation. *J. Cereb. Blood Flow Metab. 31*, 1051–1063.

Yasuno, Y., Makita, S., Sutoh, Y., Itoh, M., and Yatagai, T. (2002). Birefringence imaging of human skin by polarization-sensitive spectral interferometric optical coherence tomography. *Opt. Lett. 27*, 1803.

Yu, L., Nguyen, E., Liu, G., Choi, B., and Chen, Z. (2010). Spectral Doppler optical coherence tomography imaging of localized ischemic stroke in a mouse model. *J. Biomed. Opt. 15*, 066006–066006–6.

Zhao, Y., Chen, Z., Saxer, C., Xiang, S., de Boer, J.F., and Nelson, J.S. (2000). Phase-resolved optical coherence tomography and optical Doppler tomography for imaging blood flow in human skin with fast scanning speed and high velocity sensitivity. *Opt. Lett. 25*, 114.

11 Advanced Optical Imaging in the Study of Acute and Chronic Response to Implanted Neural Interfaces

Cristin G. Welle and Daniel X. Hammer

CONTENTS

INTRODUCTION

Implanted medical devices that interface with the nervous system are currently used to diagnose and treat a wide variety of neurological and psychiatric disorders and impairments. The acceptance of these devices is likely to grow over the next decade to the extent that they are demonstrated to provide benefit to patients who are resistant to treatment by pharmaceutical or other interventions. Neurostimulation devices treat medical conditions such as chronic pain, Parkinson's disease, essential tremor, epilepsy, hearing loss, and urinary incontinence, among others. There are an estimated 800,000 implanted neurostimulation devices in patients worldwide, and the market share is expected to grow (Medtech Insight 2013). Neural recording devices, which detect neural signals, are currently marketed for epilepsy monitoring and brain mapping. Several medical devices recently approved by the U.S. Food and Drug Administration (FDA) include both stimulation and recording elements. These include a closed-loop system, the NeuroPace

Responsive Neurostimulation System for epilepsy, which detects brain electrical signals and provides stimulation to interrupt seizures. Similarly, the Inspire Upper Airway stimulation system to treat sleep apnea detects ventilatory effort and responds with stimulation of the hypoglossal nerve to open the airway. The closed-loop detection/therapy combination of neural sensing and stimulation in a single device platform has the potential to increase the therapeutic potency of future devices.

In addition to currently marketed devices, there is a robust pipeline for neural interface devices designed to treat new disease indications currently in nonclinical or clinical study evaluation. These include peripheral nerve interface devices designed to treat a diversity of disorders, including tinnitus, stroke, and inflammation (Engineer et al. 2013; Hays et al. 2013; Pavlov and Tracey 2012). Central and peripheral recording interfaces are being explored for the restoration of movement function in individuals with paralysis or amputation (Collinger et al. 2013; Simeral et al. 2011; Tan et al. 2014). Some of these devices also incorporate stimulation elements to restore functionality to intact musculature or to provide sensory feedback to the device user (Dhillon and Horch 2005; Raspopovic et al. 2014; Tan et al. 2014).

Despite the success of neural interface devices in bringing medically necessary therapies to patients, incontrovertible evidence suggests that implanted devices elicit a response from the neuroinflammatory system. The magnitude of this response can vary depending on the device properties and implantation region, but the lack of biocompatibility can limit device performance. This is particularly true for neural recording devices, whose effectiveness can be more sensitive to the inflammatory response in the surrounding tissue than stimulating devices. For example, central or peripheral neural interfaces used to detect neural signals to control prosthetic or robotic assist devices often demonstrate a loss in detection capability that is ascribed, at least in part, to the inflammation of the surrounding neural tissue (Kim et al. 2008; Rousche and Normann 1998; Simeral et al. 2011).

The dynamics of the innate tissue response and attempts to mitigate the response to improve device performance have primarily been studied in animal models. These investigations typically rely on postmortem histological assessments and occasionally include in vivo electrical impedance spectroscopy. While this line of research has yielded important insights regarding the identity and temporal expression of molecules involved in neuroinflammation, the techniques used have inherent limitations and have not yet presented a complete picture of neuroinflammatory dynamics. For instance, histological techniques can be applied only to a single, postmortem timepoint, and the extracted and processed tissue may be disrupted or distorted by the processing methodology. In addition, electrical impedance spectroscopy is affected by numerous factors in vivo, beyond just those involved in tissue response, making interpretation of the data challenging. For example, changes in insulation or electrode contact metallization can also cause changes in impedance that are difficult to dissociate from those caused by cellular factors.

Advanced optical imaging techniques allow in vivo longitudinal assessment of biological responses without the limitations described here. These assessments can be paired with electrophysiological recordings or impedance spectroscopy to provide a multimodal evaluation of device performance. In addition, the use of genetic manipulations in animal models can allow for simultaneous optical assessment of both tissue structural changes and functional dynamics of neural circuitry. Utilization of a multiparametric optical approach provides unparalleled access to ongoing structural and functional correlations and their impact on device performance.

Advanced imaging techniques have been successfully applied to the study of normal neural structure and function. For example, a great deal has been learned about experience-dependent neural plasticity through the resolution and quantification of dendritic spine and axonal bouton lifetime and turnover (Holtmaat et al. 2005; Keck et al. 2008). Insight into the complex interplay between the protective and destructive roles of microglia in the central nervous system (CNS), as well as a new understanding of their role in surveillance, formerly thought to be limited, has also been gained only with time-lapsed, in vivo optical imaging (Davalos et al. 2005; Nimmerjahn et al. 2005). The neurovascular unit and its role in the response to ischemia is another active area of exploration for the development of stroke treatment and recovery strategies. Investigation of specific targets critical to initiation and progression of other diseases (e.g., amyloid plaques for Alzheimer's disease) has only just begun (Gomez-Nicola and Perry 2015). Likewise, the biological response to neural implants has, until now, gone relatively unexplored.

Recently, a limited number of studies have begun to demonstrate the advantages of advanced imaging methodology for evaluating implanted neural electrodes. Acute studies of cortical regions in the mouse show reorientation of microglia processes toward the penetrating electrode tips (Kozai et al. 2012). Chronic studies of epidural surface electrodes, which do not penetrate into the brain, have revealed vascular remodeling over the course of weeks and spontaneously emerging vascular microbleeds (Schendel et al. 2013). In contrast, label-free imaging of the dense capillary network surrounding a single implanted electrode shank shows relative stability of the vascular network, implying that the device configuration may have implications for vascular integrity near implanted devices (Hammer et al. 2014).

The expanded use of in vivo imaging modalities in the context of neural interface devices can address key questions with regard to device safety and effectiveness through direct observation of experimental manipulation. For example, the degree of cellular atrophy, particularly for neurons, in very close proximity to the electrode has been debated in literature. This question is difficult to answer with histology due to tissue disruption inherent to the technique, but can be easily addressed with chronic, in vivo imaging of regions surrounding a brain implant. Other questions that could be answered with in vivo imaging techniques include the timing of the neuroinflammatory response, particularly with respect to microglia and astrocytes, and the correlation of the tissue response with the observed decline of high-quality electrophysiological signals. In addition, the changes in vascularization and blood flow dynamics over time and with respect to different electrode morphologies can be meaningfully investigated with these techniques. Understanding cellular atrophy, neuroinflammation, hypoxia and vascular remodeling, and the correlation with device parameters and performance can provide the insights needed to dramatically improve neural interface performance, and bring the next-generation technology to benefit patients in need of neurological or psychiatric intervention.

In this chapter we begin by exploring the knowledge gained to date from histological studies of the response to implanted electrodes (Response to Implanted Electrodes Characterized by Histology section). A survey of in vivo optical imaging studies in animal models follows (In Vivo Optical Neural Imaging section), with particular emphasis on two-photon laser scanning microscopy (TPLSM), which has been the workhorse for neurophotonics for the last 20 years. Finally, we review what has been reported to date in terms of acute and long-term imaging of implanted, penetrating electrodes (Implanted Electrode Injury section). As neural interface devices become more prevalent and more sophisticated, the need to understand mechanistic interactions between brain and device only grows in urgency and importance.

RESPONSE TO IMPLANTED ELECTRODES CHARACTERIZED BY HISTOLOGY

The brain's response to implanted electrodes has been characterized through histological and biochemical evaluation of explanted brain tissues. Well over half a century of research using these techniques has provided a thorough description of cellular, vascular, and biochemical changes precipitated by chronic electrode implantation. However, inherent limitations in the histological and biochemical approaches have left gaps in our understanding of the brain's response to microelectrode implantation. Perhaps for this reason, research to date has failed to identify a solution to entirely mitigate the negative aspects of tissue response in a way that allows for long-term, high-fidelity cortical recordings from implanted microelectrodes. A brief summary of the principles of the neuroinflammatory tissue response known to date is provided in the following. More comprehensive reviews can be found elsewhere (Jorfi et al. 2015; Polikov et al. 2005).

CELLULAR, VASCULAR, AND MOLECULAR COMPONENTS
OF THE BRAIN'S RESPONSE TO IMPLANTED ELECTRODES

Implantation of microelectrodes elicits a robust cellular response, including a broad spectrum of resident neural and glial cells, along with nonnative infiltrating cell types from the surrounding vasculature. The acute response to electrode insertion is particularly well characterized, with later timepoints infrequently appearing in published reports. Within the first 24 h following electrode implantation, there is a reported loss of neuronal cell bodies up to 50–100 μm from the electrode shank (Biran et al. 2005). Loss of neurons has also been reported at 16 weeks (Potter et al. 2012). A progressive death of dendritic processes, in conjunction with axonal demyelination, has been observed out to 16 weeks post-implantation (Collias and Manuelidis 1957; McConnell et al. 2009; Winslow et al. 2010).

Resident glial cells actively respond to acute injury caused by insertion, with microglia initiating the response within the first several hours (Davalos et al. 2005; Kozai et al. 2012). Microglial reactivity, as measured by ED-1, reduces in the subsequent weeks (Potter et al. 2012), and microglial density declines from the initial reaction and then plateaus after 1–2 months (McConnell et al. 2009). At this point, microglia are primarily localized in a ring surrounding the electrode track (Potter et al. 2012; Saxena et al. 2013; Winslow and Tresco 2010), or adhered to the electrode itself (Biran et al. 2005). Astrocytes respond to acute injury more slowly, with a clear response developing within several days following insertion. The reactivity continues to develop and increase over the following 1–2 months (Biran et al. 2005; McConnell et al. 2009; Szarowski et al. 2003; Turner et al. 1999), but stabilizes after 3 months (Saxena et al. 2013; Winslow and Tresco 2010), at which point it is primarily localized to the first several hundred microns from the electrode interface. Resident fibroblasts also accumulate at the electrode interface at acute timepoints (Collias and Manuelidis 1957).

Nonresident cell types, primarily circulating macrophages, infiltrating myeloid cells, neutrophils, and leukocytes have also been detected in the electrode tissue response (Kim et al. 2004; Saxena et al. 2013; Woolley et al. 2013). Nonresident cell types have received less attention, and the details of their numbers, timing of arrival, and persistence are not comprehensively described. Yet, nonresident circulating macrophages and myeloid cells play a role in eliciting and producing inflammatory reactions in affected brain tissues (Burda and Sofroniew 2014; Ransohoff and Cardona 2010). Some studies have suggested that they may be one of the primary cellular components of the brain's tissue response, and some evidence indicates that a majority of Iba1+ microglia cells are actually nonresident macrophages (Ravikumar et al. 2014). Although very long-term studies of the tissue response are rare, such investigations have suggested that the cellular response remains stable up to 6 months post-implantation (Collias and Manuelidis 1957; Rousche and Normann 1998).

Cellular reactivity is also accompanied by changes to vascular structure and integrity. Vascular structural changes, such as proliferation of capillaries, can occur at timepoints as early as 7 days (Collias and Manuelidis 1957). Leakage of the blood–brain barrier has been consistently demonstrated to result from implanted electrodes, and markers of blood–brain barrier leakage such as hemosiderin-laden macrophages or IgG accumulation have been shown to persist for up to 16 weeks (McConnell et al. 2009; Saxena et al. 2013; Winslow and Tresco 2010). Additionally, Saxena et al. found correlations between blood–brain barrier leakage, electrode performance, and neurotoxic/proinflammatory markers, providing evidence that blood–brain barrier leakage may be intimately related to cellular responses and even neural activity (Saxena et al. 2013). Evidence for long-term aberrations in vascular performance also comes from the postmortem analysis of human brain tissues, which showed markers of vascular damage up to 10 months following electrode extraction (Liu et al. 2012). The temporal relationships between cellular and vascular reactivity are demonstrated in Figure 11.1.

Observed cellular and vascular changes are closely associated with increases in neuroinflammatory molecules in local proximity to the electrode. ED-1 expressing microglia, part of the brain's immune response, have been shown to release inflammatory cytokines (Biran et al. 2005). Upregulation of pre-inflammatory matrix metalloproteases, such as MMP-9, which is associated with blood–brain barrier permeability has been shown up to 4 months post-implantation (Saxena et al. 2013). Evidence of reactive oxygen species, which are implicated in the breakdown of the blood–brain barrier and can be neurotoxic (Haorah et al. 2007; Potter et al. 2013; Sarker et al. 2000), has been demonstrated for 4 weeks following electrode insertion (Potter et al. 2013).

Some aspects of the cellular and vascular tissue response can be mitigated, at least in the acute phase, through pharmaceutical interventions. Anti-inflammatory agents, including antibiotics and dexamethasone, have been shown to reduce gliosis and improve electrical recording signal quality. However, evidence is lacking for the success of this strategy beyond 2 months post-implantation

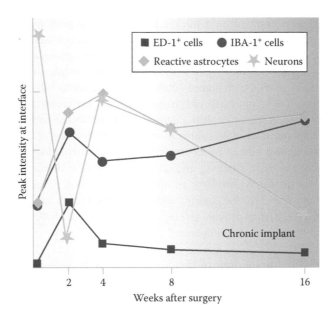

FIGURE 11.1 Cortical device implantation results in increased astrocytic and microglial response. ED-1 is a biomarker for activated microglia and macrophages and IBA-1 is a biomarker for total microglia and macrophages. Trauma-induced/blood-derived events are reflected by a yellow gradient while central nervous system–derived events are shown as a brown gradient. (Adapted from Potter, K.A., Buck, A.C., Self, W.K., and Capadona, J.R., Stab injury and device implantation within the brain results in inversely multiphasic neuroinflammatory and neurodegenerative responses, *J. Neural Eng.*, 9(4), 046020, 2012. Reproduced with permission from IOP Publishing. With permission.)

(Rennaker et al. 2007; Spataro et al. 2005; Zhong and Bellamkonda 2007). Antioxidants, such as resveratrol and curcumin, may also mitigate neuronal cell body loss within the first several weeks and reduce blood–brain barrier leakage (Potter et al. 2013, 2014). Paradoxically, these agents can also produce proinflammatory results. For instance, dexamethasone may increase vascular and microglial reactivity (Spataro et al. 2005), and resveratrol increases glial reactivity (Potter et al. 2013).

LIMITATIONS OF CONVENTIONAL HISTOLOGICAL TECHNIQUES

As summarized earlier, the current understanding of the brain's response to electrode implantation reveals a complex and interrelated set of cellular and molecular reactions, but falls short of identifying a key mitigation strategy for improving implanted electrode performance. This is due in part to the complexity of the response and the evolving nature of the tissue reactivity over time post-implantation. However, inherent limitations in the histological and biochemical approaches applied to this topic stymie efforts to understand the dynamic nature of the brain's response to foreign objects. Limitations in access to spatial and temporal information prevent a complete understanding of the response kinetics. In addition, difficulties in obtaining and correlating multimodal information sources describing structural, physiological, and biochemical alterations simultaneously have hampered efforts to draw strong correlations between observed phenomena.

Histological techniques are limited in describing spatial relationships by several factors. First, extraction of the implanted electrode normally removes local tissue as well, disrupting the device/tissue interface. Cellular adhesions to explanted electrodes have been widely reported (Azemi et al. 2011; Barrese et al. 2013; Rousche and Normann 1998; Turner et al. 1999; Woolley et al. 2011), presumably indicating the presence of corresponding gaps in the remaining tissue. Woolley and colleagues pioneered a technique to surmount this issue, called device-capture histology, in which immunohistochemistry is performed on a thicker block of tissue with the electrode still in place (Woolley et al. 2011, 2013). Application of device-capture histology has created useful representations of tissue immediately surrounding the electrode and brought insight into the relationship of microglia and nonparenchymal cell types with respect to the distance from the cortical surface. However, implementation of this technique is currently restricted to "shank" style multielectrode arrays due to the limitations of antibody penetration into thicker tissue slices, and the technique is laborious. In addition, even device-capture histology requires tissue sectioning, preventing volumetric analysis of the spatial interrelationships of cellular and molecular components. The loss of three-dimensional information makes understanding vascular morphological changes particularly difficult, requiring labor-intensive stitching and reconstruction of tortuous capillary vessels across tissue sections. Nonuniform tissue shrinkage resulting from fixation processes further complicates this task. Recent developments in tissue clearing and imaging protocols, including techniques such as CLARITY, Scale, and SeeDB, may provide an elegant solution to the challenges of tissue sectioning, but these have not yet been applied to implanted microelectrode arrays (Chung et al. 2013; Hama et al. 2011; Ke et al. 2013).

In addition to the spatial limitations of histological methods, because the techniques require tissue extraction, they are inherently limited to a single timepoint. This severely limits the ability to describe and understand nonstatic processes. The neuroinflammatory response of the brain is widely recognized as biphasic, with acute and chronic phases. However, recent evidence suggests that it may actually be multiphasic, and perhaps fluctuating (Harris et al. 2011; McConnell et al. 2009; Potter et al. 2012). Given the interrelated cellular, vascular, and molecular responses during inflammation, it is not a stretch to suggest that inflammation may be cyclic or involve repeated cascades of linked events; something impossible to verify with a single timepoint per animal. Incomplete understanding of the temporal elements of the tissue response may have hampered previous efforts at pharmacological intervention. The initial approach has been to target interventions according to time post electrode insertion; a more effective strategy might be to target based on the milieu of inflammatory molecules present at a given point in time. Thus, real-time monitoring of the temporal response could improve the effectiveness of therapeutic strategies.

A lack of spatial and temporal resolution in observing the evolving tissue response has left crucial elements of the response dynamics unexplored. Processes such as cellular migration, death, and proliferation are difficult to track with histology. Particularly unresolved is the issue of neuronal apoptosis and remodeling. There has not been a broad consensus yet on neuronal survival. Loss of neurons immediately surrounding the electrode may be due to apoptosis (Biran et al. 2005; McConnell et al. 2009) or displacement of cell bodies by other cell types (Edell et al. 1992; Liu et al. 1999; Woolley et al. 2013). If apoptosis does occur, little is known regarding the timing. Others have reported loss or displacement of neural branches and axonal demyelination (Collias and Manuelidis 1957; McConnell et al. 2009; Winslow and Tresco 2010). Again, differentiating between dendritic atrophy and displacement due to other cellular elements, and observing the timing of those events, has not been possible with current techniques. Given that the goal of implanted electrodes is to interface electrically with neural structures, and this requires physical proximity to excitable membranes, understanding changes to neuronal structure is crucial to identifying strategies to improve device performance. Another example of a missing element in current knowledge regarding device implantation is the response of the vasculature to implanted electrodes. Although capillary sprouting has been observed (Collias and Manuelidis 1957), a comprehensive picture of vascular dynamics has not been described to date. This has important implications for metabolism and tissue hypoxia, which are critically related to neurovascular coupling. Finally, both astrocytes and microglia have been shown to have a diverse set of responses, and perhaps distinct subpopulations during injury. Understanding these dynamics has required the use of in vivo measurement techniques, including imaging (Bardehle et al. 2013; Roth et al. 2013).

THE NEED FOR DYNAMIC IN VIVO INVESTIGATIONS OF BRAIN INJURY RESPONSE

To understand the factors involved in implanted recording electrode failure, a multimodal correlation between measurements of the electrode and tissue response is key. Thus far, correlations between histological findings and electrophysiological recordings have been mixed. Leakage of the blood–brain barrier was found to correspond with electrode recording signal-to-noise, while astrocyte and microglia/macrophage markers were not (Saxena et al. 2013). Liu and colleagues found a positive correlation between the presence of large diameter neurons around a given electrode and the quality of neural recordings from that electrode (Liu et al. 1999). The formation of the glial scar has been proposed as a cause of changes in device performance (Edell et al. 1992). However, several histological studies have noted an apparent contradiction between the timing of robust neuroinflammatory processes (peak around 1 month) and the disruption of neural signal recording (6–12 months) (Barrese et al. 2013; McConnell et al. 2009; Saxena et al. 2013). Because histological samples are extracted from the terminal timepoint for a given animal, direct comparison between ongoing changes in electrophysiological recording or stimulating performance and neuroinflammatory responses is difficult. In addition, tissue damage resulting from electrode extraction or shrinking/expanding resulting from histological processing can make even general spatial correlations between individual recording contacts and the local environment challenging (Hama et al. 2011; Woolley et al. 2011).

IN VIVO OPTICAL NEURAL IMAGING

Because of the limitations in histological methods outlined in the previous section—primarily collection of only a single timepoint from each animal, but also other application-specific issues related to the neural-electrode interface—the use of in vivo optical imaging techniques to study the acute and chronic biological response to implanted electrodes is increasing. Before discussing recent imaging studies for that application in the next section, in this section we will provide an overview of imaging methodology and a survey of nonclinical, in vivo optical imaging investigations of neurons, neural pathways, neurological function, and neural injury and disease.

ANIMAL MODELS, ANATOMY, AND PHYSIOLOGY

In vivo optical imaging techniques require an optically transparent interface through which photons can be directed to and collected from the tissue sample. A majority of longitudinal two-photon and optical coherence tomography (OCT) investigations take place in the neocortex of rodent or nonhuman primate animal models, due the tractability of accessing neural tissue with only limited intervention. However, acute and chronic imaging studies have also observed tissues in the spinal cord and recently, peripheral nervous system (Brodnick et al. 2014).

The use of the rodent model has numerous advantages for in vivo advanced imaging modalities. Mouse and rat brains are roughly 10 and 15 mm in diameter, while the cortical layers extend ~1.0 and 1.5 mm deep, respectively (Figure 11.2). The cortex is without sulci or gyri, which keeps the major structural features aligned with the tissue's surface contour, and the dura is relatively thin, allowing for uniform optical access. Cortical structures are arranged in a laminar structure, allowing for visual discrimination between cortical layers with many imaging modalities. These layers are interconnected vertically, forming structures referred to as "columns" (Mountcastle 1997). In the mouse, imaging modalities with deeper penetration profiles, such as TPLSM and photoacoustic microscopy, can image over the depth of a cortical column, throughout the cortical gray matter. As an example, Figure 11.3 shows a TPLSM depth stack (3-D view) of a transgenic mouse (Thy1-EYFP-H) with sparsely labeled layer 5 pyramidal neurons. Optical access for rats does not extend through the entire cortex, although this can be obtained with MRI-based imaging methods. In addition, deeper brain regions have been optically imaged either through the removal of superficial tissue, such as cortex, or with penetrating lenses (Andermann et al. 2013; Ghosh et al. 2011; Mizrahi et al. 2004; Ziv et al. 2013).

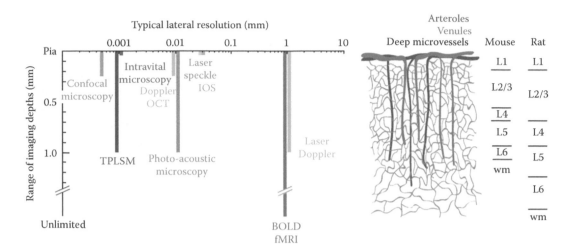

FIGURE 11.2 Comparison of penetration depth and lateral resolution for various imaging modalities with respect to rodent cortical layers and vasculature. (Reproduced with permission from Macmillan Publishers Ltd. *J. Cerebr. Flow Metabol.*, Shih, A.Y, Driscoll, J.D., Drew, P.J., Nishimura, N., Schaffer, C.B., and Kleinfeld, D., Two-photon microscopy as a tool to study blood flow and neurovascular coupling in the rodent brain, 32(7), 1277–1309, Copyright 2012. With permission.)

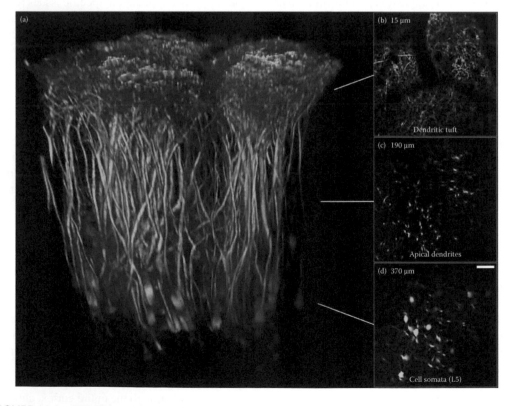

FIGURE 11.3 TPLSM stack of layer 5 pyramidal neurons (350 × 350 μm lateral field of view, 2 μm z-steps). (a) 3-D isometric view, and single z-slices of the (b) dendritic tuft, (c) apical dendrites, and (d) cell somata. Approximate depth of z-slice is indicated. Scale bar (b–d) = 50 μm.

SURGICAL PREPARATION

Owing to the relatively thin murine skull, optical access to cortical tissue can be gained directly through the skull (Wang and Hurst 2007; Yoder and Kleinfeld 2002), but more often than not imaging is accomplished either by thinning or complete removal of the skull, accompanied by window implantation (Drew et al. 2010; Holtmaat et al. 2009). For acute studies, the thinned skull preparation causes less perturbation of the underlying tissue (Xu et al. 2007). However, bone regrowth, even when a window is implanted, often limits imaging to a few months (Drew et al. 2010). For longitudinal studies designed to last months to years, a craniotomy and window implantation may be a better strategy. However, an initial tissue response, including inflammation, needs to be accounted for in the analysis. With this preparation, we found significant inflammation and vascular growth and remodeling confined to the superficial layers beneath the window (up to ~200 µm). Figure 11.4 illustrates vasodilation and vessel growth in this region, which, for the most part, stabilizes by the end of the first month after surgery and window implantation (Hammer et al. 2014). The capillary network in the underlying layers >200 µm does not appear to be significantly altered by the surgery.

Recent advances in surgical preparations for imaging small animal models in neuroscience have focused on implanted microscopes. The primary advantage of these devices is in imaging awake, unanaesthetized, freely behaving animals. For neurovascular coupling, this is especially important because anesthesia is known to suppress the vascular response (Drew et al. 2011). The microscopes have evolved from a bench-top, fiber-tethered approach (Flusberg et al. 2008) to fully integrated optical designs with some performance parameters (resolution and field of view) approaching traditional fluorescence microscopes (Ghosh et al. 2011). The integrated optics approach harnesses advances in consumer electronics and optics and can be mass produced at very low cost.

OPTICAL IMAGING MODES AND METHODS

The most common optical imaging modalities applied to in vivo neuroscience animal investigations are confocal, fluorescence, two-photon, optical coherence, and photoacoustic microscopy, as well as diffuse tomography and near-infrared spectroscopy (NIRS). Figure 11.2 shows a comparison of penetration depth and lateral resolution for various imaging modalities with respect to rodent cortical layers and vasculature. Many of these techniques are covered elsewhere in this book, in particular Chapters 2 through 5 on NIRS, and Chapters 8 through 10 on optical coherence microscopy. Several other earlier works also contain excellent reviews of optical neuroimaging (Frostig 2009; Holtmaat and Svoboda 2009; Kleinfeld et al. 2011; Shih et al. 2012; Svoboda and Yasuda 2006).

Confocal microscopy was invented in 1957 by Minsky (1961) and continues to be a mainstay for in vivo animal imaging. Confocal imaging enhances contrast by rejection of light scattered from adjacent voxels (axial or lateral) with a pinhole placed at a focal (i.e., detector) plane conjugate to the object plane. A confocal setup typically forms the core upon which most microscopes provide further capabilities with fluorescence hardware. Fluorescence microscopes excite extrinsic and intrinsic fluorophores labeling cells or organelles to emit red-shifted light (Masters 2010). Fluorescence spectra contain a wealth of molecular information and the spatial, temporal, and spectral characteristics of this technique have been exploited for various uses in neuroimaging. However, because a great deal of visible light is required to excite the fluorophores, particularly when intrinsic targets are used, the method suffers from poor penetration and lower sensitivity and generally requires a confocal arrangement for moderate axial sectioning. Approximately 25 years ago, Denk and Webb first applied the two-photon effect to microscopy, where two near-infrared photons are absorbed to emit visible light at half the wavelength using high intensity ultrashort

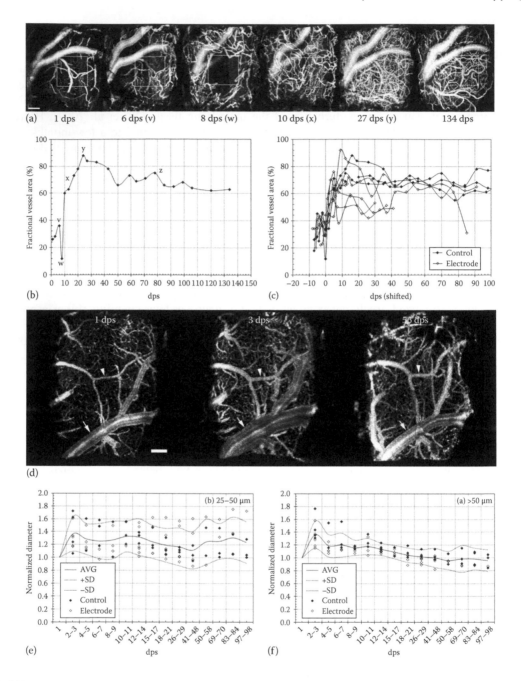

FIGURE 11.4 Massive vessel dilation and vascularization are observed in the superficial layers after surgery and window implantation. (a) In the plane 0–100 μm below the window, vascular remodeling and growth take place. (b) Corresponding vessel area measurements (fractional %) over the time course of several months in the center of the window (ROI indicated in (a)). (c) Similar trends were observed for all mice with no differences between control and electrode groups. (d) In the plane 100–200 μm below the window vessel dilation peaks in the first few days after surgery. The large vessel (arrow) returns to normal diameter within a few months while dilation in the smaller vessel (arrowhead) persists. Vessel diameter measurements from 10 mice for smaller. (e) and larger (f) vessels showed no difference between control and electrode groups. (Adapted from Hammer, D.X., Lozzi, A., Abliz, E., Greenbaum, N., Agrawal, A., Krauthamer, V., and Welle, C.G., Longitudinal vascular dynamics following cranial window and electrode implantation measured with speckle variance optical coherence angiography, *Biomed. Opt. Express*, 5(8), 2823, 2014. Reproduced with permission from Optical Society of America. With permission.)

pulsed lasers (Denk et al. 1990). Two-photon (and later three-photon and multiphoton) microscopy has several advantages over traditional fluorescence techniques, including better optical sectioning (without a pinhole) because the nonlinear effect is confined to regions of high light intensity, better penetration because tissue has a lower scattering coefficient for near-infrared light used for excitation, and less damaging because longer wavelength light has lower energy. TPLSM is the predominant imaging technique used in neurophotonics laboratories, though photoacoustic microscopy is gaining traction because of its significantly higher penetration capability (Wang 2009; Wang and Hu 2012).

While TPLSM decouples axial resolution (and sectioning) from the optical depth of focus via a nonlinear fluorescence process, OCT does the same via low coherence interferometry. The axial resolution and depth sectioning are inversely proportional to the source bandwidth. Moreover, OCT provides access to the intensity and phase of the reflected light, which provides additional useful information that can be exploited, particularly for flow quantification. OCT is predominantly used for ophthalmology, but several groups have shown its promise in neuroscience applications (Liu et al. 2012; Mahmud et al. 2013; Srinivasan et al. 2012a).

For measurement of blood flow, custom methods have been developed for TPLSM that sequentially scan the image beam across and along a vessel of interest (Kleinfeld et al. 1998). Streaks associated with individual red blood cells (RBC) from these line scans yield flow velocity rates (mm/s) as well as RBC flux (pL/s, or RBC counts/s). Except for slow dilatory responses that can be captured with image frame rates of several Hz, line scanning is typically required to accurately calculate evoked blood velocity changes that result from stimulation (Drew et al. 2011; Shih et al. 2012). Line scanning can also be performed on labeled cells during stimulation (Kleinfeld et al. 2011). While TPLSM flow precision is exceptional, the line scan technique requires custom scans for each cell/vessel or a set of cells/vessels in an image plane. Despite intelligent scanning methods that have enabled interrogation of a moderate number of cells in the field of view (Valmianski et al. 2010), it remains a single-point, nonimaging approach. Despite this constraint, some investigators have used line scans of vessels laterally separated by ~0.5 mm to gain a better understanding of coherence in the vasomotor response (Drew et al. 2011). TPLSM flow methods also require injection of dye (e.g., dextran) into vasculature to provide contrast.

OCT-based angiography (OCA) provides visualization of vessels by the detection of speckle- (i.e., intensity) and phase-based changes associated with flowing blood. Figure 11.5 shows examples of cross-sectional and en face images, obtained with OCA, of the capillary network in mouse motor cortex surrounding an implanted electrode. Because intrinsic contrast is used to detect flow, no dye injection is required. Methods have also been developed to quantify vessel and capillary flow rates, using the high-speed capabilities of OCT (Srinivasan et al. 2010, 2012b; Yousefi et al. 2013). The flow detection capabilities of OCT have allowed direct quantification of wide-field tissue perfusion via the generation of cortical flow maps ($\mu L/mm^2/min$) (Srinivasan et al. 2013). In a manner similar to TPLSM line scanning, the dynamic range of OCT flow detection can be adjusted to match known vessel flow rates, though the dynamic range is generally smaller for OCT compared to TPLSM. Thus multiple methods are often required to capture both extremely low flow in capillaries (<1 mm/s) and moderate flow in penetrating arterioles and ascending venules (~5–10 mm/s in mice). Also, in contrast to TPLSM, OCT cannot visualize pooled blood because the intrinsic signature relies on motion and therefore provides no means to investigate the effects of extravasation. Most previous OCT techniques do not resolve individual RBC passage through capillaries, although some recent progress has been made to configure OCT to rapidly map metrics that are proportional to RBC flux (Lee et al. 2014). For the investigation of the neurovascular unit, TPLSM and OCT may prove complementary.

It should be clear that many imaging methods have been demonstrated for use in neuroscience, each with advantages and disadvantages and each suited to different applications. Other emerging in vivo optical imaging techniques that have been applied to neural applications include intrinsic signal optical imaging (ISOI) (Frostig and Chen-Bee 2009), TPLSM with longer wavelengths (1300 nm)

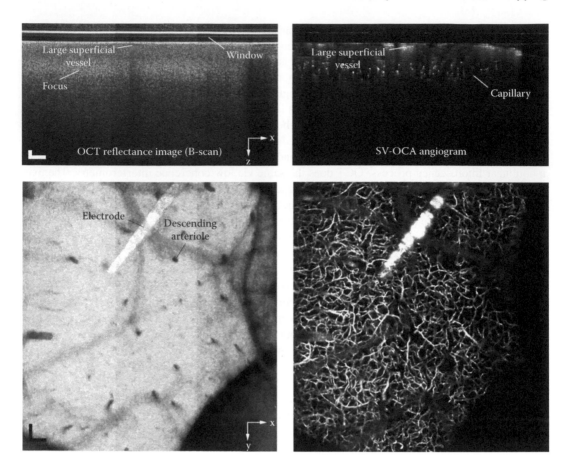

FIGURE 11.5   OCT images of single-shank electrode (Neuronexus Inc.) penetrating into mouse cortex visualized through craniotomy window. Top row shows cross-sectional images. Focus is set near the electrode tip ~350 μm below bottom of window. Bottom row shows en face maximum intensity projections (MIP) of 100 μm thick z-stack just above the electrode tip. Left column shows standard OCT reflectance processing. Right column shows intrinsic contrast OCA processing to visualize capillary network.

for deep tissue probing (Kobat et al. 2009), three-photon microscopy (Horton et al. 2013), and light sheet microscopy (Keller et al. 2015; Prevedel et al. 2014).

Along with imaging instrumentation and methodology, development of labeling methods for in vivo cellular and molecular targets has advanced considerably in the last 15 years. While targeting intrinsic constituents (e.g., reflectance variations arising from optical scattering, endogenous fluorophores) is desirable from the stand-point of reduced experimental complexity and perturbation of the animal model, intrinsic contrast imaging rarely has the specificity necessary for optical imaging due to scattering, low fluorescence cross section of endogenous fluorophores, and other factors. Prior to 2000, the canonical approach to labeling cells was via organic dye injection and uptake into the target cell. For vascular imaging, large molecule dyes such as dextran that cannot penetrate the vessel wall are injected into the tail vein of rodents. Investigators have also designed artery-specific dyes and red-shifted dyes to resolve deeper flow (Kobat et al. 2009; Shen et al. 2012). Around the turn of the century, animals genetically modified to express green fluorescent protein (GFP) or other spectral variants were introduced for in vivo tracking of specific cells and organelles (Feng et al. 2000; Miyawaki 2005). The fluorescent proteins are heritable, enable high specificity and sparse labeling (important to understand single cell behavior in cortical columns with hundreds of thousands of neurons), and can be engineered for a wide range of functional activity (e.g., calcium indicators

[Heim et al. 2007; Mank et al. 2008]) and spectral emission. The last two features are especially advantageous for simultaneous imaging of the complex, multifaceted interactions between neurons, support cells, and vasculature, both for structural and functional endpoints. This advancement in biomarkers has profoundly changed the way animal models are imaged to explore normal function and disease dysfunction, particularly in the field of neuroscience.

Functional optical imaging, which resolves physiological activity in addition to anatomy or structure, has a long history with applications in cellular physiology and neuroscience (Cohen et al. 1968; Tasaki et al. 1968). Because of the brain's architecture of coordinated networks, imaging can provide vastly more information about function than single electrode recordings or even electrode arrays. Functional optical imaging typically involves recording changes in intrinsic absorption and scattering, or more commonly, fluorescent biomarkers. Voltage-sensitive dyes (VSD) rapidly change absorption or emission spectra in response to changes in membrane potential and therefore can be used to optically detect action potentials or other fast cellular dynamics (Civillico 2011). Fluorescent probes based upon calcium binding to proteins are the most commonly used activity-dependent probe. While the application of these methods in neuroscience is immensely important, they have yet to be used extensively to image the biological response to implanted electrodes. Therefore, VSD, calcium imaging, and other functional imaging techniques are not described extensively in this chapter.

SURVEY OF OPTICAL IMAGING APPLIED TO NEURAL STRUCTURE, FUNCTION, AND DISEASE

Optical imaging (compared to MRI, for example) is particularly well suited to investigate brain function and neurological disease in small animal (or primate) models because it provides cellular and subcellular (i.e., micron-scale) resolution within a relatively controlled environment. Moreover, various techniques have been invented to probe these structural targets and their underlying functional dynamics (generation and propagation of action potentials, blood flow and perfusion, membrane ion flux, enzymatic activities, cell migration, cell morphological changes, etc.). This section provides a survey of some key advances in neuroscience aided by optical microscopy.

Neuronal Structure and Function

Soon after the demonstration of functional imaging of single neurons (Svoboda et al. 1997), early TPLSM studies began to explore changes in substructures, including dendritic spines and axonal boutons (Grutzendler et al. 2002; Lendvai et al. 2000; Trachtenberg et al. 2002). Structural plasticity refers collectively to changes to axonal and dendritic arborization, changes to synaptic structures (lifetime and turnover of dendritic spines and axonal boutons), and synaptogenesis (as opposed to neurotransmitter modulation of synaptic strength) (Holtmaat and Svoboda 2009). Time-lapsed optical microscopy reversed earlier observations using static measurements and conventional labeling (Golgi method) and found neural arborization to be relatively stable over long periods of time on a macroscale in response to experiences such as learning. However, synaptic structures were found to have a larger role in structural plasticity. There is no consensus on the magnitude of this role because dendritic spine and axonal bouton lifetime and turnover appear to be highly dependent on cell type and region (Paola et al. 2006). Also, the relationship between spine size and stability (transience vs. persistence) is not well understood. Spine growth occurs rapidly in the immediate postnatal period and is followed by a longer period of spine pruning and stabilization driven by normal sensory experience (Holtmaat et al. 2005; Zuo et al. 2005). In adults, the majority of spines are stable and yet there is much evidence that the ~10% change in total persistent spines (a fraction of the new persistent spines observed over 1 month following whisker trimming in the mouse barrel cortex) can be attributed to experience-dependent plasticity in the adult neocortex (Holtmaat et al. 2006; Keck et al. 2008). Synapse formation is relatively slow and there are several times more nonsynaptic contacts than synapses formed. Two-thirds of new spines make contact with multiple synaptic boutons and thus dendritic spine turnover is approximately twice bouton turnover. This may also indicate that the mechanisms of dendritic spine and axonal bouton formation are not closely coupled. While many questions remain about structural plasticity, in vivo optical microscopy of genetically labeled neocortical neurons has enabled development of models to describe structural changes that are regulated by sensory experience (Holtmaat and Svoboda 2009).

Neural function is most often explored using calcium indicators, introduced either with multicell bolus loading (Garaschuk et al. 2006) or by genetic encoding (Kotlikoff 2007; Mank et al. 2008). Calcium imaging can be used to explore synaptic plasticity or spiking activity (Helmchen 2009). Calcium imaging with 3-D line scanning can also be used to rapidly explore volumetric population activity and local network dynamics (Göbel and Helmchen 2007). Investigations of local population activity can examine the extent of stimulation responses and network-mediated modulation of signal transmission. Functional studies can be aided either by multimodal approaches that use adjunctive structural imaging (i.e., intrinsic signal optical imaging [Frostig and Chen-Bee 2009]) or by recent progress mapping the mouse brain connectome (Oh et al. 2014).

Imaging Nonneuronal Cells

Optical imaging is also contributing to understanding the function of support cells (macrophages, microglia, astrocytes) in the brain parenchyma, especially during neurodegeneration and recovery (Helmchen and Kleinfeld 2008). Microglia are constantly active within the CNS, playing several functional roles including surveillance, synaptic pruning or remodeling, modulation of neural activity, phagocytosis, and proliferation for self-renewal to maintain homeostasis or in response to injury or neurodegenerative diseases (Gomez-Nicola and Perry 2015). Originally thought to be quiescent, the perpetual activity of microglia in a surveillance role, extending and retracting its processes while maintaining a fixed distribution and size of soma within a mosaic organization, was revealed by in vivo optical imaging (Davalos et al. 2005; Nimmerjahn et al. 2005). It is now understood that microglial processes make short duration (~5 min) contacts with synapses approximately every hour in the normal rodent brain. During injury (ischemia), the contact duration is extended and there is some evidence that microglia play a role in synapse turnover (Wake et al. 2009). In their phagocytotic role, microglia participate in removal of amyloid beta plaque in Alzheimer's disease, but are less effective in clearing misfolded proteins in prion disease (Gomez-Nicola and Perry 2015). During chronic neurodegeneration with systemic inflammation, microglia proliferate, and with astrocytes, activate and contribute to neuronal damage. The exact inflammatory phenotype of microglia in neurodegenerative diseases such as Alzheimer's, Parkinson's, and Huntington's is not completely known, but evidence is building, which indicates inflammation is causal in the initiation and progression of these diseases. The overall role of microglia in neural health is thus mixed, with some functions benign and others deleterious.

Neurovascular Coupling

Owing to the massive metabolic demand of the brain and the critical link between function and perfusion, neurovascular coupling (hyperemia) is an active area of research. Neuronal activity is spatially and temporally linked to a hemodynamic response, which includes immediate central dilation and delayed surround constriction (and flow decrease), and involves a complex cascade involving neuronal processes, astroglial cell endfeet, and the arteriole vessel wall smooth muscle (Attwell et al. 2010; Devor et al. 2008). Recent investigations have sought to identify the neural signals, patterns, and pathways that lead to vasoactivity, the role of intermediate support cells in the signaling cascade, and the neurovascular coupling changes associated with disease (primarily stroke) and the ability to restore function after injury (Shih et al. 2012).

Multichannel TPLSM is especially well suited for the investigation of neurovascular coupling because, unlike OCT that has difficulty resolving individual neurons with intrinsic contrast, all the key structures of the neurovascular unit (neurons, support cells, capillaries) can be resolved and simultaneously examined. Kleinfeld et al. used a two-channel approach and custom line scanning to simultaneously image neurons and astrocytes (with the Ca^{2+} reporter Oregon green bapta-1-AM) and record RBC velocity in capillaries (labeled with fluorescein-dextran) during electrical stimulation of the rat hindlimb (Kleinfeld et al. 2011). They found rapid neuronal Ca^{2+} transients and flow changes synchronous with stimulation. Spontaneous neuronal activity also occurred separately. Calcium transients in astrocytes occurred on much slower timescales, apparently without any temporal correspondence to stimulation. Kleinfeld et al. used the same dual-channel approach to simultaneously image blood flow and arterial smooth muscle activation using GCaMP (Kleinfeld et al. 2011). The multichannel approach allows moderate examination of the spatiotemporal environment involved in neurovascular coupling and the functional and metabolic response to stimulation or degradation from disease and age.

Blood oxygenation at the single capillary scale can also be measured with two-photon microscopy. Sakadzic et al. used a two-photon enhanced phosphorescent O_2 nanoprobe, PtP-C343, to measure oxygen partial pressure (pO_2) in vessels and tissue (Sakadžić et al. 2010). Because phosphorescent lifetime (inversely proportional to pO_2), rather than intensity, was measured, the low phosphorescence cross section (local probe concentration) was not a limiting factor in determining oxygenation. They observed a decrease in arteriole pO_2 with depth over a range of ~50–250 μm below the cortical surface. In tissue oxygenation experiments during normoxia and hypoxia (induced with 30 s respiratory arrest), tissue pO_2 during hypoxia decreased more quickly and recovered more slowly with increasing distance from the nearest artery. This investigation demonstrates the potential of exploring functional and metabolic measures relatively deep in the brain with high resolution. It may serve as an adjunctive method to other optical methods for cerebral blood flow (CBF) and oxygenation measurement, such as laser speckle imaging and blood oxygenation level dependent functional magnetic resonance imaging (BOLD fMRI).

Stroke

Optical imaging methods have aided in the understanding of the spatiotemporal characteristics of stroke. Many studies have focused on the peri-infarct region (penumbra) to aid development of treatment strategies targeting salvageable tissue (Shih et al. 2013). Continual (i.e., time-lapse) imaging, both on acute and chronic timescales, provides dynamic information not afforded by histologic techniques. For stroke, this includes structural visualization of dendritic and vascular injury but also recovery (e.g., reperfusion of ischemic regions).

Primary models of stroke include Rose Bengal photothrombosis (Watson et al. 1985), in which controlled cortical regions are exposed to laser irradiation after absorptive dye injection, and middle cerebral artery occlusions (MCAO), in which flow in a primary cerebral artery is halted, either temporarily and reversibly with a silicon coated filament, or permanently by ligature. Models of penumbra include endothelin-1-induced focal ischemia, which is designed to moderately and reversibly reduce blood flow via binding sites in the cerebral vessel endothelium in a manner similar to tissue infarction (Macrae et al. 1993). Photothrombosis provides a more controllable lesion size, even targeting individual vessels, and some, though idiosyncratic, reperfusion to examine recovery. MCAO can cause slightly variable stroke core damage size but more reproducible reperfusion. Microhemorrhages induced by femtosecond laser ablation are spatially confined by clotting and cause a slight decrease in nearby capillary flow from mild compression without large-scale changes in dendrite morphology (Rosidi et al. 2011). Sustained microglia/macrophage response, including increased density and directed processes, indicates injury mechanism is primarily inflammatory and not neurodegenerative.

Dendrites exhibit pervasive blebbing in the ischemic core where blood flow is reduced in the vast majority of vessels (10% flow). Extravasation and the accompanying tissue response (e.g., release of glutamate or serum-derived factors) does not appear to be a dendritic damage mechanism in stroke (Zhang and Murphy 2007). The size of the clotted area is correlated to the degeneration speed. Clotted regions >0.2 mm^2 reach damage threshold within 30 min, while a smaller area ~0.1 mm^2

can take up to 3 h (Zhang and Murphy 2007). In the peri-infarct region, both dendrites and vessels orient ~90° with respect to the core rim in the en face plane, yielding a slanted appearance in cross section (Brown et al. 2007). Capillary density significantly increases in the weeks after stroke, while only a slight increase in capillary flow is observed. In the border region, the distance between damaged dendrites and capillaries (6 μm diameter) and small vessels (29 μm diameter) is 84 and 40 μm, respectively.

Excitatory synapses in the brain terminate at dendritic extensions or spines, and considerable focus has been directed toward quantitation of dynamic spine characteristics and the structural and functional relationship between capillaries and spines during injury. The average distance between a spine and capillary in the uninjured mouse brain is ~13 μm, and the average spine is supplied by ~100 red blood cells (RBC)/s (Zhang et al. 2005). Moderate ischemia (50% flow) doesn't affect spines within 5 h while severe ischemia (10% flow) causes loss of spines within 10 min. Reperfusion within an hour is sufficient for spine restoration. While spines in the peri-infarct region have not been examined longitudinally because of the challenge in tracking spines during the dendritic/vascular remodeling described in the previous paragraph, spine density and turnover have been examined over acute timescales (up to 6 h) at 1, 2, and 6 weeks after stroke (Zhang and Murphy 2007). Spine density decreased in the first week and returned to control level after 6 weeks. In a normal animal, spine turnover is very small, on the order of 0.25% gained or lost in a given region. In the peri-infarct region (<500 μm), spine turnover peaks at 1 week after stroke, with spines gained or lost in the range 0.5%–1.5%, and rates do not return to normal after 6 weeks (Figure 11.6).

These imaging results, combined with functional measures (intrinsic optical imaging with hind- and fore-leg stimulations) have provided a clearer picture of the stroke core and peri-infarct transition to normal tissue (Zhang and Murphy 2007). The core is characterized by a lack of function, dendrite degeneration, and flow absence. In a ring approximately 80 μm around the core (the "nouveau" penumbra), function and flow are absent but there is little spine loss. In a ring 300 μm outside the core and nouveau penumbra is the classical penumbra, where function is absent, flow is partial, but spines are normal.

More recently, attention has been paid to understanding the generation and/or remodeling of compensatory sensory pathways during stroke response and recovery. Rapid alteration in sensory processing in the penumbra on the order of minutes may be attributed to the altered function of existing pathways, rather than adaptations (Murphy and Corbett 2009). There is also evidence of the redistribution of activity to the contralateral hemisphere (Mohajerani et al. 2011).

Acute (filament middle cerebral artery occlusion, fMCAO) and chronic (distal middle cerebral artery occlusion, dMCAO) mouse models of stroke were investigated in a multiparametric OCA study (Srinivasan et al. 2013). Tissue infarction, vessel diameter, capillary density, capillary perfusion, and absolute CBF were measured from the reflectance, angiography, and power Doppler flow outputs of a spectral-domain OCT (SD-OCT) system and correlated to histology during experimental stroke and recovery. The OCT images showed development of persistent scattering changes in regions that exhibited nonperfusion during the acute stroke phase, which correlated with altered cellular morphology and indicated permanent infarction (stroke core). The OCT reflectance changes after acute stroke were accompanied by MCA dilation and anterior cerebral artery (ACA) constriction and flow deficit (Figure 11.7). For chronic stroke, MCA flow deficit, MCA branch flow reversal, dilation of MCA and ACA arteries and collaterals, and capillary remodeling occurred in a combined active and passive response to the focal hypoperfusion.

Optical coherence angiography (OCA) has the potential to complement two-photon microscopy studies of neurovascular coupling. While subcellular resolution of neurons remains somewhat elusive (Srinivasan et al. 2012a), OCA can resolve vasculature without extrinsic dyes and quantify flow over wide fields with depth penetration and sectioning similar to TPLSM. Other techniques such as laser speckle imaging have also been applied for chronic vascular imaging during ischemic stroke and recovery, though without depth sectioning capability (Schrandt and Shams Kazmi 2015).

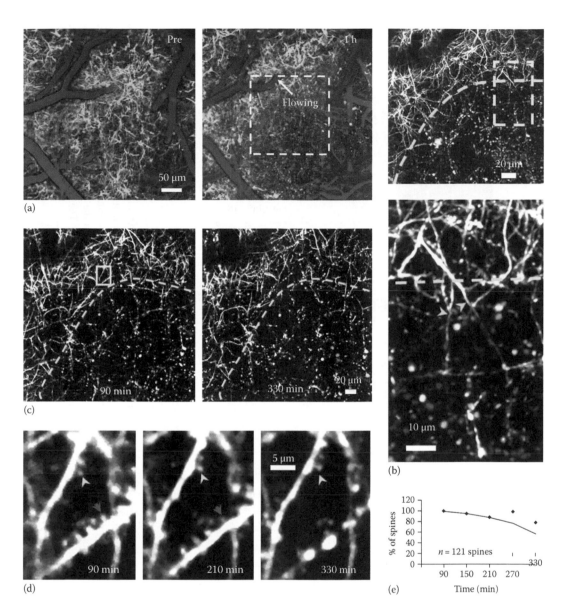

FIGURE 11.6 Localized Stroke Damage to Dendrites and Maintenance of Stable and Sharp Borders for Dendritic Damage. (a) Low-magnification image of the vascular and dendritic structure before and 1 h after photoactivation of RB. (b) Image showing localized stroke damage to dendrites. Top panel, a magnified view of the white boxed region in (a) showing dendritic structure (a substack Z-projection) 90 min after photoactivation of RB; bottom panel, a magnified view of the yellow-boxed region in top panel showing individual dendrites (arrowheads) that were more damaged and beaded when they projected into an ischemic zone. (c) A magnified view of the yellow-boxed region in (a) showing dendritic structure (a substack Z-projection) at 90 and 330 min after RB photoactivation. Note that the sharp border (blue line) indicates a sharp transition from a relatively normal dendritic structure to a beaded and swollen structure. (d) Further magnified view (yellow box in [c]) showing a stable spine (blue arrowhead) and a lost spine (red arrowhead) 5.5 h after stroke near the border region. (e) Spine number expressed as a percentage of the number present at the timepoint 90 min after RB photoactivation (it took time to locate the stroke edge, and data prior to 90 min were not obtained for this region). (From Zhang, S. and Murphy, T.H., Imaging the impact of cortical microcirculation on synaptic structure and sensory-evoked hemodynamic responses in vivo, *PLoS Biol.*, 5(5), e119, 2007. Reproduced with permission from the authors and Public Library of Science. With permission.)

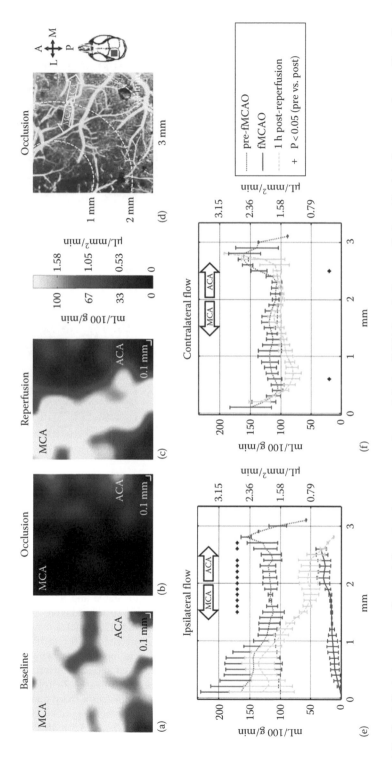

FIGURE 11.7 Spatially heterogeneous flow and diameter changes were observed during fMCAO and after reperfusion. Absolute flow maps show uniform flow at baseline (a), preferentially reduced MCA flow during occlusion (b), and restored MCA flow, with a persistent ACA flow deficit, after reperfusion (c). (d) An angiogram from the same animal during occlusion shows capillary nonperfusion in the lateral portion of the cranial window, presumably delineating the region destined for infarction, as suggested by Figure 11.6. (e and f) Regional blood flow was estimated as a function of distance between the MCA and ACA supplied territories. Flow was restored to baseline values on the MCA side after reperfusion, while a flow deficit persisted on the MCA side after reperfusion, while a flow deficit persisted on the ACA side (e). These changes were not mirrored by the contralateral hemisphere (f).   *(Continued)*

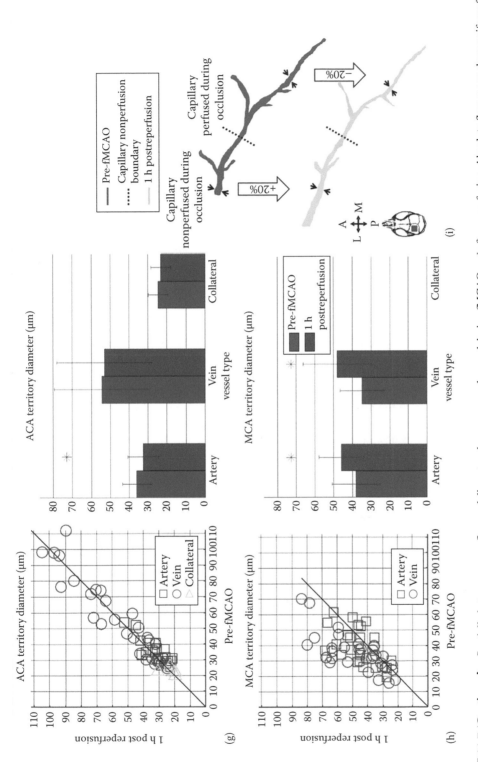

FIGURE 11.7 (Continued) Spatially heterogeneous flow and diameter changes were observed during fMCAO and after reperfusion. Absolute flow maps show uniform flow at baseline (g and h) Consistent with these results, vessels preferentially dilated in the MCA region, and constricted in the ACA region after reperfusion. (i) Remarkably, even along a single artery, both dilation and constriction were observed, depending on the earlier presence of nearby capillary non-perfusion. (From Srinivasan, V.J., Mandeville, E.T., Can, A., Blasi, F., Climov, M., Daneshmand, A., Lee, J.H. et al., Multiparametric, longitudinal optical coherence tomography imaging reveals acute injury and chronic recovery in experimental ischemic stroke, *PLoS One*, 8(8), e71478, 2013. Reproduced with permission from the authors and Public Library of Science. With permission.)

Other Diseases

The mechanisms and features of neurodegenerative diseases such as Alzheimer's are beginning to be uncovered with the help of optical imaging techniques (Tsai et al. 2004). Investigations of other neurological conditions, including Parkinson's, prion disease, developmental disorders, schizophrenia, and epilepsy, could also certainly benefit from the dynamic information acquired with real-time in vivo optical microscopy in appropriate animal models. These fields of study within neuroscience, like the studies of brain-electrode interactions described in the next section, are nascent areas of exploration, and can be guided by the rapid progress made in optical imaging over the past 20 years.

IMPLANTED ELECTRODE INJURY

Relatively few studies have used in vivo imaging to explore the response of local tissue to implanted electrodes. This may be due in part to the inherent technical challenges of this approach, or to a preference for larger animal models for electrode implantation, where optical imaging would be more challenging. However, the proliferation of available mouse lines with fluorescent markers of specific cellular populations, along with advances in imaging modalities with regard to resolution and penetration depth, now provides ample opportunity for the application of these techniques to understanding the biological response to chronically implanted electrodes.

NEURAL ELECTRODE TYPES

Penetrating and surface electrode arrays come in various standard designs and most manufacturers offer custom arrangements suited for specific applications. Figure 11.8 shows a few representative examples of research-grade, commercially available microelectrode arrays from various manufacturers. Penetrating electrode arrays (Figure 11.8a) can record action potentials from single neurons, along with local field potentials derived from membrane voltage changes from local populations of neurons. Depending on the array configuration, signals from groups of neurons within a single cortical layer, or through a cortical column, can be obtained simultaneously. Surface arrays (Figure 11.8b) are less invasive by design and, due to the cortical architecture, record only population-based neural signals, typically called microencephalography (μECoG).

ACUTE AND CHRONIC RESPONSE TO IMPLANTED PENETRATING ELECTRODE

Advanced optical imaging techniques have recently been applied to acute studies of electrode insertion. Multiphoton microscopy used to determine the effect of penetrating electrode insertion in cortex on vasculature demonstrated that rupture of the vasculature predicted worse outcomes for vascular integrity (Kozai et al. 2010). Avoidance of the rupture of major penetrating vessels reduces vascular damage by 82.8% in deep cortical layers and 73.7% in superficial cortical regions. A subsequent study by Kozai and colleagues applied multiphoton microscopy to the neuroinflammatory process observed in the first few hours following microelectrode insertion (Kozai et al. 2012). CX3CR1-GFP transgenic mice with fluorescent expression in microglia were imaged in the 6 h following electrode insertion and demonstrated reorientation of microglial processes toward the electrode at a rate of 1.6 μm/min. This process was most robust within the first 30–45 min following insertion and involved microglia within 100 μm of the electrode surface. Reorientation was not accompanied by movement of the microglia somata, and microglia at distances greater than 200 μm did not appear to leave the ramified state to reorient. Apparent loss of vascular perfusion immediately adjacent to the electrode was observed within the first 7 h post-implantation. Taken together, these results suggest variable damage thresholds depending on vascular integrity, and a local neuroinflammatory response within the first few hours post-insertion. However, this work leaves open the question of long-term neurovascular responses following chronic microelectrode implantation.

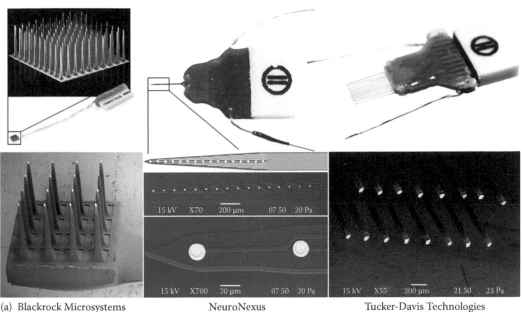

(a) Blackrock Microsystems   NeuroNexus   Tucker-Davis Technologies

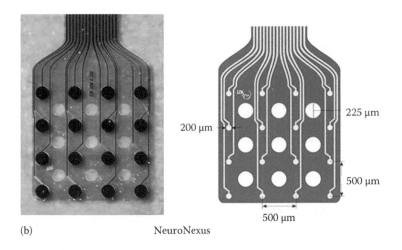

(b)   NeuroNexus

FIGURE 11.8 Photographs, mechanical and wiring diagrams, and SEM images of a representative selection of commercially available research-grade (a) penetrating and (b) surface (μECoG) electrode arrays from various manufacturers. (Courtesy of Blackrock Microsystems, Salt Lake City, UT; NeuroNexus, Ann Arbor, MI; Tucker-Davis Technologies, Alachua, FL.)

An alternate approach uses transparent, nonpenetrating cortical surface electrodes to allow longitudinal observation of vascular remodeling (Schendel et al. 2013). Intravascular dye injection allowed for fluorescence microscopy imaging of vascular dynamics for 1–2 months post device implantation. Hypervascularization was observed within 24 days post-surgery and was often accompanied by microhemorrhages, both of which then stabilized. Recent technological advances in graphene electrode technology have improved the clarity of the electrode contacts, allowing for greater optical access through surface electrodes. Acute studies show fluorescent imaging of functional neural circuitry in tissue slices (Kuzum et al. 2014) and OCT angiography of vasculature below the electrode contacts (Park et al. 2014).

Recent work has applied OCT to the longitudinal analysis of vascular responses to an implanted penetrating, single-shank, microelectrode array over a period of several months (Hammer et al. 2014).

Surface vasculature shows a stereotypic response, with vasodilation occurring 2–3 days post-surgery, and subsiding in large diameter (>50 μm) but not smaller diameter (<50 μm) vessels in the following days. Dilation is followed by hypervascularization at 3–7 days post-surgery, and stabilizing, or in some cases subsiding, by around 20–30 days post-surgery (Figure 11.4). Interestingly, this remodeling is not correlated with penetrating electrode insertion, but instead seems to be related to the surgical procedure of the implantation of a glass window over the brain surface. Moreover, quantification of the dense capillary network directly surrounding the penetrating electrode revealed remarkable stability over the course of many months. These results indicate that single-shank, "Michigan"-style microelectrode arrays do not induce major vascular modifications on a chronic timescale.

With OCA, in addition to structural angiography from the axial planes in the region around an electrode (Figure 11.9), we are also able to quantify flow rates in capillaries, based upon a technique using the autocorrelation function introduced by Srinivasan et al. (2012b). Figure 11.10 shows the structural angiography and flow quantification from one animal with a stable capillary network. We summed composite OCA images at two focal depths, 50 μm above and below the electrode. The electrode in both cases was within the ~100 μm depth of focus of the objective. Only two focal depths are collected (in contrast to typical TPLSM scanning protocols), because many adjacent B-scans must be collected to resolve capillary flow (increasing substantially the scan time), and also because OCA has inherent depth sectioning capability. We summed eight different regions by maximum intensity projection with respect to focus (f) and electrode (e), three of which are shown in Figure 11.9. For a single penetrating electrode in a wild-type mouse, we found very little evidence of capillary remodeling in the region immediately surrounding the electrode and very little variability in flow rates (Figure 11.10) for >6 months post-implantation.

TPLSM and OCA can be combined to provide complementary information on the biological response to implanted electrodes in the mouse cortex. We examined acute (<24 h) and longer-term (weeks) changes in motor cortex from single-shank (Figure 11.11) and four-shank (Figure 11.12) electrode implantation with both TPLSM and OCA in the same transgenic mice (Thy1-EYFP-H). Similar to the result in the wild-type mouse observed in Figure 11.10, the single-shank electrodes seemed to elicit minimal capillary remodeling. Individual capillary segments (arrowheads in middle row) could be tracked and identified over long time periods, showing remarkable stability in fine capillary features. The TPLSM images also revealed presumptive mechanical disruption of dendrites superficial to the plunging electrode, as individual dendrites appear stretched over the electrode

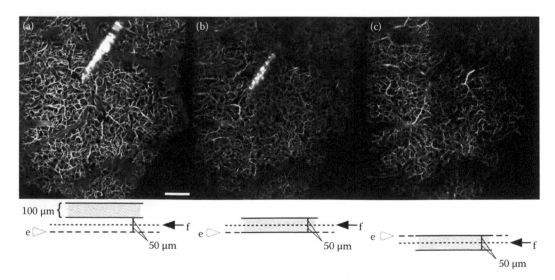

FIGURE 11.9 Capillary network at three slices with respect to electrode (e) and focus (f). Layer thickness is 100 μm. (a) Slice 2, (b) slice 4, and (c) slice 8. Scale bar = 200 μm.

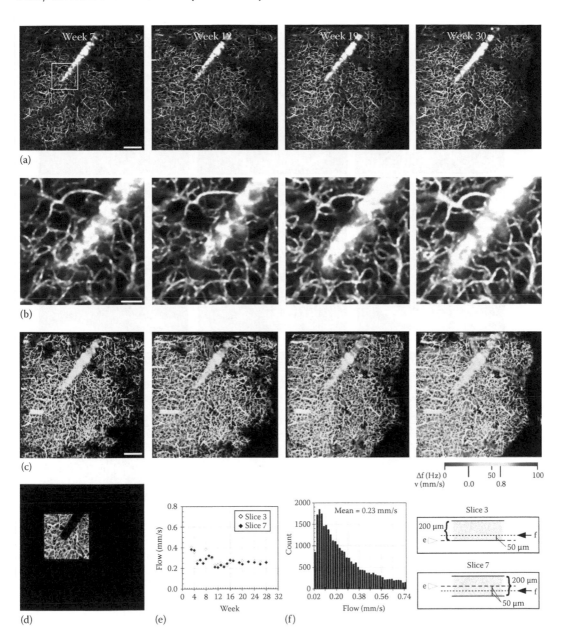

FIGURE 11.10 Minimal capillary remodeling and flow differences were observed in region around electrode. (a) Capillary network visualized in slice 3 (thickness = 200 μm) at four timepoints over the first 7 months. Scale bar = 200 μm. (b) Magnified region (300 × 300 μm) around electrode shows very little vascular remodeling. Scale bar = 50 μm. (c) Flow maps show relatively consistent capillary flow. Scale bar = 200 μm. (d) 500 × 500 μm region of interest excluding electrode used to measure flow. (e) Mean flow in region-of-interest (ROI) for slices 3 and 7 measured 0.28 mm/s over the first 7 months. (f) Example histogram of flow measured in ROI for slice 3 in week 12.

surface (arrowheads in bottom row). Over the next 2 weeks, recovery of typical dendritic structures is apparent, and the presence of the electrode is less distinguishable from the cortical surface.

Upon insertion of a four-shank electrode, we observed a large acute, persistent tissue response at 24 h post-implantation (Figure 11.12). The flow in local capillaries is eliminated and neural processes are fragmented within several hundred microns of the outer boundaries of the electrode array. Observed reductions in capillary flow vacillated in severity over time. After a few days, some

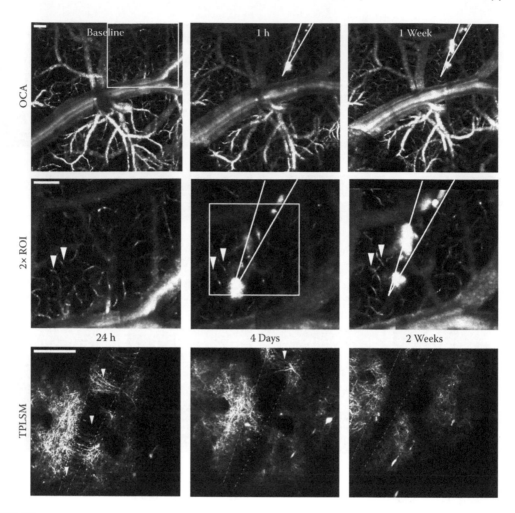

FIGURE 11.11 OCA (MIP, slice 4: thickness = 100 μm) and TPLSM images (MIP, thickness = 100 μm) of mouse motor cortex with single-shank electrode through craniotomy window. Top and middle rows show OCA at baseline before electrode implantation and at one acute and one longer-term timepoint. Middle row is 2× zoom of ROI indicated by box in first panel of first row. Approximate location of electrode is indicated. Arrowheads in zoomed region indicate two stable capillary segments (of many). Bottom row shows TPLSM images at three time-points in the first 2 weeks (approximate location for TPLSM scans is shown in middle OCA panel and approximate location of electrode is indicated by dashed lines). Early mechanical stress on the dendrites from plunging electrode is evident, which appears predominantly at the 24 h timepoint (arrowheads). Scale bar = 100 μm.

capillary flow returned before it began to degrade again over the course of a month, after which it slowly recovered. No TPLSM signal was detected after 96 h. The large changes observed between 1 and 24 h post-insertion do not necessarily contradict the results of Kozai et al. described earlier. Clearly, the damage resultant from multi-shank electrode insertion is multiphasic and multifactorial, and observations at multiple timepoints within the first few hours and for the first few months are necessary to fully characterize the complex electrode-tissue interaction.

From the preceding discussion it is clear that we have a very incomplete understanding of the complex interactions that occur at the brain-electrode interface. In vivo optical imaging will continue to provide qualitative and quantitative data to increase our understanding of the causes of long-term electrode signal degradation. Multimodal imaging approaches, such as the combination of OCA and TPLSM, provide powerful tools to understand complex interactions between molecular, cellular and tissue-level damage mechanisms over chronic durations. For example, Figure 11.13 shows composite TPLSM and

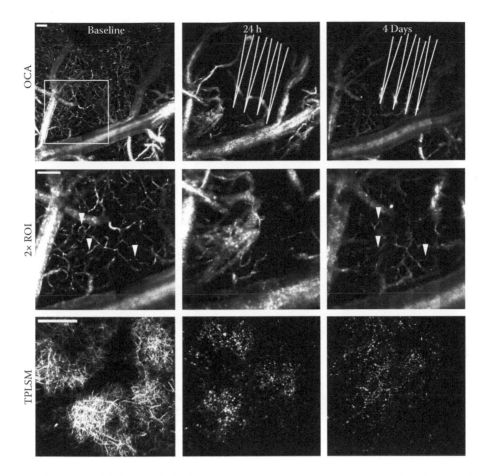

FIGURE 11.12 OCA (MIP, slice 4: thickness = 100 μm) and TPLSM images (MIP, thickness = 100 μm) of mouse motor cortex with four-shank electrode through craniotomy window. Panel arrangement is same as Figure 11.11. Zoomed regions for OCA are in the lower left (indicated in first panel). 24 h timepoint shows massive disruption of capillary network around the electrodes. After 4 days, flow returns to the capillary network. Arrowheads in zoomed region indicate three capillary segments (of many) that reperfuse. TPLSM images are from regions near electrodes. Note loss of dendrite density after 24 h. Scale bar = 100 μm.

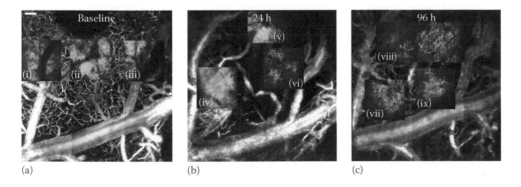

FIGURE 11.13 Composite image of capillary (OCA) and dendritic (TPLSM, green) networks for (a) baseline, (b) 24 h, and (c) 96 h after insertion of a 4-shank electrode. Regions of lower fluorescent signal and abnormal dendritic appearance correspond to regions where no flow is present immediately surrounding electrode shanks. Note especially the change in appearance of the dendrites across the transition zone in the 96 h TPLSM overlay (viii). Both OCA and TPLSM are MIP slices (thickness = 100 μm). Scale bar = 100 μm.

OCA images, overlaid from the same lateral and depth coordinates in the mouse cortex with implanted electrodes, indicating temporal correlation of vascular occlusion and neuronal degeneration. Advanced imaging methods such as these will allow tracking of all neurovascular dynamics in live animals on the micrometer scale, which can provide insight into complex damage reactions, point to short-term biomarkers predictive of long-term device performance, and potentially aid development of more reliable neural devices and better outcomes for the patients that depend on them.

ACKNOWLEDGMENTS

This work was funded by an interagency agreement with DARPA (RE-NET Program, Microsystems Technology Office) and grants from the FDA Office of the Chief Scientist, Critical Path Initiative, and Medical Counter Measures Initiative. We thank everyone in the Neural Implant Program that contributed to this work, especially Do-Hyun Kim, who built the two-photon microscope; Adam Boretsky and Andrea Lozzi, who spent countless hours and tireless work performing surgeries, imaging the animals, and analyzing and producing results; and Gene Civillico, for helpful discussions and thoughtful edits during manuscript preparation. We also thank Victor Krauthamer for motivating the work and providing constant and steady guidance, direction, and support.

The mention of commercial products, their sources, or their use in connection with material reported herein is not to be construed as either an actual or implied endorsement of such products by the US Department of Health and Human Services.

REFERENCES

Andermann, M. L., N. B. Gilfoy, G. J. Goldey, R. N. S. Sachdev, M. Wölfel, D. A. McCormick, R. C. Reid, and M. J. Levene. 2013. Chronic cellular imaging of entire cortical columns in awake mice using microprisms. *Neuron* 80 (4): 900–913. doi:10.1016/j.neuron.2013.07.052.

Attwell, D., A. M. Buchan, S. Charpak, M. Lauritzen, B. A. Macvicar, and E. A. Newman. 2010. Glial and neuronal control of brain blood flow. *Nature* 468 (7321): 232–243. doi:10.1038/nature09613.

Azemi, E., C. F. Lagenaur, and X. T. Cui. 2011. The surface immobilization of the neural adhesion molecule L1 on neural probes and its effect on neuronal density and gliosis at the probe/tissue interface. *Biomaterials* 32 (3): 681–692. doi:10.1016/j.biomaterials.2010.09.033.

Bardehle, S., M. Krüger, F. Buggenthin, J. Schwausch, J. Ninkovic, H. Clevers, H. J. Snippert et al. 2013. Live imaging of astrocyte responses to acute injury reveals selective juxtavascular proliferation. *Nature Neuroscience* 16 (5): 580–586. doi:10.1038/nn.3371.

Barrese, J. C., N. Rao, K. Paroo, C. Triebwasser, C. Vargas-Irwin, L. Franquemont, and J. P. Donoghue. 2013. Failure mode analysis of silicon-based intracortical microelectrode arrays in non-human primates. *Journal of Neural Engineering* 10 (6): 066014. doi:10.1088/1741-2560/10/6/066014.

Biran, R., D. C. Martin, and P. A. Tresco. 2005. Neuronal cell loss accompanies the brain tissue response to chronically implanted silicon microelectrode arrays. *Experimental Neurology* 195 (1): 115–126. doi:10.1016/j.expneurol.2005.04.020.

Brodnick, S. K., M. R. Hayat, S. Kapur, T. J. Richner, M. W. Nonte, K. W. Eliceiri, L. Krugner-Higby, J. C. Williams, and S. O. Poore. 2014. A chronic window imaging device for the investigation of in vivo peripheral nerves. *Conference Proceedings: Annual International Conference of the IEEE Engineering in Medicine and Biology Society* 2014: 1985–1988. doi:10.1109/EMBC.2014.6944003.

Brown, C. E., P. Li, J. D. Boyd, K. R. Delaney, and T. H. Murphy. 2007. Extensive turnover of dendritic spines and vascular remodeling in cortical tissues recovering from stroke. *The Journal of Neuroscience* 27 (15): 4101–4109. doi:10.1523/JNEUROSCI.4295-06.2007.

Burda, J. E. and M. V. Sofroniew. 2014. Reactive gliosis and the multicellular response to CNS damage and disease. *Neuron* 81 (2): 229–248. doi:10.1016/j.neuron.2013.12.034.

Chung, K., J. Wallace, S.-Y. Kim, S. Kalyanasundaram, A. S. Andalman, T. J. Davidson, J. J. Mirzabekov et al. 2013. Structural and molecular interrogation of intact biological systems. *Nature* 497 (7449): 332–337. doi:10.1038/nature12107.

Civillico, E. F. 2011. Voltage-sensitive dye imaging of cortical function in vivo. In *Neuronal Network Analysis*, eds. T. Fellin and M. Halassa, 67: 83–99. Totowa, NJ: Humana Press. http://link.springer.com/10.1007/7657_2011_2.

Cohen, L. B., R. D. Keynes, and B. Hille. 1968. Light scattering and birefringence changes during nerve activity. *Nature* 218 (5140): 438–441. doi:10.1038/218438a0.

Collias, J. C. and E. E. Manuelidis. 1957. Histopathological changes produced by implanted electrodes in cat brains; comparison with histopathological changes in human and experimental puncture wounds. *Journal of Neurosurgery* 14 (3): 302–328. doi:10.3171/jns.1957.14.3.0302.

Collinger, J. L., B. Wodlinger, J. E. Downey, W. Wang, E. C. Tyler-Kabara, D. J. Weber, A. J. C. McMorland, M. Velliste, M. L. Boninger, and A. B. Schwartz. 2013. High-performance neuroprosthetic control by an individual with tetraplegia. *Lancet (London, England)* 381 (9866): 557–564. doi:10.1016/S0140-6736(12)61816-9.

Davalos, D., J. Grutzendler, G. Yang, J. V. Kim, Y. Zuo, S. Jung, D. R. Littman, M. L. Dustin, and W.-B. Gan. 2005. ATP mediates rapid microglial response to local brain injury in vivo. *Nature Neuroscience* 8 (6): 752–758. doi:10.1038/nn1472.

Denk, W., J. H. Strickler, and W. W. Webb. 1990. Two-photon laser scanning fluorescence microscopy. *Science (New York, N.Y.)* 248(4951): 73–76.

Devor, A., E. M. C. Hillman, P. Tian, C. Waeber, I. C. Teng, L. Ruvinskaya, M. H. Shalinsky et al. 2008. Stimulus-induced changes in blood flow and 2-deoxyglucose uptake dissociate in ipsilateral somatosensory cortex. *The Journal of Neuroscience* 28 (53): 14347–14357. doi:10.1523/JNEUROSCI.4307-08.2008.

Dhillon, G. S. and K. W. Horch. 2005. Direct neural sensory feedback and control of a prosthetic arm. *IEEE Transactions on Neural Systems and Rehabilitation Engineering* 13 (4): 468–472. doi:10.1109/TNSRE.2005.856072.

Drew, P. J., A. Y. Shih, J. D. Driscoll, P. M. Knutsen, P. Blinder, D. Davalos, K. Akassoglou, P. S. Tsai, and D. Kleinfeld. 2010. Chronic optical access through a polished and reinforced thinned skull. *Nature Methods* 7 (12): 981–984. doi:10.1038/nmeth.1530.

Drew, P. J., A. Y. Shih, and D. Kleinfeld. 2011. Fluctuating and sensory-induced vasodynamics in rodent cortex extend arteriole capacity. *Proceedings of the National Academy of Sciences of the United States of America* 108 (20): 8473–8478. doi:10.1073/pnas.1100428108.

Edell, D. J., V. V. Toi, V. M. McNeil, and L. D. Clark. 1992. Factors influencing the biocompatibility of insertable silicon microshafts in cerebral cortex. *IEEE Transactions on Bio-Medical Engineering* 39 (6): 635–643. doi:10.1109/10.141202.

Engineer, N. D., A. R. Møller, and M. P. Kilgard. 2013. Directing neural plasticity to understand and treat tinnitus. *Hearing Research* 295 (January): 58–66. doi:10.1016/j.heares.2012.10.001.

Feng, G., R. H. Mellor, M. Bernstein, C. Keller-Peck, Q. T. Nguyen, M. Wallace, J. M. Nerbonne, J. W. Lichtman, and J. R. Sanes. 2000. Imaging neuronal subsets in transgenic mice expressing multiple spectral variants of GFP. *Neuron* 28(1): 41–51.

Flusberg, B. A., A. Nimmerjahn, E. D. Cocker, E. A. Mukamel, R. P. J. Barretto, T. H. Ko, L. D. Burns, J. C. Jung, and M. J. Schnitzer. 2008. High-speed, miniaturized fluorescence microscopy in freely moving mice. *Nature Methods* 5 (11): 935–938. doi:10.1038/nmeth.1256.

Frostig, R. D. 2009. *In Vivo Optical Imaging of Brain Function*, 2nd ed. Frontiers in Neuroscience. Boca Raton, FL: CRC Press.

Frostig, R. D. and C. H. Chen-Bee. 2009. Visualizing adult cortical plasticity using intrinsic signal optical imaging. In *In Vivo Optical Imaging of Brain Function*, ed. R. D. Frostig. Frontiers in Neuroscience. Boca Raton, FL: CRC Press. http://www.ncbi.nlm.nih.gov/books/NBK20227/.

Garaschuk, O., R.-I. Milos, and A. Konnerth. 2006. Targeted bulk-loading of fluorescent indicators for two-photon brain imaging in vivo: Article: Nature protocols. *Nature Protocols* 1 (1): 380–386. doi:10.1038/nprot.2006.58.

Ghosh, K. K., L. D. Burns, E. D. Cocker, A. Nimmerjahn, Y. Ziv, A. E. Gamal, and M. J. Schnitzer. 2011. Miniaturized integration of a fluorescence microscope. *Nature Methods* 8 (10): 871–878. doi:10.1038/nmeth.1694.

Göbel, W. and F. Helmchen. 2007. In vivo calcium imaging of neural network function. *Physiology* 22 (6): 358–365. doi:10.1152/physiol.00032.2007.

Gomez-Nicola, D. and V. H. Perry. 2015. Microglial dynamics and role in the healthy and diseased brain a paradigm of functional plasticity. *The Neuroscientist* 21(2): 169–184.

Grutzendler, J., N. Kasthuri, and W.-B. Gan. 2002. Long-term dendritic spine stability in the adult cortex. *Nature* 420 (6917): 812–816. doi:10.1038/nature01276.

Hama, H., H. Kurokawa, H. Kawano, R. Ando, T. Shimogori, H. Noda, K. Fukami, A. Sakaue-Sawano, and A. Miyawaki. 2011. Scale: A chemical approach for fluorescence imaging and reconstruction of transparent mouse brain. *Nature Neuroscience* 14 (11): 1481–1488. doi:10.1038/nn.2928.

Hammer, D. X., A. Lozzi, E. Abliz, N. Greenbaum, A. Agrawal, V. Krauthamer, and C. G. Welle. 2014. Longitudinal vascular dynamics following cranial window and electrode implantation measured with speckle variance optical coherence angiography. *Biomedical Optics Express* 5 (8): 2823–2836. doi:10.1364/BOE.5.002823.

Haorah, J., S. H. Ramirez, K. Schall, D. Smith, R. Pandya, and Y. Persidsky. 2007. Oxidative stress activates protein tyrosine kinase and matrix metalloproteinases leading to blood–brain barrier dysfunction. *Journal of Neurochemistry* 101 (2): 566–576. doi:10.1111/j.1471-4159.2006.04393.x.

Harris, J. P., A. E. Hess, S. J. Rowan, C. Weder, C. A. Zorman, D. J. Tyler, and J. R. Capadona. 2011. *In vivo* deployment of mechanically adaptive nanocomposites for intracortical microelectrodes. *Journal of Neural Engineering* 8 (4): 046010. doi:10.1088/1741-2560/8/4/046010.

Hays, S. A., R. L. Rennaker, and M. P. Kilgard. 2013. Targeting plasticity with vagus nerve stimulation to treat neurological disease. *Progress in Brain Research* 207: 275–299. doi:10.1016/B978-0-444-63327-9.00010-2.

Heim, N., O. Garaschuk, M. W. Friedrich, M. Mank, R. I. Milos, Y. Kovalchuk, A. Konnerth, and O. Griesbeck. 2007. Improved calcium imaging in transgenic mice expressing a troponin C–based biosensor. *Nature Methods* 4 (2): 127–129. doi:10.1038/nmeth1009.

Helmchen, F. 2009. Two-photon functional imaging of neuronal activity. In *In Vivo Optical Imaging of Brain Function*, ed. R. D. Frostig, 2nd ed. *Frontiers in Neuroscience*. Boca Raton, FL: CRC Press. http://www.ncbi.nlm.nih.gov/books/NBK20230/.

Helmchen, F. and D. Kleinfeld. 2008. Chapter 10. In vivo measurements of blood flow and glial cell function with two-photon laser-scanning microscopy. *Methods in Enzymology* 444: 231–254. doi:10.1016/S0076-6879(08)02810-3.

Holtmaat, A., T. Bonhoeffer, D. K. Chow, J. Chuckowree, V. D. Paola, S. B. Hofer, M. Hübener et al. 2009. Long-term, high-resolution imaging in the mouse neocortex through a chronic cranial window. *Nature Protocols* 4 (8): 1128–1144. doi:10.1038/nprot.2009.89.

Holtmaat, A. and K. Svoboda. 2009. Experience-dependent structural synaptic plasticity in the mammalian brain. *Nature Reviews Neuroscience* 10 (9): 647–658. doi:10.1038/nrn2699.

Holtmaat, A., L. Wilbrecht, G. W. Knott, E. Welker, and K. Svoboda. 2006. Experience-dependent and cell-type-specific spine growth in the neocortex. *Nature* 441 (7096): 979–983. doi:10.1038/nature04783.

Holtmaat, A. J. G. D., J. T. Trachtenberg, L. Wilbrecht, G. M. Shepherd, X. Zhang, G. W. Knott, and K. Svoboda. 2005. Transient and persistent dendritic spines in the neocortex in vivo. *Neuron* 45 (2): 279–291. doi:10.1016/j.neuron.2005.01.003.

Horton, N. G., K. Wang, D. Kobat, C. G. Clark, F. W. Wise, C. B. Schaffer, and C. Xu. 2013. In vivo three-photon microscopy of subcortical structures within an intact mouse brain. *Nature Photonics* 7 (3): 205–209. doi:10.1038/nphoton.2012.336.

Insight, M. 2013. U.S. neurostimulation devices market. *Medical Market and Technology Report*. Bridgewater, NJ: Elsevier Business Intelligence.

Jorfi, M., J. L. Skousen, C. Weder, and J. R. Capadona. 2015. Progress towards biocompatible intracortical microelectrodes for neural interfacing applications. *Journal of Neural Engineering* 12 (1): 011001. doi:10.1088/1741-2560/12/1/011001.

Ke, M.-T., S. Fujimoto, and T. Imai. 2013. SeeDB: A simple and morphology-preserving optical clearing agent for neuronal circuit reconstruction. *Nature Neuroscience* 16 (8): 1154–1161. doi:10.1038/nn.3447.

Keck, T., T. D. Mrsic-Flogel, M. V. Afonso, U. T. Eysel, T. Bonhoeffer, and M. Hübener. 2008. Massive restructuring of neuronal circuits during functional reorganization of adult visual cortex. *Nature Neuroscience* 11 (10): 1162–1167. doi:10.1038/nn.2181.

Keller, P. J., M. B. Ahrens, and J. Freeman. 2015. Light-sheet imaging for systems neuroscience. *Nature Methods* 12 (1): 27–29. doi:10.1038/nmeth.3214.

Kim, S.-P., J. D. Simeral, L. R. Hochberg, J. P. Donoghue, and M. J. Black. 2008. Neural control of computer cursor velocity by decoding motor cortical spiking activity in humans with tetraplegia. *Journal of Neural Engineering* 5 (4): 455–476. doi:10.1088/1741-2560/5/4/010.

Kim, Y.-T., R. W. Hitchcock, M. J. Bridge, and P. A. Tresco. 2004. Chronic response of adult rat brain tissue to implants anchored to the skull. *Biomaterials* 25(12): 2229–2237.

Kleinfeld, D., P. Blinder, P. J. Drew, J. D. Driscoll, A. Muller, P. S. Tsai, and A. Y. Shih. 2011. A guide to delineate the logic of neurovascular signaling in the brain. *Frontiers in Neuroenergetics* 3 (April): 1. doi:10.3389/fnene.2011.00001.

Kleinfeld, D., P. P. Mitra, F. Helmchen, and W. Denk. 1998. Fluctuations and stimulus-induced changes in blood flow observed in individual capillaries in layers 2 through 4 of rat neocortex. *Proceedings of the National Academy of Sciences of the United States of America* 95(26): 15741–15746.

Kobat, D., M. E. Durst, N. Nishimura, A. W. Wong, C. B. Schaffer, and C. Xu. 2009. Deep tissue multiphoton microscopy using longer wavelength excitation. *Optics Express* 17 (16): 13354–13364. doi:10.1364/OE.17.013354.

Kotlikoff, M. I. 2007. Genetically encoded Ca^{2+} indicators: Using genetics and molecular design to understand complex physiology. *The Journal of Physiology* 578 (Pt 1): 55–67. doi:10.1113/jphysiol.2006.120212.

Kozai, T. D. Y., T. C. Marzullo, F. Hooi, N. B. Langhals, A. K. Majewska, E. B. Brown, and D. R. Kipke. 2010. Reduction of neurovascular damage resulting from microelectrode insertion into the cerebral cortex using *in vivo* two-photon mapping. *Journal of Neural Engineering* 7 (4): 046011. doi:10.1088/1741-2560/7/4/046011.

Kozai, T. D. Y., A. L. Vazquez, C. L. Weaver, S.-G. Kim, and X. T. Cui. 2012. *In vivo* two-photon microscopy reveals immediate microglial reaction to implantation of microelectrode through extension of processes. *Journal of Neural Engineering* 9 (6): 066001. doi:10.1088/1741-2560/9/6/066001.

Kuzum, D., H. Takano, E. Shim, J. C. Reed, H. Juul, A. G. Richardson, J. de Vries et al. 2014. Transparent and flexible low noise graphene electrodes for simultaneous electrophysiology and neuroimaging. *Nature Communications* 5: 5259. doi:10.1038/ncomms6259.

Lee, J., J. Y. Jiang, W. Wu, F. Lesage, and D. A. Boas. 2014. Statistical intensity variation analysis for rapid volumetric imaging of capillary network flux. *Biomedical Optics Express* 5 (4): 1160. doi:10.1364/BOE.5.001160.

Lendvai, B., E. A. Stern, B. Chen, and K. Svoboda. 2000. Experience-dependent plasticity of dendritic spines in the developing rat barrel cortex in vivo. *Nature* 404 (6780): 876–881. doi:10.1038/35009107.

Liu, G., A. J. Lin, B. J. Tromberg, and Z. Chen. 2012. A comparison of doppler optical coherence tomography methods. *Biomedical Optics Express* 3(10): 2669–2680.

Liu, X., D. B. McCreery, R. R. Carter, L. A. Bullara, T. G. H. Yuen, and W. F. Agnew. 1999. Stability of the interface between neural tissue and chronically implanted intracortical microelectrodes. *IEEE Transactions on Rehabilitation Engineering* 7(3): 315–326.

Macrae, I. M., M. J. Robinson, D. I. Graham, J. L. Reid, and J. McCulloch. 1993. Endothelin-L-induced reductions in cerebral blood flow: Dose dependency, time course, and neuropathological consequences. *Journal of Cerebral Blood Flow and Metabolism* 13276–284.

Mahmud, M. S., D. W. Cadotte, B. Vuong, C. Sun, T. W. H. Luk, A. Mariampillai, and V. X. D. Yang. 2013. Review of speckle and phase variance optical coherence tomography to visualize microvascular networks. *Journal of Biomedical Optics* 18 (5): 050901–050901. doi:10.1117/1.JBO.18.5.050901.

Mank, M., A. F. Santos, S. Direnberger, T. D. Mrsic-Flogel, S. B. Hofer, V. Stein, T. Hendel et al. 2008. A genetically encoded calcium indicator for chronic in vivo two-photon imaging. *Nature Methods* 5 (9): 805–811. doi:10.1038/nmeth.1243.

Masters, B. R. 2010. The development of fluorescence microscopy. In *Encyclopedia of Life Sciences*, ed. John Wiley & Sons, Ltd. Chichester, U.K.: John Wiley & Sons, Ltd. http://doi.wiley.com/10.1002/9780470015902.a0022093.

McConnell, G. C., H. D. Rees, A. I. Levey, C.-A. Gutekunst, R. E. Gross, and R. V. Bellamkonda. 2009. Implanted neural electrodes cause chronic, local inflammation that is correlated with local neurodegeneration. *Journal of Neural Engineering* 6 (5): 056003. doi:10.1088/1741-2560/6/5/056003.

Minsky, M. 1961. Microscopy apparatus. http://www.google.com/patents/US3013467.

Miyawaki, A. 2005. Innovations in the imaging of brain functions using fluorescent proteins. *Neuron* 48 (2): 189–199. doi:10.1016/j.neuron.2005.10.003.

Mizrahi, A., J. C. Crowley, E. Shtoyerman, and L. C. Katz. 2004. High-resolution in vivo imaging of hippocampal dendrites and spines. *The Journal of Neuroscience* 24 (13): 3147–3151. doi:10.1523/JNEUROSCI.5218-03.2004.

Mohajerani, M. H., K. Aminoltejari, and T. H. Murphy. 2011. Targeted mini-strokes produce changes in interhemispheric sensory signal processing that are indicative of disinhibition within minutes. *Proceedings of the National Academy of Sciences* 108 (22): E183–E191. doi:10.1073/pnas.1101914108.

Mountcastle, V. B. 1997. The columnar organization of the neocortex. *Brain: A Journal of Neurology* 120(Pt 4): 701–722.

Murphy, T. H. and D. Corbett. 2009. Plasticity during Stroke recovery: From synapse to behaviour. *Nature Reviews Neuroscience* 10 (12): 861–872. doi:10.1038/nrn2735.

Nimmerjahn, A., F. Kirchhoff, and F. Helmchen. 2005. Resting microglial cells are highly dynamic surveillants of brain parenchyma in vivo. *Science (New York, N.Y.)* 308 (5726): 1314–1318. doi:10.1126/science.1110647.

Oh, S. W., J. A. Harris, L. Ng, B. Winslow, N. Cain, S. Mihalas, Q. Wang et al. 2014. A mesoscale connectome of the mouse brain. *Nature* 508 (7495): 207–214. doi:10.1038/nature13186.

Paola, D., A. H. Vincenzo, G. Knott, S. Song, L. Wilbrecht, P. Caroni, and K. Svoboda. 2006. Cell type-specific structural plasticity of axonal branches and boutons in the adult neocortex. *Neuron* 49 (6): 861–875. doi:10.1016/j.neuron.2006.02.017.

Park, D.-W., A. A. Schendel, S. Mikael, S. K. Brodnick, T. J. Richner, J. P. Ness, M. R. Hayat et al. 2014. Graphene-based carbon-layered electrode array technology for neural imaging and optogenetic applications. *Nature Communications* 5: 5258. doi:10.1038/ncomms6258.

Pavlov, V. A. and K. J. Tracey. 2012. The vagus nerve and the inflammatory reflex—Linking immunity and metabolism. *Nature Reviews. Endocrinology* 8 (12): 743–754. doi:10.1038/nrendo.2012.189.

Polikov, V. S., P. A. Tresco, and W. M. Reichert. 2005. Response of brain tissue to chronically implanted neural electrodes. *Journal of Neuroscience Methods* 148 (1): 1–18. doi:10.1016/j.jneumeth.2005.08.015.

Potter, K. A., A. C. Buck, W. K. Self, M. E. Callanan, S. Sunil, and J. R. Capadona. 2013. The effect of resveratrol on neurodegeneration and blood brain barrier stability surrounding intracortical microelectrodes. *Biomaterials* 34 (29): 7001–7015. doi:10.1016/j.biomaterials.2013.05.035.

Potter, K. A., A. C. Buck, W. K. Self, and J. R. Capadona. 2012. Stab injury and device implantation within the brain results in inversely multiphasic neuroinflammatory and neurodegenerative responses. *Journal of Neural Engineering* 9 (4): 046020. doi:10.1088/1741-2560/9/4/046020.

Potter, K. A., M. Jorfi, K. T. Householder, E. Johan Foster, C. Weder, and J. R. Capadona. 2014. Curcumin-releasing mechanically adaptive intracortical implants improve the proximal neuronal density and blood-brain barrier stability. *Acta Biomaterialia* 10 (5): 2209–2222. doi:10.1016/j. actbio.2014.01.018.

Prevedel, R., Y.-G. Yoon, M. Hoffmann, N. Pak, G. Wetzstein, S. Kato, T. Schrödel et al. 2014. Simultaneous whole-animal 3D imaging of neuronal activity using light-field microscopy. *Nature Methods* 11 (7): 727–730. doi:10.1038/nmeth.2964.

Ransohoff, R. M. and A. E. Cardona. 2010. The myeloid cells of the central nervous system parenchyma. *Nature* 468 (7321): 253–262. doi:10.1038/nature09615.

Raspopovic, S., M. Capogrosso, F. M. Petrini, M. Bonizzato, J. Rigosa, G. D. Pino, and J. Carpaneto. et al. 2014. Restoring natural sensory feedback in real-time bidirectional hand prostheses. *Science Translational Medicine* 6(222): 222ra19–222ra19. doi: 10.1126/scitranslmed.3006820.

Ravikumar, M., S. Sunil, J. Black, D. S. Barkauskas, A. Y. Haung, R. H. Miller, S. M. Selkirk, and J. R. Capadona. 2014. The roles of blood-derived macrophages and resident microglia in the neuroinflammatory response to implanted intracortical microelectrodes. *Biomaterials* 35 (28): 8049–8064. doi:10.1016/j. biomaterials.2014.05.084.

Rennaker, R. L., J. Miller, H. Tang, and D. A. Wilson. 2007. Minocycline increases quality and longevity of chronic neural recordings. *Journal of Neural Engineering* 4 (2): L1. doi: 10.1088/1741-2560/4/2/L01.

Rosidi, N. L., J. Zhou, S. Pattanaik, P. Wang, W. Jin, M. Brophy, W. L. Olbricht, N. Nishimura, and C. B. Schaffer. 2011. Cortical microhemorrhages cause local inflammation but do not trigger widespread dendrite degeneration. *PLoS One* 6 (10): e26612. doi:10.1371/journal.pone.0026612.

Roth, T. L., D. Nayak, T. Atanasijevic, A. P. Koretsky, L. L. Latour, and D. B. McGavern. 2013. Transcranial amelioration of inflammation and cell death after brain injury. *Nature* 505 (7482): 223–228. doi:10.1038/ nature12808.

Rousche, P. J. and Normann, R. A. 1998. Chronic recording capability of the utah intracortical electrode array in cat sensory cortex. *Journal of Neuroscience Methods* 82 (1): 1–15.

Sakadžić, S., E. Roussakis, M. A. Yaseen, E. T. Mandeville, V. J. Srinivasan, K. Arai, S. Ruvinskaya et al. 2010. Two-photon high-resolution measurement of partial pressure of oxygen in cerebral vasculature and tissue. *Nature Methods* 7 (9): 755–759. doi:10.1038/nmeth.1490.

Sarker, M. H., D.-E. Hu, and P. A. Fraser. 2000. Acute effects of bradykinin on cerebral microvascular permeability in the anaesthetized rat. *The Journal of Physiology* 528 (1): 177–187. doi:10.1111/j.1469-7793.2000.00177.x.

Saxena, T., L. Karumbaiah, E. A. Gaupp, R. Patkar, K. Patil, M. Betancur, G. B. Stanley, and R. V. Bellamkonda. 2013. The impact of chronic blood–brain barrier breach on intracortical electrode function. *Biomaterials* 34 (20): 4703–4713. doi:10.1016/j.biomaterials.2013.03.007.

Schendel, A. A., S. Thongpang, S. K. Brodnick, T. J. Richner, B. D. B. Lindevig, L. Krugner-Higby, and J. C. Williams. 2013. A cranial window imaging method for monitoring vascular growth around chronically implanted micro-ECoG devices. *Journal of Neuroscience Methods* 218 (1): 121–130. doi:10.1016/j. jneumeth.2013.06.001.

Schrandt, C. J., S. M. Shams Kazmi, T. A. Jones, and A. K. Dunn. 2015. Chronic monitoring of vascular progression after ischemic stroke using multiexposure speckle imaging and two-photon fluorescence microscopy. *Journal of Cerebral Blood Flow and Metabolism* 35 (6): 933–942. doi:10.1038/jcbfm.2015.26.

Shen, Z., Z. Lu, P. Y. Chhatbar, P. O'Herron, and P. Kara. 2012. An artery-specific fluorescent dye for studying neurovascular coupling. *Nature Methods* 9 (3): 273–276. doi:10.1038/nmeth.1857.

Shih, A. Y., P. Blinder, P. S. Tsai, B. Friedman, G. Stanley, P. D. Lyden, and D. Kleinfeld. 2013. The smallest stroke: Occlusion of one penetrating vessel leads to infarction and a cognitive deficit. *Nature Neuroscience* 16 (1): 55–63. doi:10.1038/nn.3278.

Shih, A. Y., J. D. Driscoll, P. J. Drew, N. Nishimura, C. B. Schaffer, and D. Kleinfeld. 2012. Two-photon microscopy as a tool to study blood flow and neurovascular coupling in the rodent brain. *Journal of Cerebral Blood Flow and Metabolism* 32 (7): 1277–1309. doi:10.1038/jcbfm.2011.196.

Simeral, J. D., S.-P. Kim, M. J. Black, J. P. Donoghue, and L. R. Hochberg. 2011. Neural control of cursor trajectory and click by a human with tetraplegia 1000 days after implant of an intracortical microelectrode array. *Journal of Neural Engineering* 8 (2): 025027. doi:10.1088/1741-2560/8/2/025027.

Spataro, L., J. Dilgen, S. Retterer, A. J. Spence, M. Isaacson, J. N. Turner, and W. Shain. 2005. Dexamethasone treatment reduces astroglia responses to inserted neuroprosthetic devices in rat neocortex. *Experimental Neurology* 194 (2): 289–300. doi:10.1016/j.expneurol.2004.08.037.

Srinivasan, V. J., E. T. Mandeville, A. Can, F. Blasi, M. Climov, A. Daneshmand, J. H. Lee et al. 2013. Multiparametric, longitudinal optical coherence tomography imaging reveals acute injury and chronic recovery in experimental ischemic stroke. *PLoS One* 8 (8): e71478. doi:10.1371/journal.pone.0071478.

Srinivasan, V. J., H. Radhakrishnan, J. Y. Jiang, S. Barry, and A. E. Cable. 2012a. Optical coherence microscopy for deep tissue imaging of the cerebral cortex with intrinsic contrast. *Optics Express* 20 (3): 2220–2239. doi:10.1364/OE.20.002220.

Srinivasan, V. J., H. Radhakrishnan, E. H. Lo, E. T. Mandeville, J. Y. Jiang, S. Barry, and Alex E. Cable. 2012b. OCT Methods for Capillary Velocimetry. *Biomedical Optics Express* 3 (3): 612–629. doi:10.1364/BOE.3.000612.

Srinivasan, V. J., S. Sakadi, I. Gorczynska, S. Ruvinskaya, W. Wu, J. G. Fujimoto, and D. A. Boas. 2010. Quantitative cerebral blood flow with optical coherence tomography. *Optics Express* 18 (3): 2477–2494. doi:10.1364/OE.18.002477.

Svoboda, K., W. Denk, D. Kleinfeld, and D. W. Tank. 1997. In vivo dendritic calcium dynamics in neocortical pyramidal neurons. *Nature* 385 (6612): 161–165. doi:10.1038/385161a0.

Svoboda, K. and R. Yasuda. 2006. Principles of two-photon excitation microscopy and its applications to neuroscience. *Neuron* 50 (6): 823–839. doi:10.1016/j.neuron.2006.05.019.

Szarowski, D. H., M. D. Andersen, S. Retterer, A. J. Spence, M. Isaacson, H. G. Craighead, J. N. Turner, and W. Shain. 2003. Brain responses to micro-machined silicon devices. *Brain Research* 983 (1): 23–35.

Tan, D. W., M. A. Schiefer, M. W. Keith, J. R. Anderson, J. Tyler, and D. J. Tyler. 2014. A neural interface provides long-term stable natural touch perception. *Science Translational Medicine* 6 (257): 257ra138–257ra138. doi:10.1126/scitranslmed.3008669.

Tasaki, I., A. Watanabe, R. Sandlin, and L. Carnay. 1968. Changes in fluorescence, turbidity, and birefringence associated with nerve excitation. *Proceedings of the National Academy of Sciences of the United States of America* 61 (3): 883–888.

Trachtenberg, J. T., B. E. Chen, G. W. Knott, G. Feng, J. R. Sanes, E. Welker, and K. Svoboda. 2002. Long-term in vivo imaging of experience-dependent synaptic plasticity in adult cortex. *Nature* 420 (6917): 788–794. doi:10.1038/nature01273.

Tsai, J., J. Grutzendler, K. Duff, and W.-B. Gan. 2004. Fibrillar amyloid deposition leads to local synaptic abnormalities and breakage of neuronal branches. *Nature Neuroscience* 7 (11): 1181–1183. doi:10.1038/nn1335.

Turner, J. N., W. Shain, D. H. Szarowski, M. Andersen, S. Martins, M. Isaacson, and H. Craighead. 1999. Cerebral astrocyte response to micromachined silicon implants. *Experimental Neurology* 156 (1): 33–49. doi:10.1006/exnr.1998.6983.

Valmianski, I., A. Y. Shih, J. D. Driscoll, D. W. Matthews, Y. Freund, and D. Kleinfeld. 2010. Automatic identification of fluorescently labeled brain cells for rapid functional imaging. *Journal of Neurophysiology* 104 (3): 1803–1811. doi:10.1152/jn.00484.2010.

Wake, H., A. J. Moorhouse, S. Jinno, S. Kohsaka, and J. Nabekura. 2009. Resting microglia directly monitor the functional state of synapses in vivo and determine the fate of ischemic terminals. *The Journal of Neuroscience* 29 (13): 3974–3980. doi:10.1523/JNEUROSCI.4363-08.2009.

Wang, L. V. 2009. Multiscale photoacoustic microscopy and computed tomography. *Nature Photonics* 3 (9). 503–509. doi:10.1038/nphoton.2009.157.

Wang, L. V. and S. Hu. 2012. Photoacoustic tomography: In vivo imaging from organelles to organs. *Science (New York, N.Y.)* 335 (6075): 1458–1462. doi:10.1126/science.1216210.

Wang, R. K. and S. Hurst. 2007. Mapping of cerebro-vascular blood perfusion in mice with skin and skull intact by optical micro-angiography at 1.3 µm wavelength. *Optics Express* 15 (18): 11402–11412. doi:10.1364/OE.15.011402.

Watson, B. D., W. Dalton Dietrich, R. Busto, M. S. Wachtel, and M. D. Ginsberg. 1985. Induction of reproducible brain infarction by photochemically initiated thrombosis. *Annals of Neurology* 17 (5): 497–504. doi:10.1002/ana.410170513.

Winslow, B. D., M. B. Christensen, W.-K. Yang, F. Solzbacher, and P. A. Tresco. 2010. A comparison of the tissue response to chronically implanted parylene-C-coated and uncoated planar silicon microelectrode arrays in rat cortex. *Biomaterials* 31 (35): 9163–9172. doi:10.1016/j.biomaterials.2010.05.050.

Winslow, B. D. and P. A. Tresco. 2010. Quantitative analysis of the tissue response to chronically implanted microwire electrodes in rat cortex. *Biomaterials* 31 (7): 1558–1567. doi:10.1016/j.biomaterials.2009.11.049.

Woolley, A. J., H. A. Desai, and K. J. Otto. 2013. Chronic intracortical microelectrode arrays induce non-uniform, depth-related tissue responses. *Journal of Neural Engineering* 10 (2): 026007. doi:10.1088/1741-2560/10/2/026007.

Woolley, A. J., H. A. Desai, M. A. Steckbeck, N. K. Patel, and K. J. Otto. 2011. In situ characterization of the brain–microdevice interface using device capture histology. *Journal of Neuroscience Methods* 201 (1): 67–77. doi:10.1016/j.jneumeth.2011.07.012.

Xu, H.-T., F. Pan, G. Yang, and W.-B. Gan. 2007. Choice of cranial window type for in vivo imaging affects dendritic spine turnover in the cortex. *Nature Neuroscience* 10 (5): 549–551. doi:10.1038/nn1883.

Yoder, E. J. and D. Kleinfeld. 2002. Cortical imaging through the intact mouse skull using two-photon excitation laser scanning microscopy. *Microscopy Research and Technique* 56 (4): 304–305. doi:10.1002/jemt.10002.

Yousefi, S., J. Qin, and R. K. Wang. 2013. Super-resolution spectral estimation of optical micro-angiography for quantifying blood flow within microcirculatory tissue beds in vivo. *Biomedical Optics Express* 4 (7): 1214. doi:10.1364/BOE.4.001214.

Zhang, S., J. Boyd, K. Delaney, and T. H. Murphy. 2005. Rapid reversible changes in dendritic spine structure in vivo gated by the degree of ischemia. *The Journal of Neuroscience* 25 (22): 5333–5338. doi:10.1523/JNEUROSCI.1085-05.2005.

Zhang, S. and T. H. Murphy. 2007. Imaging the impact of cortical microcirculation on synaptic structure and sensory-evoked hemodynamic responses in vivo. *PLoS Biology* 5 (5): e119. doi:10.1371/journal.pbio.0050119.

Zhong, Y. and R. V. Bellamkonda. 2007. Dexamethasone-coated neural probes elicit attenuated inflammatory response and neuronal loss compared to uncoated neural probes. *Brain Research* 1148 (May): 15–27. doi:10.1016/j.brainres.2007.02.024.

Ziv, Y., L. D. Burns, E. D. Cocker, E. O. Hamel, K. K. Ghosh, L. J. Kitch, A. El Gamal, and M. J. Schnitzer. 2013. Long-term dynamics of CA1 hippocampal place codes. *Nature Neuroscience* 16 (3): 264–266. doi:10.1038/nn.3329.

Zuo, Y., G. Yang, E. Kwon, and W.-B. Gan. 2005. Long-term sensory deprivation prevents dendritic spine loss in primary somatosensory cortex. *Nature* 436 (7048): 261–265. doi:10.1038/nature03715.

12 Photoacoustic Neuroimaging

Lihong V. Wang, Jun Xia, and Junjie Yao

CONTENTS

INTRODUCTION TO PHOTOACOUSTIC NEUROIMAGING

The photoacoustic effect, first discovered by Alexander Bell in 1880 (Bell 1880), refers to the formation of sound waves following light absorption in an object. The conversion of optical energy to acoustic energy allows visualization of deep tissue optical absorption with acoustically defined spatial resolution. Starting in the 1990s, with the advent of short-pulsed lasers and high-sensitivity ultrasound transducers, the photoacoustic effect began to be utilized for biomedical imaging (Kruger 1994, Karabutov et al. 1996, Oraevsky et al. 1997, Wang et al. 2002). In 2003, Wang et al. (2003a,b) reported the first functional photoacoustic tomography (PAT), which imaged hemodynamic response noninvasively in the rat brain. Since then, the field has been growing rapidly, and PAT is now becoming an important neuroimaging modality (Wang 2009, Hu and Wang 2010, Wang and Hu 2012, Xia and Wang 2013).

Similar to other commonly used neuroimaging modalities, such as functional magnetic resonance imaging (fMRI) and functional near-infrared spectroscopy, PAT images brain functions through the neurovascular coupling effect, which refers to the close correlation between local neural activity and changes in cerebral hemodynamics (Attwell and Iadecola 2002). Because photoacoustic signals originate from optical absorption, PAT readily benefits from endogenous oxyhemoglobin (HbO$_2$) and deoxyhemoglobin (HbR) contrasts. The two forms of hemoglobin possess different optical absorption coefficients (Horecker 1943), and their concentrations can be quantified through spectral inversion. Thus, PAT provides label-free functional brain imaging of oxygen saturation (sO$_2$) and total hemoglobin concentration (C$_{Hb}$). In comparison, the blood-oxygen-level dependent signal in fMRI mainly originates from HbR (Steinbrink et al. 2006), and thus fMRI cannot distinguish between increased blood oxygenation and decreased blood volume (Stein et al. 2009). Compared with purely optical imaging modalities, PAT breaks through the optical diffusion limit, allowing high-resolution imaging of cortical vasculature through an intact scalp. Moreover, the imaging depth and spatial resolution of PAT are fully scalable across both optical and ultrasonic dimensions, providing an unprecedented opportunity to bridge the gap between microscopic and macroscopic neuroimaging. PAT also benefits from advances in exogenous contrast agents, such as organic dyes, nanoparticles, and reporter genes, allowing exploration of molecular pathways underlying neurological disorders (Hammoud et al. 2007, Kim et al. 2010).

In this chapter, we review the advances in photoacoustic neuroimaging. The second section introduces the principle of PAT and describes various photoacoustic neuroimaging systems. The third section highlights representative photoacoustic neuroimaging studies, and the last section discusses further improvements in photoacoustic neuroimaging. It should be noted that PAT also has applications in many other biomedical areas, including oncology (Mallidi et al. 2011), dermatology (Favazza et al. 2011a,b), and cardiology (Taruttis et al. 2012, Wang et al. 2012). Interested readers can refer to Kim et al. (2010), Beard (2011), Mallidi et al. (2011), Wang and Hu (2012), and Xia and Wang (2014) for recent reviews of PAT.

PHOTOACOUSTIC NEUROIMAGING SYSTEMS

PRINCIPLE OF PHOTOACOUSTIC TOMOGRAPHY

As mentioned previously, photoacoustic signals originate from optical absorption. Once an object absorbs light, the absorbed optical energy is converted into heat, which generates a temperature rise. Thermoelastic expansion then takes place, resulting in the emission of acoustic waves. Following a short laser pulse excitation, the initial photoacoustic pressure can be written as

$$p_0(\vec{r}) = \Gamma \mu_a F(\vec{r}), \tag{12.1}$$

where
  $\Gamma(\vec{r})$ is the Gruneisen parameter
  μ_a is the absorption coefficient
  $F(\vec{r})$ is the local fluence (Wang 2008)

Once the initial pressure is generated, the acoustic wave propagates through the tissue and is detected by a single ultrasonic transducer or an array of transducers.

For photoacoustic microscopy (PAM) (Yao and Wang 2013), which uses a focused single-element ultrasonic transducer to raster scan the tissue, the distribution of initial pressure can be approximated by simply back projecting the temporal signal along the acoustic axis. For photoacoustic computed tomography (PACT) (Xu and Wang 2006), whose receiving elements are unfocused and thus have a large acceptance angle, the initial pressure can be reconstructed only by merging data from all transducer elements.

Over the past decade, numerous PACT image reconstruction algorithms have been proposed (Cox and Beard 2005, Xu and Wang 2005, Burgholzer et al. 2007). As this chapter focuses on photoacoustic neuroimaging, we will only briefly discuss the two most commonly used photoacoustic image reconstruction algorithms—the universal back projection (UBP) algorithm (Xu and Wang 2005) and the time reversal (TR) algorithm (Xu and Wang 2004, Yulia et al. 2008). Reviews of PACT image reconstruction are available in the following references: Xu and Wang (2006) and Mark et al. (2009). The UBP algorithm is exact for three scanning geometries, that is, spherical, planar, and cylindrical, and is achieved by back projecting the filtered pressure data onto a collection of concentric spherical surfaces that are centered at each transducer location. Because all the back projecting indexes can be precalculated and stored, UBP can be done in tens of milliseconds (Yuan et al. 2013) and is thus commonly used when computational time is critical. The TR algorithm is realized by solving the wave propagation model backward, using the measured data as the boundary condition. While computationally more intensive, the TR algorithm can be applied to arbitrarily closed surfaces and can incorporate acoustic heterogeneities. This is particularly useful when imaging the brain of large animals and humans, where the acoustic distortion from the skull is nonnegligible (Huang et al. 2012a,b).

PHOTOACOUSTIC MICROSCOPY

Based on the spatial resolution and targeted imaging depth, PAM generally falls into two groups: optical-resolution PAM (OR-PAM) (Maslov et al. 2008), with a lateral resolution ranging from submicrometer to a few micrometers and an imaging depth less than one transport mean free path (TMFP, typically <0.6 mm in the brain), and acoustic-resolution PAM (AR-PAM) (Maslov et al. 2005), with a lateral resolution on the level of tens of micrometers and an imaging depth of 1–10 TMFPs (typically 0.6–6 mm in the brain). In the following, we will discuss each PAM technology, their imaging characteristics, and representative applications in neuroscience.

Optical-Resolution PAM

OR-PAM targets an imaging depth within one TMFP, which is similar to the thickness of a mouse cortex (Maslov et al. 2008, Hu et al. 2011). As shown in Figure 12.1a, in OR-PAM, the excitation laser beam is tightly focused by an objective lens into the tissue, and a single-element ultrasonic transducer detects the resultant photoacoustic signals. Although the ultrasonic transducer is often

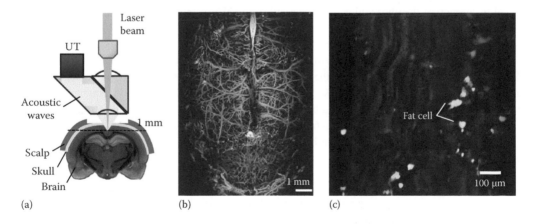

FIGURE 12.1 OR-PAM of the brain. (a) Schematic of second-generation OR-PAM (2G-OR-PAM), where the lateral resolution is determined by the optical focusing. UT, ultrasonic transducer. (b) OR-PAM of mouse cortical vasculature with the scalp removed but the skull intact (Hu et al. 2009). The optical lateral resolution of ~3 μm allows imaging cortical blood vessels on the capillary level, with a penetration depth of less than 1 mm. (c) OR-PAM image of an unsectioned sciatic nerve at 121 0 nm (Matthews et al. 2014). The lipids in myelin provided the image contrast. The wavy fibrous structure in the image is caused by bundles of myelin coated axons that form the nerve fascicles. The bright round structures are probably surrounding fat cells. (Adapted from Hu, S. et al., *J. Biomed. Optics*, 14(4), 3, 2009; Matthews, T.P. et al., *J. Biomed. Optics*, 19(1), 16004, 2014. With permission.)

confocally aligned with the objective lens for optimum detection sensitivity, the optical focusing is generally more than 10 times tighter in diameter than the acoustic focusing, and thus the lateral resolution of OR-PAM is determined by the optical focusing. The axial resolution of OR-PAM is determined by the detection bandwidth of the ultrasonic transducer, which is chosen to match the acoustic path length, taking into account frequency-dependent acoustic attenuation. With adjustments to the numerical aperture (NA) of the optical objective lens and/or the excitation wavelength, OR-PAM has achieved lateral resolutions ranging from 220 nm to 5 µm, with imaging depths ranging from 100 µm to 1.2 mm in chicken breast tissue (Yao and Wang 2013).

Capitalizing on its high spatial resolution, OR-PAM can provide high-quality mouse cortical vasculature images, with the scalp removed but skull intact, as shown in Figure 12.1b (Hu et al. 2009). The cortical vasculature of a mouse brain is well resolved, down to capillaries (Hu et al. 2009). Therefore, OR-PAM is suited for monitoring cortical hemodynamics as surrogates of the underlying neural activities. OR-PAM can also image neurons by using lipids in the myelin as the endogenous imaging contrast. As shown in Figure 12.1c, the wavy fibrous structure in the OR-PAM image of an unsectioned sciatic nerve is caused by bundles of myelin-coated axons that form the nerve fascicles (Matthews et al. 2014). The bright round structures are most likely surrounding fat cells. By using OR-PAM with appropriate contrast agents, it is potentially feasible to study single neuron activities, such as action potential propagation, neural transmitter release, and communications between synapses.

Acoustic-Resolution PAM

AR-PAM targets imaging depths in the quasidiffusive regime, which is more than one TMFP but less than ten TMFPs (Maslov et al. 2005, Zhang et al. 2006). Such an imaging depth is ideal for small animal whole-brain imaging, since the entire mouse brain is about 5–7 mm thick. Similar to OR-PAM, AR-PAM focuses the laser pulses to an area in tissue that coincides with the focal spot of a wideband ultrasonic detector (Figure 12.2a). However, unlike OR-PAM, the laser focus in AR-PAM is intentionally tuned wider than the ultrasonic focal spot, so that the entire volume of the ultrasonic focal zone is adequately illuminated. In this case, the resolution does not closely depend on the tissue's optical scattering characteristics because it is not the optical focusing ability but the ultrasonic focusing that determines the resolution at depths within a few TMFPs. By adjusting the central frequency of the ultrasonic transducer and/or the NA of the acoustic focusing lens, the lateral resolution of AR-PAM can be scaled. As in OR-PAM, the axial resolution of AR-PAM is determined by the bandwidth of the ultrasonic transducer.

Considering the acoustic properties of the mouse scalp and skull, Stein et al. specially engineered AR-PAM for noninvasive imaging of the major mouse cortical vessels with both the scalp and skull intact (Figure 12.2b). By using a focused ultrasonic transducer with a central frequency of 20 MHz and bandwidth of 90%, they achieved a lateral resolution of 70 µm and an axial resolution of 27 µm, with an imaging depth of more than 3.6 mm in the mouse brain (Stein et al. 2009). Significantly, AR-PAM has achieved high-resolution brain imaging beyond the optical diffusion limit with total noninvasiveness.

Scalability of PAM

The key imaging parameters of PAM are highly scalable. By adjusting both the optical illumination and acoustic detection configurations, PAM can scale in spatial resolution, penetration depth, imaging speed, and detection sensitivity over a wide range. Specifically, the lateral resolution of OR-PAM can be improved by either increasing the objective NA or using a shorter excitation wavelength, with the maximum imaging depth scaled accordingly. For example, with a wavelength of 532 nm and an NA of 1.23 implemented with a water immersion objective, subwavelength PAM (SW-PAM) achieved a lateral resolution of 220 nm with a maximum imaging depth of 100 µm (Zhang et al. 2010). In comparison, the lateral resolution of AR-PAM can be scaled by varying the acoustic central frequency and the acoustic lens NA. For example, by using a 5 MHz ultrasonic transducer, deep-penetration photoacoustic macroscopy (PAMac) relaxes the lateral resolution to

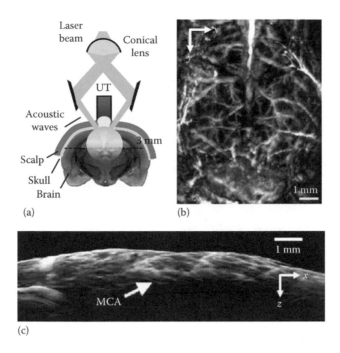

FIGURE 12.2 AR-PAM of the brain. (a) Schematic of the dark-field AR-PAM, where the lateral resolution is determined by the diffraction-limited acoustic focusing. UT, ultrasonic transducer. (b) AR-PAM of the cortical vasculature in a living adult mouse with both the scalp and skull intact (Stein et al. 2009). A penetration depth of ~3 mm can be achieved with an acoustic-diffraction-limited resolution of ~70 μm. (c) x–z projection of the cortical vasculature image by AR-PAM, showing the deep penetration. MCA, middle cerebral artery. (Adapted from Stein, E.W. et al., *J. Biomed. Optics*, 14, 020502, 2009. With permission.)

560 μm and extends the maximum imaging depth to a few centimeters (Song and Wang 2007, 2008). Nevertheless, since the naked eye can already discern features with 50 μm resolution, PAMac is generally not classified as microscopy. Further, there is actually no clear boundary between OR-PAM and AR-PAM, and the transition from OR-PAM to AR-PAM is rather continuous. As optical focusing becomes gradually inefficient with increasing imaging depth, OR-PAM eventually transitions into AR-PAM (Xing et al. 2013).

PHOTOACOUSTIC COMPUTED TOMOGRAPHY

PACT is another major implementation of PAT. Compared to PAM, PACT usually targets at deeper imaging depths with lower spatial resolution. In the following, we will discuss various PACT systems, based on their detection geometry.

Circular-View PACT

A circular-view PACT system provides cross-sectional images of the brain. The first-generation circular-view PACT systems used a single-element transducer that scanned around the object's head (Wang et al. 2003a,b). Because of the mechanical scanning and limited pulse repetition rate of the laser (10 Hz), those systems took several minutes for a 2D scan and several hours for a 3D scan. The imaging speed of a circular-view PACT has now been greatly improved by using a full-ring transducer array (Gamelin et al. 2008, 2009, Li et al. 2010). The system shown in Figure 12.3a is based on a 512-element full-ring transducer array with 50 mm diameter and 5 MHz central frequency (Gamelin et al. 2009). Each element in the array is curved in elevation (Figure 12.3a) to produce a mechanical focal depth of 19 mm. The combined foci of all elements ensures selectivity in elevation.

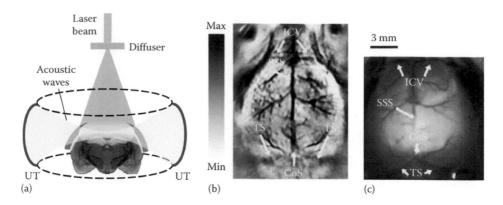

FIGURE 12.3 Circular-view PACT of the brain. (a) Schematic of a circular-view PACT system. UT, ultrasonic transducer. (b) *In vivo* PACT image of the cortical vasculature of an adult mouse with both the scalp and skull intact. (c) Photograph of the cortical vasculature corresponding to (c) with scalp removed. CoS, confluence of sinuses; ICV, inferior cerebral vein; SSS, superior sagittal sinus; TS, transverse sinus. (Reproduced from Nasiriavanaki, M. et al., *Proc. Natl. Acad. Sci. USA*, 111(1), 21, 2014. With permission.)

Figure 12.3b is an *in vivo* photoacoustic image of the cortical vasculature of a 5–6-week-old mouse (~20 g) with both the scalp and skull intact (Nasiriavanaki et al. 2014). It can be seen that the vasculature structures agree well with the open scalp photograph (Figure 12.3c) acquired after the experiment.

The axial resolution of a circular-view PACT system is primarily determined by the bandwidth of the ultrasonic transducer (Xu and Wang 2003). For a wide bandwidth transducer, the axial resolution approximates to $0.8\lambda_c$, where λ_c is the acoustic wavelength at the high cutoff frequency. The lateral resolution is inversely related to the transducer's aperture and the distance between the imaging point and the scanning center (Xu 2009). For the full-ring array system shown in Figure 12.3, the axial resolution is 100 µm, while the lateral resolution ranges from 100 to 150 µm within the 20 mm diameter field of view (Xia et al. 2011).

Planar-View PACT

A planar-view PACT system detects photoacoustic signals along a 2D plane. To ensure a large detection angle, each transducer element should have a small aperture—ideally smaller than the acoustic wavelength. In this regard, a Fabry–Perot interferometer (FPI) is advantageous over a piezoelectric transducer, as its detector size is defined optically by the focal spot of the probe beam (Zhang et al. 2008, Laufer et al. 2009). The FPI sensor is transparent and can be placed on the surface of the animal's head without blocking the excitation laser. Figure 12.4a shows a schematic of an FPI-based PACT system. The excitation laser illuminates the animal brain from the top, while the 1550 nm probe beam is focused and raster scans over the surface of the sensor to map the distribution of the photoacoustic waves arriving at the sensing film. This configuration is equivalent to scanning a single-element piezoelectric transducer with an active area equaling the size of the focal spot. Figure 12.4c is an *ex vivo* image of a mouse brain with an intact scalp and skull. The scanning speed of the system is mainly limited by the pulse repetition rate (10 Hz) of the excitation laser. To minimize motion artifacts due to the long acquisition time (15 min), the mouse was scarified before the experiment (Laufer et al. 2009).

Similar to circular-view PACT, the axial resolution of a planar-view PACT is spatially invariant and determined mainly by the bandwidth and central frequency of the sensor. The lateral resolution depends on the detection aperture, the effective acoustic element size, and the bandwidth of the sensor (Zhang et al. 2008). The system shown in Figure 12.4 achieved 50 µm axial and 50–100 µm lateral resolutions (Zhang et al. 2008).

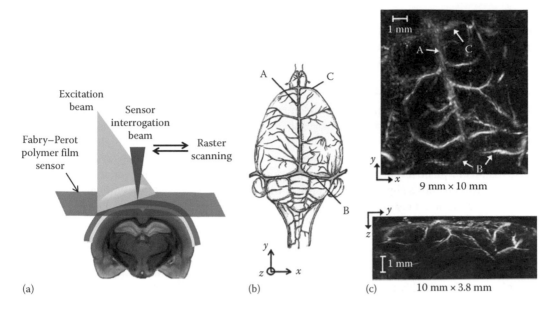

FIGURE 12.4   (a) Experimental arrangement of a planar-view PACT system based on FPI. The Fabry–Perot film sensor was placed on the surface of the mouse head. A focused 1550 nm laser beam was raster scanned over the surface of the sensor in order to map the distribution of the photoacoustic waves arriving at the sensor head. (b) Schematic of superficial cerebral vascular anatomy: A, superior sagittal sinus; B, transverse sinus; C, inferior cerebral vein. (c) x–y (top) and y–z (bottom) maximum intensity projections of the 3D photoacoustic image. (Reproduced from Laufer, J. et al., *Appl. Opt.*, 48(10), D299, 2009. With permission.)

Spherical-View PACT

Compared to planar- and circular-view PACT, a spherical-view PACT system can provide nearly isotropic spatial resolution and is ideal for volumetric imaging (Kruger et al. 2009, Xiang et al. 2013). Figure 12.5a shows a schematic of a hemispherical-view PACT system with 128 elements (Kruger et al. 2009). Each transducer element has a central frequency of 5 MHz and an active area of 3 mm in diameter. To ensure sufficient spatial sampling, the array had to rotate by 16 steps during each volumetric image acquisition, which took 1.6 s as limited by the 10 Hz laser repetition rate.

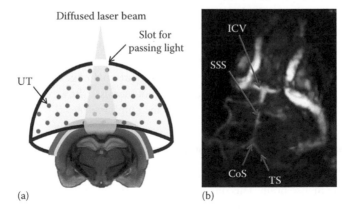

FIGURE 12.5   (a) Schematic of a spherical-view PACT system with 128 elements. (b) Photoacoustic image of mouse brain, acquired using the system in (a). CoS, confluence of sinuses; ICV, inferior cerebral vein; SSS, superior sagittal sinus; TS, transverse sinus. (Adapted from http://www.endrainc.com/. With permission; Courtesy of Dr. Sam Gambhir, Stanford University, Stanford, CA.)

The spatial resolution was measured to range from 190 to 270 μm, depending on the orientation and location of the object. Figure 12.5b is an image of a mouse brain, acquired using the hemispherical-view PACT system. Major cortical vessels, such as the confluence of sinuses (CoS), inferior cerebral vein (ICV), superior sagittal sinus (SSS), and transverse sinus (TS), can be clearly identified.

Comparison of Different PACT Scanning Geometries

Of all the three scanning geometries discussed previously, circular-view PACT is the only one that can acquire a cross-sectional image at each laser shot and is thus particularly useful in studying fast hemodynamics in the brain. A downside of circular-view PACT is the poor elevational resolution. For instance, the elevational resolution in the 512-element full-ring array system is 10 times worse than its in-plane resolution (1 mm vs. 0.1 mm) (Xia et al. 2011). However, the poor elevation resolution also enables sampling of the entire cortical layer of a mouse brain without 3D scanning. Further, the elevational resolution has been improved to 0.6 mm using additional elevational scanning and 3D image reconstruction (Xia et al. 2011). The planar- and spherical-view PACT systems can provide volumetric images with relatively uniform resolutions along all three axes and can be used to study the vascular structures of the brain. However, these systems can produce an image only after a complete 3D scan, making real-time imaging challenging. Advanced image reconstruction algorithms, such as highly constrained back projection (Kruger et al. 2009), have been proposed to address this issue.

PHOTOACOUSTIC NEUROIMAGING STUDIES

In the following, we highlight representative photoacoustic neuroimaging studies. Photoacoustic neuroimaging of small animals, that is, mice and rats, using endogenous contrasts will be discussed first, followed by imaging with exogenous contrasts, and then studies performed on large animals and an adult human skull.

SINGLE-CELL PHOTOACOUSTIC FLOW OXIMETRY IN THE MOUSE BRAIN

By using hemoglobin as an endogenous oxygen tracer, OR-PAM is capable of studying oxygen metabolism in the brain. Recently, by integrating fine spatial and temporal scales, single-cell photoacoustic flowoxigraphy, a new implementation of OR-PAM, is capable of imaging oxygen release from single red blood cells (RBCs) *in vivo* (Wang et al. 2013). As shown in Figure 12.6, by fast line scanning (20 Hz) along a capillary with two wavelength excitations, photoacoustic flowoxigraphy can simultaneously measure multiple hemodynamic parameters that are required to quantify the oxygen release rate by RBCs, such as sO_2 (Figure 12.6a) and RBC flow speed (Figure 12.6b). The oxygen release rate is closely related to the local oxygen metabolism of neural cells. Experimental results show that photoacoustic flowoxigraphy can image the coupling between neural activity and oxygen delivery in response to different physiological challenges, such as visual stimulation (Figure 12.6c) and acute systematic hypoglycemia. Photoacoustic flowoxigraphy can be extremely useful in understanding how the brain is powered at the single-cell level.

REAL-TIME PHOTOACOUSTIC TOMOGRAPHY OF CORTICAL HEMODYNAMICS IN THE MOUSE BRAIN

A full-ring array PACT system allows real-time monitoring of the hemodynamics within the entire cerebral cortex of the mouse brain. This imaging capability was first demonstrated by Li et al. (2010), who monitored the wash-in process of a contrast dye in a Swiss Webster mouse brain. After the control image was acquired (Figure 12.7a), about 0.1 mL of Evans blue dye at 3% concentration was administered through tail vein injection. The entire cortical region was then continuously imaged at 620 nm (the peak absorption wavelength for Evans blue) at a temporal resolution of 1.6 s/frame. Since the optical absorption by blood is weak at 620 nm (Li and Wang 2009), the control image

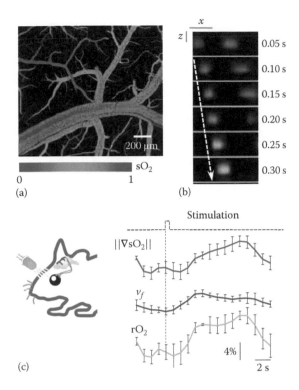

FIGURE 12.6 Single-cell label-free PAM of oxygen metabolism *in vivo* (Wang et al. 2013). (a) sO_2 mapping of the brain vasculature. The yellow dashed box indicates a capillary chosen for oxygen unloading measurement. (b) Single-cell oxygen unloading was measured by fast line scanning along a capillary with two wavelength excitations. Blood flows from left to right. The dashed arrow follows the trajectory of a single flowing RBC. Scale bars: $x = 10$ µm, $z = 30$ µm. (c) OR-PAM of neuron–RBC coupling in the mouse visual cortex. The eye of a mouse was stimulated by a flashing LED (left), and transient responses to a single visual stimulation were monitored. Clear increases were observed in the magnitude of the sO_2 gradient ($\|\nabla sO_2\|$), blood flow speed (v_f), and oxygen unloading rate (rO_2). (Adapted from Wang, L. et al., *Proc. Natl. Acad. Sci. USA*, 110(15), 5759, 2013. With permission.)

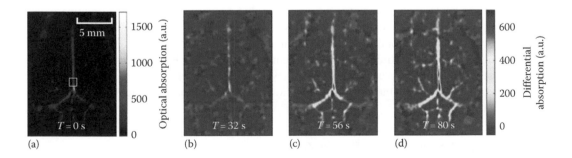

FIGURE 12.7 Noninvasive imaging of the wash-in process of Evans blue in cortical vasculature. (a) Photoacoustic image obtained before the dye injection. (b–d) PA image difference relative to (a) at time 32, 56, and 80 s, respectively. All these images were obtained at 620 nm wavelength. (Reproduced from Li, C. et al., *J. Biomed. Optics*, 15(1), 010509-010501-010503, 2010. With permission.)

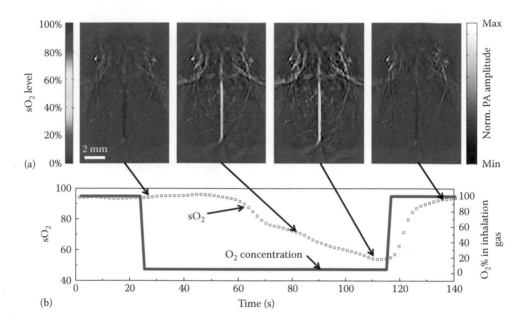

FIGURE 12.8   (a) Photoacoustic images of cortex vasculature acquired noninvasively at 650 nm laser wavelength. (b) Average sO_2 level within the superior sagittal sinus. (Adapted from Xia, J. et al., *Opt. Lett.*, 38(15), 2800, 2013. With permission.)

(Figure 12.7a) misses many fine structures. With the increase in the optical absorption by Evans blue during the wash-in process, more vascular structures appeared, as shown in Figure 12.7b through d.

Using the same system, Xia et al. imaged the cerebral hemodynamics in response to different oxygenation concentrations in the inhalation gas (Xia et al. 2013). A ratio-based sO_2 calculation method was developed to circumvent the estimation of local fluence. During the experiment, two lasers, both operating at 10 Hz, alternately illuminated the object, with a short delay of 75 μs. Therefore, at each imaging frame, the two images (acquired at 650 and 750 nm) were captured at nearly the same time. The pixel-by-pixel amplitude ratio of the two images was utilized for the sO_2 calculation. Compared to the absolute photoacoustic amplitude, the signal ratio is less sensitive to differences in light attenuation at different wavelengths and thus provides a more accurate estimation of the sO_2 level. Figure 12.8a shows photoacoustic images acquired at 650 nm. The sO_2 values within the SSS were superposed on each image. Because 650 nm is an HbR-sensitive wavelength, an increase in signal amplitude indicates a drop in the sO_2 level. Figure 12.8b shows the average sO_2 value within the superior sagittal sinus. It can be seen that a complete transition from hyperoxia to hypoxia took 85 s, while the subsequent recovery to hyperoxia took only 25 s. This observation is consistent with other brain challenge studies (Stein et al. 2009).

The imaging speed (1.6 s/frame) of the 512-element circular-view PACT is mainly limited by its custom-made 64-channel data acquisition (DAQ) system. DAQ systems with hundreds of channels are now commercially available and will soon allow high-speed neuroimaging at the laser's pulse repetition rate.

PHOTOACOUSTIC TOMOGRAPHY OF RESTING-STATE FUNCTIONAL CONNECTIVITY IN THE MOUSE BRAIN

Resting-state functional connectivity imaging is an emerging neural research approach that aims to identify low-frequency, spontaneous cerebral hemodynamic fluctuations and their associated functional connections (Fox and Raichle 2007, Deco et al. 2011, Buckner et al. 2013). Recent research finds that, in a healthy brain, resting-state hemodynamic fluctuations correlate interhemispherically between bilaterally homologous regions, as well as intrahemispherically within the same functional regions. In patients with brain disorders, such as Alzheimer's, correlation between certain regions was found to be greatly reduced (Deco et al. 2011, Bero et al. 2012). The task-free nature of resting-state imaging makes it an appealing approach for animal studies. However, while functional connectivity magnetic resonance imaging has been successfully used in humans, it cannot be easily applied to mice, a species with the widest variety of neurological disease models (Cazzin and Ring 2010). The reason for this is in mouse brain MRI, the ratio of air space to tissue volume is much higher than that of human MRI. The high ratio causes strong local magnetic field gradients, which results in image distortion and signal losses (Guilfoyle et al. 2013). Moreover, the ultra-high magnetic field used in small animal imaging may stimulate the peripheral nerves and affect the normal brain function (Schaefer et al. 2000).

Using the full-ring array PACT system, Nasiriavanaki et al. successfully achieved high-resolution functional connectivity imaging of the mouse brain (Nasiriavanaki et al. 2014). The experiment was performed noninvasively on 3–4-month-old male Swiss Webster mice. To extract the resting-state signal, the reconstructed temporal photoacoustic images were processed through spatial smoothing, mean pixel value subtraction, temporal filtering (to the functional connectivity frequency band), and global regression. A seed-based algorithm was then used to derive the functional connectivity maps. The results, shown in Figure 12.9a through c, clearly indicate bilateral correlations in eight main regions as well as several subregions. A unique advantage of PACT is that the functional connectivity maps are automatically coregistered with high-resolution cortical vascular images, allowing to pinpoint the exact location of neural activity. In addition to the seed-based approach, a parcellation algorithm was developed to generate the functional divisions in a data-driven manner. The resulting parcellation map (Figure 12.9d) agrees well with the standard functional brain atlas (Figure 12.9c). This study indicates that functional connectivity PACT can be used as a noninvasive tool for the diagnosis and therapeutic monitoring of brain diseases.

STIMULATION-BASED FUNCTIONAL PHOTOACOUSTIC NEUROIMAGING

In 2003, PAT was developed for the first time to study the brain response to stimulation (Wang et al. 2003a,b). The experiment was performed on adult Sprague-Dawley rats, and hemodynamic changes in the whisker-barrel cortex were observed. Using the full-ring array PACT system, Yao et al. refined the study in 2013 and imaged both hemodynamic and metabolic responses to forepaw stimulation (Yao et al. 2013). Metabolic imaging was enabled by a fluorescent 2-deoxyglucose analog (2-NBDG). Two wavelengths were used: 478 nm, the peak absorption wavelength for 2-NBDG, and 570 nm, the isosbestic wavelength for Hb and HbO_2 absorption. Thirty minutes after the tail vein injection of 2-NBDG, the mouse was stimulated through needle electrodes inserted under the skin of the forepaws. Each paw was stimulated twice, each for 3 min with a 3 min rest period between stimulations. As shown in Figure 12.10, the stimulation caused an increase in both hemodynamics and glucose metabolism, indicating close coupling between oxygen metabolism and glucose metabolism in the brain. It can also be seen that the glucose response area was more homogenous and confined within the somatosensory region, while the hemodynamic response area had a clear vascular pattern and spread wider than the somatosensory. This is because the glucose response is a focal activity induced by neurons within the stimulated site, while the hemoglobin response may spread into neighboring vessels of the same vascular network due to blood flow (Yao et al. 2013). As a nonradioactive imaging approach, 2-NBDG–PACT has great potential to be used as a fluorodeoxy-glucose positron emission tomography (FDG-PET) alternative for studying the cerebral metabolism and its correlation with neurological diseases (Baxter et al. 1985, Bremner et al. 1997).

Vascular response to electrical stimulation can also be studied at the microscopic level using OR-PAM (Tsytsarev et al. 2011). The experiment was also performed on a Swiss Webster mouse. To ensure light focus on the cortical surface, a cranial opening (4×4 mm^2) was made using a dental drill.

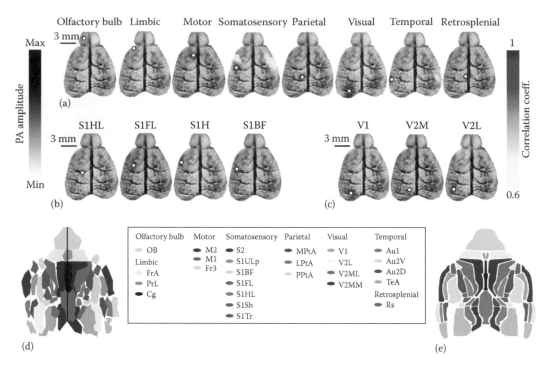

FIGURE 12.9 Functional connectivity maps in a live mouse brain acquired noninvasively by PACT. Correlation maps of (a) the eight main functional regions, (b) the four subregions of the somatosensory cortex, and (c) the three subregions of the visual cortex. White circles indicate the seed regions. (d) Parcellation map of the mouse in (a–c). (e) Corresponding functional regions from the Paxinos histological atlas. The regions and their subregions indicated in the atlas are as follows: Au, auditory cortex; Au1, primary auditory cortex; Au2D, secondary auditory—dorsal area; Au2V, secondary auditory—ventral area; Cg, cingulate; Fr3, frontal cortex area 3; FrA, frontal association; LPtA, lateral parietal association; M1, primary motor cortex; M2, secondary motor cortex; M, motor cortex; MPtA, medial parietal association; OB, olfactory bulb; P, parietal region; PPtA, posterior parietal association; PrL, prelimbic; RS, retrosplenial area; S1ULp, primary somatosensory—upper lips region; S1BF, primary somatosensory—barrel field; S1FL, primary somatosensory—forelimb region; S1HL, primary somatosensory cortex—hindlimb region; S1Sh, primary somatosensory—shoulder region; S1Tr, primary somatosensory cortex—trunk region; S2, secondary somatosensory; TeA, temporal association cortex; V1, primary visual cortex; V2, secondary visual cortex; V2MM, secondary visual cortex—mediomedial region; V2ML, secondary visual cortex—mediolateral region; and V2L, secondary visual cortex—lateral region. (Reproduced from Nasiriavanaki, M. et al., *Proc. Natl. Acad. Sci. USA*, 111(1), 21, 2014. With permission.)

The exposed dura mater surface was then cleaned with artificial cerebrospinal fluid. Direct electrical stimulation was induced through a 10 μm diameter tungsten electrode, which was placed into the cortex to a depth of 0.1–0.2 mm through the opening (Tsytsarev et al. 2011). Before stimulation, the sO_2 within a small region of interest around the tip of the microelectrode (Figure 12.11a) was quantified using two wavelengths (570 and 578 nm). B-scan imaging was then performed along the dashed line in Figure 12.11a for hemodynamic monitoring at various stimulation intensities. Figure 12.11b and c demonstrates the vasoconstriction and vasodilatation of the same microvessel (V1 in Figure 12.11a) in response to electrical stimulations at different current levels. It can be seen that the 100 μA stimulation induced vasoconstriction, while the 150 μA stimulation caused vasodilation. It was also found that 107 μA was the critical stimulation intensity corresponding to the transition from vasoconstriction to vasodilatations, indicating both coexistence and competition between the two responses. Because the vascular response was caused by reactions in cortical neurons, smooth muscle cells, and astrocytes, this study indicates that OR-PAM is a promising tool for *in vivo* studies of neurovascular coupling at the microscopic level.

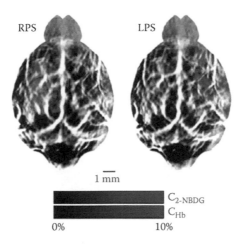

FIGURE 12.10 Overlaid images showing the relative changes of 2-NBDG concentration ($C_{2\text{-NBDG}}$, shown in blue) and total C_{Hb} (shown in red), superimposed on the resting-state image at 570 nm (shown in gray). RPS: right paw stimulation, LPS, left paw stimulation. (Reproduced from Yao, J. et al., *NeuroImage*, 64, 257, 2013. With permission.)

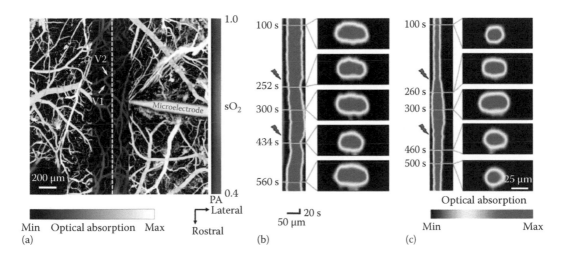

FIGURE 12.11 (a) Superimposed open-skull photoacoustic images of the mouse cortical microvasculature. The maximum-amplitude projection image acquired at 570 nm is shown in gray scale, and the vessel-by-vessel hemoglobin sO_2 mapping of a smaller region calculated from dual-wavelength measurements is shown in color scale. (b and c) B-scan monitoring of vasoconstriction and vasodilatation induced by direct electrical stimulations at (b) 100 µA and (c) 150 µA. In each panel, the left column is the time course of the change in vessel diameter. The right column is the vessel cross-sectional image at different time points, indicated by the green lines. The red lightning symbol indicates the onset of the stimulation. (Reproduced from Tsytsarev, V. et al., *J. Biomed. Optics*, 16, 076002, 2011. With permission.)

PHOTOACOUSTIC TOMOGRAPHY OF MOUSE CEREBRAL EDEMA INDUCED BY COLD INJURY

Besides hemoglobin, water is another intrinsic contrast for photoacoustic imaging. It has a peak absorption coefficient around 975 nm, with a full width at half maximum of 920–1040 nm (Xu et al. 2010). Taking advantage of this fact, Xu et al. imaged mouse cerebral edema using the full-ring array PACT system (Xu et al. 2011). After a control experiment on the healthy brain, cold injury was induced by placing an aluminum tube filled with liquid nitrogen on the mouse scalp for 30 s. The mouse was then imaged 12, 24, and 36 h after the cold injury, using two wavelengths (610 and 975 nm), to track the changes in both blood volume and water distribution. Figure 12.12 shows the brain images acquired *in vivo* by the PACT system and subsequently acquired *ex vivo* by MRI. All images were acquired approximately 1 mm beneath the brain surface. The 610 nm images (Figure 12.12a and b) indicate the decreased amount of blood in the vessel during the first 24 h after the cold injury, while the 975 nm images (Figure 12.12c and d) show an accumulation of water after the cold injury. These observations are consistent with brain cold injury studies (Frei et al. 1973). Immediately after the PACT experiment, the mouse was

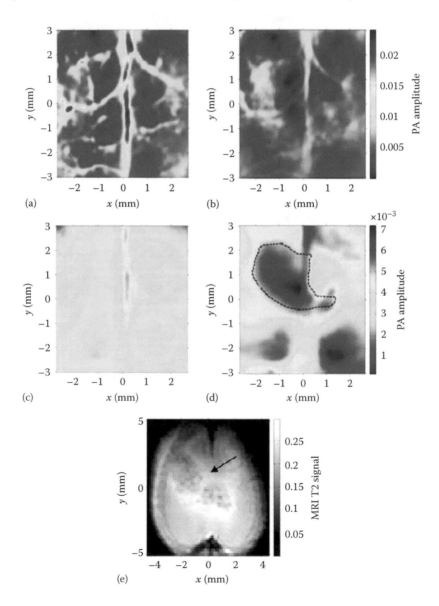

FIGURE 12.12 Transcranial photoacoustic images of the mouse brain acquired noninvasively before cold injury (a, c) and 24 h after cold injury (b, d). The upper and lower row images were acquired at 610 and 975 nm wavelengths, respectively. The dashed line area in (d) outlines the contour area according to 60% of the maximum water signal acquired at 975 nm. (e) MRI image of the mouse cortex, taken immediately after the mouse, was sacrificed. The edema is indicated by the arrow. (Reproduced from Xu, Z. et al., *J. Biomed. Optics*, 16(6), 066020, 2011. With permission.)

sacrificed and scanned in a small-animal MRI scanner. The T2-based MRI image (Figure 12.12e) agrees well with the PACT image in Figure 12.12d.

Other endogenous chromophores, such as lipid (Allen 2012) and cytochrome (Zhang et al. 2013), can also be imaged using PAT and may enable structural brain imaging. In a pilot study, Lou et al. demonstrated that different brain structures, such as the striatum, hippocampus, ventricles, and cerebellum, can be clearly imaged in an exposed mouse brain, possibly based on the lipid and cytochrome contrasts (Lou et al. 2014).

MOLECULAR PHOTOACOUSTIC NEUROIMAGING

PAT also benefits from the wide choice of optical probes for molecular brain imaging. In 2008, Li et al. reported that PAT can provide simultaneous functional and molecular imaging of a brain tumor, with a high spatial resolution (Li et al. 2008). The molecular probe used in the study was a near-infrared dye named IRDye800-NHS (Li-Cor, Inc.), conjugated with a cyclic peptide, cyclo(Lys-Arg-Gly-Asp-Phe) [c(KRGDf) for short]. IRDye800-c(KRGDf) targets $\alpha_v\beta_3$ integrin, which is overexpressed in the U87 glioblastoma tumor. The experiment was performed on a nude mouse with implanted human U87 glioblastoma tumor cells in the brain. To ensure blood–brain barrier penetration, IRDye800-c(KRGDf) was administered through the tail vein with mannitol. Twenty hours after the injection, the mouse was imaged using a single-element circular-view PACT system with four wavelengths: 764, 784, 804, and 824 nm. These four wavelengths were chosen for spectral separation of Hb, HbO_2, and IRDye800-c(KRGDf). Functional images of sO_2 and C_{Hb} are shown in Figure 12.13a and b, respectively. The tumor region (red arrow, Figure 12.13a) expresses lower sO_2 than the surrounding normal tissue, indicating hypoxic tumor vasculature. Higher C_{Hb} can be observed around the same region due to tumor angiogenesis. Figure 12.13c is a composite image with the segmented molecular distribution superposed on the structural image of the brain cortex. It can be seen that the contrast agent mainly accumulates at the tumor region. The study indicates that PAT can noninvasively evaluate the neoangiogenesis and the invasive potential of brain tumors.

In addition to organic dyes, nanoparticles have also been widely used in photoacoustic neuroimaging for vascular contrast enhancement (Wang et al. 2004, Yang et al. 2007), cancer detection (Ray et al. 2011), drug delivery (Yuan et al. 2014), and multimodality imaging (Kircher et al. 2012). However, while nanoparticles normally have higher molar absorption coefficients and better wavelength tunability than organic dyes, the nanotoxicity of these materials is still unclear (Gwinn and Vallyathan 2006, Lovell et al. 2011).

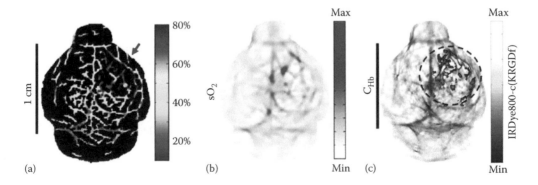

FIGURE 12.13 *In vivo* molecular imaging of a nude mouse brain with a U87 glioblastoma xenograft. (a) sO_2 image. The arrow indicates the tumor region. (b) C_{Hb} image. (c) Composite of segmented molecular image of IRDye800-c(KRGDf) and structural image. (Reproduced from Li, M.-L. et al., *Proc. IEEE*, 96(3), 481, 2008. With permission.)

PHOTOACOUSTIC NEUROIMAGING OF LARGE ANIMALS

Beyond small-animal neuroimaging, PAT has also been successfully applied to large animals, such as sheep (Petrov et al. 2005, 2012) and monkeys (Yang and Wang 2008, Jo et al. 2012, Yang 2014). The sheep has a thick scalp and skull (4 and 6 mm, respectively), which are comparable to that of humans (4 and 8–10 mm, respectively) (Petrov et al. 2012), while the monkey brain is structurally similar to the human brain. Thus, these studies indicate possible translation of PAT to human brain imaging.

The effects of a thick scalp and skull on photoacoustic neuroimaging were first studied by Yang and Wang (2008). The experiment was performed on a fresh head sample of a 20-month-old rhesus monkey, whose skull thickness was approximately 2 mm, similar to that of a human infant. A single-element circular-view PACT system was employed in the study. Since the ultrasonic signals propagating through a thick skull are less distorted at lower frequencies (0.5–1 MHz) (Fry 1977), the experiment utilized a 1 MHz central frequency transducer with 60% bandwidth and 12 mm diameter element size. In comparison, most of the mouse brain studies were performed using transducers with 5 MHz or higher central frequencies. A Nd:YAG laser operating at 1064 nm was used as the excitation source. The experiment was first performed with intact scalp and skull bone. The result (Figure 12.14a) shows hardly any brain features. The scalp and the skull bone were then removed in sequence, and the resulting images are shown in Figure 12.14b and c, respectively. The photoacoustic image (Figure 12.14b) through the exposed skull bone shows high correlation (0.7) with the photoacoustic image (Figure 12.14c) of the exposed brain, indicating that the effect of the skull bone alone on PACT is not significant. The presence of the scalp significantly degraded the photoacoustic image (Figure 12.14a), probably because (1) the scalp and dura mater generated photoacoustic signals that clutter the cortex signal, and (2) they severely limited the laser penetration, and thus reduced the image's SNR (Yang and Wang 2008). However, major cortical vessels can still be

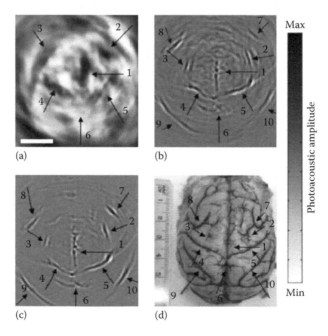

FIGURE 12.14 (a) PAT of a rhesus monkey brain with intact scalp and skull bone. Scale bar: 10 mm. (b) PAT of the monkey brain with the skull bone exposed. (c) PAT of the exposed monkey brain. (d) Photograph of the exposed monkey brain. The blood vessels that can be identified from photoacoustic images have matching numbers. (Reproduced from Yang, X. and Wang, L.V., *J. Biomed. Optics*, 13(4), 044009, 2008. With permission.)

identified in Figure 12.14a, indicating that PAT can be used for cerebral blood oxygenation monitoring in large animals or human infants.

PHOTOACOUSTIC TOMOGRAPHY THROUGH A WHOLE ADULT HUMAN SKULL

A human skull poses more challenges for photoacoustic neuroimaging, as it is thicker (7–11 mm thickness) and strongly attenuates both light and ultrasound, significantly degrading the image's SNR. In addition, the skull also introduces severe acoustic signal aberration associated with acoustic wave reflection and refraction within the skull. In principle, if the photoacoustic signals possess sufficient SNR, acoustic distortions can be corrected through advanced image reconstruction algorithms (Huang et al. 2012a,b). Thus, it is essential to improve the light transmittance through the skull, which subsequently increases the SNR of photoacoustic signals.

To resolve this issue, Nie et al. designed a photon recycler to reflect backscattered light back to the skull (Nie et al. 2012). The photon recycler was made of plastic polyvinyl coated with a mixture of titanium white pigment and epoxy. As shown in Figure 12.15a, the photon recycler can significantly improve the light delivery efficiency. The experiment was performed on a single-element circular-view PACT system. Similar to the monkey brain experiment, a low-frequency 1 MHz transducer was employed in the study. The photon recycler was first tested on a phantom experiment, and the SNR was found to increase by 2.4 times, from 6.3 to 15. The authors then imaged a 3-month-old canine brain (Figure 12.15c) positioned in the cortical area of the human skull. As expected, the photoacoustic image (Figure 12.15d) was strongly distorted, and the brain features were barely identifiable. Interestingly, the control image with only the skull present (Figure 12.15e) showed similar features, indicating that the majority of the signals came from the skull. By subtracting the control image from Figure 12.15d, the authors obtained a differential image (Figure 12.15f) with blurred brain features. Because the high-frequency signals were strongly attenuated through the skull, the image was further high-pass filtered to enhance the high-frequency components. The filtered image

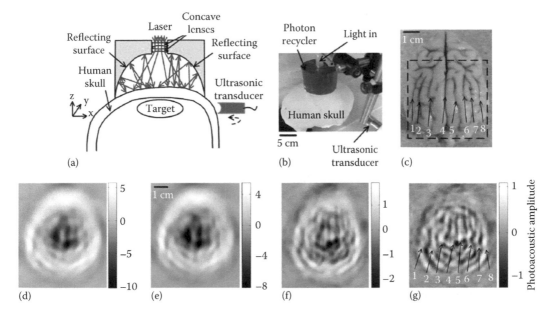

FIGURE 12.15 (a) Schematic of the photon recycler reflecting backscattered photons back to the human skull for PAT. (b) Photograph of the PAT setup with photon recycler. (c) Photograph of the canine brain cortex. (d) PAT image of a canine brain acquired through the human skull. (e) PAT image of the human skull only. (f) Differential image of (d) and (e). (g) Image from (f) after high-pass filtering. (Reproduced from Nie, L. et al., *J. Biomed. Optics*, 17(11), 110506, 2012. With permission.)

(Figure 12.15g) clearly shows the eight main features, marked by numbered arrows in the photograph of the canine brain (Figure 12.15c). As many functional brain studies are based on differential analysis, this study indicates that PACT can potentially be used for monitoring the functional activities of the human cerebral cortex. It should be noted that the images in Figure 12.15 were simply reconstructed using the universal back-projection algorithm. The image quality can be significantly improved by using advanced reconstruction algorithms that account for acoustic distortion from the skull (Huang et al. 2012a,b).

Besides the photon recycler, other light delivery schemes, such as light delivery through the nasal or oral cavities, can also be explored to image structures of the deep brain. Photoacoustic endoscopy (Yang et al. 2012) through the nasal cavity may be implemented as well and can potentially avoid strong acoustic and light distortions from the relatively thick frontal and parietal bones.

CONCLUSION AND OUTLOOK

Since the first demonstration of functional photoacoustic neuroimaging in 2003 (Wang et al. 2003a,b), the last decade has seen an explosion in the development and use of PAT for neuroimaging (Wang 2009, Hu and Wang 2010, Xia and Wang 2013). Neurovascular, functional, metabolic, and molecular brain images at multiple spatial scales have been demonstrated on various animal models. Translation to human brain imaging is also underway. As a maturing technology, PAT still needs further improvements to realize its full potential in neuroimaging. In terms of imaging speed, the bottleneck is the pulse repetition rate of high-energy pulsed lasers. The state of the art PAT system has a frame rate of 20 Hz (Needles et al. 2013), corresponding to the pulse repetition rate of commonly used pulsed lasers. With advances in laser techniques and DAQ systems, hundred-Hz or even kilo-Hz pulse-repetition-rate lasers may soon provide sufficient pulse energy for PAT imaging. Such a high frame rate will allow in the monitoring of fast neurovascular dynamics or neural firing. In terms of quantitative neuroimaging, advanced data processing algorithms are needed to account for variations in local fluence (Shao et al. 2011, Cox et al. 2012, Guo et al. 2012, Xia et al. 2013). Fast wavelength tuning lasers (Wang et al. 2012) are also essential for monitoring fast functional changes in the brain. As with optical heterogeneities, acoustic heterogeneities also need to be considered, especially when imaging large animals or humans, where the skull effect cannot be neglected (Huang et al. 2012a,b).

There are also several emerging neuroimaging directions that have yet not been demonstrated in PAT. For instance, imaging of awake animals is drawing increasing interest in neuroscience because it avoids the depression of neuron activities during anesthesia and allows correlating neural activities with behavior (Dombeck et al. 2007, Greenberg et al. 2008). To implement this in PAT, existing animal mounting schemes should be modified to introduce head restraint mechanisms (Dombeck et al. 2007). Alternatively, PAT systems can be further miniaturized to be mounted on the animal's head without jeopardizing the animal's normal activity (Helmchen et al. 2001). Direct imaging of neural activities is another important neural imaging direction. PAT can potentially achieve this using voltage-sensitive dyes or calcium-sensitive dyes, whose spectral properties change when a neuron fires an action potential (Knöpfel et al. 2006, Perron et al. 2009).

In conclusion, with its unique combination of optical absorption sensitivity and ultrasonic imaging depth and resolution scalability, PAT promises to play an important role in neuroimaging.

ACKNOWLEDGMENTS

The authors appreciate Prof. James Ballard's close reading of the manuscript. This work was sponsored in part by National Institutes of Health (NIH) grants DP1 EB016986 (NIH Director's Pioneer Award), R01 CA186567 (NIH Director's Transformative Research Award), R01 EB016963, R01 EB010049, and R01 CA159959. L. V. Wang has a financial interest in MicroPhotoAcoustics, Inc., and Endra, Inc., which, however, did not support this work.

REFERENCES

Allen, T. J., A. Hall, A. P. Dhillon, J. S. Owen, and P. C. Beard. (2012). Spectroscopic photoacoustic imaging of lipid-rich plaques in the human aorta in the 740 to 1400 nm wavelength range. *Journal of Biomedical Optics* **17**(6): 0612091–06120910.

Attwell, D. and C. Iadecola. (2002). The neural basis of functional brain imaging signals. *Trends in Neurosciences* **25**(12): 621–625.

Baxter, L. R., Jr, M. E. Phelps, J. C. Mazziotta et al. (1985). Cerebral metabolic rates for glucose in mood disorders: Studies with positron emission tomography and fluorodeoxyglucose f 18. *Archives of General Psychiatry* **42**(5): 441–447.

Beard, P. (2011). Biomedical photoacoustic imaging. *Interface Focus* **1**(4): 602–631.

Bell, A. G. (1880). On the production and reproduction of sound by light. *American Journal of Science* **20**: 305–325.

Bero, A. W., A. Q. Bauer, F. R. Stewart, B. R. White, J. R. Cirrito, M. E. Raichle, J. P. Culver, and D. M. Holtzman. (2012). Bidirectional relationship between functional connectivity and amyloid-Î² deposition in mouse brain. *The Journal of Neuroscience* **32**(13): 4334–4340.

Bremner, J., R. B. Innis, C. K. Ng et al. (1997). POsitron emission tomography measurement of cerebral metabolic correlates of yohimbine administration in combat-related posttraumatic stress disorder. *Archives of General Psychiatry* **54**(3): 246–254.

Buckner, R. L., F. M. Krienen, and B. T. T. Yeo. (2013). Opportunities and limitations of intrinsic functional connectivity MRI. *Nature Neuroscience* **16**(7): 832–837.

Burgholzer, P., G. J. Matt, M. Haltmeier, and G. N. Paltauf. (2007). Exact and approximative imaging methods for photoacoustic tomography using an arbitrary detection surface. *Physical Review E* **75**(4): 046706.

Cazzin, C. and C. J. A. Ring. (2010). Recent advances in the manipulation of murine gene expression and its utility for the study of human neurological disease. *Biochimica et Biophysica Acta (BBA)—Molecular Basis of Disease* **1802**(10): 796–807.

Cox, B., J. G. Laufer, S. R. Arridge, and P. C. Beard. (2012). Quantitative spectroscopic photoacoustic imaging: A review. *Journal of Biomedical Optics* **17**(6): 061202.

Cox, B. T. and P. C. Beard. (2005). Fast calculation of pulsed photoacoustic fields in fluids using k-space methods. *Journal of the Acoustical Society of America* **117**: 3616–3627.

Deco, G., V. K. Jirsa, and A. R. McIntosh. (2011). Emerging concepts for the dynamical organization of resting-state activity in the brain. *Nature Reviews Neuroscience* **12**(1): 43–56.

Dombeck, D. A., A. N. Khabbaz, F. Collman, T. L. Adelman, and D. W. Tank. (2007). Imaging large-scale neural activity with cellular resolution in awake, mobile mice. *Neuron* **56**(1): 43–57.

Favazza, C. P., L. A. Cornelius, and L. H. V. Wang. (2011a). In vivo functional photoacoustic microscopy of cutaneous microvasculature in human skin. *Journal of Biomedical Optics* **16**(2): 026004.

Favazza, C. P., O. Jassim, L. A. Cornelius, and L. H. V. Wang. (2011b). In vivo photoacoustic microscopy of human cutaneous microvasculature and a nevus. *Journal of Biomedical Optics* **16**(1): 016015.

Fox, M. D. and M. E. Raichle. (2007). Spontaneous fluctuations in brain activity observed with functional magnetic resonance imaging. *Nature Reviews Neuroscience* **8**(9): 700–711.

Frei, H. J., T. Wallenfang, W. Poll, H. J. Reulen, R. Schubert, and M. Brock. (1973). Regional cerebral blood flow and regional metabolism in cold induced oedema. *Acta Neurochirurgica* **29**(1): 15–28.

Fry, F. J. (1977). Transkull transmission of an intense focused ultrasonic beam. *Ultrasound in Medicine and Biology* **3**(2–3): 179–184.

Gamelin, J., A. Aguirre, A. Maurudis, F. Huang, D. Castillo, L. V. Wang, and Q. Zhu. (2008). Curved array photoacoustic tomographic system for small animal imaging. *Journal of Biomedical Optics* **13**(2): 10.

Gamelin, J., A. Maurudis, A. Aguirre, F. Huang, P. Guo, L. V. Wang, and Q. Zhu. (2009). A real-time photoacoustic tomography system for small animals. *Optics Express* **17**(13): 10489–10498.

Greenberg, D. S., A. R. Houweling, and J. N. Kerr. (2008). Population imaging of ongoing neuronal activity in the visual cortex of awake rats. *Nature Neuroscience* **11**(7): 749–751.

Guilfoyle, D. N., S. V. Gerum, J. L. Sanchez, A. Balla, H. Sershen, D. C. Javitt, and M. J. Hoptman. (2013). Functional connectivity fMRI in mouse brain at 7T using isoflurane. *Journal of Neuroscience Methods* **214**(2): 144–148.

Guo, Z., C. Favazza, A. Garcia-Uribe, and L. V. Wang. (2012). Quantitative photoacoustic microscopy of optical absorption coefficients from acoustic spectra in the optical diffusive regime. *Journal of Biomedical Optics* **17**(6): 066011–066011.

Gwinn, M. R. and V. Vallyathan (2006). Nanoparticles: Health effects—Pros and cons. *Environmental Health Perspectives* **114**(12): 1818–1825.

Hammoud, D. A., J. M. Hoffman, and M. G. Pomper. (2007). Molecular neuroimaging: From conventional to emerging techniques 1. *Radiology* **245**(1): 21–42.

Helmchen, F., M. S. Fee, D. W. Tank, and W. Denk. (2001). A miniature head-mounted two-photon microscope: High-resolution brain imaging in freely moving animals. *Neuron* **31**(6): 903–912.

Horecker, B. L. (1943). The absorption spectra of hemoglobin and its derivatives in the visible and near infra-red regions. *Journal of Biological Chemistry* **148**(1): 173–183.

Hu, S., K. Maslov, V. Tsytsarev, and L. V. Wang. (2009). Functional transcranial brain imaging by optical-resolution photoacoustic microscopy. *Journal of Biomedical Optics* **14**(4): 3.

Hu, S., K. Maslov, and L. V. Wang. (2011). Second-generation optical-resolution photoacoustic microscopy with improved sensitivity and speed. *Optics Letters* **36**(7): 1134–1136.

Hu, S. and L. V. Wang. (2010). Neurovascular photoacoustic tomography. *Frontiers in Neuroenergetics* **2**: 10.

Huang, C., L. Nie, R. W. Schoonover, Z. Guo, C. O. Schirra, M. A. Anastasio, and L. V. Wang. (2012a). Aberration correction for transcranial photoacoustic tomography of primates employing adjunct image data. *Journal of Biomedical Optics* **17**(6): 0660161–0660168.

Huang, C., L. Nie, R. W. Schoonover, L. V. Wang, and M. A. Anastasio. (2012b). Photoacoustic com-puted tomography correcting for heterogeneity and attenuation. *Journal of Biomedical Optics* **17**(6): 061211–061211.

Jo, J., H. Zhang, P. D. Cheney, and X. Yang. (2012). Photoacoustic detection of functional responses in the motor cortex of awake behaving monkey during forelimb movement. *Journal of Biomedical Optics* **17**(11): 110503–110503.

Karabutov, A., N. B. Podymova, and V. S. Letokhov. (1996). Time-resolved laser optoacoustic tomography of inhomogeneous media. *Applied Physics B* **63**(6): 545–563.

Kim, C., C. Favazza, and L. H. V. Wang. (2010). In vivo photoacoustic tomography of chemicals: High-resolution functional and molecular optical imaging at new depths. *Chemical Reviews* **110**(5): 2756–2782.

Kircher, M. F., A. de la Zerda, J. V. Jokerst et al. (2012). A brain tumor molecular imaging strategy using a new triple-modality MRI-photoacoustic-Raman nanoparticle. *Nature Medicine* **18**(5): 829–834.

Knöpfel, T., J. Díez-García, and W. Akemann. (2006). Optical probing of neuronal circuit dynamics: Genetically encoded versus classical fluorescent sensors. *Trends in Neurosciences* **29**(3): 160–166.

Kruger, R., D. Reinecke, G. Kruger, M. Thornton, P. Picot, T. Morgan, K. Stantz, and C. Mistretta. (2009). HYPR-spectral photoacoustic CT for preclinical imaging. *Proceedings of SPIE* **7177**: 71770F.

Kruger, R. A. (1994). Photoacoustic ultrasound. *Medical Physics* **21**(1): 127–131.

Laufer, J., E. Zhang, G. Raivich, and P. Beard. (2009). Three-dimensional noninvasive imaging of the vas-culature in the mouse brain using a high resolution photoacoustic scanner. *Applied Optics* **48**(10): D299–D306.

Li, C., A. Aguirre, J. Gamelin, A. Maurudis, Q. Zhu, and L. V. Wang. (2010). Real-time photoacous-tic tomography of cortical hemodynamics in small animals. *Journal of Biomedical Optics* **15**(1): 010509-010501-010503.

Li, C. and L. V. Wang. (2009). Photoacoustic tomography and sensing in biomedicine. *Physics in Medicine and Biology* **54**(19): R59–R97.

Li, M.-L., O. Jung-Taek, X. Xueyi, K. Geng, W. Wei, L. Chun, G. Lungu, G. Stoica, and L. V. Wang. (2008). Simultaneous molecular and hypoxia imaging of brain tumors in vivo using spectroscopic photoacoustic tomography. *Proceedings of the IEEE* **96**(3): 481–489.

Lou, Y., J. Xia, and L. V. Wang. (2014). Mouse brain imaging using photoacoustic computed tomography. *Proceedings of SPIE* **8943**: 894340.

Lovell, J. F., C. S. Jin, E. Huynh, H. Jin, C. Kim, J. L. Rubinstein, W. C. W. Chan, W. Cao, L. V. Wang, and G. Zheng. (2011). Porphysome nanovesicles generated by porphyrin bilayers for use as multimodal bio-photonic contrast agents. *Nature Materials* **10**(4): 324–332.

Mallidi, S., G. P. Luke, and S. Emelianov. (2011). Photoacoustic imaging in cancer detection, diagnosis, and treatment guidance. *Trends in Biotechnology* **29**: 213–221

Mark, A., K. Peter, and K. Leonid. (2009). On reconstruction formulas and algorithms for the thermoacoustic tomography. In L. V. Wang, ed., *Photoacoustic Imaging and Spectroscopy*, CRC Press: Boca Raton, FL, pp. 89–101.

Maslov, K., G. Stoica, and L. V. Wang. (2005). In vivo dark-field reflection-mode photoacoustic microscopy. *Optics Letters* **30**(6): 625–627.

Maslov, K., H. F. Zhang, S. Hu, and L. V. Wang. (2008). Optical-resolution photoacoustic microscopy for in vivo imaging of single capillaries. *Optics Letters* **33**(9): 929–931.

Matthews, T. P., C. Zhang, D. K. Yao, K. Maslov, and L. H. V. Wang. (2014). Label-free photoacoustic micros-copy of peripheral nerves. *Journal of Biomedical Optics* **19**(1): 16004.

Nasiriavanaki, M., J. Xia, H. Wan, A. Q. Bauer, J. P. Culver, and L. V. Wang. (2014). High-resolution photoacoustic tomography of resting-state functional connectivity in the mouse brain. *Proceedings of the National Academy of Sciences of the United States of America* **111**(1): 21–26.

Needles, A., A. Heinmiller, J. Sun, C. Theodoropoulos, D. Bates, D. Hirson, M. Yin, and F. S. Foster. (2013). Development and initial application of a fully integrated photoacoustic micro-ultrasound system. *IEEE Transactions on Ultrasonics, Ferroelectrics and Frequency Control* **60**(5): 888–897.

Nie, L., X. Cai, K. Maslov, A. Garcia-Uribe, M. A. Anastasio, and L. V. Wang. (2012). Photoacoustic tomography through a whole adult human skull with a photon recycler. *Journal of Biomedical Optics* **17**(11): 110506–110506.

Oraevsky, A. A., S. L. Jacques, and F. K. Tittel. (1997). Measurement of tissue optical properties by time-resolved detection of laser-induced transient stress. *Applied Optics* **36**(1): 402–415.

Perron, A., H. Mutoh, W. Akemann, S. G. Gautam, D. Dimitrov, Y. Iwamoto, and T. Knöpfel. (2009). Second and third generation voltage-sensitive fluorescent proteins for monitoring membrane potential. *Frontiers in Molecular Neuroscience* **2**: 5.

Petrov, I. Y., Y. Petrov, D. S. Prough, I. Cicenaite, D. J. Deyo, and R. O. Esenaliev. (2012). Optoacoustic monitoring of cerebral venous blood oxygenation though intact scalp in large animals. *Optics Express* **20**(4): 4159–4167.

Petrov, Y. Y., D. S. Prough, D. J. Deyo, M. Klasing, M. Motamedi, and R. O. Esenaliev. (2005). Optoacoustic, noninvasive, real-time, continuous monitoring of cerebral blood oxygenation: An in vivo study in sheep. *Anesthesiology* **102**(1): 69–75.

Ray, A., X. Wang, Y.-E. Lee et al. (2011). Targeted blue nanoparticles as photoacoustic contrast agent for brain tumor delineation. *Nano Research* **4**(11): 1163–1173.

Schaefer, D. J., J. D. Bourland, and J. A. Nyenhuis. (2000). Review of patient safety in time-varying gradient fields. *Journal of Magnetic Resonance Imaging* **12**(1): 20–29.

Shao, P., B. Cox, and R. J. Zemp. (2011). Estimating optical absorption, scattering, and Grueneisen distributions with multiple-illumination photoacoustic tomography. *Applied Optics* **50**(19): 3145–3154.

Song, K. H. and L. V. Wang. (2007). Deep reflection-mode photoacoustic imaging of biological tissue. *Journal of Biomedical Optics* **12**(6): 060503.

Song, K. H. and L. V. Wang. (2008). Noninvasive photoacoustic imaging of the thoracic cavity and the kidney in small and large animals. *Medical Physics* **35**(10): 4524–4529.

Stein, E. W., K. Maslov, and L. V. Wang. (2009). Noninvasive, in vivo imaging of blood-oxygenation dynamics within the mouse brain using photoacoustic microscopy. *Journal of Biomedical Optics* **14**: 020502.

Steinbrink, J., A. Villringer, F. Kempf, D. Haux, S. Boden, and H. Obrig. (2006). Illuminating the BOLD signal: Combined fMRI-fNIRS studies. *Magnetic Resonance Imaging* **24**(4): 495–505.

Taruttis, A., J. Claussen, D. Razansky, and V. Ntziachristos. (2012). Motion clustering for deblurring multispectral optoacoustic tomography images of the mouse heart. *Journal of Biomedical Optics* **17**(1): 016009.

Tsytsarev, V., S. Hu, J. Yao, K. Maslov, D. L. Barbour, and L. V. Wang. (2011). Photoacoustic microscopy of microvascular responses to cortical electrical stimulation. *Journal of Biomedical Optics* **16**: 076002.

Wang, L., K. Maslov, and L. V. Wang. (2013). Single-cell label-free photoacoustic flowoxigraphy in vivo. *Proceedings of the National Academy of Sciences of the United States of America* **110**(15): 5759–5764.

Wang, L. D., K. Maslov, W. X. Xing, A. Garcia-Uribe, and L. H. V. Wang. (2012). Video-rate functional photoacoustic microscopy at depths. *Journal of Biomedical Optics* **17**(10): 106007.

Wang, L. V. (2008). Tutorial on photoacoustic microscopy and computed tomography. *IEEE Journal of Selected Topics in Quantum Electronics* **14**(1): 171–179.

Wang, L. V. and S. Hu. (2012). Photoacoustic tomography: In vivo imaging from organelles to organs. *Science* **335**(6075): 1458–1462.

Wang, X. (2009). Functional and molecular photoacoustic tomography of small-animal brains. In L. V. Wang, ed., *Photoacoustic Imaging and Spectroscopy*, CRC Press: Boca Raton, FL, pp. 251–262.

Wang, X., Y. Pang, G. Ku, G. Stoica, and L. V. Wang. (2003a). Three-dimensional laser-induced photoacoustic tomography of mouse brain with the skin and skull intact. *Optics Letters* **28**(19): 1739–1741.

Wang, X., Y. Pang, G. Ku, X. Xie, G. Stoica, and L. V. Wang. (2003b). Noninvasive laser-induced photoacoustic tomography for structural and functional in vivo imaging of the brain. *Nature Biotechnology* **21**(7): 803–806.

Wang, X. D., Y. Xu, M. H. Xu, S. Yokoo, E. S. Fry, and L. H. V. Wang. (2002). Photoacoustic tomography of biological tissues with high cross-section resolution: Reconstruction and experiment. *Medical Physics* **29**(12): 2799–2805.

Wang, Y., X. Xie, X. Wang, G. Ku, K. L. Gill, D. P. O'Neal, G. Stoica, and L. V. Wang. (2004). Photoacoustic tomography of a nanoshell contrast agent in the in vivo rat brain. *Nano Letters* **4**(9): 1689–1692.

Xia, J., A. Danielli, Y. Liu, L. Wang, K. Maslov, and L. V. Wang. (2013). Calibration-free quantification of absolute oxygen saturation based on the dynamics of photoacoustic signals. *Optics Letters* **38**(15): 2800–2803.

Xia, J., Z. Guo, K. Maslov, A. Aguirre, Q. Zhu, C. Percival, and L. V. Wang. (2011). Three-dimensional photoacoustic tomography based on the focal-line concept. *Journal of Biomedical Optics* **16**(9): 090505.

Xia, J. and L. Wang. (2013). Photoacoustic tomography of the brain. In S. J. Madsen, *Optical Methods and Instrumentation in Brain Imaging and Therapy*, Springer: New York, Vol. 3, pp. 137–156.

Xia, J. and L. Wang. (2014). Small-animal whole-body photoacoustic tomography: A review. *IEEE Transactions on Biomedical Engineering* **61**(5): 1380–1389.

Xiang, L., B. Wang, L. Ji, and H. Jiang. (2013). 4-D photoacoustic tomography. *Scientific Reports* **3**(1113): 1–8.

Xing, W., L. Wang, K. Maslov, and L. V. Wang. (2013). Integrated optical- and acoustic-resolution photoacoustic microscopy based on an optical fiber bundle. *Optics Letters* **38**(1): 52–54.

Xu, M. (2009). Analysis of spatial resolution in photoacoustic tomography. In L. V. Wang, ed., *Photoacoustic Imaging and Spectroscopy*, CRC Press: Boca Raton, FL, pp. 47–60.

Xu, M. and L. V. Wang. (2005). Universal back-projection algorithm for photoacoustic computed tomography. *Physical Review E* **71**(1): 016706.

Xu, M. H. and L. H. V. Wang. (2006). Photoacoustic imaging in biomedicine. *Review of Scientific Instruments* **77**(4): 22.

Xu, M. H. and L. V. Wang. (2003). Analytic explanation of spatial resolution related to bandwidth and detector aperture size in thermoacoustic or photoacoustic reconstruction. *Physical Review E* **67**(5): 15.

Xu, Y. and L. V. Wang. (2004). Time reversal and its application to tomography with diffracting sources. *Physical Review Letters* **92**(3): 033902.

Xu, Z., C. H. Li, and L. V. Wang. (2010). Photoacoustic tomography of water in phantoms and tissue. *Journal of Biomedical Optics* **15**(3): 036019.

Xu, Z., Q. Zhu, and L. V. Wang. (2011). In vivo photoacoustic tomography of mouse cerebral edema induced by cold injury. *Journal of Biomedical Optics* **16**(6): 066020.

Yang, J. M., C. Favazza, R. M. Chen, J. J. Yao, X. Cai, K. Maslov, Q. F. Zhou, K. K. Shung, and L. H. V. Wang. (2012). Simultaneous functional photoacoustic and ultrasonic endoscopy of internal organs in vivo. *Nature Medicine* **18**(8): 1297+.

Yang, X. (2014). Photoacoustic imaging of brain cortex in rhesus macaques. *The Journal of the Acoustical Society of America* **135**(4): 2209–2209.

Yang, X. and L. V. Wang. (2008). Monkey brain cortex imaging by photoacoustic tomography. *Journal of Biomedical Optics* **13**(4): 044009.

Yang, X. M., S. E. Skrabalak, Z. Y. Li, Y. N. Xia, and L. H. V. Wang. (2007). Photoacoustic tomography of a rat cerebral cortex in vivo with au nanocages as an optical contrast agent. *Nano Letters* **7**(12): 3798–3802.

Yao, J. and L. V. Wang. (2013). Photoacoustic microscopy. *Laser & Photonics Reviews* **7**(5): 758–778.

Yao, J., J. Xia, K. I. Maslov, M. Nasiriavanaki, V. Tsytsarev, A. V. Demchenko, and L. V. Wang. (2013). Noninvasive photoacoustic computed tomography of mouse brain metabolism in vivo. *NeuroImage* **64**: 257–266.

Yuan, H., C. M. Wilson, J. Xia, S. L. Doyle, S. Li, A. M. Fales, Y. Liu, E. Ozaki, K. Mulfaul, and G. Hanna. (2014). Plasmonics-enhanced and optically modulated delivery of gold nanostars into brain tumor. *Nanoscale* **6**(8): 4078–4082.

Yuan, J., G. Xu, Y. Yu, Y. Zhou, P. L. Carson, X. Wang, and X. Liu. (2013). Real-time photoacoustic and ultrasound dual-modality imaging system facilitated with graphics processing unit and code parallel optimization. *Journal of Biomedical Optics* **18**(8): 086001–086001.

Yulia, H., K. Peter, and N. Linh. (2008). Reconstruction and time reversal in thermoacoustic tomography in acoustically homogeneous and inhomogeneous media. *Inverse Problems* **24**(5): 055006.

Zhang, C., K. Maslov, and L. V. Wang. (2010). Subwavelength-resolution label-free photoacoustic microscopy of optical absorption in vivo. *Optics Letters* **35**(19): 3195–3197.

Zhang, C., Y. S. Zhang, D. K. Yao, Y. N. Xia, and L. H. V. Wang. (2013). Label-free photoacoustic microscopy of cytochromes. *Journal of Biomedical Optics* **18**(2): 20504.

Zhang, E., J. Laufer, and P. Beard. (2008). Backward-mode multiwavelength photoacoustic scanner using a planar Fabry–Perot polymer film ultrasound sensor for high-resolution three-dimensional imaging of biological tissues. *Applied Optics* **47**(4): 561–577.

Zhang, H. F., K. Maslov, G. Stoica, and L. V. Wang. (2006). Functional photoacoustic microscopy for high-resolution and noninvasive in vivo imaging. *Nature Biotechnology* **24**(7): 848–851.

13 Perfusion Brain Mapping in the Treatment of Acute Stroke

Michael J. Alexander and Paula Eboli

CONTENTS

INTRODUCTION

Computed tomography perfusion (CTP) imaging is an effective imaging technique increasingly being used in the setting of acute cerebral ischemia and stroke evaluation. It allows a rapid evaluation of brain circulation and aids in the detection of acute ischemic lesions as well as areas with decreased perfusion.

Its interpretation can be challenging, and controversy exists when interpreting the results and deciding which patients are likely to benefit from early reperfusion (Allmendinger et al. 2012, Maija et al. 2013).

Essentially, CTP imaging provides a qualitative as well as quantitative evaluation of cerebral hemodynamics by providing perfusion color maps of cerebral blood volume (CBV), cerebral blood flow (CBF), and mean transit time (MTT), based on the central volume principle that CBF = CBV/MTT. These color maps are used to determine well-perfused, ischemic, and necrotic areas (Hoeffner et al. 2004, Maija et al. 2013).

Standard head computed tomography (CT) remains the initial diagnostic test in patients with stroke-related symptoms and is mainly used to triage and identify patients who are candidates for IV TPA therapy. CT angiography (CTA) and CT perfusion (CTP) are usually performed after the initial head CT adding just a few minutes to the process.

Computed tomography angiography (CTA) evaluates cerebral vasculature and aids in diagnosing vessel occlusion due to the presence of thrombus or stenosis, while CTP allows the identification of infarcted and/or ischemic tissue.

CT PERFUSION BASIC PRINCIPLES

CTP, as mentioned, is based on the principle that CBF = CBV/MTT. These values are obtained by injecting iodine contrast and obtaining its concentration and attenuation values when passing through the brain vasculature. Contrast agent time–concentration curves are generated in an arterial region of interest, and after a complex mathematical process, CBF, CBV, and MTT values are obtained (Hoeffner et al. 2004).

Based on imaging voxels, CBV refers to the total blood volume, including both intravascular and intraparenchymal blood contained within the voxel unit. CBF, on the other hand, refers to the total volume of blood moving through the voxel in a given unit of time (Allmendinger et al. 2012). Ito et al. (2004) created a database of normal human CBF, CBV, cerebral oxygen extraction fraction, and cerebral metabolic rate of oxygen measured by positron emission tomography and determined the overall mean values for CBF = 44.4 ± −6.5 mL 100 mL(−1) min(−1) and CBV = 3.8 ± −0.7 mL 100 mL(−1) for cerebral cortical regions (Ito et al. 2004).

The MTT on the other hand is the average transit time measured in seconds that the contrast takes to pass through a given volume of brain. Time to peak (TTP) is usually included in most CTPs and is the time measured in seconds that the contrast takes from its injection to maximal enhancement (Allmendinger et al. 2012).

VALIDATION OF CT PERFUSION

In 2002, Eastwood et al. did a CTP scanning pilot study in patients with acute middle cerebral artery (MCA) stroke and found the area of acute stroke to have statistically significant decreases in both mean CBF and CBV values and statistically significant increase in MTT (Eastwood et al. 2002). Around that same time, Wintermark et al., when studying the prognostic accuracy of CBF measurement by CTP, assumed the ischemic cerebral area to be the area encompassed by the penumbra as well as the core infarct and defined it as the area with 34% or more decrease in CBF compared to normal areas (Wintermark et al. 2002a,b). Within this area, the areas with CBF less than 2.5 mL per 100 g were called core infarct and those with CBF greater than 2.5 mL per 100 g were considered penumbra. They also compared the results obtained with CTP with those obtained with delayed Diffusion weighted imaging – magnetic resonance (DWI-MR) examinations and finally concluded that CTP allows accurate prediction of the extent of penumbra as well as the actual core infarct size (Wintermark et al. 2002a,b).

Subsequently, Wintermark et al. compared admission CTP studies with diffusion and perfusion-weighted (PW) magnetic resonance imaging (MRI) images in patients with acute stroke. They included thirteen patients with acute ischemic strokes. These patients underwent on admission both CTP and DWI/PWI MRI. They compared the size of ischemic area, core infarct, and penumbra on CTP with the affected area shown on DWI and calculated CBV, CVF, MTT, and TTP from PWI. They found that the infarct size on CTP corresponded to the infarct size found on DWI, and they also found that the ischemic lesion, both penumbra and core infarct, correlated well with the size of the elevated MTT map calculated from PWI, concluding that CTP and DWI/PWI MRI are equally effective to identify ischemic lesions and cerebral penumbra in acute ischemic stroke (Wintermark et al. 2002a,b).

More recently in 2013, Bivard et al. studied 183 patients who underwent CTP for stroke within 6 h of onset, followed by a 24 h MRI with DWI and PWI. Using these data, they developed a prob-ability-based model, which was used to identify thresholds parameters, and found that penumbra was best defined by a TTP greater than 5 s and ischemic core stroke by a decrease in CBF of 50% (Bivard et al. 2014).

Despite all these calculations and interpretations, there is a significant difference on CTP maps within different vendors. This is usually multifactorial and can be related to scan parameters, post-processing method, and type of algorithm used (Benson et al. 2015).

CLINICAL APPLICATIONS

Neurological deficits due to cerebral ischemia can be transient or permanent, and they can improve or worsen depending on the duration and severity of decreased blood flow to the affected area.

In general, neurological dysfunction occurs after CBF falls below 18–20 mL/100 g of tissue per minute and infarction occurs after CBF drops to less than 10 mL/100 mg of tissue per minute.

But when CBF is between 10 and 20 mL/100 g of tissue per minute, neuronal death occurs after several minutes to hours (Latchaw et al. 2003). Therefore, theoretically, the tissue in this intermediate zone can be salvaged when blood flow is restored.

From a functional standpoint, ischemic penumbra is defined as the tissue surrounding an area of recent infarction that is functionally impaired but potentially viable, and from a molecular perspective, genes that induce apoptosis are expressed (Donnan and Davis 2002). Treatment is therefore aimed to increase blood flow to this area to maximize brain tissue survival and improve clinical outcomes.

When using CTP, we assume that core infarct is the area with a matched perfusion deficit defined by decreased CBF and CBV and increased MTT and TTP, while ischemic penumbra refers to a mismatched perfusion area where CBV is preserved, CBF is diminished, and MTT and TTP are prolonged.

Therefore, CTP can help us identify those patients with areas of ischemic penumbra that will potentially benefit from thrombolytic therapies assuming that the penumbra area corresponds to salvageable tissue at risk for infarction (Hoeffner et al. 2004). Recently new trials have demonstrated high-quality evidence on the efficacy of mechanical thrombectomy in patients with large vessel occlusion. They showed improved clinical outcomes, especially in those patients where good recanalization was achieved, TICI2b–TICI3, and treatment was started within 6 h of stroke onset. On the other hand, those with partial recanalization, TICI 2a, did not do as well, and at this moment, there are not enough data to know the effectiveness of treatment when started after 6 h (Powers et al. 2015). All these studies required baseline noncontrast CT or MRI and CTA or magnetic resonance angiography (MRA) evidence of large vessel occlusion.

Two studies, EXTEND-IA and SWIFT PRIME, included CTP parameters. EXTEND-IA used CTP to identify potentially salvageable brain tissue; therefore, ischemic penumbra area had to have a time to maximum (T_{max}) greater than 6 s and ischemic core and a CBF less than 30% compared to that of normal tissue (Campbell et al. 2015, Saver et al. 2015). The mismatch ratio (penumbra/core infarct) had to be greater than 1.2, the absolute mismatch volume greater than 10 mL, and the ischemic core less than 70 mL (Powers et al. 2015). SWIFT PRIME on the other hand, for the first 71 patients, included patients with a mismatch ratio greater than 1.8, ischemic penumbra greater than 15 mL, and an infarct core area of less than 50 mL (Powers et al. 2015, Saver et al. 2015).

Since none of the five trials was designed to validate the use of CTP in the patients' selection, the 2015 American Heart Association/American Stroke (AHA/ASA) Guidelines for the early management of patients with acute ischemic stroke regarding endovascular treatment makes no formal recommendation with respect to the use of CTP and recommends the use of Alberta stroke program early computerized tomography score (ASPECTS) score greater or equal to 6 as the inclusion criteria (Powers et al. 2015).

CASE EXAMPLES

CASE #1

A 71-year-old man was awoke, aphasic, and hemiplegic, with NIHSS 24, 6 h ago, the last known time of normal function. CT showed a left dense MCA sign, and CTA showed two separate clots, one on the left internal carotid artery (ICA) and the other on the left MCA (Figures 13.1 and 13.2).

CTP showed relative mismatch between the CBF and CBV studies, indicating a potential penumbra area (Figures 13.3 and 13.4), and diagnostic cerebral angiogram showed left carotid occlusion (Figures 13.5 and 13.6).

Revascularization was done with Penumbra ADAPT clot aspiration technique within 21 min of femoral puncture. TICI 2b reperfusion was achieved (Figure 13.7). Postop CT showed a small lacunar infarct (Figure 13.8).

The patient's speech and strength returned to normal in the recovery room; he returned to light work 2 weeks later and normal job functions 4 weeks later.

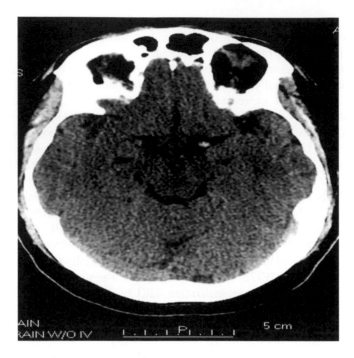

FIGURE 13.1   Head CT shows left dense MCA sign.

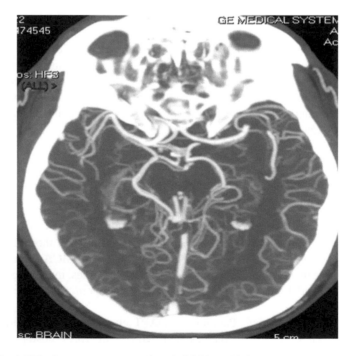

FIGURE 13.2   Head CTA shows two separate clots, left ICA and left MCA.

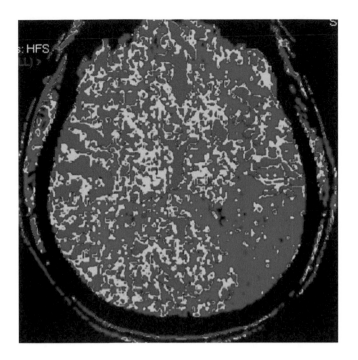

FIGURE 13.3 CTp CBF.

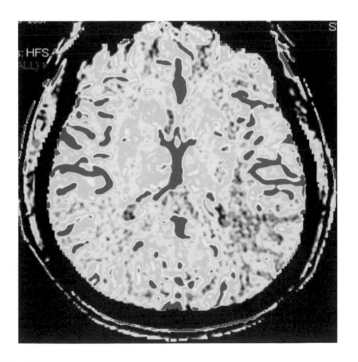

FIGURE 13.4 CTp CBV.

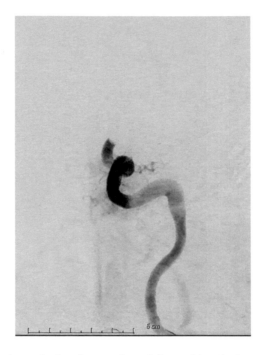

FIGURE 13.5   AP diagnostic cerebral angiogram shows left carotid occlusion.

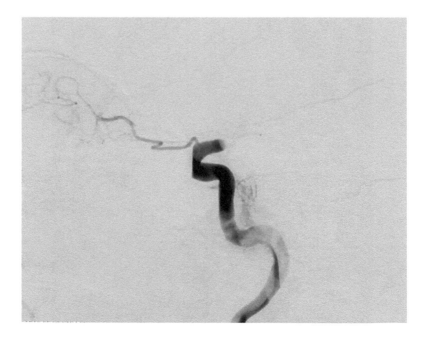

FIGURE 13.6   Lateral diagnostic cerebral angiogram showed left carotid occlusion.

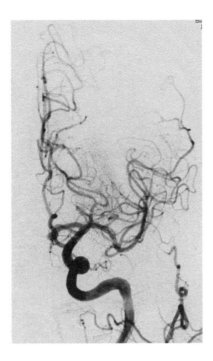

FIGURE 13.7 Postthrombectomy cerebral angiogram shows TICI 2b reperfusion.

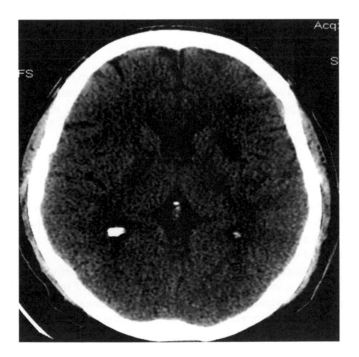

FIGURE 13.8 Postop CT shows small lacunar infarct.

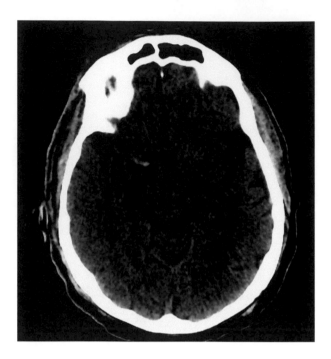

FIGURE 13.9   Head CT shows right dense MCA sign.

Case #2

A 68-year-old man, NIHSS 18 on admission, received IV tPA at 2 h and was then helicoptered in from outside the hospital. At 4 h he had no neurologic change, and CT showed a right dense MCA sign (Figure 13.9), and CTA demonstrated right MCA occlusion (Figure 13.10).

CTP showed CBF/CBV mismatch (Figure 13.11), and diagnostic cerebral angiogram showed a carotid T occlusion (Figures 13.12 and 13.13).

The ADAPT technique was used, two passes were attempted, and finally the clot was removed. TICI 3 revascularization was achieved (Figures 13.14 and 13.15) 14 min from groin puncture.

NIHSS decreased to 3 postprocedure, but the patient did have a residual left arm weakness (mRS 2); therefore, TICI 3 does not always correlate to mRS 0. Postop MRI showed a small basal ganglia stroke (Figure 13.16).

CONTROVERSIES

Immediately after the administration of TPA, a vascular imaging should be obtained to sort out those patients who would benefit from further interventional therapy. CTP compared to MRI is readily available and takes only a few minutes, but how well the ischemic stroke lesions diagnosed on CTP correlate with MRI is still debatable.

As mentioned, Wintermark et al. compared CTP and MRI on the admission on 13 patients with ischemic stroke and found a strong correlation between the size of the infarct on CTP and DWI abnormalities. Similarly, Eastwood et al. correlated early CTP images with brain MRI on 14 patients and found that low CBV on CTP correlated most highly with the extent of MR diffusion abnormality and suggested that low CBV may be used to detect infarcted tissue similarly to DWI (Wintermark et al. 2002a,b, Eastwood et al. 2003).

Nonetheless, there are some cases where decreased CVB on CTP does not correlate with an infarcted area on DWI MRI images (Figures 13.17 through 13.19). The reason is not yet clear, and further studies including more patients will be necessary to address this issue.

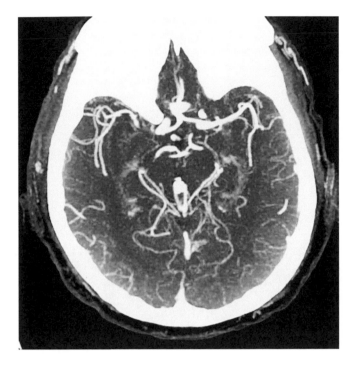

FIGURE 13.10 CTA shows right MCA occlusion.

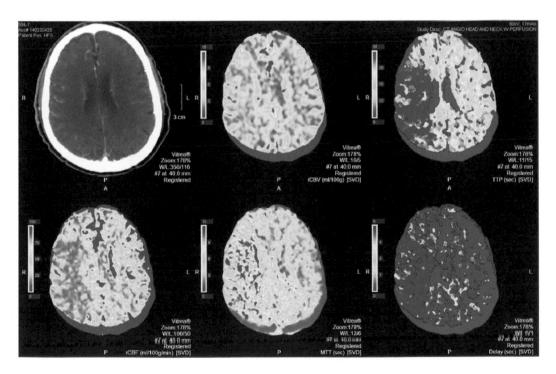

FIGURE 13.11 CTP shows CBF/CBV mismatch.

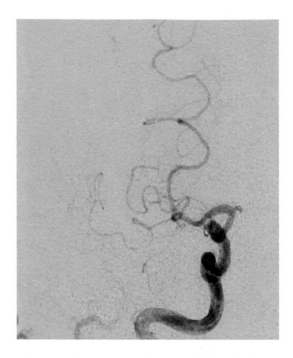

FIGURE 13.12   AP diagnostic cerebral angiogram shows carotid T occlusion.

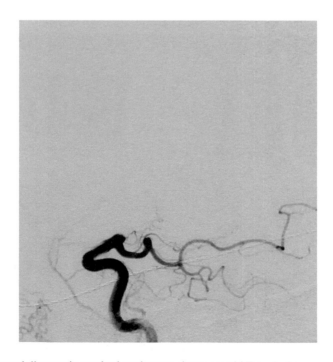

FIGURE 13.13   Lateral diagnostic cerebral angiogram shows carotid T occlusion.

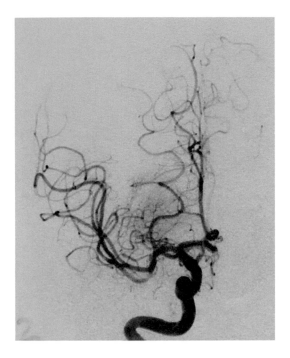

FIGURE 13.14   AP postthrombectomy diagnostic cerebral angiogram shows TICI 3 revascularization.

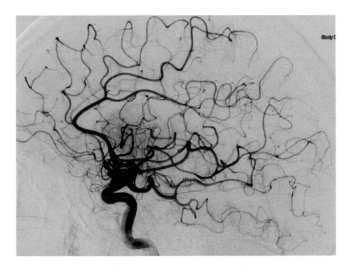

FIGURE 13.15   Lateral postthrombectomy diagnostic cerebral angiogram shows TICI 3 revascularization.

CONCLUSION

Till date, the role of CTP to identify patients who will benefit from endovascular therapy is still a matter of debate. Further studies are needed to avoid excluding from treatment those patients who have ASPECTS >6 but a matched CTP defect that may potentially benefit from an interventional procedure (see Figures 13.20 through 13.23) and to address the use of CTP in stroke cases out of the 6 h window.

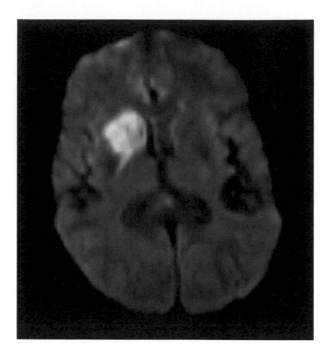

FIGURE 13.16   Postop MRI shows a small basal ganglia stroke.

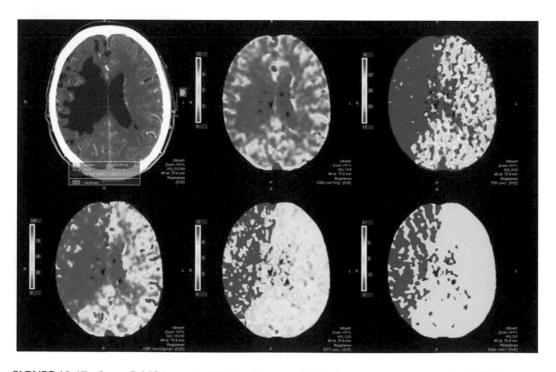

FIGURE 13.17   Large R MCA matched defect. Decreased CVB and increased MTT, TTP and CBF.

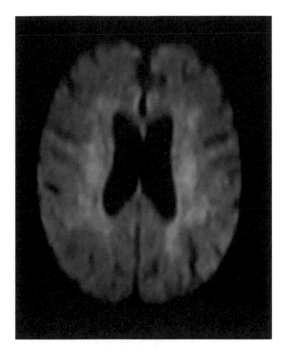

FIGURE 13.18 DWI shows no diffusion restricted lesion.

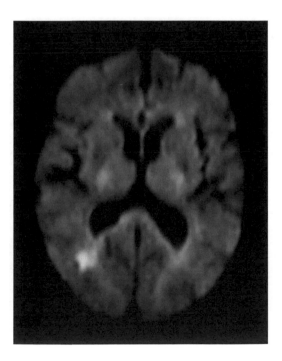

FIGURE 13.19 DWI shows small diffusion restricted lesion.

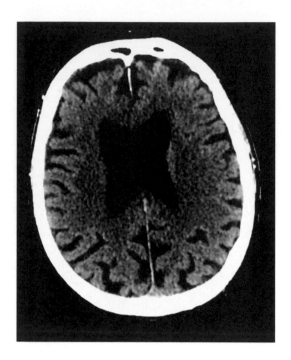

FIGURE 13.20 Head CT shows no obvious ischemic lesion.

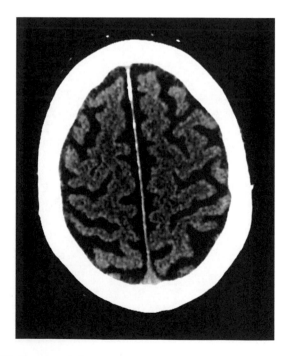

FIGURE 13.21 Head CT shows no obvious ischemic lesion.

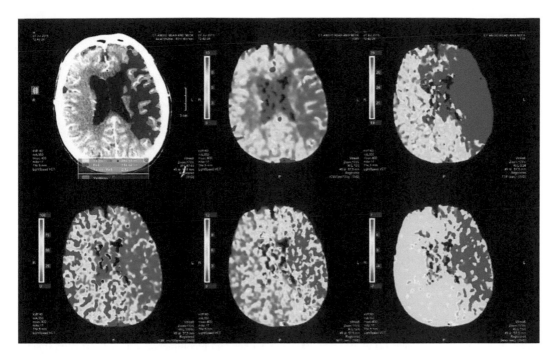

FIGURE 13.22 CTp shows large matched L MCA distribution lesion.

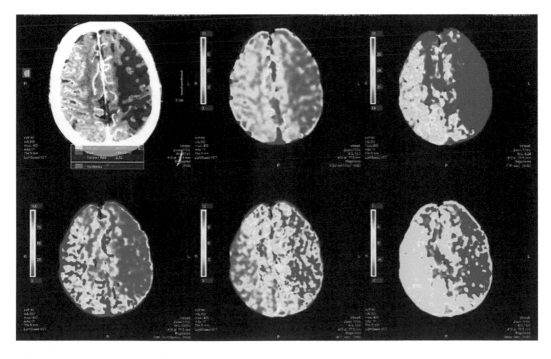

FIGURE 13.23 CTp shows large matched L MCA distribution lesion.

REFERENCES

Allmendinger AM, Tang ER, Lui YW et al. January 2012. Imaging of stroke: Part 1, Perfusion CT—Overview of imaging technique, interpretation pearls, and common pitfalls. *AJR Am J Roentgenol.* 198(1):52–62.

Benson J, Payabvash S, Salazar P et al. April 2015. Comparison of CT perfusion summary maps to early diffusion-weighted images in suspected acute middle cerebral artery stroke. *Eur J Radiol.* 84(4):682–689.

Bivard A, Levi C, Krishnamurthy V et al. December 2014. Defining acute ischemic stroke tissue pathophysiology with whole brain CT perfusion. *J Neuroradiol.* 41(5):307–315.

Campbell BC, Mitchell PJ, Kleinig TJ et al. March 12, 2015. EXTEND-IA investigators. Endovascular therapy for ischemic stroke with perfusion-imaging selection. *N Engl J Med.* 372(11):1009–1018.

Donnan GA, Davis SM. 2002. Neuroimaging, the ischemic penumbra, and selection of patients for acute stroke therapy. *Lancet Neurol.* 1:417–425.

Eastwood JD, Lev MH, Azhari T et al. 2002. CT perfusion scanning with deconvolution analysis: Pilot study in patients with acute middle cerebral artery stroke. *Radiology* 222:227–236.

Eastwood JD, Lev MH, Wintermark M et al. October 2003. Correlation of early dynamic CT perfusion imaging with whole-brain MR diffusion and perfusion imaging in acute hemispheric stroke. *AJNR Am J Neuroradiol.* 24(9):1869–1875.

Hoeffner EG, Case I, Jain R et al. 2004. Cerebral perfusion CT: Technique and clinical applications. *Radiology* 231:632–644.

Ito H, Kanno I, Kato C et al. May 2004. Database of normal human cerebral blood flow, cerebral blood volume, cerebral oxygen extraction fraction and cerebral metabolic rate of oxygen measured by positron emission tomography with 15O-labelled carbon dioxide or water, carbon monoxide and oxygen: A multicentre study in Japan. *Eur J Nucl Med Mol Imaging* 31(5):635–643.

Latchaw RE, Yonas H, Hunter GJ et al. April 2003. Guidelines and recommendations for perfusion imaging in cerebral ischemia: A scientific statement for healthcare professionals by the writing group on perfusion imaging, from the Council on Cardiovascular Radiology of the American Heart Association. *Stroke* 34(4):1084–1104.

Maija R, Gaida K, Karlis K et al. April 24, 2013. Perfusion computed tomography relative threshold values in definition of acute stroke lesions. *Acta Radiol Short Rep.* 2(3):2047981613486099.

Powers WJ, Derdeyn CP, Biller J et al. June 29, 2015. 2015 AHA/ASA focused update of the 2013 guidelines for the early management of patients with acute ischemic stroke regarding endovascular treatment: A guideline for healthcare professionals from the American Heart Association/American Stroke Association. *Stroke* 46:3020–3035.

Saver JL, Goyal M, Bonafe A et al. June 11 2015. SWIFT PRIME Investigators. Stent-retriever thrombectomy after intravenous t-PA vs. t-PA alone in stroke. *N Engl J Med.* 372(24):2285–2295.

Wintermark M, Reichhart M, Cuisenaire O et al. August 2002a. Comparison of admission perfusion computed tomography and qualitative diffusion- and perfusion-weighted magnetic resonance imaging in acute stroke patients. *Stroke* 33(8):2025–2031.

Wintermark M, Reichhart M, Thiran JP et al. 2002b. Prognostic accuracy of cerebral blood flow measurement by perfusion computed tomography, at the time of emergency room admission, in acute stroke patients. *Ann Neurol.* 51:417–432

14 Mapping the Injured Brain
Structure and Function Following Traumatic Brain Injury

Chandler Sours, Jiachen Zhuo, and Rao P. Gullapalli

CONTENTS

TRAUMATIC BRAIN INJURY

BACKGROUND

Traumatic brain injury (TBI) is a leading cause of death and lifelong disability throughout developed nations. TBI is the result of an external mechanical force applied to the skull leading to temporary and permanent impairments, functional disability, or psychosocial maladjustment (Steyerberg et al., 2008). The Centers for Disease Control and Prevention (CDC) estimates that 1.7 million people sustain TBI each year resulting in 275,000 hospitalizations and 52,000 deaths, equal to about one-third (30.5%) of all injury-related deaths in the United States (Faul et al., 2010). However, of those who survive this traumatic event, many are left with physical, cognitive, and psychosocial deficits, resulting in a limited ability to return to work and a reduced quality of life. As a consequence of these injuries, patients and family members are left with a large emotional and financial burden, and

273

the nation is left with a vast fiscal burden, with direct and indirect costs estimated to range from $62 to $78 billion annually (Centers for Disease Control and Prevention (CDC), 2013).

TYPES OF TBI

Current clinical practice categorizes TBI in multiple ways including level of severity, level of consciousness, or mental status following head injury. Perhaps the most frequently used classification system is the Glasgow Coma Scale (GCS) that divides patients into mild (GCS 13–15), moderate (GCS 9–12), and severe TBI (GCS < 8) (Teasdale and Jennett, 1974). The GCS is generally used to classify patients because of its ease of use and relatively consistent correlations with the Glasgow Outcome Scale (GOS) and the Disability Rating Scale (Thornhill et al., 2000). However, due to the heterogeneous nature of the injury, there is a large discrepancy in patient outcomes despite similar GCS scores, suggesting that new classification systems need to be actively explored (Thornhill et al., 2000).

Despite variable long-term outcomes across the range of severities based on the GCS classification, the majority of TBI cases, approximately 75%, are diagnosed as mild TBI (mTBI) (Centers for Disease Control and Prevention, 2003). Furthermore, these mild injuries can be further divided based on mechanism of injury into civilian or military blast injuries, each of which has unique features (Magnuson et al., 2012). The leading causes of civilian injuries, in order of prevalence include falls, unintentional blunt trauma, motor vehicle accidents, and assaults (CDC webpage). While civilian mTBI generally results from a direct impact to the brain, military blast mTBI can occur with or without a direct insult to the head, but is caused by overpressurization shock waves originating from high-energy explosives. It is the propagation of this shock wave that travels through the body into the head that is proposed to result in a portion of military mTBI in the absence of a direct impact (Chapman and Diaz-Arrastia, 2014). Further complicating matters, the latest research points to a cumulative effect of multiple concussive or subconcussive events. Referred to as chronic traumatic encephalopathy (CTE), this form of neurodegeneration results from repetitive subconcussive events and results in progressive memory loss and ultimately dementia similar to Alzheimer's disease (Lakhan and Kirchgessner, 2012).

BIOLOGY

The primary injury to the brain is a result of sudden acceleration, deceleration, and/or rotational forces. The acceleration and deceleration of the brain can cause cortical contusions (bruises of the brain tissue) and hemorrhages (bleeding) when the brain hits the skull. The rotational forces may also cause deeper cerebral lesions when white matter (WM) axons are stretched and damaged, as well as shearing injuries at the gray matter (GM) and WM interface (Ommaya et al., 2002). This disperse shearing injury is referred to as diffuse or traumatic axonal injury (DAI or TAI, respectively). Following the initial impact, the injury is propagated throughout the brain through various biomolecular and cellular pathways, which results in widespread degeneration of neurons and glial cells. However, this primary damage can lead to myriad molecular events, referred to as the secondary injury, including cellular edema, inflammation, vascular dysregulation, ischemia, disruption in plasma membrane and neurotransmitter release, mitochondrial dysfunction, production of reactive oxygen species, altered anaerobic metabolism, and lactic acidosis. Together, these secondary injury pathways result in both necrotic and apoptotic cell death (Lye and Shores, 2000; Nemetz et al., 1999). Therefore, secondary injuries likely contribute to long-term cognitive impairment and are ultimately the deciding factors in patient recovery.

POSTCONCUSSIVE SYMPTOMS

It has been estimated that there are approximately 5.3 million people in the United States living with long-term disability from TBI (Centers for Disease Control and Prevention (CDC), 2013).

Furthermore, it is believed that a significant fraction (~40%) of mTBI patients will remain impaired for at least 3 months, and a substantial subgroup of these patients will show deficits up to 1 year after injury leading to lost productivity and resultant socioeconomic consequences (Alves et al., 1993; Bazarian et al., 1999; Centers for Disease Control and Prevention, 2003). Postconcussive symptoms include neuropsychological symptoms such as difficulty with socializing, depression, and anxiety, cognitive symptoms such as deficits in attention, executive function and working memory, reduced information processing speed, as well as somatic symptoms such as headaches, chronic pain, sensory perception disorders, and language difficulty (Chaumet et al., 2008; Hillary et al., 2010; Immonen et al., 2009; Johansson et al., 2009; Makley et al., 2008; McAllister et al., 2006; McDowell et al., 1997; Menzel 2008; Miotto et al., 2010; Nampiaparampil 2008). Furthermore, at 1 year post injury, 82% of mTBI patients continue to report the presence of at least one symptom (McMahon et al., 2014). This suggests that across the spectrum of mild to severe injuries, both structural damage and its related functional disruptions caused by trauma produce persistent post-concussive symptoms.

CURRENT CLINICAL IMAGING FOR TRAUMA

In the clinical setting, computed tomography (CT) is the primary diagnostic tool in the evaluation of TBI. CT is used to triage patients who arrive at the hospital with a suspected head injury based on the presence of intracranial injuries, focal injuries, and/or the presence of diffuse pattern of parenchymal injury known as diffuse axonal injury (DAI) (Cihangiroglu et al., 2002; Zimmerman, 1999). While an admittance CT does provide immediate information to the patient's clinical team, it should be noted that the clinical presentation often does not match the presence of abnormalities seen on this initial CT. This may be due to the continued evolution of intracranial injuries or due to the fact that the CT scanning often does not have the sensitivity to detect DAI.

CONVENTIONAL CLINICAL MR IMAGING

A conventional MRI for trauma-related injuries often consists of T1-weighted images to assess the presence of focal intracranial damage; T2-weighted images to determine the extent of contusions and hemorrhages; fluid attenuated inversion recovery (FLAIR) images to determine the parenchymal integrity, cortical surface lesion, brain stem, and ventricular hemorrhage; and a proton-density-weighted image to assess WM abnormalities. In recent years, the use of susceptibility-weighted imaging (SWI), a 3D technique that is more sensitive to microhemorrhages resulting from DAI, has greatly aided the clinical evaluation of trauma (Ashwal et al., 2006; Haacke et al., 2004; Reichenbach et al., 1997). For instance, the extent of SWI lesions has been shown to correlate with initial GCS, length of coma as well as the presence of neurologic impairments in memory and attention in pediatric TBI (Tong et al., 2004). Furthermore, the clinical utility of SWI in mTBI populations has been demonstrated; there is a noted correlation between the aggregate SWI lesion volume and measures of clinical severity (Benson et al., 2012). Conventional MRI has proven superior to CT imaging for the detection of DAI and has greatly improved the assessment of intracranial trauma. However, both CT imaging and traditional MRI have limited sensitivity in detecting and characterizing DAI in the acute trauma setting (Parizel et al., 1998; Wilson et al., 1988). Thus, additional research in the evaluation of trauma using advanced neuroimaging techniques is warranted.

In recent years, several advanced imaging techniques have been introduced that can probe the microstructural and cerebrovascular changes that in turn lead to modifications in cellular metabolism and consequently alterations in cortical and subcortical function. There is currently great interest in the field to apply advanced imaging techniques during the acute stage of injury to improve the prognostic ability of neuroimaging to determine long-term outcomes.

ADVANCED IMAGING

ADVANCED STRUCTURAL MR IMAGING

Following TBI, the secondary injury cascade ultimately results in necrotic or apoptotic cell death that contributes to measurable cerebral atrophy due to neuronal loss. Structural imaging using the high-resolution T1-weighted magnetization prepared-rapid acquisition gradient echo (T1-MPRAGE) sequence is used to characterize the time course of these cerebral and ventricular volume changes. Cerebral atrophy appears to be a continuous process becoming increasingly prominent in the chronic stages of injury in cortical and subcortical regions (Warner et al., 2010a,b). Whole-brain and regional volume loss has been shown to correlate with injury severity, neurocognitive performance, long-term disability, and learning (Levine et al., 2008; Warner et al., 2010a,b). However, while this cerebral atrophy is often visually apparent in severe TBI populations, it is often very subtle in mTBI populations suggesting that more sensitive techniques to assess structural damage may provide additional information in mTBI populations.

COMBINING STRUCTURE AND FUNCTION

It can be assumed that the structural damage to neurons, glia, and astrocytes induced by both the primary and secondary injuries associated with head injuries will likely disrupt the communication between large scale neural networks supporting both sensory and higher order cognitive processes. The management of each of the postconcussive symptoms individually provides a challenge to patients and clinicians; however, many patients experience a unique combination of symptoms making the creation of treatment plans exceedingly difficult for clinicians. However, while the structural and functional damage produced by the trauma is diverse, it is likely that the patients suffering from similar symptoms will share similarities in the location and extent of injury paving the way for individualized medicine. The following discussion highlights the current techniques available to measure the subtle structural injuries and consequential functional damage induced by TBI.

STRUCTURE: DIFFUSION MRI IN THE BRAIN

Diffusion MRI has long been used as an important tool in studying neurological disorders because it provides *in vivo* measurements of tissue microstructure changes that are unable to be detected using conventional CT or MR imaging. Diffusion tensor imaging (DTI) entails measuring water diffusion in at least six directions to obtain an appropriate representation of a diffusion tensor, describing the preferential diffusion direction and an ellipsoidal diffusivity profile. Briefly speaking, common measurements in DTI include mean diffusivity (MD) and apparent diffusion coefficient, which both measure the average magnitude of diffusivity, and fractional anisotropy (FA), which measures the disproportion of diffusion along the three principal axes of the diffusion ellipsoid. The diffusion properties of different brain tissues (GM, WM, CSF) exhibit unique structural properties. Diffusion in CSF is similar to free diffusion in water, so MD is extremely high ($\sim 3 \times 10^{-3}$ mm^2/s) and FA is almost 0. Diffusion in GM is generally nondirectional or isotropic because it is mostly composed of neurons and glial cells (FA < 0.2). On the other hand, in WM, diffusion is highly anisotropic due to the myelinated axons, which restrict water diffusion in the direction perpendicular to the axon. Therefore, axial diffusivity (AD, diffusion measured along the axon) can be as much as seven times the radial diffusivity (RD, diffusion measured perpendicular to the axon averaged across two axes) (FA $\sim$ 0.45–0.8) (Song et al., 2003). Figure 14.1 shows an example of water diffusion in WM axons and the effect of axon membrane injury to water diffusion, which leads to increased RD and reduced FA.

Intact axons have high anisotropy while damaged axons have reduced anisotropy. Neurodevelopmental studies in both human and animal models have demonstrated increases in FA throughout the early stages of brain development (to adolescence), likely due to the formation

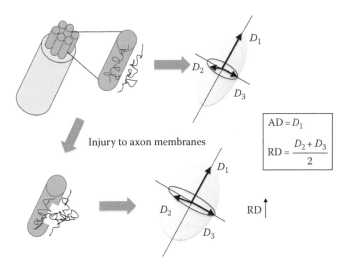

FIGURE 14.1 Water diffusion in white matter axons and the effect of axon membrane injury to water diffusion. Due to the collisions with axon membranes, water molecules travel less distance perpendicular to the axon directions than parallel to the axons. When there is injury to axon membranes, water diffusion along the radial direction will increase. D_1, D_2, and D_3 are the diffusion coefficients along the three principal axes of the diffusion tensor. AD, axial diffusivity; RD, radial diffusivity.

and maturation of myelin. This FA value plateaus in adulthood but starts to decline in older age (after age 60) likely because of the loss of myelin integrity associated with aging (Lebel and Beaulieu, 2011; Pfefferbaum et al., 2005; Zhang et al., 2007).

DTI is commonly used to study alterations in WM regions because FA has been proven to be especially sensitive in detecting subtle WM microstructure changes. Damage to WM axons can also cause reduced neurotransmission in the brain, altered communication within the neural network, and, therefore, reduced cognitive function. Changes in FA have been shown to correlate with clinical presentation and cognitive performance in patients suffering from multiple neurological disorders such as Alzheimer's disease (Bozzali et al., 2002), multiple sclerosis (Schmierer et al., 2004), and Parkinson's disease (Karagulle Kendi et al., 2008). More specifically, the AD is believed to be associated with axonal integrity, while the RD is believed to be associated with the integrity of myelin (Mac Donald et al., 2011; Sidaros et al., 2008; Song et al., 2003).

While DTI is widely used to assess damage in WM regions, it has limited sensitivity in assessing alterations in GM regions or lesion regions with relatively isotropic diffusion. Recently, the ability of diffusional kurtosis imaging (DKI) to study the heterogeneity of the microenvironment of a tissue has gained significant attention. DTI assumes the underlying water diffusion in tissues and follows Gaussian distribution, which is really only the case in a uniform diffusion environment. The heterogeneous diffusion environment due to cellular membranes gives rise to deviation of the water diffusion from the Gaussian distribution, mathematically termed kurtosis (K) (Figure 14.2). In DKI, both the Gaussian and non-Gaussian water diffusion are included in the estimation model (Jensen et al., 2005; Lu et al., 2006). The Gaussian portion of the model provides more accurate estimation of DTI parameters (Veraart et al., 2011). The non-Gaussian portion of the model provides measures of kurtosis, which is high if the underlying tissue complexity/heterogeneity is higher (e.g., in GM or WM), and lower otherwise (e.g., in CSF) (Jensen and Helpern 2010). WM regions generally have higher mean kurtosis (MK) (~1.2) due to the densely packed axonal membranes/myelin sheath. GM regions have median MK values (~0.7), which are relatively high (unlike the low FA in the GM) and provide a tool to probe GM tissue microstructure changes. MK has been shown to be a more sensitive marker in tumor grading (Raab et al., 2010; Van Cauter et al., 2012), Alzheimer's disease (Falangola et al., 2013; Gong et al., 2013), Parkinson's disease (Wang et al., 2011), attention deficit

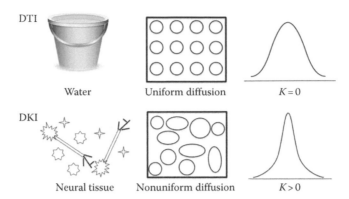

FIGURE 14.2 Diagram representing Gaussian and non-Gaussian diffusion displacement in different diffusion environments. DTI, diffusion tensor imaging; DKI, diffusion kurtosis imaging.

hyperactivity disorder (ADHD) (Helpern et al., 2011), and TBI (Grossman et al., 2012; Zhuo et al., 2012). In an animal model of TBI, our group has proven DKI to be sensitive to reactive astrogliosis (Zhuo et al., 2012). For example, increased MK was observed in the rats that had increased glial fibrillary acidic protein staining in the cortex contralateral to the site of injury (Zhuo et al., 2012). MK changes were observed in the absence of MD or FA changes, suggesting that MK may be more sensitive to subtle tissue microstructure changes in GM regions.

FUNCTION: FUNCTIONAL MRI IN THE BRAIN

Functional MRI (fMRI) is a valuable tool as it can identify the deficits in neural networks associated with various cognitive processes. Specifically, fMRI provides an indirect measure of large-scale neural activation and is based on the MR signal differences between deoxygenated blood and oxygenated blood. When individual neurons that are recruited for a given task produce an axon potential, there is an increase in freshly oxygenated blood to the local tissue to keep up with the increased neuronal demand. This change in the ratio of deoxygenated blood to oxygenated blood in the activated region causes a change in the tissue signal as the local tissue changes from a predominantly paramagnetic state to diamagnetic state. It is this MR signal change that is measured in fMRI and is called the blood-oxygen-level-dependent (BOLD) signal (Figure 14.3a). Currently fMRI data are acquired in one of two ways using a task-based fMRI paradigm and a resting-state fMRI (rs-fMRI) paradigm (Figure 14.3b and c).

Using a task-based paradigm, participants are instructed to perform a specific task while in the scanner. Depending on the specific analysis desired, task-based paradigms can be performed using either a block design (Figure 14.3b) or an event-related design. With task-based fMRI, researchers are able to indirectly measure which regions of the brain are recruited to perform a specific task based on which regions of the brain demonstrate increased BOLD signal, during task performance compared to resting conditions.

Using a resting-state paradigm, it is possible to examine functional brain networks to measure the interaction between global brain regions in disparate locations. In this method, participants are instructed to rest in the MRI scanner and are not required to participate in a task. Referred to as resting state functional connectivity (rs-FC), this method measures the strength of functional interactions between brain regions based on temporal correlations between small fluctuations in the BOLD signal (Biswal et al., 1995; Sporns 2011; van den Heuvel and Hulshoff, Pol 2010) (Figure 14.3c). While historically these small fluctuations in BOLD signal were thought to be signal noise, it was noted that regions recruited to perform specific tasks displayed similar temporal patterns of fluctuations during resting conditions (Biswal et al., 1995). Analysis of functional connectivity can be

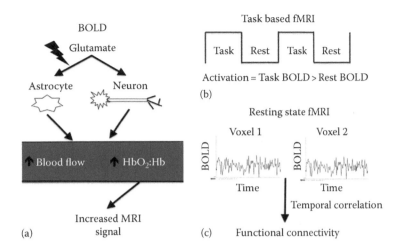

FIGURE 14.3 Diagram representing functional MRI. (a) Representation of the BOLD signal as an indirect measure of neuronal function. (b) Example of a block-design task-based fMRI paradigm. (c) Example of a seed-based analysis of functional connectivity derived from rs-fMRI.

performed using a hypothesis seed-based method or a data-driven independent component analysis method (Calhoun et al., 2009). Regardless of which method is selected, resting-state networks are consistently replicated across studies and often include networks that are associated with sensory systems (auditory, visual, somatosensory, and motor) as well as networks associated with higher-order cognitive processes (Raichle, 2010). Understanding the differences in neural network communications related to postconcussive symptoms among TBI patients will provide valuable information on the precise mechanisms of these deficits.

APPLICATION OF DIFFUSION-WEIGHTED IMAGING TO TBI

WHY IS DIFFUSION-WEIGHTED IMAGING USEFUL IN TBI?

TBI causes focal damage to brain tissue, diffuse injury to the axons from the shearing forces exerted on the brain, and disperse cellular changes due to a multitude of secondary injury mechanisms. Therefore, it is critical to be able to characterize these structural changes through *in vivo* imaging markers to aid patient management and prediction of long-term recovery.

Much of what has been learned regarding structural alterations following trauma have come from experimental models of TBI in rodent studies. Pathophysiology results from these studies suggest initially the axons swell up in response to injury due to the loss of integrity of ionic transport channels located on the axon. While some swelling resolves, extensive unresolved swelling often results in broken axons with terminal axon bulbs. Accompanying damage may also involve loss of the integrity of the myelin sheath known as demyelination. This demyelination often progresses over time which results in reduced axonal integrity in the chronic stages of injury. In addition to direct axonal damage, the injury also results in a transient increase in numbers of astrocytes and microglial cells (Chen et al., 2003). The atypical increase in the number of astrocytes in a region due to the death of nearby neurons is referred to as astrogliosis. Reactive astrogliosis are believed to play essential roles in preserving healthy neurons and minimizing inflammation within the surrounding brain tissue (Myer et al., 2006).

In additional to axonal damage and reactive astrogliosis, cerebral edema, which is an excess accumulation of water in the intracellular and/or extracellular space of the brain, also typically occurs following TBI. Intracellular or cytotoxic edema typically happens immediately following traumatic injury due to a change in the cellular metabolism resulting in inadequate functioning of the sodium and potassium ion channels in the cell membrane. Extracellular or vasogenic edema develops more slowly over time due to a failure of the blood-brain barrier (a barrier that separates circulating blood and CSF and maintains the integrity of CSF). The development of cerebral edema often also results in compressive forces on surrounding brain tissue that has been shown to elevate intracranial pressure and reduce cerebral blood flow, causing further damage to the brain (Greve and Zink, 2009).

DIFFUSION TENSOR MRI APPLICATIONS IN TBI

As previously mentioned, DAI represents the most common form of injury resulting from TBI and constitutes nearly half of all TBI-related injuries (Arfanakis et al., 2002). This axonal injury is also a powerful predictor of morbidity and mortality (Greve and Zink, 2009). However, both CT and traditional MRI have limited sensitivity in detecting and characterizing DAI in the acute trauma setting (Parizel et al., 1998; Wilson et al., 1988). In recent years, diffusion MRI has quickly gained interest in the TBI community. The various diffusion parameters are quickly developing into new imaging biomarkers with potential prognostic value for predicting patient outcomes in both mildly and severely injured patients. Early studies investigating diffusion MRI in TBI populations have demonstrated an increased sensitivity in detecting DAI-related lesions (Ezaki et al., 2006). In addition, whole-brain MD measurements demonstrate diagnostic value by being able to independently indicate TBI in spite of normal conventional MRI findings (Shanmuganathan et al., 2004). Figure 14.4 shows an example of a moderately injured patient with loss of consciousness and altered mental status and seizures who presented with normal conventional MRI and CT. Diffusion tensor imaging (DTI) of the patient reflects disrupted fiber tracks from the corpus callosum, as well as increased whole-brain MD as compared to normal controls, indicating disturbed homeostasis of water diffusion in the brain.

Recent studies using DTI tend to focus primarily on alterations in the WM microstructure associated with DAI due to the great sensitivity of FA and axial/radial diffusivity in detecting axonal injury. Region of interest (ROI) analysis based on either anatomical regions or tractography-based tract regions

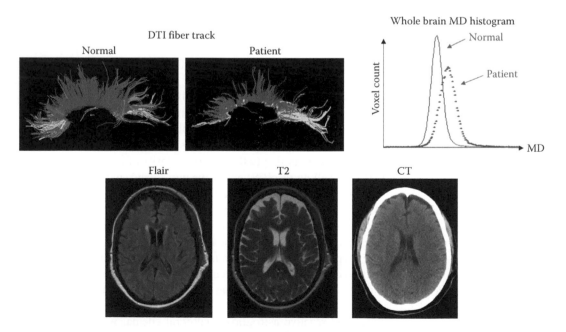

FIGURE 14.4 An example of a moderately injured patient with loss of consciousness and altered mental status and seizures who presented with normal conventional MRI and CT. DTI of the patient reflects disrupted fiber tracks from the corpus callosum as well as increased whole-brain MD as compared to normal controls.

Symptom improving   Symptom worsening

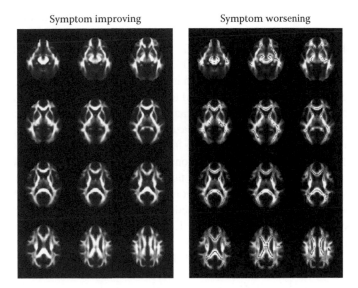

FIGURE 14.5   Diffuse acute FA reduction in mTBI patients ($n = 14$) who had eventual worsening symptoms at 6 months compared to a control group ($n = 30$), while no changes in FA were noted in a group of patients ($n = 11$) who had eventual improvement of symptom severity. Results based on TBSS analysis thresholded at $p < 0.001$.

indicate that commonly damaged regions following TBI include the corpus callosum (Bazarian et al., 2007; Kumar et al., 2009; Mayer et al., 2010; Warner et al., 2010a), internal capsule (Arfanakis et al., 2002; Bazarian et al., 2007), and cingulum bundles (Mac Donald et al., 2011). More recently, whole-brain analysis, such as tract-based spatial statistics (TBSS) (Smith et al., 2006), revealed more widely spread WM abnormalities extending to the superior and inferior longitudinal fasciculus, corona radiate, frontal and temporal lobes, etc. (Messe et al., 2011; Yuh et al., 2013), which is consistent with the diffuse nature of the injury. Alterations in DTI parameters following TBI, most strongly FA, have been widely demonstrated, correlating with injury severity (Benson et al., 2007) and/or functional outcomes after TBI, including neuropsychological scores (Kraus et al., 2007; Niogi et al., 2008), GOS (Sidaros et al., 2008), and postconcussive symptoms (Wilde et al., 2008). More importantly, acute DTI abnormalities have shown prognostic values in predicting outcomes in severe TBI (Betz et al., 2012; Shanmuganathan et al., 2004). Even in mTBI, FA reduction in at least one region in the brain was shown to be the most robust predictor for 6-month outcome, over conventional MRI and/or any other clinical, demographic/socioeconomic characteristics (Yuh et al., 2013). Figure 14.5 demonstrates diffuse acute FA reduction in mTBI patients ($n = 14$) who had eventual worsening symptoms at 6 months compared to a control group ($n = 30$), while no changes in FA were noted in a group of patients ($n = 11$) who had eventual improvement of symptom severity.

TBI is an evolving dynamic injury, which results in a series of diffusion properties that change progressively throughout various phases of injury. The most consistent DTI findings have been the reduced FA and the increased MD at chronic stages of injury (6 months or longer), regardless of injury severity. Such changes are mostly attributed to increased AD, which is consistent with axonal damage and cell death. These changes in FA also correlate with injury severity (Benson et al., 2007), cognitive deficits (Kraus et al., 2007; Niogi et al., 2008), impairment of learning and memory (Salmond et al., 2006), and functional outcome (Wilde et al., 2006). On the other hand, increased RD, which is an indication of irreversible myelin damage, tends to be more often observed in severely injured than mildly injured TBI patients (Kraus et al., 2007; Newcombe et al., 2007; Sidaros et al., 2008)

DTI findings at acute (within 1 week) or semiacute (within 3 months) stages after injury are more heterogeneous. In moderate to severe TBI, nearly all DTI studies reported increased MD and reduced FA even at the acute stage, which are correlated with clinical or neuropsychological measures (Benson et al., 2007; Betz et al., 2012; Perez et al., 2014). This may be attributed to more severe cellular and axonal damage, while in mTBI, increased (Bazarian et al., 2007; Chu et al., 2010; Wilde et al., 2008), reduced (Arfanakis et al., 2002; Betz et al., 2012; Inglese et al., 2005), bidirectional (Kou and Vandevord 2014), or no changes in FA values (Ilvesmaki et al., 2014) have all been reported. Nevertheless, an overall trend for mTBI tends to be that at very early stages post injury, FA is increased, with an associated reduction in MD and/or RD. This indicates an inflammatory response such as axonal swelling or cytotoxic edema (Bazarian et al., 2007; Chu et al., 2010; Wilde et al., 2008). Increased FA and decreased RD are also found to correlate with severity of postconcussion symptoms suggesting a link between altered axonal integrity and long-term outcome (Wilde et al., 2008). The FA increase is followed by a partial normalization, which may either indicate recovery or just a transient stage before further FA reduction. For example, Mayer et al. (2010) reported acute FA increase at 12 days post injury, which returned to normal level at 3–5 months, associated with a reduction of TBI symptoms (Mayer et al., 2010). Ling and colleagues reported a very similar change with increased FA up to 21 days post injury, yet a recovery at 4 months (Ling et al., 2012). On the other hand, Ilvesmaki et al. (2014) reported no DTI changes in 75 mTBI patients either at acute or 1 month post injury.

Depending on the exact timing of imaging post injury and the heterogeneous injury progression for each individual patient, DTI changes during this dynamic phase can indeed be highly variable. Most studies have employed group analysis that report on abnormalities in regions that are consistently found across all or most patients. Some studies characterized injured brain regions by evaluating abnormal FA voxels or regions (e.g., z-score of 2 or more compared to control values) for each individual patient. Oftentimes, the analysis was relaxed to include any abnormal FA values regardless of the direction of changes. For example, Yuh and colleagues found the criteria that more than one ROI with reduced FA at acute stage being a reliable predictor for 6 month patient outcome in mTBI (Yuh et al., 2013). Kou et al. reported that abnormal FA voxels (FA lesion load) within 24 h after injury are correlated with patient neurocognitive data, which is also associated with an increase of these patients' serum neuronal and glial biomarkers (2013). In a sports-related concussion study, Bazarian et al. found a significantly higher number of pre- and postseason FA and MD changes in athletes than controls, with an even higher voxel number change in a concussed athlete (Bazarian et al., 2012). However the pathologic changes underlying the bidirectional FA/MD changes are not well understood.

Despite all the efforts reported in previous studies, the exact time course of the primary and secondary injuries associated with TBI and the effect of diffusion parameters remain unclear. The heterogeneous nature of TBI injuries necessitate more longitudinal studies with larger patient size and better controlled imaging time points in order to better characterize the injuries. In addition, individual patient-oriented analysis is also needed with larger sample sizes to better understand the prognostic values of the DTI measures. As TBI may result in many underlying pathological changes in relation to axons, neurons, and glial cells that affect DTI parameters, studies in animal models of TBI are needed to fully understand the pathological changes underlying these DTI parameter changes. However, the heterogeneous nature of human TBI makes this line of research difficult.

Diffusional Kurtosis MRI Applications in TBI

DKI is a relatively new addition to the DTI model and provides an independent measure for diffusion heterogeneity. High MK values have been associated with reactive gliosis in the animal model of TBI at the subacute stage post injury (Zhuo et al., 2012). This is due to the complex cellular structure of astrocyte/microglial that gives rise to the overall diffusion heterogeneity. In human studies of chronic mTBI, reduced MK was found in the thalamus and internal capsule, indicating a reduction of diffusion heterogeneity that may suggest a degenerative process leading to neuronal shrinkage and changes in axonal and myelin density (Grossman et al., 2012). Furthermore, reduced MK in the thalamus and internal capsule demonstrate positive correlations with cognitive measures of attention and processing speed. In a longitudinal study of a group of 24 mTBI patients followed through acute (<10 days), subacute (1 month), and chronic (6 months) stages post injury, reduced MK and radial kurtosis (RK) in the anterior internal capsule were reported across the three visits (Stokum et al., 2015). Interestingly, increased (improvement) in MK and RK were correlated with patients' cognitive improvements in multiple regions, including thalamus, internal capsule, and corpus callosum, which demonstrate that DKI may be sensitive in tracking pathophysiological changes associated with mTBI and can serve as a complimentary tool to conventional DTI in evaluating longitudinal changes following TBI.

While DKI applications in TBI are still scarce, its usage in other neuroimaging studies has shed some light into the potential underlying cellular structures for the kurtosis parameters. For example, higher kurtosis values have been found in high-grade than low-grade gliomas, which may be attributed to increased cellular density, decreased cell size, and increased complexity of the intracellular microenvironment (Raab et al., 2010; Van Cauter et al., 2012). Reduced kurtosis values have also been found in Alzheimer's disease, suggesting a loss of neuron cell bodies, synapses and dendrites, and increased extracellular space (Falangola et al., 2013; Gong et al., 2013).

While the DKI method is relatively new, it is quite promising as it allows for the *in vivo* analysis of microstructural changes in both GM and WM, which may provide additional insight into understanding the cellular process following TBI. Figure 14.6 represents a case study of a severe TBI patient scanned at 8 and 33 days post injury. In this case the contrast differences between the MK map and the contrast provided by the MD and FA map is quite noticeable. There is edema around lesion (yellow arrow) acutely, which resolved at 1 month. Restricted diffusion at the center of the lesion is manifested as extremely low MK (blue arrow). At 1 month, lesion area looks homogeneous on SWI and FLAIR but had both high and low MD and MK components (red arrows). Focal area with restricted diffusion also shows reduced MK. High MK ring around the lesion may be indicative of glial scars formed around damaged tissues (green arrows).

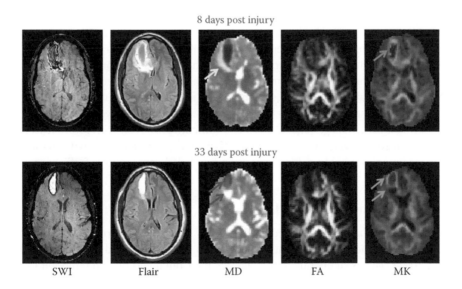

8 days post injury

33 days post injury

| SWI | Flair | MD | FA | MK |

FIGURE 14.6 Case study of 41-year-old male (GCS 4) with significant postconcussive symptoms. In this case different contrast that the MK map brought out than MD and FA can be appreciated. There is edema around the lesion (yellow arrow) acutely, which resolves at 1 month. Restricted diffusion at the center of the lesion manifest as extremely low MK (blue arrow). At 1 month, the lesion area looks homogeneous on SWI and FLAIR but had both high and low MD (red arrows) and MK components (green arrows). Focal area with restricted diffusion also showed reduced MK.

SUMMARY

DTI has provided great insight into WM microstructural damage following TBI, even when conventional MRI or CT fails to show any anatomical abnormalities, and has exhibited prognostic values for long-term outcomes. DKI further extends the ability of DTI to probe into GM microstructural changes. However the nature of the heterogeneity in TBI and study populations, inconsistency in imaging timing, and image acquisition methodologies have caused great discrepancies in findings among studies. Well-controlled longitudinal studies are in great need to better characterize the time course of microstructural changes following TBI and to identify potential imaging markers with prognostic values.

APPLICATION OF FUNCTIONAL MR IMAGING TO TBI

WHY IS FUNCTIONAL MR IMAGING USEFUL IN TBI?

As previously noted, both CT and traditional MR imaging have limited sensitivity in detecting and characterizing subtle DAI in the acute trauma setting (Parizel et al., 1998; Wilson et al., 1988). The previous section regarding diffusion-weighted imaging provided an extensive review of the invaluable use of DTI and DKI in diagnosing and predicting recovery in trauma patients. Despite advances in diffusion-weighted imaging at detecting alterations in tissue microstructural changes caused by these diffuse injuries, there is still great variability in chronic functional outcome causing clinicians to struggle with the long-term prognosis for trauma patients. Since there is a great deal of research noting altered structural integrity of WM tracts in the brain due to the DAI associated with TBI, it stands to reason that the communication between disperse functional regions in the brain will likely be disrupted as well. Furthermore, there may be functional alterations in network communication that contribute to poor cognitive and emotional outcomes that are unable to be directly measured in terms of structural damage at the current sensitivity of DTI and DKI.

Determining a direct link between structural connectivity as measured by DTI and functional connectivity measured with rs-fMRI has been a challenge to researchers in the past decade (Hagmann et al., 2010; Honey et al., 2009). Yet great strides have been made in recent years with multiple groups finding associations between altered structural integrity of WM tracts and altered functional communication within and between neural networks. This altered communication can be measured using either task-based or rs-fMRI. A further complication to this research is the great heterogeneity within the TBI population, including inconsistent injury mechanism and location as well as variables in patient demographics such as age, education, and sex. Everyone's injury is unique, making generalizations across patients exceedingly difficult. However, with future research combining structural connectivity and functional connectivity measures, the field will be one step closer to individualized medicine being able to treat trauma-induced damage to specific neural networks.

TASK-BASED fMRI APPLICATIONS IN TBI

While TBI is a diffuse injury, task-based fMRI studies have focused primarily on tasks related to executive functioning and cognition that are symptoms often associated with frontal lobe damage (McAllister, 2011). This is likely due to the fact that the frontal and medial temporal lobes (MTLs) are highly susceptible to damage from TBI due to their proximity to the osseous protrusions and depressions. For example, multiple task-based fMRI studies on patients with mTBI have shown alterations in BOLD responses during tasks designed to probe spatial memory (Slobounov et al., 2011), working memory (Christodoulou et al., 2001; McAllister et al., 2001; Sanchez-Carrion et al., 2008; Turner and Levine, 2008), executive function (Scheibel et al., 2003; Soeda et al., 2005), and sustained attention (Maruishi et al., 2007). Whereas some research groups have reported increased activation of task-positive regions following mTBI (McAllister et al., 1999, 2001; Smits et al., 2009), others have observed reduced activation (Chen et al., 2012; Slobounov et al., 2011; Witt et al., 2010). While results regarding increased or decreased activations and cluster size appear contradictory, this may be related to study design differences that include inconsistencies in time since injury and the wide range in difficulty of tasks chosen between studies. For example, McAllister et al. (2001) found that in an auditory N-back task, chronic mTBI patients had increased activations during a moderate working memory load but had decreased activations during a high working memory load suggesting that mTBI populations have differing compensatory mechanisms based on differences in task difficulty or cognitive load. Therefore, the impact of cognitive load on alterations in neural processing leading to variable BOLD responses may result in contrasting conclusions regarding the cognitive effects following TBI.

While the use of task-based fMRI is difficult to perform during the acute stages of severe TBI, research in the chronic stage has led to many new insights into severe TBI. Kasahara and colleagues found that chronic TBI patients had reduced performance and reduced activations during a working memory task indicating a failure of TBI patients to adequately activate the parietal regions of the task-positive network (TPN) (Kasahara et al., 2011). However, other groups have found over-recruitment of task-positive regions. Groups have argued that this increased activation should be interpreted as altered engagement of regions that are recruited by the patients at a lower cognitive load than in controls (Turner et al., 2012); however, others have asserted that increased activation results from the allocation of "latent resources" (Hillary, 2011). Furthermore, there is controversy in the field regarding the extent of functional recovery of mTBI patients in the chronic stage of injury. Some groups have provided evidence of recovery (Mayer et al., 2011), whereas others have noted persistent functional alterations in the chronic stage (Witt et al., 2010).

These findings highlight the critical question as to whether TBI patients demonstrate increased activation within brain regions used to perform a given task due to reduced efficiency of neural communication or if TBI patients demonstrate reduced activation of brain regions recruited to perform a task because they are unable to recruit the needed neural resources. A third possibility is that these patients may recruit new or novel brain regions that are not recruited in healthy populations to perform the tasks effectively. This over-recruitment may represent a form of compensatory processing to overcome deficiencies in neural recruitment. On the other hand, reduced activation may be due to compromised structural integrity of WM tracts within specific neural networks and recruitment of novel regions could represent functional reorganization.

Although advances have been made in understanding changes in brain systems following TBI with fMRI through examining fMRI activation patterns during task performance, critical gaps remain in the field's understanding of trauma-related functional alterations suggesting further research is needed to clarify the question.

An additional area of active research concerns the deactivation of the default mode network (DMN), a group of regions that are known to display reduced activation during task-related activities while demonstrating increased activity during rest conditions (Greicius et al., 2003; Raichle et al., 2001). It has been proposed that the failure to successfully deactivate the DMN during goal-directed behavior results in an interference in cognitive processing due to the intrusion of internal thoughts. This failure to deactivate these regions manifests itself as a lapse in attention and reduced behavioral performance (Sonuga-Barke and Castellanos, 2007). Further research suggests the balance between activation of the TPN and deactivation of the DMN is associated with reduced cognitive performance in healthy controls (Sala-Llonch et al., 2012; Weissman et al., 2006) and is disrupted in aging (Prakash et al., 2012; Sambataro et al., 2010), sleep deprivation (De Havas et al., 2012), schizophrenia (Woodward et al., 2011), and ADHD (Castellanos et al., 2009). Most importantly, groups have reported reduced deactivation of DMN regions during task execution in severe or mixed TBI populations (Bonnelle et al., 2011; Sharp et al., 2011), as well as mTBI patients in the subacute stage (Mayer et al., 2012).

In addition, a groundbreaking work on a severe TBI population has recently linked this failure to deactivate the DMN with structural damage within WM tracts connecting the right anterior insula as measured using DTI (Bonnelle et al., 2012). This is particularly intriguing considering that the proposed role of the salience network, which includes the insula, is to detect external sensory inputs and modulate the balance between internal focus represented by activation of the DMN and external attention represented by activation of TPN (Menon and Uddin, 2010; Seeley et al., 2007). These findings provide an invaluable link between altered structural integrity and reduced functional activation on a network level.

Resting-State fMRI Applications in TBI

Disparities between the task-based fMRI studies may be due to variability in the severity of injury in sample populations, inconsistency in time since injury, and the wide range in difficulty of tasks chosen between studies. More importantly, task-evoked functional BOLD techniques have limited clinical utility during the acute stage on a TBI patient when the patient may not be cognitively capable of performing a task. A proposed solution is the use of resting-state BOLD techniques that are both independent of cognitive load and easily administered during all stages of TBI. The ability of rs-fMRI to acquire functional information in the acute stage of TBI, when task-evoked functional BOLD techniques may be limited, is of great clinical value in assessing a TBI population.

Due to the rotational forces exerted on the brain during trauma, corpus callosum, the WM region connecting the right and left hemispheres, is particularly vulnerable to damage from TBI. Since this region is often damaged following TBI, it is not surprising that multiple groups have demonstrated reduced interhemispheric functional connectivity (IH-FC) among TBI participants following severe TBI (Marquez de la Plata et al., 2011), mTBI (Sours et al., 2015), and sports-related mTBI (Slobounov et al., 2011). As shown in Figure 14.7a, data from our group demonstrate a specific reduction in IH-FC between the right and left dorsolateral prefrontal cortex (DLPFC) in both patients at the subacute stage with high ($n = 26$) and low ($n = 15$) numbers of postconcussive symptoms compared to a control group ($n = 30$). Furthermore, this study noted an association between reduced IH-FC in the DLPFC and reduced performance on an assessment of mathematical processing (Sours et al., 2015). Damage to the corpus callosum as measured by magnetic resonance spectroscopy (1H-MRS) was found to be associated with reduced IH-FC in a population of mTBI participants (Johnson et al., 2012a) further emphasizing the link between structural damage and altered functional communication.

TBI has been shown to reduce functional connectivity within multiple neural networks, including the motor network (Kasahara et al., 2010), TPN (Hillary et al., 2011; Mayer et al., 2011), and the DMN (Bonnelle et al., 2011; Hillary et al., 2011; Johnson et al., 2012b; Mayer et al., 2011; Sharp et al., 2011; Zhou et al., 2012). While it is generally assumed that reduced structural connectivity is associated with reduced functional connectivity, recent research in TBI suggests that this is not always the case. For instance, multiple groups have shown increased resting-state functional connectivity between the thalamus and cortex (Tang et al., 2011; Zhou et al., 2014). Similar to previous reports, our group has recently shown that in a population of 41 mTBI patients at approximately 1 month post injury, there is increased rs-FC between the thalamus and cortical regions, including the right insula, right primary somatosensory cortex, and right somatosensory association area, in mTBI patients who reported a high number of postconcussive symptoms compared to those with low symptoms Figure 14.7b (Sours et al., 2015). Since many of the postconcussive symptoms are somatic in nature, it represents an intriguing possibility that increased thalamocortical connectivity that may contribute to those symptoms.

In addition, similar to the evidence of a disruption in the balance between the DMN and TPN based on altered activation and deactivation patterns during task execution, groups have also noted increased resting-state functional connectivity (representative of a reduced anticorrelation) following mTBI between the DMN and TPN (Mayer et al., 2011; Sours et al., 2013), perhaps due to damage to inhibitory circuits. It has been suggested that TPN and DMN operate in an antagonistic fashion (Fox et al., 2005). While great strides have been made in investigating resting-state

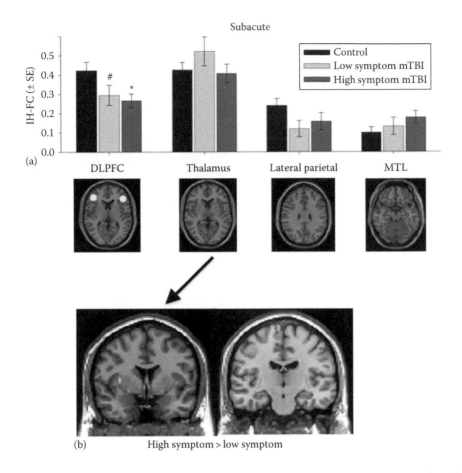

(a)

(b) High symptom > low symptom

FIGURE 14.7 (a) ROI analysis of group differences in IH-FC within the DLPFC, thalamus, lateral parietal lobe, and MTL in the subacute stage. Significance assessed using separate one-way ANCOVA controlling for age. Fisher's least significant difference test was used to correct for multiple comparisons. # $p < 0.10$, * $p < 0.05$. (b) Visualization of the between group functional connectivity maps for the left thalamus ROI. Warm colors represent regions of increased resting-state functional connectivity in the high symptom mTBI group compared to the low symptom mTBI group. Functional connectivity maps were thresholded at voxel-wise p-value of 0.001 (uncorrected) and cluster extent threshold of p-value of 0.05 using a family-wise error correction for multiple comparisons.

functional connectivity following injury, this may provide a limited understanding of the alterations in neural processing. Although more easily acquired, these data may paint a limited view of the potential changes observed following mTBI, especially in patients with subtle cognitive deficits. Since mTBI is such a subtle injury, deficits in neural communication may only become apparent when the patient is cognitive taxed.

To investigate this notion, groups have begun to investigate functional connectivity measures not only during resting conditions but also during task execution (Bonnelle et al., 2011; Caeyenberghs et al., 2012). Preliminary data from our group suggest that following injury, mTBI patients in the subacute and chronic stages demonstrate increased functional connectivity (a reduced anticorrelation) between the task-positive and task-negative regions during the 2-back working memory paradigm compared to a control group (Figure 14.8). Furthermore, analysis investigating this alteration in network communication using the graph theory measure of modularity suggests that chronic mTBI patients demonstrate reduced segregation of these two networks compared to control participants.

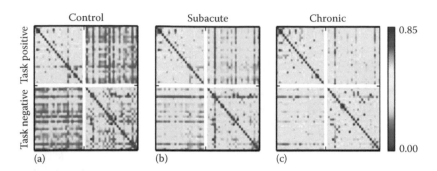

FIGURE 14.8 Functional connectivity matrices illustrating both within- and between-network connectivity values during the 2-back condition. The white boundaries separate the task-positive and task-negative networks. Regions selected from the activation and deactivation patterns from the performance of the 2-back task in the control group. (a) Controls ($n = 19$). (b) Subacute ($n = 30$). (c) Chronic ($n = 21$). (Courtesy of J. Kinnison.)

SUMMARY

The measurement of unique aspects of large-scale neural networks, including resting-state functional connectivity and alterations in the recruitment and segregation of network communication during cognitively demanding tasks has provided researchers with indispensable information characterizing neural networks that may be disrupted following TBI. Furthermore, great strides have been made determining how these alterations are associated with cognitive deficits and postconcussive symptoms among TBI patients. Multiple avenues of active investigation have been brought to light from these studies with the ultimate goal of improving clinicians' ability to diagnose and predict patient functional outcome.

CONCLUSION

TBI often causes permanent physical, cognitive, and psychosocial problems, resulting in a limited ability to return to work and reduced quality of life for these patients. Furthermore, these injuries result in both financial and emotional burden on the nation. Through recent advancements in noninvasive techniques described in this chapter that are able to detect trauma-induced structural and functional damage, great strides in understanding the pathophysiology of TBI have been made. However, the ability to integrate information from structural and functional imaging modalities is needed to provide a complete picture of this complex injury. Multiparametric approaches using the most sensitive, advanced imaging techniques that can measure cerebrovascular, biophysical, and metabolic changes in conjunction with neurocognitive assessments are needed. This will result in a more complete understanding of the neurodegenerative processes, leading the way for the development of innovative therapeutic interventions, while at the same time identifying novel imaging biomarkers, all of which are needed to actively manage consequences of TBI.

REFERENCES

Alves W, Macciocchi SN, Barth JT. 1993. Postconcussive symptoms after uncomplicated mild head injury. *Journal Head Trauma and Rehabilitation* 8(3):48–59.

Arfanakis K, Haughton VM, Carew JD, Rogers BP, Dempsey RJ, Meyerand ME. 2002. Diffusion tensor MR imaging in diffuse axonal injury. *American Journal of Neuroradiology* 23(5):794–802.

Ashwal S, Holshouser BA, Tong KA. 2006. Use of advanced neuroimaging techniques in the evaluation of pediatric traumatic brain injury. *Developmental Neuroscience* 28(4–5):309–326.

Bazarian JJ, Wong T, Harris M, Leahey N, Mookerjee S, Dombovy M. 1999. Epidemiology and predictors of post-concussive syndrome after minor head injury in an emergency population. *Brain Injury* 13(3):173–189.

Bazarian JJ, Zhong J, Blyth B, Zhu T, Kavcic V, Peterson D. 2007. Diffusion tensor imaging detects clinically important axonal damage after mild traumatic brain injury: A pilot study. *Journal of Neurotrauma* 24(9):1447–1459.

Bazarian JJ, Zhu T, Blyth B, Borrino A, Zhong J. 2012. Subject-specific changes in brain white matter on diffusion tensor imaging after sports-related concussion. *Magnetic Resonance Imaging* 30(2):171–180.

Benson RR, Gattu R, Sewick B, Kou Z, Zakariah N, Cavanaugh JM, Haacke EM. 2012. Detection of hemorrhagic and axonal pathology in mild traumatic brain injury using advanced MRI: Implications for neurorehabilitation. *NeuroRehabilitation* 31(3):261–279.

Benson RR, Meda SA, Vasudevan S, Kou Z, Govindarajan KA, Hanks RA, Millis SR, Makki M, Latif Z, Coplin W et al. 2007. Global white matter analysis of diffusion tensor images is predictive of injury severity in traumatic brain injury. *Journal of Neurotrauma* 24(3):446–459.

Betz J, Zhuo J, Roy A, Shanmuganathan K, Gullapalli RP. 2012. Prognostic value of diffusion tensor imaging parameters in severe traumatic brain injury. *Journal of Neurotrauma* 29(7):1292–1305.

Biswal B, Yetkin FZ, Haughton VM, Hyde JS. 1995. Functional connectivity in the motor cortex of resting human brain using echo-planar MRI. *Magnetic Resonance in Medicine: Official Journal of the Society of Magnetic Resonance in Medicine/Society of Magnetic Resonance in Medicine* 34(4):537–541.

Bonnelle V, Ham TE, Leech R, Kinnunen KM, Mehta MA, Greenwood RJ, Sharp DJ. 2012. Salience network integrity predicts default mode network function after traumatic brain injury. *Proceedings of the National Academy of Sciences of the United States of America* 109(12):4690–4695.

Bonnelle V, Leech R, Kinnunen KM, Ham TE, Beckmann CF, De Boissezon X, Greenwood RJ, Sharp DJ. 2011. Default mode network connectivity predicts sustained attention deficits after traumatic brain injury. *The Journal of Neuroscience: The Official Journal of the Society for Neuroscience* 31(38):13442–13451.

Bozzali M, Falini A, Franceschi M, Cercignani M, Zuffi M, Scotti G, Comi G, Filippi M. 2002. White matter damage in Alzheimer's disease assessed in vivo using diffusion tensor magnetic resonance imaging. *Journal of Neurology, Neurosurgery, and Psychiatry* 72(6):742–746.

Caeyenberghs K, Leemans A, Heitger MH, Leunissen I, Dhollander T, Sunaert S, Dupont P, Swinnen SP. 2012. Graph analysis of functional brain networks for cognitive control of action in traumatic brain injury. *Brain: A Journal of Neurology* 135(Pt 4):1293–1307.

Calhoun VD, Liu J, Adali T. 2009. A review of group ICA for fMRI data and ICA for joint inference of imaging, genetic, and ERP data. *NeuroImage* 45(1 Suppl):S163–S172.

Castellanos FX, Kelly C, Milham MP. 2009. The restless brain: Attention-deficit hyperactivity disorder, resting-state functional connectivity, and intrasubject variability. *Canadian Journal of Psychiatry. Revue Canadienne De Psychiatrie* 54(10):665–672.

Centers for Disease Control and Prevention. 2003. Report to Congress on mild traumatic brain injury in the United States: Steps to prevent a serious public health problem. CDC, Atlanta, GA.

Centers for Disease Control and Prevention (CDC). 2013. CDC grand rounds: Reducing severe traumatic brain injury in the United States. *Morbidity and Mortality Weekly Report* 62(27):549–552.

Chapman JC, Diaz-Arrastia R. 2014. Military traumatic brain injury: A review. *Alzheimer's & Dementia: The Journal of the Alzheimer's Association* 10(3 Suppl):S97–S104.

Chaumet G, Quera-Salva MA, Macleod A, Hartley S, Taillard J, Sagaspe P, Mazaux JM, Azouvi P, Joseph PA, Guilleminault C et al. 2008. Is there a link between alertness and fatigue in patients with traumatic brain injury? *Neurology* 71(20):1609–1613.

Chen CJ, Wu CH, Liao YP, Hsu HL, Tseng YC, Liu HL, Chiu WT. 2012. Working memory in patients with mild traumatic brain injury: Functional MR imaging analysis. *Radiology* 264(3):844–851.

Chen S, Pickard JD, Harris NG. 2003. Time course of cellular pathology after controlled cortical impact injury. *Experimental Neurology* 182(1):87–102.

Christodoulou C, DeLuca J, Ricker JH, Madigan NK, Bly BM, Lange G, Kalnin AJ, Liu WC, Steffener J, Diamond BJ et al. 2001. Functional magnetic resonance imaging of working memory impairment after traumatic brain injury. *Journal of Neurology, Neurosurgery, and Psychiatry* 71(2):161–168.

Chu Z, Wilde EA, Hunter JV, McCauley SR, Bigler ED, Troyanskaya M, Yallampalli R, Chia JM, Levin HS. 2010. Voxel-based analysis of diffusion tensor imaging in mild traumatic brain injury in adolescents. *American Journal of Neuroradiology* 31(2):340–346.

Cihangiroglu M, Ramsey RG, Dohrmann GJ. 2002. Brain injury: Analysis of imaging modalities. *Neurological Research* 24(1):7–18.

De Havas JA, Parimal S, Soon CS, Chee MW. 2012. Sleep deprivation reduces default mode network connectivity and anti-correlation during rest and task performance. *NeuroImage* 59(2):1745–1751.

Ezaki Y, Tsutsumi K, Morikawa M, Nagata I. 2006. Role of diffusion-weighted magnetic resonance imaging in diffuse axonal injury. *Acta Radiologica (Stockholm, Sweden: 1987)* 47(7):733–740.

Falangola MF, Jensen JH, Tabesh A, Hu C, Deardorff RL, Babb JS, Ferris S, Helpern JA. 2013. Non-Gaussian diffusion MRI assessment of brain microstructure in mild cognitive impairment and Alzheimer's disease. *Magnetic Resonance Imaging* 31(6):840–846.

Faul M, Xu L, Wald M, Coronado V. 2010. *Traumatic Brain Injury in the United States: Emergency Department Visits, Hospitalizations, and Deaths*. Centers for Disease Control and Prevention, Atlanta, GA.

Fox MD, Snyder AZ, Vincent JL, Corbetta M, Van Essen DC, Raichle ME. 2005. The human brain is intrinsically organized into dynamic, anticorrelated functional networks. *Proceedings of the National Academy of Sciences of the United States of America* 102(27):9673–9678.

Gong NJ, Wong CS, Chan CC, Leung LM, Chu YC. 2013. Correlations between microstructural alterations and severity of cognitive deficiency in Alzheimer's disease and mild cognitive impairment: A diffusional kurtosis imaging study. *Magnetic Resonance Imaging* 31(5):688–694.

Greicius MD, Krasnow B, Reiss AL, Menon V. 2003. Functional connectivity in the resting brain: A network analysis of the default mode hypothesis. *Proceedings of the National Academy of Sciences of the United States of America* 100(1):253–258.

Greve MW, Zink BJ. 2009. Pathophysiology of traumatic brain injury. *The Mount Sinai Journal of Medicine, New York* 76(2):97–104.

Grossman EJ, Ge Y, Jensen JH, Babb JS, Miles L, Reaume J, Silver JM, Grossman RI, Inglese M. 2012. Thalamus and cognitive impairment in mild traumatic brain injury: A diffusional kurtosis imaging study. *Journal of Neurotrauma* 29(13):2318–2327.

Haacke EM, Xu Y, Cheng YC, Reichenbach JR. 2004. Susceptibility weighted imaging (SWI). *Magnetic Resonance in Medicine: Official Journal of the Society of Magnetic Resonance in Medicine/Society of Magnetic Resonance in Medicine* 52(3):612–618.

Hagmann P, Cammoun L, Gigandet X, Gerhard S, Grant PE, Wedeen V, Meuli R, Thiran JP, Honey CJ, Sporns O. 2010. MR connectomics: Principles and challenges. *Journal of Neuroscience Methods* 194(1):34–45.

Helpern JA, Adisetiyo V, Falangola MF, Hu C, Di Martino A, Williams K, Castellanos FX, Jensen JH. 2011. Preliminary evidence of altered gray and white matter microstructural development in the frontal lobe of adolescents with attention-deficit hyperactivity disorder: A diffusional kurtosis imaging study. *Journal of Magnetic Resonance Imaging* 33(1):17–23.

Hillary FG. 2011. Determining the nature of prefrontal cortex recruitment after traumatic brain injury: A response to Turner. *Frontiers in Systems Neuroscience* 5:24.

Hillary FG, Genova HM, Medaglia JD, Fitzpatrick NM, Chiou KS, Wardecker BM, Franklin RG, Jr, Wang J, DeLuca J. 2010. The nature of processing speed deficits in traumatic brain injury: Is less brain more? *Brain Imaging and Behavior* 4(2):141–154.

Hillary FG, Slocomb J, Hills EC, Fitzpatrick NM, Medaglia JD, Wang J, Good DC, Wylie GR. 2011. Changes in resting connectivity during recovery from severe traumatic brain injury. *International Journal of Psychophysiology: Official Journal of the International Organization of Psychophysiology* 82(1):115–123.

Honey CJ, Sporns O, Cammoun L, Gigandet X, Thiran JP, Meuli R, Hagmann P. 2009. Predicting human resting-state functional connectivity from structural connectivity. *Proceedings of the National Academy of Sciences of the United States of America* 106(6):2035–2040.

Ilvesmaki T, Luoto TM, Hakulinen U, Brander A, Ryymin P, Eskola H, Iverson GL, Ohman J. 2014. Acute mild traumatic brain injury is not associated with white matter change on diffusion tensor imaging. *Brain: A Journal of Neurology* 137(Pt 7): 1876–1882.

Immonen RJ, Kharatishvili I, Grohn H, Pitkanen A, Grohn OH. 2009. Quantitative MRI predicts long-term structural and functional outcome after experimental traumatic brain injury. *NeuroImage* 45(1):1–9.

Inglese M, Makani S, Johnson G, Cohen BA, Silver JA, Gonen O, Grossman RI. 2005. Diffuse axonal injury in mild traumatic brain injury: A diffusion tensor imaging study. *Journal of Neurosurgery* 103(2):298–303.

Jensen JH, Helpern JA. 2010. MRI quantification of non-Gaussian water diffusion by kurtosis analysis. *NMR in Biomedicine* 23(7):698–710.

Jensen JH, Helpern JA, Ramani A, Lu H, Kaczynski K. 2005. Diffusional kurtosis imaging: The quantification of non-Gaussian water diffusion by means of magnetic resonance imaging. *Magnetic Resonance in Medicine: Official Journal of the Society of Magnetic Resonance in Medicine/Society of Magnetic Resonance in Medicine* 53(6):1432–1440.

Johansson B, Berglund P, Ronnback L. 2009. Mental fatigue and impaired information processing after mild and moderate traumatic brain injury. *Brain Injury* 23(13–14):1027–1040.

Johnson B, Zhang K, Gay M, Horovitz S, Hallett M, Sebastianelli W, Slobounov S. 2012b. Alteration of brain default network in subacute phase of injury in concussed individuals: Resting-state fMRI study. *NeuroImage* 59(1):511–518.

Johnson B, Zhang K, Gay M, Neuberger T, Horovitz S, Hallett M, Sebastianelli W, Slobounov S. 2012a. Metabolic alterations in corpus callosum may compromise brain functional connectivity in MTBI patients: An 1 H-MRS study. *Neuroscience Letters* 509:5–8.

Karagulle Kendi AT, Lehericy S, Luciana M, Ugurbil K, Tuite P. 2008. Altered diffusion in the frontal lobe in Parkinson disease. *American Journal of Neuroradiology* 29(3):501–505.

Kasahara M, Menon DK, Salmond CH, Outtrim JG, Tavares JV, Carpenter TA, Pickard JD, Sahakian BJ, Stamatakis EA. 2011. Traumatic brain injury alters the functional brain network mediating working memory. *Brain Injury* 25(12):1170–1187.

Kasahara M, Menon DK, Salmond CH, Outtrim JG, Taylor Tavares JV, Carpenter TA, Pickard JD, Sahakian BJ, Stamatakis EA. 2010. Altered functional connectivity in the motor network after traumatic brain injury. *Neurology* 75(2):168–176.

Kou Z, Vandevord PJ. 2014. Traumatic white matter injury and glial activation: From basic science to clinics. *Glia* 62(11):1831–1855

Kraus MF, Susmaras T, Caughlin BP, Walker CJ, Sweeney JA, Little DM. 2007. White matter integrity and cognition in chronic traumatic brain injury: A diffusion tensor imaging study. *Brain: A Journal of Neurology* 130(Pt 10):2508–2519.

Kumar R, Husain M, Gupta RK, Hasan KM, Haris M, Agarwal AK, Pandey CM, Narayana PA. 2009. Serial changes in the white matter diffusion tensor imaging metrics in moderate traumatic brain injury and correlation with neuro-cognitive function. *Journal of Neurotrauma* 26(4):481–495.

Lakhan SE, Kirchgessner A. 2012. Chronic traumatic encephalopathy: The dangers of getting "dinged". *SpringerPlus* 1:2-1801-1-2.

Lebel C, Beaulieu C. 2011. Longitudinal development of human brain wiring continues from childhood into adulthood. *The Journal of Neuroscience: The Official Journal of the Society for Neuroscience* 31(30):10937–10947.

Levine B, Kovacevic N, Nica EI, Cheung G, Gao F, Schwartz ML, Black SE. 2008. The Toronto traumatic brain injury study: Injury severity and quantified MRI. *Neurology* 70(10):771–778.

Ling J, Merideth F, Caprihan A, Pena A, Teshiba T, Mayer AR. 2012. Head injury or head motion? Assessment and quantification of motion artifacts in diffusion tensor imaging studies. *Human Brain Mapping* 33(1):50–62.

Lu H, Jensen JH, Ramani A, Helpern JA. 2006. Three-dimensional characterization of non-Gaussian water diffusion in humans using diffusion kurtosis imaging. *NMR in Biomedicine* 19(2):236–247.

Lye TC, Shores EA. 2000. Traumatic brain injury as a risk factor for Alzheimer's disease: A review. *Neuropsychology Review* 10(2):115–129.

Mac Donald CL, Johnson AM, Cooper D, Nelson EC, Werner NJ, Shimony JS, Snyder AZ, Raichle ME, Witherow JR, Fang R et al. 2011. Detection of blast-related traumatic brain injury in U.S. military personnel. *The New England Journal of Medicine* 364(22):2091–2100.

Magnuson J, Leonessa F, Ling GS. 2012. Neuropathology of explosive blast traumatic brain injury. *Current Neurology and Neuroscience Reports* 12(5):570–579.

Makley MJ, English JB, Drubach DA, Kreuz AJ, Celnik PA, Tarwater PM. 2008. Prevalence of sleep disturbance in closed head injury patients in a rehabilitation unit. *Neurorehabilitation and Neural Repair* 22(4):341–347.

Marquez de la Plata CD, Garces J, Shokri Kojori E, Grinnan J, Krishnan K, Pidikiti R, Spence J, Devous MDS, Moore C, McColl R et al. 2011. Deficits in functional connectivity of hippocampal and frontal lobe circuits after traumatic axonal injury. *Archives of Neurology* 68(1):74–84.

Maruishi M, Miyatani M, Nakao T, Muranaka H. 2007. Compensatory cortical activation during performance of an attention task by patients with diffuse axonal injury: A functional magnetic resonance imaging study. *Journal of Neurology, Neurosurgery, and Psychiatry* 78(2):168–173.

Mayer AR, Ling J, Mannell MV, Gasparovic C, Phillips JP, Doezema D, Reichard R, Yeo RA. 2010. A prospective diffusion tensor imaging study in mild traumatic brain injury. *Neurology* 74(8):643–650.

Mayer AR, Mannell MV, Ling J, Gasparovic C, Yeo RA. 2011. Functional connectivity in mild traumatic brain injury. *Human Brain Mapping* 32(11):1825–1835.

Mayer AR, Yang Z, Yeo RA, Pena A, Ling JM, Mannell MV, Stippler M, Mojtahed K. 2012. A functional MRI study of multimodal selective attention following mild traumatic brain injury. *Brain Imaging and Behavior* 6(2):343–354.

McAllister TW. 2011. Neurobiological consequences of traumatic brain injury. *Dialogues in Clinical Neuroscience* 13(3):287–300.

McAllister TW, Flashman LA, McDonald BC, Saykin AJ. 2006. Mechanisms of working memory dysfunction after mild and moderate TBI: Evidence from functional MRI and neurogenetics. *Journal of Neurotrauma* 23(10):1450–1467.

McAllister TW, Sparling MB, Flashman LA, Guerin SJ, Mamourian AC, Saykin AJ. 2001. Differential working memory load effects after mild traumatic brain injury. *NeuroImage* 14(5):1004–1012.

McAllister TW, Saykin AJ, Flashman LA, Sparling MB, Johnson SC, Guerin SJ, Mamourian AC, Weaver JB, Yanofsky N. 1999. Brain activation during working memory 1 month after mild traumatic brain injury: A functional MRI study. *Neurology* 53(6):1300–1308.

McDowell S, Whyte J, D'Esposito M. 1997. Working memory impairments in traumatic brain injury: Evidence from a dual-task paradigm. *Neuropsychologia* 35(10):1341–1353.

McMahon P, Hricik A, Yue JK, Puccio AM, Inoue T, Lingsma HF, Beers SR, Gordon WA, Valadka AB, Manley GT et al. 2014. Symptomatology and functional outcome in mild traumatic brain injury: Results from the prospective TRACK-TBI study. *Journal of Neurotrauma* 31(1):26–33.

Menon V, Uddin LQ. 2010. Saliency, switching, attention and control: A network model of insula function. *Brain Structure & Function* 214(5–6):655–667.

Menzel JC. 2008. Depression in the elderly after traumatic brain injury: A systematic review. *Brain Injury* 22(5):375–380.

Messe A, Caplain S, Paradot G, Garrigue D, Mineo JF, Soto Ares G, Ducreux D, Vignaud F, Rozec G, Desal H et al. 2011. Diffusion tensor imaging and white matter lesions at the subacute stage in mild traumatic brain injury with persistent neurobehavioral impairment. *Human Brain Mapping* 32(6):999–1011.

Miotto EC, Cinalli FZ, Serrao VT, Benute GG, Lucia MC, Scaff M. 2010. Cognitive deficits in patients with mild to moderate traumatic brain injury. *Arquivos De Neuro-Psiquiatria* 68(6):862–868.

Myer DJ, Gurkoff GG, Lee SM, Hovda DA, Sofroniew MV. 2006. Essential protective roles of reactive astrocytes in traumatic brain injury. *Brain: A Journal of Neurology* 129(Pt 10):2761–2772.

Nampiaparampil DE. 2008. Prevalence of chronic pain after traumatic brain injury: A systematic review. *The Journal of the American Medical Association* 300(6):711–719.

Nemetz PN, Leibson C, Naessens JM, Beard M, Kokmen E, Annegers JF, Kurland LT. 1999. Traumatic brain injury and time to onset of Alzheimer's disease: A population-based study. *American Journal of Epidemiology* 149(1):32–40.

Newcombe VF, Williams GB, Nortje J, Bradley PG, Harding SG, Smielewski P, Coles JP, Maiya B, Gillard JH, Hutchinson PJ et al. 2007. Analysis of acute traumatic axonal injury using diffusion tensor imaging. *British Journal of Neurosurgery* 21(4):340–348.

Niogi SN, Mukherjee P, Ghajar J, Johnson C, Kolster RA, Sarkar R, Lee H, Meeker M, Zimmerman RD, Manley GT et al. 2008. Extent of microstructural white matter injury in postconcussive syndrome correlates with impaired cognitive reaction time: A 3T diffusion tensor imaging study of mild traumatic brain injury. *American Journal of Neuroradiology* 29(5):967–973.

Ommaya AK, Goldsmith W, Thibault L. 2002. Biomechanics and neuropathology of adult and paediatric head injury. *British Journal of Neurosurgery* 16(3):220–242.

Parizel PM, Ozsarlak, Van Goethem JW, van den Hauwe L, Dillen C, Verlooy J, Cosyns P, De Schepper AM. 1998. Imaging findings in diffuse axonal injury after closed head trauma. *European Radiology* 8(6):960–965.

Perez AM, Adler J, Kulkarni N, Strain JF, Womack KB, Diaz-Arrastia R, Marquez de la Plata CD. 2014. Longitudinal white matter changes after traumatic axonal injury. *Journal of Neurotrauma*. 31(17):1478–1485

Pfefferbaum A, Adalsteinsson E, Sullivan EV. 2005. Frontal circuitry degradation marks healthy adult aging: Evidence from diffusion tensor imaging. *NeuroImage* 26(3):891–899.

Prakash RS, Heo S, Voss MW, Patterson B, Kramer AF. 2012. Age-related differences in cortical recruitment and suppression: Implications for cognitive performance. *Behavioural Brain Research* 230(1):192–200.

Raab P, Hattingen E, Franz K, Zanella FE, Lanfermann H. 2010. Cerebral gliomas: Diffusional kurtosis imaging analysis of microstructural differences. *Radiology* 254(3):876–881.

Raichle ME. 2010. Two views of brain function. *Trends in Cognitive Sciences* 14(4):180–190.

Raichle ME, MacLeod AM, Snyder AZ, Powers WJ, Gusnard DA, Shulman GL. 2001. A default mode of brain function. *Proceedings of the National Academy of Sciences of the United States of America* 98(2):676–682.

Reichenbach JR, Venkatesan R, Schillinger DJ, Kido DK, Haacke EM. 1997. Small vessels in the human brain: MR venography with deoxyhemoglobin as an intrinsic contrast agent. *Radiology* 204(1):272–277.

Sala-Llonch R, Pena-Gomez C, Arenaza-Urquijo EM, Vidal-Pineiro D, Bargallo N, Junque C, Bartres-Faz D. 2012. Brain connectivity during resting state and subsequent working memory task predicts behavioural performance. *Cortex; A Journal Devoted to the Study of the Nervous System and Behavior* 48(9):1187–1196.

Salmond CH, Menon DK, Chatfield DA, Williams GB, Pena A, Sahakian BJ, Pickard JD. 2006. Diffusion tensor imaging in chronic head injury survivors: Correlations with learning and memory indices. *NeuroImage* 29(1):117–124.

Sambataro F, Murty VP, Callicott JH, Tan HY, Das S, Weinberger DR, Mattay VS. 2010. Age-related alterations in default mode network: Impact on working memory performance. *Neurobiology of Aging* 31(5):839–852.

Sanchez-Carrion R, Gomez PV, Junque C, Fernandez-Espejo D, Falcon C, Bargallo N, Roig-Rovira T, Ensenat-Cantallops A, Bernabeu M. 2008. Frontal hypoactivation on functional magnetic resonance imaging in working memory after severe diffuse traumatic brain injury. *Journal of Neurotrauma* 25(5):479–494.

Scheibel RS, Pearson DA, Faria LP, Kotrla KJ, Aylward E, Bachevalier J, Levin HS. 2003. An fMRI study of executive functioning after severe diffuse TBI. *Brain Injury* 17(11):919–930.

Schmierer K, Altmann DR, Kassim N, Kitzler H, Kerskens CM, Doege CA, Aktas O, Lunemann JD, Miller DH, Zipp F et al. 2004. Progressive change in primary progressive multiple sclerosis normal-appearing white matter: A serial diffusion magnetic resonance imaging study. *Multiple Sclerosis (Houndmills, Basingstoke, England)* 10(2):182–187.

Seeley WW, Menon V, Schatzberg AF, Keller J, Glover GH, Kenna H, Reiss AL, Greicius MD. 2007. Dissociable intrinsic connectivity networks for salience processing and executive control. *The Journal of Neuroscience: The Official Journal of the Society for Neuroscience* 27(9):2349–2356.

Shanmuganathan K, Gullapalli RP, Mirvis SE, Roys S, Murthy P. 2004. Whole-brain apparent diffusion coefficient in traumatic brain injury: Correlation with Glasgow coma scale score. *American Journal of Neuroradiology* 25(4):539–544.

Sharp DJ, Beckmann CF, Greenwood R, Kinnunen KM, Bonnelle V, De Boissezon X, Powell JH, Counsell SJ, Patel MC, Leech R. 2011. Default mode network functional and structural connectivity after traumatic brain injury. *Brain: A Journal of Neurology* 134(Pt 8):2233–2247.

Sidaros A, Engberg AW, Sidaros K, Liptrot MG, Herning M, Petersen P, Paulson OB, Jernigan TL, Rostrup E. 2008. Diffusion tensor imaging during recovery from severe traumatic brain injury and relation to clinical outcome: A longitudinal study. *Brain: A Journal of Neurology* 131(Pt 2):559–572.

Slobounov SM, Gay M, Zhang K, Johnson B, Pennell D, Sebastianelli W, Horovitz S, Hallett M. 2011. Alteration of brain functional network at rest and in response to YMCA physical stress test in concussed athletes: RsFMRI study. *NeuroImage* 55(4):1716–1727.

Smith SM, Jenkinson M, Johansen-Berg H, Rueckert D, Nichols TE, Mackay CE, Watkins KE, Ciccarelli O, Cader MZ, Matthews PM et al. 2006. Tract-based spatial statistics: Voxelwise analysis of multi-subject diffusion data. *NeuroImage* 31(4):1487–1505.

Smits M, Dippel DW, Houston GC, Wielopolski PA, Koudstaal PJ, Hunink MG, van der Lugt A. 2009. Postconcussion syndrome after minor head injury: Brain activation of working memory and attention. *Human Brain Mapping* 30(9):2789–2803.

Soeda A, Nakashima T, Okumura A, Kuwata K, Shinoda J, Iwama T. 2005. Cognitive impairment after traumatic brain injury: A functional magnetic resonance imaging study using the stroop task. *Neuroradiology* 47(7):501–506.

Song SK, Sun SW, Ju WK, Lin SJ, Cross AH, Neufeld AH. 2003. Diffusion tensor imaging detects and differentiates axon and myelin degeneration in mouse optic nerve after retinal ischemia. *NeuroImage* 20(3):1714–1722.

Sonuga-Barke EJ, Castellanos FX. 2007. Spontaneous attentional fluctuations in impaired states and pathological conditions: A neurobiological hypothesis. *Neuroscience and Biobehavioral Reviews* 31(7):977–986.

Sours C, Rosenberg J, Kane R, Roys S, Zhuo J, Shanmuganathan K, Gullapalli RP. 2015. Associations between interhemispheric functional connectivity and the automated neuropsychological assessment metrics (ANAM) in civilian mild TBI. *Brain Imaging and Behavior* 9(2):190–203.

Sours C, Zhuo J, Janowich J, Aarabi B, Shanmuganathan K, Gullapalli RP. 2013. Default mode network interference in mild traumatic brain injury—A pilot resting state study. *Brain Research* 1537:201–215.

Sporns O. 2011. The human connectome: A complex network. *Annals of the New York Academy of Sciences* 1224:109–125.

Steyerberg EW, Mushkudiani N, Perel P, Butcher I, Lu J, McHugh GS, Murray GD, Marmarou A, Roberts I, Habbema JD et al. 2008. Predicting outcome after traumatic brain injury: Development and international validation of prognostic scores based on admission characteristics. *PLoS Medicine* 5(8):e165; discussion e165.

Stokum JA, Sours C, Zhuo J, Kane R, Shanmuganathan K, Gullapalli RP. 2015. A longitudinal evaluation of diffusion kurtosis imaging in patients with mild traumatic brain injury. *Brain Injury* 29(1):47–57.

Tang L, Ge Y, Sodickson DK, Miles L, Zhou Y, Reaume J, Grossman RI. 2011. Thalamic resting-state functional networks: Disruption in patients with mild traumatic brain injury. *Radiology* 260(3):831–840.

Teasdale G, Jennett B. 1974. Assessment of coma and impaired consciousness. A practical scale. *Lancet* 2(7872):81–84.

Thornhill S, Teasdale GM, Murray GD, McEwen J, Roy CW, Penny KI. 2000. Disability in young people and adults one year after head injury: Prospective cohort study. *BMJ (Clinical Research Ed.)* 320(7250):1631–1635.

Tong KA, Ashwal S, Holshouser BA, Nickerson JP, Wall CJ, Shutter LA, Osterdock RJ, Haacke EM, Kido D. 2004. Diffuse axonal injury in children: Clinical correlation with hemorrhagic lesions. *Annals of Neurology* 56(1):36–50.

Turner GR, Levine B. 2008. Augmented neural activity during executive control processing following diffuse axonal injury. *Neurology* 71(11):812–818.

Turner GR, McIntosh AR, Levine B. 2012. Dissecting altered functional engagement in TBI and other patient groups through connectivity analysis: One goal, many paths (A response to Hillary). *Frontiers in Systems Neuroscience* 6:10.

Van Cauter S, Veraart J, Sijbers J, Peeters RR, Himmelreich U, De Keyzer F, Van Gool SW, Van Calenbergh F, De Vleeschouwer S, Van Hecke W et al. 2012. Gliomas: Diffusion kurtosis MR imaging in grading. *Radiology* 263(2):492–501.

van den Heuvel MP, Hulshoff Pol HE. 2010. Exploring the brain network: A review on resting-state fMRI functional connectivity. *European Neuropsychopharmacology: The Journal of the European College of Neuropsychopharmacology* 20(8):519–534.

Veraart J, Poot DH, Van Hecke W, Blockx I, Van der Linden A, Verhoye M, Sijbers J. 2011. More accurate estimation of diffusion tensor parameters using diffusion kurtosis imaging. *Magnetic Resonance in Medicine: Official Journal of the Society of Magnetic Resonance in Medicine/Society of Magnetic Resonance in Medicine* 65(1):138–145.

Wang JJ, Lin WY, Lu CS, Weng YH, Ng SH, Wang CH, Liu HL, Hsieh RH, Wan YL, Wai YY. 2011. Parkinson disease: Diagnostic utility of diffusion kurtosis imaging. *Radiology* 261(1):210–217.

Warner MA, Marquez de la Plata C, Spence J, Wang JY, Harper C, Moore C, Devous M, Diaz-Arrastia R. 2010a. Assessing spatial relationships between axonal integrity, regional brain volumes, and neuropsychological outcomes after traumatic axonal injury. *Journal of Neurotrauma* 27(12):2121–2130.

Warner MA, Youn TS, Davis T, Chandra A, Marquez de la Plata C, Moore C, Harper C, Madden CJ, Spence J, McColl R et al. 2010b. Regionally selective atrophy after traumatic axonal injury. *Archives of Neurology* 67(11):1336–1344.

Weissman DH, Roberts KC, Visscher KM, Woldorff MG. 2006. The neural bases of momentary lapses in attention. *Nature Neuroscience* 9(7):971–978.

Wilde EA, Chu Z, Bigler ED, Hunter JV, Fearing MA, Hanten G, Newsome MR, Scheibel RS, Li X, Levin HS. 2006. Diffusion tensor imaging in the corpus callosum in children after moderate to severe traumatic brain injury. *Journal of Neurotrauma* 23(10):1412–1426.

Wilde EA, McCauley SR, Hunter JV, Bigler ED, Chu Z, Wang ZJ, Hanten GR, Troyanskaya M, Yallampalli R, Li X et al. 2008. Diffusion tensor imaging of acute mild traumatic brain injury in adolescents. *Neurology* 70(12):948–955.

Wilson JT, Wiedmann KD, Hadley DM, Condon B, Teasdale G, Brooks DN. 1988. Early and late magnetic resonance imaging and neuropsychological outcome after head injury. *Journal of Neurology, Neurosurgery, and Psychiatry* 51(3):391–396.

Witt ST, Lovejoy DW, Pearlson GD, Stevens MC. 2010. Decreased prefrontal cortex activity in mild traumatic brain injury during performance of an auditory oddball task. *Brain Imaging and Behavior* 4(3–4):232–247.

Woodward ND, Rogers B, Heckers S. 2011. Functional resting-state networks are differentially affected in schizophrenia. *Schizophrenia Research* 130(1–3):86–93.

Yuh EL, Mukherjee P, Lingsma HF, Yue JK, Ferguson AR, Gordon WA, Valadka AB, Schnyer DM, Okonkwo DO, Maas AI et al. 2013. Magnetic resonance imaging improves 3-month outcome prediction in mild traumatic brain injury. *Annals of Neurology* 73(2):224–235.

Zhang Y, Schuff N, Jahng GH, Bayne W, Mori S, Schad L, Mueller S, Du AT, Kramer JH, Yaffe K et al. 2007. Diffusion tensor imaging of cingulum fibers in mild cognitive impairment and Alzheimer disease. *Neurology* 68(1):13–19.

Zhou Y, Lui YW, Zuo XN, Milham MP, Reaume J, Grossman RI, Ge Y. 2014. Characterization of thalamo-cortical association using amplitude and connectivity of functional MRI in mild traumatic brain injury. *Journal of Magnetic Resonance Imaging* 39(6):1558–1568.

Zhou Y, Milham MP, Lui YW, Miles L, Reaume J, Sodickson DK, Grossman RI, Ge Y. 2012. Default-mode network disruption in mild traumatic brain injury. *Radiology* 265(3):882–892.

Zhuo J, Xu S, Proctor JL, Mullins RJ, Simon JZ, Fiskum G, Gullapalli RP. 2012. Diffusion kurtosis as an in vivo imaging marker for reactive astrogliosis in traumatic brain injury. *NeuroImage* 59(1):467–477.

Zimmerman, R.A. 1999. Craniocerebral trauma. In: SH Lee, KCVG Rao, RA Zimmerman (eds.), *Cranial MRI and CT*, 4th edn. McGraw-Hill, New York, pp. 413–452.

Section III

Intraoperative Neuroimaging

15 UV-Based Imaging Technologies for Intraoperative Brain Mapping

Babak Kateb, Frank Boehm, Ray Chu, Sam Chang, Keith Black, and Shouleh Nikzad

CONTENTS

INTRODUCTION

There is a strong and rapidly increasing trend in medicine toward the development of more compact, less invasive, smarter, and highly efficacious diagnostic and therapeutic devices and systems. These emerging technologies will be of particular benefit as relates to the diagnostic imaging and mapping of the human brain. Due to the invasiveness of malignant gliomas, it is challenging to adequately differentiate their boundaries from surrounding healthy brain tissues.

Complete tumor resection is a critical factor that ultimately impacts patient outcomes, their quality of life, and survival. The capacity for precise intraoperative imaging feedback, which can accurately distinguish between tumors and healthy brain tissues, is currently lacking. This deficit drives the development of novel intraoperative imaging strategies that might have the potential to significantly enhance the surgical precision with which tumors are removed. This chapter will survey and articulate the latest advances and applications associated with one of these approaches, via an exploration of ultraviolet (UV)-based intraoperative brain imaging and mapping technologies.

BRIEF SURVEY OF PHOTONICS IN INTRAOPERATIVE BRAIN IMAGING AND SPECTROSCOPY

A number of photonics-based strategies have been applied to the diagnostic and intraoperative imaging of the brain. Various fluorescent dyes (Kabuto, et al., 1997; MacDonald et al., 1998;

Kremer et al., 2000; Lin et al., 2000; Zhang et al., 2009), laser-induced fluorescence of photosensitizers (Poon et al., 1992; Tsai et al., 1993; Hebeda et al., 1998; Fisher et al., 2013; Kitai et al., 2014), and time-resolved laser-induced fluorescence spectroscopy (TR-LIFS) (Thompson et al., 2001a,b, 2002; Butte et al., 2010, 2011; Haidar et al., 2015) have been employed to delineate the peripheries of cancerous tumors in the brain. However, the interrogation of photosensitizers is contingent on the injection of specific photonically sensitive compounds, and the time required for the acquisition of data from time-resolved laser-induced spectroscopy may be quite protracted. The utility of TR-LIFS, by which fluorescence was induced via a pulsed (3 ns) nitrogen laser (337 nm) through a fiber-optic catheter, was shown to be contingent on both the wavelength and the species of tissue being imaged (Thompson et al., 2002).

The demarcation of brain tumors using laser-induced fluorescence attenuation spectroscopy remains in its nascent clinical stages; hence, it is not yet available for intraoperative use (Shehada et al., 2000). Confocal microscopy was utilized in conjunction with femtosecond laser pulses and photomultiplier tubes to enable time-resolved single-photon counting detection. This strategy employed differences in the fluorescence lifetimes between breast cancer (Hs578T) and healthy (Hs578Bst) cells in vitro, through the quantitative assessment of free to enzyme-bound ratios of reduced nicotinamide adenine dinucleotide hydrogenase (NADH) coenzymes, as related to their concentration and conformations (Yu and Heikal, 2009).

INTRAOPERATIVE IMAGING FOR DIAGNOSTICS

Here we give a brief example of imaging as a diagnostic tool with the use of infrared (IR) light. Although emission and detection mechanisms are different for the UV and IR, as will be described later in this chapter, this example might serve as an analogy to demonstrate the utility of leveraging imaging technologies, such as near IR or UV, as diagnostic tools.

A novel, rapid, and noninvasive neuroimaging technique known as intraoperative thermal imaging (ITI) exhibited the potential ability to define the boundaries of primary and metastatic brain tumors. Consequently, brain tumor resections might be significantly enhanced via the instantaneous anatomical and pathophysiological data that it provides. A thermal profile of a cerebral cortex resident tumor and adjunctive healthy cerebral neocortical tissues were imaged using a ThermaCAM P60 (TCP60) IR camera (FLIR Systems). The TCP60 was capable of immediately interrogating and analyzing the IR emission spectrum of a reasonably large sized clinically relevant field, with a spatial resolution of 7.5–13 μm and a thermal sensitivity of 0.06°C.

However, ITI is restricted to measuring heat profiles at the surfaces of entities with limited penetration, which translates to the requirement of the full exposure of a lesion. Nevertheless, these data may serve as a useful intraoperative means for the scrutiny of surgical fields to demarcate neoplasms. In a case study conducted by Kateb et al. (2009), intraoperative ITI data generated the distinct demarcation (location and dimensional periphery) of the tumor of a patient with metastatic intracortical melanoma, via contrasts in temperature (~3.3°C), between the core of the tumor (36.4°C) and healthy tissue (33.1°C), which was confirmed by pre-resection MR, CT, and ultrasound.

Through the use of IR thermal imaging, Papaioannou et al. endeavored to differentiate brain tumors from healthy brain tissue in fourteen Wistar rats, each of which was injected with 100,000 C6 glioma cells. These rats subsequently underwent bilateral craniotomies at ~2 weeks following the introduction of the malignant cells. A shortwave (~3.4–5 μm) Inframetrics thermal camera (model PM-290), which employed a cryogenically cooled 256 × 256 platinum silicide focal plane array, was then used to image the open external brain tissues. The attained spatial resolution was ~100 μm, with a thermal sensitivity of >0.1°C at 30°C. Based on the measured thermal variances between the tumor sites and normal brain tissues, the results revealed that the malignancies were cooler than healthy brain tissue (Konerding and Steinberg, 1988), with core temperatures at a low of ~28.7°C and high of ~38°C. This was counterintuitive, as tumors possess higher metabolic rates and are dense, albeit somewhat leaky vascularization, which should induce hyperthermia. However, it was surmised that the tumors might indeed have inherently reduced metabolic rates and be less vascularized in contrast to intensely microvascularized healthy brain tissues (Papaioannou et al., 2002).

ENDOSCOPIC FLUORESCENCE LIFETIME IMAGING MICROSCOPY

The application of an endoscopic fluorescence lifetime imaging microscopy (FLIM) system for the intraoperative diagnosis of glioblastoma multiforme (GBM) was described by Sun et al. (2010). GBM comprises the most widespread and aggressive type of human brain tumor, which accounts for 52% of all brain tumor cases, with a median survival rate of 12–15 months. Current optimal therapies include maximal safe surgical resection, with successive adjuvant chemoradiotherapy. A number of studies have revealed that the degree of surgical resection is the primary factor in overall survival (Andreou et al., 2001; Brandes et al., 2008; Sanai and Berger, 2008; Robins et al., 2009). A prototypical FLIM system that consisted of a gated (down to ~0.2 ns) intensifier imaging system with a fiber bundle (Ø0.5 mm) housed 10,000 fibers with a gradient index lens objective 0.5 NA, and 4 mm field of view, which enabled intraoperative access to the surgical field. Trials involving three patients who underwent craniotomy for tumor resection confirmed that the FLIM system had the capacity to distinguish tumors from the healthy cortex. The fluorescence of NADH (at 460 ± 25 nm wavelength band emission) in GBM was observed to be less intense (35% less, p-value <0.05) and exhibited increased longevity ($^{\tau}$CBM-Amean = 1.59 ± 0.24 ns) than that in healthy cortex tissues ($^{\tau}$CBM-Amean = 1.28 ± 0.04 ns—p-value <0.005) (Sun et al., 2010).

LASER SPECTROSCOPY AND LASER-INDUCED FLUORESCENCE

It was demonstrated by Pascu et al. that an in vitro comparison of the autofluorescence spectra of pairs of brain tumor tissues and normal tissues from the same patient, which were induced by UV and visible light laser irradiation, allowed them to be identified and differentiated. The laser source comprised a frequency-doubled tunable dye laser that was pumped by a nitrogen-pulsed laser, which included emission at 337.1 nm, a laser beam bandwidth of <0.1 nm, a pulse full time width (FTW) of 1 ns, an energy/pulse of 50 µJ, a peak power per pulse of 500 kW, and a continuously adjustable pulse repetition rate of 1 and 10 pulses per second (pps) (Danaila and Pascu, 1999, 2001).

This group suggested that although the measured autofluorescence spectra of tumor samples were quite close to those of normal tissues, several distinguishing factors set them apart. First, for each pair of tumor/healthy brain tissue samples, the peak autofluorescence was typically shifted between 10 and 20 nm, from the normal to tumorous tissue (Figure 15.1). Second, the overall fluorescence intensity between the samples also varied, in the range from 15%–30%. Finally, deviations in the ratios of certain fluorescence intensity peaks between the brain tumor and healthy tissue samples were found to range from 10% to 40% (Pascu et al., 2009).

Interestingly, from the clinical intraoperative perspective, the fluorescence of various anesthetics that might be present during brain surgery was also investigated, which might impact overall fluorescence measurements. Some of these include sodium thiopental, pancuronium, fentanyl, suxamethonium, and sevoflurane. It was found that subsequent to excitation, sodium thiopental, pancuronium, and fentanyl exhibited no fluorescence emission. In contrast, suxamethonium-influenced tryptophan side-chain chromophores, compared to tryptophan fluorescence were quenched by sevoflurane, which reversibly increased NADH fluorescence (Ramanujam et al., 1994; Utzinger et al., 1999). Since anesthetics are dispersed within both malignant and healthy brain tissues, they affect the autofluorescence spectrums of both. This might be advantageous if their concentration within tumors is elevated, as this may serve to more precisely demarcate tumor/healthy brain tissue interfaces.

An investigation into the utility of intraoperative laser-induced fluorescence spectroscopy to demarcate tumor margins was undertaken by Poon et al., who employed intracerebral glioma rat models for the work. The intravenous introduction of a fluorescent dye (chloro-aluminum phthalocyanine tetrasulfonate [ClAlPcS4]) (68 nm fluorescence peak in contrast to 47 nm for normal

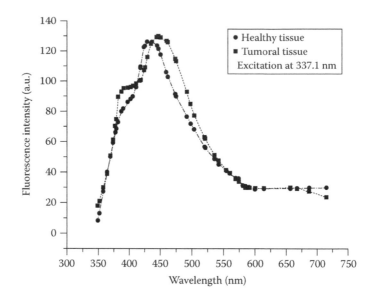

FIGURE 15.1 Autofluorescence emission of healthy brain tissue and malignant brain tumor tissue subsequent to excitation at 337.1 nm, showing a 15 nm peak shift between the two. (Reproduced from Pascu, A. et al., *Anat. Rec.* (*Hoboken*), 292(12), 2013, December 2009. With permission.)

brain tissue) was performed 24 h prior to tumor removal, which took place under an operating microscope. Tumor resection was performed on one group of rats (G1-9 individuals) under the guidance of laser-induced fluorescence measurements, whereas for the second group (G2-10 individuals), it was conducted by visual evaluation. It was revealed that the G1 rats retained a considerably lower residual tumor mass than did the G2 rats (G1 = 0.5 ± 0.2 mm³) (G2 = 13.7 ± 4.0 mm³). Further, at 2 weeks postresection, three of the G1 rats were tumor-free, whereas there were none from G2. The acquisition of spatially resolved spectra using a fiber-optic probe required five seconds, whereupon a contrast ratio of 40:1 was observed for glioma/normal brain fluorescence signals. These data verified that higher-resolution tumor margins could be distinguished via an intraoperative laser-induced fluorescence system under an operating microscope when compared with unaided visual assessment (Poon et al., 1992).

Photoacoustic Tomography

Photoacoustic tomography (PAT) comprises an emerging hybrid strategy for the generation of high-resolution/high-contrast imagery of biological tissues. This technology utilizes the acoustic detection of optical absorption signals, which are derived from either physiologically ambient chromophores (e.g., oxyhemoglobin and deoxyhemoglobin) or introduced contrast agents (e.g., organic dyes or nanoparticles). The acquisition of high-resolution images with sharp contrast (spanning both optically ballistic and diffusive domains) is made possible due to the reduced propensity of ultrasound to scatter within tissues, in contrast to photonic waves (Xia et al., 2014). Multiscale/multicontrast imagery of dynamic biological structures may thus be achieved, ranging from cellular organelles to entire organs, and supplies useful data pertaining to anatomy, functionality, metabolism, biomolecules, and genetics.

The challenges posed by the high degree of photonic scattering within biological tissues may be surmounted through the application of the photoacoustic effect, whereby the molecular absorption of photons generates thermally induced pressure spikes and ultrasonic waves, which are captured by acoustic detection devices and translated to imagery. The spatial resolution of images may be adjusted and maintained as a depth-to-resolution ratio, which is typically1/200 of the selected imaging depth, demonstrated to attain a maximum 8.4 cm (in chicken breast tissue) to date (Favazza et al., 2010; Wang and Hu, 2012; Xia et al., 2014).

With regard to the imaging of the brain, Nasiriavanaki et al. developed a PAT system to noninvasively image the functional connectivity of the resting state mouse brain, which facilitated an extensive field of view and enhanced spatial resolution. The observation and demarcation of eight functional areas of the brain (olfactory bulb, limbic, parietal, somatosensory, retrosplenial, visual, motor, and temporal regions, and several subregions), and hence, correlations between them, was made possible. Upon sequentially exposing the mice to hyperoxic and hypoxic conditions, robust and faint functional interrelations, respectively, were revealed between these domains, with the concurrent acquisition of vascular images (Nasiriavanaki et al., 2014). This work demonstrated that PAT possesses the capacity to noninvasively elucidate the functional connectivity of the mouse brain, which may in future also be applicable to correlating detailed functionalities between various domains within the human brain. Although PAT shows much promise, a number of constraints remain toward the realization of its full potential. These include limitations on imaging depth, imaging speed, and the measurement of localized fluence distribution. Advanced contrast agents are also being investigated toward the augmentation of resolution within deeper tissues (Xia et al., 2014).

QUANTUM DOT OPTICAL IMAGING

Quantum dots are semiconductor nanoparticles that exhibit strong potential in biomedical imaging applications, inclusive of the brain; however, they can elicit cytotoxicity in human cells. This situation may be remediated to a certain degree when these nanometric entities are coated with biocompatible substances. The synthesis of biocompatible dextrin-coated cadmium sulfide nanoparticles (CdS-Dx/QDs) was undertaken by Reyes-Esparza et al., who, as part of their investigation, observed their transport, distribution, and activity in the brain. In vitro studies revealed that they were not toxic to HepG2 or HeLa cells in low concentrations (<1 μg/mL), and the uptake of CdS-Dx/QDs by these cells generated robust fluorescence.

In animal models, it was verified that the Ø3–5 nm CdS-Dx/QDs crossed the blood–brain barrier (BBB) via the detection of low level fluorescence in all layers of the cerebral cortex, with the most intense fluorescence occurring in the inner granular and multiform cell layers, with lower levels in the multiform cell layer. Even more intense fluorescence was observed in the plexiform layer. The CdS-Dx/QDs were activated using a 488 nm laser, with their responses detectable at 515 nm. With additional refinement and further enhanced biocompatibility, targeted QDs may facilitate the detailed imaging and demarcation of distinct regions of the brain (Reyes-Esparza et al., 2015).

Li et al. developed novel dual-modality nanometric entities (Gd-DOTA-Ag$_2$S QDs), which they referred to as Gd-Ag2S nanoprobes. This biocompatible platform exploited the synergistic advantages of Gd (deep tissue magnetic resonance imaging) combined with Ag$_2$S quantum dots (high signal-to-noise ratio and high spatiotemporal resolution imaging). The brain tumor of a mouse model was clearly demarcated in situ, and a successful intraoperative resection was accomplished via the precise fluorescence imaging of the injected Gd-Ag2S nanoprobes (Li et al., 2015).

MECHANISMS AND SIGNATURES IN THE UV AND ITS APPLICATION IN INTRAOPERATIVE BRAIN MAPPING: CLINICAL TRIAL AT CEDARS-SINAI MEDICAL CENTER

Kateb et al. described in a recent patent application (Kateb et al., 2014) that the surface of the brain of a patient was intraoperatively scrutinized under UV illumination in vivo for the excision of a malignant tumor, which integrated a Generation I UV/visible camera (Figure 15.2) with a back-illuminated silicon imaging detector; employing either a charge-coupled device (CCD) under

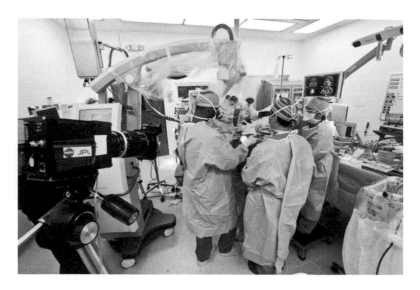

FIGURE 15.2 Neurosurgeons/researchers (Drs. Black, Chu, and Kateb) from Cedars-Sinai Medical Center and the Maxine Dunitz Neurosurgical Institute utilizing an adapted UV camera from NASA's JPL, California Institute of Technology, to demarcate brain tumor tissue. JPL's UV imaging technology was based on the delta doping technology described in this chapter and provided high sensitivity and stability across the entire UV spectral range.

exposure to UV light, images captured by the camera could be subsequently processed to investigate the demarcation and differentiation of tumors from healthy tissue.

These capacities were enabled by distinguishing variations in the autofluorescence of NADH that exists between cancerous and healthy cells. As tumor cells have amplified energy requirements to fuel higher metabolic activity, NADH is upregulated in cancerous tissues, which facilitates its detection to enable the differentiation of tumors and healthy brain tissues with a UV optical camera. It remains under investigation, however, whether NADH autofluorescence is indeed intensified across all types of cancerous cells/tissues in a linear manner, as opposed to healthy tissues. Multiple complex factors encompassing localized tumor environments (e.g., necrotic and hypoxic/anoxic domains, metabolic profiles, the presence of viable cells) may influence the ultimate autofluorescence of tumor resident NADH.

The autofluorescence of the NADH coenzyme possesses two constituent absorption peak excitations, which are enabled by the dual chromophores that it contains, namely, adenine (260 nm) and dihydronicotinamide (340 nm) (Hull et al., 2001), with an emission maximum at 480 nm. The conformational structures (folded or unfolded) of NADH may be significantly influenced by temperature and solvents, which is of significance as efficient fluorescence resonance energy transfer (FRET) between the two chromophores (at ~3 ps time scales), following photonic stimulation, occurs only when it is in the folded configuration. The unfolding of the NADH molecule ensues when it is exposed to methanol (MeOH) for instance, which diminishes the FRET between the NADH chromophores (Ladokhin and Brand, 1995; Rover et al., 1998; Heiner et al., 2013).

Advanced UV Technologies for Intraoperative Brain Mapping

UV light spans the wavelength range from ~100 to 400 nm, which is subdivided into several domains, encompassing UVC (100–260 nm), UVB (260–315 nm), and UVA (315–400 nm) (Sliney and Wolbarsht, 1980). Since it was found that UV could activate fluorescence in NADPH, this phenomenon was exploited (initially in the 1970s) for brain slice imaging (Lipton, 1973) and, subsequently, in vivo imaging (Jobsis et al., 1971; Rosenthal and Jobsis, 1971); however, a drawback that emerged was the phototoxicity of UVB and UVA (Kim et al., 2015). Mitochondrial flavoprotein fluorescence elicited by UV came to be employed to produce functional signals, both in vitro (Shuttleworth et al., 2003; Brennan et al., 2006) and in vivo (Shibuki et al., 2003; Reinert et al., 2004; Husson et al., 2007). These flavoprotein fluorescence signals possess longevity that is comparable to optical intrinsic signal (OIS) imaging; albeit they have been observed to exhibit improved signal-to-noise ratios with higher spatial localization (Tohmi et al., 2006; Husson et al., 2007).

Rapid advances in the materials/nanomaterials sciences, synthesis and fabrication processes, and further reductions in the dimensions of electronic components to nanometric realms will likely lead to the emergence of sophisticated UV imaging technologies that will enable the development of facilitative UV-based intraoperative brain imaging applications. These methods involve simple modifications to conventional imaging apparatus (excitation and emission filters, dichroic), but have not, to date, been implemented for intraoperative imaging.

There are a number of silicon-based UV imaging technologies that use ion implantation, chemisorption, silicon-on-oxide (SOI) technologies, as well as silicon-on-sapphire technologies. Here we discuss one of the more sensitive and stable forms of the technologies that have been developed for space and consumer applications.

Advances in imaging technologies have enabled the development of much more compact cameras with higher-resolution, high-sensitivity, and high-stability UV cameras based on NASA's Jet Propulsion Laboratory (JPL)-invented 2D doping of silicon arrays. Future plans include the integration of next-generation .5 megapixel cameras into intraoperative microscopes and endoscopes.

DELTA-DOPING AND SUPERLATTICE DOPING IN SILICON ARRAYS FOR UV IMAGING

Delta doping technology, as applied to back-illuminated silicon detectors and imaging arrays, was invented and developed at NASA's JPL, California Institute of Technology. This technology was originally developed for astronomy and planetary sciences; however, its potential for applications in other fields was soon realized. The more recent invention of superlattice-doped silicon arrays (Hoenk et al., 2014a) has been applied to machine vision, X-ray detection, deep UV detection, and near UV detection with high efficiency, ultrahigh stability, and uniformity.

Solid-state silicon imaging detectors such as CCDs or CMOS arrays are typically illuminated through their front surfaces; however, gate structures, encompassing polysilicon electrodes, patterned oxides, and metallic interconnects reduce quantum efficiency (QE) and resolution through the assimilation and dispersion of ancillary light. Furthermore, because UV is absorbed within the first few nanometers of material, front-illuminated silicon arrays are blind to it.

Back-illuminated CCDs were developed where the introduction of light into the device is inverted and illuminated from the opposite side of the silicon wafer. This configuration necessitates the removal of the low-resistivity substrate and the passivation of the exposed silicon surface. Substrate thinning and surface passivation are still the required prerequisite steps for high-resistivity substrates in the preparation of high QE CCDs, which have the capacity for UV imaging. The use of molecular beam epitaxy (MBE), as in delta doping and superlattice doping, provides an optimal interfacial passivation strategy for the elimination of QE hysteresis in back-illuminated UV imaging arrays at the wafer level (Nikzad et al., 1994a,c; Hoenk et al., 1992a,b). Prior to MBE processing, silicon surfaces are atomically cleaned, to allow the growth of 2–3 nm single silicon crystals contained in a single atomic sheet, 30% of which is comprised of a monolayer of dopants (Hoenk et al., 2014a,b; Nikzad, 2014; Nikzad et al., 2015).

Delta-doped arrays and superlattice-doped silicon arrays consistently demonstrate 100% internal QEs and are devoid of hysteresis (Hoenk, 1992a,b, 2014a,b; Nikzad et al., 1994a, 2012). The response of the device may be further enhanced and tailored for specific regions of the spectrum (Nikzad et al., 1994a,b, 2012; Nikzad, 2014). In recent years, unprecedented QE has been demonstrated by the JPL team through the combination of superlattice doping or delta doping with custom precision coatings using atomic layer deposition (Hamden et al., 2011; Nikzad et al., 2012; Hoenk et al., 2014a,b).

UV Laser–Based Fluorescence Spectroscopy

It remains challenging to achieve total tumor resection in patients with GBM due to its infiltrative nature. A further challenge relates to the differentiation of morphologically similar malignant domains from neighboring normal brain tissue. Haj-Hosseini et al. developed a hand held optical touch pointer in conjunction with a fluorescence spectroscopy system to intraoperatively discriminate between GBM and healthy brain tissue. This group had previously exploited 5-aminolevulinic acid (5-ALA)–induced fluorescence and MR spectroscopy to conduct stereotactic biopsies of human glioblastomas (Pålsson et al., 2003). Further, they developed a prototypical optic-based fluorescence spectroscopy system for guided neurosurgical tumor resection (Ilias et al., 2007; Haj-Hosseini et al., 2008), which enabled the detection of fluorescence peaks associated with malignant tumor tissue. This was facilitated subsequent to the introduction of a relatively low dose of 5-ALA (5 mg/kg), which traversed the leaky BBB of the tumor in contrast to the intact BBB in healthy brain tissue (Haj-Hosseini et al., 2010). Once within malignant tissue, 5-ALA is converted to the fluorescent protoporphyrin IX (PpIX) marker (Moan et al., 2001; Novotny and Stummer, 2003), which comprises a natural substance involved with the heme cycle that is quickly removed from the body. PpIX absorbs laser light at 405 nm and reemits fluorescence at peaks of 635 and 704 nm.

The components of the optical touch pointer system included a near-UV laser module (maximal 405 nm excitation light and power of 50 nW), which could function in either pulsed or continuous (dual) mode. For the photonic readout, a 2048 element CCD with accommodation for a 240–850 nm wavelength range at 3 nm resolution was employed. For back-reflected light suppression, a long pass cutoff filter (450 nm) was affixed to the front of the spectrometer detector slit. The excitation light was conveyed to the tissue through a handheld fiber-optic probe that housed a single central transmitting fiber (Ø0.64 mm), which was circumscribed by nine receiver fibers (Ø0.24 mm) that were matched on the opposite end of the cable to the detector slit of the spectrometer (Figure 15.3)

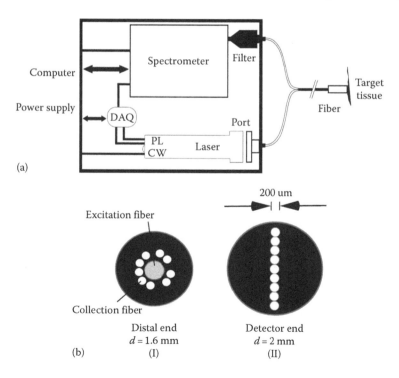

FIGURE 15.3 (a) Hardware design of pulsed fiber-optic system. (b) Configuration of fiber probe (I), probe tip, and detector end (II). (Reproduced from Haj-Hosseini, N. et al., *Lasers Surg. Med.*, 42(1), 9, January 2010. With permission.)

A micropositioner was mounted at the interface of the laser and fiber probe to facilitate their alignment (Haj-Hosseini et al., 2010).

The administration of 5 mg/kg of 5-ALA to patients who were to undergo surgical resection for presumed glioblastoma, commenced 2 h prior to the procedure. Subsequent measurements conducted with healthy white or gray brain tissues detected no presence of PpIX, whereas GBM-affected areas showed distinct indications of PpIX, which verified malignant cell uptake. The pulse duration and laser power settings that enabled distinguishable fluorescence signals, following probe contact with brain tissue, were 0.4 s and 5 mW, respectively, imparting an irradiance of 1.78 W/cm^2 at a dose of 0.71 J/cm^2 per pulse. To negate the potential for thermally induced tissue damage, tissue temperature increases were quantified using the optic and thermal parameters of brain tissue (Sturesson and Andersson-Engels, 1996; Schwarzmaier et al., 2002) via Monte Carlo, for light absorption and propagation, in conjunction with the COMSOL Multiphysics 3.4 program to simulate heat transfer (Yaroslavsky et al., 2002). It was estimated that the maximal temperature increase per laser pulse (5 mW, 0.4 s) was 0.48°C (Haj-Hosseini et al., 2010).

Optimally, this laser-based fluorescence spectroscopy system had the capacity to exclusively discern malignant brain tumor tissue from healthy brain tissue in ~1 s per measurement point. Thus, it was not anticipated that for this particular application, any clinical complication would arise as the result of laser heating, as the pulsed modulation kept the tissue energy exposure to a minimum through the synchronization of laser light delivery and fluorescence acquisition. Future studies have the aim of correlating the intensity of PpIX fluorescence with the degree of malignancy (Haj-Hosseini et al., 2010).

CONCLUSIONS AND FUTURE DIRECTIONS

UV imaging is a promising technology for intraoperative mapping if it is employed in conjunction with other neurophotonic techniques as a component of real-time multimodal imaging systems. Therefore, UV, IR, and other neurophotonic imaging techniques might be used for real-time intraoperative brain mapping, which is urgently required to enhance the resection of brain tumors and epileptic regions of the brain. These neurophotonic techniques enable the prediction and adjustment of brain shifts and tumor resection margins during surgery.

New generations of intraoperative microscopes will not only be equipped with advanced neurophotonic technologies/imaging and arrays, but also have the ability to learn from surgeries and communicate with each other through machine learning, cloud computing, and artificial intelligence, which will further facilitate predictive modeling and pattern recognition.

REFERENCES

Andreou AG, Kalayjian ZK, Apsel A, Pouliquen P, Athale RA, Simonis G, Reedy R. Silicon on sapphire CMOS for optoelectronic microsystems. *IEEE Circuits Syst Mag*, 2001; 1(3): 22–30.

Brandes A, Tosoni A, Franceschi E, Reni M, Gatta G, Vecht C. Glioblastoma in adults. *Crit Rev Oncol Hematol*, 2008; 67:139–152.

Brennan AM, Connor JA, Shuttleworth CW. NAD(P)H fluorescence transients after synaptic activity in brain slices: Predominant role of mitochondrial function. *J Cereb Blood Flow Metab*, 2006; 26:1389–1406.

Butte PV, Fang Q, Jo JA, Yong WH, Pikul BK, Black KL, Marcu L. Intraoperative delineation of primary brain tumors using time-resolved fluorescence spectroscopy. *J Biomed Opt*, March–April 2010; 15(2):027008.

Butte PV, Mamelak AN, Nuno M, Bannykh SI, Black KL, Marcu L. Fluorescence lifetime spectroscopy for guided therapy of brain tumors. *Neuroimage*, January 2011; 54(Suppl 1):S125–S135.

Danaila L, Pascu ML. Photodynamic therapy of the cerebral malignant tumours. *Rom J Biophys*, 1999; 10:145–152.

Danaila L, Pascu ML. 2001, *Lasers in Neurosurgery*. Bucharest, Romania: Academia Romana.

Favazza C, Maslov K, Cornelius L, Wang LV. 2010. In vivo functional human imaging using photoacoustic microscopy: Response to ischemic and thermal stimuli. *Photons Plus Ultrasound: Imaging and Sensing 2010*, San Francisco, CA, pp. 75640Z–75640Z-6.

Fisher CJ, Niu CJ, Lai B, Chen Y, Kuta V, Lilge LD. Modulation of PPIX synthesis and accumulation in various normal and glioma cell lines by modification of the cellular signaling and temperature. *Lasers Surg Med*, September 2013; 45(7):460–468.

Haidar DA, Leh B, Zanello M, Siebert R. Spectral and lifetime domain measurements of rat brain tumors. *Biomed Opt Express*, March 11, 2015; 6(4):1219–1233.

Haj-Hosseini N, Andersson-Engels S, Wårdell K. 2008. Evaluation of a fiber-optic based pulsed laser system for fluorescence spectroscopy. *14th Nordic-Baltic Conference on Biomedical Engineering and Medical Physics*, Riga, Latvia, pp. 363–366.

Haj-Hosseini N, Richter J, Andersson-Engels S, Wårdell K. Optical touch pointer for fluorescence guided glioblastoma resection using 5-aminolevulinic acid. *Lasers Surg Med*, January 2010; 42(1):9–14.

Hamden E, Greer F, Blacksberg J, Hoenk ME, Nikzad S, Schiminovich D. Anti-reflection coatings for use in UV detector design. *Appl Opt*, 2011; 50:4180–4188.

Hebeda KM, Saarnak AE, Olivo M, Sterenborg HJ, Wolbers JG. 5-Aminolevulinic acid induced endogenous porphyrin fluorescence in 9L and C6 brain tumours and in the normal rat brain. *Acta Neurochir (Wien)*, 1998; 140(5):503–512; discussion 512–513.

Heiner Z, Roland T, Léonard J, Haacke S, Groma GI. 2013. Ultrafast absorption kinetics of NADH in folded and unfolded conformations. *XVIIIth International Conference on Ultrafast Phenomena*. Published by EDP Sciences, Vol. 41, Lausanne, Switzerland, pp. 1–3, 07003.

Hoenk ME, Grunthaner PJ, Grunthaner FJ, Terhune RW, Fattahi M, Tseng HF. Growth of a delta-doped silicon layer by molecular beam epitaxy on a charge-coupled device for reflection limited quantum efficiency. *Appl Phys Lett*, 1992a; 61:1084–1086.

Hoenk ME, Grunthaner PJ, Grunthaner FJ, Terhune RW, Fattahi M. August 1992b. Epitaxial growth of p+ silicon on a backside-thinned CCD for enhanced UV response. *Proceedings of the SPIE 1656, High-Resolution Sensors and Hybrid Systems*, San Jose, CA, p. 488.

Hoenk ME, Nikzad S, Carver AG, Jones TJ, Hennessy J, Jewell AD, Sgro J, Tsur S, McClish M, Farrell R. July 30, 2014a. Superlattice-doped imaging detectors: Progress and prospects. *Proceedings of the SPIE. 915413, High Energy, Optical, and Infrared Detectors for Astronomy VI*, Montreal, Quebec, Canada, June 24, 2014.

Hoenk ME, Nikzad S, Jones TJ, Greer F, Carver AG. September 2014b. Delta-doping at wafer level for high throughput, high yield fabrication of silicon imaging arrays, US Patent 2014/8828852.

Hull RV, Conger PS 3rd, Hoobler RJ. Conformation of NADH studied by fluorescence excitation transfer spectroscopy. *Biophys Chem*, March 15, 2001; 90(1):9–16.

Husson TR, Mallik AK, Zhang JX, Issa NP. Functional imaging of primary visual cortex using flavoprotein autofluorescence. *J Neurosci*, 2007; 27:8665–8675.

Ilias MA, Richter J, Westermark F, Brantmark M, Andersson-Engels S, Wårdell K. Evaluation of a fiber-optic fluorescence spectroscopy system to assist neurosurgical tumor resections. *Proc SPIE*, 2007; 6631:66310W01–66310W08.

Jobsis FF, O'Connor M, Vitale A, Vreman H. Intracellular redox changes in functioning cerebral cortex. I. Metabolic effects of epileptiform activity. *J Neurophysiol*, 1971; 34:735–749.

Kabuto M, Kubota T, Kobayashi H, Nakagawa T, Ishii H, Takeuchi H, Kitai R, Kodera T. Experimental and clinical study of detection of glioma at surgery using fluorescent imaging by a surgical microscope after fluorescein administration. *Neurol Res*, February 1997; 19(1):9–16.

Kateb B, Black KL, Nikzad S. September 25, 2014. UV imaging for intraoperative tumor delineation, US Patent 2014/0288433.

Kateb B, Yamamoto V, Yu C, Grundfest W, Gruen JP. Infrared thermal imaging: A review of the literature and case report. *Neuroimage*, August 2009; 47(Suppl 2):T154–T162.

Kim K, Park H, Lim KM. Phototoxicity: Its mechanism and animal alternative test methods. *Toxicol Res*, June 2015; 31(2):97–104.

Kitai R, Takeuchi H, Miyoshi N, Andriana B, Neishi H, Hashimoto N, Kikuta K. Determining the tumor-cell density required for macroscopic observation of 5-ALA- induced fluorescence of protoporphyrin IX in cultured glioma cells and clinical cases. *No Shinkei Geka*, June 2014; 42(6):531–536, Article in Japanese.

Konerding MA, Steinberg F. Computerized infrared thermographic and ultrastructural studies of xenotransplanted human tumors on nude mice. *Thermology*, 1988; 3:7–14.

Kremer P, Wunder A, Sinn H, Haase T, Rheinwald M, Zillmann U, Albert FK, Kunze S. Laser-induced fluorescence detection of malignant gliomas using fluorescein-labeled serum albumin: Experimental and preliminary clinical results. *Neurol Res*, July 2000; 22(5):481–489.

Ladokhin AS, Brand L. Evidence for an excited-state reaction contributing to NADH fluorescence. *J Fluoresc*, March 1995; 5(1):99–106.

Lesser M. Antireflection coatings for silicon charge-coupled devices. *Opt Eng*, 1987; 26:911–915.

Li C, Cao L, Zhang Y, Yi P, Wang M, Tan B, Deng Z, Wu D, Wang Q. Preoperative detection and intraoperative visualization of brain tumors for more precise surgery: A new dual-modality MRI and NIR nanoprobe. *Small*, September 16, 2015; 11(35):4517–4525.

Lin WC, Toms SA, Motamedi M, Jansen ED, Mahadevan-Jansen A. Brain tumor demarcation using optical spectroscopy: An in vitro study. *J Biomed Opt*, April 2000; 5(2):214–220.

Lipton P. Effects of membrane depolarization on nicotinamide nucleotide fluorescence in brain slices. *Biochem J*, 1973; 136:999–1009.

MacDonald TJ, Tabrizi P, Shimada H, Zlokovic BV, Laug WE. Detection of brain tumor invasion and micrometastasis in vivo by expression of enhanced green fluorescent protein. *Neurosurgery*, December 1998; 43(6):1437–1442; discussion 1442–1443.

Moan J, Van Den Akker JTHM, Juzenas P, Ma LW, Angell-Petersen E, Gadmar ØB, Iani V. On the basis for tumor selectivity in the 5-aminolevulinic acid-induced synthesis of protoporphyrin IX. *J Porphyr Phthalocya*, 2001; 5(2):170–176.

Nasiriavanaki M, Xia J, Wan H, Bauer AQ, Culver JP, Wang LV. High-resolution photoacoustic tomography of resting-state functional connectivity in the mouse brain. *Proc Natl Acad Sci U S A*, January 7, 2014; 111:21–26.

Nikzad S. 2014. High-performance silicon imagers and their applications in astrophysics, medicine and other fields, in *High Performance Silicon Imaging: Fundamentals and Applications of CMOS and CCD Image Sensors*, D. Durini, ed. Woodhead Publishing, Cambridge, U.K., pp. 411–438.

Nikzad S, Hoenk ME, Greer F et al. Delta doped electron multiplied CCD with absolute quantum efficiency over 50% in the near to far ultraviolet range for single photon counting applications. *Appl Opt*, 2012; 51:365.

Nikzad S, Hoenk ME, Grunthaner P, Terhune RW, Grunthaner FJ. Delta-doped CCDs for enhanced UV performance. *SPIE X-Ray UV Detectors*, 1994a; 2278:138–142.

Nikzad S, Hoenk ME, Grunthaner PJ, Grunthaner FJ, Terhune RW, Winzenread R, Fattahi M, Tseng HF, Lesser M. March 1994b. Delta-doped CCDs: High QE with long-term stability at UV and visible wavelengths. *Proceedings of SPIE*, 2198. *Instrumentation in Astronomy VIII*, San Jose, CA, p. 907.

Nikzad S, Hoenk ME, Grunthaner PJ, Grunthaner FJ, Terhune RW, Winzenread R, Fattahi M, Tseng HF, Lesser M. July 22, 1994c. Delta-doped CCDs as stable, high-sensitivity, high resolution UV imagers. *Proceedings of the SPIE 2217, Aerial Surveillance Sensing Including Obscured and Underground Object Detection*, Orlando, FL, p. 355.

Nikzad S, Jewell AD, Carver A, Maki J, Hoenk ME, Bell LD. 2015. Digital imaging for planetary exploration, in *Handbook of Digital Imaging*, Michael, K. ed. Wiley Books, Hobokin, NJ.

Novotny A, Stummer W. 5-Aminolevulinic acid and the blood-brain barrier—A review. *Med Laser Appl*, 2003; 18(1):36–40.

Pålsson S, Backlund EO, Eriksson O, Lundberg P, Smedby Ő, Sturnegk P, Wårdell K, Svanberg K, Andersson-Engels S. 2003. ALA-PpIX fluorescence and MR spectroscopy in connection with stereotactic biopsy of human glioblastomas, PhD thesis. Lund, Sweden: Lund Institute of Technology.

Papaioannou T, Thompson RC, Kateb B, Sorokoumov O, Grundfest WS, Black KL. May 13, 2002. Biomedical diagnostic, guidance, and surgical-assist systems IV. *Proceedings of the SPIE 4615*, Vol. 46, Bellingham WA, p. 32.

Pascu A, Romanitan MO, Delgado JM, Danaila L, Pascu ML. Laser-induced autofluorescence measurements on brain tissues. *Anat Rec (Hoboken)*, December 2009; 292(12):2013–2022.

Poon WS, Schomacker KT, Deutsch TF, Martuza RL. Laser-induced fluorescence: Experimental intraoperative delineation of tumor resection margins. *J Neurosurg*, April 1992; 76(4):679–686.

Ramanujam N, Mitchell MF, Mahadevan A, Warren S, Thomsen S, Silva E, Richards-Kortum R. In vivo diagnosis of cervical intraepithelial neoplasia using 337-nm-excited laser-induced fluorescence. *Proc Natl Acad Sci USA*, 1994; 91: 10193–10197.

Reinert KC, Dunbar RL, Gao W, Chen G, Ebner TJ. Flavoprotein autofluorescence imaging of neuronal activation in the cerebellar cortex in vivo. *J Neurophysiol*, 2004; 92:199–211.

Reyes-Esparza J, Martínez-Mena A, Gutiérrez-Sancha I, Rodríguez-Fragoso P, de la Cruz GG, Mondragón R, Rodríguez-Fragoso L. Synthesis, characterization and biocompatibility of cadmium sulfide nanoparticles capped with dextrin for in vivo and in vitro imaging application. *J Nanobiotechnology*, November 17, 2015; 13:83.

Robins HI, Lassman AB, Khuntia D. Therapeutic advances in malignant glioma: Current status and future prospects. *Neuroimaging Clin North Am*, 2009; 19:647–656.

Rosenthal M, Jobsis FF. Intracellular redox changes in functioning cerebral cortex. II. Effects of direct cortical stimulation. *J Neurophysiol*, 1971; 34:750–762.

Rover Júnior L, Fernandes JC, de Oliveira Neto G, Kubota LT, Katekawa E, Serrano SH. Study of NADH stability using ultraviolet-visible spectrophotometric analysis and factorial design. *Anal Biochem*, June 15, 1998; 260(1):50–55.

Sanai N, Berger MS. Glioma extent of resection and its impact on patient outcome. *Neurosurgery*, 2008; 62:753–764.

Schwarzmaier H-J, Eickmeyer F, Fiedler VU, Ulrich F. Basic principles of laser induced interstitial thermo-therapy in brain tumors. *Med Laser Appl*, 2002; 17(2):147–158.

Shehada RE, Marmarelis VZ, Mansour HN, Grundfest WS. Laser induced fluorescence attenuation spectroscopy: Detection of hypoxia. *IEEE Trans Biomed Eng*, March 2000; 47(3):301–312.

Shibuki K, Hishida R, Murakami H, Kudoh M, Kawaguchi T, Watanabe M, Watanabe S, Kouuchi T, Tanaka R. Dynamic imaging of somatosensory cortical activity in the rat visualized by flavoprotein autofluorescence. *J Physiol*, 2003; 549:919–927.

Shuttleworth CW, Brennan AM, Connor JA. NAD(P)H Fluorescence imaging of postsynaptic neuronal activation in murine hippocampal slices. *J Neurosci*, 2003; 23:3196–3208.

Sliney DH, Wolbarsht ML. 1980. *Safety with Lasers and Other Optical Sources: A Comprehensive Handbook.* New York: Plenum Press.

Sturesson C, Andersson-Engels S. Mathematical modeling of dynamic cooling and pre-heating, used to increase the depth of selective damage to blood vessels in laser treatment of port wine stains. *Phys Med Biol*, 1996; 41(3):413–428.

Sun Y, Hatami N, Yee M, Phipps J, Elson DS, Gorin F, Schrot RJ, Marcu L. Fluorescence lifetime imaging microscopy for brain tumor image-guided surgery. *J Biomed Opt*, September–October 2010; 15(5):056022.

Thompson RC, Black KL, Kateb B, Marcu L. September 2001a. Time-resolved laser-induced fluorescence spectroscopy for detection of experimental brain tumors. *Congress of Neurological Surgeons*, San Diego, CA, oral/poster presentation.

Thompson RC, Black KL, Kateb B, Marcu L. Detection of brain tumors using time resolved laser-induced fluorescence spectroscopy. *Proc SPIE*, 2002; 8(4613):8–12.

Thompson RC, Papaioannov T, Kateb B, Black KL. April 2001b. Application of laser spectroscopy and thermal imaging for detection of brain tumors. American Association of Neurological Surgeons. Toronto, Ontario, Canada, abstract/poster.

Tohmi M, Kitaura H, Komagata S, Kudoh M, Shibuki K. Enduring critical period plasticity visualized by transcranial flavoprotein imaging in mouse primary visual cortex. *J Neurosci*, 2006; 26:11775–11785.

Tsai JC, Kao MC, Hsiao YY. Fluorospectral study of the rat brain and glioma in vivo. *Lasers Surg Med*, 1993; 13(3):321–331.

Utzinger U, Trujillo EV, Atkinson EN, Mitchell MF, Cantor SB, Richards-Kortum R. Performance estimation of diagnostic tests for cervical precancer based on fluorescence spectroscopy: Effects of tissue type, sample size, population, and signal-to-noise ratio. *IEEE Trans Biomed Eng*, 1999; 46:1293–1303.

Wang LV, Hu S. Photoacoustic tomography: In vivo imaging from organelles to organs. *Science*, March 23, 2012; 335(6075):1458–1462.

Xia J, Yao J, Wang LV. Photoacoustic tomography: Principles and advances. *Electromagn Waves (Camb)*, 2014; 147:1–22.

Yaroslavsky AN, Schulze PC, Yaroslavsky IV, Schober R, Ulrich F, Schwarzmaier HJ. Optical properties of selected native and coagulated human brain tissues in vitro in the visible and near infrared spectral range. *Phys Med Biol*, 2002; 47(12):2059–2073.

Yu Q, Heikal AA. Two-photon autofluorescence dynamics imaging reveals sensitivity of intracellular NADH concentration and conformation to cell physiology at the single-cell level. *J Photochem Photobiol B*, April 2, 2009; 95(1):46–57.

Zhang X, Zheng X, Jiang F, Zhang ZG, Katakowski M, Chopp M. Dual-color fluorescence imaging in a nude mouse orthotopic glioma model. *J Neurosci Methods*, July 30, 2009; 181(2):178–185.

16 Visible Light Optical Imaging and Spectroscopy during Neurosurgery

Sameer Allahabadi, Caroline Jia, and Neal Prakash

CONTENTS

INTRODUCTION

The current, widely used intraoperative neurosurgical brain mapping techniques include electrical stimulation mapping (ESM), electrocorticography (ECoG), microelectrode recordings (MER), and also, less commonly, intraoperative magnetic resonance imaging (MRI). Intraoperative brain mapping with optical imaging using visible light is a technique that first emerged in 1992 (Haglund et al., 1992), but is still not a commonly used technique. Refinements in optical imaging show promise for improved surgical outcomes: (1) from more precise delineation of lesions leading to more complete resections and (2) through improved intraoperative mapping of critical brain functions leading to less postoperative morbidity. Intraoperative optical imaging can utilize either ultraviolet, visible, or near infrared light to examine brain structure and function. Each spectral range has unique *in vivo* imaging substrates, imaging devices, and advantages and limitations. This chapter reviews intraoperative optical imaging and spectroscopy (iOIS), which allows mapping of perfusion and oxygenation changes. iOIS shows potential for (1) precise, cost-effective delineation of epileptic foci, gliomas, and cerebral arteriovenous malformations (AVM) and (2) functional mapping of language and sensorimotor cortices.

Intraoperative brain imaging techniques have revolutionized neurosurgery by allowing surgeons to update real-time information for navigational systems, to monitor tumor resections, to adjust the

TABLE 16.1

Comparison of Utility of Intraoperative Brain Imaging Techniques

Functional Mapping	Tumor Delineation
Electrocorticography	Direct visualization
Microelectrode recordings	Fluorescent imaging
Electrocortical stimulation mapping	Contrast-enhanced ultrasound
(Functional) Magnetic resonance imaging and spectroscopy	
Intraoperative optical imaging and spectroscopy	

approach to intracranial lesions, and to guide functional and drug or cell delivery procedures. Use of intraoperative brain imaging can help avoid inadvertent injury of important anatomic and vascular structures. In addition, complications such as ischemia or hemorrhage can be detected early during the procedures. Intraoperative brain imaging is particularly useful for ensuring that biopsies yield diagnostic tissue, for assessing the completeness of tumor resection, and for creating a more detailed map of the brain. Preoperative and intraoperative functional mapping techniques help surgeons prevent injury to critical brain regions, such as those in the motor and language cortex which could lead to permanent neurological impairment if resected. Theoretically, improvements in current intraoperative techniques would yield enhanced maps, which should increase the success of intraoperative surgeries and reduce unexpected injuries, although data proving improved outcomes are limited. Three current standards for intraoperative imaging are ESM, ECoG, and MER. iOIS is a fairly straightforward technique that has slowly been developed in several independent institutions (Table 16.1).

ELECTROCORTICAL STIMULATION MAPPING

ESM is performed in different stages. The surgeon must open the skull and expose the brain; then a small electrical probe is used to correlate specific tissues with the systemic functions. This allows the surgeon to delineate areas of critical brain function. ESM requires the use of a stimulator device; the stimulation parameters consist of the individual pulse type, pulse width, frequency of stimulation, applied intensity, and the stimulation probe being used to deliver the electric current. For brain mapping, the application of short pulse trains with frequencies of 25–60 Hz has been established as the paradigm. The most commonly applied frequencies are 50 Hz in Europe and 60 Hz in North America. The safest and most reliable stimulator for brain mapping purposes is the constant-current stimulator; they deliver the preset current independently from the impedance of the cortical or subcortical surface. The constant-current stimulator surpasses the constant-voltage stimulators because the latter's delivered current also depends on the impedance. Although ESM is still the golden standard for intraoperative brain mapping, its major drawback is its potential harm on the language cortex. However, when ESM is combined with other preoperative techniques, it is easier to differentiate between the networks within the brain that are indispensable and those that are not (Mandonnet et al., 2010).

ELECTROCORTICOGRAPHY

ECoG records spectral changes in various frequency bands due to normal cortical function during overt or imagined motor activity. ECoG not only provides clinical recordings for epilepsy monitoring on an unparalleled spatiotemporal scale (Kuruvilla and Flink, 2003) but also resolves task-associated spectral changes in high-frequency bands that reflect local cortical activity (Crone et al., 1998; Szurhaj et al., 2005; Leuthardt et al., 2007; Miller et al., 2007, 2010). The greatest methodological challenge involved in activity-based ECoG sensorimotor mapping is having the patient awake and reliably cooperating with a cued series of specific body movements repeatedly (Su and Ojemann, 2013).

MICROELECTRODE RECORDINGS

Intraoperative MER has been introduced to improve the target localization during stereotactic surgery (Lozano et al., 1995; Benazzouz et al., 2002; Krack et al., 2003). This is an invasive technique that can identify a small target population of neurons deep within the brain. Although MER reveals the electrical activity and patterns of different brain substructures, it is difficult to determine whether MER has a positive effect on the clinical outcome and whether the influence that MER has had on the choice of the target has been beneficial. These questions, however, are very difficult to answer since the clinical outcome is determined by many more factors than the use of MER alone.

INTRAOPERATIVE MAGNETIC RESONANCE IMAGING

During surgery, the brain shift phenomenon, after lesion and cerebrospinal fluid removal, challenges the surgeon's ability to complete the resection. Moreover, after resection of most the tumor, it is difficult to differentiate between tumor and normal tissue due to infarction, edema, or iatrogenic contusion (Pamir et al., 2013). Intraoperative (i) MRI is capable of identifying tumor tissues that are not visible to the unaided eye. IMRI is performed using a system combining an MRI scanner and an optical tracking system; this allows for the magnetic field to be limited to the area between the magnets, allowing for radiofrequency shielding to be feasible. Using an ultra-low-field iMRI, one study concluded that iMRI was useful in guiding for lesions in the eloquent area that did not require complete removal, but maximum resection (Livne et al.).

Magnetic resonance spectroscopy (MRS) utilizes the fact that different chemical compositions of tissue contain molecules that resonate at different frequencies (Roder et al., 2014). MRS has great potential in delineating residual tumor from normal brain tissue when performed using 3 Tesla iMRI, which is particularly important near the eloquent cortex. Pamir et al. noted that only around one-third of the T2-hyperintense areas near the resected tumor actually contained residual tumor (Pamir et al., 2013). Thus, it has been recommended that intraoperative MR scanners include MRS in its tumor imaging protocols to ensure that the minimal amount of tissue is removed.

fMRI is a functional neuroimaging technique that is an alternative to ESM and is becoming more effective in facilitating localization and margin determination of extra-axial lesion resections (Livne et al.). It also has several potential advantages, particularly for clinical trials, as it is a noninvasive imaging technique that does not require the injection of contrast agent or radiation exposure and thus can be repeated many times during a longitudinal study. fMRI has relatively high spatial and reasonable temporal resolution and can be acquired in the same session as structural magnetic resonance imaging. Perhaps most importantly, fMRI may provide useful information about the functional integrity of brain networks supporting memory and other cognitive domains, including the neural correlates of specific behavioral events, such as successful versus failed memory formation (Sperling et al., 2010).

A challenge for ifMRI is that a patient under general anesthesia either cannot perform a motor or cognitive task, or has very weak or absent functional signals for sensory stimuli used to generate the fMRI signal. Advances in awake anesthesia methods have allowed for the creation of awake intraoperative fMRI (aifMRI) protocols to delineate the motor, language, and other cognitive areas (Lu et al., 2013). However, there still remain challenges of maintaining patient alertness during the procedure, as well as potentially altered hemodynamic responses due to the anesthetic drug. Thus, although aifMRI is feasible, the effectiveness of this technique relative to other intraoperative mapping techniques is still unclear.

ULTRASOUND

Ultrasound has been known to provide direct visualization of vascularization patterns and tissue resistance for other organs besides the brain. In terms of intraoperative brain surgery, the use of ultrasound has created high resolution images before tumor resection, but its images lack any information in terms of the perfusion of the lesion. Thus, the use of contrast-enhanced ultrasound (CEUS) began as a novel field to provide a real-time direct view onto the vascularization and flow patterns of the brain (Prada et al., 2014). Prada et al. found that with 71 different brain lesions iCEUS was useful in illustrating not only the borders of the tumor, but also the tumor's perfusion patterns. iCEUS aided the surgeons to identify the vascular peduncles correctly, helping the lesion resection process and maximizing the amount of damaged tissue removed (Table 16.2).

OPTICAL IMAGING AND SPECTROSCOPY

Optical imaging and spectroscopy (OIS) techniques can allow for a detailed, noninvasive view of brain structure and function. OIS techniques can use infrared, near-infrared, visible, and ultraviolet light to create detailed images of organs, body tissues, and cells. Each portion of the light spectra has different major absorptive moieties and functional signals *in vivo* (Table 16.3).

INFRARED LIGHT

Infrared imaging has been utilized in animals and humans for an assessment of brain temperature gradients. Initial studies, in which a low-resolution infrared camera was used to image the brain intraoperatively, showed that the temperature of primary brain tumors was lower than that of the surrounding normal tissue (Gorbach et al., 2003). The infrared camera has an array of recording elements that are sensitive to infrared light. Because infrared light is emitted in proportion to absolute temperature, the infrared light detected can be converted to temperature map. This grey scale image can then be played on the video monitor screen; black and white areas represent the lowest temperatures and highest temperatures in the field respectively. Recently, improved infrared cameras that have a finer spatial resolution and greater temperature sensitivity than ones used previously are employed (Kateb et al., 2009). Consequently, as brain tumors induce changes in cerebral blood flow (CBF) in the cortex, they can be made visible with infrared imaging during cranial surgery.

NEAR-INFRARED LIGHT

The largest changes in cortical reflectance in mammals come from the absorption of visible and near-infrared light by hemoglobin. These changes provide an indirect index of underlying neural activity, as neuronal firing increases perfusion to the active area. Fox and Raichle showed neural activity within humans results in both an increase in blood volume and a net increase in hemoglobin oxygenation (Fox and Raichle, 1986). Both of these changes can be visualized with OIS. At isosbestic points where oxyhemoglobin (HbO) and deoxyhemoglobin (HbR) absorption are equal (~422, 452, 500, 530, 548, 570, 586, and 797 nm), decreases in reflection correspond to increases in blood volume, and vice versa. This has been verified by a variety of techniques, including intravascular dye imaging and optical spectroscopy (Narayan et al., 1995; Malonek and Grinvald, 1996). At points where absorption of HbR is dominant, the signal is more sensitive to oxygen extraction. Typically, visible wavelengths between 600 and 660 nm are used, or near-infrared light below 800 nm can also be applied. Imaging at wavelengths that are sensitive to oxygen extraction produces maps that are more spatially correlated with underlying neuronal activity compared to wavelengths at isosbestic points that are influenced by blood volume changes (Frostig et al., 1990; Grinvald et al., 1991; Hodge et al., 1997; Vanzetta and Grinvald, 1999). This is because changes in oxidative metabolism are more tightly coupled to electrical activity than perfusion-related responses (Vanzetta and Grinvald, 1999). Imaging at 600–660 nm also offers the greatest signal-to-noise ratio and, when the imaging is focused beneath the cortical surface, emphasizes blood vessel artifacts the least (Hodge et al., 1997).

Functional near infrared light spectroscopy (fNIRS), a noninvasive technology, measures neurological changes in the hemodynamic state of the brain up to the outer 5–8 mm of the

TABLE 16.2

Advantages and Limitations of Various Intraoperative Brain Mapping Techniques

Technique	Resolution	Advantages	Limitations
Electrical stimulation mapping	Millimeter	Delineates vital tissues using small electrical probes; Can map the motor, sensory, and visual functions of the brain.	Potential harm to language cortex; Induction of transient functional responses; Does not directly identify the noncritical functional parts of the brain
Electrocorticography	Centimeter	Can provide clinical recordings for epilepsy monitoring; Resolves task-associated spectral changes in high-frequency bands; Contains high signal-to-noise ratio and is less susceptible to artifacts; Reveals functional connectivity which advances understanding of large-scale cortical processes; Identifies the noncritical functional regions in the brain	Requires the patient to be awake and cooperate with series of repetitive body movements; Very invasive procedure; Inaccurate across cortical regions with varying depth profiles
Microelectrode readings	Micrometer, but limited field of view	Direct visualization of target centers; Potential for routine movement disorder surgery	Difficult to accurately determine position of microelectrode tip; Impossible to systematically map three-dimensional spatial organization of multiple brain regions, particularly deep structures; Requires special equipment and expertise; Increases the total intraoperative time to perform the recordings.; Procedures must be performed under local anesthetics
Functional magnetic resonance imaging	Millimeter	Expensive; Noncontact/noninvasive	Results may contradict ESM; Pathological features of the tumor influence the accuracy; Images are not as high resolution as typical MRI due a low magnetic field
Magnetic resonance spectroscopy	Centimeter	Capable of identifying tumors in T2-hyperintense areas after tumor resection; Fast and can be included in features of iMRI; Consistent with histological results	Had to accommodate for noise in 10% of patients (Roder et al., 2014)
Contrast enhanced ultrasound	Millimeter	Fast, safe, and relatively inexpensive; Provides dynamic and continuous real-time imaging and characterization of brain lesions; Allows for visualization of afferent and efferent vessels and hyperperfused areas which helps to tailor the tumor removal approach	Individual imaging sections can only analyze one portion of the lesion at a time; Needs to integrate information from B-mode and color Doppler imaging; Ultrasound contrast agent must be precisely timed in order to accurately calculate the transit time
Optical imaging and spectroscopy	Micrometer	Noncontact/noninvasive; Relatively inexpensive	Surface imaging only; Equipment, methodology, and analysis not standardized

TABLE 16.3

Absorptive Moieties *In Vivo* across the Light Spectrum

Moieties with Functional Signal are Listed below in Bold

Ultraviolet	Visible	Near-Infrared	Infrared
Oxidized and Reduced Nicotinamide adenine dinucleotide (NAD+/NADH)			
Tryptophan		**Deoxygenated hemoglobin**	
Tyrosine		**Oxygenated hemoglobin**	
DNA			
Urocanic acid			
Flavins			
Collagen			
		Cytochrome C oxidase	**Heat**
			Lipids
Water			
Elastin			
Melanin			

brain cortex (Boas et al., 2004). Most fNIRS employ continuous wave systems and time-resolved systems (Scholkmann et al., 2014). Thus, fNIRS is relatively inexpensive compared to fMRI techniques and contains less noise, as it is not dependent on the time of flight. Utilizing a cap containing flexible fiber optic cables, fNIRS ejects low levels of nonionizing red (690 nm) and near-infrared (830 nm) light into the brain (Huppert et al., 2013). The fiber optic cable records to the detector the amount of light the brain absorbs at each wavelength. These light intensities can then be converted to associated levels of oxygenated and deoxygenated hemoglobin. fNIRS has the advantage over fMRI because it does not limit the position of the patient and allows for better testing of the motor cortex. In clinical settings, fNIRS is a reliable technique in group tests and is improving in being reliable in single patients as well. Some difficulties impeding single patient testing include the superficial tissue that light has to penetrate through to reach the brain, systemic physiological changes which could alter brain signals, and imitations in technology that lead to a problematic signal-to-noise ratio (Scholkmann et al., 2014).

Ultraviolet Light

In the 1970s, it was discovered that ultraviolet light excites NAD(P)H fluorescence in brain slices (Lipton, 1973) and *in vivo* (Jobsis et al., 1971; Rosenthal and Jobsis, 1971). The main disadvantage of ultraviolet light is its phototoxicity. More recently, mitochondrial flavoprotein fluorescence, excited with blue light, has been used to generate functional signals, both *in vitro* (Shuttleworth et al., 2003; Brennan et al., 2006) and *in vivo* (Shibuki et al., 2003; Reinert et al., 2004; Husson et al., 2007). Flavoprotein fluorescence signals have a similar time course to hemodynamic OIS signals, but they reportedly have greater signal-to-noise ratio and are more spatially localized (Tohmi et al., 2006; Husson et al., 2007). These methods involve simple modifications to conventional imaging apparatus (excitation and emission filters, dichroic) (Toga et al., 1995; Brennan et al., 2006).

Numerous exogenous fluorescent agents, such as 5-aminolevulinic acid and fluorescein, have also been used to help improve visualization and guidance during gross total resection of various brain tumors (Valdes et al., 2016). Targeted fluorescent dyes typically are given intravenously and taken up preferentially by cancer cells. Intraoperatively, they are excited by ultraviolet light and emit light in the visible spectrum, enabling delineation of the brain tumor. Ongoing evolution of the various dyes and technologies will continue to improve the sensitivity and specificity of this technique as well as the potential toxicities from the dye and tissue damage from ultraviolet light exposure.

TABLE 16.4

Advantages and Limitations of Optical Imaging with Different Light Spectra

	Advantages	Limitations
Infrared light	Does not require altering surgical tumor removal techniques	Lack of software that could autodelineate the tumor margin
	Does not require use of chemical contrast medium because warm blood is clearly differentiated from normal cortical surface	Higher resolution camera than what we used in this case; report is needed for the future studies
	Operative procedure does not have to be interrupted for imaging	Less than 0.5 mm IR penetration depth (Kateb et al., 2009)
		Nonstandardized analysis
Near infrared	Includes both electronic and vibrational spectroscopy	Incomplete knowledge of how light propagates inside the head
	Can be applied to various types of materials	
Visible light	Creates HbO, HbR, and HbT maps	Nonstandardized analysis
	Spatial resolution as high as 50–100 μm	
Ultraviolet light	Detects NAD+ and NADH fluorescence	Phototoxicity
	Mitochondrial flavoprotein fluorescence, used to generate functional signals, have greater signal-to-noise ratio and is more spatially localized	Nonstandardized analysis
		Limited depth penetration

VISIBLE LIGHT

Hemoglobin is the major source of light absorption in the visible spectrum of exposed living brain. In the green spectrum, HbO and HbR have identical absorption at four different isosbestic points; hence maps with green-light OIS are largely maps of total-hemoglobin (HbT) changes. In the red spectrum, HbR absorbs light much more than HbO, hence maps with red-light OIS are largely maps of HbR changes (Pouratian et al., 2002). The final maps are generated by comparing pictures of the brain at rest versus when stimulated. OIS exploits a trait that is inherent to vascularized tissues, especially in the cerebral cortex—when illuminated with light, the active cortex and its associated vasculature exhibit changes in light reflectance relative to inactive areas. Hybrid techniques such as 2DOS (Prakash et al., 2007; Sheth et al., 2009) and hyperspectral imaging (Mori et al., 2014) may offer an ideal balance between spatial, spectral, and temporal resolution (Table 16.4).

HISTORY OF VISIBLE LIGHT OPTICAL IMAGING AND SPECTROSCOPY

The history of OIS with visible light began with voltage-sensitive dyes (VSDs). The first VSDs applied to invertebrate preparations showed that it was possible to optically record fast membrane potential changes like action potentials in single neurons (Tasaki et al., 1968; Salzberg et al., 1973). These pioneering recordings inspired further research in the next decade, most of which has targeted measurements of the intact brain during its processing of sensory information. The first *in vivo* VSD recordings of sensory processing in the vertebrate brain were obtained in the optic tectum of the frog (Grinvald et al., 1984) and subsequently in the rat somatosensory and visual cortex (Orbach et al., 1985). These early investigations reported complications due to small signal amplitudes, large contaminating hemodynamic noise, phototoxicity, and limitations in detector and computer technology. The hemodynamic "noise" was then recognized to be a potential mapping signal in a groundbreaking paper (Grinvald et al., 1986).

Since 1986, activity-dependent light reflectance patterns, occurring without dyes or tracers and have characteristic spatiotemporal features, were recognized as "*intrinsic signals*" (Prakash and Frostig, 2005; Prakash et al., 2007, 2009). Two main sources of intrinsic signals exist in the visible light spectrum (Frostig et al., 1990; Malonek and Grinvald, 1996; Mayhew et al., 2000; Sheth et al., 2004).

With green-yellow light (~500–599 nm), HbT is the dominant source of the intrinsic signal. This signal is typically a monophasic response peaking in the capillary bed 3–5 s post stimulation, representing increased HbT from functional hyperemia (Frostig et al., 1990; Malonek and Grinvald, 1996; Mayhew et al., 2000; Sheth et al., 2004). With red light (~600–699 nm), HbR is the dominant source of the intrinsic signal. This signal is typically a biphasic response in the capillary bed, with an early phase peaking 0.5–1.5 s post stimulation representing increased HbR (Frostig et al., 1990; Malonek and Grinvald, 1996; Mayhew et al., 2000; Sheth et al., 2004) via increased cerebral metabolic rate of oxygen (Thompson et al., 2003), followed by a late phase peaking 3–5 s poststimulation from functional hyperemia.

INTRAOPERATIVE OPTICAL IMAGING OF INTRINSIC SIGNALS

In 1992, the first reported case of consented intraoperative optical imaging with visible light added another technique available to the neurosurgical arsenal (Haglund et al., 1992). The selection of the appropriate intraoperative imaging tool depends on the neurosurgical procedure performed (Lunsford et al., 1996). iOIS is a less harmful and more precise way of mapping brain function and imaging brain tumors, such as glioblastomas (GBMs). iOIS is a brain mapping technique that can visualize brain compartments with micrometer and millisecond resolution (Prakash and Frostig, 2005; Prakash et al., 2007, 2009) OIS requires an optically visible brain, and as such, it has been used primarily in animals for neurovascular research (Grinvald et al., 1986, 1991; Frostig et al., 1990; Prakash et al., 1996, 2000, 2009; Pouratian et al., 2002) but also in humans intraoperatively (Haglund et al., 1992; Toga et al., 1995; Cannestra et al., 1996, 1998a,b, 2000, 2001, 2004, ; Pouratian et al., 2000; Shoham and Grinvald, 2001; Sato et al., 2002, 2005; Schwartz et al., 2004; Nariai et al., 2005). Specifically, iOIS has been demonstrated to be a potentially useful neurosurgical tool for both functional brain mapping (Haglund et al., 1992; Toga et al., 1995; Cannestra et al., 2000, 2001; Lin et al., 2001; Nariai et al., 2005; Schwartz, 2005) and lesion delineation (Cannestra et al., 2004; Toms et al., 2005; Popescu and Toms, 2006). iOIS maps are a unique tool as they can correlate extremely well with ESM. Studies have shown the ability to regularly produce optical maps that are adjacent to vascular lesions. iOIS offers a resolution as high as 50–100 μm as demonstrated by numerous studies, potentially allowing for a much finer delineation of eloquence. iOIS is performed in the operating room by attaching a charge-coupled device (CCD) camera and optical filter(s) to the operating scope. In essence, iOIS improves the neurosurgeon's eyesight by creating functional maps and delineating cortical lesions, by extracting the most relevant information from reflected light arriving in the operating scope. iOIS has been slowly evolving, as technical challenges, including high degrees of movement and noise in the operating room environment, are being overcome.

KEY ADVANCEMENTS IN VISIBLE LIGHT INTRAOPERATIVE OPTICAL IMAGING AND SPECTROSCOPY

A recent study performed by Sobottka et al. in 41 patients with tumor lesions have found that prolonged stimulation and rest periods of 30 s allow for complications, including noise and moment of cortex to be reduced (Sobottka et al., 2013a). Additionally, the effects of brain movement have also been accounted for by algorithms that calculate the differences in pixel gradients between that of the reference image and that of the deformed image. Robust maps, with a lower signal-to-ratio, can thus, be generated. Furthermore, Sobottka et al. has demonstrated the flexibility of optical intraoperative imaging by applying it for the first time, in the human visual cortex (Sobottka et al., 2013b).

iOIS can be used as an alternative to other methods for the identification of sensory cortex areas and offers the added benefit of a micrometer map of functional activity. It has great potential for visualizing and monitoring additional specific functional brain areas, such as the visual, motor, and speech cortex. iOIS may also add more information regarding neurovascular function and dysfunction, than electric activity alone. It is a noncontact imaging technique and introduces no potentially harmful compounds. Moreover, mapping is relatively rapid (Pouratian et al., 2000). Further *in vitro* studies may also broaden our understanding of neuronal–glial optical signals (Pal et al., 2013) and better assist intraoperative pathological diagnosis in the future. Further studies are needed comparing intraoperative imaging techniques and examination of their impacts on surgical outcomes.

TWO-DIMENSIONAL VISIBLE LIGHT OPTICAL SPECTROSCOPIC METHODS

Newer variants in OIS include two-dimensional optical spectroscopy (2DOS) (Prakash et al., 2007; Sheth et al., 2009) and hyperspectral imaging (Mori et al., 2014). These hybrid techniques optimize spatial, spectral, and temporal resolutions for potentially improved intraoperative imaging. They show promise for impacting neurosurgical procedures by being capable of allowing noncontact, rapid delineation of brain function and/or tumor delineation (Figures 16.1 and 16.2).

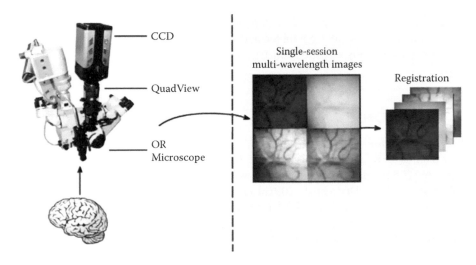

FIGURE 16.1 Setup for i2DOS. With single wavelength intraoperative optical imaging, a charge-coupled device (CCD) is mounted to the standard operating room microscope with a single optical filter. 2DOS is performed by adding a series of beam splitters with four different optical filters (QuadView); simultaneous acquisition of four images from four different wavelengths of visible light is possible. The four subimages can then be registered and analyzed to calculate maps of blood volume and blood oxygenation (Prakash et al., 2007, 2009; Sheth et al., 2009). (Data from Prakash, N. et al., *NeuroImage*, 47, T116, 2007.)

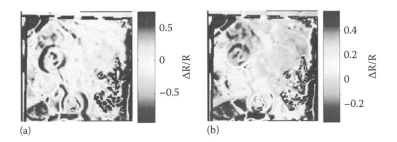

FIGURE 16.2 Example of somatosensory maps of finger stimulation with i2DOS. (a) the HbT map shows maximal perfusion changes along venules and arterioles in the upper left portion of the image, which matched the location of the finger mapped by ECoG (under electrode "2"). (b) the HbR map shows maximal deoxygenation changes in a similar location, but more distributed in the capillary bed. (The diameter of the numbered electrodes is 5 mm.)

CONCLUSIONS

OIS offers several potential advantages over the current gold standard for intraoperative mapping, ESM. The spatial resolution of iOIS is much greater than that of ESM. Resection within 1 cm of essential areas identified by ESM increases the likelihood of postoperative deficits (Haglund et al., 1992), suggesting the limit of resolution of ESM is on the order of 1 cm. This limited resolution does not ensure resection of as much pathological tissue as possible. iOIS, on the other hand, can potentially provide a more detailed functional map of the exposed cortex, with consequently more precise resections. The iOIS setup requires no modification of the traditional operating room configuration and does not require any major modification to neurosurgical techniques, as has been necessary for the implementation of iMRI. Furthermore, the camera used for optical imaging and spectroscopy can be mounted atop a second operating microscope so that it can be moved into and out of the sterile operating field as needed, minimizing surgical interference. iOIS is also advantageous because it is not in contact with the brain during surgery. Since iOIS only detects changes in light reflectance, it does not require any contact with potentially normal tissue that the surgeon may not want to disturb. Moreover, OIS equipment is affordable since it only requires four major components: camera, light source, computer, and a spectroscopic filter. iOIS dramatically improves a neurosurgeon's eyesight by creating functional and lesion maps from images from the surgical microscope.

FUTURE DIRECTIONS

Many novel methods for visible light intraoperative imaging such as 2DOS, multispectral imaging, and laser imaging show great promise for the future. Laser-based techniques, specifically, laser speckle imaging (LSI) has excellent spatiotemporal resolution for surface imaging of cerebral blood flow (CBF) (Briers, 2001; Dunn et al., 2001; Liu et al., 2005). OCT (Giese et al., 2006) and optical Doppler tomography (Chen et al., 1999) have exquisite three-dimensional spatial resolution. OCT has been validated for intraoperative detection of residual tumor during the resection of human gliomas (Giese et al., 2006; Finke et al., 2012). Newer imaging technologies may allow recordings across the ultraviolet to infrared spectrum, such that changes in NADH, NAD+, oxyhemoglobin, deoxyhemoglobin, water, and heat can all be performed simultaneously, potentially allowing superior functional mapping and tumor delineation.

REFERENCES

Benazzouz, A., Breit, S., Koudsie, A., Pollak, P., Krack, P., and Benabid, A. L. 2002. Intraoperative microrecordings of the subthalamic nucleus in Parkinson's disease. *Mov Disord* 17(Suppl 3):S145–S149.

Boas, D. A., Dale, A. M., and Franceschini, M. A. 2004. Diffuse optical imaging of brain activation: Approaches to optimizing image sensitivity, resolution, and accuracy. *Neuroimage* 23(Suppl 1):S275–S288. doi: 10.1016/j.neuroimage.2004.07.011.

Brennan, A. M., Connor, J. A., and Shuttleworth, C. W. 2006. NAD(P)H fluorescence transients after synaptic activity in brain slices: Predominant role of mitochondrial function. *J Cereb Blood Flow Metab* 26(11):1389–1406. doi: 10.1038/sj.jcbfm.9600292.

Briers, J. D. 2001. Laser Doppler, speckle and related techniques for blood perfusion mapping and imaging. *Physiol Meas* 22(4):R35–R66.

Cannestra, A. F., Black, K. L., Martin, N. A., Cloughesy, T., Burton, J. S., Rubinstein, E., Woods, R. P., and Toga, A. W. 1998a. Topographical and temporal specificity of human intraoperative optical intrinsic signals. *Neuroreport* 9(11):2557–2563.

Cannestra, A. F., Blood, A. J., Black, K. L., and Toga, A. W. 1996. The evolution of optical signals in human and rodent cortex. *Neuroimage* 3(3 Pt 1):202–208. doi: 10.1006/nimg.1996.0022.

Cannestra, A. F., Bookheimer, S. Y., Pouratian, N., O'Farrell, A., Sicotte, N., Martin, N. A., Becker, D., Rubino, G., and Toga, A. W. 2000. Temporal and topographical characterization of language cortices using intraoperative optical intrinsic signals. *Neuroimage* 12(1):41–54. doi: 10.1006/nimg.2000.0597.

Cannestra, A. F., Pouratian, N., Bookheimer, S. Y., Martin, N. A., Beckerand, D. P., and Toga, A. W. 2001. Temporal spatial differences observed by functional MRI and human intraoperative optical imaging. *Cereb Cortex* 11(8):773–782.

Cannestra, A. F., Pouratian, N., Forage, J., Bookheimer, S. Y., Martin, N. A., and Toga, A. W. 2004. Functional magnetic resonance imaging and optical imaging for dominant-hemisphere perisylvian arteriovenous malformations. *Neurosurgery* 55(4):804–812; discussion 812-4.

Cannestra, A. F., Pouratian, N., Shomer, M. H., and Toga, A. W. 1998b. Refractory periods observed by intrinsic signal and fluorescent dye imaging. *J Neurophysiol* 80(3):1522–1532.

Chen, Z., Zhao, Y., Srinivas, S. M., Nelson, J. S., Prakash, N., and Frostig, R. D. 1999. Optical Doppler tomography. *IEEE J Sel Top Quantum Electron* 5(4):1134–1142.

Crone, N. E., Miglioretti, D. L., Gordon, B., and Lesser, R. P. 1998. Functional mapping of human sensorimotor cortex with electrocorticographic spectral analysis. II. Event-related synchronization in the gamma band. *Brain* 121(12):2301–2315.

Dunn, A. K., Bolay, H., Moskowitz, M. A., and Boas, D. A. 2001. Dynamic imaging of cerebral blood flow using laser speckle. *J Cereb Blood Flow Metab* 21(3):195–201. doi: 10.1097/00004647-200103000-00002.

Finke, M., Kantelhardt, S., Schlaefer, A., Bruder, R., Lankenau, E., Giese, A., and Schweikard, A. 2012. Automatic scanning of large tissue areas in neurosurgery using optical coherence tomography. *Int J Med Robot* 8(3):327–336. doi: 10.1002/rcs.1425.

Fox, P. T. and Raichle, M. E. 1986. Focal physiological uncoupling of cerebral blood flow and oxidative metabolism during somatosensory stimulation in human subjects. *Proc Natl Acad Sci U S A* 83(4):1140–1144.

Frostig, R. D., Lieke, E. E., Ts'o, D. Y., and Grinvald, A. 1990. Cortical functional architecture and local coupling between neuronal activity and the microcirculation revealed by in vivo high-resolution optical imaging of intrinsic signals. *Proc Natl Acad Sci U S A* 87(16):6082–6086.

Giese, A., Böhringer, H. J., Leppert, J., Kantelhardt, S. R., Lankenau, E., Koch, P., Birngruber, R., and Hüttmann, G. 2006. Non-invasive intraoperative optical coherence tomography of the resection cavity during surgery of intrinsic brain tumors. Paper read at *Biomedical Optics 2006, Photonic Therapeutics and Diagnostics II, SPIE*, San Jose, CA.

Gorbach, A. M., Heiss, J., Kufta, C., Sato, S., Fedio, P., Kammerer, W. A., Solomon, J., and Oldfield, E. H. 2003. Intraoperative infrared functional imaging of human brain. *Ann Neurol* 54(3):297–309. doi: 10.1002/ana.10646.

Grinvald, A., Anglister, L., Freeman, J. A., Hildesheim, R., and Manker, A. 1984. Real-time optical imaging of naturally evoked electrical activity in intact frog brain. *Nature* 308(5962):848–850.

Grinvald, A., Frostig, R. D., Siegel, R. M., and Bartfeld, E. 1991. High-resolution optical imaging of functional brain architecture in the awake monkey. *Proc Natl Acad Sci U S A* 88(24):11559–11563.

Grinvald, A., Lieke, E., Frostig, R. D., Gilbert, C. D., and Wiesel, T. N. 1986. Functional architecture of cortex revealed by optical imaging of intrinsic signals. *Nature* 324(6095):361–364. doi: 10.1038/324361a0.

Haglund, M. M., Ojemann, G. A., and Hochman, D. W. 1992. Optical imaging of epileptiform and functional activity in human cerebral cortex. *Nature* 358(6388):668–671. doi: 10.1038/358668a0.

Hodge, C. J., Jr., Stevens, R. T., Newman, H., Merola, J., and Chu, C. 1997. Identification of functioning cortex using cortical optical imaging. *Neurosurgery* 41(5):1137–1144; discussion 1144–1145.

Huppert, T., Schmidt, B., Beluk, N., Furman, J., and Sparto, P. 2013. Measurement of brain activation during an upright stepping reaction task using functional near-infrared spectroscopy. *Hum Brain Mapp* 34(11):2817–2828. doi: 10.1002/hbm.22106.

Husson, T. R., Mallik, A. K., Zhang, J. X., and Issa, N. P. 2007. Functional imaging of primary visual cortex using flavoprotein autofluorescence. *J Neurosci* 27(32):8665–8675. doi: 10.1523/jneurosci.2156-07.2007.

Jobsis, F. F., O'Connor, M., Vitale, A., and Vreman, H. 1971. Intracellular redox changes in functioning cerebral cortex. I. Metabolic effects of epileptiform activity. *J Neurophysiol* 34(5):735–749.

Kateb, B., Yamamoto, V., Yu, C., Grundfest, W., and Gruen, J. P. 2009. Infrared thermal imaging: A review of the literature and case report. *NeuroImage* 47(Suppl 2):T154–T162. doi: 10.1016/j.neuroimage.2009.03.043.

Krack, P., Batir, A., Van Blercom, N., Chabardes, S., Fraix, V., Ardouin, C., Koudsie, A. et al. 2003. Five-year follow-up of bilateral stimulation of the subthalamic nucleus in advanced Parkinson's disease. *N Engl J Med* 349(20):1925–1934. doi: 10.1056/NEJMoa035275.

Kuruvilla, A. and Flink, R. 2003. Intraoperative electrocorticography in epilepsy surgery: Useful or not? *Seizure* 12(8):577–584. doi: 10.1016/S1059-1311(03)00095-5.

Leuthardt, E. C., Miller, K., Anderson, N. R., Schalk, G., Dowling, J., Miller, J., Moran, D. W., and Ojemann, J. G. 2007. Electrocorticographic frequency alteration mapping: A clinical technique for mapping the motor cortex. *Neurosurgery* 60(4):260–271. doi: 10.1227/01.NEU.0000255413.70807.6E.

Lin, W. C., Toms, S. A., Johnson, M., Jansen, E. D., and Mahadevan-Jansen, A. 2001. In vivo brain tumor demarcation using optical spectroscopy. *Photochem Photobiol* 73(4):396–402.

Lipton, P. 1973. Effects of membrane depolarization on nicotinamide nucleotide fluorescence in brain slices. *Biochem J* 136(4):999–1009.

Liu, Q., Wang, Z., and Luo, Q. 2005. Temporal clustering analysis of cerebral blood flow activation maps measured by laser speckle contrast imaging. *J Biomed Opt* 10(2):024019. doi: 10.1117/1.1891105.

Livne, O., Harhel, R., Hadani, M., Spiegelmann, R., Feldman, Z., and Cohen, Z. R. 2014. Intraoperative magnetic resonance imaging for resection of intra-axial brain lesions: A decade of experience using low-field magnetic resonance imaging, Polestar N-10, 20, 30 systems. *World Neurosurg* 82(5):770–776. doi: 10.1016/j.wneu.2014.02.004.

Lozano, A. M., Hutchison, W. D., and Dostrovsky, J. O. 1995. Microelectrode monitoring of cortical and subcortical structures during stereotactic surgery. *Acta Neurochir Suppl* 64:30–34.

Lu, J.-F., Zhang, H., Wu, J.-S., Yao, C.-J., Zhuang, D. X., Qiu, T.-M., Jia, W.-B., Mao, Y., and Zhou, L. F. 2013. "Awake" intraoperative functional MRI (ai-fMRI) for mapping the eloquent cortex: Is it possible in awake craniotomy? *NeuroImage Clin* 2:132–142. doi: 10.1016/j.nicl.2012.12.002.

Lunsford, L. D., Kondziolka, D., and Bissonette, D. J. 1996. Intraoperative imaging of the brain. *Stereotact Funct Neurosurg* 66(1–3):58–64.

Malonek, D. and Grinvald, A. 1996. Interactions between electrical activity and cortical microcirculation revealed by imaging spectroscopy: Implications for functional brain mapping. *Science* 272(5261):551–554.

Mandonnet, E., Winkler, P. A., and Duffau, H. 2010. Direct electrical stimulation as an input gate into brain functional networks: Principles, advantages and limitations. *Acta Neurochir (Wien)* 152(2):185–193. doi: 10.1007/s00701-009-0469-0.

Mayhew, J., Johnston, D., Berwick, J., Jones, M., Coffey, P., and Zheng, Y. 2000. Spectroscopic analysis of neural activity in brain: Increased oxygen consumption following activation of barrel cortex. *Neuroimage* 12(6):664–675. doi: 10.1006/nimg.2000.0656.

Miller, K. J., Leuthardt, E. C., Schalk, G., Rao, R. P. N., Anderson, N. R., Moran, D. W., Miller, J. W., and Ojemann, J. G. 2007. Spectral changes in cortical surface potentials during motor movement. *J Neurosci* 27(9):2424–2432.

Miller, K. J., Schalk, G., Fetz, E. E., den Nijs, M., Ojemann, J. G., and Rao, R. P. N. 2010. Cortical activity during motor execution, motor imagery, and imagery-based online feedback. *Proc Natl Acad Sci* 107(9):4430–4435. doi: 10.1073/pnas.0913697107.

Mori, M., Chiba, T., Nakamizo, A., Kumashiro, R., Murata, M., Akahoshi, T., Tomikawa, M. et al. 2014. Intraoperative visualization of cerebral oxygenation using hyperspectral image data: A two-dimensional mapping method. *Int J Comput Assist Radiol Surg* 9(6):1059–1072. doi: 10.1007/s11548-014-0989-9.

Narayan, S. M., Esfahani, P., Blood, A. J., Sikkens, L., and Toga, A. W. 1995. Functional increases in cerebral blood volume over somatosensory cortex. *J Cereb Blood Flow Metab* 15(5):754–765. doi: 10.1038/jcbfm.1995.99.

Nariai, T., Sato, K., Hirakawa, K., Ohta, Y., Tanaka, Y., Ishiwata, K., Ishii, K., Kamino, K., and Ohno, K. 2005. Imaging of somatotopic representation of sensory cortex with intrinsic optical signals as guides for brain tumor surgery. *J Neurosurg* 103(3):414–423. doi: 10.3171/jns.2005.103.3.0414.

Orbach, H. S., Cohen, L. B., and Grinvald, A. 1985. Optical mapping of electrical activity in rat somatosensory and visual cortex. *J Neurosci* 5(7):1886–1895.

Pal, I., Nyitrai, G., Kardos, J., and Heja, L. 2013. Neuronal and astroglial correlates underlying spatiotemporal intrinsic optical signal in the rat hippocampal slice. *PLoS One* 8(3):e57694. doi: 10.1371/journal.pone.0057694.

Pamir, M. N., Özduman, K., Yıldız, E., Sav, A., and Dinçer, A. 2013. Intraoperative magnetic resonance spectroscopy for identification of residual tumor during low-grade glioma surgery. *J Neurosurg* 118(6):1191–1198. doi: 10.3171/2013.1.JNS111561.

Popescu, M. A. and Toms, S. A. 2006. In vivo optical imaging using quantum dots for the management of brain tumors. *Expert Rev Mol Diagn* 6(6):879–890. doi: 10.1586/14737159.6.6.879.

Pouratian, N., Bookheimer, S. Y., O'Farrell, A. M., Sicotte, N. L., Cannestra, A. F., Becker, D., and Toga, A. W. 2000. Optical imaging of bilingual cortical representations. Case report. *J Neurosurg* 93(4):676–681. doi: 10.3171/jns.2000.93.4.0676.

Pouratian, N., Bookheimer, S. Y., Rex, D. E., Martin, N. A., and Toga, A. W. 2002. Utility of preoperative functional magnetic resonance imaging for identifying language cortices in patients with vascular malformations. *J Neurosurg* 97(1):21–32. doi: 10.3171/jns.2002.97.1.0021.

Prada, F., Perin, A., Martegani, A., Aiani, L., Solbiati, L., Lamperti, M., Casali, C. et al. 2014. Intraoperative contrast-enhanced ultrasound for brain tumor surgery. *Neurosurgery* 74(5):542–552; discussion 552. doi: 10.1227/neu.0000000000000301.

Prakash, N., Biag, J. D., Sheth, S. A., Mitsuyama, S., Theriot, J., Ramachandra, C., and Toga, A. W. 2007. Temporal profiles and 2-dimensional oxy-, deoxy-, and total-hemoglobin somatosensory maps in rat versus mouse cortex. *Neuroimage* 37(Suppl 1):S27–S36. doi: 10.1016/j.neuroimage.2007.04.063.

Prakash, N., Cohen-Cory, S., and Frostig, R. D. 1996. RAPID and opposite effects of BDNF and NGF on the functional organization of the adult cortex in vivo. *Nature* 381(6584):702–706. doi: 10.1038/381702a0.

Prakash, N. and Frostig, R. D. 2005. What has intrinsic signal optical imaging taught us about NGF-induced rapid plasticity in adult cortex and its relationship to the cholinergic system? *Mol Imaging Biol* 7(1):14–21. doi: 10.1007/s11307-005-0956-5.

Prakash, N., Vanderhaeghen, P., Cohen-Cory, S., Frisen, J., Flanagan, J. G., and Frostig, R. D. 2000. Malformation of the functional organization of somatosensory cortex in adult ephrin-A5 knock-out mice revealed by in vivo functional imaging. *J Neurosci* 20(15):5841–5847.

Prakash, N., Uhlemann, F., Sheth, S. A., Bookheimer, S., Martin, N., and Toga, A. W. 2009. Current trends in intraoperative optical imaging for functional brain mapping and delineation of lesions of language cortex. *NeuroImage* 47(Suppl 2):T116–T126. doi: http://dx.doi.org/10.1016/j.neuroimage.2008.07.066.

Reinert, K. C., Dunbar, R. L., Gao, W., Chen, G., and Ebner, T. J. 2004. Flavoprotein autofluorescence imaging of neuronal activation in the cerebellar cortex in vivo. *J Neurophysiol* 92(1):199–211. doi: 10.1152/jn.01275.2003.

Roder, C., Skardelly, M., Ramina, K. F., Beschorner, R., Honneger, J., Nagele, T., Tatagiba, M. S., Ernemann, U., and Bisdas, S. 2014. Spectroscopy imaging in intraoperative MR suite: Tissue characterization and optimization of tumor resection. *Int J Comput Assist Radiol Surg* 9(4):551–559. doi: 10.1007/s11548-013-0952-1.

Rosenthal, M. and Jobsis, F. F. 1971. Intracellular redox changes in functioning cerebral cortex. II. Effects of direct cortical stimulation. *J Neurophysiol* 34(5):750–762.

Salzberg, B. M., Davila, H. V., and Cohen, L. B. 1973. Optical recording of impulses in individual neurones of an invertebrate central nervous system. *Nature* 246(5434):508–509.

Sato, K., Nariai, T., Sasaki, S., Yazawa, I., Mochida, H., Miyakawa, N., Momose-Sato, Y. et al. 2002. Intraoperative intrinsic optical imaging of neuronal activity from subdivisions of the human primary somatosensory cortex. *Cereb Cortex* 12(3):269–280.

Sato, K., Nariai, T., Tanaka, Y., Maehara, T., Miyakawa, N., Sasaki, S., Momose-Sato, Y., and Ohno, K. 2005. Functional representation of the finger and face in the human somatosensory cortex: Intraoperative intrinsic optical imaging. *Neuroimage* 25(4):1292–1301. doi: 10.1016/j.neuroimage.2004.12.049.

Scholkmann, F., Kleiser, S., Metz, A. J., Zimmermann, R., Pavia, J. M., Wolf, U., and Wolf, M. 2014. A review on continuous wave functional near-infrared spectroscopy and imaging instrumentation and methodology. *NeuroImage* 85(Part 1):6–27. doi: 10.1016/j.neuroimage.2013.05.004.

Schwartz, T. H. 2005. The application of optical recording of intrinsic signals to simultaneously acquire functional, pathological and localizing information and its potential role in neurosurgery. *Stereotact Funct Neurosurg* 83(1):36–44. doi: 10.1159/000085025.

Schwartz, T. H., Chen, L. M., Friedman, R. M., Spencer, D. D., and Roe, A. W. 2004. Intraoperative optical imaging of human face cortical topography: A case study. *Neuroreport* 15(9):1527–1531.

Sheth, S. A., Nemoto, M., Guiou, M., Walker, M., Pouratian, N., Hageman, N., and Toga, A. W. 2004. Columnar specificity of microvascular oxygenation and volume responses: Implications for functional brain mapping. *J Neurosci* 24(3):634–641. doi: 10.1523/jneurosci.4526-03.2004.

Sheth, S. A., Prakash, N., Guiou, M., and Toga, A. W. 2009. Validation and visualization of two-dimensional optical spectroscopic imaging of cerebral hemodynamics. *NeuroImage* 47(Suppl 2):T36–T43. doi: 10.1016/j.neuroimage.2008.09.060.

Shibuki, K., Hishida, R., Murakami, H., Kudoh, M., Kawaguchi, T., Watanabe, M., Watanabe, S., Kouuchi, T., and Tanaka, R. 2003. Dynamic imaging of somatosensory cortical activity in the rat visualized by flavoprotein autofluorescence. *J Physiol* 549(Pt 3):919–927. doi: 10.1113/jphysiol.2003.040709.

Shoham, D. and A. Grinvald. 2001. The cortical representation of the hand in macaque and human area S-I: High resolution optical imaging. *J Neurosci* 21(17):6820–6835.

Shuttleworth, C. W., Brennan, A. M., and Connor, J. A. 2003. NAD(P)H fluorescence imaging of postsynaptic neuronal activation in murine hippocampal slices. *J Neurosci* 23(8):3196–3208.

Sobottka, S. B., Meyer, T., Kirsch, M., Koch, E., Steinmeier, R., Morgenstern, U., and Schackert, G. 2013a. Intraoperative optical imaging of intrinsic signals: A reliable method for visualizing stimulated functional brain areas during surgery. *J Neurosurg* 119(4):853–863. doi: 10.3171/2013.5.jns122155.

Sobottka, S. B., Meyer, T., Kirsch, M., Reiss, G., Koch, E., Morgenstern, U., and Schackert, G. 2013b. Assessment of visual function during brain surgery near the visual cortex by intraoperative optical imaging. *Biomed Tech (Berl)* 58(3):249–256. doi: 10.1515/bmt-2012-0074.

Sperling, R. A., Dickerson, B. C., Pihlajamaki, M., Vannini, P., LaViolette, P. S., Vitolo, O. V., Hedden, T. et al. 2010. Functional alterations in memory networks in early Alzheimer's disease. *Neuromolecular Med* 12(1):27–43. doi: 10.1007/s12017-009-8109-7.

Su, D. K. and Ojemann, J. G. 2013. Electrocorticographic sensorimotor mapping. *Clin Neurophysiol* 124(6):1044–1048. doi: 10.1016/j.clinph.2013.02.114.

Szurhaj, W., Bourriez, J.-L., Kahane, P., Chauvel, P., Mauguière, F., and Derambure, P. 2005. Intracerebral study of gamma rhythm reactivity in the sensorimotor cortex. *Eur J Neurosci* 21(5):1223–1235.

Tasaki, I., Watanabe, A., Sandlin, R., and Carnay, L. 1968. Changes in fluorescence, turbidity, and birefringence associated with nerve excitation. *Proc Natl Acad Sci U S A* 61(3):883–888.

Thompson, J. K., Peterson, M. R., and Freeman, R. D. 2003. Single-neuron activity and tissue oxygenation in the cerebral cortex. *Science* 299(5609):1070–1072. doi: 10.1126/science.1079220.

Toga, A. W., Cannestra, A. F., and Black, K. L. 1995. The temporal/spatial evolution of optical signals in human cortex. *Cereb Cortex* 5(6):561–565.

Tohmi, M., Kitaura, H., Komagata, S., Kudoh, M., and Shibuki, K. 2006. Enduring critical period plasticity visualized by transcranial flavoprotein imaging in mouse primary visual cortex. *J Neurosci* 26(45):11775–11785. doi: 10.1523/jneurosci.1643-06.2006.

Toms, S. A., Lin, W. C., Weil, R. J., Johnson, M. D., Jansen, E. D., and Mahadevan-Jansen, A. 2005. Intraoperative optical spectroscopy identifies infiltrating glioma margins with high sensitivity. *Neurosurgery* 57(Suppl 4):382–391; discussion 382–391.

Valdes, P. A., Roberts, D. W., Lu, F. K., and Golby, A. 2016. Optical technologies for intraoperative neurosurgical guidance. *Neurosurg Focus* 40(3):E8. doi: 10.3171/2015.12.focus15550.

Vanzetta, I. and Grinvald, A. 1999. Increased cortical oxidative metabolism due to sensory stimulation: Implications for functional brain imaging. *Science* 286(5444):1555–1558.

17 Intraoperative Infrared Optical Imaging in Neurosurgery

Michael E. Wolf, Richard P. Menger, Osama Ahmed, Shahdad Sherkat, and Babak Kateb

CONTENTS

INTRODUCTION

Over the last century, the advancement of photonics has enabled the generation and manipulation of light toward the creation of new technologies that would be perceived as unfathomable to previous generations. Cutting-edge technologies such as lasers, fiber optics, and cameras have not only been integrated into numerous consumer and commercial technology platforms but have also aided surgical technologies.

Infrared (IR) technologies are rapidly gaining popularity in the interdisciplinary fields of neuroscience, neurology, and neurosurgery. Neurosurgery is a field that is dominated by vision, and thus, imaging in the light spectra beyond the visible essentially expands the visual perception of the surgeon. This chapter presents an overview of the current literature on intraoperative IR optical imaging techniques used in neurosurgery, with particular emphasis on tumor, vascular, and functional applications.

INFRARED TECHNOLOGY

IR light is a form of electromagnetic radiation with a wavelength that is greater than visible light and less than that of radio waves. Although not visible to the naked eye, we often perceive IR photons in the mid and long wavelength as the heat that radiates from objects. This region is known as the "thermal infrared," and its wavelength lies between 8,000 and 14,000 nm. Thermal IR is the region where intraoperative thermal imaging (ITI) is employed, and its importance will be discussed in the upcoming sections. Another region of IR light, known as near-infrared (NIR), at 700–2500 nm, also has the capacity to facilitate valuable applications in medical imaging (Kateb et al. 2009).

THERMAL IMAGING

Human body temperature is influenced by a multitude of internal biological factors; hence, any deviation from normal may provide insights into pathophysiological anomalies. Temperature changes in the body may arise from inflammation, carcinogenesis, or neuropathology and can be locally or systematically expressed (Nola et al. 2012). The first recorded use of thermobiological assessment dates back to 480 BC in the writings of Hippocrates (Hippocrates and Adams 1939). It was surmised that underlying organ pathologies might be understood by spreading mud over the patient's body and observing the areas that would dry the most quickly. The development of IR imaging technologies has allowed for the evolution of thermal evaluation from a conviction to a subject of scientific inquiry.

As discussed earlier, thermal IR refers to the portion of the electromagnetic spectra with a wavelength of 8–14 μm. According to Planck's law, all objects will emit energy as photons at their prescribed wavelength. The Stefan–Boltzmann law states that the amount of radiation emitted is proportionate to the fourth power of the object's thermodynamic temperature. Thermal imaging detects this IR radiation and converts the acquired data into a visual image that is detectable by the human eye (Ring and Ammer 2012). This method is completely noninvasive and nonionizing and captures images in real time.

The initial use of diagnostic thermal imaging was known as human thermography, which was introduced in 1956 when Robert Lawson observed a higher temperature on the skin over breast cancer than the skin over normal breast tissue (Lawson 1956). Thermography had a brief spark in the field of medicine but subsequent studies failed to demonstrate its ability to effectively screen for breast cancer (Hitchcock et al. 1968; Feig et al. 1977). Subpar performance in clinical practice, combined with the high cost of this technology, eventually led to waning interest in this diagnostic modality.

The twenty-first century, however, has seen a resurgence of interest in thermography due to lower costs and improved performance through newer and improved technologies that employ high-speed computers, powerful imaging processors, enhanced algorithms, and high-tech cameras (Jones 1998; Arora et al. 2008). Along with its initial use in oncology, thermographic technology has spread to serve as an imaging tool in the fields of orthopedics, pain management, dentistry, veterinary medicine, neurology, and neurosurgery (Diakides et al. 2012). Generally, the term thermography is used for nonoperative diagnostics of the skin overlaying the structure of interest, while ITI is the terminology employed for surgical procedures that involve thermal imaging technologies. ITI has recently come to prominence in the field of neurosurgery where it has exploited the thermal properties of CNS afflictions and has shown promise as an imaging aid during tumor resection.

EARLY NEUROSURGICAL EXPERIMENTS USING ITI

The earliest use of ITI of the brain was for localizing vascular structures during the resection of cerebral arteriovenous malformations (AVMs) (Okudera et al. 1994, 1998; Ecker et al. 2002; Watson et al. 2002).

The Ecker study, conducted at the Mayo Clinic in Rochester, Minnesota, used ITI in AVM resections in addition to exploring the technology's potential in other neurosurgical cases such as epilepsy, functional mapping of the cerebral cortex, and the differentiation of brain tumors. IR imaging was able to accurately locate seizure foci in 7 of 9 patients enrolled into the study, identify the focus of seizures upon stimulation in 4 of 9 patients, and reveal distinct thermal footprints for 14 of 16 brain tumors. A concurrent study using similar technologies had shown this technology to be effective in the detection of brain tumors in rats (Papaioannou et al. 2002).

Another pioneer of ITI technology was Dr. Alexander Gorbach who proposed neurologic applications of this science in 1993. In a 2003 study, Gorbach and his team used ITI to detect the exposed cerebral cortex of 21 patients (Gorbach et al. 2003). This was done during language and motor tasks, as well as during the stimulation of the contralateral median nerve. The detected IR functional localizations were compared with fMRI, which concluded that it could provide adequate information on spatial and temporal patterns of sensory, motor, and language representations. The following year, Gorbach published on ITI for the imaging of brain tumors and, akin to Ecker and Papaioannou, Gorbach was able to distinguish brain tumors via their abnormal perfusion and vascular dynamics. In these ITI tumor studies, cortical surfaces were recorded prior to, during, and following the excision of tumorous masses in 34 patients (21 primary brain tumors, 10 brain metastases, 1 falx meningioma, 1 cavernous angioma, and 1 radiation necrosis–astrocytosis) (Gorbach et al. 2004). Metastatic tumors exhibited higher temperatures than surrounding healthy tissues, while gliomas exhibited lower temperatures than their surroundings. These early findings demonstrated the necessity for the thorough investigation of various tumor thermal profiles if ITI was to have a future in the surgical environment.

OPTICAL IMAGING USING NEAR-INFRARED LIGHT

The Near Infrared (NIR) region is the region of IR light that is closest to visible light with a wavelength of 750–1300 nm. The use of NIR light for imaging purposes arises from the basis that, from between 650 and 900 nm, there exists a region known as the "infrared window," where the absorption of light by biological tissues is minimal and the light is able to diffuse through several centimeters of the medium (Vo-Dinh 2003). In this wavelength range, the primary limitation for light penetration is due to mitochondrial membrane scattering, as well as the scattering initiated by other vesicles that are present in the tissue (Beauvoit et al. 1995). The ability of NIR light to penetrate the skull was first analyzed by Professor Frans Jöbsis, who utilized a spectroscopy technique known as near-infrared spectroscopy (NIRS) to noninvasively monitor the status of cerebral oxidative metabolism (Jöbsis-Vander Vliet 1977). Jobsis demonstrated that NIR light could be applied and detected through the intact scalp to monitor hemoglobin changes associated with the hemodynamic response. Within the brain, this hemodynamic response, brought on due to increased regional brain activity, is referred to as neurovascular coupling.

The main and obvious limitation of NIRS was the lack of spatial data. It did not take long for investigators to overcome this limitation by combining multiple NIRS measurements consisting of multiple clusters of sources and detectors to pinpoint and localize brain emanating signals to form useful images of the brain (Maki et al. 1995). In the realm of noninvasive imaging, NIRS technology essentially served as the seed that sprouted into a tree of possibilities. The discovery of NIRS has led to imaging and tomography techniques such as diffuse optical tomography (DOT) and optical coherence tomography (OCT). DOT functions by extrapolating NIRS data with multiple measurement points to create three-dimensional computed tomographic images of the brain

(Hielscher et al. 2002). OCT, on the other hand, is an interferometry technique that employs NIR light to scan cross-sectional images at various depths to create two- or three-dimensional images (Fujimoto et al. 2000). These technologies have shown to be a valuable method for noninvasive neuroscience discovery and medical applications, and their uses range from cognitive brain mapping to disease detection.

Optical Coherence Tomography

Similar to NIRS and DOT, the research on OCT has been largely constrained to nonoperative imaging. Recently, however, it has also been implemented as an intraoperative tool in neurosurgery. OCT is a low coherence interferometry technique that operates by applying NIR light to the tissue and collecting reflected light while rejecting the multiple scattered photons, which make up the majority of the light–tissue interactions. This method, based on the echolocation principal, has earned it the nickname "optical ultrasound." It is noninvasive, nonionizing, does not require sample preparation, is economical, and is easy to use. OCT provides highly detailed resolution in the micrometer range that is decoupled in depth and transverse resolution (Huang et al. 1991). This technology has been advanced in speed and sensitivity through spectral encoding (Boer et al. 2003; Wojtkowski et al. 2002) and continues to push boundaries as the field of photonics advances. Akin to NIRS, OCT can utilize various approaches such as time domain, spectral domain, time-encoded domain, and frequency domain. A linear scan creates a two-dimensional image, while an area scan achieves a three-dimensional image (Dubois et al. 2002). These are known as confocal OCT and full field OCT (FF-OCT), respectively.

Although OCT is inadequate for whole brain imaging, it has the ability to perform "optical biopsies" in which small samples of tissue may be imaged during a surgical procedure without having to resect or process the specimen (Boppart 2003). The resolution is approaching that of histology without the burden of the required downtime for histological processing. At its current level of development, OCT cannot fully replace histological samples; thus it serves as a complimentary tool in normal/cancerous tissue differentiation.

OCT has revealed a wide array of other uses including integration into neuroendoscopies to add further imaging capabilities, particularly for opaque conditions such as ventricular systems (Böhringer et al. 2006a). Another study showed the ability of a spectral-domain configured OCT to differentiate brain tumor regions ex vivo (Böhringer et al. 2006b). The researchers imaged gliomas in a mouse model as well as in human brain tumor biopsies. The study utilized both spectral domain-OCT and time-domain OCT, but the latter proved to be ineffective at depths of greater than 1 mm. The spectral-domain OCT provided a highly detailed microstructure and differentiation of the solid tumor, infiltration zone, and normal brain tissue, which uncovered the potential of OCT as a rapid intraoperative imaging technique in tumor delineation and neurosurgical resection guidance. The use of OCT in brain tumor delineation was further supported in rat glioma models (Zhang et al. 2011). Another recent clinical study done on eighteen human glioma samples demonstrated FF-OCT as a viable method for the identification of tissue architectures, encompassing microcysts, microcalcifications, and gliomal tumor cells with good correlations to histological examinations (Assayag et al. 2013).

Near-Infrared Fluorescence Imaging

The utilization of the IR window by NIRS and NIRS-derived technologies has influenced the development of intraoperative fluorescent imaging modalities that function in this photonic range. The NIR region has a limited level of tissue absorbance, scattering, and autofluorescence from endogenous chromophores and thus provides a neutral background that is ideally suited for fluorescent imaging. There are, however, two main drawbacks to this technology: the injection of exogenous contrasting agents and the necessity of cameras and imaging technologies to "see" the emitted NIR light. While there exist fluorophores that emit light in the visible spectrum, which can be seen with the naked eye, the higher signal-to-noise ratio and enhanced resolution of NIR fluorescent imaging, compared to visible light fluorescence, as well as its relatively inexpensive cost, outweighs this inconvenience. Three currently used intraoperative NIR imaging systems are the SPY system™, Photodynamic Eye™, and FLARE™. Adding to this growing armamentarium, researchers have developed hands-free wireless goggles that allow surgeons to observe both visible and NIR light illuminated from tumor specific fluorophores (Liu et al. 2011). The primary NIR fluorophores used for intraoperative neurosurgical studies have thus far been those of the cyanine family.

NEUROSURGICAL ONCOLOGY

TUMOR THERMAL PROFILE

The variations in tumor properties, particularly between different tumor types, make their thermodynamic assessment impossible to generalize. Different tumor cells may display a range of thermal properties from hypo- to hyperthermic, and some have even been documented to exhibit mixtures of both higher and lower temperatures (Gorbach et al. 2004).

The two main factors that cause a hyperthermic tumor profile are the inflammatory response and the process of carcinogenesis (Nola et al. 2012). The injury of tissues by bacteria, toxins, trauma, or cancers initiates localized inflammatory responses by which biochemicals such as cytokines, histamines, and bradykinins are released into the affected area, thus causing tissue swelling, hyperemia, and hyperthermia (Lawrence et al. 2002). Carcinogenesis leads to a process that promotes tumor growth and ensures a cell's survival. Tumors require increased blood flow in order to support their high metabolic demand and nutrient needs, and they achieve this through angiogenesis, or by promoting the secretion of vasodilators such as nitric oxide and increasing the dilation of existing blood vessels (Nola et al. 2012).

The causes of hypothermic tumor profiles in neurologic tissues arise from different sets of circumstances. One potential factor is the occurrence of edema in the surrounding environment due to the increase in white matter volume caused by tumor growth (Ecker et al. 2002; Gorbach et al. 2004). This edema also leads to the "disconnection effect" where the cortex overlying the intracranial tumor has decreased metabolism. These factors may not directly decrease the tumor temperature; however, the reduction in perfusion and cerebral blood flow as a repercussion of these factors may cause a decrease in the volume of warm blood that is supplied to the tumor. Morphology may also play a role in this process as it has also been shown that additional heat is lost via convection from extruding tumors with larger surface areas (Tepper et al. 2013). A protruding shape is typical in transplanted tumors in animal models (Papaioannou et al. 2002; Xie et al. 2004; Song et al. 2007), which have been shown to exhibit a hypothermic tumor profile. The greater surface area also leads to necrosis at the center of the tumor (Tepper and Gannot 2014), which will inevitably magnify the temperature reduction. In reality, tumors possess a combination of all the factors described in this section. It is the overall net effect of these factors that will determine the hypothermic or hyperthermic nature of the tumor.

ADVANCED NEUROSURGICAL INTRAOPERATIVE THERMAL IMAGING FOR TUMOR RESECTION

Early studies by Ecker and Gorbach served as a catalyst for other ITI research on brain tumors. Subsequent studies investigated the procedure with greater detail, and were aided with more advanced technologies. Newer and improved detector materials augmented thermal sensitivity and spatial resolution, while smart imaging algorithms enhanced the computation and deciphering of detected signals (Mital and Scott 2007; Arora et al. 2008). In a prominent case report, thermal imaging was used intraoperatively on a human patient with a metastatic intracortical melanoma (Kateb et al. 2009). The surgeons used a TCP60 IR camera that provided a high spatial resolution of 7.5–13 μm and a thermal sensitivity of 0.06°C to assess the thermal profile of the tumor and its surrounding tissue. ITI was successful in revealing a distinct demarcation of the tumor via significant temperature differences. The tumor core presented a temperature of 3.1°C greater than the surrounding normal tissue. It was estimated that 83.6% of the brain tumor was removed, for near total resection, and 16.4% of the residuals could be visualized using ITI. Magnetic resonance imaging (MRI) and computed tomography (CT) were used pre-resection to image the tumor location and dimensions. Ultrasound and light microscopy were used intraoperatively to confirm the tumor depth and position, and to observe the extent of its resection. Finally, MRI was employed to image the tumor post-resection. These studies demonstrated ITI's value when used in conjunction with the gold standard imaging techniques. ITI provided adjuvant imaging quickly, intraoperatively, and noninvasively.

Since this 2009 report, analogous studies have been conducted confirming the usefulness of ITI. In 2013, a time-resolved thermography technique was employed to successfully highlight tumorous regions intraoperatively (Hollmach et al. 2013). The following year, a different team went a step further where they conducted the procedure with five different subjects (Kastek et al. 2014). The uniqueness of this study was the variation in tumor type and degree: three gliomas, one meningioma, and one metastatic tumor of lung cancer origin. The three gliomas had different extents of malignancy (G2–G4) and thus presented their own variations in metabolic activity and temperature exclusive of interpatient variability. One important observation from this study was the revelation of temperature decreases during the section of blood vessels despite the use of bipolar coagulation. Tumor vascularity, in this study, was determined to play a prominent role in the tumor's temperature and thermal reading but was addressed by conducting the surgery in a more methodical manner. Nevertheless, this experiment reached the same conclusion as previous experiments: thermal imaging comprises a feasible technique for the intraoperative delineation of brain tumors.

TUMOR PAINT

Tumor paint is a new and radical brain tumor imaging agent that was initially discovered by pediatric oncologist James Olson and his team at the Fred Hutchinson Cancer Research Center and Seattle Children's Hospital Research Institute. Tumor paint comprises a molecule that consists of a NIR fluorophore that is conjugated with chlorotoxin (CTX), a peptide found in the venom of the Israeli deathstalker scorpion (*Leiurus quinquestriatus*). The CTX portion preferentially binds to glioma cells (Soroceanu et al. 1998) and other tumors of neuroectodermal origin (Lyons et al. 2002) in contrast to normal brain cells. There have been no reports of toxicity in the use of CTX during in vivo studies (Stroud et al. 2011).

In Veiseh's 2007 study, CTX-Cy5.5 Tumor Paint was used for tumor delineation in transgenic and xenograft mice with prostate cancer, sarcoma, intestinal cancer, glioma, and medulloblastoma. Cy5.5 is a commonly used cyanine fluorophore that emits photons in the NIR spectrum. This tumor paint was successful in detecting tumor foci and metastases both in vitro and during simulated surgical operating conditions. It had been hypothesized that the specific binding of CTX is facilitated by matrix metalloproteinase-2 (MMP-2), and this was made evident by the reduction in tumor paint binding subsequent to the administration of a pharmacologic MMP-2 blocker.

A newly discovered tumor paint is Tumor Paint BLZ-100, which uses indocyanine green (ICG) as the conjugate NIR fluorophore. In one study, researchers implanted human glioma cells in mouse brains and injected Tumor Paint BLZ-100 48 hours prior to the removal of the brain for imaging (Butte et al. 2014). The investigators employed a novel high-resolution, two-in-one charge-coupled device camera system to capture both the visible light and the NIR fluorescence emitted by the tumor paint upon stimulation by NIR pulses. The camera imaging system was unique in its ability to detect and visualize the ICG down to 50 nmol concentrations while maintaining a high signal-to-noise ratio. The BLZ-100 had a high affinity for the glioma and thus required the minimal use of a contrast agent. The research on tumor paint has garnered much publicity, both within the medical community and beyond; however, further cross-sectional studies with mice and human subjects are required to properly evaluate the efficacy of tumor paint in comparison with other intraoperative contrast agents.

RGD–Cyanine

Arginylglycylaspartic acid (RGD) is a tripeptide composed of L-arginine, glycine, and L-aspartic acid that exhibits specific adhesion to αvβ3 integrins. These αvβ3 integrins are overexpressed in many tumor cell lines thus the use of RGD ligands may increase the internalization of fluorophores into tumor cells (Ye and Chen 2011). RGD and its associated conjugates are diminutive in size, giving them easier access to tumor tissues. These compounds also pose minimal risk of immune response, and their synthesis is inexpensive and simple (Wang et al. 2013). RGD may be decorated with imaging moieties while maintaining biological interaction with specific receptors.

Dr. Chen, in 2004, described RGD-Cy5.5 as a contrasting agent in vitro, in vivo, and ex vivo, to illuminate integrin-expressing tumor cells in a U87MG glioblastoma xenograft. Doses were given thirty minutes to twenty-four hours postinjection. Intermediate doses of 0.5 nmol produced improved tumor contrasting than high doses at 3 nmol. This was thought to be due to the partial self-inhibition of receptor-specific tumor uptake at higher doses. Low doses of 0.1 nmol were insufficient due to significant background noise. The uptake of RGD-Cy5.5 into tumors was blocked by competitive inhibition of injected unlabeled c(RGDyK). A later study using a U87MG glioblastoma xenograft tested the binding affinities of Cy5.5-conjugated RGD monomers, dimers, and tetramers (Cheng et al. 2005). All three peptide-dye conjugates displayed tumor uptake, both in vitro and in vivo however the tetramer displayed the highest level of uptake and tumor-to-normal tissue ratio thirty minutes to four hours postinjection, with the dimer being the second best. RGD multimers were also found to be successful for tumor imaging using Cy-7 as the fluorophore (Wu et al. 2006).

Indocyanine Green

ICG is a water-soluble cyanine dye with an IR range excitation and emission peak of 800 nm and 850 nm, respectively (Kirchherr et al. 2009) that functions by binding to human serum proteins which are then taken up by tumor cells via enhanced permeability and retention (EPR) effects (Kosaka et al. 2011). ICG is FDA approved and its low-negative reaction profile and lack of radiation make it a significantly safer imaging method than most (Polom et al. 2011). In its current iteration, the maximum depth of ICG visibility is reported to be at 21 mm (Pleijhuis et al. 2011), thus, superficial tumors are ideal for ICG. ICG has shown to be able to distinguish between low-grade and high-grade tumors; low-grade tumors have a clearance of approximately five minutes, while high grade tumors exhibited a sequestering phenomenon in which ICG fluorescence was maximized at ten minutes following injection (Haglund et al. 1996).

While the ICG technology was initially used for vascular surgeries, it has recently been implemented in tumor delineation. A 2011study used ICG videoangiography to analyze its potential for the surgical identification of high-grade gliomas, low-grade gliomas, meningiomas, metastatic tumors, and hemangioblastomas (Kim et al. 2011). Kim and colleagues used an operative microscope with NIR light sources as well as NIR filter pass bands for detection. The microscope had an attached video camera, which was used to image the vascular fluorescence of the various tumors. Due to its quick clearance rate, ICG was administered intraoperatively via intravenous injection. Prior to resection, ICG videoangiography provided information on the pathology-induced changes of the tumoral and surrounding peritumoral brain circulation. Postresection, ICG videoangiography provided an examination of potential damage to healthy vessels close to the tumor mass.

Not surprisingly, further research experiments using ICG videoangiography for brain tumor imaging, has been done with hemangioblastomas (Tamura et al. 2012; Hojo et al. 2014). Not only do hemangioblastomas exhibit marked differences in vascularity, which are easily detected by ICG videoangiography, but their presence at superficial dural lesions make them a particularly amenable candidate for this depth-limited technology (Tamura et al. 2012). ICG videoangiography may be used to evaluate vessels of various diameters while the current gold standard technique, digital subtraction angiography (DSA), is more expensive, has a longer latency for results, and is not appropriate for small-vessel use (Raabe et al. 2003).

PRESENT APPLICATIONS OF INFRARED INTRAOPERATIVE OPTICAL IMAGING

NEUROVASCULAR SURGERY

The current gold standard for the diagnosis of aneurysms and the evaluation of treated aneurysms is DSA, whereby images are compiled prior to and following the injection of contrast agents to "subtract" noncontrasted tissues. This requires access to the arterial systems of a patient, generally through the femoral artery. This procedure is not without risks; however, complication rates range from 0.4% to 2.6% (Raabe et al. 2005). Complications may include local site hematoma, dissection, stroke, aneurysm rupture, or hemorrhage. Furthermore, this procedure dramatically increases operative time through repositioning, cannulation, and imaging of the patient through cumbersome equipment. A new and separate puncture incision requires the preoperative coordination of neurointerventionalists, radiology technicians, and nursing. It also requires a completely different set of technologies in the operating room, including biplanar fluoroscopy.

In contrast to DSA, during the intraoperative use of ICG-VA, a dye is administered to the patient by the anesthesia team, and vascular flow is assessed by video technologies (Raabe et al. 2003). ICG-VA requires the use of microscope-related technology with high resolution contrasted imaging. The microscope is appropriately calibrated by the operating surgeon to coordinate the field needed for investigation. After the surgeon visually confirms the operative field, the microscope is switched to a setting that changes the light source to a NIR light that encompasses the ICG absorption band. Optical filters block out any other light sources seen through the microscope to the operating surgeon, such that only the vascular structures that are actively being filled with blood are visible via ICG-VG illumination. This allows for direct feedback for the operative surgeon regarding the patency of vessels (illumination), the occlusion of aneurysm (no illumination), or the removal of an AVM (no illumination). The use of these real-time techniques allows for the immediate implementation of operative strategies (Raabe et al. 2005).

Clinical assessment tools and operative feedback is vitally important, and without these emerging technologies or similar intraoperative evaluations, 2%–8% of aneurysms undergoing clipping showed a residual neck or nidus (Raabe et al. 2005). The 2%–8% of patients are either subject to undergoing an additional procedure for definitive treatment of their aneurysm or must suffer the risk of a re-bleed or re-rupture from an untreated portion of an aneurysm. Even more critical is that postoperative angiography had a 4%–12% incidence of inadvertent occlusion of a normal artery via an aneurysm clip (Raabe et al. 2005).

Approximately 90% of the vessel clip abnormalities seen by intraoperative angiography are detected by ICG-VA. The complication rate for ICG-VA is low, documented as 0.05%–0.2%, presenting generally with histamine-flushing-type reactions, including tachycardia, hypotension, and skin lesions. The clinical utility here cannot be overstated. Due to the time constraints and delay of intraoperative DSA, clip adjustment for a DSA determined inadvertent clipping of a normal artery still had a stroke rate of 33%; utilization of ICG-VA can decrease the time to treatment considerably (Raabe et al. 2005).

The limitations of ICG-VA center on its lack of three-dimensional interpretations and the constrained perimeters of the microscopic field. Planes that lie beyond those available in the operative microscope cannot be fully appreciated during the ICG-VA interrogation. Anatomy beyond the direct exposed field cannot be viewed (Raabe et al. 2003). Residual necks located behind the aneurysm are thereby opacified in appearance and therefore missed during an ICG-VA screening. Complex aneurysms with atherosclerosis and partial thrombosis would be similarly obscured (Raabe et al. 2003).

In another promising application for this technology, extracranial–intracranial bypass procedures, ICG-VA serves to provide the rapid and reliable evaluation of newly created arterial constructs. Cerebral revascularization is vital in the treatment of moyamoya disease and stroke and as an adjuvant in complex aneurysm surgery. It is critical to leave the operating room with a proven patient vessel, as the new vascular construct (superficial temporal artery to middle cerebral artery anastomosis) must be viable. The inspections of anastomoses by surgeons without radiographic evaluations have a documented failure rate of up to 10% (Raabe et al. 2003). The addition of ICG-VA has 100% sensitivity for the detection of intraoperative graft occlusions (Connolly et al. 2012). This is due to the fact that most constructed grafts tend to be of the high flow variety, thereby increasing the reliability of the ICG-VA technology.

EMERGENCY NEUROSURGERY

The inherent frustration of neurosurgery and neurological diagnosis is the cryptic nature of the disease process; it occurs inside the body. As such, clinicians have historically relied upon the pearls of clinical examination to localize injuries and subsequently treatments thereof. The advent of new technologies has allowed the clinician to have improved access to the disease process via the proxy of detailed imaging. However, these imaging modalities come at a significant financial cost, and the burdens of both radiation exposure and time. IR technology is developing as a direct clinical adjuvant to the diagnosis of neurosurgical disease during emergent situations.

Approximately 1.5 million people annually seek treatment in an emergency room for the treatment and diagnosis of head trauma in the United States (Zisakis et al. 2013). Cohorts of individuals present to hospital systems without advanced imaging technologies in rural areas or underserved areas. Oftentimes, clinicians must decide to treat empirically or outsource care to a tertiary center at considerable cost and burden to the patient. The use of NIR technology may be employed to detect intracerebral hematomas in rapid and reliable format (Zisakis et al. 2013). This is a noninvasive technology with an ease of applicability that can aid the medical professional in the evaluation of intracranial hematomas. The patient may then be triaged appropriately for additional imaging or conservative management.

NIR technologies exploit the different properties of intravascular and extravascular blood (Zisakis et al. 2013). Extravascular blood typically has a higher concentration of hemoglobin than the brain tissue that immediately surrounds the clot. As hemoglobin absorbs NIR radiation the extravascular blood may, under these conditions, be visually differentiated from intravascular blood and the surrounding tissue (Zisakis et al. 2013).

This technology has also proven useful for clinical assessment following traumatic cranial injury. With the diagnosis of initial intracranial trauma, serial imaging is frequently weighed against the burden of radiation exposure and cost—frequently delaying treatment pending clinical evidence of interval progression. NIRS, however, allows for the noninvasive determination of delayed hematoma in patients with trauma injury. It does so without the radiation exposure, patient transport, and cost of the standard modality, CT imaging. Up to 16% of patients with traumatic brain injury develop a delayed hematoma (Gopinath et al. 1995), this can occur within seventy-two hours following the inciting trauma. Even more promising is the early detection afforded to patients through the use of NIRS. Gopinath et al. noted that the use of this technology illustrated an increased size of the hematoma prior to a clinical exam or CT scan in 24 of 27 patients. This allowed for an even earlier diagnosis and improved treatment than the current modalities that are utilized in neurosurgical intensive care units (clinical exam and CT scan).

Traumatically injured patients require early triage and proper diagnosis for maximum treatment and proper management. The use of NIR technology does not replace the standard of care regarding neuroaxis imaging or complex trauma management, but shows expanding promise to augment current standards of care.

PEDIATRIC NEUROSURGERY

The risks of diagnostic imaging radiation exposure are most significant within the pediatric population. Again, NIRS has tremendous applicability in this particular patient cohort as it greatly reduces the radiation burden. Already, NIR technologies are demonstrating utility within pediatric neurosurgical care. In one pediatric intensive care unit, NIRS had the sensitivity to detect a intracranial hemorrhage of 1.0 with a specificity of 0.8 (Salonia et al. 2012). All four regions of the brain (frontal, temporal, parietal, and occipital) are able to be evaluated (Salonia et al. 2012) for the presence of intracranial hemorrhage without the anesthesia that may be required for pediatric patients to complete traditional imaging modalities.

FUNCTIONAL NEUROSURGERY

The utilization of NIR technology in neurosurgery has proven useful in functional neurosurgery as well. Deep brain stimulation employs the use of preoperative imaging and intraoperative monitoring with microelectrode recordings to confirm, enhance precision, and improve outcomes. Unfortunately, microelectrode recording does not come without its disadvantages, including increased operative time and risk of hemorrhage.

One of the earliest utilizations of NIR imaging in functional neurosurgery was described by Giller in 2000, and again in 2009. A fiber optic probe was devised with a blunt tip that had a light source in the middle with surround light-returning fibers and assembled with a portable spectrometer. The blunt tip profile of the probe was passed through intracerebral structures. As the probe is passed through different tissues, scattered light is returned at different wavelengths, denoting the different anatomic structures. In pallidotomies and thalamic procedures, many structures were consistently identified with the NIR probe. The studies, however, showed less reliable with intrathalamic structures. As the probe passed through a sulcus, for example, cerebrospinal fluid caused a dip in the NIR image return.

THE FUTURE OF NEAR-INFRARED FLUORESCENCE IMAGING

Future research in this technology requires a focus on overcoming some of the limitations of current NIR organic dyes. ICG, the only clinically approved NIR fluorophore to date, has a quantum yield of only 0.01 in an aqueous environment and a rapid clearance rate (Pauli et al. 2009). When seeking to identify an ideal contrasting agent, many issues need to be addressed such as toxicity, quantum yield, photostability, spectral bandwidths, solubility, molecular targeting, ease of synthesis, and ability to functionalize and conjugate. For brain studies, NIR fluorophores face the additional hurdle of having to traverse the blood brain barrier. Newly developed NIR fluorescent dyes such as squaraines, phthalocyanines, porphyrin derivatives, and boron-dipyrromethene analogs have made much progress in meeting the necessary criteria and, thus, are potential candidates for future intraoperative brain imaging studies (Luo et al. 2011).

As seen with tumor paint and RGD conjugates, NIR fluorophores attached to cancer-targeting ligands may allow for greater binding specificity and thus require smaller concentrations of agents, as well as providing greater resolution. These cancer-targeted ligands may be comprised of small organic molecules, proteins, peptides, antibodies, DNA, siRNA, or aptamers. Therefore, many combinations can be made with various ligands and NIR fluorophores.

The new era of nanomedicine will also constitute a highly influential factor in the development of NIR imaging. Nanoparticles may improve drug bioavailability by serving as drug delivery agents, and they have also been effectively used as contrasting agents for ultrasound and MRI (Kateb and Heiss 2013). Nanoparticles may bind directly to fluorophores, encapsulate fluorophores, or provide their own NIR luminescence. Glycol chitosan nanoparticles labeled with the NIR dye Cy5.5 have already been successfully used for in vivo noninvasive optical imaging (Na et al. 2011). Nanoparticles, however, face many of the same challenges as other imaging agents. For example, quantum dots have generated a great deal of interest for their photostability and brilliant NIR emission, but their large dimensions can impede their efficient clearance from renal systems, thus exhibiting long-term toxicity (Pelley et al. 2009). Engineered nanoparticles and the addition of strong affinity-based functional groups may enable the creation of far more specific and less toxic molecules.

CONCLUSIONS

This chapter has summarized optical image-guided surgery technologies that employ IR, NIR, and/or ultraviolet ranges of light. These emerging technologies are gaining acceptance throughout the subspecialties of neurosurgery as they are cost effective, versatile, and allow for a continuum to bridge the uncertainty between preoperative and postoperative imaging. These technologies are particularly well suited for vascular and tumor imaging, and thus, research efforts have concentrated on these applications. Further, these technologies have potential for applications in other CNS disorders.

The core intent of these imaging methods is to maximize optical resolution while minimizing toxicity. As there is a constructive competition among these imaging modalities and overlap in their applications, a wealth of comparative studies are open to further exploration. The key to optimizing our understanding of anatomic pathology may well be the collaboration of multiple techniques, and many of these technologies are compatible with each other. Additionally, the integration of these imaging technologies with advanced computers, algorithms, and other quantitative systems may offer a positive shift from subjective to objective diagnosis in clinical care.

REFERENCES

Adams, F. 1939. *The Genuine Work of Hippocrates* (1 ed.), Vol. 1. Baltimore, MD: Williams and Wilkins.

Arora, N., D. Martins, D. Ruggerio, E. Tousimis, A.J. Swistel, M.P. Osborne, and R.M. Simmons. 2008. Effectiveness of a noninvasive digital infrared thermal imaging system in the detection of breast cancer. *The American Journal of Surgery* 194 (4): 523–526. doi:10.1016/j.amjsurg.2008.06.015.

Assayag, O., K. Grieve, B. Devaux, F. Harms, J. Pallud, F. Chretien, C. Boccara, and P. Varletb. April 2013. Imaging of non-tumorous and tumorous human brain tissues with full-field optical coherence tomography. *NeuroImage Clinical* 2: 549–557. doi:10.1016/j.nicl.2013.04.005.

Beauvoit, B., S.M. Evans, T.W. Jenkins, E.E. Miller, and B. Chance. 1995. Correlation between the light scattering and the mitochondrial content of normal tissues and transplantable rodent tumors. *Analytic Biochemistry* 226 (1): 167–174.

Boer, D., F. Johannes, C. Barry, B. Hyle Park, M.C. Pierce, G.J. Tearney, and B.E. Bouma. 2003. Improved signal-to-noise ratio in spectral-domain compared with time-domain optical coherence tomography. *Optic Letters* 28 (21): 2067–2069. doi:10.1364/OL.28.002067.

Böhringer, H.J., D. Boller, J. Leppert, U. Knopp, E. Lankenau, E. Reusche, G. Huttemann, and A. Giese. 2006a. Time-domain and spectral-domain optical coherence tomography in the analysis of brain tumor tissue. *Lasers in Surgery and in Medicine* 38 (6): 588–597.

Böhringer, H.J., E. Lankenau, V. Rohde, G. Hüttmann, and A. Giese. 2006b. Optical coherence tomography for experimental neuroendoscopy. *Minimally Invasive Neurosurgery* 49 (5): 269–275.

Boppart, S.A. 2003. Optical coherence tomography: Technology and applications for neuroimaging. *Psychophysiology* 40 (4): 529–541.

Butte, P.V., A. Mamelak, J. Parrish-Novak, D. Drazin, F. Shweikeh, P.R. Gang alum, A. Chesn okova, J. Ljubimova, and K. Black. 2014. Near-infrared imaging of brain tumors using the tumor paint BLZ-100 to achieve near-complete resection of brain tumors. *Neurosurgical Focus* 36 (2): 1. doi:10.3171/2013.11. FOCUS13497.

Cheng, Z., Z. Xiong, S.S. Gambhir, and X. Chen. 2005. Near-infrared fluorescent RGD peptides for optical imaging of integrin alphavbeta3 expression in living mice. *Bioconjugate Chemistry* 16 (6): 1433–1441.

Connolly, E.S., A.A. Rabinstein, J. Ricardo Carhuapoma, C.P. Derdeyn, J. Dion, R.T. Higashida, B.L. Hoh et al. 2012. Guidelines for the management of aneurysmal subarachnoid hemorrhage a guideline for healthcare professionals from the American Heart Association/American Stroke Association. *Stroke* 43 (6): 1711–1737. doi:10.1161/STR.0b013e3182587839.

Diakides, M., J.D. Bronzino, and D.R. Peterson. 2012. *Medical Infrared Imaging: Principles and Practices*. Boca Raton, FL: CRC Press.

Dubois, A., L. Vabre, A.-C. Boccara, and E. Beaurepaire. 2002. High-resolution full-field optical coherence tomography with a Linnik microscope. *Applied Optics* 41 (4): 805–812. doi:10.1364/AO.41.000805.

Ecker, R.D., S.J. Goerss, F.B. Meyer, A.A. Cohen-Gadol, J.W. Britton, and J.A. Levine. 2002. Vision of the future: Initial experience with intraoperative real-time high-resolution dynamic infrared imaging. *Journal of Neurosurgery* 97 (6): 1460–1471.

Feig, S.A., G.S. Shaber, G.F. Schwartz, A. Patchefsky, H.I. Libshitz, J. Edeiken, R. Nerlinger, R.F. Curley, and J.D. Wallace. 1977. Thermography, mammography, and clinical examination in breast cancer screening. Review of 16,000 studies. *Radiology* 122 (1): 123–127.

Fujimoto, J.G., C. Pitris, S.A. Boppart, and M.E. Brezinski. 2000. Optical coherence tomography: An emerging technology for biomedical imaging and optical biopsy. *Neoplasia (New York, NY)* 2 (1–2): 9–25.

Giller, C.A., M. Johns, and H. Liu. 2000. Use of an intracranial near-infrared probe for localization during stereotactic surgery for movement disorders. *Journal of Neurosurgery* 93 (3): 498–505. doi:10.3171/jns.2000.93.3.0498

Giller, C.A., H. Liu, D.C. German, D. Kashyap, and R.B. Dewey. 2009. Use of an intracranial near-infrared probe for localization during functional neurosurgical procedures: Further experience. *Journal of Neurosurgery* 110 (2): 263–273. doi:10.3171/2008.8.JNS08728

Gopinath, S.P., C.S. Robertson, C.F. Contant, R.K. Narayan, R.G. Grossman, and B. Chance. 1995. Early detection of delayed traumatic intracranial hematomas using near-infrared spectroscopy. *Journal of Neurosurgery* 83 (3): 438–444. doi:10.3171/jns.1995.83.3.0438.

Gorbach, A.M. 1993. Infrared imaging of brain function. *Advances in Experimental Medicine and Biology* 333: 95–123.

Gorbach, A.M., J.D. Heiss, L. Kopylev, and E.H. Oldfield. 2004. Intraoperative infrared imaging of brain tumors. *Jounral of Neurosurgery* 101 (6): 960–969. doi:10.3171/jns.2004.101.6.0960.

Gorbach, A.M., J.D. Heiss, C. Kufta, S. Sato, P. Fedio, W.A. Kammerer, J. Solomon, and E.H. Oldfield. 2003. Intraoperative infrared functional imaging of human brain. *Annals of Neurology* 54 (3): 297–309. doi:10.1002/ana.10646.

Haglund, M.M., M.S. Berger, and D.W. Hochman. 1996. Enhanced optical imaging of human gliomas and tumor margins. *Neurosurgery* 38 (2): 308–317.

Hielscher, A.H., A.Y. Bluestone, G.S. Abdoulaev, A.D. Klose, J. Lasker, M. Stewart, U. Netz, and J. Beuthan. 2002. Near-infrared diffuse optical tomography. *Disease Markers* 18 (5–6): 313–337. doi:10.1155/2002/164252.

Hitchcock, C.R., D.F. Hickok, J. Soucheray, T. Moulton, and R.C. Baker. 1968. Thermography in mass screening for occult breast cancer. *The Journal of the American Medicine Association* 204 (6): 419–422.

Hojo, M., Y. Arakawa, T. Funaki, K. Yoshida, T. Kikuchi, Y. Takagi, Y. Araki et al. 2014. Usefulness of tumor blood flow imaging by intraoperative indocyanine green videoangiography in hemangioblastoma surgery. *World Neurosurgery* 82 (3–4): e495–e501. doi:10.1016/j.wneu.2013.02.009.

Hollmach, J., N. Hoffmann, C. Schnabel, S. Küchler, S. Sobottka, M. Kirsch, G. Schackert, E. Koch, and G. Steiner. 2013. Highly sensitive time-resolved thermography and multivariate image analysis of the cerebral cortex for intrasurgical diagnostics. SPIE Digital Library Proceedings. SPIE 8565, Photonic Therapeutics and Diagnostics (IX). Vol. 856550. doi:10.1117/12.2002342.

Huang, D., E.A. Swanson, C.P. Lin, J.S. Schuman, W.G. Stinson, W. Chang, M.R. Hee et al. 1991. Optical coherence tomography. *Science* 254 (5035): 1178–1181.

Jöbsis-Vander Vliet, F.F. 1977. Noninvasive, infrared monitoring of cerebral and myocardial oxygen sufficiency and circulatory parameters. *Science* 198 (4323): 1264–1267. doi:10.1126/science.929199.

Jones, B.F. 1998. A reappraisal of the use of infrared thermal image analysis in medicine. *IEEE Transactions on Medical Imaging* 17 (6): 1019–1027. doi:10.1109/42.746635.

Kastek, M., T. Piatkowski, H. Polakowski, Z. Czernicki, J. Bogucki, and M. Zębala. 2014. Intraoperative application of thermal camera for the assessment of during surgical resection or biopsy of human's brain tumors. SPIE Proceedings SPIE 9105, Thermosense: Thermal Infrared Applications (XXXVI). Vol. 910508, Baltimore, MD. doi:10.1117/12.2050306.

Kateb, B. and J.D. Heiss. 2013. *The Textbook of Nanoneuroscience and Nanoneurosurgery*. CRC Press/Taylor & Francis Group, Boca Raton, FL. https://www.crcpress.com/product/isbn/9781439849415.

Kateb, B., V. Yamamoto, C. Yu, W. Grundfest, and J.P. Gruen. August 2009. Infrared thermal imaging: A review of the literature and case report. *NeuroImage* 47: T154–T162. doi:10.1016/j. neuroimage.2009.03.043.

Kim, E.H., J.M. Cho, J.H. Chang, S.H. Kim, and K.S. Lee. 2011. Application of intraoperative indocyanine green videoangiography to brain tumor surgery. *Acta Neurochirurgica* 153 (7): 1487–1495. doi:10.1007/s00701-011-1046-x.

Kirchherr, A.K., A. Briel, and K. Mader. 2009. Stabilization of indocyanine green by encapsulation within micellar systems. *Molecular Pharmaceutics* 6 (2): 480–491. doi:10.1021/mp8001649.

Kosaka, N., M. Mitsunaga, M.R. Longmire, P.L. Choyke, and H. Kobayashi. 2011. Near Infrared fluorescence-guided real-time endoscopic detection of peritoneal ovarian cancer nodules using intravenously injected indocyanine green. *International Journal of Cancer* 129 (7): 1671–1677. doi:10.1002/ijc.26113.

Lawrence, T., D.A. Willoughby, and D.W. Gilroy. 2002. Anti-inflammatory lipid mediators and insights into the resolution of inflammation. *Nature Reviews Immunology* 2 (10): 787–795.

Lawson, R. 1956. Implications of surface temperatures in the diagnosis of breast cancer. *Canadian Medical Association Journal* 75 (4): 309–311.

Liu, Y., A.Q. Bauer, W. Akers, G. Sudlow, K. Liang, D. Shen, M. Berezin, J.P. Culver, and S. Achilefua. 2011. Hands-free, wireless goggles for near-infrared fluorescence and real-time image-guided surgery. *Surgery* 149 (5): 689–698. doi:10.1016/j.surg.2011.02.007.

Luo, S., E. Zhang, S. Yongping, T. Cheng, and C. Shi. 2011. A review of NIR dyes in cancer targeting and imaging. *Biomaterials* 2 (29): 7127–7138. doi:10.1016/j.biomaterials.2011.06.024.

Lyons, S., J. O'Neal, and H. Sontheimer. 2002. Chlorotoxin, a scorpion-derived peptide, specifically binds to gliomas and tumors of neuroectodermal origin. *Glia* 39 (2): 162–173. doi:10.1002/glia.10083.

Maki, A., Y. Yamashita, Y. Ito, E. Watanabe, Yoshiaki, M. and H. Koizumi. 1995. Spatial and temporal analysis of human motor activity using noninvasive NIR topography. *Medical Physics* 22 (12): 1997–2005. doi:10.1118/1.597496.

Mital, M. and Scott, M. 2007. Thermal detection of embedded tumors using infrared imaging. *Journal of Biomechanical Engineering* 129 (1): 33–39.

Na, J.H., K. Heebeom, S. Lee, H.M. Kyung, P. Kyeongsoon, Y. Heon, L.S. Hoon et al. 2011. Real-time and non-invasive optical imaging of tumor-targeting glycol chitosan nanoparticles in various tumor models. *Biomaterials* 32 (22): 5252–5261. doi: 10.1016/j.biomaterials.2011.03.076.

Ng, Y.P., N.K.K. King, K.R. Wan, E. Wang, and I. Ng. 2013. Uses and limitations of indocyanine green videoangiography for flow analysis in arteriovenous malformation surgery. *Journal of Clinical Neuroscience* 20 (2): 224–232. doi:10.1016/j.jocn.2011.12.038.

Nola, I.A., K. Gotovac, and D. Kolaric. 2012. Thermography in biomedicine—Specific requirements. In *IEEE Explore Digital Library*, Croatia, pp. 355–357. doi:10.1109/ELMAR.2005.193735.

Okudera, H., K. Shigeaki, and T. Toshihide. 1994. Intraoperative regional and functional thermography during resection of cerebral arteriovenous malformation. *Neurosurgery* 34 (6): 1065–1067; discussion 1067.

Okudera, H., S. Kobayashi, T. Takemae, H. Nagashima, S. Muraoka, and T. Takizawa. 1998. Intraoperative non-invasive infrared imaging during resection of large arteriovenous malformations. *Journal of Clinical Neuroscience* 5 (Supplement): 39–41. doi:10.1016/S0967-5868(98)90009-1.

Papaioannou, T., R.C. Thompson, B. Kateb, O. Sorokoumov, W.S. Grundfest, and K L. Black. 2002. Thermal imaging of brain tumors in a rat glioma model. In: I. Vo-Dinh, D.A. Benaron, and W.S. Grundfest, eds., *Proceedings of SPIE 4615, Biomedical Diagnostic, Guidance, and Surgical-Assist Systems IV*, San Jose, CA, pp. 32–35. doi:10.1117/12.466653.

Pauli, J., T. Vag, R. Haag, M. Spieles, M. Wenzel, W.A. Kaiser, U. Resch-Genger, and I. Hilger. 2009. An in vitro characterization study of new near infrared dyes for molecular imaging. *European Journal of Medicinal Chemistry* 44 (9): 3496–3503. doi:10.1016/j.ejmech.2009.01.019.

Pelley, J.L., A.S. Daar, and M.A. Saner. 2009. State of academic knowledge on toxicity and biological fate of quantum dots. *Toxicological Sciences: An Official Journal of the Society of Toxicology* 112 (2): 276–296. doi:10.1093/toxsci/kfp188.

Pleijhuis, R.G., G.C. Langhout, W. Helfrich, G. Themelis, A. Sarantopoulos, L.M. Crane, N.J. Harlaar, J.S. de Jong, and V. van Dam, and G.M. Ntziachristos. 2011. Near-infrared fluorescence (NIRF) imaging in breast-conserving surgery: Assessing intraoperative techniques in tissue-simulating breast phantoms. *European Surgical of Oncology* 37 (1): 32–39. doi:10.1016/j.ejso.2010.10.006.

Polom, K., D. Murawa, R. Young-soo, P. Nowaczyk, M. Hünerbein, and P. Murawa. 2011. Current trends and emerging future of indocyanine green usage in surgery and oncology: A literature review. *Cancer* 117 (21): 4812–4822. doi:10.1002/cncr.26087.

Raabe, A., J. Beck, R. Gerlach, M. Zimmermann, and V. Seifert. 2003. Near-infrared indocyanine green video angiography: A new method for intraoperative assessment of vascular flow. *Neurosurgery* 52 (1): 132–139; discussion 139. doi:10.1097/00006123-200301000-00017.

Raabe, A., P. Nakaji, J. Beck, L.J. Kim, F.P.K. Hsu, J.D. Kamerman, V. Seifert, and R.F. Spetzler. 2005. Prospective evaluation of surgical microscope—Integrated intraoperative near-infrared indocyanine green videoangiography during aneurysm surgery. *Journal of Neurosurgery* 103 (6): 982–989. doi:10.3171/jns.2005.103.6.0982.

Ring, E.F.J. and K. Ammer. 2012. Infrared thermal imaging in medicine. *Physiological Measurement* 33 (3): R33–R46. doi:10.1088/0967-3334/33/3/R33.

Salonia, R., M.J. Bell, P.M. Kochanek, and R.P. Berger. 2012. The utility of near infrared spectroscopy in detecting intracranial hemorrhage in children. *Journal of Neurotrauma* 29 (6): 1047–1053. doi:10.1089/neu.2011.1890.

Song, C., Appleyard, V., Murray, K., Frank, T., Sibbett, W., Cuschieri, A., and Thompson, A. 2007. Thermographic assessment of tumor growth in mouse xenografts. *International Journal of Cancer* 121 (5): 1055–1058.

Soroceanu, L., Gillespie, Y., Khazaeli, M.B., and Sontheimer, H. 1998. Use of chlorotoxin for targeting of primary brain tumors. *Cancer Research* 58 (21): 4871–4879.

Stroud, M.R., S.J. Hansen, and J.M. Olson. 2011. In vivo bio-imaging using chlorotoxin-based conjugates. *Current Pharmaceutical Design* 17 (38): 4362–4371.

Tamura, Y., Y. Hirota, S. Miyata, Y. Yamada, A. Tucker, and T. Kuroiwa. 2012. The use of intraoperative near-infrared indocyanine green videoangiography in the microscopic resection of hemangioblastomas. *Acta Neurochirurgica* 154 (8): 1407–1412; discussion 1412. doi:10.1007/s00701-012-1421-2.

Tepper, M. and I. Gannot. March 2014. Parametric study of different contributors to tumor thermal profile. *SPIE Digital Library* VII: 89400–89400. doi:10.1117/12.2040745.

Tepper, M., A. Shoval, O. Hoffer, H. Confino, M. Schmidt, I. Kelson, Y. Keisari, and I. Gannot. 2013. Thermographic investigation of tumor size, and its correlation to tumor relative temperature, in mice with transplantable solid breast carcinoma. *Journal of Biomedical Optics* 18 (11): 111410–111410. doi:10.1117/1.JBO.18.11.111410.

Veiseh, M., P. Gabikian, S.B. Bahrami, O. Veiseh, M. Zhang, R.C. Hackman, A.C. Ravanpay et al. 2007. Tumor paint: A chlorotoxin:Cy5.5 bioconjugate for intraoperative visualization of cancer foci. *Cancer Research* 67 (14): 6882–6888.

Vo-Dinh, T. 2003. *Biomedical Photonics Handbook*, 1st edn. Oak Ridge, TN: CRC Press.

Wang, F., Y. Li, Y. Shen, A. Wang, S. Wang, and T. Xie. 2013. The functions and applications of RGD in tumor therapy and tissue engineering. *International Journal of Molecular Sciences* 14 (7): 13447–13462. doi:10.3390/ijms140713447.

Watson, J.C., A.M. Gorback, R.M. Pluta, R. Rak, J.D. Heiss, and E.H. Oldfield. 2002. Real-time detection of vascular occlusion and reperfusion of the brain during surgery by using infrared imaging. *Journal of Neurosurgery* 96 (5): 918–923.

Wojtkowski, M., R. Leitgeb, A. Kowalczyk, T. Bajraszewski, and A.F. Fercher. 2002. In vivo human retinal imaging by fourier domain optical coherence tomography. *Journal of Biomedical Optics* 7 (3): 457–463.

Wu, Y., W. Cai, and X. Chen. 2006. Near-infrared fluorescence imaging of tumor integrin alpha v beta 3 expression with Cy7-labeled RGD multimers. *Molecular Imaging and Biology* 8 (4): 226–236. doi:10.1007/s11307-006-0041-8.

Xie, W., P. McCahon, K. Jakobsen, and C. Parish. 2004. Evaluation of the ability of digital infrared imaging to detect vascular changes in experimental animal tumours. *International Journal of Cancer* 108 (5): 790–794. doi:10.1002/ijc.11618.

Ye, Y. and X. Chen. 2011. Integrin targeting for tumor optical imaging. *Theranostics* 1: 102–126.

Zhang, K., Y. Huang, G. Pradilla, B. Tyler, and J.U. Kang. 2011. Real-time intraoperative full-range complex FD-OCT guided cerebral blood vessel identification and brain tumor resection in neurosurgery. In *SPIE Proceedings*, Vol. 7883, San Francisco, CA. doi:10.1117/12.874190.

Zisakis, A.K., V. Varsos, and A. Exadaktylos. November 2013. What is new and innovative in emergency neurosurgery? Emerging diagnostic technologies provide better care and influence outcome: A specialist review. *Emergency Medicine International* 2013: e568960. doi:10.1155/2013/568960.

18 Autofluorescence-Guided Resection of Intracranial Tumor
Past, Present, and Future

*Fartash Vasefi, Zhaojun Nie, David S. Kittle,
Chirag G. Patil, and Pramod Butte*

CONTENTS

INTRODUCTION

In the United States, the incidence rate of malignant brain tumors was 7.3 per 100,000 persons per year during 2004–2008 according to the National Cancer Institute (2012). Although this rate is quite low compared to other cancers (e.g., 45 per 100,000 persons per year for colorectal cancer, according to Centers for Disease Control and Prevention, 2012), primary brain tumors, especially malignant gliomas, have a devastating effect on patients' lives. The current 18-month survival rate of glioblastoma patients treated with surgery for biopsy only, partial resection, and complete resection ranges from 15% to 34%, making glioma one of the most aggressive and lethal tumors. Although chemotherapy and radiotherapy are used to treat glioblastoma, surgical treatment remains the most effective (Ducray et al. 2010; Fazekas 1977; Robins et al. 2009). Additionally, the extent of tumor resection has been shown to be the most important factor for longer survival (Berger 1994; Byar et al. 1983; Sanai and Berger 2008). Due to the infiltrating characteristics of malignant gliomas, complete resection is difficult to achieve without removing healthy, functional, normal tissue. Several technologies, such as stereotactic image-guided surgery based on preoperative MRI scans, intraoperative ultrasound (US) (Sosna et al. 2005), and intraoperative magnetic resonance imaging (iMRI) (Elias et al. 2007), aid the surgeons to ensure the near-complete resection of the tumor. However, these techniques have their limitations. Image-guided stereotactic surgery suffers from "brain shift," in which the accurate registration of the tumor based on preoperative MRI is lost due to movement of the brain after the craniotomy (Kubben et al. 2011). iMRI requires a large initial investment and special surgical tools. Due to space constrains, it also limits patient positioning, making it hard to reach certain areas. Intraoperative US has very low resolution. The only reliable method for intraoperative tissue diagnosis is the "frozen section," but this time-consuming process can only be performed a limited number of times. Therefore, newer technologies are needed to aid the surgeon in achieving near-complete resection while avoiding damage to nearby eloquent areas.

Technologies using minimally invasive optical techniques, called optical biopsies, have been developed for intraoperative diagnosis in a number of applications (Butte et al. 2014; Pogue et al. 2010; Vasefi et al. 2016). Optical biopsy interrogates either the cellular structure such as optical coherence tomography (OCT) (Giese et al. 2006) or biochemical composition (fluorescence [Butte et al. 2011], Raman [Zhou et al. 2012], and diffuse reflectance [DR] spectroscopy [Yaroslavsky et al. 2002]) of biological tissue based on the interactions between light and tissue. Additionally, technologies such as photoacoustic tomography (Staley et al. 2010) and fluorescent-based imaging (Pogue et al. 2010) and optical wavefront engineering (Horstmeyer et al. 2015; Ruan et al. 2015) are also being investigated for their potential in identifying tissues intraoperatively. In recent studies, Jermyn et al. have implemented Raman spectroscopy for intraoperative detection of brain cancer (Jermyn et al. 2015). This Raman-based probe technique is able to detect brain tumors from grade 2 to 4 from *in vivo* study. It is able to distinguish cancer-cell-invaded brain from normal brain with sensitivity of 93% and a specificity of 91% from 17 patients and 161 optical measurements using leave-one-out statistical analysis. In addition, OCT has been used for intraoperative detection of brain tumors via *ex vivo* tissue analysis (Kut et al. 2015). This study is able to construct a color-coded map to distinguish cancer from normal tissue with high sensitivity (92%–100%) and specificity (80%–100%) from 37 patients (data from 16 patients are used for classification training). Both techniques promise near real-time data acquisition and analysis. With recent developments in fiber-optic probe and endoscopic techniques, it is feasible to translate these modalities from benchtop to bedside (Zysk et al. 2007). The light illumination and collection can be realized using a fiber probe/endoscope for the *in vivo* study without significant interference to the surgery. The optical biopsy has several advantages, making it compatible for clinical studies: (1) it is noninvasive or minimally invasive, (2) it provides rapid measurements that can lead to real-time diagnosis, (3) it is able to detect small tumors with high spatial resolution (Wang and Song 2012), and (4) it is able to detect precancerous conditions according to spectral characteristics that are associated with molecular changes (Prasad 2003). Exogenous contrast agents such as ICG attached to a tumor-specific moiety such as antibodies or peptides (e.g., chlorotoxin) are also being recently investigated (Butte et al. 2014; Veiseh et al. 2007).

In principle, the optical biopsy modalities that provide endogenous biochemical and morphological features are preferred in intraoperative applications over those that rely on exogenous contrast agents. Autofluorescence spectroscopy and imaging techniques are common non-contrast-agent-based modalities that can be used in intraoperative applications (Richards-Kortum and Sevick-Muraca 1996). Autofluorescence signals from the endogenous fluorophores in biological tissue have demonstrated high correlation with the tissue's biological composition and cellular structures. Time-resolved fluorescence spectroscopy (TRFS) measures dynamic fluorescence signals such as fluorescence lifetime, in addition to fluorescence spectrum (Lakowicz 2006). TRFS is particularly favored for *in vivo* diagnostic applications because fluorescence lifetime is independent of intensity variation artifacts common with *in vivo* measurements. However, TRFS can still be quite sensitive to microenvironment changes such as pH, ionization, and temperature (Fang et al. 2004).

In this chapter we introduce TRFS, which has already demonstrated strong potentials in characterizing brain tumors from previous outcomes (Butte et al. 2011). TRFS utilizes an ultrashort pulsed laser to excite the proteins and metabolites in the tissue and records the corresponding fluorescence intensity decay. It has been shown that, based on the fluorescence spectrum and the decay characteristics at various color bands, tumorous tissue can be characterized from the normal brain tissue in real time. Although TRFS is a single-point detection system, it can be incorporated with standard neuronavigation systems, which act as a 3D GPS (Figure 18.1), enabling the TRFS system to provide a 3D map as the probe is swept over different areas of the brain.

Such advances will help in translating this technology in the clinical setting and allowing neurosurgeons to ensure near-complete resection while saving the uninvolved brain tissue. In addition,

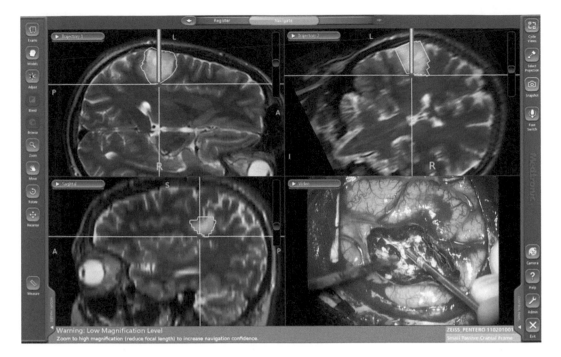

FIGURE 18.1 TRFS measurement registered by neuronavigation.

it will reduce the number of frozen section biopsies and subsequent time-consuming analysis, consequently reducing surgical time.

ENDOGENOUS FLUORESCENCE IN BIOLOGICAL TISSUE

There are several endogenous fluorophores that include amino acids (tyrosine and tryptophan), structural proteins (elastin and collagen), and enzyme cofactors (e.g., reduced nicotinamide adenine dinucleotide [NADH] and flavin-adenine dinucleotide [FAD]), which exhibit great potential to monitor tumor environments. The changes in the fluorescence signature from these fluorophores in the altered biochemical state and metabolism in tumors can be utilized for tissue characterizations. For example, an increase in NADH concentration is found in tissues with higher metabolic activities. NADH exhibits fluorescence emission in the spectral range of 400–700 nm when excited with UV radiation, whereas the oxidized form (NAD+) is nonfluorescent. In addition, NADH fluorescence lifetime is either short or long depending on its binding with proteins. The ratio change of free to bound NADH in tumors results in changes of the average lifetime of NADH (Skala et al. 2007). Thus, the intensity and lifetime of NADH become potential indicators to monitor the tumor's environment. Most endogenous fluorophores can be excited by UV-VIS light (280–450 nm) and emit fluorescence in the range of 370–700 nm, as shown in Figure 18.2. The fluorescence lifetime values of endogenous fluorophores are in the range of 0.5–6 ns. Detailed fluorescent properties of endogenous fluorophores are summarized in Table 18.1.

The TRFS system is based on natural fluorescence properties of brain tissue (Butte et al. 2005, 2010, 2011). Some of the main fluorophores in the brain such as NADH, FAD, lipopigments, and porphyrins have been extensively studied by Lin et al. (2001, 2009) (Figure 18.2). In general, it has been observed that the tumors have lower fluorescence emission when compared to normal tissue while exciting in the UV range at 355 nm wavelength. One such approach demonstrated that the ratio of NADH/FAD fluorescence emission may play a significant factor in identifying brain tumors (Liu et al. 2011).

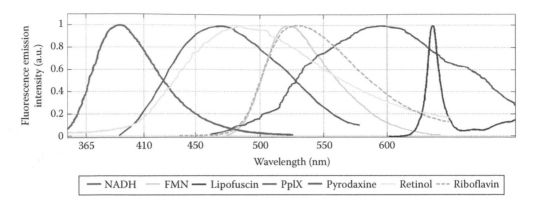

FIGURE 18.2   Natural fluorophores of biological tissue.

TABLE 18.1
Summary of Endogenous Fluorophores' Properties

Fluorophores	Excitation Peak (nm)	Emission Peak (nm)	Lifetime (ns)
Amino acid			
Tryptophan	280	350	0.5–2.8
Tyrosine	275	300	3.6
Phenylalanine	260	280	6.4
Enzymes/coenzyme			
NADH	340	450	0.4–2.8
FAD/FMN	370/440	510–520	2.3–5
Vitamin			
Pyridoxine	332, 340	400	
Pyridoxamine	335	400	0.11
Lipids			
Phospholipids	436	540, 560	
Lipofuscin	340–395	540–640, 430–460	1.5
Ceroid	340–395	540–640, 430–460	2
Lipopigments	340–395	430–460, 540	1.5

FLUORESCENCE MEASUREMENT METHODS

Fluorescence spectroscopy can be used either to investigate the variation in the intensity of fluorescence emissions (steady-state) or to detect the fluorescence decay (time-resolved). Both steady-state fluorescence spectroscopy and TRFS techniques have been investigated for their potential as an intraoperative diagnostic tool. Steady-state fluorescence spectroscopy measures the fluorescence intensity in a broadband spectral region, whereas TRFS measures the fluorescence decay or fluorescence dynamics, in addition to the intensity measurement. TRFS is considered a more robust tool for optical biopsy, as the fluorescence lifetime is independent of the emission and quantum yield, but is sensitive to changes in the tumor microenvironment. This property can be used to differentiate the fluorescence components with overlapped fluorescence emission spectra. In other words, it becomes a sensitive tool to probe the changes in microenvironment (e.g., pH, ionization, and temperature). Thus, TRFS can provide additional information compared to steady-state fluorescence techniques.

One advantage of the steady-state fluorescence spectroscopy is its relatively simple instrumentation, which usually includes a narrow band light source (laser or LED) and a grating-based spectrometer to collect the fluorescence emission. Alternately, TRFS requires a more complex instrumentation, including more sophisticated lasers, high-speed sensors, and electronics in picosecond speed range where most biological tissues demonstrate fluorescence decay. A short-pulsed laser is generally used as an excitation light in the time-domain measurement, whereas a modulated light source is used in frequency-domain measurement. To date, several time-domain applications have been used in clinical studies based on different data acquisition methods.

1. *Time-correlated single-photon counting (TCSPC)-based detection*: TCSPC is the most common technique to detect fluorescence signal of a fluorophore. In the TCSPC system, a pulsed laser with high repetition rate (e.g., 80 MHz) is used to excite the fluorophore, and only the single photon that reaches the detector first is captured after each excitation. A histogram of the photons with respect to their arrival time is generated after multiple excitations; thus, the profiles of fluorescence decay can be retrieved from this histogram. This method has high sensitivity and resolution, such that it can detect the weak autofluorescence signals from biological tissues. It has been extensively used in cellular or *ex vivo* tissue studies. However, the data acquisition time is relatively long, which limits its application as an intraoperative tool in surgery.

2. *Time-gated detection*: Time-gated detection provides an alternative method to detect the weak fluorescence signal in time-domain. In this method, the sample is excited by a short-pulsed laser. The fluorescence emission at different times along the fluorescence decay are detected by an imaging detector with a fixed gating time (e.g., 200 ps) and then are used to reconstruct the decay curve in time-domain. Time-gated, intensified charge-coupled device is the typical device used in such measurements. This method is suitable to obtain the image for a large spatial area. However, the temporal resolution and the sensitivity are relatively low.

3. *Streak camera–based detection*: Streak camera is a fast photo-detector that uses the time of flight to measure light intensity changes as a function of time. It operates by dispersing the photoelectrons across an imaging screen. Therefore, the photons arriving at the streak camera at different times are detected at different spatial positions at the sensor area. By using streak camera, multiple decay curves can be measured simultaneously with high temporal resolution of several picoseconds. However, the sensitivity is still lower than that of TCSPC.

4. *Pulse-sampling-based detection*: With the development of the high-speed digitizer, it is possible to acquire short fluorescence decay in picosecond or nanosecond temporal resolution. After the sample is excited by a laser pulse, fluorescence decay signal is detected by a point detector such as photomultiplier tube (PMT) or avalanche photodiode and then digitized by a high-speed digitizer. The fluorescence decay signal can be acquired directly after each laser excitation in real time. However, the temporal resolution of fluorescence signal is limited by the sampling rate of the digitizer

Considering the feasibility of these methods toward clinical studies, the advantages and disadvantages of each method are summarized in Table 18.2. In summary, the TCSPC method can provide the best sensitivity and temporal resolution. However, long acquisition time and high instrumentation costs limit its broad applications in clinical studies. On the contrary, time-gated- and pulse-sampling-based methods are much more feasible for clinical applications due to their fast data acquisition speed and affordable instrumentation costs.

Both single-point spectroscopy (Fang et al. 2004; Yuan et al. 2008, 2009) and multispectral and hyperspectral imaging system (Nie et al. 2013; Shrestha et al. 2010; Sun et al. 2009) have been investigated for their potential use in tissue diagnosis. In single-point measurements, a fiber-optic

TABLE 18.2
Comparison of Time-Resolved Fluorescence Measurement Methods

Method	Advantage	Disadvantage	Prices
TCSPC	High sensitivity High temporal resolution Low system error	Slow data acquisition	$$$$
Time-gated	Fast data acquisition Simultaneous acquisition for imaging	Low sensitivity	$$
Pulse-sampling	Fast data acquisition High intensity	Limited time resolution Large system error	$
Streak camera	High resolution Fast data acquisition	Low sensitivity	$$$$

probe is used to deliver laser light and collect fluorescence signal, which makes the TRFS much more compatible with clinical studies. In imaging measurements, a single fiber or fiber bundles are used to collect the image in the wide-field-based fluorescence lifetime imaging microscopy (FLIM) (Elson et al. 2007; Sun et al. 2009), whereas high-speed scanning modules such as galvanometer mirrors are used in scanning-based FLIM (Shrestha et al. 2010). Recently, video-frequency-based fluorescence imaging systems were reported to provide near real-time fluorescence lifetime monitor (Cheng et al. 2013; Schwartz et al. 1997).

CURRENT TRFS SYSTEMS

As shown in Figure 18.3, the TRFS system uses a 355 nm laser pulse (400 ps, 5 uJ/pulse) to excite the tissue fluorescence, a custom demuxer to split the spectral channels into 6 discrete wavelength bands from 360 to 700 nm, a custom fiber delay unit to delay individual color bands before being combined onto the photomultiplier tube (MCP-PMT, 80 ps rise time), a MCP-PMT to detect fluorescence, and a digitizer to sample the fluorescence lifetime measurements. The fluorescence decay was estimated at each color band using Laguerre deconvolution. The probe consisted of nonsolarizing silica/silica step index fibers of 0.12 numerical apertures (NA), as shown in Figure 18.3.

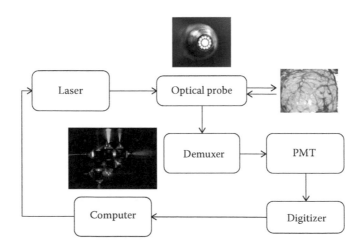

FIGURE 18.3 TRFS system diagram.

It had a central excitation fiber of 600 µm core diameter, surrounded by a collection ring of twelve 200 µm core diameter fibers. The collection fibers were beveled at a 10° angle in order to improve excitation/collection overlap for small tissue-to-probe distances.

Alternatively, the TRFS system may be combined with DR measurements to make the TRFS-DR system, as illustrated in Figure 18.4a (Nie 2014). The integrated system consists of four modules: the TRF subsystem, the DR subsystem, the dual-modality fiber-optic probe, and the control unit. The TRF subsystem uses an acousto-optic-tunable-filter (AOTF)-based spectrometer to collect dynamic fluorescence decay signals and uses a grating-based spectrometer to collect steady-state fluorescence spectra. The DR subsystem collects spatially resolved DR spectra in a small tissue volume. The four meter length clinical sterilizable fiber-optic probe is used to deliver illumination

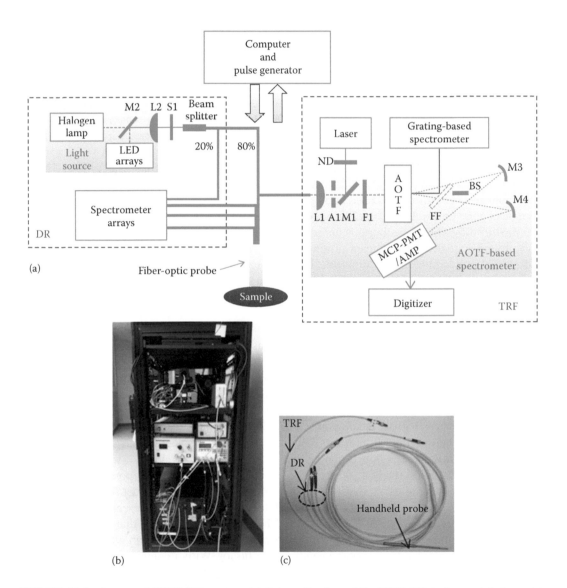

FIGURE 18.4 Integrated TRFS-DR system. (a) Schematic view of the TRFS-DR instrument that consists of four modules: the TRF subsystem, the DR subsystem, the dual-mode clinical fiber probe, and the control unit. (b) TRFS-DR instrument housed in a mobile cart, making the system compatible with clinical studies. (c) Dual-modality clinical sterilizable fiber-optic probe that is used to collect TRF and DR signals from the same tissue volume.

light and to collect TRF and DR signals. Photographs of TRFS-DR instrument housed in a mobile cart and the dual-modality fiber-optic probe are shown in Figure 18.4b and c, respectively. During clinical studies, autofluorescence and DR signals are collected in sequence within 2 s.

CLINICAL STUDIES FOR BRAIN TUMORS

Both steady-state and time-resolved fluorescence techniques have been extensively investigated for brain tumor diagnosis in the past decades. A brief review of fluorescence applications of brain tumor studies is presented in this section.

For steady-state autofluorescence spectroscopy, NADH and FAD are the two main endogenous fluorophores that were investigated in brain tissue. Usually, the emission intensity ratio of these two fluorophores is used to discriminate tumor from normal tissue. Chung et al. reported in an *ex vivo* study that gliomas had lower autofluorescent intensity than normal brain tissue when using different excitation lights in the spectral range of 360–490 nm. They also found that both fluorescent intensities of NADH and FAD were lower in all kinds of tumors than in normal tissue, but the ratio of NADH to FAD fluorescence intensity was not significantly different between the tumors and normal tissue. Another *in vivo* brain study has been investigated by Lin et al. (2001) using the combination of fluorescence and DR spectroscopy. This study obtained a sensitivity of 100% and a specificity of 76% when discriminating brain tumors from normal tissue. Croce et al. (2003) also investigated the autofluorescence properties of neoplastic and normal brain tissue using *ex vivo* tissue sections and homogenates. It was found that the fluorescence intensity ratio of 520 and 470 nm in glioblastoma is higher than that in normal tissue by using a 360 nm excitation light. This indicates an oxidized FAD and lipopigment increase in the neoplastic tissue compared to the normal sample. Lately, Lin et al. (2001) also used multiple excitation wavelengths (337, 360, and 440 nm) to differentiate pediatric neoplastic and epileptogenic tissue from normal tissue. Their study found statistically significant differences (P value <0.05) between neoplastic brain tissue and normal gray/white matter. Moreover, Liu et al. (2011) aimed to analyze the redox ratio, the intensity ratio of FAD and NADH, for brain tumors in a rat brain model. The results suggested that the fluorescence intensity ratio of different wavelengths can be used as a feature for brain tissue discrimination.

The initial TRF spectroscopy study of *ex vivo* brain tumors was reported by Marcu et al. (2004), in which the fluorescence lifetime of glioblastoma was found longer than that in normal white matter and cortex. In a later *ex vivo* study, the fluorescence lifetime features of low-grade and high-grade gliomas had been investigated by Yong et al. (2006). This study found the low-grade gliomas had faster decay than high-grade gliomas at the emission wavelength of 460 nm. Another *ex vivo* study by Butte et al. (2005) analyzed meningiomas using the TRF spectroscopy. Their results showed that the meningiomas have different fluorescence spectral intensities and lifetime values than normal dura and cerebral cortex tissue. By using both spectral and time-resolved features, the meningiomas can be distinguished with a high sensitivity and specificity of 89% and 100%, respectively. Lately, two *in vivo* clinical studies were reported by Butte et al. (2010, 2011). These studies showed that the low-grade gliomas had distinct fluorescence emission characteristics from the normal cortex and white matter, whereas fluorescence characteristics varied with different high-grade gliomas types. Their study also showed that both spectral and time-resolved features varied not only with the tumor phenotypes, but also with the location of the examination spot due to the heterogeneity of the tumor. With a combination of spectral and lifetime parameters, the low-grade gliomas could be distinguished from other tissue types with 100% sensitivity and 98% specificity. On the contrary, high-grade gliomas were distinguishable with 47% sensitivity and 94% specificity. Sun et al. (2010) demonstrated the feasibility of using FLIM in glioblastoma multiform diagnosis using a fiber bundle and a gated intensifier imaging system. The clinical results showed that FLIM was able to differentiate tumor from normal cortex at the emission wavelength of 460 nm.

TRFS CLINICAL STUDIES

We report a recent clinical trial using TRFS intraoperatively in patients (N = 9) undergoing surgical resection of glioblastoma as a pilot study. The patients were scheduled for routine surgical removal of brain tumor and underwent the planned operation. The TRFS system described in "Current TRFS System" (Figure 18.3) was used during the procedure. The UV light was delivered to the tumor site using a custom-made fiber-optic probe. This fiber-optic probe was positioned above the areas of interest and the brain tissue was spectroscopically interrogated. A biopsy was performed at the site investigated by the TRFS and histopathological analysis was performed to confirm the diagnosis. The histopathological diagnosis was used as a standard to validate our results. A classification algorithm based on linear discriminant analysis (LDA) was developed and the fluorescence emission data collected from the patients was used as a preliminary training set. A total number of 35 measurements were recorded, including normal cortex (NC, N = 10), normal white matter (WM, N = 12), and glioblastoma (GBM, N = 13). The training set for the classification algorithm included the measurements with histopathological analysis confirmation as a single tissue type (mixed tissues such as various levels of infiltration were excluded). Additional normal cortex measurements (N = 8) from other noninfiltrating tumor surgeries such as meningioma were also acquired, but no biopsies were performed. These measurements were deemed normal as there was no evidence of glioblastoma.

Examples of classification results to differentiate between normal cortex, white matter, and glioma during *in vivo* brain tumor surgery are shown in Figure 18.5. The LDA classification has been implemented by three classifier sets, in which the output of the classifier will be either of the training groups. With the NC classifier, for example, we group WM and GBM tissue measurements as the "not normal cortex" group. Then the NC classifier will be trained based on NC and Not NC training sets. The same subclassifiers have been developed for WM and GBM where NC+GBM and NC+WM have been regrouped as Not WM and Not GBM, respectively. Figure 18.5a–c shows the subclassifiers' power to discriminate between the training groups. All three classifiers can be integrated in a three-dimensional scatterogram. Figure 18.5d shows the classification scatterogram for normal cortex (n = 10, blue), normal white matter (n = 12, green), and glioma (n = 13, red) from 9 patients.

Figure 18.6 shows the classification test on 8 normal cortex measurements in the meningioma case study. None of the normal cortex measurements are confirmed by biopsy. However, due to the meningioma developmental anatomy, the chance of getting white matter or glioma at the normal cortex areas is almost impossible.

MULTIMODALITY OPTICAL BIOPSY FOR CLINICAL STUDIES

Even though each aforementioned modality has the potential to differentiate tumor from normal tissue, neither of them can be used as a "standard" tool. Improvements in measurement and data analysis are still required to achieve the desired performance for clinical applications. On the other hand, information acquired from multimodality optical biopsies may be able to complement the drawback of each single modality. A combination of steady-state fluorescence spectroscopy and TRFS has been used in several clinical studies. By combining these two modalities, the fluorescence features (e.g., metabolic status, cellular structures) and optical properties (e.g., scattering size and density, hemoglobin absorption) can be used to investigate the measured biological tissue. This will help us to study the differences between various tissue types. Lin and his coworkers demonstrated the feasibility of dual modality in determining the brain tumor types (Lin et al., 2001; Valdés et al., 2011; Yadav et al., 2013). These studies illustrate that it is possible to achieve higher contrast for tissue diagnosis using multimodality optical biopsy. Moreover, several clinical instruments that combine spatially resolved DR spectroscopy and steady-state fluorescence spectroscopy have been proposed by several groups. Optical properties can be calculated using the DR measurement. Based on the optical properties of biological tissue, it is possible to estimate the distortion of fluorescence excitation light and emission light, thus retrieving the intrinsic fluorescence signals. Several methods have been developed to solve this problem.

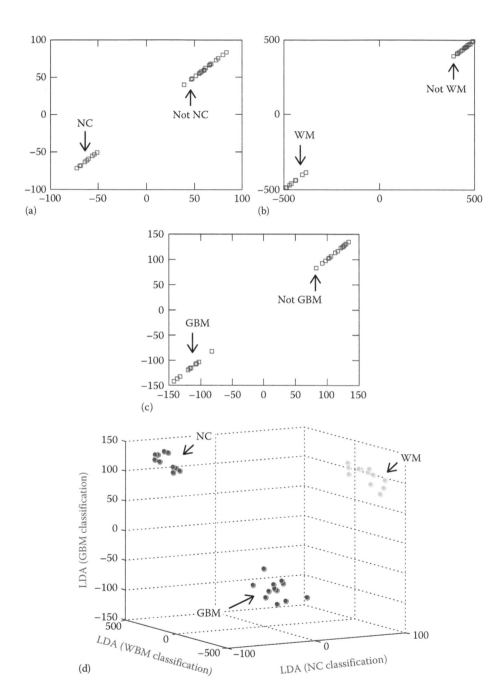

FIGURE 18.5 TRFS classification output for differentiating normal cortex, white matter, and GBM tissue types. (a) Normal cortex classification, (b) white matter classification, (c) GBM classification, and (d) three dimensional scatterogram of tissue type using three subclassifiers shown in (a)–(c). (Data from Vasefi, D.S., et al. 2016. Butte. Intraoperative optical biopsy for brain tumors using spectro-lifetime properties of intrinsic fluorophores. *Proc. SPIE*. 9711, 97111T.)

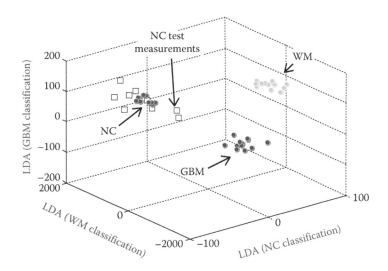

FIGURE 18.6   Meningioma case study with normal cortex measurements (N = 8).

REFERENCES

Berger, M.S. Malignant astrocytomas: Surgical aspects. *Seminars in Oncology* 21 (1994): 172–185.

Butte, P.V., Q. Fang, J.A. Jo, W.H. Yong, B.K. Pikul, K.L. Black, and L. Marcu. Intraoperative delineation of primary brain tumors using time-resolved fluorescence spectroscopy. *Journal of Biomedical Optics* 15(2) (2010): 027008–027008.

Butte, P.V., A. Mamelak, J. Parrish-Novak, D. Drazin, F. Shweikeh, P.R. Gangalum, A. Chesnokova, J.Y. Ljubimova, and K. Black. Near-infrared imaging of brain tumors using the Tumor Paint BLZ-100 to achieve near-complete resection of brain tumors. *Neurosurgical Focus* 36(2) (2014): E1.

Butte, P.V., A.N. Mamelak, M. Nuno, S.I. Bannykh, K.L. Black, and L. Marcu. Fluorescence lifetime spectroscopy for guided therapy of brain tumors. *Neuroimage* 54 (2011): S125–S135.

Butte, P.V., B.K. Pikul, A. Hever, W.H. Yong, K.L. Black, and L. Marcu. Diagnosis of meningioma by time-resolved fluorescence spectroscopy. *Journal of Biomedical Optics* 10(6) (2005): 064026–064026.

Byar, D.P., S.B. Green, T.A. Strike, and M.D. Walker. Prognostic factors for malignant glioma. *Oncology of the Nervous System* 12 (1983): 379–395.

Centers for Disease Control and Prevention. Colorectal cancer rates by state, 2012. http://www.cdc.gov/cancer/colorectal/statistics/state.htm, accessed January 23, 2017.

Cheng, S., J. Jesus Rico-Jimenez, J. Jabbour, B. Malik, K.C. Maitland, J. Wright, Y.-S. Lisa Cheng, and J.A. Jo. Flexible endoscope for continuous in vivo multispectral fluorescence lifetime imaging. *Optics Letters* 38(9) (2013): 1515–1517.

Chorvat Jr, D. and A. Chorvatova. Multi-wavelength fluorescence lifetime spectroscopy: A new approach to the study of endogenous fluorescence in living cells and tissues. *Laser Physics Letters* 6(3) (2009): 175.

Chung, Y.G., J.A. Schwartz, C.M. Gardner, R.E. Sawaya, and S.L. Jacques. Diagnostic potential of laser-induced autofluorescence emission in brain tissue. *Journal of Korean Medical Science* 12 (1997): 135–142.

Croce, A.C., S. Fiorani, D. Locatelli, R. Nano, M. Ceroni, F. Tancioni, E. Giombelli, E. Benericetti, and G. Bottiroli. Diagnostic potential of autofluorescence for an assisted intraoperative delineation of glioblastoma resection margins. *Photochemistry and Photobiology* 77(3) (2003): 309–318.

Ducray, F., G. Dutertre, D. Ricard, E. Gontier, A. Idbaih, and C. Massard. Advances in adults' gliomas biology, imaging and treatment. *Bulletin du Cancer* 97(1) (2010): 17–36.

Elias, J.W., K.-M. Fu, and R.C. Frysinger. Cortical and subcortical brain shift during stereotactic procedures. *Journal of Neurosurgery* 107(5) (2007): 983–988.

Elson, D.S., J.A. Jo, and L. Marcu. Miniaturized side-viewing imaging probe for fluorescence lifetime imaging (FLIM): Validation with fluorescence dyes, tissue structural proteins and tissue specimens. *New Journal of Physics* 9(5) (2007): 127.

Eyüpoglu, I.Y., M. Buchfelder, and N.E. Savaskan. Surgical resection of malignant gliomas—Role in optimizing patient outcome. *Nature Reviews Neurology* 9(3) (2013): 141–151.

Fang, Q., T. Papaioannou, J.A. Jo, R. Vaitha, K. Shastry, and L. Marcu. Time-domain laser-induced fluorescence spectroscopy apparatus for clinical diagnostics. *Review of Scientific Instruments* 75(1) (2004): 151–162.

Fazekas, J.T. Treatment of grades I and II brain astrocytomas. The role of radiotherapy. *International Journal of Radiation Oncology, Biology and Physics* 2 (1977): 661–666.

Giese, A., H.J. Böhringer, J. Leppert, S.R. Kantelhardt, E. Lankenau, P. Koch, R. Birngruber, and G. Hüttmann. Non-invasive intraoperative optical coherence tomography of the resection cavity during surgery of intrinsic brain tumors, *Proc. SPIE 6078, 60782Z, Photonic Therapeutics and Diagnostics II*, January 21, 2006, San Jose, CA, 2006.

Horstmeyer, R., H. Ruan, and C. Yang. Guidestar-assisted wavefront-shaping methods for focusing light into biological tissue. *Nature Photonics* 9(9) (2015): 563–571.

Jermyn, M., K. Mok, J. Mercier, J. Desroches, J. Pichette, K. Saint-Arnaud, L. Bernstein, M.-C. Guiot, K. Petrecca, and F. Leblond. Intraoperative brain cancer detection with Raman spectroscopy in humans. *Science Translational Medicine* 7(274) (2015): 274ra19–274ra19.

Kubben, P.L., K.J. ter Meulen, O. EMG Schijns, M.P. ter Laak-Poort, J.J. van Overbeeke, and H. van Santbrink. Intraoperative MRI-guided resection of glioblastoma multiforme: A systematic review. *The Lancet Oncology* 12(11) (2011): 1062–1070.

Kut, C., K.L. Chaichana, J. Xi, S.M. Raza, X. Ye, E.R. McVeigh, F.J. Rodriguez, A. Quiñones-Hinojosa, and X. Li. Detection of human brain cancer infiltration ex vivo and in vivo using quantitative optical coherence tomography. *Science Translational Medicine* 7(292) (2015): 292ra100–292ra100.

Lakowicz, J.R. *Principles of Fluorescence Spectroscopy*, 3rd edn. Springer, New York, 2006.

Lin, W.-C., D.I. Sandberg, S. Bhatia, M. Johnson, G. Morrison, and J. Ragheb. Optical spectroscopy for in-vitro differentiation of pediatric neoplastic and epileptogenic brain lesions. *Journal of Biomedical Optics* 14(1) (2009): 014028–014028.

Lin, W.-C., S.A. Toms, M. Johnson, E.D. Jansen, and A. Mahadevan-Jansen. Vivo brain tumor demarcation using optical spectroscopy. *Photochemistry and Photobiology* 73(4) (2001): 396–402.

Liu, Q., G. Grant, J. Li, Y. Zhang, F. Hu, S. Li, C. Wilson, K. Chen, D. Bigner, and T. Vo-Dinh. Compact point-detection fluorescence spectroscopy system for quantifying intrinsic fluorescence redox ratio in brain cancer diagnostics. *Journal of Biomedical Optics* 16(3) (2011): 037004–037004.

Marcu, L., J.A. Jo, P.V. Butte, W.H. Yong, B.K. Pikul, K.L. Black, and R.C. Thompson. Fluorescence lifetime spectroscopy of glioblastoma multiforme. *Photochemistry and Photobiology* 80(1) (2004): 98–103.

National Cancer Institute. CBTRUS statistical report: Primary brain and central nervous system tumors diagnosed in the United States in 2004–2008, 2012.

Nie, Z. Integration of time-resolved fluorescence and diffuse reflectance spectroscopy for intraoperative detection of brain tumor margin, PhD thesis, McMaster University, Hamilton, Ontario, Canada, 2014.

Nie, Z., A. Ran, J.E. Hayward, T.J. Farrell, and F. Qiyin. Hyperspectral fluorescence lifetime imaging for optical biopsy. *Journal of Biomedical Optics* 18(9) (2013): 096001.

Pogue, B.W., S.L. Gibbs-Strauss, P. Valdés, K.S. Samkoe, D.W. Roberts, and K.D. Paulsen. Review of neurosurgical fluorescence imaging methodologies. *Journal of Selected Topics in Quantum Electronics, IEEE* 16(3) (2010): 493–505.

Prasad, P.N. *Introduction to Biophotonics*, John Wiley & Sons, New York, 2003.

Ramanujam, N. Fluorescence spectroscopy of neoplastic and non-neoplastic tissues. *Neoplasia* 2(1) (2000): 89–117.

Richards-Kortum, R. and E. Sevick-Muraca. Quantitative optical spectroscopy for tissue diagnosis. *Annual Review of Physical Chemistry* 47(1) (1996): 555–606.

Robins, H.I., A.B. Lassman, and D. Khuntia. Therapeutic advances in malignant glioma: Current status and future prospects. *Neuroimaging Clinics of North America* 19(4) (2009): 647–656.

Ruan, H., M. Jang, and C. Yang. Optical focusing inside scattering media with time-reversed ultrasound microbubble encoded light. *Nature Communications* 6 (2015): 8968.

Sanai, N. and M.S. Berger. Glioma extent of resection and its impact on patient outcome. *Neurosurgery* 62 (2008): 753–764; discussion 264–756.

Shrestha, S., B.E. Applegate, J. Park, X. Xiao, P. Pande, and J.A. Jo. High-speed multispectral fluorescence lifetime imaging implementation for in vivo applications. *Optics Letters* 35(15) (2010): 2558–2560.

Skala, M.C., K.M. Riching, A. Gendron-Fitzpatrick, J. Eickhoff, K.W. Eliceiri, J.G. White, and N. Ramanujam. In vivo multiphoton microscopy of NADH and FAD redox states, fluorescence lifetimes, and cellular morphology in precancerous epithelia. *Proceedings of the National Academy of Sciences* 104(49) (2007): 19494–19499.

Sosna, J., M.M. Barth, J.B. Kruskal, and R.A. Kane. Intraoperative sonography for neurosurgery. *Journal of Ultrasound in Medicine* 24(12) (2005): 1671–1682.

Staley, J., P. Grogan, A.K. Samadi, H. Cui, M.S. Cohen, and X. Yang. Growth of melanoma brain tumors monitored by photoacoustic microscopy. *Journal of Biomedical Optics* 15(4) (2010): 040510–040510.

Sun, Y., N. Hatami, M. Yee, J. Phipps, D.S. Elson, F. Gorin, R.J. Schrot, and L. Marcu. Fluorescence lifetime imaging microscopy for brain tumor image-guided surgery. *Journal of Biomedical Optics* 15(5) (2010): 056022–056022.

Sun, Y., J. Phipps, D.S. Elson, H. Stoy, S. Tinling, J. Meier, B. Poirier, F.S. Chuang, D.G. Farwell, and L. Marcu. Fluorescence lifetime imaging microscopy: In vivo application to diagnosis of oral carcinoma. *Optics Letters* 34(13) (2009): 2081–2083.

Valdés, P.A., A. Kim, F. Leblond, O.M. Conde, B.T. Harris, K.D. Paulsen, B.C. Wilson, and D.W. Roberts. Combined fluorescence and reflectance spectroscopy for in vivo quantification of cancer biomarkers in low-and high-grade glioma surgery. *Journal of Biomedical Optics* 16(11) (2011): 116007–11600714.

Vasefi, F., D.S. Kittle, Z. Nie, C. Falcone, C.G. Patil, R.M. Chu, A.N. Mamelak, K.L. Black, and P.V. Butte. Intraoperative optical biopsy for brain tumors using spectro-lifetime properties of intrinsic fluorophores. In: *Proc. SPIE. Imaging, Manipulation, and Analysis of Biomolecules, Cells, and Tissues IX 9711, 97111T*, February 13, 2016, San Francisco, CA, 2016.

Vasefi, F., N. MacKinnon, D.L. Farkas, and B. Kateb. Review of the potential of optical technologies for cancer diagnosis in neurosurgery: A step toward intraoperative neurophotonics. *Neurophoton* 4(1) (December 2016): 011010.

Veiseh, M., P. Gabikian, S. Bahrami, O. Veiseh, M. Zhang, R.C. Hackman, A.C. Ravanpay et al. Tumor paint: A chlorotoxin: Cy5. 5 bioconjugate for intraoperative visualization of cancer foci. *Cancer Research* 67(14) (2007): 6882–6888.

Wagnieres, G.A., W.M. Star, and B.C. Wilson. In vivo fluorescence spectroscopy and imaging for oncological applications. *Photochemistry and Photobiology* 68(5) (1998): 603–632.

Wang, L.V. and H. Song. Photoacoustic tomography: In vivo imaging from organelles to organs. *Science* 335(6075) (2012): 1458–1462.

Yadav, N., S. Bhatia, J. Ragheb, R. Mehta, P. Jayakar, W. Yong, and W.-C. Lin. In vivo detection of epileptic brain tissue using static fluorescence and diffuse reflectance spectroscopy. *Journal of Biomedical Optics* 18(2) (2013): 027006–027006.

Yaroslavsky, A.N., P.C. Schulze, I.V. Yaroslavsky, R. Schober, F. Ulrich, and H.J. Schwarzmaier. Optical properties of selected native and coagulated human brain tissues in vitro in the visible and near infrared spectral range. *Physics in Medicine and Biology* 47(12) (2002): 2059.

Yong, W.H., P.V. Butte, B.K. Pikul, J.A. Jo, Q. Fang, T. Papaioannou, K.L. Black, and L. Marcu. Distinction of brain tissue, low grade and high grade glioma with time-resolved fluorescence spectroscopy. *Frontiers in Bioscience: A Journal and Virtual Library* 11 (2006): 1255.

Yuan, Y., J.-Y. Hwang, M. Krishnamoorthy, K. Ye, Y. Zhang, J. Ning, R.C. Wang, M.J. Deen, and Q. Fang. High-throughput acousto-optic-tunable-filter-based time-resolved fluorescence spectrometer for optical biopsy. *Optics Letters* 34(7) (2009): 1132–1134.

Yuan, Y., T. Papaioannou, and Q. Fang. Single-shot acquisition of time-resolved fluorescence spectra using a multiple delay optical fiber bundle. *Optics Letters* 33(8) (2008): 791–793.

Zhou, Y., C.-H. Liu, Y. Sun, P. Yang, S. Boydston-White, Y. Liu, and R.R. Alfano. Human brain cancer studied by resonance Raman spectroscopy. *Journal of Biomedical Optics* 17(11) (2012): 116021–116021.

Zysk, A.M., F.T. Nguyen, A.L. Oldenburg, D.L. Marks, and S.A. Boppart. Optical coherence tomography: A review of clinical development from bench to bedside. *Journal of Biomedical Optics* 12(5) (2007): 051403–051403.

19 Spectro-Temporal Autofluorescence Contrast– Based Imaging for Brain Tumor Margin Detection and Biobanking

Asael Papour, Zach Taylor, Linda Liau, William H. Yong, Oscar Stafsudd, and Warren Grundfest

CONTENTS

INTRODUCTION

BRAIN CANCER

Brain tumors are the most complicated and challenging in treatment. Life expectancy is very low, and conventional treatment such as chemotherapy is mostly ineffective due to the blood–brain barrier. In most cases, treatment will involve physical intervention and surgery that will reduce the major part of the tumor (Burger et al. 1985; Brandes et al. 2008). Currently, the standard margin detection methods for tumors are visual and tactile feedback during surgical resection with frozen section microscopy. The efficacy of this approach varies widely based on the skill of the surgeon and is subjected to sampling errors. However, complete resection is nearly impossible because gliomas, and glioblastoma multiforme (GBM) in particular, are diffusely infiltrating with irregular, indistinct margins (Wallner et al. 1989; Hess 1999). This problem is an example of an unmet clinical need, calling for an intraoperative tool that can aid surgeons in detecting margins of abnormal tissue, and give higher confidence levels in complete resections. Many approaches have been explored and abandoned mainly due to high costs, extensively prolonged operational time, and complexity of use (Lunsford et al. 1996). In this chapter, we will introduce a new technology that is capable of imaging contrast and differentiating tissues in real time, based on indigenous tissue fluorophores' response to optical excitation.

In Translational Research

Detecting biochemical and molecular markers, which have been established in spectroscopic bio-medical research, can be utilized not only to offer video-rate contrast imaging intraoperatively but also to provide a refuge in tissue screening limitations in brain repositories. Tissue repositories play a salient role in translational research; providing a reserve of high quality tumor specimens, facilitating innovative research, and enabling advance treatments for diseases and cancers (Khanna et al. 2010). To date, there are more than 300 million tissue samples that are stored in biobanks in the United States alone, however, major number of researchers have trouble finding samples of sufficient quality (Baker 2012). Currently, pathologists rely on visual assessment of tissues prior to distribution to researchers. This calls for a system that can assist pathologists in prioritizing samples to provide rapid and accurate assessments of surface structure and complexity, tissue type, and tissue viability in regards to DNA and RNA extraction.

As a first step toward better abnormal tissue margin detection and advancing translational research, this chapter presents a robust method and system that brings innovation from multidisciplinary university research to undertake real-world topics.

THEORY

Ultraviolet (UV) radiation is part of the electromagnetic spectrum, between the visible and x-ray regions, that spans from extreme UV at 10 nm (nanometers, 10^{-9} m) to visible violet at 400 nm. Most of the UV spectra, extreme to middle UV (10–300 nm), is absorbed by the atmosphere and does not propagate to long distances. UV radiation is used in many applications in black-light activated technologies and serves different purposes at various wavelength bands; near UV 300–400 nm is used in security measures and counterfeit detection, middle UV 200–300 nm used in disinfection devices, and far UV 100–200 nm that is absorbed by oxygen and does not propagate well in air. The middle UV has more potential in damaging tissue and can cause severe damage to DNA (Sinha and Hader 2002). Devices that use this wavelength band exhibit harmful radiation levels and use it to denature proteins to effectively kill bacteria. Excimer lasers often operate in the far UV and are used in the semiconductor industry for photolithography and in medicine for eye surgeries (LASIK), ablating material without heating the tissue.

The technologies that are based on the less energetic near-UV light explore the light–material interaction to detect chemicals that otherwise are left invisible. By illuminating near-UV light on ID cards, for example, a hidden pattern will glow (fluoresce) and confirm if genuine. The emitted light is the fluorescence response of the added chemicals (fluorophores) that are visible to the naked eye under UV illumination. Some fluorophores emit broadband light which appears to be toned white, while others are engineered to fluoresce in a narrow regime of the spectrum and show a pronounced color.

Proteins in our body include metabolic, structural, and others, each revealing a unique auto-fluorescence spectrum that can be identified and mapped within the tissue or individual cells. Looking at intensities and color distribution (spectrum) of the emitted light using amplitude fluorescence technique, much can be learned on cellular organization and mechanisms, tissue morphology and structure (Ramanujam 2000). This technology has shown good results in imaging materials, however, it encountered many difficulties in tissue: Differentiation between endogenous fluorophores is less accurate in cluttered environment and is sensitive to scattering and absorption and misrepresents the true content of the tissue.

In order to overcome these shortcomings, another aspect of fluorescence emission has been investigated and shown to have improved detection capabilities especially in high variability and clutter-dominated tissue. By looking at the fluorescence intensity change in time, after a short pulse of UV light, the time dependent behavior of the emitted signal can be measured and analyzed. This technique can characterize tissue with higher sensitivity and specificity (80%–90%) (Tadrous et al. 2003).

Unlike phosphorescence materials, in which a diminishing observable "glow" can be seen for a relatively long time (milliseconds to hours), fluorescence materials tend to emit light in a much shorter time scale (pico-, nano-, and microseconds). Fluorescence emission is a process that "allows" the molecule to release the energy of the absorbed UV light and relax back to a stable state with lower energy. This is done by emitting light at a longer wavelength (often visible).

Fluorescence lifetime imaging microscopy (FLIM) is a research technique that extracts the autofluorescence lifetime from the fluorescence intensity decay at each point on the tissue sample and delineates tissue constituents based on variations in decay time. This is accomplished by exciting the sample with a very short (picosecond) pulse of a UV laser and detecting the autofluorescence decay coefficient at a range of wavelengths (Marcu et al. 2004; Lakowicz 2006; Sun et al. 2010).

The fluorescence decay rate is governed by the specific molecular energy levels structure: Slow relaxation mechanisms will reveal longer fluorescence decay emission, while fast energy relaxation, which is an increase in probability for relaxation, will reveal a short fluorescence decay behavior. This added observation, of energy levels configuration and relaxation rate, is a detailed indicator of molecular and chemical structure that results in greater detection specificity (Lakowicz 2006). Energy levels and relaxation processes can also be sensitive to environmental changes—acidity (pH), pressure, temperature, and proximity of other molecules or proteins—and cause temporal and spectral changes to the emitted fluorescence signals. These changes are important and can serve as an indicator for tissue health and metabolic processes (Lakowicz 2006).

Traditional time-resolved fluorescence imaging techniques, such as FLIM, measure the intensity decay profile point by point consecutively in time. This method records the fluorescence decay profile in steps to recreate a theoretical model that can fit and estimate the decay shape characteristics. Fluorescence decay is a statistical phenomenon that can be closely estimated and described mathematically by an exponential decay function. The radiative rate of decay is governed by the parameter τ (Greek, Tau) that represents the fluorescence lifetime of a material with a simple single exponential behavior (Equation 19.1). The initial amplitude is 1 (normalized function) for simplification.

$$F(t) = \exp(-\tau/T) \tag{19.1}$$

A theoretical line can be drawn to fit the data points in a shape of an exponential decay function to estimate and extract the characteristic lifetime coefficient (Figure 19.1).

Decay rates are dependent on the biochemical composition of the material, and significant contrast can be generated using maps of decay times. Time-resolved fluorescence is superior to amplitude fluorescence, since the fluorophores' lifetime is not related to fluorescence intensity and does not depend on factors that would significantly degrade amplitude-only measurements such as concentration, absorption, and scattering by tissue surface geometry.

However, this technique has not transcended the research and development stage into the clinical level due to several limiting factors. Data processing of the large recorded datasets is computationally intensive and uses methods of lifetime extraction, for example, via orthogonal basis calculation (Laguerre polynomials), to identify fluorophore decay times and distributions (Wagnieres et al. 1998; Maarek et al. 2000; De Veld et al. 2005; Jo et al. 2005; Phipps et al. 2012). This powerful approach is slow and prone to significant errors as signal-to-noise ratio (SNR) drops (Jo et al. 2005). Moreover, the solution is not unique and requires setting a priori assumptions that can oversimplify the complex nature of the decay behavior (Lakowicz 2006). There exists no robust, objective manner

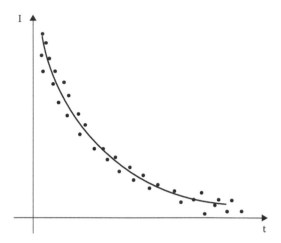

FIGURE 19.1 Conventional fluorescence lifetime measurement: Intensity measurements of the signal at each time point (dots) and a curve fit that approximates the exponential decay (solid curve).

to choose the number of terms in the formula, and different numbers of terms can all produce good fits but provide completely different interpretations of the results. Another limiting factor is found in the image acquisition routine; low signal levels require a longer measurement time in order to produce better fitting results; hence, a typical acquisition time for a large field of view (>0.5 mm) including processing time is in the order of a few minutes per lifetime image (Jo et al. 2005). To overcome these limitations, many systems are based on single-point spectroscopy and photon-counting techniques or small field-of-view optical fiber bundle. Point spectroscopy, for example, requires mechanical scanning to create a 2D image and has hindered the clinical acceptance of this technique.

SPECTRO-TEMPORAL AUTOFLUORESCENCE CONTRAST IMAGING

Imaging Theory

A new approach has been developed in pursuit of a wide-field, video-rate capability for autofluorescence signatures imaging. Since traditional calculations of decay rate are used to create a contrast-based color map, a new paradigm utilizes the differences in lifetime signatures and converts them directly to a contrast image, without any lifetime calculations (Jiang et al. 2011; Papour et al. 2013). By controlling alternative illumination and processing methods, a shift from the traditional short-pulsed UV laser scheme allows for an intelligent measurement that encapsulates the lifetime property without the extraction of the decay coefficient using inexpensive light-emitting diodes (LEDs). This system combines a nonconventional illumination and acquisition scheme with a normalization algorithm to produce direct contrast–based images of fluorescence signatures based on lifetime differences. By using a new illumination and acquisition technique, a tremendous gain in total acquisition and processing time is achieved, in contrast with traditional FLIM that uses the shortest illumination pulse possible to brute force a fluorescence lifetime decay coefficient measurement.

Raw fluorescence signals are acquired as images and undergo processing to relative lifetime maps. The maps show contrast in areas with different structure based on the underlying constituents of the tissue. Variance in constituent concentration and location are seen as changes in contrast, similar to the way x-rays create a contrast based on relative changes in density. Constituents such as fat, collagen, elastin, and nicotinamide adenine dinucleotide + hydrogen (NADH) dominate the lifetime map, with the level of expression dependent on the detection wavelength, due to their unique fluorescence signature. These maps can show normal and abnormal distributions of constituents and lead to a better understanding of the tissue complexity and biochemical structure. As a result, the contrast maps are a sensitive indication for tumor boundaries and other abnormalities, including the integrity of the tissue sample.

Understanding and visualizing chemical and morphological changes on the tissue's surface has the potential to delineate tissue abnormalities and accelerate tissue sorting by allowing a degree of identification that is currently unavailable in brain surgeries and biobanking. Moreover, analysis of relationships between the different spectral lifetime maps could potentially enable quantitative deduction of the tumor's stage (Yong 2006).

SIGNAL PROCESSING

The image processing procedure is not computationally intensive; however, it encapsulates a key capability of the system. Two fluorescence images are acquired and then normalized by element-wise matrix division. The resulting values represent the fluorescence decay rate. Higher values represent longer fluorescence lifetimes, while lower values represent shorter lifetimes. Each pixel acquires different information from the tissue surface, thus it can have a different final value. The difference between pixels' values is the mechanism of contrast in the final relative lifetime map, and can be optimized by acquisition parameters that involve both numerical and physical illumination manipulation. The imaging methodology encompasses acquisition of only two images; the first image is taken during excitation and is referred to as the "calibration image," while the second image is taken during the fluorescence decay and referred to as the "decay image" (Figure 19.2).

The calibration image is acquired during the LEDs' constant illumination for ~10 ns and then saved. This image holds important information of the fluorescence yield and locality. By looking at the steady-state intensity, we can eliminate scattering and absorption factors and ultimately be sensitive to the lifetime factor alone. Next, the LEDs are turned off and the second image is acquired during the fluorescence decay for a period of ~10 ns, depending on the tissue type. The decay image is recording the unique decay signatures of the tissue at each location (pixel). In this acquisition system, all of the pixels are acquired in parallel using conventional imaging optics.

In the processing stage, the decay image is divided by the calibration image to create a relative lifetime decay map: Each pixel's intensity value in the decay image is divided with the corresponding pixel in the calibration image. This results in a final relative lifetime image (normalized image) (Figure 19.3). This method creates contrast by taking only two images: one during illumination and the second during the decay period. By dividing the decay signature with its corresponding steady-state fluorescence yield, a quotient that correlates only to lifetime is accomplished.

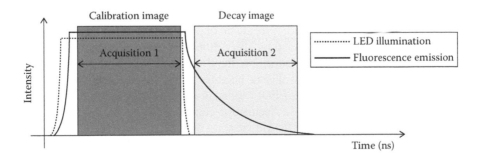

FIGURE 19.2 Fluorescence relative lifetime acquisition algorithm; two measurements are sufficient in extracting the differences in fluorescence lifetimes of a sample. A calibration image is acquired during LED constant illumination and then saved. The LED is turned off and the camera captures the second image during the fluorescence decay.

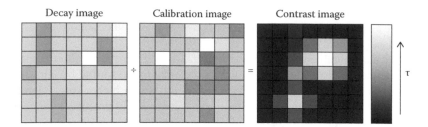

FIGURE 19.3 Fluorescence relative lifetime processing algorithm; the decay image is divided by the calibration image: each pixel's intensity value in the decay image is divided with the corresponding pixel in the calibration image. This results in a final relative lifetime image (normalized image). This image is characterized in high signal-to-noise ratio since each pixel is integrating the intensity of the fluorescence signal during the acquisition period. In this methodology, the fluorophore lifetime and the integrated signals are correlated to produce relative lifetime information where bright pixels represent longer lifetimes.

Variation in lifetime on the tissue surface will reveal image contrast of the different corresponding locations. A larger quotient means a long relative lifetime while a smaller quotient represents a shorter lifetime.

This procedure results in pronounced contrast images with high SNR due to the acquisition metrics; each pixel is integrating the intensity of the fluorescence signal during the acquisition periods, and allows for the total imaging time to be very short, less than 1 second.

This measurement can be repeated in many different excitation and collection wavelengths using various LEDs and band-pass filters to divide the spectrum and isolate individual constituents for better contrast and increased specificity. The ability to create multispectral imaging can also increase sensitivity to find higher-order correlations of normal versus abnormal tissue locations.

Finding lifetime information in a relative type measurement method is similar to x-ray, and magnetic resonance imaging (MRI), where contrast is the result of local changes in density or concentration, respectively. This new modality displays optical variation in tissue by morphological and density changes of the constituents in the top layers of the tissue, to reveal features that are not visible to the naked eye.

The analogy to x-ray and MRI serves as a good comparison to modalities in which maps of relative measurements are used. These modalities do not supply absolute measurements and yet serve as a powerful tool for diagnostics. The fast autofluorescence contrast imaging modality is compatible with clinical settings, since the short acquisition time eliminates blur and signal degradation due to natural movements, pulse, breathing, etc. In comparison to conventional FLIM systems, the relative measurement system yields a hundred fold decrease in acquisition time and 10,000 fold decrease in processing time. This novel system also eliminates common problems of FLIM such as damage to the tissue and photobleaching due to the high peak power of lasers by using LED technology with long flat-top pulses and low peak power. Since processing for image generation requires only simple mathematical operations, it is possible to process an entire image with good contrast at video rate with the minimal processing power found on every mobile computer platform. Since this method does not require fluorophore lifetimes extraction, the illumination pulse widths can vary from 1 to 20 ns depending on tissue type. This avoids the need for an expensive, bulky, pulsed UV laser and permits the use of inexpensive, compact LEDs (Kennedy et al. 2008; Jiang et al. 2011; Papour 2013; Papour et al. 2013; Sherman et al. 2013). Acquisition can be completed in less than 1 second while processing time is ~5 ms for a 100 mm² field of view at ~35 µm resolution (defined by image intensifier and lens specifications).

SYSTEM DESIGN

A key component of the system is the imager, a camera that is capable of recording nanosecond events at high rates (Figure 19.4). This unique capability arises from decades of development in light

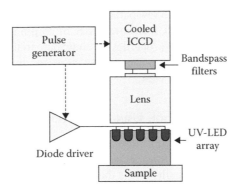

FIGURE 19.4 Schematics of the fluorescence acquisition system. The optical pulse and the camera are synchronized by the pulse generator to capture the images. A circular array of light-emitting diodes ensures uniform illumination over a large field of view of the lens.

intensifying technology. Understanding this technology is a crucial step in harnessing an efficient manner to which imaging autofluorescence signals at video rate can be achieved. Intensified camera or intensified charge-coupled device (iCCD) is a camera type that has an intensifier unit in front of the CCD. This module operates as the "shutter" and as an amplifier, capable of increasing the sensitivity of the device in low light conditions. A photocathode film captures the incident image light at visible wavelengths (400–700 nm) in front of the camera, and generates photoelectrons by means of the photoelectric effect. The emitted electrons are drawn inside the camera into an intensifier tube by an applied electric field. The intensifier tube is comprised of a microchannel plate (MCP) that operates under high voltage to accelerate electrons and multiply them in a chain reaction that amplifies the signal to a thousand fold. After passing through the MCP, the electrons collide on a green phosphorous layer to generate photons to be collected on the CCD (for this reason night vision often projects a green image).

In nanosecond exposure conditions, high voltage across the MCP is switched by an electronic circuit to operate as a shutter and amplifies the incoming image. Although a small signal can be collected on the CCD in one short exposure (of few nanoseconds), in conditions of a repetitive signal, data can be timely sampled and accumulated on the CCD to generate an image that otherwise would not be detected. In a conventional CCD camera, such a fast shutter capability is not possible nor is the amplification of the signal due to intrinsic limitations of imaging sensor technology.

GATE CHARACTERISTICS (SHUTTER EXPOSURE AND LIMITATIONS)

The intensifier's efficiency in passing an optical signal is a complex function of the device assembly and the operational settings. The gating mechanism is capable of transmitting the optical signals in a gate period as short as 100 ps, however at the cost of signal intensity reduction due to inefficiencies of the image intensifier. The efficiency is defined by the percentage intensity of an incoming signal that passes through the intensifier. Longer gate times result in significantly better transmission efficiency and serve as another factor for using long gate times.

Traditional FLIM requires short (subnanosecond) gate times in order to achieve a good resolution decay curve. With low gating efficiency, however, it will take longer to achieve an adequate image. In comparison to the contrast measurement system, operation of considerably longer gate times of 10 ns is used. A high image transmission efficiency of almost a hundred percent is reached. This further reduces the total image acquisition time and optimizes the performance of the system. These benefits serve as an additional advantage when using long shutter times in this system to achieve real-time imaging.

The resolution of the camera is set by a few parameters, and unlike a regular CCD, iCCD cameras are usually inferior in resolution. The MCP channel's size of 25 µm in diameter and further restrictions of the phosphorous layer and other properties all contribute to the degradation of the observed image in the camera's final output. Resolution can be increased by choosing magnifying optics, however, at the cost of a smaller field of view.

ILLUMINATION AND CONTRAST GENERATION USING LED TECHNOLOGY

Another key advantage of the system architecture is the circular array of LEDs about the lens aperture. LEDs can easily illuminate a large tissue area with good uniformity without the need for complicated optics. They often include a built-in lens that makes the setup very compact, unlike lasers that require fiber optics to guide the light to the tissue and a tailored optics solution to uniformly illuminate the tissue surface. Beyond LEDs' evident cost effectiveness and robustness, a major consideration in choosing them over lasers is that they do not require the same regulatory compliance that lasers must acquire for use in medical devices.

Looking at the illumination methodology, and specifically the LED operation, enables a deeper aspect of this technology to reveal the crux of the system. Analyzing the optical signals and processing methodologies shows the direct contrast mechanism without performing traditional deconvolution or lifetime calculation of any sort. Analytical simulation of two fluorophores with similar lifetime decay coefficients (2 and 2.5 ns) is performed (Figure 19.5a) with the illumination profile of an LED (Figure 19.5b). The LED has a noticeable decay profile that is a result of its operational mechanism and cannot be completely eliminated. The simulated LED shows a worst case scenario with a decay coefficient larger than the investigated fluorophores (3 ns). The simulation results (Figure 19.5c) show fluorescence signals emitted (convolved signals) by the fluorophores during the LED illumination. The contrast itself arises from the difference of the two curves in Figure 19.5c and is shown in Figure 19.5d. Contrast generation in this system is a result of direct differentiation between response behaviors of fluorophores and is achievable even with "slow" LEDs with long illumination profiles. By controlling the LED pulse shape and decay characteristics, a considerable gain in contrast can be achieved; however, this is outside the scope of this book. This unique direct contrast mechanism allows differentiation of autofluorescence signals in real time with relatively inexpensive parts and an extremely small footprint.

SENSITIVITY

The system has proven the ability to distinguish between fluorophores with less than 6% difference in lifetime (Jiang et al. 2011). In imaging tissue, clutter and noise dominate the lifetime map and limit the detection rate for small abnormalities. Factoring in noise and a decrease in resolution, sensitivity will decrease but will depend on the type of tissue. Continuing study of the system's performance and large statistical analysis is undergoing and will reveal the detection rate and sensitivity of the system in different tissue types.

DESIGN AND LIMITATIONS

The system is designed to achieve good contrast in tissue and is relying on a certain autofluorescence lifetime regime in order to do so. The system can best distinguish autofluorescence decay constants between 0.5 and 20 ns. Most abundant fluorophores in tissue have autofluorescence in this regime. Identifying fluorophores like flavin adenine dinucleotide (FAD) and nicotinamide adenine dinucleotide (NAD), which have multiple subnanosecond decay factors, is not easily detectable (Islam et al. 2013). Moreover, proteins that are found in tissue in comparably minute quantities would not contribute to a macroscopic contrast map (above 1 cm²).

Maximum permissible exposure (MPE) is a limit defined by the American National Standard (Z136.1-2007) and sets the safety limits for radiation by definitions of exposure parameters including

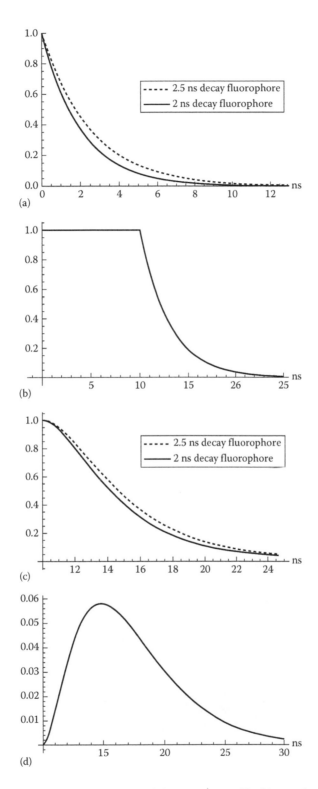

FIGURE 19.5 (a) Pure fluorescence signals, (b) LED illumination profile, (c) convolved fluorescence signals, and (d) contrast measure.

duration and peak power for the different wavelengths. The contrast-based autofluorescence imaging system utilizes the near-UV region (300–400 nm) for illumination of biological specimen without using lasers. The exposure and safety limits for patients and personnel using this LED-based technology allow for comfortable conditions and safe exposures that enable the performance of experiments without harming biological tissues.

BRAIN IMAGING AND RESULTS

Figure 19.6a displays gray scale image of dura mater sample that was analyzed for proof of concept experiment. Figure 19.6b shows the autofluorescence lifetime contrast image of the sample; brighter pixels represent longer lifetime. Clear distinction between different regions in the tissue

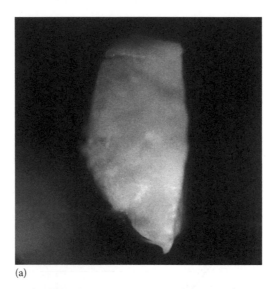

(a)

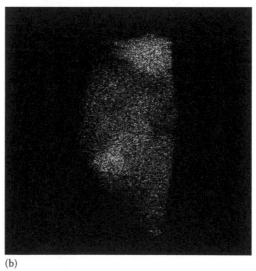

(b)

FIGURE 19.6 (a) Dura mater sample under UV illumination. Brighter pixels correspond to high intensity fluorescence emission. (b) Dura mater sample after processing: Brighter pixels correspond to longer lifetimes. The bright pixels in Figure 19.6a do not correspond to the bright pixels in this image since the algorithm normalizes the intensity and eliminates intensity bias; thus, the process image shows only lifetime information.

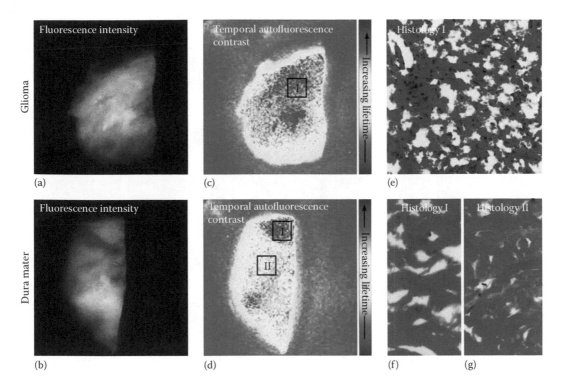

FIGURE 19.7 Brain tissue results and analysis. (a, b) Fluorescence intensity images of the "calibration image." (c, d) Custom color map of relative lifetimes: red represents longer lifetime. (e) Histology results of the marked region in (c), showing malignant tissue composition; glioma tissue was relatively uniform. (f, g) Histology results of regions I and II, respectively, in (d).

is visible in the contrast image and suggests variable tissue morphology. Further analysis and contrast optimization, including a glioma sample, are given in Figure 19.7: Fluorescence intensity (steady-state) images of the glioma sample (a) and dura mater (b) correspond to brighter pixels. The autofluorescence contrast images in (c) and (d) are a color map representation, where shorter relative lifetimes are yellow (and blue) and longer relative lifetimes are red. These lifetime images display contrast that is not detectable in the gray scale images (a) and (b), showing clear structure corresponding to tissue morphology. Figure 19.7e shows a low to moderate cellularity area of neoplastic astrocytes from a glioblastoma patient. The histology of region I is fairly uniform which correlates well with the relative uniformity of the lifetime fluorescence image. Figure 19.7f and g shows histology from two distinct regions of a sample of dura mater. Region I is a relatively cellular area showing multiple nuclei. Region II was relatively devoid of fibroblastic nuclei and consisted mostly of dense collagen. Pathological results correlate with the autofluorescence contrast map and further confirm the system's ability to detect tissue abnormalities. The heterogeneous imaging signals arising from regions I and II may reflect the relative presence or absence of fibroblasts in these areas.

HEAD AND NECK SQUAMOUS CELL CARCINOMA

Versatility of the system has enabled imaging of various tissue samples including head and neck squamous cell carcinoma (HNSCC). Figure 19.8, the color images show top view of three freshly dissected pieces of tissue from a single head and neck surgery. The color image on top left shows two cancerous tissues from the same tumor mass which was cut in half to reveal inner structure. The small pale tissue waged between the two cancerous tissues is a noncancerous tissue, and was added

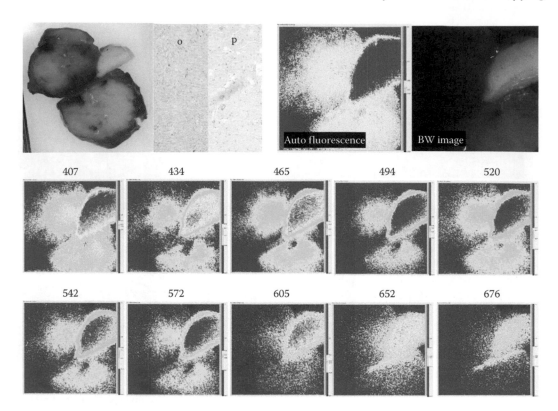

FIGURE 19.8 Head and neck squamous cell carcinoma (HNSCC) tissues from a single patient. Top left: Normal (middle) and malignant tissue samples with the corresponding histology. Top left: Black and white image at room illumination, and fluorescence intensity image. The 10 images in the bottom two rows represent colored fluorescence lifetime images acquired at different wavelengths in nanometers, from 407 to 676 nm. Relative lifetimes in each image are represented in colors from the longest to the shortest in the following order: Red, yellow, green, cyan, and blue.

in order to better show the capability of the system in distinguishing normal and abnormal tissues in HNSCC samples. In addition to the autofluorescence contrast and black and white image of the field of view (top right), a series of color-coded contrast images are shown. These images represent spectral acquisition; each was acquired at a different wavelength band using band-pass filters at the time of imaging. The number above each picture represents the central wavelength of the band-pass filter. The color code represents relative autofluorescence lifetimes: longer lifetimes are red and shorter lifetimes are green and yellow. Validation of the results in microscopy images of hematoxylin and eosin (H&E) stains shows evidence of a high-grade squamous cell carcinoma (o) of the halved cancerous tissue, with the normal tissue stained region (p) for comparison.

CONCLUSIONS

Years of research and development in fluorescence lifetime imaging technology have advanced our understanding of biological systems and processes. Despite the advancements, this modality has not yet adapted from the bench to the clinical setting. Among the challenges that impede clinical acceptance is the requirement for real-time imaging that can provide usable contrast and differentiate tissues based on morphological and metabolic changes.

The spectro-temporal autofluorescence system has the ability to create real-time, wide-field contrast images based on relative fluorescence lifetime signatures. This system holds the potential to differentiate and detect tissues in biorepositories (*ex vivo*) and as an intraoperative device in clinical setting (*in vivo*). The rapid imaging rate facilitates implementation of FLIM modality in intraoperative setting and can enable cancer margin delineation and accurate scission, to achieve better patient outcome. Those capabilities coupled with modular, low-cost components bring closer the realization of the implementation of this system in a clinical setting. This technology is just one example of new advances in biophotonics that enable scientists, engineers, and clinicians to advance technologies from the bench to the bedside with continuing decrease in complexity and development times.

ACKNOWLEDGMENT

This research was supported by the DoD Telemedicine & Advanced Technology Research Center (TATRC), grant# W81XWH-12-2-0075.

REFERENCES

Baker, M. 2012. Biorepositories: Building better biobanks. *Nature* 486 (7401): 141–146. doi:10.1038/486141a.

Brandes, A. A., A. Tosoni, E. Franceschi, M. Reni, G. Gatta, and C. Vecht. 2008. Glioblastoma in adults. *Critical Reviews in Oncology/Hematology* 67 (2): 139–152. doi:10.1016/j.critrevonc.2008.02.005.

Burger, P. C., F. Stephen Vogel, S. B. Green, and T. A. Strike. 1985. Glioblastoma multiforme and anaplastic astrocytoma pathologic criteria and prognostic implications. *Cancer* 56 (5): 1106–1111. doi:10.1002/1097-0142(19850901)56:5<1106::AID-CNCR2820560525>3.0.CO;2-2.

De Veld, D. C. G., M. J. H. Witjes, H. J. C. M. Sterenborg, and J. L. N. Roodenburg. 2005. The status of in vivo autofluorescence spectroscopy and imaging for oral oncology. *Oral Oncology* 41 (2): 117–131. doi:10.1016/j.oraloncology.2004.07.007.

Hess, K. R. 1999. Extent of resection as a prognostic variable in the treatment of gliomas. *Journal of Neuro-Oncology* 42 (3): 227–231. doi:10.1023/A:1006118018770.

Islam, M., M. Honma, T. Nakabayashi, M. Kinjo, and N. Ohta. 2013. pH dependence of the fluorescence lifetime of FAD in solution and in cells. *International Journal of Molecular Sciences* 14 (1): 1952–1963. doi:10.3390/ijms14011952.

Jiang, P.-C., W. S. Grundfest, and O. M. Stafsudd. 2011. Quasi-real-time fluorescence imaging with lifetime dependent contrast. *Journal of Biomedical Optics* 16 (8): 086001–086001. doi:10.1117/1.3609229.

Jo, J. A., Q. Fang, and L. Marcu. 2005. Ultrafast method for the analysis of fluorescence lifetime imaging microscopy data based on the Laguerre expansion technique. *IEEE Journal of Quantum Electronics* 11 (4): 835–845. doi:10.1109/JSTQE.2005.857685.

Kennedy, G. T., D. S. Elson, J. D. Hares, I. Munro, V. Poher, P. M. W. French, and M. A. A. Neil. 2008. Fluorescence lifetime imaging using light emitting diodes. *Journal of Physics D: Applied Physics* 41 (9): 094012. doi:10.1088/0022-3727/41/9/094012.

Khanna, A., S. S. Reddy, M. Jan, T. Swapnil, T. Vasudev Rao, S. A. V. Rao, A. L. Naik, S. Totey, and N. K. Venkataramana. 2010. Establishment of a brain tumor tissue repository in India: Maintaining quality standards. *Journal of Stem Cells* 5 (2): 89–101.

Lakowicz, J. R., ed. 2006. Time-domain lifetime measurements. In *Principles of Fluorescence Spectroscopy*, pp. 97–155. Springer, New York. http://www.springerlink.com/content/t04768544418v407/abstract/.

Lunsford, L. D., D. Kondziolka, and D. J. Bissonette. 1996. Intraoperative imaging of the brain. *Stereotactic and Functional Neurosurgery* 66 (1–3): 58–64.

Maarek, J.-M. I., L. Marcu, W. J. Snyder, and W. S. Grundfest. 2000. Time-resolved fluorescence spectra of arterial fluorescent compounds: Reconstruction with the Laguerre expansion technique. *Photochemistry and Photobiology* 71 (2): 178–187. doi:10.1562/0031-8655(2000)0710178TRFSOA2.0.CO2.

Marcu, L., J. A. Jo, P. V. Butte, W. H. Yong, B. K. Pikul, K. L. Black, and R. C. Thompson. 2004. Fluorescence lifetime spectroscopy of glioblastoma multiforme. *Photochemistry and Photobiology* 80 (1): 98–103. doi:10.1111/j.1751-1097.2004.tb00055.x.

Papour, A. 2013. *Analysis and Optimization of a Lifetime Fluorescence System to Detect Structural Protein Signatures in Varying Host Mediums for Rapid Biomedical Imaging*. University of California, Los Angeles, CA. http://gradworks.umi.com/15/30/1530877.html.

Papour, A., Z. Taylor, A. Sherman, D. Sanchez, G. Lucey, L. Liau, O. Stafsudd, W. Yong, and W. Grundfest. 2013. Optical imaging for brain tissue characterization using relative fluorescence lifetime imaging. *Journal of Biomedical Optics* 18 (6): 060504–060504. doi:10.1117/1.JBO.18.6.060504.

Phipps, J. E., Y. Sun, M. C. Fishbein, and L. Marcu. 2012. A fluorescence lifetime imaging classification method to investigate the collagen to lipid ratio in fibrous caps of atherosclerotic plaque. *Lasers in Surgery and Medicine* 44 (7): 564–571. doi:10.1002/lsm.22059.

Ramanujam, N. 2000. Fluorescence spectroscopy of neoplastic and non-neoplastic tissues. *Neoplasia (New York, N.Y.)* 2 (1–2): 89–117.

Sherman, A. J., A. Papour, S. Bhargava, M. A. R. St. John, W. H. Yong, Z. Taylor, W. S. Grundfest, and O. M. Stafsudd. March 2013. Normalized fluorescence lifetime imaging for tumor identification and margin delineation, *Proc SPIE 8572 Advanced Biomedical and Clinical Diagnostic Systems XI* 85721H–85721H. doi:10.1117/12.2013414.

Sinha, R. P. and D.-P. Häder. 2002. UV-induced DNA damage and repair: A review. *Photochemical and Photobiological Sciences* 1 (4): 225–236.

Sun, Y., N. Hatami, M. Yee, J. Phipps, D. S. Elson, F. Gorin, R. J. Schrot, and L. Marcu. 2010. Fluorescence lifetime imaging microscopy for brain tumor image-guided surgery. *Journal of Biomedical Optics* 15 (5): 056022. doi:10.1117/1.3486612.

Tadrous, P. J., J. Siegel, P. M. W. French, S. Shousha, E.-N. Lalani, and G. W. H. Stamp. 2003. Fluorescence lifetime imaging of unstained tissues: Early results in human breast cancer. *The Journal of Pathology* 199 (3): 309–317. doi:10.1002/path.1286.

Wagnieres, G. A., W. M. Star, and B. C. Wilson. 1998. In vivo fluorescence spectroscopy and imaging for oncological applications. *Photochemistry and Photobiology* 68 (5): 603–632. doi:10.1111/j.1751-1097.1998.tb02521.x.

Wallner, K. E., J. H. Galicich, G. Krol, E. Arbit, and M. G. Malkin. 1989. Patterns of failure following treatment for glioblastoma multiforme and anaplastic astrocytoma. *International Journal of Radiation Oncology, Biology, and Physics* 16 (6): 1405–1409. doi:10.1016/0360-3016(89)90941-3.

Yong, W. 2006. Distinction of brain tissue, low grade and high grade glioma with time-resolved fluorescence spectroscopy. *Frontiers in Bioscience* 11 (1): 1255. doi:10.2741/1878.

20 Optical Coherence Tomography and Quantitative Optical Imaging of Brain Cancer

Carmen Kut, Jordina Rincon-Torroella,
Alfredo Quinones-Hinojosa, and Xingde Li

CONTENTS

INTRODUCTION

Surgery is an essential component of brain cancer management. Glioma is the most common adult primary brain cancer, with inevitable recurrence and finite survival times (Stupp et al. 2005; McGirt et al. 2008, 2009; Chaichana et al. 2010a,b, 2012). While glioma treatment consists of a combination of surgery, radiation, and chemotherapy, surgery remains the first-line therapy and the most common initial step for glioma treatment. Recent studies have consistently demonstrated the significant long-term benefits of safe, maximum cancer resection, including improved overall surgical and delayed cancer recurrence for both low-grade and high-grade gliomas (Karim et al. 1996; Sanford et al. 2002; Laws et al. 2003; Stummer et al. 2006; McGirt et al. 2008, 2009; Smith et al. 2008; Chaichana et al. 2011, 2012, 2013a,b, 2014; Sanai et al. 2011; Jakola et al. 2012; Markert 2012; Chaichana and Quinones-Hinojosa 2013, 2014; Almeida et al. 2015). Currently, the surgical paradigm is to resect as much cancer as possible, while avoiding new neurological deficits and preserving critical neurological functions such as motor function and speech.

As recent studies have elucidated, the surgical extent of resection (EOR) is critical in prolonging progression-free and overall survivals for high-grade glioma patients (Laws et al. 2003; Stummer et al. 2006; McGirt et al. 2009; Chaichana et al. 2011, 2012, 2013a,b, 2014; Sanai et al. 2011; Chaichana and Quinones-Hinojosa 2013, 2014; Almeida et al. 2015). Sanai et al. conducted a systematic study with 500 glioblastoma (i.e., GBM, or high-grade glioma) patients and reported significant improvements in survival for patients with ≥78% EOR (Sanai et al. 2011); furthermore, GBM patients with gross total resection had a 1.6 fold increase in survival when compared with patients with subtotal resection. Similarly, Chaichana et al. reported a 70% EOR threshold that is associated with increased progression-free and overall survivals and that every 5% increase in resection decreases the risk of death by 5.2% for glioblastoma patients (Chaichana et al. 2013a). Finally, the same trend can be found in low-grade glioma patients. Of the patients with low-grade glioma, 50%–75% die from recurrence or progression to a malignant (i.e., higher grade) glioma (Keles et al. 2001). Maximal EOR in low-grade glioma is also significantly associated with improved survival and delayed recurrence (Karim et al. 1996; Sanford et al. 2002; McGirt et al. 2008; Smith et al. 2008; Jakola et al. 2012; Markert 2012). Berger et al. demonstrated that patients with a residual tumor volume of less than 10 cm^3 have a progression-free survival when compared with patients with >10 cm^3 residual tumor volume (50 vs. 30 months) (Berger et al. 1994); similarly, McGirt et al. have reported that gross total resection is associated with increased overall survival and progression-free survival (when compared with subtotal resection); similarly, Smith et al. reports that a >90% EOR (vs. <90% EOR) is associated with over 50% increased overall survival at the 5–8-year time points (McGirt et al. 2008; Smith et al. 2008).

In summary, surgery is critical to brain cancer management and there is a clear clinical need to maximize cancer resection, which delays recurrence, improves response to adjuvant therapies (such as chemotherapy and radiation treatments), and increases survival. The tools used in the operating room (OR) are evolving, and in this review, we discuss the present and future adjuvant tools that will aid the surgeon in maximizing resection while preserving maximal function, so that the patients can enjoy the survival benefits in increased resection and achieve a high quality of life.

ADVANTAGES AND LIMITATIONS OF EXISTING TECHNOLOGIES FOR GUIDING BRAIN CANCER RESECTION

Intra-operative cancer versus noncancer tissue differentiation is very difficult to achieve by the naked eye and even with the use of the operating microscope. To maximize the surgical EOR, it is necessary to develop an effective and reliable imaging technology, which can provide continuous, real-time, and high-resolution intra-operative guidance. As a result, various image-guided technologies have been introduced into the OR to address this clinical challenge, including magnetic resonance imaging (MRI) and computed tomography (CT), ultrasound (US), surgical microscopy, and navigational platforms, as well as more recent and experimental technologies including 5-aminolevulinic acid (5-ALA) fluorescence and Raman spectroscopy (Figure 20.1). All of these imaging technologies have important contributions and provide complementary information to the surgeon; nevertheless, each of these technologies has their unique advantages and disadvantages and do not completely resolve the important clinical need to perform maximal safe cancer resections.

As a standard of care, MRI has been conventionally used to diagnose, characterize, and monitor gliomas. MRI images are obtained preoperatively to examine important anatomical information of brain cancer and establish the surgical plan (Hammoud et al. 1996; Young et al. 2015). Additionally, the preoperative MRI may be used for intra-operative orientation and localization when co-registered with neurosurgical navigation platforms in the OR (e.g., Brainlab, StealthStation, and Vector Vision). MRI is indispensable for brain cancer diagnoses, management strategy, and follow-up due to its large field of view (which enables whole-brain imaging) and its capacity to provide 3D anatomical details (which enables presurgical planning). Nevertheless, MRI has limited resolution (3–26 mm^3) and suffers from poor biological specificity; in other words, it does not differentiate cancer directly. Rather, MRI T$_2$-weighted signals detect abnormality via an increase in water and contrast enhancement, which occurs when there is a breakdown of blood–brain barrier as a result of either cancer or a number of other factors including inflammation and neurodegenerative diseases (Upadhyay and Waldman 2014). In addition, the lack of real-time data acquisition makes guidance with preoperative MRI sensitive to brain shift and positional errors during the surgery (Spivak and Pirouzmand 2005). On the other hand, intra-operative MRI allows for repeated image acquisition during the surgery. Although intra-operative MRI reduces the effects of brain shift and positional errors while it is being performed intra-operatively, it is very expensive (i.e., several million dollars per unit) and time consuming (i.e., about 50 minutes), which also means more time in the OR and

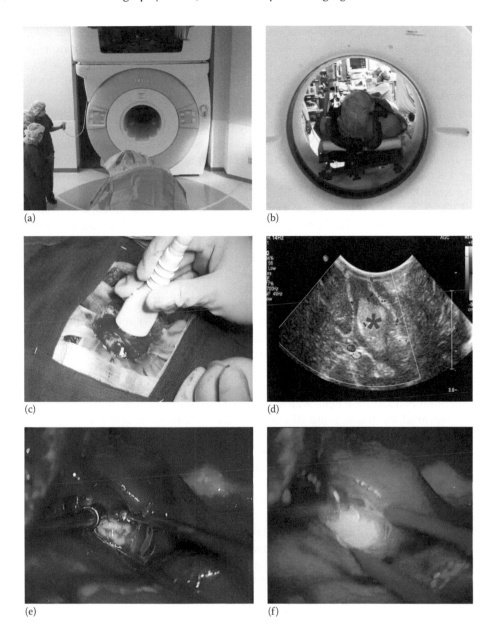

FIGURE 20.1 Many existing imaging technologies have been introduced to maximize brain cancer resection while preserving normal bran tissues. This figure illustrates several of those technologies. (a) Technicians introducing the intra-operative MRI into the OR. The surgery is stopped and the anesthetized patient stays at the surgical bed while monitoring continues. The imaging procedure takes approximately 50 min. (b) Intra-operative CT scanner in the OR. The intra-operative CT scans for brain cancer patients can be completed at any point during surgery (in this case, immediately before the surgery starts) but does not provide real-time feedback of brain motions once surgery begins. The head holder and bed need to be CT compatible. (c) Use of intra-operative US in a surgery for brain metastases. (d) The US elucidates the tumor core (*) located subcortically and surrounded by blood vessels that can be visualized with Doppler. However, it is generally more difficult to visualize a primary brain tumor with US, especially in the infiltrative areas. (e) Intra-operative visualization under the operative microscope during the removal of a high-grade glioma. (f) Fluorescence detection of glioblastoma with an orally administered prodrug 5-ALA under microscopic visualization (Quinones-Hinojosa 2012 Present). The cancerous tissue is fluorescent since 5-ALA leads to the accumulation of fluorescent porphyrins in malignant gliomas. 5-ALA has heterogeneous uptake in the tumor and therefore has greater contrast in tumor core and high-grade gliomas.

more costs associated with the procedure. Thus, while MRI is critical for brain cancer detection and surgical planning, it cannot provide continuous, real-time operative guidance and must require other imaging adjuncts to provide complementary information. Similarly, CT scanning is occasionally used in the clinic to detect brain cancer, but it is significantly less sensitive than MRI with limited visualization of both the tumor and the normal anatomy (Zhang et al. 2015). US is an affordable and noninvasive imaging modality for brain tumor detection. It is optimal for needle placements of brain biopsies (Zhang et al. 2015), as US has a relatively large field of view and can provide 3D continuous guidance during surgery. In addition, US helps surgeons to detect tumor anatomical structures and also to detect and preserve surrounding blood vessels, preventing subsequent bleeding or stroke. However, US lacks the sufficient resolution (~150–250 μm) and contrast to discriminate infiltrated tumor borders effectively (Rygh et al. 2008; Selbekk et al. 2013). The fluorescent dye indocyanine green (ICG) has been used to visualize peritumoral and intratumoral vascular structures (Jusué-Torres et al. 2013); however, ICG binds to plasma proteins in the blood stream and is nonspecific for tumoral vessel detection. Routinely used imaging technologies are essential to brain cancer management, even though they have not resolved the important clinical problem in detecting and maximizing resection of brain tumor in the OR.

New experimental imaging technologies such as 5-ALA fluorescence and Raman spectroscopy have also been introduced into the OR. While these techniques have not received FDA approval, they have shown early promise in clinical studies (Stummer et al. 2013; Jermyn et al. 2015). The most popular method in fluorescence imaging of brain tumor involves the use of an orally administered drug, 5-ALA (Quinones-Hinojosa 2012-Present; Stummer et al. 2013). 5-ALA is a prodrug in the heme biosynthetic pathways, which eventually leads to the accumulation of fluorescent porphyrins in malignant gliomas (Stummer et al. 2006; Quinones-Hinojosa 2012-Present). Thus, 5-ALA visualizes brain tumors under blue light, which leads to more complete brain tumor resection and consequently improved survival in phase III clinical trial (Stummer et al. 2006). However, the 5-ALA uptake varies throughout brain tumor areas, especially the locations with lower metabolism, for example, infiltrative cancer areas and low-grade gliomas (Stummer et al. 2006; Valdes et al. 2011; Roberts et al. 2012). Thus, 5-ALA has excellent specificity (92%–100%) but relatively lower sensitivity (47%–84%) (Stummer et al. 2013). On the other hand, Raman spectroscopy is a relatively new technology with the very first clinical feasibility study conducted in 2015 with 17 patients (Jermyn et al. 2015). Raman spectroscopy characterizes tissues based on their molecular signatures, and early studies have shown excellent results in detecting tumor cell populations in mice and humans. It is optimal for spot detection during brain cancer resections; however, since Raman spectroscopy has a restricted field of view (0.00000025 mm^2 area to 1 mm^3 volume per spot) and a relatively slow scanning rate (1 second per spot), it cannot detect large cancer volumes within the surgical time frame (Ji et al. 2013; Jermyn et al. 2015).

In contrast, optical coherence tomography (OCT) is a noninvasive, label-free, and cost-effective technology that can provide continuous, real-time imaging of 3D tissue anatomy at micron-level resolution in the OR. This chapter discusses new OCT findings in neurosurgery (Boppart et al. 1998; Bohringer et al. 2006, 2009; Kut et al. 2012; Assayag et al. 2013a; Kantelhardt et al. 2013; Xi et al. 2013; Kut et al. 2015a,b), and how optical property maps can be derived from OCT imaging data to provide direct visual cues for surgeons to detect brain cancer infiltration (Kut et al. 2015a,b).

PRINCIPLES OF OPTICAL COHERENCE TOMOGRAPHY

OCT is an optical analog of US, except that OCT uses a near-infrared light rather than sound waves, and does not require the use of any matching medium, for example, gels. OCT is capable of noncontact imaging, that is, acquiring images a couple of centimeters above the tissue surface (Li et al. 2005; Kut et al. 2015a), which can reduce the risks of infection. In addition, OCT provides cross-sectional images of tissue volume at approximately several μm resolution. The image contrast of

OCT relies upon tissue optical properties (e.g., absorption and scattering of light) and does not require the use of contrast agents.

OCT is an established medical technique for ophthalmological applications, that is, to image the anterior segment of the eye and the retina (Fujimoto 2003). In addition, OCT has been used in a variety of clinical applications (Fujimoto 2003; Böhringer et al. 2006; Zhang et al. 2013) and is currently FDA approved for ophthalmic, gastrointestinal, and intravascular applications. Furthermore, there are several studies in the application of OCT for brain cancer detection and surgical guidance (Boppart et al. 1998; Levitz et al. 2004; Bizheva et al. 2005; Bohringer et al. 2009; Finke et al. 2012; Assayag et al. 2013a; Kantelhardt et al. 2013). For example, Boppart et al. demonstrated the first OCT imaging of metastatic brain tumor specimens (intracortical melanoma) (Boppart et al. 1998); since then, other studies have extended OCT studies to cross-sectional imaging of primary human brain cancers (Bizheva et al. 2005; Böhringer et al. 2006a,b, 2009; Assayag et al. 2013b). Bizheva et al. performed 2D OCT imaging of formalin-fixed human tissue specimens such as meningioma and ganglioglioma (Bizheva et al. 2005), and Assayag et al. performed *en face* studies on human brain cancer specimens (Assayag et al. 2013b). Finally, Bohringer et al. (2006a,b, 2009) conducted proof-of-concept studies of *ex vivo* and *in vivo* primary brain cancers (solely with attenuation data).

Built upon the results of previous group, a recent study imaged and systematically compared cancer with noncancer brain tissues in a controlled environment using freshly resected *ex vivo* human tissues and an *in vivo* murine model implanted with human brain cancer (Kut et al. 2015a). In addition, this study proposed a method that could quantitatively analyze and display OCT imaging data and illustrated the use of a color-coded optical property map that detected brain cancer with direct visual cues with excellent sensitivity and specificity (Kut et al. 2015a). The subsequent sections detail the techniques, methodology, and results of that study and discuss the implications and potential of OCT imaging in brain cancer management.

OCT IMAGING SYSTEM PARAMETERS

To design an OCT imaging system for brain cancer applications, several parameters must be considered. First, the imaging system must be high-speed to enable relatively large volumetric imaging in real time. This can be achieved by using >100 kHz swept-source infrared laser, >0.5GS/second digitizer, and a graphics processing unit (GPU) video card to process a sustained data stream at >1 GB/second. Second, the OCT imaging algorithms should be optimized to match the data processing speed with the image acquisition rates (which is determined by the OCT hardware). Third, a light source with a higher wavelength (e.g., 1300 nm rather than 800 nm) should be considered for a greater imaging depth. Finally, the OCT probe should be small (~5–15 mm in diameter), lightweight, and easy-to-use for neurosurgical applications. An OCT imaging probe with the desired specifications (e.g., resolution, working distance, and probe diameter/size) can be achieved by carefully choosing the right combination of lenses and other probe parameters. Figure 20.2 illustrates an example of a swept-source OCT imaging system with a high-speed swept-source infrared laser and a handheld imaging probe (Kut et al. 2015a).

OPTICAL PROPERTY QUANTIFICATION AND MAPPING

Like US, OCT acquires cross-sectional images that are also known as B-scan images. These images represent planar slices into the tissue volume, and each cross-sectional frame consists of a series of neighboring A-lines. Each A-line, in turn, represents the OCT intensity data as a function of depth along a specific x–y coordinate.

The OCT intensity data can be quantitatively analyzed to yield important optical properties of human tissues. For example, the OCT intensity has a decay profile (along depth) that is similar to a decaying exponential. This profile is dependent on the optical attenuation

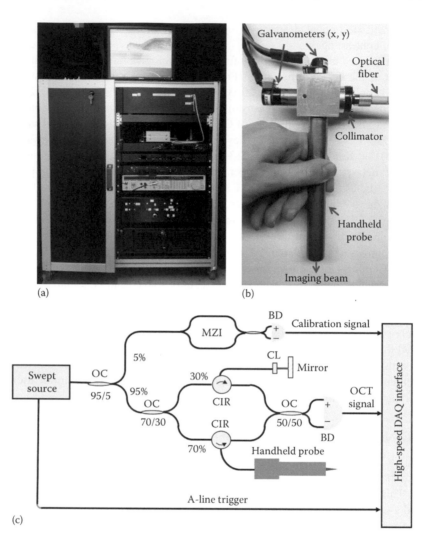

FIGURE 20.2 Example of an OCT imaging system. This is a swept-source OCT system with a 1300 nm, Fourier-domain mode locking swept fiber laser (a) and a handheld imaging probe that scans in both *x* and *y* directions (and collects data along the *z* direction) for volumetric samples (b). The OCT schematic (c) describes system components including BD, balanced detector; CIR, circulator; CL, collimating lens; DAQ, data acquisition; MZI, Mach–Zehnder Interferometer; and OC, optical coupler. (Reprinted from Kut, C. et al., *Sci. Transl. Med.*, 7(292), 292ra100, 2015. With permission.)

(i.e., absorption and scattering) of the tissue, as determined by the following equation (Cobb et al. 2006; Xi et al. 2013):

$$I(z) = k\sqrt{\mu_{bs}}\, e^{-\mu_t z} h(z), \tag{20.1}$$

where

I(z) is the OCT intensity profile along depth z (in millimeters)

k is the system constant

μ_t is the optical attenuation of the tissue sample (which is the sum of tissue scattering and absorption)

μ_{bs} is the backscattering coefficient

h(z) is the geometric factor of the imaging beam

Due to the presence of h(z) in Equation 20.1, the OCT profile is not purely exponential due to the depth-dependent effects of the beam profile. To remove this influence, the OCT intensity signal from the tissue sample can be cross-referenced with equivalent data from a silicon oxide phantom with known optical properties (Xi et al. 2013).

The reference phantom is constructed with silicon oxide nanoparticles embedded within a gelatin phantom. The gelatin simulates the mechanical properties, while the silicon oxide nanoparticles simulate the optical properties of human tissues. Since the nanoparticles' concentration and diameters are known, the attenuation coefficient ($\mu_{t,p}$) of the reference phantom can be approximated based on Mie theory predictions (Xi et al. 2013). Thus, the attenuation coefficient ($\mu_{t,b}$) of the biological tissue can be obtained as follows:

$$\frac{I_b(z)}{I_p(z)} = \sqrt{\mu_{bs}^b / \mu_{bs}^p}\, e^{-(\mu_{t,b}-\mu_{t,p})*z}, \tag{20.2}$$

where
  $I_b(z)$ and $I_p(z)$ are the OCT intensity profiles of the biological sample and the reference phantom, respectively
  μ_{bs}^b and μ_{bs}^p are the backscattering coefficients of the sample and phantom, respectively
  z is the imaging depth

Thus, the attenuation coefficient of the biological tissue ($\mu_{t,b}$) can be quickly calculated by first taking the logarithm of Equation 20.2 and then performing a simple linear fit. This method has been shown to provide reliable data without the influence of depth-dependent effects.

Using this method, an attenuation value can be computed along each A-line (i.e., at each x–y coordinate). However, a sliding window algorithm should be adopted in actual practice for speed considerations (Kut et al. 2015a). A sliding window algorithm includes two parameters—the window size and the step size; thus, attenuation values will be computed by laterally averaging all A-lines within a specific window. For example, assuming that (1) a cross-sectional image (i.e., B-scan) consists of 2000 A-lines laterally, (2) the window size is around 330 A-lines, and (3) the step size is 33 A-lines, attenuations will be calculated with 52 values per B-scan (i.e., averaged for windows with A-line number 1 to 330, 331 to 660, … , 1671 to 2000). The sliding window algorithm is advantageous because it speeds up the fitting time (e.g., fitting 52 values instead of 2000 values in the aforementioned example), and it suppresses speckle noise due to averaging. Since the brain tissue is relatively homogeneous, we can afford to use a larger window size to suppress speckle noise without compressing image quality.

Finally, additional algorithms are desired to overcome intra-operative constraints such as brain motion and the presence of blood (Kut et al. 2015a). Brain motion can be addressed by detecting the tissue surface at each time point, while blood pooling on top of the brain surface can be diluted by saline flush and excluded from analysis. Finally, intact blood vessels can be detected and visualized using Doppler algorithms.

Using the aforementioned algorithms, the OCT data will be used to reconstruct a 3D human tissue volume; overlaid on top (of the tissue volume) will be an optical property map consisting of color-coded attenuation values (Kut et al. 2015a). Three-dimensional visualization of brain tissues can provide spatial orientation of brain tissues. Furthermore, it provides direct visual cues for surgeons to detect cancer versus noncancer regions. Figure 20.3 illustrates an example of this visualization method with OCT imaging data obtained from a freshly resected brain tissue obtained from a glioma patient. In this illustration, the color red represents low optical attenuation (which is associated with brain cancer) and the color green represents high optical attenuation (which is associated with noncancer white matter). The following sections detail the methods, the physiological causes behind the attenuation differences, as well as the statistical analysis used to establish a diagnostic threshold (cancer vs. noncancer).

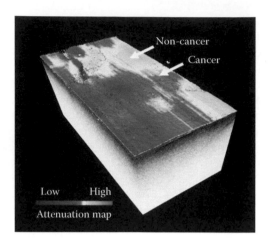

FIGURE 20.3 OCT can distinguish cancer (red) from noncancer (green) brain tissues. This image includes a color-coded optical property map overlaid on the 3D OCT dataset collected from a brain cancer patient. The data can be collected in real time and without the need of any injected contrast agents. Scale bars: 1 mm.

PHYSIOLOGICAL EXPLANATION OF THE ATTENUATION DIFFERENCES

For most systemic cancers, the optical attenuation is higher in cancer (compared with noncancer) because it has a higher cell density and nuclear-to-cytoplasmic ratio (Wax et al. 2003; McLaughlin et al. 2010).

In brain cancer, however, optical attenuation is influenced by additional factors. It is well known that noncancer white matter has high attenuation due to its myelin content (Amberger et al. 1998; Binder and Berger 2002; Alaminos et al. 2005). In addition, brain cancer cells are known to invade along white matter by breaking down and decreasing the expression of myelin (Amberger et al. 1998; Bevilacqua et al. 1999; Binder and Berger 2002; Alaminos et al. 2005; Bohringer et al. 2009; van der Meer et al. 2010). Noncancer gray matter, on the other hand, does not have myelin content. Thus, initial results have demonstrated that brain cancer has a lower overall attenuation when compared with surrounding noncancer white matter (with myelin content) but has a high attenuation when compared with surrounding noncancer gray matter (without myelin content) (Kut et al. 2015a). The following section describes the statistical analyses behind these results.

SYSTEMATIC ANALYSIS OF HUMAN BRAIN CANCER

Initial OCT results were collected from fresh surgically resected, *ex vivo* brain tissues of 32 consented patients (15 patients with high-grade glioma, 12 patients with low-grade glioma, and 5 control patients with noncancer brain lesions) (Kut et al. 2015a). During imaging, tissue volumes were imaged with OCT and carefully marked with India ink for subsequent correlation with histology. The first 16 patients entered into a training dataset to establish a diagnostic attenuation threshold (to distinguish cancer from noncancer), while the final 16 patients entered into a validation dataset to establish the OCT detection sensitivity and specificity for high- and low-grade glioma patients (Kut et al. 2015a).

In the training dataset, we find that noncancer white matter has significantly higher attenuation (6.2 ± 0.8 mm^{-1}, n = 5) when compared with infiltrated zones of both high-grade (3.5 ± 1.6 mm^{-1}, n = 9) and low-grade (2.7 ± 1.0 mm^{-1}, n = 2) glioma patients (Kut et al. 2015a). To establish a diagnostic attenuation threshold, we look for the attenuation value that would yield the highest sensitivity while maintaining at least 80% specificity; using separate analyses, this attenuation threshold was determined to be 5.5 mm^{-1} for both high- and low-grade glioma patients (Kut et al. 2015a) (Figure 20.4).

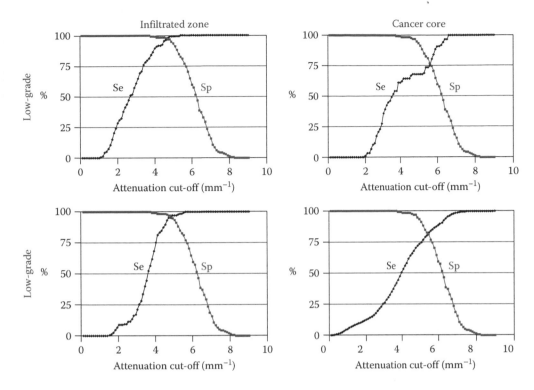

FIGURE 20.4 Diagnostic sensitivity (Se) and specificity (Sp) versus optical attenuation for infiltrated zone and cancer core in high- and low-grade patients. At 5.5 mm^{-1} threshold, max sensitivity was achieved with ≥80% specificity. (Modified and reprinted from Kut, C. et al., *Sci. Transl. Med.*, 7(292), 292ra100, 2015. With permission.)

Using this attenuation threshold value, we determined that OCT has a detection sensitivity of 92% and a specificity of 100% for high-grade glioma tissues from the validation dataset (Kut et al. 2015a). Similarly, OCT has a detection sensitivity of 100% and a specificity of 80% for low-grade glioma tissues (Kut et al. 2015a).

In addition to *ex vivo* human studies, the attenuation threshold (at 5.5 mm^{-1}) has also been validated in *in vivo* murine models that are implanted with patient-derived high-grade brain cancer (i.e., glioblastoma, or grade IV glioma) (Kut et al. 2015a). After implantation, the mice were operated for tumor removal, following the animal care standards, and the resection cavity was studied with the OCT technique before and after tumor resection. Initial OCT results correlated well with histology, which confirmed the accuracy of optical property maps in detecting brain cancer *in vivo* (Figure 20.5) (Kut et al. 2015a). Through studies on freshly resected human *ex vivo* tissues as well as *in vivo* rodent models implanted with patient-derived brain cancer (Kut et al. 2015a), this technology has demonstrated tremendous potential for future applications in the OR, where surgeons will be able to use OCT in the future (in conjunction with other technologies) to maximize EOR in a safe, expeditious, and cost-effective matter.

CONCLUSIONS AND FUTURE DIRECTIONS

To summarize, this review explores the potential of a new quantitative, optical mapping method to detect brain cancer using OCT. Current literature has shown robust evidence that maximizing the EOR of human brain tumors leads to long-term benefits, such as improved survival and delayed recurrence. While some intra-operative brain cancer detection tools are already available in the OR

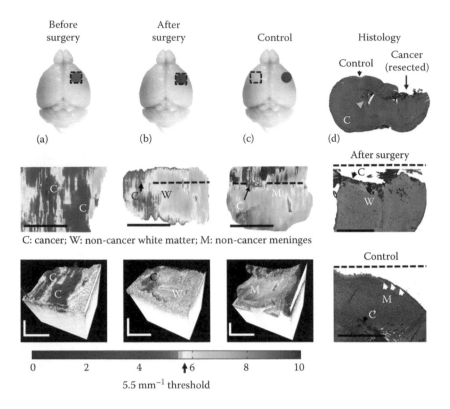

C: cancer; W: non-cancer white matter; M: non-cancer meninges

0   2   4   ↑6   8   10
5.5 mm⁻¹ threshold

FIGURE 20.5   In vivo OCT imaging and *en face* optical attenuation mapping on a murine model of human GBM at the brain cancer site before (a) and after (b) surgery and at normal brain surface (c). OCT results are confirmed with corresponding histology as shown in (d), which are 2D cross-sectional histological sections perpendicular to the optical attenuation map along the dotted lines. Scale bars: 1 mm. (Modified and reprinted from Kut, C. et al., *Sci. Transl. Med.*, 7(292), 292ra100, 2015. With permission.)

with good imaging technologies (such as MRI, CT, US, and fluorescence), these modalities have limitations in the ability to provide quantitative, real-time, and/or 3D continuous guidance in the OR with high resolution and contrast.

OCT can bring tremendous value to brain cancer management when used in conjunction with existing imaging technologies. OCT adds unique advantages as it provides real-time, continuous intra-operative guidance and is able to image volumetric tissue samples quickly and efficiently at micron-level resolution. Furthermore, quantitative optical mapping techniques using OCT data can provide surgeons with color-coded visual cues in the detection of cancer versus brain cancer surgery at high sensitivity and specificity. However, OCT cannot provide molecular characteristics of brain cancer and, as a result, can benefit from complementary surface imaging techniques such as 5-ALA and Raman spectroscopy. In addition, OCT does not provide a whole-brain field of view (unlike MRI or CT); as a result, it should be integrated into a surgical guidance system (e.g., Polaris or Brainlab) that can map the 3D position of the OCT probe, relative to the brain resection cavity, in real time.

So far, the feasibility of this technology has been demonstrated in freshly resected *ex vivo human* and *in vivo* rodent studies with patient-derived xenografts. For eventual translational into the OR, OCT can be integrated into existing imaging technologies for maximum value. Macroscopically, OCT can be integrated with surgical guidance systems (e.g., surgical microscopes or marker-based tracking systems) to spatially differentiate cancer from noncancer throughout the resection cavity.

Microscopically, spot detection techniques (e.g., Raman spectroscopy or multiphoton fluorescence) can further improve OCT detection accuracy by providing complementary molecular information at critical zones. Finally, additional *in vivo* human clinical studies need to be performed to evaluate the utility of OCT in the OR, to integrate OCT with other imaging modalities, and to overcome any intra-operative constraints. With these advances, OCT can lead to an enabling technology for guiding brain surgery rapidly and efficiently, increasing the EOR for brain cancer, and thus improving patient survival. Furthermore, the same technology could also be used for tumor detection in other clinical applications.

REFERENCES

Alaminos, M., V. Dávalos et al. (2005). EMP3, a myelin-related gene located in the critical 19q13. 3 region, is epigenetically silenced and exhibits features of a candidate tumor suppressor in glioma and neuroblastoma. *Cancer Research* **65**(7): 2565–2571.

Almeida, J. P., K. L. Chaichana et al. (2015). The value of extent of resection of glioblastomas: Clinical evidence and current approach. *Current Neurology and Neuroscience Reports* **15**(2): 1–13.

Amberger, V. R., T. Hensel et al. (1998). Spreading and migration of human glioma and rat C6 cells on central nervous system myelin in vitro is correlated with tumor malignancy and involves a metalloproteolytic activity. *Cancer Research* **58**(1): 149–158.

Assayag, O., M. Antoine et al. (2013a). Large field, high resolution full-field optical coherence tomography: A pre-clinical study of human breast tissue and cancer assessment. *TCRT Express* **1**(1): 21–34.

Assayag, O., K. Grieve et al. (2013b). Imaging of non tumorous and tumorous human brain tissue with full-field optical coherence tomography. *NeuroImage: Clinical* **2**: 549–557.

Berger, M. S., A. V. Deliganis et al. (1994). The effect of extent of resection on recurrence in patients with low grade cerebral hemisphere gliomas. *Cancer* **74**(6): 1784–1791.

Bevilacqua, F., D. Piguet et al. (1999). In vivo local determination of tissue optical properties: Applications to human brain. *Applied Optics* **38**(22): 4939–4950.

Binder, D. K. and M. S. Berger (2002). Proteases and the biology of glioma invasion. *Journal of Neuro-Oncology* **56**(2): 149–158.

Bizheva, K., W. Drexler et al. (2005). Imaging ex vivo healthy and pathological human brain tissue with ultra-high-resolution optical coherence tomography. *Journal of Biomedical Optics* **10**(1): 011006–0110067.

Bohringer, H. J., E. Lankenau et al. (2006b). Optical coherence tomography for experimental neuroendoscopy. *Minimally Invasive Neurosurgery: MIN* **49**(5): 269–275.

Bohringer, H. J., E. Lankenau et al. (2009). Imaging of human brain tumor tissue by near-infrared laser coherence tomography. *Acta Neurochirurgica* **151**(5): 507–517; discussion 517.

Böhringer, H., D. Boller et al. (2006a). Time-domain and spectral-domain optical coherence tomography in the analysis of brain tumor tissue. *Lasers in Surgery and Medicine* **38**(6): 588–597.

Boppart, S. A., M. E. Brezinski et al. (1998). Optical coherence tomography for neurosurgical imaging of human intracortical melanoma. *Neurosurgery* **43**(4): 834–841.

Chaichana, K. L. and A. Quinones-Hinojosa (2013). Neuro-oncology: Paediatric brain tumours—When to operate?. *Nature Reviews Neurology* **9**(7): 362–364.

Chaichana, K. L. and A. Quinones-Hinojosa (2014). The need to continually redefine the goals of surgery for glioblastoma. *Neuro-Oncology* **16**(4): 611–612.

Chaichana, K. L., E. E. Cabrera-Aldana et al. (2014). When gross total resection of a glioblastoma is possible, how much resection should be achieved?. *World Neurosurgery* **82**(1): e257–e265.

Chaichana, K. L., K. K. Chaichana et al. (2011). Surgical outcomes for older patients with glioblastoma multiforme: Preoperative factors associated with decreased survival. Clinical article. *Journal of Neurosurgery* **114**(3): 587–594.

Chaichana, K. L., I. Jusue-Torres et al. (2013a). Establishing percent resection and residual volume thresholds affecting survival and recurrence for patients with newly diagnosed intracranial glioblastoma. *Neuro-Oncology*.

Chaichana, K. L., M. J. McGirt et al. (2010a). Recurrence and malignant degeneration after resection of adult hemispheric low-grade gliomas. *Journal of Neurosurgery* **112**(1): 10–17.

Chaichana, K. L., C. Pendleton et al. (2013b). Multi-institutional validation of a preoperative scoring system which predicts survival for patients with glioblastoma. *Journal of Clinical Neuroscience* **20**(10): 1422–1426.

Chaichana, K. L., P. Zadnik et al. (2012). Multiple resections for patients with glioblastoma: Prolonging survival. *Journal of Neurosurgery* **118**(4): 812–820.

Chaichana, K., S. Parker et al. (2010b). A proposed classification system that projects outcomes based on preoperative variables for adult patients with glioblastoma multiforme. *Journal of Neurosurgery* **112**(5): 997–1004.

Cobb, M. J., Y. Chen et al. (2006). Non-invasive imaging of carcinogen-induced early neoplasia using ultrahigh-resolution optical coherence tomography. *Cancer Biomarkers: Section A of Disease Markers* **2**(3–4): 163–173.

Finke, M., S. Kantelhardt et al. (2012). Automatic scanning of large tissue areas in neurosurgery using optical coherence tomography. *The International Journal of Medical Robotics and Computer Assisted Surgery* **8**(3): 327–336.

Fujimoto, J. G. (2003). Optical coherence tomography for ultrahigh resolution in vivo imaging. *Nature Biotechnology* **21**(11): 1361–1367.

Hammoud, M. A., R. Sawaya et al. (1996). Prognostic significance of preoperative MRI scans in glioblastoma multiforme. *Journal of Neuro-Oncology* **27**(1): 65–73.

Jakola, A. S., K. S. Myrmel et al. (2012). Comparison of a strategy favoring early surgical resection vs a strategy favoring watchful waiting in low-grade gliomas. *JAMA* **308**(18): 1881–1888.

Jermyn, M., K. Mok et al. (2015). Intraoperative brain cancer detection with Raman spectroscopy in humans. *Science Translational Medicine* **7**(274): 274ra219–274ra219.

Ji, M., D. A. Orringer et al. (2013). Rapid, label-free detection of brain tumors with stimulated Raman scattering microscopy. *Science Translational Medicine* **5**(201): 201ra119–201ra119.

Jusué-Torres, I., R. Navarro-Ramírez et al. (2013). Indocyanine green for vessel identification and preservation before dural opening for parasagittal lesions. *Neurosurgery* **73**: ons145.

Kantelhardt, S., M. Finke et al. (2013). Evaluation of a completely robotized neurosurgical operating microscope. *Neurosurgery* **72**: A19–A26.

Karim, A. B., B. Maat et al. (1996). A randomized trial on dose-response in radiation therapy of low-grade cerebral glioma: European Organization for Research and Treatment of Cancer (EORTC) Study 22844. *International Journal of Radiation Oncology, Biology, and Physics* **36**(3): 549–556.

Keles, G. E., K. R. Lamborn et al. (2001). Low-grade hemispheric gliomas in adults: A critical review of extent of resection as a factor influencing outcome. *Journal of Neurosurgery* **95**(5): 735–745.

Kut, C., K. L. Chaichana et al. (2015a). Detection of human brain cancer infiltration ex vivo and in vivo using quantitative optical coherence tomography. *Science Translational Medicine* **7**(292): 292ra100.

Kut, C., S. Raza et al. (2012). High resolution optical guidance of glioma margin identification—Optimizing extent of resection. Poster presented at *2012 Society for Neuro-Oncology Annual Meeting*, Washington, DC.

Kut, C., J. Xi et al. (2013). Application of optical coherence tomography in brain cancer: Detecting glioma invasion from non-neoplastic white matter in human ex vivo samples. Oral presentation presented at *2013 SPIE Annual Meeting*, San Francisco, CA.

Kut, C., J. Xi et al. (2015b). Real-time, label-free optical property mapping for detecting glioma invasion with SSOCT for potential guidance of surgical intervention. Accepted for oral presentation at *2015 SPIE Annual Meeting*, San Diego, CA.

Laws, E., M. E. Shaffrey et al. (2003). *Surgical Management of Intracranial Gliomas—Does Radical Resection Improve Outcome? Acta Neurochirurgica Supplement* **85**: 47–53.

Levitz, D., L. Thrane et al. (2004). Determination of optical scattering properties of highly-scattering media in optical coherence tomography images. *Optics Express* **12**(2): 249–259.

Li, X., S. Martin et al. (2005). High-resolution optical coherence tomographic imaging of osteoarthritic cartilage during open knee surgery. *Arthritis Research & Therapy* **7**(2): R318.

Markert, J. M. (2012). THe role of early resection vs biopsy in the management of low-grade gliomas. *JAMA* **308**(18): 1918–1919.

McGirt, M. J., K. L. Chaichana et al. (2008). Extent of surgical resection is independently associated with survival in patients with hemispheric infiltrating low-grade gliomas. *Neurosurgery* **63**(4): 700–707; author reply 707–708.

McGirt, M. J., K. L. Chaichana et al. (2009). Independent association of extent of resection with survival in patients with malignant brain astrocytoma. *Journal of Neurosurgery* **110**(1): 156–162.

McLaughlin, R. A., L. Scolaro et al. (2010). Parametric imaging of cancer with optical coherence tomography. *Journal of Biomedical Optics* **15**(4): 046029.

Quinones-Hinojosa, A. (2012-Present). Aminolevulinic acid hydrochloride in identifying brain tumor stem cells in patients undergoing surgery for anaplastic astrocytoma or glioblastoma. In: ClinicalTrials.gov [Internet]. Bethesda, MD: National Library of Medicine (US). Available from: http://www.cancer.gov/about-cancer/treatment/clinical-trials/search/view?cdrid=721825&version=HealthProfessional of the record NLM Identifier: NCT01502605.

Roberts, D. W., P. A. Valdes et al. (2012). Adjuncts for maximizing resection: 5-aminolevuinic acid. *Clinical Neurosurgery* **59**: 75–78.

Rygh, O. M., T. Selbekk et al. (2008). Comparison of navigated 3D ultrasound findings with histopathology in subsequent phases of glioblastoma resection. *Acta Neurochirurgica* **150**(10): 1033–1042.

Sanai, N., M. Y. Polley et al. (2011). An extent of resection threshold for newly diagnosed glioblastomas. *Journal of Neurosurgery* **115**(1): 3–8.

Sanford, A., L. Kun et al. (2002). Low-grade gliomas of childhood: Impact of surgical resection. A report from the Children's Oncology Group. *Journal of Neurosurgery* **96**: 427–428.

Selbekk, T., A. S. Jakola et al. (2013). Ultrasound imaging in neurosurgery: Approaches to minimize surgically induced image artefacts for improved resection control. *Acta Neurochirurgica* **155**(6): 973–980.

Smith, J. S., E. F. Chang et al. (2008). Role of extent of resection in the long-term outcome of low-grade hemispheric gliomas. *Journal of Clinical Oncology: Official Journal of the American Society of Clinical Oncology* **26**(8): 1338–1345.

Spivak, C. J. and F. Pirouzmand (2005). Comparison of the reliability of brain lesion localization when using traditional and stereotactic image-guided techniques: A prospective study. *Journal of Neurosurgery* **103**(3): 424–427.

Stummer, W., U. Pichlmeier et al. (2006). Fluorescence-guided surgery with 5-aminolevulinic acid for resection of malignant glioma: A randomised controlled multicentre phase III trial. *The Lancet Oncology* **7**(5): 392–401.

Stummer, W., J.-C. Tonn et al. (2013). 5-ALA-derived tumor fluorescence: The diagnostic accuracy of visible fluorescence qualities as corroborated by spectrometry and histology and post-operative imaging. *Neurosurgery* **74**(3): 310–320.

Stupp, R., W. P. Mason et al. (2005). Radiotherapy plus concomitant and adjuvant temozolomide for glioblastoma. *The New England Journal of Medicine* **352**(10): 987–996.

Upadhyay, N. and A. Waldman (2014). Conventional MRI evaluation of gliomas. *The British Journal of Radiology* **84**(2): S107–S111.

Valdes, P. A., A. Kim et al. (2011). Delta-aminolevulinic acid-induced protoporphyrin IX concentration correlates with histopathologic markers of malignancy in human gliomas: The need for quantitative fluorescence-guided resection to identify regions of increasing malignancy. *Neuro-Oncology* **13**(8): 846–856.

van der Meer, F. J., D. J. Faber et al. (2010). Apoptosis- and necrosis-induced changes in light attenuation measured by optical coherence tomography. *Lasers in Medical Science* **25**(2): 259–267.

Wax, A., C. Yang et al. (2003). In situ detection of neoplastic transformation and chemopreventive effects in rat esophagus epithelium using angle-resolved low-coherence interferometry. *Cancer Research* **63**(13): 3556–3559.

Xi, J., Y. Chen et al. (2013). Characterizing optical properties of nano contrast agents by using cross-referencing OCT imaging. *Biomedical Optics Express* **4**(6): 842–851.

Young, R. M., A. Jamshidi et al. (2015). Current trends in the surgical management and treatment of adult glioblastoma. *Annals of Translational Medicine* **3**(9): 121.

Zhang, Y., Y. Chen et al. (2013). Visible and near-infrared spectroscopy for distinguishing malignant tumor tissue from benign tumor and normal breast tissues in vitro. *Journal of Biomedical Optics* **18**(7): 077003–077003.

Zhang, Z. Z., L. B. Shields et al. (2015). The art of intraoperative glioma identification. *Frontiers in Oncology* **5**: 175.

21 Intraoperative Optical Guidance for Neurosurgery

Chia-Pin Liang, Cha-Min Tang, and Yu Chen

CONTENTS

MOTIVATION

Navigating instruments accurately and safely to deep brain targets significantly impacts the therapeutic outcome of stereotactic neurosurgery. A fundamental limitation of needle-based stereotactic neurosurgeries, such as biopsy (Colbassani et al., 1988; Owen and Linskey, 2009), catheterization (Roitberg et al., 2001; Venkataramana et al., 2010), and electrode placement (Limousin et al., 1998; Starr, 2002), is that they are blind procedures in which the operator does not have real-time feedback as to what lies immediately ahead of the advancing instrument. This results in several problems:

1. Numerous medium-sized blood vessels that cannot be detected by computed tomography (CT) and magnetic resonance imaging (MRI) are at risk of being lacerated by the advancing probe. Cerebral hemorrhage is one of the leading complications of needle-based

stereotactic neurosurgery. Severe hemorrhage requires immediate surgical evacuation; otherwise, increased intracranial pressure will lead to coma and death (Binder et al., 2003, 2005; Sawin et al., 1998). Even with emergency treatments, however, patients may still suffer from neurological deficit or paralysis. The overall hemorrhage rate of stereotactic brain biopsy is as high as 59.8%, and the risk of symptomatic hemorrhage is 8.8% (Kulkarni et al., 1998). Biopsy of deep-seated lesion (basal ganglion and thalamus) is associated with two-fold higher hemorrhage risk than other regions (Kongkham et al., 2008). One study reported that 3 in 12 cases of biopsy in the basal ganglion and thalamus developed the hematoma requiring surgical removal and two of these three patients resulted in permanent hemiparesis (Nishihara et al., 2010). On the other hand, hemorrhage is also one of the most worried complications of deep brain stimulation (DBS) surgery. DBS is a surgical approach for managing certain neurological and psychiatric disorders (Skidmore et al., 2006; Mohr, 2008) by implanting therapeutic electrodes to deep brain targets. The most widely accepted condition is Parkinson's disease (PD) (Andrews et al., 2010; Flora et al., 2010). The symptomatic hemorrhage rate associated with the insertion of sharp electrodes is 2%–5% (Terao et al., 2003; Gorgulho et al., 2005; Deuschl et al., 2006; Sansur et al., 2007; Voges et al., 2007; Bronstein et al., 2010) and two to eight fold higher for hypertensive patients (Terao et al., 2003; Gorgulho et al., 2005; Sansur et al., 2007). The severe consequence includes permanent paralysis and death (Hariz, 2002; Binder et al., 2003, 2005). The major problem is that operators do not have real-time imaging feedback of the vessels lying ahead of advancing needles. Therefore, there is a critical need to detect the at-risk vessels sitting in front of the advancing instrument.

2. The discrepancy between the actual position of the brain target and its preoperatively determined coordinate caused by inaccuracy of imaging navigation, intraoperative positioning issues, and loss of cerebrospinal fluid can lead to false biopsy (Hall, 1998), failure of surgery (Hariz, 2002), and severe adverse events (Lang et al., 2006). Despite the preoperative CT or MRI guidance, loss of cerebrospinal fluid once the cranium is open, inaccuracy of imaging navigation, and brain shift due to positioning can lead to marked discrepancies in stereotactic coordinate up to 1 cm (Hirabayashi et al., 2002). In a surgery targeting the deep brain nuclei in size of few mm to cm, the offset could result in failed surgery. In DBS surgery, for example, considering how close critical pathways travel next to each other in these nuclei, it is not surprising that even an mm shift of the electrodes can introduce unwanted side effects, such as muscle twitching, unpleasant sensation, and acute depression (Funkiewiez et al., 2001). Also, along with the development of gene or cell therapies for neurological disorders, precise delivery of therapeutic agents into a specific brain region becomes more and more important. For example, in the glial-cell-derived neurotrophic factor trials for PD (34 patients), misplaced infusion catheters led to severe adverse events and required surgical repositioning in two patients and complete removal in another (Lang et al., 2006). Currently, microelectrode recording (MER) is the gold standard for compensating the stereotactic discrepancy. Deep brain nuclei are typically identified by their characteristic neuron firing patterns (Gross et al., 2006). However, refinement mapping comes at cost. Inserting the electrode for mapping in one track requires up to 2 hours, and typically 3 or more tracks are necessary to locate the target. Therefore, guiding therapeutic instrument efficiently and accurately to the target is still an unmet challenge.

3. Connectomic surgery that can restore many brain diseases by stimulating specific neural fiber tracts (Greenberg et al., 2008; Malone et al., 2009; Henderson, 2012) also requires precise intraoperative guidance to locate the pathology-related fiber tracts. Electrode misplacement by a few mm generates no therapeutic effect at all (Henderson, 2012). However, there is no effective intraoperative guidance tool to locate these tracts. MER cannot acquire signal from the fiber tracts. Deviation from the tractographically defined target, such as dentatorubrothalamic tract, can lead to poor tremor control, as shown in a patient with

bilateral DBS for essential tremor (Henderson, 2012). In theory, diffusion tensor imaging MRI can locate these fiber tracts, but in reality, intraoperative MRI (iMRI) tractography still suffers from poor signal-to-noise ratio, poor spatial resolution, and fiber orientation ambiguity at crossing points (Henderson, 2012). Therefore, there is a critical need to develop an intraoperative guidance tool that can target the pathologically related tracts, and it is also important to develop an imaging platform for studying the connectome in micron-scale.

Optical coherence tomography (OCT) is an elegant solution to address the challenges of stereotactic neurosurgery targeting brain nuclei (Jafri et al., 2005), avoiding hemorrhage (Liang et al., 2011), and locating neural fibers (Wang et al., 2011). Jafri et al. (2005) inserted a side-viewing OCT probe into human brain ex vivo following the surgical trajectory toward DBS targets. Operators can easily know whether the probe has reached or passed the targets by reading the structural landmarks. The same group also demonstrated that the OCT probe can guide drug delivery and tissue dissection aiming at deep brain targets in small rodents in vivo (Jafri et al., 2009). Liang et al. (2011) demonstrated that a forward-imaging OCT probe can identify at-risk blood vessels sitting in front of advancing instruments in sheep brain in vivo and detect important tissue micro-landmarks for surgical guidance. Furthermore, polarization-sensitive OCT can separate the white matter (WM) track from gray matter (GM) based on their birefringence contrast and can also measure the orientation of fiber tracks (Wang et al.). Thus, an OCT system that is equipped with three imaging protocols—structure, Doppler, and polarization for detecting nuclei, avoiding hemorrhage, and targeting specific fiber tracts—will be an evolutionary tool for intraoperative neurosurgery guidance. In the following sections, we will summarize some recent efforts in developing translational optical device for intraoperative neurosurgery guidance.

SIDE-VIEWING OCT CATHETER FOR NEUROSURGERY GUIDANCE

SIDE-VIEWING OCT PROBE

Jafri et al. first propose to use a side-viewing OCT catheter (Figure 21.1) for neurosurgery guidance (Jafri et al., 2005). The probe consists of a single mode fiber, a microfocusing lens, and a reflective mirror. The focusing lens determines the lateral resolution, and the fold mirror deflects the laser beam to the side of the probe for circumferential scanning. By rotating and pulling back the fiber probe, a 3D volumetric image can be obtained.

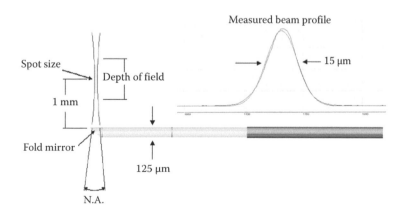

FIGURE 21.1 Side-viewing OCT catheter. (Reprinted from Jafri, M.S. et al., *J. Biomed. Opt.*, 10, 051603, 2005. With permission.)

Ex Vivo Human Brain OCT Imaging

Figure 21.2d shows that the probe was inserted into a human brain ex vivo following a surgical trajectory. A 3D volumetric image was obtained. The cross-sectional images from putamen (PUT), WM lamina, and globus pallidus interna (GPi) are shown in Figure 21.2b, c, and e, respectively. The images clearly demonstrate the potential of using OCT for neurosurgery guidance. In the PUT (Figure 21.2b), OCT reveals the distribution of the scattered WM tracts (bright, round structure) within the GM nuclei (dark, homogeneous background). It also detects the homogeneous and shining WM lamina that separates different nuclei (Figure 21.2c). OCT is capable of identifying the

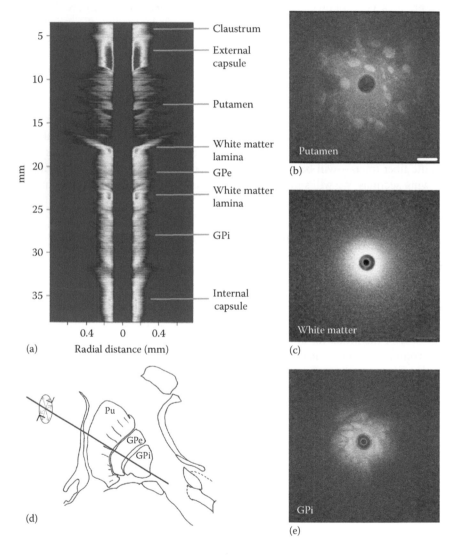

FIGURE 21.2 Structures within the striatum can be clearly identified by OCT. (a) L-mode projection of OCT displays the absolute averaged intensity of backscattered light and depth of penetration as the probe is linearly advanced. (b) Individual standard OCT image of the PUT obtained during a single 360° rotation of the beam. The oval-shaped bright spots are cross sections of the WM striations in the PUT. (c) WM is characterized by strong backscattering and poor penetration. (d) Schematic of the OCT track through the striatum. (e) GPi has the appearance of closely packed WM bundles separated by thin GM bands. Scale bar, 0.5 mm. (Reprinted from Jafri, M.S. et al., *J. Biomed. Opt.*, 10, 051603, 2005. With permission.)

surgical target, GPi, which consists of densely packed WM tracks (Figure 21.2e). Most importantly, OCT can determine whether the probe reaches the target by L-mode imaging, which is a longitudinal sectioning image along the probe insertion trajectory (Figure 21.2a). The L-mode image shows that there are shining WM lamina and internal capsule (IC) surrounding the target, GPi. Differentiating the landmarks surrounding the target is critical for verifying whether the probe reaches its intended target. Currently there is no other imaging modality that is capable of detecting the micro-landmarks due to the limited spatial resolution or contrast.

IN VIVO SMALL RODENT BRAIN SURGERY GUIDED BY OCT

Jafri et al. further demonstrated that the side-viewing OCT probe can be used for guiding deep brain tissue dissection and injection in vivo (Jafri et al., 2009). Figure 21.3 shows the injection of adeno-associated virus carrying enhanced green fluorescent protein (AAV2-EGFP) to either the dentate subiculum or CA1 region of a rat hippocampus. In vivo OCT image at the injection site is shown on the right (Figure 21.3b, d, and f), and a postmortem photomicrograph showing the fluorescent cells superimposed on the corresponding brightfield image is on the left (Figure 21.3a, c and e). The circle to the left of the probe is the injection needle. The location of the transfected cells exactly matches the OCT image of the needle tip with respect to the structure of the dentate (Figure 21.3a). Similarly, the tissue structure and needle position shown in the OCT images of the subiculum (Figure 21.3c) and CA1 region (Figure 21.3e) correlate with the location of the transfected cells (Figure 21.3d and f, respectively). These experiments show that OCT image confirms the location of the instrument at the time of intervention and also monitors the injection that cannot otherwise be carried out.

FORWARD-VIEWING OCT PROBE FOR NEUROSURGERY GUIDANCE*

INTRODUCTION

Although OCT imaging has great potential to transform neurosurgery guidance, it will not be possible to bring this technology to the bedside without a translational device. An effective image guidance probe for stereotactic procedures in the brain needs to satisfy three criteria: it must have a thin needlelike geometry (O.D. < 1 mm), forward-imaging ability to detect tissue landmarks in front, and the ability to detect blood vessels. Numerous forward-imaging probes have been developed for OCT (Boppart et al., 1997; Sergeev et al., 1997; Pan et al., 2001, 2003; Xie et al., 2003; Zara et al., 2003; Jain et al., 2004; Liu et al., 2004; Cobb et al., 2005; Wang et al., 2005, 2007; Zara and Patterson, 2006; Munce et al., 2008; Takahashi et al., 2008; Aguirre et al., 2010; Fleming et al., 2010), but their distal-end designs (>1 mm) are too large to fit into surgical instruments. As discussed previously, Jafri et al. used a thin side-viewing OCT probe to perform imaging-guided DBS (Jafri et al., 2005) and deep brain dissection and injection (Jafri et al., 2009). The weakness of the side-viewing probe is that it cannot visualize the at-risk vessels sitting in front of the advancing probe and thus cannot reduce the risk of hemorrhage.

Another novel design using a paired-angle-rotation scanning probe (Wu et al., 2006; Han et al., 2008) achieves forward-imaging needle-based (O.D. = 820 μm) endoscopy by using two counter-rotating gradient-index (GRIN) lenses. Yet another group developed an elegant forward-imaging probe by relaying the scanning beam using a long GRIN rod lens, while keeping the endoscope itself stationary (Xie et al., 2006). This design provides fast dynamic focus tracking, which can perform high-quality noncontact in vivo 3D imaging (Xie et al., 2009). Two different dynamic ranges of GRIN rod lenses have been reported: 72 and 108 dB (Xie et al., 2006; Yu et al., 2009). In order to optimize the dynamic range, 8-deg-beveled glass windows and antireflection coating were applied to reduce the reflection from the end surface (Xie et al., 2009). Previously reported GRIN rod lens–based probes are primarily for laryngoscopy or laparoscopy applications with probe sizes ranging from 2.7 to 4.58 mm (Xie et al., 2006, 2009; Guo et al., 2009; Yu et al., 2009). To our knowledge,

---

* Source: Liang, C.P. et al., Opt. Express, 19, 26283, 2011. With permission.

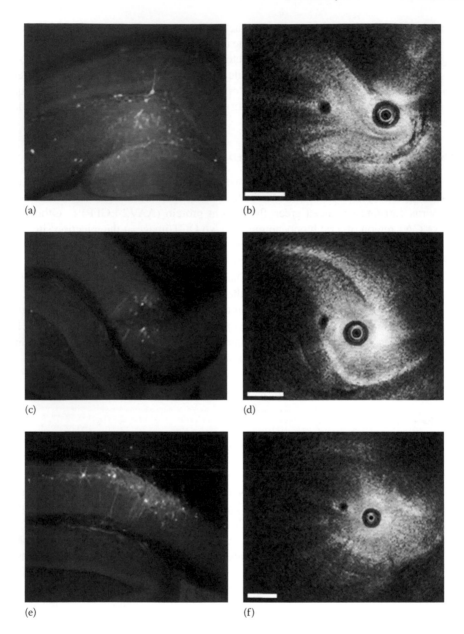

FIGURE 21.3 Precise targeting of viral vector delivery using OCT. Photomicrographs of AAV2-EGFP expression and brightfield images in rat hippocampal dentate gyrus (a), subiculum (c), and CA1 region (e). Corresponding OCT images show real-time structural images of dentate gyrus (b), subiculum (d), and CA1 region (f). The dark circle to the left of the probe is the injection needle. Note the correlation between the position of the injection needle and the gene expression. Scale bars, 0.5 mm. (Reprinted from Jafri, M.S. et al., *J. Neurosci. Methods*, 176, 85, 2009. With permission.)

Doppler capability of the probes with O.D. smaller than 1 mm has not yet been demonstrated. This may be due to the difficulty of actuating the optics in such limited space. The limitations of existing approaches for OCT-guided stereotactic brain procedures motivated us to develop a GRIN rod lens approach that can satisfy the thin forward-imaging, needle-type criteria, and the optical Doppler imaging criteria.

Liang et al. demonstrated that the GRIN-rod-lens design can be miniaturized to a needle-imaging probe (O.D. = 740 µm) (Liang et al., 2011). High-speed (100 frames/s) and high-sensitivity (>90 dB) OCT imaging was achieved by using this needle probe. The stationary GRIN-rod-lens also provides high-quality Doppler optical coherence tomography (DOCT) imaging with 41 dB velocity dynamic range (VDR) and ±17 µm/s velocity resolution. The effectiveness and robustness of the system was demonstrated in the studies of sheep brain in vivo and human brain ex vivo.

FORWARD-IMAGING OCT PROBE

Figure 21.4a shows the schematic of the OCT and handheld needle imaging device. The swept-source OCT system (Li et al., 2009; Yuan et al., 2009, 2010) utilizes a wavelength-swept laser as its light source. Spectrum bandwidth of the laser is 100 nm centered at 1325 nm (Thorlabs, SL1325-P16). The wavelength-swept frequency is 16 kHz with 12 mW output power; therefore, for 160 axial lines images, the frame rate is equivalent to 100 frames per second. A Mach–Zehnder interferometer (MZI) receives 3% of the laser output power and uses it to generate a clock signal with uniformly spaced optical frequency to trigger the sampling of the OCT signal in the analog-to-digital (A/D) converter. The sample and reference arms of a Michelson interferometer receive equal portions of the remaining 97% of the laser power. The galvanometer scanning mirrors deflect the sample arm light into the GRIN rod imaging needle (NSG America, LRL-050-P400) through a low numerical aperture (NA) lens. Careful NA matching between GRIN lens (NA 0.084) and coupling lens ensures maximum coupling efficiency. The distal end of the GRIN needle receives the scanning laser spot from the proximal end and relays the backscattered signal back to the interferometer. The diameter of the GRIN lens is 0.5 mm and the outer diameter, including the stainless steel tubing, is

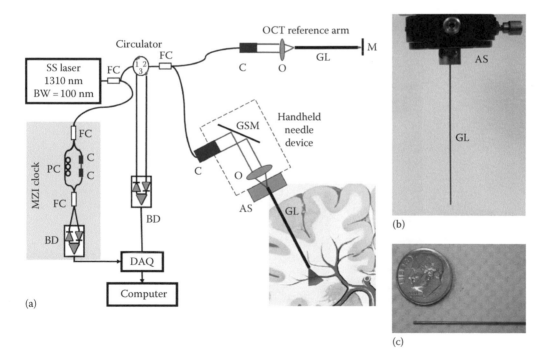

FIGURE 21.4 Schematic and pictures of the handheld OCT device. (a) Schematic of the handheld OCT system. FC, fiber coupler; PC, polarization controller; C, collimator; BD, balanced detector; MZI, Mach–Zehnder interferometer (frequency clocks); DAQ, data acquisition board; M, mirror; GSM, galvanometer scanning mirror; O, objective lens; AS, alignment stages; GL, GRIN lens needle. (b) GRIN needle. (c) GRIN needle placed beside a U.S. dime. (Reprinted from Liang, C.P. et al., *Opt. Express*, 19, 26283, 2011. With permission.)

0.74 mm. To reduce the risk of blood vessel laceration and mechanical resistance in the tissue, we created a blunt tip by attaching a transparent cap to the distal end of the GRIN lens. Figure 21.4b shows the picture of the GRIN needle (GL), and Figure 21.4c shows the GRIN needle placed beside a U.S. dime.

In Vivo Sheep Brain Deep Vessel Imaging

The blood vessel DOCT imaging from an anesthetized sheep by the needle probe shows the feasibility of vessel detection in vivo. Figure 21.5 shows OCT and DOCT superimposed images at different time points of a vessel as it was detected, monitored, and then compressed by the advancing probe. The OCT images in Figure 21.5 show that the speckle size in the lumen of vessels is much smaller than the speckle size of the wall, making the boundary of the vessel wall easy to identify. The colormap scale for the DOCT signal is analogous to the colormap used for ultrasound (US) Doppler signals. Red to yellow represents increasing flow speed in one direction, whereas blue to cyan represents increasing speed in the opposite direction. However, the alias rings in these images do not indicate the change of flow direction. Instead, the abrupt change of color represents higher velocity that is "wrapped." "Wrapping" occurs because the velocity exceeds the VDR. When velocities exceed this limit in either direction, they "wrap" to values at the other end of colormap (Yang et al., 2003). The superimposed image further verified the boundary of lumen and indicated that this system can provide measurement of the vessel size (~0.2 mm in diameter) that is verified by both OCT and DOCT. Figure 21.5a shows a vessel was detected in front of the probe. Figure 21.5 shows the probe was approaching the vessel. Figure 21.5b shows an image of the probe that was stopped right before hitting the vessel. Figure 21.6 shows the vessel pulsation while keeping the probe still. The periodic changes observed in the flow speed images strongly suggest that the vessel is a pulsating artery. Figure 21.5c shows the needle probe compressing the vessel. Figure 21.5 shows that, with the probe advancing gradually, the lumen of the vessel becomes more and more constricted and the

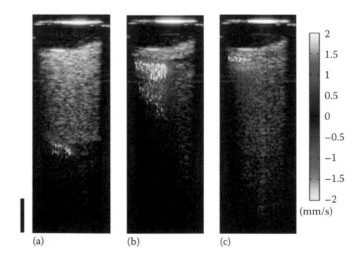

(a)   (b)   (c)

FIGURE 21.5 A high-risk case of contacting a brain vessel, showing compression of the vessel by the OCT probe. All the images are from the same vessel. The bright line on the top of the images is the junction between GRIN rod lens and the transparent cap. The front curved surface of the cap had direct contact with the tissue. The scale bar in all the figures is 0.25 mm. (a) The OCT/DOCT superimposed image shows a vessel 0.65 mm in front of the probe. (b) The superimposed image shows a vessel right in front of the probe (c) The superimposed image from shows the vessel was compressed by the probe. (Reprinted from Liang, C.P. et al., *Opt. Express*, 19, 26283, 2011. With permission.)

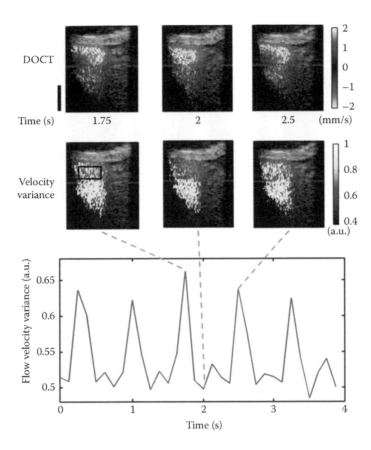

FIGURE 21.6 Quantification of pulsation by flow velocity variance. The first row shows Doppler images at different time points specified in the second row. The third row shows the corresponding velocity variance images. The black rectangle shows the ROI. The value of flow velocity variance in the plot is the average value of ROI. The dashed lines indicate the time points of the corresponding image set. The scale bar is 0.2 mm. (Reprinted from Liang, C.P. et al., *Opt. Express*, 19, 26283, 2011. With permission.)

Doppler signal also decreases accordingly. Figure 21.6 shows that the pulsation can be quantified by flow velocity variance. Flow velocity variance varies monotonically with flow velocity and has a wider measurement range. Therefore, it provides more accurate estimation of the flow speed than the aliased DOCT signal. However, it is prone to large error in regions with low SNR, so we set our ROI in a region having stable DOCT signal (the black rectangle in the variance image). Both DOCT and variance images show a dramatic difference between peaks and valleys. The larger red signal area in the DOCT images indicates higher flow velocity at the peaks. Also, the variance images show stronger signal corresponding to the higher flow velocity at peak. The average of the ROI variance signal versus time shows clear pulsation pattern. The heart rate is 80 beats per minute. The high SNR Doppler signal enables us not only to detect the vessel but also to monitor the physiological change of the blood flow.

An example of the vessel being pushed aside by the probe is shown in Figure 21.7. A vessel was detected 0.93 mm in front of the probe by its Doppler signal (Figure 21.7). As the probe approached to the vessel, the vessel was gradually pushed aside (Figure 21.7a and c) and finally out of the field of view (FOV) (Figure 21.7d). We also notice that the high blood flow speed in this large vessel made the Doppler alias rings smaller than the pixel size resulting in randomized Doppler signals. Although DOCT cannot provide quantitative flow data in this case, it still acts as a sensitive detection tool.

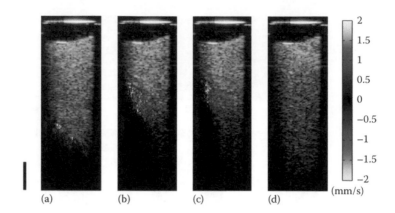

FIGURE 21.7 A low-risk case of contacting a brain vessel. The images from a to d are in sequence as the probe was advancing. The scale bar in the figure is 0.25 mm. The OCT/DOCT superimposed image of a vessel (a) 0.75 mm in front of the probe, (b) 0.35 mm in front of the probe with the vessel was being pushed aside, (c) 0.35 mm in front of the probe with the vessel was pushed further away from the probe, and (d) after the probe had passed the vessel. (Reprinted from Liang, C.P. et al., *Opt. Express*, 19, 26283, 2011. With permission.)

Ex Vivo Human Brain Imaging

The images of the human basal ganglia ex vivo illustrate the potential for neurosurgical guidance. Figure 21.8 shows a camera image of human brain tissue and the full-track reconstructed OCT image. In the camera image of brain tissue, note the white appearance of fiber tracts surrounding the GM nuclei and the striation of white fiber bundles within the PUT. The reconstructed OCT track was obtained by pushing the probe from right to left 2 mm below and parallel to the surface of the brain slice. This figure illustrates the high degree of contrast between GM and WM generated by OCT. The transition between the thin white fiber capsule, the GM PUT, and the lamina to the globus pallidus externa (GPe) can be clearly identified. These OCT-detectable structures are useful anatomic landmarks that can be

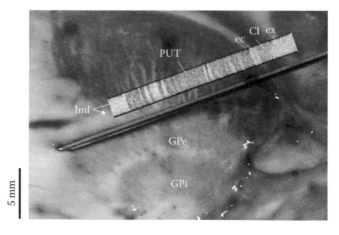

FIGURE 21.8 Full-track OCT image of a human basal ganglia. The probe on top of the brain tissue shows relative dimension and the direction of insertion (from top right to bottom left). The full-track OCT image is placed proximal to the insertion passage. Major structures are labeled: extreme capsule (ex), claustrum (Cl), external capsule (ec), PUT, lateral medullary lamina (lml), GPe, and GPi. The width of the reconstructed OCT images in this figure is expanded for better visualization. The real width is 0.44 mm. (From Liang, C.P. et al., *Opt. Express*, 19, 26283, 2011. With permission.)

used in stereotactic surgery (Jafri et al., 2005, 2009; Jeon et al., 2006). The structures identified in this passage include extreme capsule (ex), claustrum (Cl), external capsule (ec), PUT, and lateral medullary lamina (lml). The WM typically has much higher intensity with shallower penetration depth. Also, the characteristic WM striations in the PUT display high contrast and can be easily identified. Because the needle was inserted 2 mm below and parallel to the surface of the brain slice, the number and the size of the fiber tracts in the camera image are different from the fiber tracks in OCT image. Reconstruction of the full-track OCT image allows better identification of the probe location. This information can assist in the guidance of therapeutic tools to deep brain targets with micrometer precision.

DISCUSSION

The GRIN rod lens approach provides a solution to the multiple requirements for stereotactic neurosurgery. By using motion-free GRIN rod lenses as relay optics, the high-speed A-line scanning rate of the laser system can be fully exploited without compromising the needle size. This high-speed forward-imaging capability allows us to promptly avoid the at-risk vessels (Figure 21.5) and locate the tip position (Figure 21.8). Moreover, stable and high-speed A-line scanning is critical for optimizing the flow VDR (Yang et al., 2003). Unstable and slow A-line scanning rates will degrade VDR and hinder vessel detection by DOCT. Our system not only detects vessels but also monitors the physiological changes of the blood flow (Figure 21.6). Since there is no complex actuation at the distal end, this probe is one of the thinnest forward-imaging OCT probes. The diameter of the probe is a compromise between providing sufficient FOV and minimizing tissue injury. A GRIN lens diameter of 0.5 mm and FOV of 0.44 mm allow easy recognition of vessels, and the overall diameter of 0.74 mm is smaller than the inner diameter of existing stereotactic cannulae.

Vessels within the parenchyma can be easily pushed to the side by the probe (Figure 21.7), whereas vessels in deep sulci (Figure 21.5) that are tightly attached to the pia mater are not easily pushed aside by the probe. Vessels in sulci are really surface vessels that are buried by the cortical folds. This is consistent with anecdotal experiences of neurosurgeons who meticulously avoid entering cortical sulci.

Besides detecting the vessels, the probe can also monitor the pulsation (Figure 21.6) and differentiate the vessel type. These capabilities could be valuable for screening the vessels posing high risk in neurosurgery. The aliasing problem may hinder using DOCT signal to quantify blood flow; however, this problem can be solved by using velocity variance (Yang et al., 2003) or axial Kasai algorithm (Morofke et al., 2007). Also, the high-speed Fourier domain mode locking laser should be able to increase the velocity detection limit by one to two orders (Adler et al., 2007; Klein et al.).

In addition to detecting the at-risk vessels, this forward-imaging needle-type OCT probe can be an important complementary technology for current stereotaxic neurosurgery. Prior studies have demonstrated the ability of catheter-based OCT to provide information on the position of the probe relative to neighboring anatomic landmarks (Jafri et al., 2005, 2009). Although the forward-scanning probe does not provide as much FOV as with the rotating side-imaging approach (Jafri et al., 2005), the needle probe can be used to provide real-time feedback on the degree to which the brain has shifted due to the surgery-induced CSF leak. This is accomplished by establishing the probe tip relative to critical landmarks such as a prominent GM–WM junction (Figure 21.8). Since the stereotactic system provides the expected position, any deviation between the expected and actual location of a landmark provides the surgeon useful information on the degree to which the brain has shifted during surgery. Secondly, it can often provide optical signatures of specific anatomic structures (i.e., the pencils of fibers in the PUT). The use of image guidance probes in conjunction with preoperative MRI or CT images will give the surgeons greater confidence on the probe location. However, for diagnosis of other pathologies, a wider FOV may be desired. A possible way is to reconstruct a wider FOV image using manual scanning and optical tracking (Ren et al., 2009). Besides the application in stereotactic procedures, this device can potentially be applied to many other image-guided interventions and for the detection of blood flow involving difficult-to-reach structures.

CONCLUSION

The forward-imaging needle-type OCT probe is a promising complementary technology for current stereotaxic neurosurgery. With real-time OCT/DOCT forward-imaging capability, it may be possible to avoid laceration of at-risk intraparenchymal vessels and infer probe location relative to OCT-detectable landmarks. The device described here may be adapted to multiple intervention procedures in addition to stereotactic neurosurgery.

CONCURRENT MAGNETIC RESONANCE IMAGING AND OPTICAL COHERENCE TOMOGRAPHY*

INTRODUCTION

As described previously, OCT is a great complementary technology for stereotactic neurosurgery, but small FOV limits its usage. MRI is the most widely used neuroimaging tool with large FOV. Combining high-resolution OCT with large FOV MRI will allow an efficient and accurate performance of neurosurgery.

Superior tissue contrast and versatile imaging protocols of iMRI make it a promising guidance tool for various surgeries (Schulz et al., 2004; Hall and Truwit, 2005b; Larson et al., 2012). However, sending a patient into the constrictive MRI bore interrupts the surgical rhythm, and thus the number of imaging studies in one procedure is limited to two or three times. On the other hand, the iMRI system with big opening and better accessibility to the patient suffers from poor imaging quality. In order to take full advantage of MRI guidance, MRI-compatible robotic systems that can operate inside the restrictive MRI bore with near real-time MRI guidance have been developed (Tsekos et al., 2007; Sutherland et al., 2008; Yang et al., 2011). These systems are promising, but their targeting capability is still limited by the resolution of MRI. Moreover, the artifact surrounding surgical tools (such as metallic electrode or biopsy needle) obscures the important region of interest (ROI) and makes it difficult to achieve sub-mm accuracy. In order to overcome these challenges, an OCT imaging system is integrated with an MRI-compatible robot for high precision surgery. The innovative MRI-compatible robot (Yang et al., 2011) developed by Dr. Jaydev Desai's group enables us to teleoperate the needle device within the constrictive MRI bore and performs OCT under continuous MRI monitoring. The combination of macroscale MRI morphology and microscale OCT architecture in a concurrent and coregistered manner paves a new avenue to the iMRI interventions required sub-mm precision.

MRI/OCT SYSTEM

Figure 21.9a shows the picture of the integrated MRI/OCT imaging system, which includes an MRI-compatible robotic device (Yang et al., 2011). To work within a strong magnetic field (3T MRI, SIEMENS), we actuate the robot by a pneumatic cylinder instead of a conventional motor. By pressurizing or depressurizing the air chambers in pneumatic cylinder (the pressure is controlled by a pneumatic directional control valve (MPYE-5-1, FESTO) connected to an air compressor), we can move the needle forward or backward. The MRI-incompatible pressure control units, including the control valve and the air compressor, are outside the MRI room and are connected to the robot through long plastic tubes. The position of the needle is recorded by an encoder with 12.7 µm resolution for feedback control. The smallest step size that is achievable by the pneumatic robot is 1 mm, and the response time for 1 mm displacement is 0.2 second. Regarding the OCT system, the spectrum bandwidth of the swept-source OCT laser is 100 nm, centered at 1325 nm (SL1325-P16, Thorlabs). Therefore, the calculated axial resolution is 7.7/5.9 µm in air/water. The wavelength-swept frequency is 16 kHz with 12 mW output power. The OCT system (Liang et al.) sitting outside the MRI room is connected to a side-viewing OCT probe (ImageWire, St. Jude Medical) through an

---

* *Source*: Liang, C.P. et al., *J. Biomed. Opt.*, 18, 046015, 2013b. With permission.

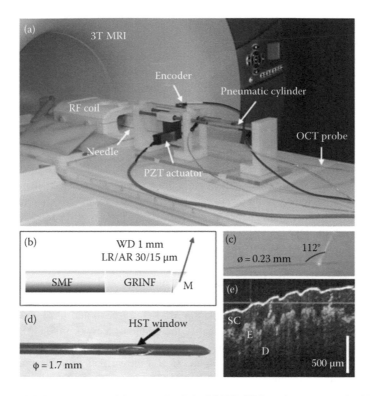

FIGURE 21.9 MRI/OCT system. (a) Photograph of the MRI/OCT imaging system. (b) Side-viewing OCT probe consisted of an SMF, a gradient index multimode fiber, and a micromirror (M). (c) Photo of the OCT probe. (d) MRI-compatible titanium needle probe for OCT imaging, with the side window covered by a transparent HST. (e) Cross-sectional OCT image of a human finger with SC, E, and D. (Reprinted from Liang, C.P. et al., *J. Biomed. Opt.*, 18, 046015, 2013b. With permission.)

8 mm long single mode fiber. The side-viewing probe consists of a single mode fiber (SMF) for light transmission, a GRIN multimode fiber for light focusing, and a micromirror (M) for beam deflection (Figure 21.9b and c). he OCT probe (O.D. = 0.23 mm) is placed in a 16 gauge (O.D. = 1.7 mm) titanium biopsy needle (in vivo). A side window is created on the needle to allow the laser beam from the OCT probe to reach the sample, and the window is encapsulated in a transparent heat shrink tube (HST) to prevent tissue collection (Figure 21.9d). The measured working distance from the outer surface of the needle probe is 0.2 mm. The HST deteriorates the transverse resolution from 20 to 30 μm and the axial resolution remains 15 μm. The sensitivity drop introduced by the HST is minimized to 1–2 dB by filling the needle with water for index matching. The OCT probe is linearly actuated forward and backward along the long axis of the needle by an MRI-compatible piezoelectric (PZT) actuator (Bending actuators, Piezo Systems) attached at the proximal end. We only use the information from the pullback direction because it is difficult to insert a flexible fiber linearly from a distal site. Although we only use half of the information, the tissue is under oversampling. With 1 mm/s needle insertion speed and 18.3 mm/s sampling rate (15 Hz frame rate and 1.22 mm scanning range), the overlap between OCT images is 94%. However, oversampling can provide us additional information, such as tissue deformation, elastography, and Doppler flow. On the other hand, spiral scanning may be more efficient, but it is also more complex and expensive to build an MRI-compatible rotary motor. Instead, if the tissue is relatively homogeneous in transverse direction, a simple PZT bender that can perform cross-sectional imaging is sufficient. Figure 21.9e shows an example of an OCT image of a human finger. The microstructures beneath the skin surface, including stratum corneum (SC), epidermis (E), and dermis (D), can be easily identified. Sweat glands, the spiraling columns in

the SC region, can also be easily identified. The PZT actuator and the OCT needle probe are moved in tandem, driven by the pneumatic actuator.

The accuracy of MRI/OCT coregistration cannot be better than the resolution of MRI (~1 mm). It may be difficult to achieve sub-mm coregistration, but the targeting precision is only limited by how accurately OCT can determine the boundary surrounding the target, and it is not limited by the resolution of MRI.

Ex Vivo Human Brain MRI/OCT Image

We image an ex vivo human basal ganglia to validate MRI/OCT for neurosurgery guidance. iMRI, which provides large FOV and direct visualization of brain targets in operation, has been applied to ste-reotactic neurosurgery, including brain tumor biopsy (Hall and Truwit, 2005a), catheterization (White et al., 2011), and DBS (Larson et al., 2012; Martin et al., 2005). However, the resolution and the contrast of MRI are not good enough to detect certain nuclei in the basal ganglia (Rezai et al., 2006). On the other hand, researchers have used OCT to locate these nuclei in human brain ex vivo (Jafri et al., 2005) and in rat brain in vivo (Jafri et al., 2009). We place a slab of frozen human brain tissue containing basal ganglia in a gelatin holder filled with saline at room temperature and wait until it is completely thawed. Then, we acquire a few preoperative T2 images (4 minutes per frame) to plan the trajectory. We moni-tor the advancement of the needle by the dynamic MRI (1 frame/s) and OCT imaging (15 frames/s) in real time (Figure 21.10e). We can easily identify the boundary between the GM (low intensity with deep penetration) and the WM (high intensity with shallow penetration) while most of structures are occult in the dynamic MRI (Figure 21.10e). Figure 21.10a shows a photograph of a brain tissue. In the high-resolution MRI image (Figure 21.10b), the transitions between the GM nucleus, PUT (white on T2 image) and its surrounding WM bundles, WM lamina (WL), and IC (dark on T2 image) are clearly visible. We reconstruct a full-track OCT image (Figure 21.10c) from the OCT video and correlate it with the large-scale MRI morphology. From each frame of the OCT video, we select a band of A-lines with 66.67 μm width (1 mm/s sampling rate, same as the needle insertion speed) from the same location. Then, these images are put together to construct Figure 21.10c. We align the preoperative MRI to the intraoperative OCT images by matching them, respectively, to the dynamic MRI images. The accuracy of the alignment is limited by the resolution of dynamic MRI images, which is 2 mm. In Figure 21.10d, we plot the normalized MRI intensity (green line), normalized OCT backscattering intensity (blue line), and normalized attenuation coefficient (red line). The normalized OCT attenuation coefficient and OCT surface scattering in Figure 21.10e are smoothed for better visualization by a moving average filter with 2.35 mm average width. In the full-track OCT image, the transitions between PUT and its surrounding WM bundles (WL and IC) are obvious and are well correlated with the MRI image. The full-track OCT shows that the GM nucleus (PUT) has lower backscattering intensity, deeper penetration depth, and thus smaller attenuation coefficient (Figure 21.10d), while myelinated WM tracks have higher backscat-tering intensity, shallower penetration depth, and larger attenuation coefficient. This result agrees well with previous studies (Jeon et al., 2006). It is worth noting that the transition from IC to thalamus is not visible in the MRI image. The thalamus is a GM nucleus containing uniformly distributed WM fibers. The mixture of WM and GM gives it a very low contrast to the surrounding WM tracks in MRI image. However, we can easily identify the boundary between IC and thalamus in OCT image. Compared to the PUT, high level of WM in the thalamus reduces the optical penetration depth and increases the atten-uation coefficient (Figure 21.10d). Also, the backscattering intensity from the thalamus is not as high as in the WM bundles. This result agrees with a previous study (Jafri et al., 2005). Figure 21.10d shows that the transitions between PUT and its surrounding tissues are obvious in both imaging modalities, but only OCT provides good contrast between IC and thalamus. Figure 21.10f shows that the attenuation coefficient of the thalamus (GM/WM mixture) is between IC (WM) and PUT (GM) and the difference is statistically significant ($p < 0.0001$). The attenuation numbers in Figure 21.10f are calculated from a 1 cm wide unsmoothed data at the location indicated by the black arrows in Figure 21.10d. This result suggests that OCT can discern the MRI-occult nuclei in the basal ganglia with a different contrast.

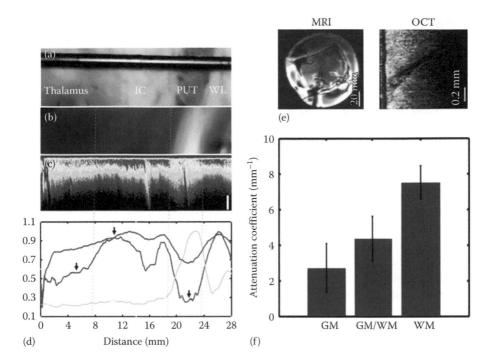

FIGURE 21.10 Coregistered MRI/OCT images of human basal ganglia ex vivo. (a) Photo of the human brain tissue containing basal ganglia. Different tissue types are visible, including WM lamina (WL), PUT, IC, and thalamus. The OCT needle probe is sitting on top of the tissue indicating the insertion track (the actual insertion track is beneath the tissue surface). (b) T2 MRI image of the brain slice. (c) Full-tack OCT with coregistered lateral position with MRI. The axial dimension is expanded for better visualization. The scale bar for the axial dimension is 0. 5 mm. (d) Plot of normalized MRI intensity (green), normalized OCT backscattering intensity (blue), and normalized OCT attenuation coefficient. (e) Concurrent dynamic MRI and OCT imaging of human brain tissue during needle insertion. (f) The attenuation coefficient of GM (PUT), GM/WM mixture (thalamus), and WM (IC). The p-values between data sets are smaller than 0.0001. (Reprinted from Liang, C.P. et al., *J. Biomed. Opt.*, 18, 046015, 2013b. With permission.)

Contrast in MRI images depends on the magnetic property of the tissue. The signal from protons with shorter T2 (e.g., protons in lipids) will decay faster than protons with longer T2 (e.g., protons in water) and will appear with lower signal intensity. Therefore, the WM, which consists of densely packed lipid membranes, has lower intensity than other tissues in T2 image. Lipids are also strong optical scatters and give bright intensity in OCT images (Jafri et al., 2005; Jeon et al., 2006; Liang et al., 2011). Densely packed axon fibers in the WM region contain high concentration of lipids and thus have strong OCT signal with shallow penetration depth. Although the source of the contrast of both imaging modalities comes from two different physical phenomena, OCT is able to produce images at a very high resolution over a small FOV. Such high resolution is currently not possible with MRI. However, the two techniques can complement each other, where MR images provide guidance to an OCT catheter to a specific location from where high-resolution images could be obtained. The combination of the two technologies enables one to span the resolution scales afforded by each of the modalities.

CONCLUSION

In conclusion, a multiscale MRI/OCT imaging system with teleoperated robot has potential to improve the accuracy and efficiency of iMRI procedures. Real-time OCT images can fill the gap between pre- and postoperative MRI images and thus allow operators to reach the target more efficiently. Moreover, high-resolution OCT imaging provides microstructural information and different contrasts that can be useful for detecting MRI-occult tissue landmarks. Therefore, combining large FOV MRI with high-resolution OCT is a promising guidance tool for high precision surgery. Future works will include the investigation of the impact of physiological motion (breathing, etc.) and bio-fouling (bleeding, etc.) on targeting accuracy under in vivo conditions.

COHERENCE-GATED DOPPLER: A FIBER SENSOR FOR PRECISE LOCALIZATION OF BLOOD FLOW*

INTRODUCTION

The needle-type forward-imaging OCT probe and MRI/OCT imaging device are promising tools that can potentially transform the neurosurgery guidance, but we also need a simpler device that can be integrated with conventional guidance tools and improve their safety.

Detecting blood vessels in front of a surgical probe in real time is an important function during interventional procedures. This may minimize the possibility of lacerating at-risk blood vessels during stereotactic neurosurgery, avoid vessels during local anesthesia procedures deep inside the body, and locate vessels during central venous/arterial cannulation. US is currently used for real-time image guidance and blood vessel localization for some of these interventional procedures. However, in some cases, such as neuraxial blockade, US guidance is especially challenging because of the complex encasement of bones that allows only a very narrow acoustic window for the US beam (Grau et al., 2002). Furthermore, the guidance efficacy in some cases is still unsatisfactory due to limited resolution (~1 mm) and the need for complex hand–eye coordination (Karakitsos et al., 2006). Optical sensors that can be integrated with surgical tools and detect blood vessels in front of instruments provide a solution to these challenges.

An effective optical sensor needs to be small, flexible, and rugged for integration with minimally invasive instruments. It also needs to have high lateral resolution (<100 μm) to detect clinically significant blood vessels. Ideally, it should also have sufficient axial detection range to detect blood vessels' millimeters ahead. Various optical technologies have been developed to monitor the blood flow. Laser Doppler flowmetry (LDF) has been used in neurosurgery (Wardell et al., 2007), dermatology (Eun, 1995), and dentistry (Kijsamanmith et al., 2011) to collect blood flow information. However, the resolution (mm–cm) of the two fiber design in a conventional LDF system is too coarse to allow location and avoidance of many vessels in the brain. Petoukhova et al. improved the axial resolution of LDF to 50 μm by using coherence gating effect (Petoukhova et al., 2001). Although this approach successfully obtained depth-resolved information from the human skin (Varghese et al., 2010), the lateral resolution is not sufficient (few mm (Petoukhova et al., 2001)) for neurosurgery. Also the two fiber design with wide separation in LDF system is not compatible with minimally invasive procedures. In contrast, DOCT creates a very small imaging spot (10 μm) with a single fiber (Chen et al., 1997). By scanning this spot, the flow information in the ROI can be mapped out with great resolution. However, in many clinical applications, it is not necessary to obtain high-resolution images of blood vessels and accept the tradeoffs associated with imaging. Often what is needed is simply to determine whether there is a blood flow at a precise location in front of a surgical probe. We developed a technology to serve in such situations.

Coherence-gated Doppler (CGD) is a real-time movement sensing technology that can be thought of as a hybrid between OCT and LDF. The system design is derived from time-domain DOCT or optical Doppler tomography (Chen et al., 1997) with several important simplifications and modifications. By converting the DOCT imaging system to the CGD sensing system, the reference phase modulator, reference path length scanning, signal digitization, and demodulation processing can be omitted. CGD only requires a simple electronic circuit to convert the Doppler beating signal to an audio signal. From the pitch and volume of the audio signal, we demonstrate that the simple CGD system enables us to differentiate tissue, vein, and artery in live animals and also demonstrate that the CGD probe can potentially predict the creation of hematomas in neurosurgery.

---

* *Source*: Liang, C.P. et al., *Biomed. Opt. Express*, 4, 760, 2013a. With permission.

CGD System and Probe

Figure 21.11a illustrates the design of the CGD system. The light source is a superluminescent diode (SLD) at 1310 nm (QPhotonics QFLD-1300-10S). The wavelength bandwidth of the light source is 3 nm, and thus the coherence length (CL) is 190 μm in water. The fiber coupler (FC) splits the photons to sample (90%) and reference arms (10%). The optical fiber circulator sends the illumination light from port 1 to port 2 and the backscattering light from port 2 to port 3. Both backscattering light from the sample and the reference mirror go to another FC (50/50), and the photons from each arm are redistributed to two output fibers. The interference fringe from output fibers goes to a dual balanced detection system, which rejects the common mode noise. After further amplification and frequency filtering

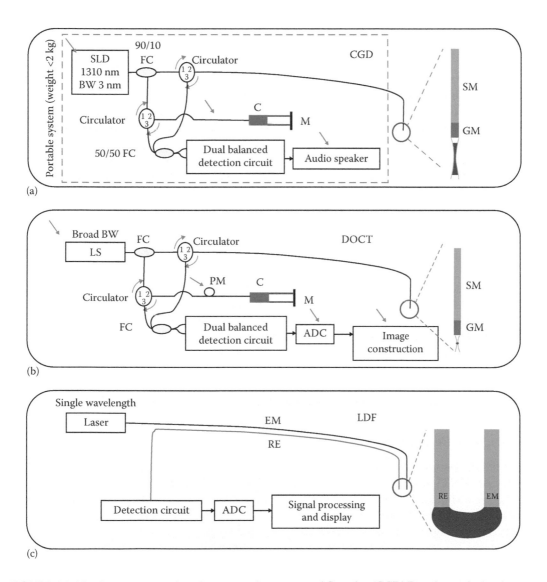

FIGURE 21.11 System comparison between coherence-gated Doppler (CGD)/Doppler optical coherence tomography (DOCT)/laser Doppler flowmetry (LDF) (a), CGD system (b), Doppler OCT system (c) LDF system. SLD, superluminescent diode; LS, light source; BW, wavelength bandwidth; FC, fiber coupler; C, collimator; M, mirror; PM, phase modulator; ADC, analog-to-digital converter; SM, single mode fiber; GM, gradient index fiber; EM, emission fiber; RE, receiver fiber. The detection volume of each system is shown in the probe drawing to the right in red. (From Liang, C.P. et al., *Biomed. Opt. Express*, 4, 760, 2013a. With permission.)

(10–20,000 Hz), the electronic signal is converted to an audio signal and broadcast by a speaker. We also collect the signal by a data acquisition (DAQ) card (National Instrument NI-6259) and process it with a computer (optional). The sampling rate of DAQ is 400 kHz. The fiber probe consists of an SMF and a GRIN multimode fiber (GM) for focusing (St. Jude Medical). The focal distance is 1.5 mm and the lateral resolution is 40 μm. Figure 21.11b and c illustrates DOCT and LDF system, respectively.

CGD Data Processing

The interference between two electric fields with Doppler frequency difference will generate a fluctuating AC signal, whose frequency is equivalent to the Doppler frequency difference. Therefore, the frequency of CGD waveform is equivalent to the Doppler frequency shift, which is linearly proportional to the moving speed. The sign of the signal does not reflect the flow direction. From the interference fringe, we can only measure the flow speed (the frequency) and the volume of moving scatters (the amplitude), but no information on the flow direction. To quantify the performance of the system, we record the interference fringe by a DAQ card (sampling rate 400 kHz) and calculate the M1 (first moment) factor, $\int \omega * P(\omega) d\omega$, which is linearly proportional to the average concentration of the scatter multiplied by the root mean square of the velocity (Rajan et al., 2009). $P(\omega)$ is the power spectrum of the interference fringe. The power spectrum of the CGD waveform is calculated by taking Fourier transform and square the amplitude. The time-domain window for Fourier transform is 0.1 second, and the integration window on frequency domain is 0–20 kHz. The M1 factor is used to characterize the linearity of speed measurement and detection volume. We also study the similarity of the time-domain signal by autocorrelation. Firstly, we square the CGD waveform signal to avoid the cancelation between positive and negative signals. Then, we use the entire signal in the 5 seconds' window to calculate the autocorrelation (Acorr) coefficient at different time shifts (0–5 seconds). We also calculate the spectrogram from the CGD waveform. The theoretical minimum resolvable frequency is 0 Hz (DC), and the maximum resolvable Doppler frequency is 15 MHz (3 dB bandwidth), which is limited by the dual balanced detector (Thorlabs PDB145C). This frequency corresponds to a flow speed of 9.75 m/s, which is well above the fastest blood flow speed in human body (~1 m/s) (Gardin et al., 1984). In practice, the maximum frequency is determined by the sampling rate of data acquisition. In our study, the sampling rate is 400 kHz and thus the maximum frequency is 200 kHz (Nyquist limit). The minimum resolvable frequency is determined by the time window of short-time Fourier transform, which is 0.1 second. Therefore, the resolution in frequency domain is 10 Hz. The minimum (10 Hz) and maximum (200 kHz) frequency correspond to Doppler velocity of 0.0065 and 130 mm/s, respectively. The signal frequencies within human audible range (<20 kHz) are displayed. A median filter is applied to improve visualization.

In Vivo Sheep Brain Blood Vessel Detection

We tested the performance of the CGD needle probe (Figure 21.12) for detecting vessels deep in the sheep brain. Figure 21.13 shows the voltage waveform and the spectrogram from tissue (Figure 21.13a), vein (Figure 21.13b), and artery (Figure 21.13c). When the CGD probe is surrounded by highly scattering brain tissues, any relative motion between the probe and the brain generates very strong Doppler signal. The spectrogram shows that the signal is strong (large volume of scatters) with focused frequency range (uniform speed). In contrast, the signal from a vein is more homogeneous in the frequency domain. The waveform of an artery shows the pulsation pattern, and the spectrogram reveals the speed variation during a pulsation cycle. When the insertion motion (tissue movement with respect to the CGD probe during insertion) signal is mixed with the artery signal, it may be challenging to differentiate them in the spectrogram. However, if we study the similarity of the time-domain signal by Acorr, we can see the clear difference between the bulk motion and the artery. We find that the difference of the frequency distribution between the different spikes in the bulk motion spectrogram generates irregular fluctuation on Acorr, which is distinctively different

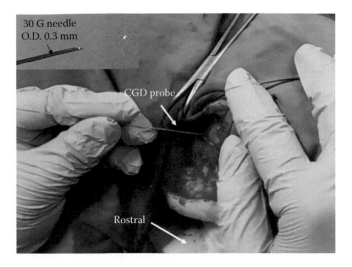

FIGURE 21.12 A CGD probe inserted into a sheep brain. (Reprinted from Liang, C.P. et al., *Biomed. Opt. Express*, 4, 760, 2013a. With permission.)

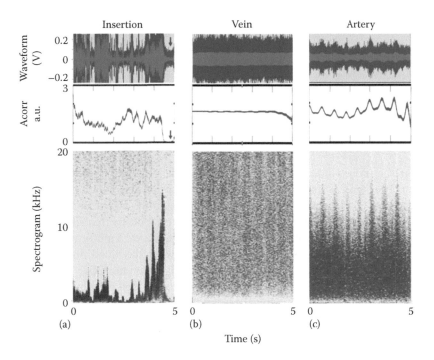

FIGURE 21.13 CGD waveform, Acorr coefficient, and spectrogram of brain vessels. (a) Inserting the probe, (b) a vein, and (c) an artery in sheep brain in vivo. The red arrows indicate the time point when the CGD probe is static relative to the brain tissue. (Reprinted from Liang, C.P. et al., *Biomed. Opt. Express*, 4, 760, 2013a. With permission.)

from the periodic Acorr of the artery. In Figure 21.13a, from 4.5 to 5 second (indicated by the red arrows), the probe is static and the motion is significantly lower than the signal in other time period. Acorr of the vein remains in a constant level due to the homogeneity of the signal.

To verify that the CGD probe can detect blood vessels in highly scattering tissues, we pushed a CGD probe toward a blood vessel in the sheep brain under US guidance. Figure 21.14a shows the

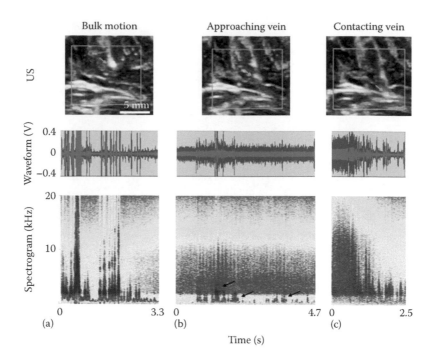

FIGURE 21.14 US imaging with the CGD flow detection in sheep brain in vivo US monitoring with Doppler detection of the CGD probe as it approaches a vessel (yellow arrows) in a sheep brain in vivo. The CGD probe tip appears as the bright white spot (red arrows). The target vein is the blue spot in the insets in top panel. (a) Insertion motion away from the vein, (b) approaching the vein, and (c) pressing on the vein resulting in the disappearance of the US Doppler signal as well as the CGD signal. (Reprinted from Liang, C.P. et al., *Biomed. Opt. Express*, 4, 760, 2013a. With permission.)

situation when the probe (indicated by red arrow) is 1–4 mm away from the vessel (indicated by yellow arrow). The signal is dominated by an insertion motion. The blood flow signal was initially detected at distance 3 mm in front of the needle. Similar to the signal from the brain vessel data (Figure 21.14a), the insertion motion has high intensity and focused frequency distribution. As the probe approaches the vessel (<1 mm), the blood flow signal shows a uniform frequency distribution (Figure 21.14b). Figure 21.14b also shows the situation when the blood flow signal is mixed with insertion motion signal (indicated by black arrows). Lastly, when the CGD probe comes in contact with the vessel, the blood flow signal initially becomes stronger, followed by a loss of the CGD flow signal. The probe is advanced until it constricts the flow and the US signal is lost; the CGD signal also attenuates.

Discussion

We have developed a thin (0.125 μm) and flexible CGD fiber probe that can detect at-risk blood vessels with real-time audio feedback up to 3 mm in front of advancing needle in brain tissue in vivo. The diffuse optics design of LDF with sample volume in the mm to cm range often includes signal from surrounding tissues and thus is not ideal for applications requiring high spatial specificity. The CGD probe with confocal optics design enables us to target a 0.4 mm capillary and differentiate blood vessels that are only 1 mm apart (data not shown; please see (Liang et al., 2013a)). On the other hand, DOCT has high spatial resolution (~10 μm), high temporal resolution (40 Hz), and wide flow speed dynamic range (7 μm/s to 52 cm/s) (Morofke et al., 2007), but it requires extensive postprocessing and an expensive system to obtain the high-resolution images. In contrast, CGD is a simple, robust, and low-cost sensing system, which provides an audio signal that is rich in content, yet is easy for the operator to interpret (Figure 21.13). Real-time temporal resolution (10 Hz) and broad electronic detection bandwidth (10–20,000 Hz) allow us to differentiate the signals from the artery, vein, and motion artifact (Figures 21.13 and 21.14).

In order to achieve the best balance between the depth sensitivity and resolution, we carefully choose the 190 μm CL. In ideal case, we hope to detect a blood vessel few millimeters away with sub-mm lateral specificity. Shorter CL provides higher resolution but also limits the depth detection range. We have used an SLD with 43 nm bandwidth at 1.325 μm center wavelength and found that it can only detect a vessel within 60 μm depth range, which is too short to prevent hemorrhage. Therefore, we further relax the detection range to 0.5 mm by increasing the CL to 0.19 mm. In the animal study, Figure 21.14 shows that it can detect a blood vessel that is 3 mm away due to longer transportation mean free path (MFP) in the brain. On the other hand, if we further relax the CL to mm level, it may further improve the depth detection range, but may also include diffuse photons from the background tissue resulting in poor spatial specificity. We have to maintain the lateral resolution at sub-mm level in order to avoid the millimeter-sized blood vessels. Therefore, a light source with CL close to the MFP of scattering tissues (~0.1 mm (Ntziachristos, 2010)) provides us the best balance between the detection range and the resolution.

CGD is best suited for situations that require multitasking rather than tasks requiring precise image guidance. If the probe is static relative to the tissue, the amplitude of audio signal is low. When the probe reaches blood vessels, a high volume of flowing scatterers generates a much louder signal, and the artery shows a clear pulsatile pattern. From the spectrum analysis (Figure 21.13), we find that the insertion motion has a narrow frequency distribution, while the movement of red blood cells in the vessels is associated with a more homogeneous frequency distribution. The audio signals from blood vessels are more like raining sound, which is clearly distinctive from the loud chirpy sounds associated with needle insertion (Figure 21.13). The pulsatile nature of arterial flow clearly distinguishes venous and arterial flow. Acorr analysis also shows the clear difference between the insertion motion, vein, and artery (Figure 21.13). The downside of the Acorr analysis is that it requires a long time window (couple seconds) that includes several heart beats and thus it cannot be implemented in real-time system. However, in real surgery, most of the instruments are advanced slowly and provide us long acquisition time. Also, since most of the bulk motion signals are from the relative movement between the probe and the tissue, stopping the probe is actually the most effective way to suppress the motion signal (red arrow in Figure 21.13a, 4.5–5 seconds). Therefore, we envision that the probe should be advanced slowly if a suspicious vessel is detected and then we can use the Acorr to analyze whether it is a pulsating artery. We will perform more animal experiments in the future to evaluate the false-positive rate.

It is worth to note that CGD may not detect a small vessel (~10 μm) at 90° angle when the probe is very close to the vessel. The MFP in brain tissue is around 0.15–0.3 mm (Jeon et al., 2006). Beyond this ballistic scattering regime, multiple scattering events effectively randomize the photon propagation direction, resulting in random incident angles to the moving blood cells. Therefore, we should be able to detect the laminar flow from any angle at the distance >0.3 mm to the blood vessel. When the probe is proximate to the vessel, however, the illumination becomes more directional and the effective incident angle depends on the scattering property of blood cells. If the diameter of the vessel is comparable to the MFP of blood (~7 μm (Smithies et al., 1998)), we will not be able to acquire the Doppler signal at 90°. For a vessel that is much larger than 7 μm, multiple scattering between blood cells will randomize the incident angle and enable us to acquire the Doppler signal from any angle. In the practical situation, we hope to stop the probe at the position >0.3 mm away from the high-risk blood vessels that are millimeters in size. Under this circumstance, we should not encounter the angle issue.

CGD probe is a great complementary technology to tomographic techniques. We can use the wide FOV tomographic imaging such as US to deliver the CGD probe to the ROI and perform precise targeting with high precision (<0.1 mm) CGD local information/feedback. Figure 21.14 shows that the CGD probe is capable of detecting blood vessels that are on average 3.4 ± 1.3 mm from the tip of the probe. With millimeters axial detection range and lateral tissue confinement, we can aim the probe to a brain vessel for tens of seconds. The signal was found to become stronger when the CGD probe contacted the blood vessel. Further advances of the CGD probe resulted in the loss of both the CGD and the US Doppler signal, and a US pattern suggestive of a hematoma is then formed in proximity to the blood vessel. On the other hand, many needle-based procedures are still performed without imaging guidance due to the limitations of technology. For example, in stereotactic neurosurgery, the skull blocks most of US signal and it is difficult to fit the US probe to the small opening along with surgical instruments. Therefore, a thin CGD probe (<30 G) that can be easily integrated with surgical instruments and provide real-time feedbacks on local blood flow in front of the probe tip has great potential to improve the safety of these "blind" procedures.

CONCLUSION

CGD provides the means for detecting blood flow with higher spatial resolution and with a smaller, more robust probe than conventional LDF at a cost of shorter depth of view. This tradeoff favors interventional procedures that require insertion of thin probes deep into tissue and where small vessels need to be detected with high precision.

SUMMARY

OCT that provides nonlabeling, real-time, and high-resolution images can potentially transform the paradigm of surgical guidance. The design and development of OCT devices for neurosurgery guidance have been presented in this chapter. A forward-imaging needle-type OCT probe that can fit into minimal invasive tools (I.D. ~1 mm), detect the at-risk blood vessels, and identify tissue micro-landmarks is a promising guidance tool to improve the safety and the accuracy of needle-based procedures, which are mostly performed without imaging feedback. However, OCT is limited by the shallow imaging depth (1–2 mm). In order to address this issue, the first OCT system that is compatible with comprehensive MRI has been developed. The multiscale and multicontrast MRI/OCT imaging system significantly improves the accuracy of iMRI from 1 to 0.01 mm. Besides the sophisticated imaging systems, a simple, thin (0.125 mm), and low-cost (1/10 cost of the OCT system) fiber sensor, called CGD that can be easily integrated with most surgical tools, has been developed to avoid intracranial hemorrhage. The transformative guidance tools pave a new avenue to better clinical outcomes.

REFERENCES

Adler, D.C., Chen, Y., Huber, R., Schmitt, J., Connolly, J., and Fujimoto, J.G. (2007). Three-dimensional endo-microscopy using optical coherence tomography. *Nature Photonics 1*, 709–716.

Aguirre, A.D., Sawinski, J., Huang, S.W., Zhou, C., Denk, W., and Fujimoto, J.G. (2010). High speed optical coherence microscopy with autofocus adjustment and a miniaturized endoscopic imaging probe. *Optical Express 18*, 4222–4239.

Andrews, C., Aviles-Olmos, I., Hariz, M., and Foltynie, T. (2010). Which patients with dystonia benefit from deep brain stimulation? A metaregression of individual patient outcomes. *Journal of Neurology Neurosurgery and Psychiatry 81*, 1383–1389.

Binder, D.K., Rau, G., and Starr, P.A. (2003). Hemorrhagic complications of microelectrode-guided deep brain stimulation. *Stereotactic and Functional Neurosurgery 80*, 28–31.

Binder, D.K., Rau, G.M., and Starr, P.A. (2005). Risk factors for hemorrhage during microelectrode-guided deep brain stimulator implantation for movement disorders. *Neurosurgery 56*, 722–732; discussion 722–732.

Boppart, S.A., Bouma, B.E., Pitris, C., Tearney, G.J., Fujimoto, J.G., and Brezinski, M.E. (1997). Forward-imaging instruments for optical coherence tomography. *Optical Letters 22*, 1618–1620.

Bronstein, J.M., Tagliati, M., Alterman, R.L., Lozano, A.M., Volkmann, J., Stefani, A., Horak, F.B., Okun, M.S., Foote, K.D., Krack, P. et al. (2010). Deep brain stimulation for Parkinson disease: An expert consensus and review of key issues. *Archives of Neurology 68*, 165.

Chen, Z.P., Milner, T.E., Srinivas, S., Wang, X.J., Malekafzali, A., vanGemert, M.J.C., and Nelson, J.S. (1997). Noninvasive imaging of in vivo blood flow velocity using optical Doppler tomography. *Optics Letters 22*, 1119–1121.

Cobb, M.J., Liu, X., and Li, X. (2005). Continuous focus tracking for real-time optical coherence tomography. *Optical Letters 30*, 1680–1682.

Colbassani, H.J., Nishio, S., Sweeney, K.M., Bakay, R.A.E., and Takei, Y. (1988). CT-assisted stereotactic brain biopsy—Value of intraoperative frozen section diagnosis. *Journal of Neurology Neurosurgery and Psychiatry 51*, 332–341.

Deuschl, G., Herzog, J., Kleiner-Fisman, G., Kubu, C., Lozano, A.M., Lyons, K.E., Rodriguez-Oroz, M.C., Tamma, F., Troster, A.I., Vitek, J.L. et al. (2006). Deep brain stimulation: Postoperative issues. *Movement Disorders 21*, S219–S237.

Eun, H.C. (1995). Evaluation of skin blood flow by laser Doppler flowmetry. *Clinical Dermatology 13*, 337–347.

Fleming, C.P., Quan, K.J., and Rollins, A.M. (2010). Toward guidance of epicardial cardiac radiofrequency ablation therapy using optical coherence tomography. *Journal of Biomedical Optics 15*, 041510.

Flora, E.D., Perera, C.L., Cameron, A.L., and Maddern, G.J. (2010). Deep brain stimulation for essential tremor: A systematic review. *Movement Disorders 25*, 1550–1559.

Funkiewiez, A., Caputo, E., Ardouin, C., Krack, P., Fraix, V., Benabid, A.L., and Pollak, P. (2001). Behavioral and mood changes associated with bilateral stimulation of the subthalamic nucleus: A consecutive series of 98 parkinsonian patients. *Neurology 56*, A274–A274.

Gardin, J.M., Burn, C.S., Childs, W.J., and Henry, W.L. (1984). Evaluation of blood-flow velocity in the ascending aorta and main pulmonary-artery of normal subjects by doppler echocardiography. *American Heart Journal 107*, 310–319.

Gorgulho, A., De Salles, A.A., Frighetto, L., and Behnke, E. (2005). Incidence of hemorrhage associated with electrophysiological studies performed using macroelectrodes and microelectrodes in functional neurosurgery. *Journal of Neurosurgery 102*, 888–896.

Grau, T., Leipold, R.W., Conradi, R., Martin, E., and Motsch, J. (2002). Efficacy of ultrasound imaging in obstetric epidural anesthesia. *Journal of Clinical Anesthesia 14*, 169–175.

Greenberg, B.D., Gabriels, L.A., Malone, D.A., Jr., Rezai, A.R., Friehs, G.M., Okun, M.S., Shapira, N.A., Foote, K.D., Cosyns, P.R., Kubu, C.S. et al. (2008). Deep brain stimulation of the ventral internal capsule/ventral striatum for obsessive-compulsive disorder: Worldwide experience. *Molecular Psychiatry 15*, 64–79.

Gross, R.E., Krack, P., Rodriguez-Oroz, M.C., Rezai, A.R., and Benabid, A.L. (2006). Electrophysiological mapping for the implantation of deep brain stimulators for Parkinson's disease and tremor. *Movement Disorders 21*, S259–S283.

Guo, S.G., Yu, L.F., Sepehr, A., Perez, J., Su, J.P., Ridgway, J.M., Vokes, D., Wong, B.J.F., and Chen, Z.P. (2009). Gradient-index lens rod based probe for office-based optical coherence tomography of the human larynx. *Journal of Biomedical Optics 14*, 014017.

Hall, W.A. (1998). The safety and efficacy of stereotactic biopsy for intracranial lesions. *Cancer 82*, 1749–1755.

Hall, W.A. and Truwit, C.L. (2005a). 1.5 T: Spectroscopy-supported brain biopsy. *Neurosurgery Clinics of North America 16*, 165.

Hall, W.A. and Truwit, C.L. (2005b). Intraoperative MR imaging. *Magnetic Resonance Imaging Clinical North America 13*, 533–543.

Han, S., Sarunic, M.V., Wu, J., Humayun, M., and Yang, C. (2008). Handheld forward-imaging needle endoscope for ophthalmic optical coherence tomography inspection. *Journal of Biomedical Optics 13*, 020505.

Hariz, M.I. (2002). Complications of deep brain stimulation surgery. *Movement Disorders 17*, S162–S166.

Henderson, J.M. (2012). Connectomic surgery: Diffusion tensor imaging (DTI) tractography as a targeting modality for surgical modulation of neural networks. *Frontiers in Integrative Neuroscience 6*, 15.

Hirabayashi, H., Tengvar, M., and Hariz, M.I. (2002). Stereotactic imaging of the pallidal target. *Movement Disorders 17*(Suppl 3), S130–134.

Jafri, M.S., Farhang, S., Tang, R.S., Desai, N., Fishman, P.S., Rohwer, R.G., Tang, C.M., and Schmitt, J.M. (2005). Optical coherence tomography in the diagnosis and treatment of neurological disorders. *Journal of Biomedical Optics 10*, 051603.

Jafri, M.S., Tang, R., and Tang, C.M. (2009). Optical coherence tomography guided neurosurgical procedures in small rodents. *Journal of Neuroscience Methods 176*, 85–95.

Jain, A., Kopa, A., Pan, Y.T., Fedder, G.K., and Xie, H.K. (2004). A two-axis electrothermal micromirror for endoscopic optical coherence tomography. *IEEE Journal of Selected Topics in Quantum Electronics 10*, 636–642.

Jeon, S.W., Shure, M.A., Baker, K.B., Huang, D., Rollins, A.M., Chahlavi, A., and Rezai, A.R. (2006). A feasibility study of optical coherence tomography for guiding deep brain probes. *Journal of Neuroscience Methods 154*, 96–101.

Karakitsos, D., Labropoulos, N., De Groot, E., Patrianakos, A.P., Kouraklis, G., Poularas, J., Samonis, G., Tsoutsos, D.A., Konstadoulakis, M.M., and Karabinis, A. (2006). Real-time ultrasound-guided catheterisation of the internal jugular vein: A prospective comparison with the landmark technique in critical care patients. *Critical Care 10*(6), R162.

Kijsamanmith, K., Timpawat, S., Vongsavan, N., and Matthews, B. (2011). Pulpal blood flow recorded from human premolar teeth with a laser Doppler flow meter using either red or infrared light. *Archives of Oral Biology 56*, 629–633.

Klein, T., Wieser, W., Eigenwillig, C.M., Biedermann, B.R., and Huber, R. (2011). Megahertz OCT for ultrawide-field retinal imaging with a 1050 nm Fourier domain mode-locked laser. *Optical Express 19*, 3044–3062.

Kongkham, P.N., Knifed, E., Tamber, M.S., and Bernstein, M. (2008). Complications in 622 cases of frame-based stereotactic biopsy, a decreasing procedure. *Canadian Journal of Neurological Sciences 35*, 79–84.

Kulkarni, A.V., Guha, A., Lozano, A., and Bernstein, M. (1998). Incidence of silent hemorrhage and delayed deterioration after stereotactic brain biopsy. *Journal of Neurosurgery 89*, 31–35.

Lang, A.E., Gill, S., Patel, N.K., Lozano, A., Nutt, J.G., Penn, R., Brooks, D.J., Hotton, G., Moro, E., Heywood, P. et al. (2006). Randomized controlled trial of intraputamenal glial cell line-derived neurotrophic factor infusion in Parkinson disease. *Annals of Neurology 59*, 459–466.

Larson, P.S., Starr, P.A., Bates, G., Tansey, L., Richardson, R.M., and Martin, A.J. (2012). An optimized system for interventional magnetic resonance imaging-guided stereotactic surgery: Preliminary evaluation of targeting accuracy. *Neurosurgery 70*, 9.

Li, Q., Onozato, M.L., Andrews, P.M., Chen, C.W., Paek, A., Naphas, R., Yuan, S.A., Jiang, J., Cable, A., and Chen, Y. (2009). Automated quantification of microstructural dimensions of the human kidney using optical coherence tomography (OCT). *Optical Express 17*, 16000–16016.

Liang, C.P., Wierwille, J., Moreira, T., Schwartzbauer, G., Jafri, M.S., Tang, C.M., and Chen, Y. (2011). A forward-imaging needle-type OCT probe for image guided stereotactic procedures. *Optics Express 19*, 26283–26294.

Liang, C.P., Wu, Y., Schmitt, J., Bigeleisen, P.E., Slavin, J., Jafri, M.S., Tang, C.M., and Chen, Y. (2013a). Coherence-gated Doppler: A fiber sensor for precise localization of blood flow. *Biomedical Optical Express 4*, 760–771.

Liang, C.P., Yang, B., Kim, I.K., Makris, G., Desai, J.P., Gullapalli, R.P., and Chen, Y. (2013b). Concurrent multiscale imaging with magnetic resonance imaging and optical coherence tomography. *Journal of Biomedical Optics 18*, 046015.

Limousin, P., Krack, P., Pollak, P., Benazzouz, A., Ardouin, C., Hoffmann, D., and Benabid, A.L. (1998). Electrical stimulation of the subthalamic nucleus in advanced Parkinson's disease. *New England Journal of Medicine 339*, 1105–1111.

Liu, X., Cobb, M.J., Chen, Y., Kimmey, M.B., and Li, X. (2004). Rapid-scanning forward-imaging miniature endoscope for real-time optical coherence tomography. *Optical Letters 29*, 1763–1765.

Malone, D.A., Jr., Dougherty, D.D., Rezai, A.R., Carpenter, L.L., Friehs, G.M., Eskandar, E.N., Rauch, S.L., Rasmussen, S.A., Machado, A.G., Kubu, C.S. et al. (2009). Deep brain stimulation of the ventral capsule/ventral striatum for treatment-resistant depression. *Biological Psychiatry 65*, 267–275.

Martin, A.J., Larson, P.S., Ostrem, J.L., Keith Sootsman, W., Talke, P., Weber, O.M., Levesque, N., Myers, J., and Starr, P.A. (2005). Placement of deep brain stimulator electrodes using real-time high-field interventional magnetic resonance imaging. *Magnetic Resonance Medicine 54*, 1107–1114.

Mohr, P. (2008). Deep brain stimulation in psychiatry. *Neuroendocrinology Letters 29*, 123–132.

Morofke, D., Kolios, M.C., Vitkin, I.A., and Yang, V.X.D. (2007). Wide dynamic range detection of bidirectional flow in Doppler optical coherence tomography using a two-dimensional Kasai estimator. *Optics Letters 32*, 253–255.

Munce, N.R., Mariampillai, A., Standish, B.A., Pop, M., Anderson, K.J., Liu, G.Y., Luk, T., Courtney, B.K., Wright, G.A., Vitkin, I.A. et al. (2008). Electrostatic forward-viewing scanning probe for Doppler optical coherence tomography using a dissipative polymer catheter. *Optics Letters 33*, 657–659.

Nishihara, M., Sasayama, T., Kudo, H., and Kohmura, E. (2010). Morbidity of stereotactic biopsy for intracranial lesions. *Kobe Journal of Medical Science 56*, E148–E153.

Ntziachristos, V. (2010). Going deeper than microscopy: The optical imaging frontier in biology. *Nature Methods 7*, 603–614.

Owen, C.M. and Linskey, M.E. (2009). Frame-based stereotaxy in a frameless era: Current capabilities, relative role, and the positive- and negative predictive values of blood through the needle. *Journal of Neuro-Oncology 93*, 139–149.

Pan, Y., Li, Z., Xie, T., and Chu, C.R. (2003). Hand-held arthroscopic optical coherence tomography for in vivo high-resolution imaging of articular cartilage. *Journal of Biomedical Optics 8*, 648–654.

Pan, Y., Xie, H., and Fedder, G.K. (2001). Endoscopic optical coherence tomography based on a microelectro-mechanical mirror. *Optics Letters 26*, 1966–1968.

Petoukhova, A.L., Steenbergen, W., and de Mul, F.F.M. (2001). Path-length distribution and path-length-resolved Doppler measurements of multiply scattered photons by use of low-coherence interferometry. *Optics Letters 26*, 1492–1494.

Rajan, V., Varghese, B., van Leeuwen, T.G., and Steenbergen, W. (2009). Review of methodological developments in laser Doppler flowmetry. *Lasers in Medical Science 24*, 269–283.

Ren, J., Wu, J., McDowell, E.J., and Yang, C. (2009). Manual-scanning optical coherence tomography probe based on position tracking. *Optics Letters 34*, 3400–3402.

Rezai, A.R., Kopell, B.H., Gross, R.E., Vitek, J.L., Sharan, A.D., Limousin, P., and Benabid, A.L. (2006). Deep brain stimulation for Parkinson's disease: Surgical issues. *Movement Disorders 21*(Suppl 14), S197–218.

Roitberg, B.Z., Khan, N., Alp, M.S., Hersonskey, T., Charbel, F.T., and Ausman, J.I. (2001). Bedside external ventricular drain placement for the treatment of acute hydrocephalus. *British Journal of Neurosurgery 15*, 324–327.

Sansur, C.A., Frysinger, R.C., Pouratian, N., Fu, K.M., Bittl, M., Oskouian, R.J., Laws, E.R., and Elias, W.J. (2007). Incidence of symptomatic hemorrhage after stereotactic electrode placement. *Journal of Neurosurgery 107*, 998–1003.

Sawin, P.D., Hitchon, P.W., Follett, K.A., and Torner, J.C. (1998). Computed imaging-assisted stereotactic brain biopsy—A risk analysis of 225 consecutive cases. *Surgical Neurology 49*, 640–649.

Schulz, T., Puccini, S., Schneider, J.P., and Kahn, T. (2004). Interventional and intraoperative MR: Review and update of techniques and clinical experience. *European Radiology 14*, 2212–2227.

Sergeev, A., Gelikonov, V., Gelikonov, G., Feldchtein, F., Kuranov, R., Gladkova, N., Shakhova, N., Snopova, L., Shakhov, A., Kuznetzova, I. et al. (1997). In vivo endoscopic OCT imaging of precancer and cancer states of human mucosa. *Optics Express 1*, 432–440.

Skidmore, F.M., Rodriguez, R.L., Fernandez, H.H., Goodman, W.K., Foote, K.D., and Okun, M.S. (2006). Lessons learned in deep brain stimulation for movement and neuropsychiatric disorders. *CNS Spectrums 11*, 521.

Smithies, D.J., Lindmo, T., Chen, Z.P., Nelson, J.S., and Milner, T.E. (1998). Signal attenuation and localization in optical coherence tomography studied by Monte Carlo simulation. *Physics in Medicine and Biology 43*, 3025–3044.

Starr, P.A. (2002). Placement of deep brain stimulators into the subthalamic nucleus or Globus pallidus internus: Technical approach. *Stereotactic and Functional Neurosurgery 79*, 118–145.

Sutherland, G.R., Latour, I., and Greer, A.D. (2008). Integrating an image-guided robot with intraoperative MRI: A review of the design and construction of neuroArm. *IEEE Engineering in Medicine and Biology Magazine 27*, 59–65.

Takahashi, Y., Watanabe, Y., and Sato, M. (2008). High speed spectral domain optical coherence tomography with forward and side-imaging probe. *Japanese Journal of Applied Physics 47*, 6540–6543.

Terao, T., Takahashi, H., Yokochi, F., Taniguchi, M., Okiyama, R., and Hamada, I. (2003). Hemorrhagic complication of stereotactic surgery in patients with movement disorders. *Journal of Neurosurgery 98*, 1241–1246.

Tsekos, N.V., Khanicheh, A., Christoforou, E., and Mavroidis, C. (2007). Magnetic resonance—Compatible robotic and mechatronics systems for image-guided interventions and rehabilitation: A review study. *Annual Review of Biomedical Engineering 9*, 351–387.

Varghese, B., Rajan, V., Van Leeuwen, T.G., and Steenbergen, W. (2010). In vivo optical path lengths and path length resolved doppler shifts of multiply scattered light. *Lasers in Surgery and Medicine 42*, 692–700.

Venkataramana, N.K., Kumar, S.K.V., Balaraju, S., Radhakrishnan, R.C., Bansal, A., Dixit, A., Rao, D.K., Das, M., Jan, M., Gupta, P.K. et al. (2010). Open-labeled study of unilateral autologous bone-marrow-derived mesenchymal stem cell transplantation in Parkinson's disease. *Translational Research 155*, 62–70.

Voges, J., Hilker, R., Botzel, K., Kiening, K.L., Moss, M., Kupsch, A., Schnitzler, A., Schneider, G.H., Steude, U., Deuschl, G. et al. (2007). Thirty days complication rate following surgery performed for deep-brain-stimulation. *Movement Disorders 22*, 1486–1489.

Wang, H., Black, A.J., Zhu, J., Stigen, T.W., Al-Qaisi, M.K., Netoff, T.I., Abosch, A., and Akkin, T. (2011). Reconstructing micrometer-scale fiber pathways in the brain: Multi-contrast optical coherence tomography based tractography. *Neuroimage 58*, 984–992.

Wang, Y., Bachman, M., Li, G.P., Guo, S., Wong, B.J., and Chen, Z. (2005). Low-voltage polymer-based scanning cantilever for in vivo optical coherence tomography. *Optics Letter 30*, 53–55.

Wang, Z., Lee, C.S., Waltzer, W.C., Liu, J., Xie, H., Yuan, Z., and Pan, Y. (2007). In vivo bladder imaging with microelectromechanical-systems-based endoscopic spectral domain optical coherence tomography. *Journal of Biomedical Optics 12*, 034009.

Wardell, K., Blomstedt, P., Richter, J., Antonsson, J., Eriksson, O., Zsigmond, P., Bergenheim, A.T., and Hariz, M.I. (2007). Intracerebral microvascular measurements during deep brain stimulation implantation using laser Doppler perfusion monitoring. *Stereotactic and Functional Neurosurgery 85*, 279–286.

White, E., Woolley, M., Bienemann, A., Johnson, D.E., Wyatt, M., Murray, G., Taylor, H., and Gill, S.S. (2011). A robust MRI-compatible system to facilitate highly accurate stereotactic administration of therapeutic agents to targets within the brain of a large animal model. *Journal of Neuroscience Methods 195*, 78–87.

Wu, J.G., Conry, M., Gu, C.H., Wang, F., Yaqoob, Z., and Yang, C.H. (2006). Paired-angle-rotation scanning optical coherence tomography forward-imaging probe. *Optics Letters 31*, 1265–1267.

Xie, T., Guo, S., Chen, Z., Mukai, D., and Brenner, M. (2006). GRIN lens rod based probe for endoscopic spectral domain optical coherence tomography with fast dynamic focus tracking. *Optics Express 14*, 3238–3246.

Xie, T., Liu, G., Kreuter, K., Mahon, S., Colt, H., Mukai, D., Peavy, G.M., Chen, Z., and Brenner, M. (2009). In vivo three-dimensional imaging of normal tissue and tumors in the rabbit pleural cavity using endoscopic swept source optical coherence tomography with thoracoscopic guidance. *Journal of Biomedical Optics 14*, Article No.: 064045.

Xie, T., Xie, H., Fedder, G.K., and Pan, Y. (2003). Endoscopic optical coherence tomography with a modified microelectromechanical systems mirror for detection of bladder cancers. *Applied Optics 42*, 6422–6426.

Yang, B., Tan, U.X., McMillan, A., Gullapalli, R., and Desai, J.P. (2011). Design and control of a 1-DOF MRI compatible pneumatically actuated robot with long transmission lines. *IEEE ASME Transactions on Mechatronics 16*, 1040–1048.

Yang, V.X.D., Gordon, M.L., Qi, B., Pekar, J., Lo, S., Seng-Yue, E., Mok, A., Wilson, B.C., and Vitkin, I.A. (2003). High speed, wide velocity dynamic range Doppler optical coherence tomography (Part I): System design, signal processing, and performance. *Optics Express 11*, 794–809.

Yu, L., Liu, G., Rubinstein, M., Saidi, A., Wong, B.J., and Chen, Z. (2009). Office-based dynamic imaging of vocal cords in awake patients with swept-source optical coherence tomography. *Journal of Biomedical Optics 14*, 064020.

Yuan, S.A., Li, Q., Jiang, J., Cable, A., and Chen, Y. (2009). Three-dimensional coregistered optical coherence tomography and line-scanning fluorescence laminar optical tomography. *Optical Letters 34*, 1615–1617.

Yuan, S.A., Roney, C.A., Wierwille, J., Chen, C.W., Xu, B.Y., Griffiths, G., Jiang, J., Ma, H.Z., Cable, A., Summers, R.M. et al. (2010). Co-registered optical coherence tomography and fluorescence molecular imaging for simultaneous morphological and molecular imaging. *Physics in Medicine and Biology 55*, 191–206.

Zara, J.M. and Patterson, P.E. (2006). Polyimide amplified piezoelectric scanning mirror for spectral domain optical coherence tomography. *Applied Physics Letters 89*, 3.

Zara, J.M., Yazdanfar, S., Rao, K.D., Izatt, J.A., and Smith, S.W. (2003). Electrostatic micromachine scanning mirror for optical coherence tomography. *Optical Letters 28*, 628–630.

Section IV

Image-Guided Intervention and Phototherapy

22 Impact of Retinal Stimulation on Neuromodulation

Deborah Zelinsky

CONTENTS

INTRODUCTION

Historically, experts have considered the retina as a sensory system, feeding information into the brain's visual cortex. However, research has now demonstrated that the retina is a bidirectional neural interface that is an actual part of the central nervous system (CNS) (Vaney 1999). Since retinal signals are processed by many regions of the brain—not just the visual cortex—the implication is that retinal stimulation can affect other physical, physiological, and psychological processes, such as motor control, biochemical activity, and cognitive abilities.

During the past decades, researchers have discovered a retinal cell type that responds to luminance (external light) levels. These photosensitive cells combine the external luminance information with signals obtained through eyesight and not only send this combined feedforward information to the brain but concurrently receive feedback signals from the body (Chen et al. 2011, 2013, Schmidt et al. 2011).

The mixture of feedforward and feedback signaling enables the retina to be used as a two-way, noninvasive portal for influencing and monitoring body functions and thought processes, largely beneath the level of consciousness. Because of the retina's critical role in brain function, therapeutic eyeglasses—an important tool in neuro-optometric rehabilitation—may be used to modify processing in a range of physical and mental health disorders. These individualized lenses can change the dynamic relationship between the mind's visual inputs and the body's internal responses by altering spatial and temporal distribution of light on the retina. The novel use of light to affect the nervous system has already been successfully applied to a range of disorders, including jaundice (Tayman et al. 2010), jet lag (Parry 2002), seasonal affective disorder (Lavoie et al. 2009), brain injury (Naeser et al. 2011, Sinclair et al. 2014), and spinal cord injury (Alilain et al. 2008, Alilain and Silver 2009). Neuro-optometry also uses light to modulate brain and body functions.

This chapter presents both the theoretical framework and empirical evidence to support the use of customized eyeglasses for altering brain function. The underlying premise is that there exists a hierarchy of separate, yet interdependent, cortical and subcortical pathways, which are linked to various visual systems. The main emphasis of this discussion is on the retina's complex connections with systems other than the conscious eyesight. Those subconscious and unconscious systems can be altered by changes in the amount, frequency, intensity, or direction of incoming light to the eye.

The following section overviews retinal function, at cortical and subcortical levels, with examples of how the mind and body adapt to environmental changes. The third section discusses retinal structure. The fourth section introduces the concept of neuromodulation, and its effects on behavior and processing. Neuromodulation is described here as the process of achieving balance between mental and physical functions at both conscious and nonconscious levels. The fifth section describes the impact that eyeglasses can have on the nervous system.

RETINAL FUNCTION

The retina connects with many systems other than eyesight. Its connections include structures in the cortex, limbic system, cerebellum, midbrain, and brainstem, all of which affect systems such as the endocrine, respiratory, circulatory, digestive, and musculoskeletal. During neuro-optometric rehabilitation, careful adjustment of light entering the retina by using lenses, prisms, and/or filters alters cellular activity. This biochemical activity triggers action potentials and graded potentials (Purves and Williams 2001), affecting overall neuronal circuitry.

The basic concept outlined in the pioneering work of A.M. Skeffington, O.D., in the mid-twentieth century, refers to a hierarchy of "Where am I?," "Where is it?," and "What is it?" pathways, culminating in an emergent concept of vision that gives meaning to sensory signals (Skeffington 1957, Skeffington 1966). Optometrist Jacob Liberman's 1990 book *Light: The Medicine of the Future* added pineal gland activation by retinal stimulation—a "How am I?" pathway to Skeffington's accepted framework (Liberman 1994). Those well-recognized retinal pathways send output signals in response to changes in the environment. Bart Krekelberg, a brain researcher at Rutgers, postulated a "When is It?" pathway for time judgment in 2003 (Krekelberg 2003)—a concept that has since been documented in the past decade (Kim et al. 2014b).

The effects of injury, disease, and stress vary from individual to individual, depending on a "Who am I?" pathway. Altered mind and body functions limit processing of the external environment. Those mental and physical changes often create symptoms in either the body's internal regulation or in the mind's planning, attention, and judgment due to disrupted, mismatched, or dysfunctional sensory circuitry. For instance, patients with brain injuries often complain of light and sound sensitivity because they cannot easily filter out external sensory stimuli. Eyeglasses will affect both external sensory input (eyesight) and internal regulation, thus influencing executive ("What should I do about it?") functions. Nontraditional types of eyeglass designs can prove to be a useful tool during rehabilitation to help balance fragile biochemical and sensory systems and enhance the rehabilitative work of other professionals.

As shown in Figure 22.1, the body's survival functions and the mind's executive functions are influenced by changes in external sensory inputs. Conversely, external attention and awareness are altered by shifts in the body and/or mind. The three domains interact.

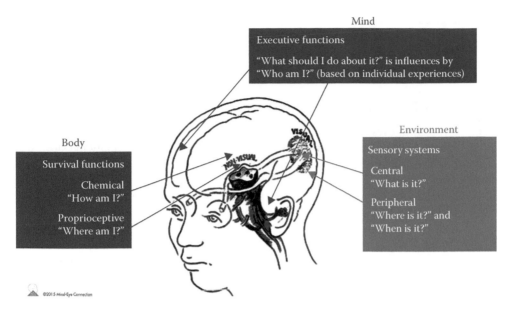

FIGURE 22.1 Three domains affected by retinal stimulation. (Courtesy of The Mind-Eye Connection, Northbrook, IL.)

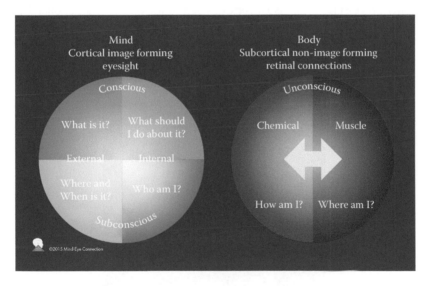

FIGURE 22.2 Mind and body, each reacts to environmental changes. (Courtesy of The Mind-Eye Connection, Northbrook, IL, Copyright 2015.)

It is useful to consider the cortical and subcortical pathways separately. The cortical pathways are described in terms of retinal input into and output from the visual cortex. The subcortical (non-image-forming) pathways are separated into neurological and biochemical systems. The multiple feedback pathways within the retina are beyond the scope of this limited chapter.

As shown in Figure 22.2, ambient processing is everything that goes on "behind the scenes" as opposed to conscious attention at a given moment. Chemical and muscle reflexes unconsciously govern "How am I?" and "Where am I?" systems at the body level, but those two concepts are also processed at a cortical level, albeit subconsciously. For example, "How am I?" subcortical would be tired, energized, etc. At the cortical level, "How am I?" would register as happy, sad, angry, etc. Meanwhile, the "Where am I?" subcortical response involves physical body balance against gravity, while, at the cortical level, mental awareness occurs, such as the realization of being in North America and on the planet Earth.

CORTICAL, IMAGE-FORMING PATHWAYS (EYESIGHT)

Unfortunately, central eyesight is deemed as the most important of all the retinal pathways, even though it is dependent on other sensory inputs. Seeing details clearly is actually the slowest retinal pathway for information processing and occurs only after conscious attention is placed on a selected target (O'Connor et al. 2002). The classic image-forming eyesight pathway from the retina to the visual cortex is only one of several visual functions. The portion of the retina that transmits clear details is not even present in newborn infants; it develops within a few months of age, after other retinal pathways are in place (Candy et al. 1998). Although the "Where is it?" (peripheral eyesight) and "What is it?" (central eyesight) pathways are connected (Yeatman et al. 2014), linkage between those cortical pathways often is not measured during eye examinations. Typically, glasses are designed to only address the clarity of surrounding targets—the "What is it?"—central system (Figure 22.3).

Input Circuitry to the Visual Cortex

As they begin the journey to the visual cortex, the vast majority of signals (approximately 90%) leaving the optic nerve travel through the lateral geniculate nucleus (LGN) of the thalamus (Goldstein 2010) for further processing in the visual cortex as part of the central and peripheral eyesight pathways. Those signals lead toward attention on a selected target. Some signals branch off at the LGN to other subcortical structures involved with neurological circuitry and others modulating chemical circuitry as discussed in "Subcortical neurological circuitry" and "Subcortical chemical circuitry."

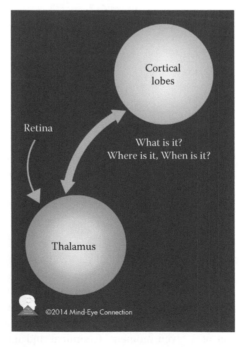

FIGURE 22.3 Cortical where?, when?, and what? retinogeniculate tracts determining space and time judgments. (Courtesy of The Mind-Eye Connection, Northbrook, IL, Copyright 2014.)

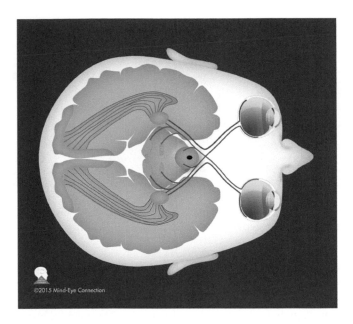

FIGURE 22.4 Input circuitry from retina to various cortical and subcortical locations. Nasal retinal fibers travel to the hypothalamus; temporal retinal fibers do not. (Courtesy of The Mind-Eye Connection, Northbrook, IL, Copyright 2015.)

Some of these subcortical structures include the intergeniculate leaflet, the superior colliculus in the midbrain, the suprachiasmatic nucleus of the hypothalamus, the pineal gland, and the habenula (part of the limbic system connecting with motor circuitry). New research also has demonstrated a direct retinal pathway in humans to the brain's pulvinar region (another portion of the thalamus) (Arcaro et al. 2015). Many of those locations send signals back to the LGN through feedback loops. Signals continue from the LGN through the optic radiations that traverse the parietal and temporal lobes.

The signal content in the optic radiations depends on the originating location of the visual signal. Signals from superior space (at or above eye level) travel through the bottom portions of each retina and head through temporal lobes, while targets below eye level send signals through the superior retina and interact with the parietal lobes. Meyer's loop in the temporal lobe was originally thought to be an anterior looping of optic radiation fibers (Jeelani et al. 2010). However, in 2015, scientists determined that Meyer's loop is a conglomeration of many sensory signals (Goga and Ture 2015) (Figure 22.4).

Retinal input into the LGN represents approximately a fifth of its total sensory input. The LGN is influenced by many other sensory signals and previous memories, as presented in Figure 22.5 as the "Who am I?" pathway. The LGN has burst modes and tonic response modes, which vary during waking or sleep states (Weyand et al. 2001, Horng et al. 2009). Exiting retinal signals in the optic nerve interact with inhibitory and excitatory feedback and feedforward signaling systems from the LGN (Guler et al. 2008, Schmidt and Kofuji 2008). In other words, a significant amount of two-way signaling occurs with "Who am I?" guiding the cortical responses to subcortical reactions. Signals from many other sensory systems also interact with retinal processing before the visual cortex becomes involved.

Output Circuitry from the Visual Cortex

The output circuitry from the visual cortex to various cortical eye fields results in quantifiable eye movements. Some of those movements are used to aim at a selected target, some to maintain balance, and others for thoughts. Optometrists can control inputs with various types of lenses and

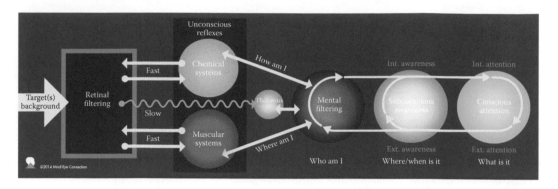

FIGURE 22.5 "Who am I?" impacts "where am I?" (brainstem) and "how am I?" (limbic system functions). (Courtesy of The Mind-Eye Connection, Northbrook, IL, Copyright 2014.)

measure changes in order to deduce processing. Knowing how a person processes incoming information is helpful when developing an individualized treatment plan for rehabilitation.

Output circuitry from the visual cortex to the eye muscles is described as part of either a ventral or a dorsal stream. Signals in those streams are governed by internal awareness, attention, and motivation. As mental priorities shift, signals travel through either the ventral or dorsal stream to the selected target and background. The ventral stream contains information regarding the selected target, and the dorsal stream contains information about the background to provide context for the details of the ventral stream. In effect, the distinction between peripheral and central eyesight—or concepts and details—depends on whether awareness and attention are on internal thoughts or external targets.

If operating properly, peripheral eyesight (background awareness) works in tandem with central eyesight (visual attention). This relationship between attention and awareness can be constricted (for instance, when looking at a sliver in a finger) or expanded (e.g., when viewing a large landscape). Many neurodegenerative diseases such as amyotrophic lateral sclerosis, multiple sclerosis (MS), and Alzheimer's disease affect this peripheral/central relationship. They can be studied by comparisons to research on glaucomatous changes in the optic nerve, which are also considered a neurodegenerative processes (Gupta and Yucel 2007).

Visual processing is not just a simple input/output mechanism. The visual cortex actually receives information from other forms of internal processing in addition to the external eyesight signals (Golomb et al. 2010) before thought-induced eye movements occur. There is feedback signaling from the visual cortex back to the LGN (Ling et al. 2015) and eventually to the retina. These *returning* retinal signals are termed retino*petal* signals, as compared to the retino*fugal* signals *exiting* the retina, described in "Input Circuitry to the Visual Cortex" section of this chapter.

A simplified viewpoint of visual output circuitry is to envision eye muscle movements as an end result after processing multisensory inputs. Eye muscles include the eyelids, pupils, and extraocular muscles. Movements can be quantified by the measurements of reaction time. In addition to cortically induced eye movements addressing eyesight, there are subcortical, reflexive eye movements induced by the brainstem and limbic system activity.

Cortical Adaptations to Change

Environmental changes trigger adaptive responses in cognition, perception, and emotions at the cortical level. Those responses are based on previous knowledge, combined with incoming sensory information. Cognitive circuitry is dependent on perceptual circuitry, which, in turn, is altered by emotional circuitry that is based on past experiences. Cortical activity is also influenced by subcortical processes and can be disrupted by brain injury or disease. In some cases, cortical processes never fully develop due to genetic mutations or other birth-related issues. However, brain plasticity allows

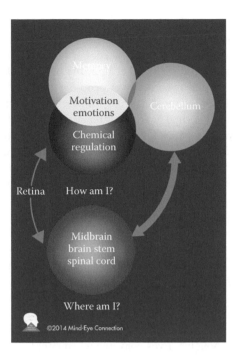

FIGURE 22.6 Retinal connections with neurological and chemical circuitry. (Courtesy of The Mind-Eye Connection, Northbrook, IL, Copyright 2014.)

for many adaptations when provided with proper stimuli. Customized eyeglasses can uniquely provide such stimuli, especially when combined with other disciplines during rehabilitation.

When describing sensory systems (such as visual and auditory), "Where is it?" and "What is it?" pathways are commonly referenced. They are often mistakenly termed "ambient and focal" processing, respectively. However, the "Where is it?" pathway is only the external portion of ambient processing. There is a second, more primitive, internal portion of ambient processing termed "Where am I?" Although the visual cortex is usually thought of as being activated by external sensory signals, prior research has determined how it is also activated by other sensory signals and/or mental imagery activities, even without the presence of actual external visual stimuli (Vetter et al. 2014). In other words, feedback information is abundant in the visual cortex, more so than feedforward information from the external environment. In autistic patients, sensory pathways are often hyperactive, and the classic eyesight pathways are often dysfunctional (Grossman et al. 2009, Tillmann et al. 2015).

Adaptation to environmental changes at the cortical level depends on such factors as motivation, propensity for risk taking, comfort, sense of security, interest, mood, appetite, anxiety level, and libido. Visual circuitry, combined with "Who am I?" internal experiences, plays an important role in assessments of time and space. Thus, cortical adaptations to change encompass many signaling processes—not simply eyesight (Figure 22.6).

SUBCORTICAL NONIMAGE-FORMING PATHWAYS

Retinal connections at an unconscious level affect chemical and muscular reactions through the retino-hypothalamic and retinotectal tracts, respectively. Research on the parvocellular ("What is it?") pathway from the ventral stream and the external magnocellular ("Where is it?") pathway from the dorsal stream show that deficiencies in those pathways are found in patients who have schizophrenia, Parkinson's disease (Altintas et al. 2008), epilepsy (van Baarsen et al. 2009), diabetes, drug addiction, autism, and Alzheimer's disease. Changes in retinal function, such as judgments in space and time, can be used as a biomarker for psychiatric disorders (Lavoie et al. 2014).

Connections build themselves on the basis of need. For instance, the auditory cortex reorganizes itself to be able to process visual motion (Shiell et al. 2014). Visual fields develop differently in deaf people than in either people who hear normally or people who have learned signing from infancy. The inferior and right visual fields are most effective for processing sign language signals (Bosworth and Dobkins 2002). People with normal hearing prefer central eyesight pathways, while those who are deaf prefer peripheral eyesight pathways. These differences are attributed to changes in *retinal* plasticity rather than to *cortical* neuroplasticity. For example, when reading was tested in children using colored filters, researchers found that those children with normal hearing did not choose the same visual filters as children who were deaf (Hollingsworth et al. 2015). Therefore, in the case of auditory sensory deprivation, neuroplasticity of the retina was again exhibited.

The effects of retinal stimulation can be quantified by measuring eye movements and pupil functions—part of which are reflexive from brainstem and limbic system activity. Other measurements are cortically induced, reflecting both conscious and subconscious thought. Signals contributing to the subcortical reactions are faster than those triggering cortical responses. For instance, convergence (aiming the eyes inward) is a measurement that has several triggers. The most common source is when a patient is engaged in the environment, consciously aiming at a selected target. A second mechanism of convergence occurs when head position is shifted. Habitual head position might be downward (creating an "eyes outward" posture), when the head tilts upwards, eyes converge. A third mechanism involves internal thoughts. Eyes converge when people are thinking about details or are stressed and diverge when they are conceptualizing or relaxed. A fourth mechanism is an internal, physiological state—eyes pull outward during sleep. Pupil reactions also can change due to external (light) or internal stimuli (fear, arousal in the mind, or drugs and chemicals in the body).

Relationships between retinal activity—more specifically the *peripheral* retina—and physical and physiological body functions have been identified in recent research—for instance, the oculocardiac reflex (Stathopoulos et al. 2012) or the adrenal glands (Kiessling et al. 2014). Impairments in retinal processing affecting predictive mechanisms have been implicated in the disrupted eye movements noted in schizophrenia (Sprenger et al. 2013). Retinal stimulation in humans has been shown to affect migraines and photophobia (Maleki et al. 2012). A study of 40,000 men during a 25 year period showed a 43% increase in open angle glaucoma in those who had gum disease. Their possible hypothesis is that toxins released from the gums travel to the eye. Patients with glaucoma also are noted as often having sleep problems attributed to loss of the retinal cells linked with circadian rhythms. Electroretinogram testing of people with seasonal affective disorder shows measureable changes in rod and cone activity (Lavoie et al. 2009).

Subcortical Neurological Circuitry

Signals that travel to the nucleus of the optic tract are used for smooth pursuit movement. Other pathways for pursuit movements exist as well, all of them interconnected (Nuding et al. 2008). Other pretectal nuclei govern reflexive eye movements and visual stability, such as optokinetic nystagmus reflexes and visual–vestibular interactions. Regions in the cortex, such as the middle, temporal, and the medial superior temporal lobes, link their signaling with subcortical structures, including parts of the accessory optic system for stable pursuit (tracking) movements (Heinen and Watamaniuk 1998) (Figure 22.7).

The Edinger–Westphal nucleus is an accessory nucleus of the oculomotor nerve and receives input for pupillary constriction from external light. This nucleus also receives internal information from the olivary pretectal nuclei, which also receive signals from a subtype of the intrinsically photosensitive retinal ganglion (ipRGC) cells. The olivary pretectal nuclei are involved in linking internal metabolism with external luminance—through the suprachiasmatic nuclei (SCN) of the hypothalamus and intergeniculate leaflet (IGL) of the thalamus (Ishikawa 2013).

That light stimulation on one part of the visual field would affect the corresponding portion of the subcortical superior colliculus make sense because the retina is mapped onto the superior colliculus

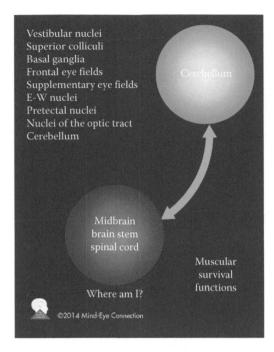

Vestibular nuclei
Superior colliculi
Basal ganglia
Frontal eye fields
Supplementary eye fields
E-W nuclei
Pretectal nuclei
Nuclei of the optic tract
Cerebellum

Cerebellum

Midbrain
brain stem
spinal cord

Muscular
survival
functions

Where am I?

©2014 Mind-Eye Connection

FIGURE 22.7 Retino-tectal tract—Where am I? (Courtesy of The Mind-Eye Connection, Northbrook, IL, Copyright 2014.)

make sense because the retina is mapped onto the superior colliculus. The superior colliculus responses are different depending on the stimulus' location (Ghose and Wallace 2014). Retinal signals travel to the top three layers (superficial superior colliculus), rather than to the bottom four layers (deep superior colliculus), which are simultaneously receiving information from other sensory systems (Ghose et al. 2014).

The subcortical superior colliculus also is involved in spatial attention (Schneider and Kastner 2009). Studies have shown that the superior colliculus/pretectal area and the visual cortical areas are each affected by changes in light (Miller et al. 1998). The superior colliculus pathway is independent of the classic cone pathway of seeing (Leh et al. 2010). Yet, the superior colliculus is still responsive to colors (Zhang et al. 2015). Sensory systems interact with the basal ganglia (Prescott et al. 2006). A clinical trial demonstrated that patients given placebo glasses were not as effectively treated as those prescribed actual glasses with prisms (Bowers et al. 2014). In other words, varying the dispersion of light onto the retina affected the patients' reactions. New research shows that prism glasses altering the "Where is it?" pathway by shifting apparent target location also have an effect on the "How am I?" chemical pathway.

Subcortical Chemical Circuitry

Chemical measurements can be made to evaluate changes in the immune system by assessing the color of the sclera, fluid in the conjunctiva, and the quantity and content of the tears. The retino-hypothalamic pathway alters adrenaline levels quickly via the hypothalamic–pituitary–adrenal axis before the slower signals from central eyesight have even focused on the target. External and internal systems come together in the retino-hypothalamic tract, where retinal changes influence body functions both muscularly and chemically (Figure 22.8).

The habenula (part of the limbic system) has direct connections with a small percentage of retinal ganglion cells and multiple connections in the brainstem. Its circuitry connects the cortex with the brainstem (Aizawa 2012) by registering changes in light. It is involved in modulation of both dopamine and serotonin systems, playing a role in sleep, depression, and schizophrenia. Dysfunction in the habenular circuit contributes to decreased REM sleep and is often linked to insomnia and depression (Aizawa et al. 2013). It is also involved in the suppression of motor control (Beretta et al. 2012).

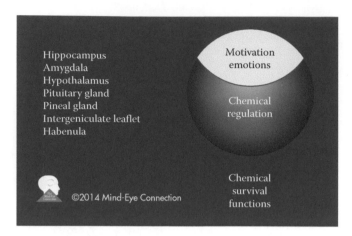

FIGURE 22.8   Retino-hypothalamic tract—How am I? (Courtesy of The Mind-Eye Connection, Northbrook, IL, Copyright 2014.)

Intergeniculate leaflet, a small section of the LGN (part of the thalamus), contributes feedforward information to the SCN regarding body metabolism and sensory conditions. It receives input from the vestibulo–visuomotor system. Thus, head movement might influence circadian rhythms (Horowitz et al. 2004, Blasiak and Lewandowski 2013, Saderi et al. 2013). "How am I?" signals from biochemical regulation are linked to "Where am I?" signals determined from the head position (Horowitz et al. 2004). Signals from the IGL to contralateral IGL are more activated in the light (Blasiak and Lewandowski 2013).

The hippocampus is one of many components that regulate adrenocortical activity at the hypothalamic level (Jacobson and Sapolsky 1991).

The amygdala is activated during eye contact. A small experiment showed that the amygdala in a cortically blind person was activated by direct gaze rather than averted gaze. The study concluded that the amygdala pathway is part of a larger network involving facial expressions (Burra et al. 2013).

The retino-hypothalamic pathway is involved at the subconscious level in circadian rhythms. While the existence of the retino-hypothalamic pathway has been known for some time, scientists have not explored how optometry can use retinal pathways to link the internal environment with external stimuli, reflecting both conscious and non-conscious pathways. Research is beginning to show how treatments using specific light wavelengths can profoundly change the response of circadian rhythms (Mure et al. 2009).

Subcortical Adaptations to Change

Pupil and eye movement reflex assessments are often used as biomarkers for neurological integrity. The relationship between the retina and the nervous system means that the eye also can be used as a noninvasive approach determining brain function. One way to assess balance in systems is with pupil measurements. The pupil is controlled by the synthesis of three photoreceptive inputs—rods, cones, and ipRGC cells—and also by nonphotoreceptive inputs such as stimulation by the autonomic nervous system (ANS). The more stressed a person is, the larger his or her pupils become.

Chemical circuitry plays a large role in visual processing. For instance, someone on pain medications having hallucinations as a side effect or someone with schizophrenia who has too much dopamine present are not fully aware of their surroundings.

The critical role that the retina performs, its interconnections with brain systems other than simple eyesight, and its relationship to disease and treatment of disease can be best understood by having some familiarity with how the retina is structured. This next section will provide a brief overview of that structure.

RETINAL STRUCTURE

The retina receives information from both external and internal sources, with the final goal of directing action. It contains pathways linking exogenous stimuli and endogenous processes, including having its own immune system (Benhar et al. 2012) and a localized circadian clock (Zele et al. 2011). At the simplest level, the eye channels light onto the retina, triggering chemical reactions in the outer retina that convert the light into electrical signals in the inner retina. Pupil responses generated by internal retinal signals are separate from responses generated by external light (Lee et al. 2014). Various diseases can be diagnosed by changes in the retinal structure. During development, the structure changes depending on need. For example, *retinal* plasticity, not *cortical* plasticity, has been found in deaf people, as mentioned in "Retinal function" (Figure 22.9).

The estimated 126 million photoreceptors in each eye (Jonas et al. 1992) receive light information and funnel signals through ten retinal layers into approximately 1,200,000 ganglion axon fibers, which exit each eye as the optic nerve (Medeiros et al. 2013). The exiting signals are not all involved in cortical eyesight. Even in blind persons, functional optic nerve fibers have been observed. Although the eyesight portions were not functioning, between 5% and 15% of the million-plus axons were still active (Cursiefen et al. 2001).

The complex filtering structure of the retina can be described in various ways—by its cell types, which serve different functions; its layers through which the signals travel; its sections, which arise from different transcription factors (Tombran-Tink and Barnstable 2008); and its thickness. Various types of chemical receptors, such as dopaminergic, serotonergic, cholinergic, and glutaminergic, are involved in signal transmission for excitatory and inhibitory signaling. For instance, when activated, dopaminergic neurons send more signals through retinal cone circuits and fewer signals through rod circuits (Witkovsky 2004). Eyeglasses can selectively stimulate various cell groups, thereby affecting informational filtering processes and, thus, output signals. Different diseases affect specific cell types, altering overall retinal function.

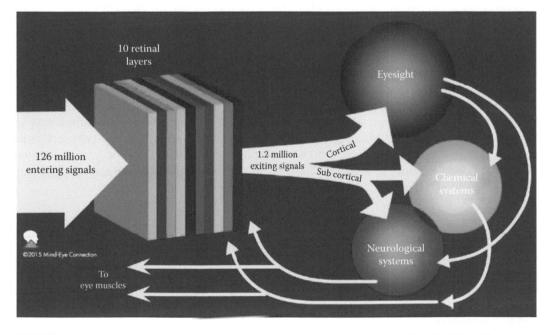

FIGURE 22.9 Retinal filtering to direct action and behavior. (Courtesy of The Mind-Eye Connection, Northbrook, IL, Copyright 2015.)

Cell Types

There are more than 50 cell types in the retina (Raviola 2002), each contributing to a different aspect of retinal processing before signals leave the eye to the brain's subcortical and cortical pathways. The main cells in the retina include the straightforward pathway—from photoreceptors to bipolar cells to ganglion cells, with horizontal and amacrine cells interspersed in between to inhibit the overflow of incoming information. There are also Mueller glial cells, which run the entire depth of the retina. This star-shaped type of glial cell (astrocyte) is found in the brain and the spinal cord and once was able to help in maintaining support structure. However, as of 2000, research has demonstrated that astrocytes are involved in metabolic transfers between extracellular environments. Disruptions in their signaling may play a possible role in neuropsychiatric disorders (Molofsky et al. 2012) and brain plasticity (Araque and Navarrete 2010, Perea and Araque 2010).

Another recent research has demonstrated that many subtypes of each kind of retinal cell exist. For instance, three types of cones, L, M, and S, respond to long, medium, and short wavelengths of light, respectively. Bipolar cells are separated into midget and diffuse general groupings, with many subtypes of those, but, functionally, they are considered to be activated by central or surrounding targets. There are ON cone bipolar cells (for each type of cone cell grouping) and OFF cone bipolar cells as well as ON and OFF rod bipolar cells. As for the inhibitory horizontal and amacrine cells, three types of horizontal cells have been identified in the human retina as of 1994 (Ahnelt and Kolb 1994), and more than 20 kinds of amacrine cells are separated into wide and narrow field classifications. The most commonly researched of these cells are the starburst amacrine and the AII cells. The AII cells link some rod and cone information before signals exit the eye, and the starburst cells are involved in directional sensitivity involved by optokinetic reflexes (Yoshida et al. 2001).

In addition to rods and cones, a third receptor (discovered in 2002) reacts to the luminance level of light. When light strikes the retina, the ipRGC cells in the ganglion layer send rudimentary eyesight signals directly to the hypothalamus (Canteras et al. 2011). This process is significantly faster than the classic eyesight pathway. Melanopsin-containing ipRGC cells are more sensitive to blue light and are more damaged in glaucoma (Bessler et al. 2010). Those cells also receive light from the rods and cones. Patients who have macular degeneration, with damage in their retinal layers, respond better to red lighting (Ishikawa 2013). The earlier-mentioned ipRGC cells receive information from both cones and rods, but show more effect with rod (peripheral retinal) information (Lall et al. 2010, Schmidt and Kofuji 2010) (Figure 22.10).

The melanopsin-containing ganglion cells have been shown to influence circuits of luminance and spatial information (Ecker et al. 2010). In 2005, the melanopsin retinal pathway was not considered a contributor to spatial vision (Schmucker et al. 2005). However, further research on mice demonstrated that the ipRGC pathway does contribute to spatial vision by mixing signals with rod and cone pathway information. The rods respond differently depending on cone information via the AII type of amacrine cell (Alam et al. 2015). There are other types of ganglion cells such as the bistratified ones whose signals travel to the koniocellular layers of the LGN (Dacey and Lee 1994). Of the 1.2 million ganglion cells, approximately 90% are parvocellular, 5% are magnocellular, and 5% are koniocellular (Freberg 2010).

Layers

Each of the cell types has a dendrite, cell body, and axon. Many scanning and electrophysiological devices are available to analyze retinal layers. If any problem exists, the abnormal cells will distort the layers, creating system breakdowns. As depicted in Figure 22.11, retinal layers filter information using both inhibitory and excitatory mechanisms, taking into account feedback and feedforward signals. Deposits of cholesterol plaques also have been found in different retinal layers (Zheng et al. 2012). In the case of sleep disorders and glaucoma, the specialized melanopsin-containing ganglion cells are involved (Gracitelli et al. 2015).

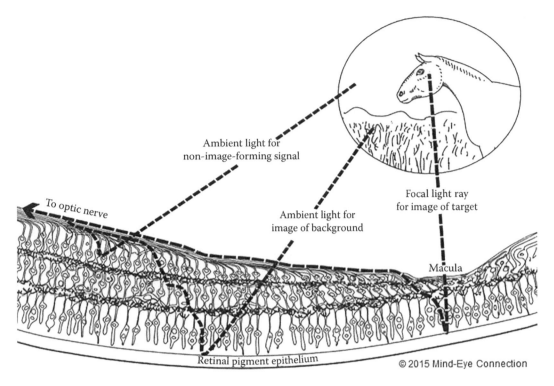

FIGURE 22.10 Three groups of retinal cells sensitive to light. (From Zelinsky, D., *Functional Magnetic Resonance Imaging: Advanced Neuroimaging Application*, InTech, 2013, p. 91. With permission.)

The retinal layers are also associated with different types of chemical transmitters. Glutamate is used through the feedforward pathway mentioned here, while GABA and glycine receptors are found in the inhibitory layers of horizontal and amacrine cells (Dutertre et al. 2012). Melatonin is produced by the retina in the dark, and dopamine is produced in the light. However, prolonged exposure to the dark lessens the effect, reducing the production of melatonin (Adler et al. 1992, Danilenko et al. 2009).

The most familiar of the retina's ten layers is the one with photoreceptors, split into the outer and inner segments of rods and cones. Photoreceptors need nourishment from another retinal layer, the retinal pigment epithelium (RPE). The constant interaction between the RPE and the rod photoreceptors is termed the *visual cycle*, triggered by lighting changes. Cones have a chemical interchange with Mueller cells. The external limiting membrane (or outer limiting membrane) separates the cell bodies of the photoreceptors from their outer and inner segments. The line of cell bodies is termed the outer nuclear layer of the retina. Oftentimes, a photoreceptor integrity line is evaluated on retinal imaging, but it is not a retinal layer. It simply represents the junction between the outer and inner segments of the photoreceptors, and its assessment is useful for determining the progress of various diseases.

A large amount of retinal processing occurs at synaptic junctions. The outer plexiform layer of the retina is where electrical signals from the photoreceptors interact with bipolar cells. This outer layer includes dendrites from bipolar and horizontal cells as well as axons from photoreceptors. The inner nuclear layer contains the cell nuclei of horizontal, bipolar, and amacrine cells. The inner plexiform layer contains axons of bipolar cells, many types of amacrine cells, and dendrites of ganglion cells (Roska et al. 2006). A web of filtered excitatory and inhibitory signals in the inner plexiform layer detects motion and suppresses eye movements (Baccus 2007). The combination of those signals affects judgments in space and time (Kim et al. 2014a, Robinson 2014) and results in the final exiting signal from the ganglion cell.

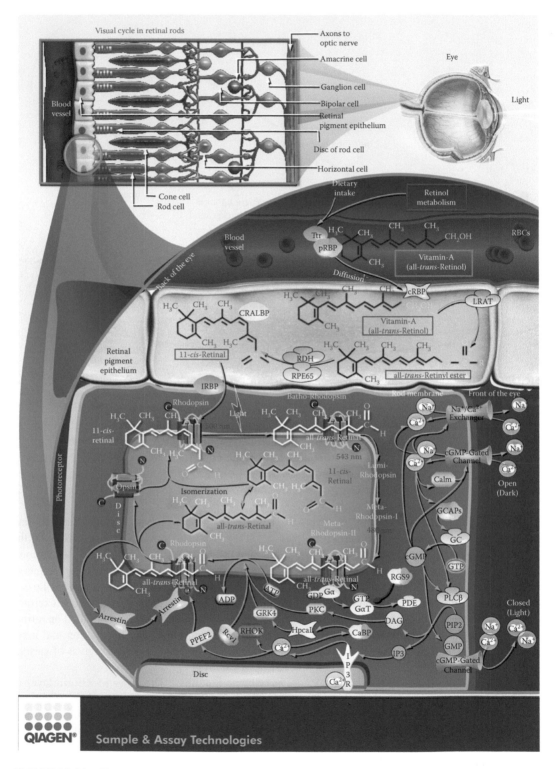

FIGURE 22.11 Visual cycle in rods © 2009 QIAGEN, all rights reserved. Qiagen testing is available for pathway specific siRNA's real-time polymerase chain reactions (PCR). (From QIAGEN, Visual cycles in rods, 2009. With permission.)

The ganglion cell layer and the axons from the ganglion cells, termed the nerve fiber layer, are next in the progression of retinal layers toward the inside of the eyeball. Loss of nerve fiber layer tissue is found to be an early biomarker for neurodegenerative diseases, such as Alzheimer's disease and glaucoma (Valenti 2011). The final layer is the porous internal limiting membrane separating the retina from the vitreous. From there, 1.2 million signals travel through the optic nerve further into the brain for processing (Figure 22.12).

SECTIONS

Retinal development arises from several different transcription factors and different portions of DNA. During gestation, the sections develop separately and then later link. Various sections of the retina react to light differently. For instance, the concentration of cone photoreceptors in the nasal retina was found to be higher than in other regions (Jonas et al. 1992). Studies as far back as 1948 identify differences in saccadic fixation abilities and melatonin production (Braendstrup 1948, Ruger et al. 2005, Johannesson et al. 2012). The ganglion cell axon diameter has been found to vary depending on its location in the retina (FitzGibbon and Taylor 2012). Studies also have shown that the nasal portion of the human retina suppresses melatonin more than the temporal portion (Visser et al. 1999). The nasal portion of the retina sends signals to the hypothalamus, while the temporal portion does not.

Differences between the inferior and superior retinal sections also have been identified. In humans, blood flow is reduced in the inferior retina when certain types of stress are applied (Harris et al. 2003). In 1985, a study demonstrated that the upper retina shows more significant contrast sensitivity than the inferior retina (Skrandies 1985). Flicker sensitivity is more closely associated with the peripheral retina (Solomon et al. 2002). In dementia, with Lewy bodies, eye movements during sleep are not normal (McCarter et al. 2013).

THICKNESS

Retinal thickness changes with either degenerative conditions or swelling. Assessments of retinal thickness have recently been developed as a simple method to follow objective changes (Choi et al. 2008, Grazioli et al. 2008). Thickness of retinal cell layers has been watched closely in glaucoma for years (Shin et al. 2014). In patients with Parkinson's disease, the parafoveal inner nuclear layer is thinner than the same layer in normal patients, and in those Parkinson's patients who also have dementia, the retinal layer is even thinner than in those with Parkinson's alone (Lee et al. 2014). In schizophrenic patients, the right nasal quadrant of the schizoaffective group is thinner than in the general schizophrenia group (Chu et al. 2012). New research is correlating retinal thickness with brain atrophy in patients with MS (Abalo-Lojo et al. 2014). In fact, studies show degeneration of the thalamus and retinal thickness in MS (Zivadinov et al. 2014).

Now that the retinal structure and function have been reviewed in detail, the remainder of this chapter will focus on neuromodulation—the continual search for homeostasis between conscious and nonconscious functions, the relationship between neuromodulation and the retina, and the overall impact of this relationship and its interplay with other brain and nervous system processes. These relationships affect human behavior, environmental response, and progression and intervention of disease processes.

NEUROMODULATION: A CONCEPTUAL FRAMEWORK

After prolonged stress, shock, injury, or disease, behavior, perception, and responses to environmental changes are frequently affected, often creating abnormal neuromodulation. One common compensatory mechanism to sensory overload is to ignore outside environmental stimuli. People have individual acceptance levels to change; how much it takes to push them over the edge varies with their "How am I?" and "Who am I?" pathways. Sensory systems interact with each other, and each

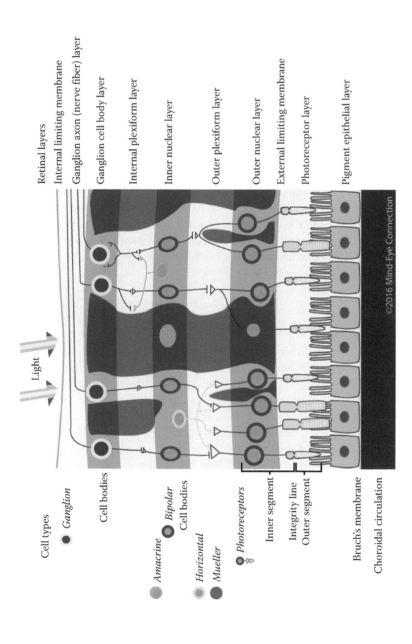

FIGURE 22.12 Retinal cells and layers. (Courtesy of The Mind-Eye Connection, Northbrook, IL, Copyright 2016.)

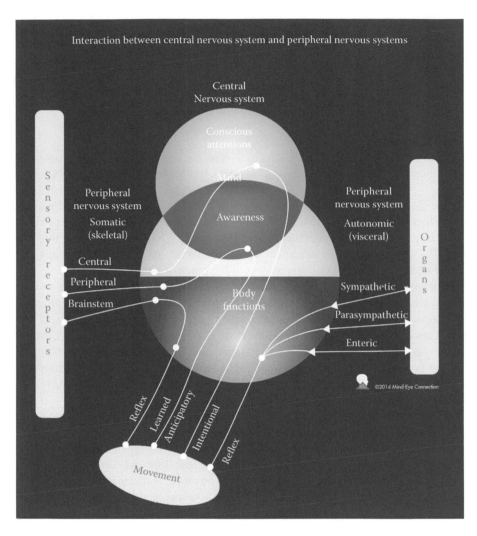

FIGURE 22.13 Internal/external interactions. (Courtesy of The Mind-Eye Connection, Northbrook, IL, Copyright 2014.)

person has an optimal load and also an upper threshold of tolerance before a breakdown is reached. Retinal stimulation via nontraditional eyeglasses or contact lenses can be extremely beneficial in helping patients regain a sense of internal comfort.

Figure 22.13 depicts how the CNS, which is composed of the brain, spinal cord, and retina, sends and receives information from the two peripheral nervous systems—the ANS and the somatosensory. The autonomic system has three portions —sympathetic, parasympathetic, and enteric.

Sensory (afferent) and motor (efferent) signals are sent to and from the CNS through both visceral and skeletal systems. The efferent pathways can be either voluntary or involuntary. The involuntary motor pathways can be either in sympathetic (fight/flight/fright) mode or parasympathetic (rest/digest mode). The enteric nervous system produces the majority of serotonin in the body and relates to the digestive system. Normal functioning of the CNS depends on the balanced interplay of both excitatory and inhibitory neurons in the body (Dutertre et al. 2012).

Cognitive reserves are limited, and in the presence of confusion or distraction from too many sensory inputs, comfort is reduced. The mind can usually "tune out" unwanted peripheral/background auditory and visual inputs and disengage eye aiming at targets in external surroundings. That selective filtering ability is often hindered when the body systems are in survival mode. Attentional

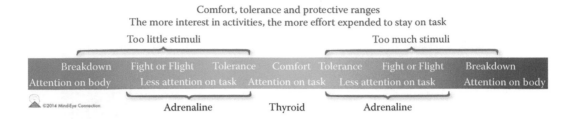

FIGURE 22.14 Comfort, tolerance, and protective mechanisms which can arise from either endogenous or exogenous sources. (Courtesy of The Mind-Eye Connection, Northbrook, IL, Copyright 2014.)

pathways are shut down or become hypersensitized in people when the internal pathways are out of balance. In other words, if the internal "How am I?" and "Where am I?" pathways are not in a range of comfort, the external perception of the "Where is it?" and "What is it?" pathways will be hindered or skewed. The degree of comfort influences actions, behavior, and attention. Typically, the mind is not aware of bodily sensations until they are out of the range of comfort.

Most patients have large ranges of comfort and tolerance, and they simply adapt to the changes. In those who are not able to adapt, such as people with brain injuries, protective mechanisms hinder cognitive processing. Something as simple as a slight tint in everyday eyeglasses might be helpful in calming a sensitive nervous system by filtering out extraneous, irritating stimuli. Comfort ranges, tolerance ranges, and protective mode form intricate circuitry that governs perception. The more comfortable a person is, the less effort is expended and the more attention paid to whatever the mind wants. If any systems are disrupted or stressed past the comfort threshold, more effort is required to deal with the outside information (Figure 22.14).

Once past tolerance ranges, the body goes into chemical and muscular protective modes. For instance, after brain injury, primitive survival reflexes emerge. Some protective mechanisms include filtering of environmental stimuli. Living near a jackhammer or flashing neon lights triggers protective signaling pathways. By comparison, when a person feels safe and comfortable, he or she can become habituated to repetitive stimuli, such as a cuckoo clock. Humans are born with certain built-in survival mechanisms (muscular reflexes), such as an asymmetrical tonic neck reflex. Those primitive survival reflexes reemerge after trauma. The role of doctors and therapists often becomes one of teaching patients how to reintegrate those reflexes. This can be achieved by movement reeducation strategies through variations of integrative manual therapy.

COMBINATION OF EXTERNAL AND INTERNAL SIGNALS

Consider what happens to adrenaline levels if a sudden movement is "caught" out of the corner of the eye, such as glimpsing an unexpected mouse running across a room. Instantly, beneath conscious control, internal chemical and muscular systems react, triggered by a moving shadow on the peripheral retina. Aiming, focusing, and classic central eyesight ("seeing") are processed *after* panic (chemical) and tension (muscular) reactions occur. Feedforward retinal signals are sent to two main processing sections in the brain—biochemical (How am I?) and muscular (Where am I?). Feedback signals to the retina from subcortical and cortical limbic structures are based on a person's experiences (Who am I?). Thus, the "How am I?" and "Where am I?" pathways are influenced by the "Who am I?" primarily beneath the conscious awareness.

Visceral systems also are affected beneath cortical control by input from temperature sensors, like sweating or shivering in response to heat or cold. Humans have built-in protective mechanisms so that, when they pass their individual comfort range and go into a tolerance range, some attention and mental energy are diverted to the discomfort or imbalance. For instance, beneath conscious awareness, some people remove a jacket without thinking about it, while others, at a conscious level, might seek a jacket to wear.

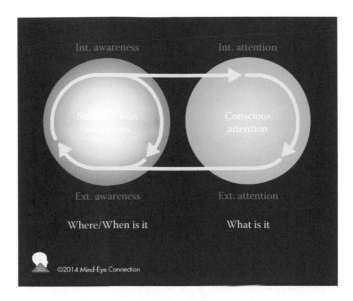

FIGURE 22.15 Awareness and attention. (Courtesy of The Mind-Eye Connection, Northbrook, IL, Copyright 2014.)

External signals have an effect on internal systems, and the altered internal signals can influence the filtering ability of external sensory systems. One example is how exposure to external noise affects retinal sensitivity (Dantsig and Diev 1986). Running with a peaceful water scene in the visual background elicits different stress chemicals than running while being chased by wild animals. People with fragile sensory integration or hypersensitive peripheral retinas visceral reactions induced daily simply by normal objects moving in their environments. These types of reactions and responses to environmental changes are part of a two-way transfer of information (Figure 22.15).

Attention and awareness can be focused on either external or internal stimuli. Eyeglasses designed to address peripheral awareness and/or body functions can affect responses to environmental changes.

When internal systems are dysregulated, external awareness (including peripheral eyesight) tends to shut down and become less sensitive to the surrounding environment. This processing beneath a conscious level varies with individual interests and energy levels. For example, a person at a fun party wanting to stay up might fight off a tired sensation. On the other hand, if the party is boring, that person might choose to surrender to the fatigue.

People differ in their experiences, temperaments, perceptions, motivations, and organizational skills. For that reason, they have different ranges of comfort before breakdowns occur. The "Who am I?" component determines whether they will activate their fight or flight mode or simply disengage from the outside environment. If a sound of a dog collar jingling is heard and the person hearing it had previous bad experiences with dogs, the memories from the cortical limbic system will "tell" the subcortical limbic system "You're not safe"—the body becomes stressed and muscles freeze or muscles run. However, the same, exact stimuli of a dog collar jingling heard by another person who has pleasant memories of dogs will create a signal indicating "This is great! Where is it?"—muscles turn toward the sound and the body relaxes.

Each system has its own comfort and stress levels. One person can eat a lot of sugary desserts and not experience a mood change; another can eat one small piece of candy and experience a sugar high. Each system can be modulated to regain stability and return to the habitual position where it feels "normal." The systems work concurrently, all vying for conscious attention. Depending on the task, sometimes the body receives attention and sometimes the mind.

The mind and body react and respond to environmental changes by activating both cortical and subcortical circuitries. Constant interaction occurs between internal and external stimuli, involving measureable shifts in eye movement. Shifting light influences the interplay between incoming signals from the outside environment (exogenous sources) and returning signals from the inside environment (endogenous sources). For instance, no single "blood pressure center" exists. Rather, networks of external and internal signals from the various nervous systems regulate blood pressure by using many factors to determine when to release hormones and retain or excrete fluids from the body. The same is true of retinal stimulation. The many visual systems use numerous factors to determine when to pay attention to changes in the external environment and when to ignore those changes.

CORTICAL INTERACTIONS WITH SUBCORTICAL CIRCUITRY

Eye movements are an observable and quantifiable result of significant brain processing at unconscious, subconscious, and conscious levels; no simple input/output system of visual acuity is at play. Instead, considerable planning and redundancy are built-in for survival. As an example, several eye field regions in various cortices connect with each other and the brainstem to govern initiation and control of eye movements (Lynch and Tian 2006) (Figure 22.16).

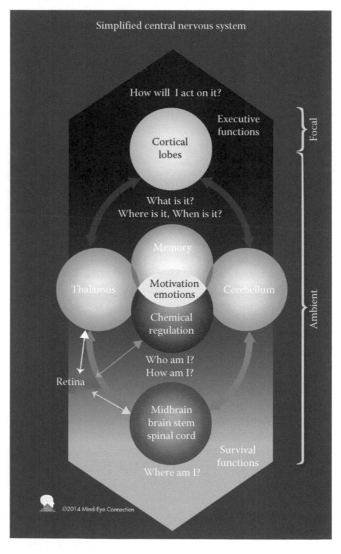

FIGURE 22.16 Retinal connections within the CNS. (Courtesy of The Mind-Eye Connection, Northbrook, IL, Copyright 2014.)

The retina is connected with the thalamus (for cortical eyesight) and the hypothalamus (for subcortical chemical circuitry) and the brainstem (for subcortical muscular circuitry).

Stress, injury, or disease can cause neurological, metabolic, and/or perceptual changes. Injury and disease disrupt electrical signaling pathways in the nervous system and alter biochemistry, affecting cognition and behavior. Diseases can occur many years after injury. Oftentimes, ocular symptoms (an end result of much processing) precede the diagnosis of CNS disorders (London et al. 2013) and separate the "Where am I?" and "How am I?" into reflex (subcortical) and subconscious cortical levels. Memory and visual perception (Khan et al. 2011) also have specific malleable circuitry. The entire field of brain plasticity is growing, as evidenced in such books as *The Brain that Changes Itself* by Norman Doidge (2010) and *The Brain's Way of Healing* by Norman Doidge (2016). Neuroplasticity is an emerging field, as is neurophotonics—the use of light to affect neurological systems.

THE IMPACT OF RETINAL STIMULATION ON NEUROMODULATION

Typically, after a brain injury, neurological systems are disrupted and comfort ranges are constricted. Signaling pathways can be rerouted, relying on the plasticity in brain wiring when creating new connections. In developing children, this circuitry can be modified to avoid dysfunctional patterns before habitual patterns become embedded.

Executive function skills develop by age 25. However, skills can be lost during aging and disease processes. These skills are based on the "Who am I?" experiences. Cognitive responses are dependent on lower-level cortical processes, such as perception and emotions, as well as on unconscious reactions at the proprioceptive and chemical levels. In fact, in addition to the right and left brain, some people discuss a top and bottom brain concept of how gray matter is organized (Kosslyn and Miller 2015). Since input circuitry to the visual cortex is routed through right, left, top (parietal), and bottom (temporal) of the cortex, therapeutic eyeglasses can be used on an individual basis to stimulate one or the other.

Neuromodulation is a continual process occurring far beneath a conscious level and involving the retina as well as the eye muscles (think of REM sleep). Even during sleep, chemical changes occur in the retinal layers. If a person sleeps with the television on, or with surrounding noise, the brain is not able to fall into as deep a sleep and restorative properties are not as good. Recent research shows that the concept of systemic tolerance is useful in pharmacological treatment of such diseases as MS and diabetes (Graham et al. 2013, Lutterotti et al. 2013). Instead of treating symptoms, researchers have achieved better results from allowing neuroplasticity of the immune system to develop antibodies.

The eye is not isolated from the other sensory, motor, emotional, and cognitive systems. As research on retinal signaling pathways continues to progress, the concept that multiple pathways exist between the eye and the body, with a great deal of activity occurring beneath the conscious awareness, will gain wider acceptance. These pathways can be both excitatory and inhibitory, involving both feedback and feedforward mechanisms. The small portion of information that reaches conscious awareness is actually a conglomeration of inputs from several senses—visual, auditory, etc. (Eagleman 2011).

Priorities continually shift among body, mind, and environment until achieving a comfortable balance. This interaction is termed homeodynamics. As stated earlier, quantifiable metrics such as eye movements, habitual eye position, pupil size, blink rate, and tear layer quality can be used to assess visual processing. Central 20/20 eyesight (classic visual acuity) is only one of several of these measurements. However, changes in perceived clarity in central eyesight can arise due to the multiple steps that occur before attention shifts to targets.

Hand–eye coordination is a good example. It is not simply sensory/motor input/output—see a ball and reach out to catch it. Rather, many other processes are involved. If they were not, people would be able to play catch endlessly. Instead, the limbic system's involvement concerning interest, fear, or pleasure, the autonomic system's activation once stress levels are reached, hunger or fatigue, and other processes all play pivotal roles.

NEURO-OPTOMETRIC APPROACHES DURING REHABILITATION

Neuro-optometry uses a noninvasive approach to brain function and has been shown helpful in the treatment of multiple types of disorders, including those that are visual and neurological. Evaluations include possible hidden dysfunctions in mind–eye connections and dysfunctions that can affect social, academic, and sports performance. When incoming visual signals are altered by particular types of filters, these pathways can be either disrupted or enhanced (Figure 22.17).

Various mechanisms are triggered by different kinds of retinal stimulation. Depending on whether the pathways are disrupted or enhanced, individualized lenses can be prescribed to assist in treating symptoms.

Major optical treatments include the following:

Lenses: They disperse light toward the edges or the center of the retina, tending to make objects appear larger or smaller by emphasizing or muting the background. This change in light primarily alters the balance between central and peripheral circuitry by having the target and background occupy different percentages of the retinal input. Depending on the type of lenses used, the treatment can enhance attention or mute peripheral distractions.

Nonyoked prisms: They can be categorized into two types: lateral and vertical. They angle light toward either nasal/temporal retinal sensors or superior/inferior receptors. The eyes will reflexively point toward the light, but the inward or outward movement prompted by these prisms will, in turn, affect different visual and postural mechanisms, affecting the placement of shoulders by shifting apparent object locations. The nasal stimulation (from base-in prism, commonly prescribed after traumatic brain injuries) affects chemical retino-hypothalamic signaling.

Yoked prisms: They angle light toward a specific section of the retinal sensors. Angling light toward one edge of the retina initially affects the body's positional sense (proprioceptive sense), because reflexive eye movements will point toward the incoming light, triggering internal postural mechanisms in the hips for stability of balance. These prisms promote a shift of the hips as the center of gravity reflexively changes. These prisms can be divided into two main categories—those that make the person comfortable, so their attention can shift to external targets, and those designed to make the person slightly uncomfortable to induce a mental reorganization. Depending on the stability of the person's sense of balance, mental attention may be shifted off of body position and onto external targets. Cortical mechanisms regarding internal perception of limb location modulate the actual temperature of the hands (Moseley et al. 2013).

Filters: They alter either spatial or temporal retinal input and affect internal processing and external perception. Tints filter out selected wavelengths of light, stimulating specific retinal cells, which then alter retinal chemistry (and thus body chemistry). Graded occlusion filters, such as Bangerters, alter the spatial components of incoming light. For instance, binasal filters reduce stimulation to the temporal retina. Neutral density filters alter the temporal components.

Mirrors: They induce a sensory mismatch in the retina between central (target) and peripheral (background). They are used in many aspects of patient treatments, such as rehabilitation of patients who have experienced strokes or visual field defects.

Alteration of corneal tear layer: When in fight or fright stage, the sympathetic nervous system is stressed and the eyes become drier. When calm, the eye and neck muscles relax and the eyes moisten as the parasympathetic system becomes dominant. By using punctual plugs to artificially moisten the eyes with the body's own tears, feedback mechanisms register moisture from tears, thereby activating the parasympathetic system.

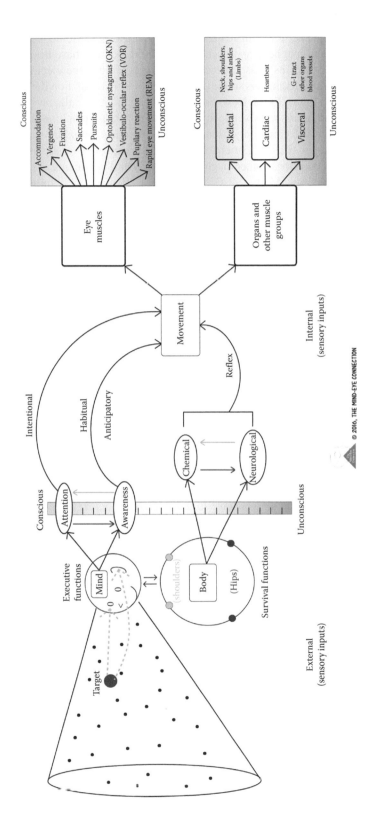

FIGURE 22.17 Effects of retinal stimulation on eye movements. (Courtesy of The Mind-Eye Connection, Northbrook, IL, Copyright 2016.)

SUMMARY OF INTERVENTIONS

A range of treatments is available to modulate nervous system responses. Methods include behavioral, electrical, pharmacologic, and magnetic approaches. Behavioral methods have been used for decades in the treatment of such diseases as autism. Electrical methods include the electroconvulsive shock therapy of the 1950s, cardiac pacemakers in the 1960s, deep brain stimulation devices in the 1990s, and vagus nerve stimulation in the 2000s. Now, since 2010, deep brain stimulation is being considered in Parkinson's treatment, using less risky versions—closed loop versus open loop (Beuter et al. 2014). Pharmacologic methods also are commonly used to control chemical signaling pathways. Examples include L-dopa for Parkinson's, ritalin for attention deficit disorder, and a variety of drugs for depression, epilepsy, and bipolar conditions. Other diagnostic methods and treatments of signaling pathways include magnetic fields, such as fMRIs, MEG, and transcranial magnetic stimulation. Combining assessment methods of "when signals are processed," such as the use of EEGs or fMRIs when signals are active, can provide a plethora of information regarding brain processing. Measurements of flicker sensitivity have been shown to chart disease progress (Falsini et al. 2000). Optogenetics—using light to modulate the nervous system—has become a new technique of brain assessment.

These approaches do not by any means exhaust the range of treatments. In addition to neuro-optometric rehabilitation methods, some optometrists use classic visual therapy activities and direct and indirect syntonic lens therapies. Available as well are some useful tools designed to enhance auditory feedback, such as the talking pen from Wayne Engineering (now known as Eye Carrot), or, from South Africa, the Sebezaphone, a self-amplification tool that enables the user to be both speaker and listener to his/her own voice, thereby enhancing language development and reading fluency (www.sebezaphone.co.za). New instruments have been developed to quantify subtle changes in eye movements, such as video pupillometers and RightEye.com's computerized testing batteries intended for patients who have sustained brain injury or have autism. An invaluable treatment instrument for any optometric office is the Germany Fusiobox (www.optomatters.com). It contains an astoundingly large array of stimuli and feedback to individualize patient treatments for lazy eyes and crossed eyes.

Since 2007, literature has been demonstrating the importance of visual/auditory linkages and Z-Bell[SM] testing (Zelinsky 2007). Just as hand–eye coordination develops with age and experience, so do eye/ear connections. Monica Gori, a researcher in Italy, has a body of work that indicates visual/auditory linkages develop rapidly until approximately age 8 (Gori et al. 2008, Gori et al. 2012, Tonelli et al. 2015).

What should be noted is not all studies have found improvements from neuro-optometric treatments. However, tests that have failed often have not considered particular variables. For instance, one study included more than 5500 children with reading impairments. They were evaluated for eye problems involving image-forming (eyesight) circuitry. Conclusion was that no optometric vision rehabilitation would be helpful. Strabismus, motor fusion, sensory fusion at a distance, refractive error, amblyopia, convergence, accommodation, or contrast sensitivity were not significantly different in those children when they were compared to children with normal reading ability (Creavin et al. 2015). However, this study did not take into account all the linkages between the eyes and other sensory systems. For instance, dyslexia involves auditory/visual interactions.

Vision is complex. The chart in Figure 22.18 separates and explains the various stages of visual development and summarizes expected responses to different optometric interventions. Nonstandard responses can provide information useful in identifying any deficient visual pathway(s) and determining the appropriate treatment or referral.

CONCLUSION

Neuromodulation via retinal stimulation is a new and promising technology that can be applied to humans today by specially trained eye care professionals. The use of eyeglasses to modulate the frequency, amount, and direction of light dispersed on the retina allows neuro-optometric rehabilitation to accelerate recovery from brain injury and provide improvement in patient comfort and tolerance

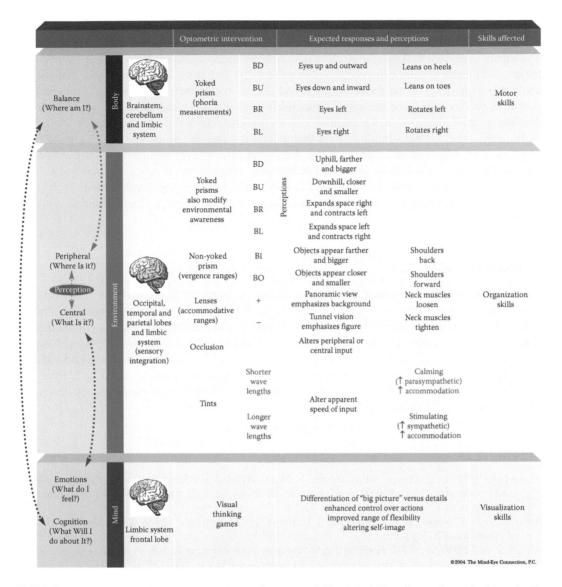

FIGURE 22.18 Intervention response chart. (Courtesy of The Mind-Eye Connection, Northbrook, IL, Copyright 2004.)

ranges to environmental changes. It can also lessen the hypersensitivity to sensory stimuli often seen in patients with brain injury, developmental disabilities, mental illnesses, and posttraumatic stress disorders. The classic use of optometric techniques to sharpen central eyesight to 20/20 impacts a patient's attention at a conscious level, with high-contrast, nonmoving targets. This approach is not sufficient in patients with neurodegenerative conditions or fragile connections between systems.

As compared to traditional behavioral and pharmacological treatments, neuro-optometric rehabilitation is safe and noninvasive. This treatment is different from classic visual therapy because it addresses the brainstem and subcortical reactions. Such therapeutic stimulation can be easily and inexpensively applied via individualized eyeglasses that cost well below other treatment options. The treatment causes no risk of life-threatening complications and no side effects. Patient compliance is good, and patients are more comfortable when they wear the glasses. This type of retinal stimulation can also be used as an adjunct to other treatments.

In the future, development of highly advanced eyeglasses that selectively stimulate retinal pathways may be possible. This tool could be used in treating patients with drug-resistant epileptic seizures or patients who have Parkinson's, Alzheimer's, or other neurodegenerative conditions as well as in patients with autism, genetic disorders, and/or mental illness. Future research may also find ways to apply this technology to metabolic disorders and alteration of gene expression. As computer/brain interfaces become more than the norms during rehabilitation, customized lenses may prove useful for developing brain plasticity and for attention training after brain injury using robotics.

In conclusion, retinal stimulation via therapeutic eyeglasses can be applied to a wide range of medical problems as an adjunct to other treatment processes to maximize rehabilitative outcomes. Doctors and scientists interested in applying these techniques are encouraged to team with neuro-optometric practitioners. Such collaboration will quickly develop a whole range of new, effective, and inexpensive therapies for an array of physical, physiological, and psychological dysfunctions.

ACKNOWLEDGMENTS

Sincere appreciation for Babak Kateb, M.D.—a true visionary. Dr. Kateb was one of the very few medical professionals who understood the tremendous diagnostic and therapeutic potential of retinal stimulation from the moment I met him. This chapter would not exist without his continuous and cheerful encouragement over the years. I would also like to acknowledge the support of Jonathan Q. Hall, Jr., O.D.—a future optometric leader—whose eye care constantly includes consideration of multiple system interactions in all of his patients. Also, appreciation goes to Dorothy Mason, Paul Zelinsky, Mara Sachs, Martha David, and Adam Boin for their abilities to create clearly descriptive illustrations and to Mike Maggio for his help with precise grammar.

REFERENCES

Abalo-Lojo, J. M., C. C. Limeres, M. A. Gomez, S. Baleato-Gonzalez, C. Cadarso-Suarez, C. Capeans-Tome, and F. Gonzalez (2014). Retinal nerve fiber layer thickness, brain atrophy, and disability in multiple sclerosis patients. *J Neuroophthalmol* **34**(1): 23–28.

Adler, J. S., D. F. Kripke, R. T. Loving, and S. L. Berga (1992). Peripheral vision suppression of melatonin. *J Pineal Res* **12**(2): 49–52.

Ahnelt, P. and H. Kolb (1994). Horizontal cells and cone photoreceptors in human retina: A Golgi-electron microscopic study of spectral connectivity. *J Comp Neurol* **343**(3): 406–427.

Aizawa, H. (2012). Habenula and the asymmetric development of the vertebrate brain. *Kaibogaku Zasshi* **87**(3): 57–58.

Aizawa, H., W. Cui, K. Tanaka, and H. Okamoto (2013). Hyperactivation of the habenula as a link between depression and sleep disturbance. *Front Hum Neurosci* **7**: 826.

Alam, N. M., C. M. Altimus, R. M. Douglas, S. Hattar, and G. T. Prusky (2015). Photoreceptor regulation of spatial visual behavior. *Invest Ophthalmol Vis Sci* **56**(3): 1842–1849.

Alilain, W. J. and J. Silver (2009). Shedding light on restoring respiratory function after spinal cord injury. *Front Mol Neurosci* **2**: 18.

Alilain, W. J., X. Li, K. P. Horn, R. Dhingra, T. E. Dick, S. Herlitze, and J. Silver (2008). Light-induced rescue of breathing after spinal cord injury. *J Neurosci* **28**(46): 11862–11870.

Altintas, O., P. Iseri, B. Ozkan, and Y. Caglar (2008). Correlation between retinal morphological and functional findings and clinical severity in Parkinson's disease. *Doc Ophthalmol* **116**(2): 137–146.

Araque, A. and M. Navarrete (2010). Glial cells in neuronal network function. *Philos Trans R Soc Lond B Biol Sci* **365**(1551): 2375–2381.

Arcaro, M. J., M. A. Pinsk, and S. Kastner (2015). The anatomical and functional organization of the human visual pulvinar. *J Neurosci* **35**(27): 9848–9871.

Baccus, S. A. (2007). Timing and computation in inner retinal circuitry. *Annu Rev Physiol* **69**: 271–290.

Benhar, I., A. London, and M. Schwartz (2012). The privileged immunity of immune privileged organs: The case of the eye. *Front Immunol* **3**: 296.

Beretta, C. A., N. Dross, J. A. Guiterrez-Triana, S. Ryu, and M. Carl (2012). Habenula circuit development: Past, present, and future. *Front Neurosci* **6**: 51.

Bessler, P., S. Klee, U. Kellner, and J. Haueisen (2010). Silent substitution stimulation of S-cone pathway and L- and M-cone pathway in glaucoma. *Invest Ophthalmol Vis Sci* **51**(1): 319–326.

Beuter, A., J. P. Lefaucheur, and J. Modolo (2014). Closed-loop cortical neuromodulation in Parkinson's disease: An alternative to deep brain stimulation?. *Clin Neurophysiol* **125**(5): 874–885.

Blasiak, T. and M. H. Lewandowski (2013). Differential firing pattern and response to lighting conditions of rat intergeniculate leaflet neurons projecting to suprachiasmatic nucleus or contralateral intergeniculate leaflet. *Neuroscience* **228**: 315–324.

Bosworth, R. G. and K. R. Dobkins (2002). Visual field asymmetries for motion processing in deaf and hearing signers. *Brain Cogn* **49**(1): 170–181.

Bowers, A. R., K. Keeney, and E. Peli (2014). Randomized crossover clinical trial of real and sham peripheral prism glasses for hemianopia. *JAMA Ophthalmol* **132**(2): 214–222.

Braendstrup, P. (1948). The functional and anatomical differences between the nasal and temporal parts of the retina; the pressure phosphenes used for the determination of the peripheral boundaries of the visual field. *Acta Ophthalmol (Copenh)* **26**(3): 351–361.

Burra, N., A. Hervais-Adelman, D. Kerzel, M. Tamietto, B. de Gelder, and A. J. Pegna (2013). Amygdala activation for eye contact despite complete cortical blindness. *J Neurosci* **33**(25): 10483–10489.

Candy, T. R., J. A. Crowell, and M. S. Banks (1998). Optical, receptoral, and retinal constraints on foveal and peripheral vision in the human neonate. *Vision Res* **38**(24): 3857–3870.

Canteras, N. S., E. R. Ribeiro-Barbosa, M. Goto, J. Cipolla-Neto, and L. W. Swanson (2011). The retinohypothalamic tract: Comparison of axonal projection patterns from four major targets. *Brain Res Rev* **65**(2): 150–183.

Chen, S. K., T. C. Badea, and S. Hattar (2011). Photoentrainment and pupillary light reflex are mediated by distinct populations of ipRGCs. *Nature* **476**(7358): 92–95.

Chen, S. K., K. S. Chew, D. S. McNeill, P. W. Keeley, J. L. Ecker, B. Q. Mao, J. Pahlberg et al. (2013). Apoptosis regulates ipRGC spacing necessary for rods and cones to drive circadian photoentrainment. *Neuron* **77**(3): 503–515.

Choi, S. S., R. J. Zawadzki, M. A. Greiner, J. S. Werner, and J. L. Keltner (2008). Fourier-domain optical coherence tomography and adaptive optics reveal nerve fiber layer loss and photoreceptor changes in a patient with optic nerve drusen. *J Neuroophthalmol* **28**(2): 120–125.

Chu, E. M., M. Kolappan, T. R. Barnes, E. M. Joyce, and M. A. Ron (2012). A window into the brain: An in vivo study of the retina in schizophrenia using optical coherence tomography. *Psychiatry Res* **203**(1): 89–94.

Creavin, A. L., R. Lingam, C. Steer, and C. Williams (2015). Ophthalmic abnormalities and reading impairment. *Pediatrics* **135**(6): 1057–1065.

Cursiefen, C., L. M. Holbach, U. Schlotzer-Schrehardt, and G. O. Naumann (2001). Persisting retinal ganglion cell axons in blind atrophic human eyes. *Graefes Arch Clin Exp Ophthalmol* **239**(2): 158–164.

Dacey, D. M. and B. B. Lee (1994). The 'blue-on' opponent pathway in primate retina originates from a distinct bistratified ganglion cell type. *Nature* **367**(6465): 731–735.

Danilenko, K. V., I. L. Plisov, A. Wirz-Justice, and M. Hebert (2009). Human retinal light sensitivity and melatonin rhythms following four days in near darkness. *Chronobiol Int* **26**(1): 93–107.

Dantsig, I. N. and A. V. Diev (1986). Light sensitivity of the central and peripheral sections of the retina in exposure to noise. *Biull Eksp Biol Med* **101**(2): 133–135.

Doidge, N. (2010). *The Brain That Changes Itself: Stories of Personal Triumph from the Frontiers of Brain Science*. Melbourne, Victoria, Australia: Scribe Publications.

Doidge, N. (2016). *The Brain's Way of Healing: Remarkable Discoveries and Recoveries from the Frontiers of Neuroplasticity*. New York: Penguin Books.

Dutertre, S., C. M. Becker, and H. Betz (2012). Inhibitory glycine receptors: An update. *J Biol Chem* **287**(48): 40216–40223.

Eagleman, D. (2011). *Incognito: The Secret Lives of the Brain*. Edinburgh, Scotland: Canongate.

Ecker, J. L., O. N. Dumitrescu, K. Y. Wong, N. M. Alam, S. K. Chen, T. LeGates, J. M. Renna, G. T. Prusky, D. M. Berson, and S. Hattar (2010). Melanopsin-expressing retinal ganglion-cell photoreceptors: Cellular diversity and role in pattern vision. *Neuron* **67**(1): 49–60.

Falsini, B., A. Fadda, G. Iarossi, M. Piccardi, D. Canu, A. Minnella, S. Serrao, and L. Scullica (2000). Retinal sensitivity to flicker modulation: Reduced by early age-related maculopathy. *Invest Ophthalmol Vis Sci* **41**(6): 1498–1506.

FitzGibbon, T. and S. F. Taylor (2012). Mean retinal ganglion cell axon diameter varies with location in the human retina. *Jpn J Ophthalmol* **56**(6): 631–637.

Freberg, L. (2010). *Discovering Biological Psychology*. Belmont, CA: Wadsworth, Cengage Learning.

Ghose, D. and M. T. Wallace (2014). Heterogeneity in the spatial receptive field architecture of multisensory neurons of the superior colliculus and its effects on multisensory integration. *Neuroscience* **256**: 147–162.

Ghose, D., A. Maier, A. Nidiffer, and M. T. Wallace (2014). Multisensory response modulation in the superficial layers of the superior colliculus. *J Neurosci* **34**(12): 4332–4344.

Goga, C. and U. Ture (2015). The anatomy of Meyer's loop revisited: Changing the anatomical paradigm of the temporal loop based on evidence from fiber microdissection. *J Neurosurg* **122**(6): 1253–1262.

Goldstein, E. B. (2010). *Sensation and Perception*. Belmont, CA: Wadsworth, Cengage Learning.

Golomb, J. D., A. Y. Nguyen-Phuc, J. A. Mazer, G. McCarthy, and M. M. Chun (2010). Attentional facilitation throughout human visual cortex lingers in retinotopic coordinates after eye movements. *J Neurosci* **30**(31): 10493–10506.

Gori, M., M. Del Viva, G. Sandini, and D. C. Burr (2008). Young children do not integrate visual and haptic form information. *Curr Biol* **18**(9): 694–698.

Gori, M., F. Tinelli, G. Sandini, G. Cioni, and D. Burr (2012). Impaired visual size-discrimination in children with movement disorders. *Neuropsychologia* **50**(8): 1838–1843.

Gracitelli, C. P., G. L. Duque-Chica, M. Roizenblatt, A. L. Moura, B. V. Nagy, G. Ragot de Melo, P. D. Borba et al. (2015). Intrinsically photosensitive retinal ganglion cell activity is associated with decreased sleep quality in patients with glaucoma. *Ophthalmology* **122**(6): 1139–1148.

Graham, J. G., X. Zhang, A. Goodman, K. Pothoven, J. Houlihan, S. Wang, R. M. Gower, X. Luo, and L. D. Shea (2013). PLG scaffold delivered antigen-specific regulatory T cells induce systemic tolerance in autoimmune diabetes. *Tissue Eng Part A* **19**(11–12): 1465–1475.

Grazioli, E., R. Zivadinov, B. Weinstock-Guttman, N. Lincoff, M. Baier, J. R. Wong, S. Hussein, J. L. Cox, D. Hojnacki, and M. Ramanathan (2008). Retinal nerve fiber layer thickness is associated with brain MRI outcomes in multiple sclerosis. *J Neurol Sci* **268**(1–2): 12–17.

Grossman, R. B., M. H. Schneps, and H. Tager-Flusberg (2009). Slipped lips: Onset asynchrony detection of auditory-visual language in autism. *J Child Psychol Psychiatry* **50**(4): 491–497.

Guler, A. D., J. L. Ecker, G. S. Lall, S. Haq, C. M. Altimus, H. W. Liao, A. R. Barnard et al. (2008). Melanopsin cells are the principal conduits for rod-cone input to non-image-forming vision. *Nature* **453**(7191): 102–105.

Gupta, N. and Y. H. Yucel (2007). Glaucoma as a neurodegenerative disease. *Curr Opin Ophthalmol* **18**(2): 110–114.

Harris, A., Y. Ishii, H. S. Chung, C. P. Jonescu-Cuypers, L. J. McCranor, L. Kagemann, and H. J. Garzozi (2003). Blood flow per unit retinal nerve-fibre tissue volume is lower in the human inferior retina. *Br J Ophthalmol* **87**(2): 184–188.

Heinen, S. J. and S. N. Watamaniuk (1998). Spatial integration in human smooth pursuit. *Vision Res* **38**(23): 3785–3794.

Hollingsworth, R. S., A. K. Ludlow, A. J. Wilkins, R. I. Calver, and P. M. Allen (2015). Visual performance and the use of colored filters in children who are deaf. *Optom Vis Sci* **92**(6): 690–699.

Horng, S., G. Kreiman, C. Ellsworth, D. Page, M. Blank, K. Millen, and M. Sur (2009). Differential gene expression in the developing lateral geniculate nucleus and medial geniculate nucleus reveals novel roles for Zic4 and Foxp2 in visual and auditory pathway development. *J Neurosci* **29**(43): 13672–13683.

Horowitz, S. S., J. H. Blanchard, and L. P. Morin (2004). Intergeniculate leaflet and ventral lateral geniculate nucleus afferent connections: An anatomical substrate for functional input from the vestibulo-visuomotor system. *J Comp Neurol* **474**(2): 227–245.

Ishikawa, H. (2013). Pupil and melanopsin photoreception. *Nihon Ganka Gakkai Zasshi* **117**(3): 246–268; discussion 269.

Jacobson, L. and R. Sapolsky (1991). The role of the hippocampus in feedback regulation of the hypothalamic-pituitary-adrenocortical axis. *Endocr Rev* **12**(2): 118–134.

Jeelani, N. U., P. Jindahra, M. S. Tamber, T. L. Poon, P. Kabasele, M. James-Galton, J. Stevens et al. (2010). 'Hemispherical asymmetry in the Meyer's Loop': A prospective study of visual-field deficits in 105 cases undergoing anterior temporal lobe resection for epilepsy. *J Neurol Neurosurg Psychiatry* **81**(9): 985–991.

Johannesson, O. I., A. G. Asgeirsson, and A. Kristjansson (2012). Saccade performance in the nasal and temporal hemifields. *Exp Brain Res* **219**(1): 107–120.

Jonas, J. B., U. Schneider, and G. O. Naumann (1992). Count and density of human retinal photoreceptors. *Graefes Arch Clin Exp Ophthalmol* **230**(6): 505–510.

Khan, Z. U., E. Martin-Montanez, and M. G. Baxter (2011). Visual perception and memory systems: From cortex to medial temporal lobe. *Cell Mol Life Sci* **68**(10): 1737–1754.

Kiessling, S., P. J. Sollars, and G. E. Pickard (2014). Light stimulates the mouse adrenal through a retinohypothalamic pathway independent of an effect on the clock in the suprachiasmatic nucleus. *PLoS One* **9**(3): e92959.

Kim, J. S., M. J. Greene, A. Zlateski, K. Lee, M. Richardson, S. C. Turaga, M. Purcaro et al. (2014a). Space-time wiring specificity supports direction selectivity in the retina. *Nature* **509**(7500): 331–336.

Kim, S. Y., F. Tassone, T. J. Simon, and S. M. Rivera (2014b). Altered neural activity in the 'when' pathway during temporal processing in fragile X premutation carriers. *Behav Brain Res* **261**: 240–248.

Kosslyn, S. M. and G. W. Miller (2015). *Top brain, Bottom Brain: Harnessing the Power of the Four Cognitive Modes*. New York: Simon & Schuster.

Krekelberg, B. (2003). Sound and vision. *Trends Cogn Sci* **7**(7): 277–279.

Lall, G. S., V. L. Revell, H. Momiji, J. Al Enezi, C. M. Altimus, A. D. Guler, C. Aguilar et al. (2010). Distinct contributions of rod, cone, and melanopsin photoreceptors to encoding irradiance. *Neuron* **66**(3): 417–428.

Lavoie, J., M. Maziade, and M. Hebert (2014). The brain through the retina: The flash electroretinogram as a tool to investigate psychiatric disorders. *Prog Neuropsychopharmacol Biol Psychiatry* **48**: 129–134.

Lavoie, M. P., R. W. Lam, G. Bouchard, A. Sasseville, M. C. Charron, A. M. Gagne, P. Tremblay, M. J. Filteau, and M. Hebert (2009). Evidence of a biological effect of light therapy on the retina of patients with seasonal affective disorder. *Biol Psychiatry* **66**(3): 253–258.

Lee, J. Y., J. M. Kim, J. Ahn, H. J. Kim, B. S. Jeon, and T. W. Kim (2014). Retinal nerve fiber layer thickness and visual hallucinations in Parkinson's Disease. *Mov Disord* **29**(1): 61–67.

Leh, S. E., A. Ptito, M. Schonwiesner, M. M. Chakravarty, and K. T. Mullen (2010). Blindsight mediated by an S-cone-independent collicular pathway: An fMRI study in hemispherectomized subjects. *J Cogn Neurosci* **22**(4): 670–682.

Liberman, J. (1994). Design technology: Light-medicine of the future. *J Healthc Des* **6**: 133–140.

Ling, S., M. S. Pratte, and F. Tong (2015). Attention alters orientation processing in the human lateral geniculate nucleus. *Nat Neurosci* **18**(4): 496–498.

London, A., I. Benhar, and M. Schwartz (2013). The retina as a window to the brain-from eye research to CNS disorders. *Nat Rev Neurol* **9**(1): 44–53.

Lutterotti, A., S. Yousef, A. Sputtek, K. H. Sturner, J. P. Stellmann, P. Breiden, S. Reinhardt et al. (2013). Antigen-specific tolerance by autologous myelin peptide-coupled cells: A phase 1 trial in multiple sclerosis. *Sci Transl Med* **5**(188): 188ra175.

Lynch, J. C. and J. R. Tian (2006). Cortico-cortical networks and cortico-subcortical loops for the higher control of eye movements. *Prog Brain Res* **151**: 461–501.

Maleki, N., L. Becerra, J. Upadhyay, R. Burstein, and D. Borsook (2012). Direct optic nerve pulvinar connections defined by diffusion MR tractography in humans: Implications for photophobia. *Hum Brain Mapp* **33**(1): 75–88.

McCarter, S. J., E. K. St Louis, and B. F. Boeve (2013). Mild cognitive impairment in rapid eye movement sleep behavior disorder: A predictor of dementia? *Sleep Med* **14**(11): 1041–1042.

Medeiros, F. A., R. Lisboa, R. N. Weinreb, J. M. Liebmann, C. Girkin, and L. M. Zangwill (2013). Retinal ganglion cell count estimates associated with early development of visual field defects in glaucoma. *Ophthalmology* **120**(4): 736–744.

Miller, A. M., W. H. Obermeyer, M. Behan, and R. M. Benca (1998). The superior colliculus-pretectum mediates the direct effects of light on sleep. *Proc Natl Acad Sci U S A* **95**(15): 8957–8962.

Molofsky, A. V., R. Krencik, E. M. Ullian, H. H. Tsai, B. Deneen, W. D. Richardson, B. A. Barres, and D. H. Rowitch (2012). Astrocytes and disease: A neurodevelopmental perspective. *Genes Dev* **26**(9): 891–907.

Moseley, G. L., A. Gallace, F. Di Pietro, C. Spence, and G. D. Iannetti (2013). Limb-specific autonomic dysfunction in complex regional pain syndrome modulated by wearing prism glasses. *Pain* **154**(11): 2463–2468.

Mure, L. S., P. L. Cornut, C. Rieux, E. Drouyer, P. Denis, C. Gronfier, and H. M. Cooper (2009). Melanopsin bistability: A fly's eye technology in the human retina. *PLoS One* **4**(6): e5991.

Naeser, M. A., A. Saltmarche, M. H. Krengel, M. R. Hamblin, and J. A. Knight (2011). Improved cognitive function after transcranial, light-emitting diode treatments in chronic, traumatic brain injury: Two case reports. *Photomed Laser Surg* **29**(5): 351–358.

Nuding, U., S. Ono, M. J. Mustari, U. Buttner, and S. Glasauer (2008). A theory of the dual pathways for smooth pursuit based on dynamic gain control. *J Neurophysiol* **99**(6): 2798–2808.

O'Connor, D. H., M. M. Fukui, M. A. Pinsk, and S. Kastner (2002). Attention modulates responses in the human lateral geniculate nucleus. *Nat Neurosci* **5**(11): 1203–1209.

Parry, B. L. (2002). Jet lag: Minimizing it's effects with critically timed bright light and melatonin administration. *J Mol Microbiol Biotechnol* **4**(5): 463–466.

Perea, G. and A. Araque (2010). GLIA modulates synaptic transmission. *Brain Res Rev* **63**(1–2): 93–102.

Prescott, T. J., F. M. Montes Gonzalez, K. Gurney, M. D. Humphries, and P. Redgrave (2006). A robot model of the basal ganglia: Behavior and intrinsic processing. *Neural Netw* **19**(1): 31–61.

Purves, D. and S. M. Williams (2001). *Neuroscience*. Sunderland, MA: Sinauer Associates.

Raviola, E. (2002). A molecular approach to retinal neural networks. *Funct Neurol* **17**(3): 115–119.

Robinson, R. (2014). Feedback, both fast and slow: How the retina deals with redundancy in space and time. *PLoS Biol* **12**(5): e1001865.

Roska, B., A. Molnar, and F. S. Werblin (2006). Parallel processing in retinal ganglion cells: How integration of space-time patterns of excitation and inhibition form the spiking output. *J Neurophysiol* **95**(6): 3810–3822.

Ruger, M., M. C. Gordijn, D. G. Beersma, B. de Vries, and S. Daan (2005). Nasal versus temporal illumination of the human retina: Effects on core body temperature, melatonin, and circadian phase. *J Biol Rhythms* **20**(1): 60–70.

Saderi, N., F. Cazarez-Marquez, F. N. Buijs, R. C. Salgado-Delgado, M. A. Guzman-Ruiz, M. del Carmen Basualdo, C. Escobar, and R. M. Buijs (2013). The NPY intergeniculate leaflet projections to the suprachiasmatic nucleus transmit metabolic conditions. *Neuroscience* **246**: 291–300.

Schmidt, T. M. and P. Kofuji (2008). Novel insights into non-image forming visual processing in the retina. *Cellscience* **5**(1): 77–83.

Schmidt, T. M. and P. Kofuji (2010). Differential cone pathway influence on intrinsically photosensitive retinal ganglion cell subtypes. *J Neurosci* **30**(48): 16262–16271.

Schmidt, T. M., M. T. Do, D. Dacey, R. Lucas, S. Hattar, and A. Matynia (2011). Melanopsin-positive intrinsically photosensitive retinal ganglion cells: From form to function. *J Neurosci* **31**(45): 16094–16101.

Schmucker, C., M. Seeliger, P. Humphries, M. Biel, and F. Schaeffel (2005). Grating acuity at different luminances in wild-type mice and in mice lacking rod or cone function. *Invest Ophthalmol Vis Sci* **46**(1): 398–407.

Schneider, K. A. and S. Kastner (2009). Effects of sustained spatial attention in the human lateral geniculate nucleus and superior colliculus. *J Neurosci* **29**(6): 1784–1795.

Shiell, M. M., F. Champoux, and R. J. Zatorre (2014). Enhancement of visual motion detection thresholds in early deaf people. *PLoS One* **9**(2): e90498.

Shin, H. Y., H. Y. Park, J. A. Choi, and C. K. Park (2014). Macular ganglion cell-inner plexiform layer thinning in patients with visual field defect that respects the vertical meridian. *Graefes Arch Clin Exp Ophthalmol* **252**(9): 1501–1507.

Sinclair, K. L., J. L. Ponsford, J. Taffe, S. W. Lockley, and S. M. Rajaratnam (2014). Randomized controlled trial of light therapy for fatigue following traumatic brain injury. *Neurorehabil Neural Repair* **28**(4): 303–313.

Skeffington, A. M. (1957). The totality of vision. *Am J Optom Arch Am Acad Optom* **34**(5): 241–255.

Skeffington, A. M. (1966). The future of optometry. *J Am Optom Assoc* **37**(9): 836–838.

Skrandies, W. (1985). Human contrast sensitivity: Regional retinal differences. *Hum Neurobiol* **4**(2): 97–99.

Solomon, S. G., P. R. Martin, A. J. White, L. Ruttiger, and B. B. Lee (2002). Modulation sensitivity of ganglion cells in peripheral retina of macaque. *Vision Res* **42**(27): 2893–2898.

Sprenger, A., P. Trillenberg, M. Nagel, J. A. Sweeney, and R. Lencer (2013). Enhanced top-down control during pursuit eye tracking in schizophrenia. *Eur Arch Psychiatry Clin Neurosci* **263**(3): 223–231.

Stathopoulos, P., M. Mezitis, G. Kostakis, and G. Rallis (2012). Iatrogenic oculocardiac reflex in a patient with head injury. *Craniomaxillofac Trauma Reconstr* **5**(4): 235–238.

Tayman, C., M. M. Tatli, S. Aydemir, and A. Karadag (2010). Overhead is superior to underneath light-emitting diode phototherapy in the treatment of neonatal jaundice: A comparative study. *J Paediatr Child Health* **46**(5): 234–237.

Tillmann, J., A. Olguin, J. Tuomainen, and J. Swettenham (2015). The effect of visual perceptual load on auditory awareness in autism spectrum disorder. *J Autism Dev Disord* **45**(10): 3297–3307.

Tombran-Tink, J. and C. J. Barnstable (2008). *Visual Transduction and Non-Visual Light Perception*. Totowa, NJ: Humana Press.

Tonelli, A., L. Brayda, and M. Gori (2015). Task-dependent calibration of auditory spatial perception through environmental visual observation. *Front Syst Neurosci* **9**: 84.

Valenti, D. A. (2011). Alzheimer's disease and glaucoma: Imaging the biomarkers of neurodegenerative disease. *Int J Alzheimers Dis* **2010**: 793931.

van Baarsen, K. M., G. L. Porro, and D. Wittebol-Post (2009). Epilepsy surgery provides new insights in retinotopic organization of optic radiations. A systematic review. *Curr Opin Ophthalmol* **20**(6): 490–494.

Vaney, D. I. (1999). Neuronal coupling in the central nervous system: Lessons from the retina. *Novartis Found Symp* **219**: 113–125; discussion 125–133.

Vetter, P., F. W. Smith, and L. Muckli (2014). Decoding sound and imagery content in early visual cortex. *Curr Biol* **24**(11): 1256–1262.

Visser, E. K., D. G. Beersma, and S. Daan (1999). Melatonin suppression by light in humans is maximal when the nasal part of the retina is illuminated. *J Biol Rhythms* **14**(2): 116–121.

Weyand, T. G., M. Boudreaux, and W. Guido (2001). Burst and tonic response modes in thalamic neurons during sleep and wakefulness. *J Neurophysiol* **85**(3): 1107–1118.

Witkovsky, P. (2004). Dopamine and retinal function. *Doc Ophthalmol* **108**(1): 17–40.

Yeatman, J. D., K. S. Weiner, F. Pestilli, A. Rokem, A. Mezer, and B. A. Wandell (2014). The vertical occipital fasciculus: A century of controversy resolved by in vivo measurements. *Proc Natl Acad Sci U S A* **111**(48): E5214–E5223.

Yoshida, K., D. Watanabe, H. Ishikane, M. Tachibana, I. Pastan, and S. Nakanishi (2001). A key role of starburst amacrine cells in originating retinal directional selectivity and optokinetic eye movement. *Neuron* **30**(3): 771–780.

Zele, A. J., B. Feigl, S. S. Smith, and E. L. Markwell (2011). The circadian response of intrinsically photosensitive retinal ganglion cells. *PLoS One* **6**(3): e17860.

Zelinsky, D. (2007). Neuro-optometric diagnosis, treatment and rehabilitation following traumatic brain injuries: A brief overview. *Phys Med Rehabil Clin North Am* **18**(1): 87–107, vi–vii.

Zelinsky, D. (2013). *Functional Magnetic Resonance Imaging: Advanced Neuroimaging Application.* InTech, p. 91.

Zhang, P., H. Zhou, W. Wen, and S. He (2015). Layer-specific response properties of the human lateral geniculate nucleus and superior colliculus. *Neuroimage* **111**: 159–166.

Zheng, W., R. E. Reem, S. Omarova, S. Huang, P. L. DiPatre, C. D. Charvet, C. A. Curcio, and I. A. Pikuleva (2012). Spatial distribution of the pathways of cholesterol homeostasis in human retina. *PLoS One* **7**(5): e37926.

Zivadinov, R., N. Bergsland, R. Cappellani, J. Hagemeier, R. Melia, E. Carl, M. G. Dwyer, N. Lincoff, B. Weinstock-Guttman, and M. Ramanathan (2014). Retinal nerve fiber layer thickness and thalamus pathology in multiple sclerosis patients. *Eur J Neurol* **21**(8): e1137–e1161.

23 FUS-Mediated Image-Guided Neuromodulation of the Brain

Seung-Schik Yoo, Wonhye Lee, and Ferenc A. Jolesz

CONTENTS

INTRODUCTION

This chapter introduces the emerging therapeutic applications of transcranial MRI-guided focused ultrasound (TcMRgFUS) as a new mode of noninvasive brain stimulation modality. Focused ultrasound (FUS), with its exquisite ability to deliver acoustic energy to the region-specific brain area with excellent depth penetration, is utilized in the implementation and demonstration of this new innovative modality. The pulsed application of low-intensity FUS to the regional brain area has shown bimodal—either excitatory or suppressive—effects on the central nervous system (CNS) with ability to stimulate the peripheral nerves. The FUS-mediated neuromodulation may be used in conjunction with procedures in other emerging fields such as the temporary disruption of the blood–brain barrier (BBB) and FUS functional neurosurgery that use the same hardware platform. The brief history, current status of research in the field, safety concerns, and their potential remedies in the context of FUS-mediated neuromodulation are discussed along with future landscapes in neurotherapeutics.

New, less invasive, or even noninvasive technologies that can deliver therapeutic effects to specific regions of interest safely and repeatedly are needed to overcome the current barriers that have hindered the development of breakthrough therapies for CNS disorders. FUS is a novel technology that has the potential for meeting these requirements.

Acoustic energy can be focused to small volumes deep into tissues, sparing the surrounding tissues of any effects. The mechanical energy deposited in the focal region can be used in a surprisingly great variety of ways and has fascinated basic and clinical neuroscientists for more than 60 years (Ballantine et al. 1960, Fry 1958, Jagannathan et al. 2009, Lele 1987, Lynn et al. 1942). During the last 10–15 years, significant technological developments by our and other groups have made the use of FUS in the brain a reality. The major issues that have hindered the use of FUS in the brain in the past have been solved, and FUS can be delivered through the intact skull in humans. FUS technology has great potential for targets throughout the body (Jolesz and McDannold 2008, Tempany et al. 2011). However, because of the invasiveness and/or significant limitations of current therapies, and

443

the critical necessity for sparing normal structures adjacent to the therapeutic target in the brain, its greatest potential is, in our view, in the brain (Colen and Jolesz 2010, Jolesz and McDannold 2014).

The creation of devices that can safely and precisely focus high-intensity ultrasound beams through the intact human skull (Aubry et al. 2003, Clement and Hynynen 2002) and the integration of these devices into a high-field MRI (Clement et al. 2005, Hynynen et al. 2004, 2006), that is, TcMRgFUS, have allowed the expectation that technology will reach its potential in clinical applications. Initial human trials with these devices are ongoing and are generating promising positive outcomes (Clement et al. 2005, Elias et al. 2013, Martin et al. 2009, McDannold et al. 2010).

Three of the fundamental medical techniques—surgery, radiation, and drugs—have significant limitations in CNS applications, and new innovative approaches can be of great value. Ultrasound energy can be noninvasively focused onto small, targeted locations deep in the brain and produce a wide range of suitable bioeffects: from temporary modification of function and structure of vascular or cellular structures to complete tissue ablation. Transcranial FUS can be used for novel therapies that have the potential for transforming a wide spectrum in basic and clinical neuroscientific therapeutic approaches. In recent years, a number of preclinical studies in humans demonstrated the feasibility and future potential of TcMRgFUS devices.

Clinical applications can have transformative potential. In neurosurgery and radiation oncology, the thermal and/or nonthermal ablation of brain tumors (McDannold et al. 2010) can replace some brain tumor surgeries and radiation surgeries. In neuro-oncology, targeted drug delivery may be used through the transiently opened BBB for chemotherapy to treat brain metastases (Hynynen et al. 2001, 2006, Kinoshita et al. 2006, McDannold et al. 2012, Treat et al. 2007). In functional neuromodulation, the direct acoustic effect or targeted delivery of neurotransmitters through the BBB can be used (Gavrilov et al. 1996, McDannold et al. 2014, Tufail et al. 2010, Yoo et al. 2011a). This chapter describes the intriguing new potential of functional neuromodulation in the brain.

BRIEF HISTORY OF FUS-MEDIATED NEUROMODULATION

For almost a century, the phenomenological evidence of ultrasound interactions with biological tissues, including ultrasonic neuromodulation, has been demonstrated. The neuromodulatory effect of ultrasound was first demonstrated in enhancing the neural activity observed from neuromuscular preparations of frogs and turtles in 1928 (Harvey et al. 1928). Three decades later, with the advancement of ultrasound physics and transducer technology, Fry et al. showed that ultrasound that was focally delivered to the lateral geniculate nucleus (LGN) of the thalamus in craniotomized cats reversibly suppressed the electroencephalographic (EEG) visual evoked potentials (VEPs) (Fry et al. 1958). Following these seminal investigations, the neuromodulatory potentials of ultrasound were examined through research on rodent brains, both *ex vivo* (Bachtold et al. 1998, Rinaldi et al. 1991) and *in vivo* (Tufail et al. 2010). In humans, the application of non-focused transcranial Doppler to the brain resulted in the elicitation of auditory sensations (Magee and Davies 1993) and mood changes among patients with chronic pain (Hameroff et al. 2013), suggesting the neuromodulatory potential of ultrasound insonification. Bystritsky et al. (2011) provided a detailed summary and review of ultrasonic functional modulation of the CNS.

RENAISSANCE OF FUS TECHNOLOGY IN THE 1990s AND 2000s

Localized transcranial delivery of acoustic energy to the neural structures located deep within the brain and the cortical areas is crucial for achieving region-specific functional brain modulation. However, in the beginning of the use of therapeutic ultrasound, this capability was not available. In 1942, Lynn and colleagues demonstrated the noninvasive destruction of focal tissue volumes inside the liver (Lynn et al. 1942) using a curved geometric shape of the ultrasound transducer. In 1962, FUS was adopted for the destruction of localized tissue areas in the CNS of craniotomized cats (Fry et al. 1955) and humans (Fry and Fry 1960). FUS later utilized the propagation of acoustic waves formed by refraction through acoustic lens placed in front of the ultrasound transducer (Lele 1962). Due to the low acoustic transmission through bones, this FUS technique has been used primarily in soft tissue applications such as ablation of tumors in the breast (Hynynen et al. 1996a), kidney (Vallancien et al. 1992), and liver (Yang et al. 1992). The high level of absorption of acoustic energy by the calcium-rich solid substance in the body gave way to the creation of the acoustic shockwave therapy in which high-energy short impulse FUS was used to break kidney stones (Chaussy and Schmiedt 1984). These earlier FUS studies, however, were conducted in the absence of modern image guidance and simply used the geometric relationship between the anatomy of interest and the FUS focal coordinates for targeting a designated treatment area.

The advancement of magnetic resonance (MR) and high-resolution ultrasound imaging technologies provided major breakthroughs in the early 1990s enabling more accurate guidance of the FUS stimulation to a localized tissue region. The first single-element FUS device for MRI-guided FUS was developed and clinically tested for the ablation of breast cancer (Cline et al. 1992, 1995, Hynynen et al. 1996b). There are now extensive clinical trials and patient care studies using these approaches (Jolesz et al. 2005). A further breakthrough was made in FUS transducer configuration in which phased-array FUS systems consisting of multiple ultrasound transducer elements, independently controlled through phase and amplitude modulation, are deployed to provide highly accurate control of focal size and aberration correction of beam distortions (Daum et al. 1998, Daum and Hynynen 1999, Hynynen et al. 2004, Thomas and Fink 1996). Several commercially available image-guided phased-array FUS systems are available in conjunction with the MR image-guidance system and now have the capability of controlling the acoustic focal size and its depth penetration under the skin (e.g., ExAblate 2000 and ExAblate Neuro, InSightec Ltd., Tirat Carmel, Israel).

Single-element or phased-array FUS systems that operate in low fundamental ultrasonic frequencies (<1 MHz; for clinical ultrasonic imagers, a higher range of 1–15 MHz is used) were developed and subsequently enabled TcMRgFUS (Figure 23.1) for overcoming the challenges of ultrasound

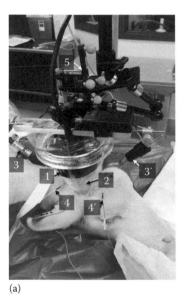 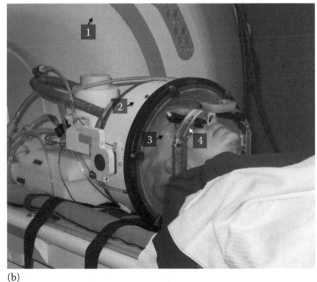

(a)   (b)

FIGURE 23.1 Examples of two different configurations of the FUS transducer system. (a) A single-element FUS transducer system for an ultrasonic neuromodulation study using a large animal model of sheep. (1) A single-element FUS transducer; (2) An acoustic impedance coupling between the transducer and the scalp using degassed water; (3, 3′) A visual guidance system for guiding the FUS focal location geometrically (Min et al. 2011b); (4, 4′) A physiological monitoring setup for measuring electroencephalography (EEG) from the head; (5) A motion tracker with infrared-sensitive markers. (b) A multielement FUS transducer system (ExAblate 4000). (1) MRI scanner for the neuroimage guidance; (2) A multielement FUS transducer system in a helmet configuration; (3) A waterproof rubber diaphragm with the degassed water between the transducer elements and head surface for acoustic coupling; (4) A head location fixation frame. (Courtesy of InSightec Ltd., Tirat Carmel, Israel.)

transcranial applications such as ultrasonic absorption, refraction, and reflection at the skull. Phased-array FUS systems having more than 1000 transducer elements have been developed in a helmet-like configuration (Song and Hynynen 2009). In applying the phased-array system, the computed tomography (CT) data of the head are used to provide information on skull structure specific to the individual and used for beam focusing by correcting for the phase/amplitude aberration introduced by the skull while MRI provides information on brain neuroanatomy and the tissue temperature using MR thermometry (Rieke and Pauly 2008). The system allowed for the transcranial application of high-intensity FUS (HIFU) and subsequent thermal brain tissue ablations, all under accurate neuroanatomical targeting. The technique was deployed for the functional neurosurgery of essential tremor (Elias et al. 2013) and neuropathic pain (Martin et al. 2009) by ablating regional structures in the thalamus.

MOTIVATION FOR USING FUS FOR NEUROMODULATION

A variety of modalities for neuromodulation techniques have been used for performing neuroscientific research and treating neurological disorders. Neuromodulation techniques provide direct and localized modulatory effects on neuronal circuits in contrast to pharmacological treatments. Invasive methods such as vagus nerve stimulation (VNS) (Ansari et al. 2007, Groves and Brown 2005, Morris and Mueller 1999), epidural cortical stimulation (EpCS) (Lefaucheur 2009), implanted electrocortical stimulation (IES) (Lefaucheur 2009), and deep brain stimulation (DBS) (Collins et al. 2010, Dowling 2008, Ellis and Stevens 2008, Grover et al. 2009, Gutman et al. 2009, Mian et al. 2010) are used, but the scope of their application is limited by the necessity for surgical procedures.

There are several noninvasive neuromodulation technologies including transcranial direct current stimulation (tDCS) (Lefaucheur 2009, Nitsche et al. 2009) and repetitive transcranial magnetic stimulation (rTMS) (Burt et al. 2002, George 2010, Kimiskidis 2010, Lipton and Pearlman 2010, Rothwell 2007). However, these techniques lack the spatial specificity and capability of reaching deep brain regions, suggesting the need for exploring a new, noninvasive technique for localized brain stimulation, including deep brain tissues. Recently proposed optogenetics techniques allow for cellular level control of the neuronal systems in the brain (Deisseroth 2011). However, the genetic modification required for the optically sensitive neurons and the limited depth penetration of the stimulatory light may hinder its immediate applications in humans.

The use of FUS for neuromodulation has recently been highlighted as a potential alternative method for overcoming the disadvantages of the currently available invasive and noninvasive brain stimulation techniques. The advantages of the transcranial FUS are that it does not require any surgical procedures or genetic modification and attains higher spatial resolution and deeper depth penetration compared to the other noninvasive methods. By using a low acoustic intensity that induces neither measurable temperature increases nor damages in the targeted tissues, the transcranial low-intensity FUS has recently been shown to reversibly modulate region-specific neural functions in a completely noninvasive manner.

REVISITING STUDIES OF FUS NEUROMODULATORY POTENTIALS IN ANIMALS

The FUS techniques that allow for functional neuromodulation of the CNS have recently been researched, primarily using animal models. We demonstrated the reversible modulation of neural excitability in region-specific brain areas in rabbits through excitation or suppression of the neural activities in the motor and the visual areas of the brain via application of low-intensity FUS sonication (Yoo et al. 2011a). The resulting neuromodulatory phenomena were examined through direct visual detection of elicited motions, electrophysiological monitoring of the elicited electromyography (EMG), and suppressed level of electroencephalography (EEG)-based VEP. The modulation of the blood-oxygenation-level-dependent (BOLD) signal level resulting from the functionally modulated brain areas was also demonstrated via functional MRI.

The functional neuromodulatory effect of sonication has also been shown in a small animal model of Sprague-Dawley (SD) rats. The neural excitatory effects of stimulation have been demonstrated by our group and others (Kim et al. 2014a, King et al. 2013, Tufail et al. 2010) through explicit motor responses such as tail movement. The suppressive effects of the FUS sonication on the epileptic EEG activity were further demonstrated in a chemically induced acute epilepsy model when the FUS sonication was applied to the thalamic area of the brain upon intraperitoneal (IP) introduction of pentylenetetrazole (PTZ) (Min et al. 2011a).

The sonications delivered to the thalamic area of healthy SD rats were shown to have modulatory effects on the extracellular level of key neurotransmitters using the microdialysis technique. The observed modulation manifested in the elevated concentration of dopamine (DA) and/or serotonin (5-HT) (Min et al. 2011b) and reduced γ-aminobutyric acid (GABA) (Yang et al. 2012). However, the stimulation did not affect the extracellular glutamate level (Yang et al. 2012). Although the assay was conducted remotely from the sonication site, the evidence suggested that sonication, indeed, modulated the level of neurotransmitters, further suggesting the potential for application of the technology in the treatment of psychiatric conditions that are mediated by aberrant neurotransmission.

It was interesting to find that the administration of sonication to the thalamus shortened the recovery time from IP ketamine/xylazine anesthesia in SD rats (Yoo et al. 2011b). This suggests the potential use of FUS in neurotherapeutics for controlling the consciousness levels in neurological disorders such as minimal consciousness states (Schiff et al. 2009). Additionally, the transcranial sonication on the rat abducens nerve (located above the base of the skull) showed the ability to control eyeball movement ipsilateral to the side of sonication (Kim et al. 2012). The stimulatory ability of FUS could potentially be used as a diagnostic and therapeutic tool with the cranial or peripheral nervous systems (PNS).

Based on the hypothesis that FUS-mediated region-specific neuromodulation would alter localized glucose metabolism in the corresponding brain region, spatial resolution of the FUS-mediated neuromodulatory areas in the brain was investigated by utilizing 2-deoxy-2[18F]fluoro-D-glucose (FDG) uptake under FUS administration to SD rats (Kim et al. 2014a). The stimulated area of the brain was visualized by FDG-positron emission topography (PET) imaging. The results showed that the spatial dimension of the neuromodulatory area was localized and approximated at full width at 90% maximum of the acoustic intensity field (Kim et al. 2013) that occupies a much smaller volume than the focal size defined by convention, that is, measured at full width at half maximum (FWHM) of the intensity field.

Recently, a study of macaque monkeys reported that FUS-mediated visuomotor behavior changes when sonication was administered to the frontal eye field (FEF) (Deffieux et al. 2013). In early 2014, a human study of ultrasonic neuromodulatory effects demonstrated suppression of the EEG activity of somatosensory evoked potentials (SEPs) generated by external median nerve stimulation and enhanced behavior modulation on two-point tactile discrimination tasks (Legon et al. 2014); however, it failed to show that FUS-related somatosensory sensation could be elicited in the absence of external tactile stimulation. Nonetheless, these results were encouraging and prompted further in-depth testing of the neuromodulatory validity and examination of the safety profile using large animal models (Lee et al. 2014).

POTENTIAL MECHANISMS BEHIND THE SCENE

The exact mechanisms underlying the FUS-mediated neuromodulation are not yet clearly elucidated on biological, cellular, or molecular levels. Understanding of the exact biological mechanisms is expected to eventually accelerate the research to control and optimize the experimental conditions such as the design of acoustic parameters for achieving the specific features of functional modulation.

Several hypotheses have been suggested to elucidate the underlying mechanism of the acoustic neuromodulation. The thermal effect was suggested for modulation of nerve conduction on the sciatic nerves extracted from the bullfrog (Colucci et al. 2009). However, successful neuromodulation using low acoustic energy in recent studies (Han et al. 2014, Kim et al. 2014a) showed that the degree of temperature change in the tissue was negligible, excluding the possibility of thermal contribution. Therefore, mechanical effects delivered via the sonication to the nervous system remained the main contributing factors responsible for the observed neuromodulatory phenomena. Kimmel and colleagues suggested that transmembrane capacitance change of the neuronal cells by exposure to FUS acoustic pressure waves may be responsible for generating the action potentials from the neuronal cells and subsequently induces brain stimulation (Krasovitski et al. 2011, Plaksin et al. 2014). Possible involvement of glial systems that have mechanoreceptors (Ostrow et al. 2011) has also been suggested for FUS-mediated neural stimulation (Kim et al. 2014).

SONICATION PARAMETERS FOR FUS-MEDIATED NEUROMODULATION

Efforts have been made to investigate the range of FUS sonication parameters for achieving efficient brain stimulations in terms of minimal acoustic intensity and energy deposition to the biological tissues (Kim et al. 2014). Based on ultrasonic stimulation of the motor cortex in the rat brain and subsequent elicitation of movement of the tail, the study showed the presence of "sweet spot" parameters among a range of FUS sonication parameters such as tone-burst duration, duty cycle, and/or sonication duration to achieve the successful neural activation with a lower acoustic energy deposition than other parameters. Interestingly, the neuronal bilayer sonophore (NBLS) model (Plaksin et al. 2014) has the potential for explaining the basis for the presence of "sweet spot" acoustic parameters. The NBLS model assumes that the neural cell membrane has piezoelectric properties, primarily through membrane capacitance changes and transmembrane ion flows. According to the model, when an acoustic pressure wave is administered to the neuronal bilayer, piezoelectric effects on the cell membrane are induced, subsequently generating the action potential and excitation of the neural cells. The use of different combinations of acoustic parameters (e.g., different sonication durations at the same duty cycle) may derive differential mechanical effects on the piezoelectric neuronal bilayer and the voltage-gated membrane potentials, conceivably contributing to the formation of a range of "sweet spot" acoustic parameters.

POSSIBLE MODULATORY EFFECTS ON WHITE MATTER TRACTS

Given the modulatory potentials of FUS on the cortical and subcortical areas of the brain, further evidence has emerged suggesting modulatory potentials of the peripheral nerve, for example, interruption of neural transmission in the peripheral (sciatic) nerves of bullfrogs (Colucci et al. 2009) and stimulation of one of the cranial nerves—the abducens nerve—in rodents (Kim et al. 2012). Therefore, it is reasonable to predict that white matter tracts of the brain, sharing features similar to the peripheral nerves, may be stimulated by FUS. The utility of modulating the neural signal transduction of the white matter tracts would create a unique opportunity for probing functional connectivity of the brain mediated by the tracts.

There is virtually no method currently available for directly assessing the function of the white matter tracts in a noninvasive fashion. Diffusion tensor MRI (DTI), through characterization of the water diffusion anisotropy of the white matter bundles, may offer macroscopic spatial orientations of the white matter tracts. However, it does not provide any means for functionally probing the degree of connectivity established by a specific white matter bundle. Selective disruption or activation of white matter tracts/bundles would provide not only foundations for examining normal brain function but also opportunities for examining aberrant neural transmission implicated in various neuropsychiatric conditions such as schizophrenia (Uhlhaas and Singer 2010) and depression (Walter et al. 2009). This potential, however, has been relatively unexplored and the subject of a debate on its efficacy because myelinated neural bundles exposed to the mechanical vibration may not generate any tangible action potentials due to rapid electrical discharge from the axonal membrane. Further study is needed to define and clarify its full potential.

SAFETY CONCERNS

A summary of safety concerns and its remedies in the context of using low-intensity FUS for neuromodulation are listed in Table 23.1. The dosimetry of exposure to the ultrasound is governed by international standards and national regulations (IEC standard 60601 part 1 and part 2).

TABLE 23.1

Safety Considerations and Remedies in the Field of FUS-Mediated Neuromodulation

Safety Concern	Potential Remedy
Potential brain tissue heating	• Conjunctional use of MR thermometry to monitor temperature changes.
	• Validation via numerical acoustic stimulation with respect to the sonication geometry and anatomy to visualize and estimate sonication propagation within the cranial cavity.
	• Use of low acoustic intensity and duty cycle, if possible.
	• Use of sufficient inter-sonication intervals.
Potential skull heating	• As for potential brain tissue heating above.
	• Increase the area of exposure to the incident sonication to distribute the acoustic energy over the wide skull surface.
	• Simulation and validation of degree of acoustic absorption via an individual-specific skull model with respect to the range of sonication frequency.
	• May accompany active/passive skin cooling by circulating cooled liquid/gel above the skin.
	• Skull base heating should be avoided using the above strategies.
Acoustic pressure-related mechanical damages to the brain	• As for potential brain tissue heating above.
	• Use of MR in conjunction with ultrasound ARFI (should improve ARFI sensitivity).
	• Validation via numerical acoustic stimulation with respect to the sonication geometry and anatomy to ensure correct sonication targeting.
	• Use of low acoustic intensity and MI.
	• Use of sufficient inter-sonication intervals.
Stimulation-related phenomena such as overexcitation of brain tissue	• Use of sufficient inter-sonication intervals to reduce the chance of neural kindling or hyper-excitatory states.
	• Limit the time of overall exposure to sonication.
	• Screen for history of seizures and neurological/psychiatric conditions and medications, as appropriate.
Subject/patient-specific related issues	• Screen for microcalcification of the brain (e.g., using CT examination) to reduce concerns for the local absorption and scattering of the acoustic beam.
	• Screen for the excessive skull thickness (>8–9 mm) as it would result in large significant attenuation of transmitted acoustic energy (reduce the stimulation efficiency) and relatively high absorption of acoustic energy at the skull (increases the change of heating).
	• Screen for potential skin irritability with respect to the acoustic coupling media to reduce the possible discomfort.
	• Appropriate screening processes should accompany the image-guidance data acquisition procedures (MRI, CT, etc.).
Other potential issues with FUS insonification	• According to the shape and location of the FUS focus in the brain, that is, treatment envelope and the size of neuromodulatory area, the proper transducer configuration should be prepared.
	• Sensory and other neurological/psychiatric phenomena associated with sonication should be communicated to subjects/patients and properly archived.
	• Patients should be informed prior to sonication that FUS may be accompanied by temporary skin sensations such as tingling or pricking of the scalp.

For example, the acoustic intensity level of 3 W/cm^2 is set as the limit for therapeutic equipment by the IEC 60601, part 2 standard (Duck 2007). For the ultrasound imager, the upper limit of the mechanical index (MI) is 1.9 according to the U.S. Food and Drug Administration (FDA) safety guideline (Duck 2007). Most of the previously reported ultrasound-mediated brain neuromodulation studies were performed within these guidelines.

There are also safety indices regarding thermal increase, for example, thermal index and bone thermal index. However, the low acoustic energy used for achieving neuromodulation is in a range far below the level that can induce a measurable temperature increase at the sonicated tissues

(Han et al. 2014, Kim et al. 2014a). Therefore, possible thermal tissue damage may be excluded as a potential hazardous factor when low-intensity ultrasound is used for FUS neuromodulation. However, care should be taken by estimating the possible temperature increase through numerical models (O'Brien 2007) before sonication or monitoring the actual thermal increase during FUS administration using MR thermometry (Rieke and Pauly 2008).

When FUS sonication is administered transcranially, the estimation of the focal region based on the stimulation is critical for accurately targeting a specific region with a desired effective acoustic intensity in the brain due to acoustic attenuation and refraction at the skull. Due to the use of low pressure and radiation force for FUS neuromodulation using low acoustic intensity, direct visualization of the acoustic pressure profile in the *in vivo* brain is not yet feasible even using the current state-of-the-art techniques of acoustic radiation force impulse (ARFI) imaging (Kaye et al. 2011). The use of MR thermometry (Rieke and Pauly 2008), which can detect moderate heating of the brain tissue, is often applied for tissue ablation using a high acoustic intensity. However, the thermal increase of the brain tissue would not be applicable to healthy human subjects. Therefore, for estimating the location and pressure profile of the FUS focal regions, numerical stimulation of acoustic wave propagation through the skull is necessary to reduce the risks and confounding factors of thermal heating (Song et al. 2012).

For transcranial application, potential pressure reverberation and standing waves should also be avoided (Song et al. 2012). In small animal models, it has been reported that the reverberation of acoustic waves inside the small skull cavity can induce complex acoustic patterns inside the rat head and stimulate multiple locations of the brain outside of the intended target (Younan et al. 2013). The study found that due to the reverberation, there was a 2.3-fold increase in I_{sppa} (or 3.6-fold increase in I_{spta}) compared to the simulations of the beam in free water. Therefore, care needs to be taken in designing acoustic parameters for the neuromodulation in small animal models. However, in large animal models or in humans, the acoustic reverberation may be negligible due to the larger size of the cranial structures (Deffieux et al. 2013). Although it is not a major issue in low-intensity application, it is also important to monitor the presence of potential skull heating in the FUS beam path and the skull base if the FUS is given close to the skull base (Connor and Hynynen 2004). The presence of microcalcification may have a slight chance of acoustic beam refraction/energy absorption and should be monitored for the similar reason.

These safety issues would be different depending on the transducer configuration, either multiarray or single element. For example, the multiarray element transducer would have an inherently increased level of risk for the standing wave formation (Song et al. 2012) but should have a sharper focal shape compared to that of the single-transducer configuration. A multiarray system can also outperform the single transducer system for sonicating deep brain structures with surgical precision. However, due to the incident angles of the insonifying beam with respect to the skull geometry, the system has a limited treatment envelope that may not cover the cortical surfaces.

POTENTIAL FUTURE APPLICATIONS

The neuromodulatory ability of FUS-mediated brain stimulation on region-specific areas in either an excitatory or suppressive manner will pave the way to development of potential future applications in the field of neuroscience and clinical medicine. Transcranial administration of focal acoustic energy to region-specific brain areas may enable a new mode of functional brain mapping that is spatially more accurate than other noninvasive methods such as TMS or tDCS and is able to reach deep brain areas. The capability of delivering acoustic energy into deep brain regions may open new possibilities for examining functional connectivity on white matter tracts and/or other important neural structures in cognition, such as the thalamus or hippocampus. However, it is important to note that the fundamental neural mechanism underpinning the efficacy of FUS is unknown and a significant amount of work remains in determining its efficacy and method of deployment.

Acoustic neuromodulation, in conjunction with the TcMRgFUS-mediated ablative therapy and neurosurgery, may also be used as a tool for the functional confirmation of the targeted ablation regions prior to conducting irreversible ablation procedures to aid in increasing the reliability of FUS neurosurgery. Based on the evidence of ultrasonic alteration on neurotransmitter concentrations in a rodent model, such as extracellular levels of DA, serotonin, and/or GABA (Yang et al. 2012), FUS has potential as a neurotherapeutic method for treating neurotransmitter-related neuropsychiatric disorders, exerting a suppressive effect on epilepsy (Min et al. 2011a), and reducing anesthesia recovery time (Yoo et al. 2011b).

It is important to note that the pulsed operation of FUS to the regional brain tissue, when used with injection of microbubble ultrasound contrast agents such as Definity or Optison, can induce temporary disruption of the BBB. The phenomenon, FUS-mediated BBB disruption (BBBd), is believed to be associated with the creation of cavitation shockwaves from bursting of microbubbles by the acoustic pressure waves (Hynynen et al. 2005). The duration of the disruption is temporary, lasting from a few minutes to hours (Aryal et al. 2014). It allows for delivery of pharmacological agents such as the anticancer medication, Herceptin (Kinoshita et al. 2006), and cisplatin (Yang and Horng 2013) that are too large to pass the BBB, or even stem cells to pass through the disrupted barrier (Burgess et al. 2011). This method was employed in a study on the development of an antibody delivery mode for AD treatments (Jordão et al. 2010) and may offer new opportunities for treatment that were not possible earlier due to the presence of the BBB. Virtually the same HIFU ablation hardware platforms can be used in neuromodulation or BBBd, operating at reduced acoustic intensity in a pulsed mode, suggesting the versatility of FUS platforms.

The new mode of noninvasive brain stimulation modality described in this chapter, TcMRgFUS, a disruptive technology, holds potential for improving upon and/or replacing existing treatments and enabling therapies that are not yet available. Among its versatile applications, CNS/PNS neuromodulation and combined use of BBBd may present mankind with the potential for great therapeutic progress.

ACKNOWLEDGMENTS

We gratefully acknowledge the editing of this chapter by Nina Geller, PhD, and the editorial support of Stephanie D. Lee and Michael Y. Park.

REFERENCES

Ansari, S., K. Chaudhri, and K. A. Al Moutaery. 2007. Vagus nerve stimulation: Indications and limitations. *Acta Neurochirurgica Supplement* 97 (Pt 2):281–286.

Aryal, M., C. D. Arvanitis, P. M. Alexander, and N. McDannold. 2014. Ultrasound-mediated blood–brain barrier disruption for targeted drug delivery in the central nervous system. *Advanced Drug Delivery Reviews* 72:94–109. doi: 10.1016/j.addr.2014.01.008.

Aubry, J. F., M. Tanter, M. Pernot, J. L. Thomas, and M. Fink. 2003. Experimental demonstration of noninvasive transskull adaptive focusing based on prior computed tomography scans. *The Journal of the Acoustical Society of America* 113 (1):84–93.

Bachtold, M. R., P. C. Rinaldi, J. P. Jones, F. Reines, and L. R. Price. 1998. Focused ultrasound modifications of neural circuit activity in a mammalian brain. *Ultrasound in Medicine & Biology* 24 (4):557–565. doi: 16/S0301-5629(98)00014-3.

Ballantine, H. T., E. Bell, and J. Manlapaz. 1960. Progress and problems in the neurological applications of focused ultrasound. *Journal of Neurosurgery* 17:858–876. doi: 10.3171/jns.1960.17.5.0858.

Burgess, A., C. A. Ayala-Grosso, M. Ganguly, J. F. Jordão, I. Aubert, and K. Hynynen. 2011. Targeted delivery of neural stem cells to the brain using MRI-guided focused ultrasound to disrupt the blood-brain barrier. *PLoS ONE* 6 (11):e27877. doi: 10.1371/journal.pone.0027877.

Burt, T., S. H. Lisanby, and H. A. Sackeim. 2002. Neuropsychiatric applications of transcranial magnetic stimulation: A meta analysis. *The International Journal of Neuropsychopharmacology* 5 (1):73–103. doi: 10.1017/S1461145702002791.

Bystritsky, A., A. S. Korb, P. K. Douglas, M. S. Cohen, W. P. Melega, A. P. Mulgaonkar, A. DeSalles, B.-K. Min, and S.-S. Yoo. 2011. A review of low-intensity focused ultrasound pulsation. *Brain Stimulation* 4 (3):125–136. doi: 16/j.brs.2011.03.007.

Chaussy, C. and E. Schmiedt. 1984. Extracorporeal shock wave lithotripsy (ESWL) for kidney stones. An alternative to surgery? *Urologic Radiology* 6 (1):80–87.

Clement, G. T. and K. Hynynen. 2002. A non-invasive method for focusing ultrasound through the human skull. *Physics in Medicine and Biology* 47 (8):1219–1236.

Clement, G. T., P. Jason White, R. L. King, N. McDannold, and K. Hynynen. 2005. A magnetic resonance imaging-compatible, large-scale array for trans-skull ultrasound surgery and therapy. *Journal of Ultrasound in Medicine* 24 (8):1117–1125.

Cline, H. E., K. Hynynen, R. D. Watkins, W. J. Adams, J. F. Schenck, R. H. Ettinger, W. R. Freund, J. P. Vetro, and F. A. Jolesz. 1995. Focused US system for MR imaging-guided tumor ablation. *Radiology* 194 (3):731–737. doi: 10.1148/radiology.194.3.7862971.

Cline, H. E., J. F. Schenck, K. Hynynen, R. D. Watkins, S. P. Souza, and F. A. Jolesz. 1992. MR-guided focused ultrasound surgery. *Journal of Computer Assisted Tomography* 16 (6):956–965.

Colen, R. R. and F. A. Jolesz. 2010. Future potential of MRI-guided focused ultrasound brain surgery. *Neuroimaging Clinics of North America* 20 (3):355–366. doi: 10.1016/j.nic.2010.05.003.

Collins, K. L., E. M. Lehmann, and P. G. Patil. 2010. Deep brain stimulation for movement disorders. *Neurobiology of Disease* 38 (3):338–345. doi: 10.1016/j.nbd.2009.11.019.

Colucci, V., G. Strichartz, F. Jolesz, N. Vykhodtseva, and K. Hynynen. 2009. Focused ultrasound effects on nerve action potential in vitro. *Ultrasound in Medicine & Biology* 35 (10):1737–1747. doi: 16/j.ultrasmedbio.2009.05.002.

Connor, C. W. and K. Hynynen. 2004. Patterns of thermal deposition in the skull during transcranial focused ultrasound surgery. *IEEE Transactions on Biomedical Engineering* 51 (10):1693–1706. doi: 10.1109/TBME.2004.831516.

Daum, D. R., M. T. Buchanan, T. Fjield, and K. Hynynen. 1998. Design and evaluation of a feedback based phased array system for ultrasound surgery. *IEEE Transactions on Ultrasonics, Ferroelectrics, and Frequency Control* 45 (2):431–438. doi: 10.1109/58.660153.

Daum, D. R. and K. Hynynen. 1999. A 256-element ultrasonic phased array system for the treatment of large volumes of deep seated tissue. *IEEE Transactions on Ultrasonics, Ferroelectrics and Frequency Control* 46 (5):1254–1268. doi: 10.1109/58.796130.

Deffieux, T., Y. Younan, N. Wattiez, M. Tanter, P. Pouget, and J.-F. Aubry. 2013. Low-intensity focused ultrasound modulates monkey visuomotor behavior. *Current Biology* 23 (23):2430–2433. doi: 10.1016/j.cub.2013.10.029.

Deisseroth, K. 2011. Optogenetics. *Nature Methods* 8 (1):26–29. doi: 10.1038/nmeth.f.324.

Dowling, J. 2008. Deep brain stimulation: Current and emerging indications. *Missouri Medicine* 105 (5):424–428.

Duck, F. A. 2007. Medical and non-medical protection standards for ultrasound and infrasound. *Progress in Biophysics & Molecular Biology* 93 (1–3):176–191. doi: 10.1016/j.pbiomolbio.2006.07.008.

Elias, W. J., D. Huss, T. Voss, J. Loomba, M. Khaled, E. Zadicario, R. C. Frysinger et al. 2013. A pilot study of focused ultrasound thalamotomy for essential tremor. *The New England Journal of Medicine* 369 (7):640–648. doi: 10.1056/NEJMoa1300962.

Ellis, T. L. and A. Stevens. 2008. Deep brain stimulation for medically refractory epilepsy. *Neurosurgical Focus* 25 (3):E11. doi: 10.3171/FOC/2008/25/9/E11.

Fry, F. J., H. W. Ades, and W. J. Fry. 1958. Production of reversible changes in the central nervous system by ultrasound. *Science* 127 (3289):83–84.

Fry, W. J. 1958. Intense ultrasound in investigations of the central nervous system. *Advances in Biological and Medical Physics* 6:281–348.

Fry, W. J., J. W. Barnard, F. J. Fry, and J. F. Brennan. 1955. Ultrasonically produced localized selective lesions in the central nervous system. *American Journal of Physical Medicine* 34 (3):413–423.

Fry, W. J. and F. J. Fry. 1960. Fundamental neurological research and human neurosurgery using intense ultrasound. *IRE Transactions on Medical Electronics* ME-7:166–181.

Gavrilov, L. R., E. M. Tsirulnikov, and I. A. Davies. 1996. Application of focused ultrasound for the stimulation of neural structures. *Ultrasound in Medicine & Biology* 22 (2):179–192.

George, M. S. 2010. Transcranial magnetic stimulation for the treatment of depression. *Expert Review of Neurotherapeutics* 10 (11):1761–1772. doi: 10.1586/ern.10.95.

Grover, P. J., E. A. C. Pereira, A. L. Green, J.-S. Brittain, S. L. F. Owen, P. Schweder, M. L. Kringelbach, P. T. G. Davies, and T. Z. Aziz. 2009. Deep brain stimulation for cluster headache. *Journal of Clinical Neuroscience* 16 (7):861–866. doi: 10.1016/j.jocn.2008.10.012.

Groves, D. A. and V. J. Brown. 2005. Vagal nerve stimulation: A review of its applications and potential mechanisms that mediate its clinical effects. *Neuroscience and Biobehavioral Reviews* 29 (3):493–500. doi: 10.1016/j.neubiorev.2005.01.004.

Gutman, D. A., P. E. Holtzheimer, T. E. J. Behrens, H. Johansen-Berg, and H. S. Mayberg. 2009. A tractography analysis of two deep brain stimulation white matter targets for depression. *Biological Psychiatry* 65 (4):276–282. doi: 10.1016/j.biopsych.2008.09.021.

Hameroff, S., M. Trakas, C. Duffield, A. Emil, M. Bagambhrini Gerace, P. Boyle, A. Lucas, Q. Amos, A. Buadu, and J. J. Badal. 2013. Transcranial ultrasound (TUS) effects on mental states: A pilot study. *Brain Stimulation* 6 (3):409–415. doi: 10.1016/j.brs.2012.05.002.

Han, H.-S., S. Y. Hwang, F. Akram, H. Jae Jeon, S. B. Nam, S. B. Jun, J. T. Kim, and T.-S. Kim. July 2–4, 2014. Neural activity modulation via ultrasound stimulation measured on multi-channel electrodes. *Proceedings of the World Congress on Engineering 2014 (WCE 2014)*, London, U.K.

Harvey, E. N., E. B. Harvey, and A. L. Loomis. 1928. Further observations on the effect of high frequency sound waves on living matter. *Biological Bulletin* 55 (6):459–469. doi: 10.2307/1536801.

Hynynen, K., G. T. Clement, N. McDannold, N. Vykhodtseva, R. King, P. J. White, S. Vitek, and F. A. Jolesz. 2004. 500 element ultrasound phased array system for noninvasive focal surgery of the brain: A preliminary rabbit study with ex vivo human skulls. *Magnetic Resonance in Medicine* 52 (1):100–107. doi: 10.1002/mrm.20118.

Hynynen, K., W. R. Freund, H. E. Cline, A. H. Chung, R. D. Watkins, J. P. Vetro, and F. A. Jolesz. 1996a. A clinical, noninvasive, MR imaging-monitored ultrasound surgery method. *Radiographics: A Review Publication of the Radiological Society of North America, Inc* 16 (1):185–195. doi: 10.1148/radiographics.16.1.185.

Hynynen, K., W. R. Freund, H. E. Cline, A. H. Chung, R. D. Watkins, J. P. Vetro, and F. A. Jolesz. 1996b. A clinical, noninvasive, MR imaging-monitored ultrasound surgery method. *Radiographics* 16 (1):185–195. doi: 10.1148/radiographics.16.1.185.

Hynynen, K., N. McDannold, N. Vykhodtseva, and F. A. Jolesz. 2001. Noninvasive MR imaging-guided focal opening of the blood-brain barrier in rabbits. *Radiology* 220 (3):640–646. doi: 10.1148/radiol.2202001804.

Hynynen, K., N. McDannold, N. A. Sheikov, F. A. Jolesz, and N. Vykhodtseva. 2005. Local and reversible blood–brain barrier disruption by noninvasive focused ultrasound at frequencies suitable for trans-skull sonications. *NeuroImage* 24 (1):12–20. doi: 10.1016/j.neuroimage.2004.06.046.

Hynynen, K., N. McDannold, N. Vykhodtseva, S. Raymond, R. Weissleder, F. A. Jolesz, and N. Sheikov. 2006. Focal disruption of the blood-brain barrier due to 260-kHz ultrasound bursts: A method for molecular imaging and targeted drug delivery. *Journal of Neurosurgery* 105 (3):445–454. doi: 10.3171/jns.2006.105.3.445.

Jagannathan, J., N. T. Sanghvi, L. A. Crum, C.-P. Yen, R. Medel, A. S. Dumont, J. P. Sheehan, L. Steiner, F. Jolesz, and N. F. Kassell. 2009. High-intensity focused ultrasound surgery of the brain: Part 1—A historical perspective with modern applications. *Neurosurgery* 64 (2):201–210; discussion 210-211. doi: 10.1227/01.NEU.0000336766.18197.8E.

Jolesz, F. A., K. Hynynen, N. McDannold, and C. Tempany. 2005. MR imaging-controlled focused ultrasound ablation: A noninvasive image-guided surgery. *Magnetic Resonance Imaging in Clinical North America* 13 (3):545–560. doi: 10.1016/j.mric.2005.04.008.

Jolesz, F. A. and N. McDannold. 2008. Current status and future potential of MRI-guided focused ultrasound surgery. *Journal of Magnetic Resonance Imaging* 27 (2):391–399. doi: 10.1002/jmri.21261.

Jolesz, F. A. and N. J. McDannold. 2014. Magnetic resonance-guided focused ultrasound: A new technology for clinical neurosciences. *Neurologic Clinics* 32 (1):253–269. doi: 10.1016/j.ncl.2013.07.008.

Jordão, J. F., C. A. Ayala-Grosso, K. Markham, Y. Huang, R. Chopra, J. McLaurin, K. Hynynen, and I. Aubert. 2010. Antibodies targeted to the brain with image-guided focused ultrasound reduces amyloid-β plaque lin the TgCRND8 mouse model of alzheimer's disease. *PLoS ONE* 5 (5):e10549. doi: 10.1371/journal.pone.0010549.

Kaye, E. A., J. Chen, and K. B. Pauly. 2011. Rapid MR-ARFI method for focal spot localization during focused ultrasound therapy. *Magnetic Resonance in Medicine* 65 (3):738–743. doi: 10.1002/mrm.22662.

Kim, H., A. Chiu, S. D. Lee, K. Fischer, and S.-S. Yoo. 2014. Focused ultrasound-mediated non-invasive brain stimulation: Examination of sonication parameters. *Brain Stimulation* 7 (5):748–756. doi: 10.1016/j.brs.2014.06.011.

Kim, H., S. D. Lee, A. Chiu, S.-S. Yoo, and S. Park. 2014a. Estimation of the spatial profile of neuromodulation and the temporal latency in motor responses induced by focused ultrasound brain stimulation. *Neuroreport* 25 (7):475–479. doi: 10.1097/WNR.0000000000000118.

Kim, H., M.-A. Park, S. Wang, A. Chiu, K. Fischer, and S.-S. Yoo. 2013. PET/CT imaging evidence of FUS-mediated (18)F-FDG uptake changes in rat brain. *Medical Physics* 40 (3):033501. doi: 10.1118/1.4789916.

Kim, H., S. J. Taghados, K. Fischer, L.-S. Maeng, S. Park, and S.-S. Yoo. 2012. Noninvasive transcranial stimulation of rat abducens nerve by focused ultrasound. *Ultrasound in Medicine & Biology* 38 (9):1568–1575. doi: 10.1016/j.ultrasmedbio.2012.04.023.

Kimiskidis, V. K. 2010. Transcranial magnetic stimulation for drug-resistant epilepsies: Rationale and clinical experience. *European Neurology* 63 (4):205–210. doi: 10.1159/000282735.

King, R. L., J. R. Brown, W. T. Newsome, and K. B. Pauly. 2013. Effective parameters for ultrasound-induced in vivo neurostimulation. *Ultrasound in Medicine & Biology* 39 (2):312–331. doi: 10.1016/j.ultrasmedbio.2012.09.009.

Kinoshita, M., N. McDannold, F. A. Jolesz, and K. Hynynen. 2006. Noninvasive localized delivery of Herceptin to the mouse brain by MRI-guided focused ultrasound-induced blood–brain barrier disruption. *Proceedings of the National Academy of Sciences* 103 (31):11719–11723. doi: 10.1073/pnas.0604318103.

Krasovitski, B., V. Frenkel, S. Shoham, and E. Kimmel. 2011. Intramembrane cavitation as a unifying mechanism for ultrasound-induced bioeffects. *Proceedings of the National Academy of Sciences* 108 (8):3258–3263. doi: 10.1073/pnas.1015771108.

Lee, W., H. Kim, S. D. Lee, M. Y. Park, and S.-S. Yoo. 2014. FUS-mediated functional neuromodulation for neurophysiologic assessment in a large animal model. *Fourth International Symposium on Focused Ultrasound 2014*, 2014/10/12/16.

Lefaucheur, J. P. 2009. Methods of therapeutic cortical stimulation. *Neurophysiologie Clinique/Clinical Neurophysiology* 39 (1):1–14. doi: 10.1016/j.neucli.2008.11.001.

Legon, W., T. F. Sato, A. Opitz, J. Mueller, A. Barbour, A. Williams, and W. J. Tyler. 2014. Transcranial focused ultrasound modulates the activity of primary somatosensory cortex in humans. *Nature Neuroscience* 17 (2):322–329 doi: 10.1038/nn.3620.

Lele, P. P. 1962. A simple method for production of trackless focal lesions with focused ultrasound: Physical factors. *The Journal of Physiology* 160 (3):494–512.

Lele, P. P., Repacholi M. H., Grandolfo M., and A. Rindi. 1987. Effects of ultrasound on "solid" mammalian tissues and tumors in vivo. In *Ultrasound*, pp. 275–306. Springer, New York.

Lipton, R. B. and S. H. Pearlman. 2010. Transcranial magnetic simulation in the treatment of migraine. *Neurotherapeutics* 7 (2):204–212. doi: 10.1016/j.nurt.2010.03.002.

Lynn, J. G., R. L. Zwemer, A. J. Chick, and A. E. Miller. 1942. A new method for the generation and use of focused ultrasound in experimental biology. *The Journal of General Physiology* 26 (2):179–193.

Magee, T. R. and A. H. Davies. 1993. Auditory phenomena during transcranial Doppler insonation of the basilar artery. *Journal of Ultrasound in Medicine* 12 (12):747–750.

Martin, E., D. Jeanmonod, A. Morel, E. Zadicario, and B. Werner. 2009. High-intensity focused ultrasound for noninvasive functional neurosurgery. *Annals of Neurology* 66 (6):858–861. doi: 10.1002/ana.21801.

McDannold, N., C. D. Arvanitis, N. Vykhodtseva, and M. S. Livingstone. 2012. Temporary disruption of the blood-brain barrier by use of ultrasound and microbubbles: Safety and efficacy evaluation in rhesus macaques. *Cancer Research* 72 (14):3652–3663. doi: 10.1158/0008-5472.CAN-12-0128.

McDannold, N., G. T. Clement, P. Black, F. Jolesz, and K. Hynynen. 2010. Transcranial magnetic resonance imaging- guided focused ultrasound surgery of brain tumors: Initial findings in 3 patients. *Neurosurgery* 66 (2):323–332; discussion 332. doi: 10.1227/01.NEU.0000360379.95800.2F.

McDannold, N., Y. Zhang, C. Power, C. Arvanitis, N. Vykhodtseva, and M. Livingstone. 2014. Targeted delivery of GABA via ultrasound-induced blood-brain barrier disruption blocks somatosensory evoked potentials. *Fourth International Symposium on Focused Ultrasound 2014*, 2014/10/14/.

Mian, M. K., M. Campos, S. A. Sheth, and E. N. Eskandar. 2010. Deep brain stimulation for obsessive-compulsive disorder: Past, present, and future. *Neurosurgical Focus* 29 (2):E10. doi: 10.3171/2010.4.FOCUS10107.

Min, B.-K., A. Bystritsky, K.-I. Jung, K. Fischer, Y. Zhang, L.-S. Maeng, S. In Park, Y.-A. Chung, F. Jolesz, and S.-S. Yoo. 2011a. Focused ultrasound-mediated suppression of chemically-induced acute epileptic EEG activity. *BMC Neuroscience* 12:23. doi: 10.1186/1471-2202-12-23.

Min, B.-K., P. S. Yang, M. Bohlke, S. Park, D. R. Vago, T. J. Maher, and S.-S. Yoo. 2011b. Focused ultrasound modulates the level of cortical neurotransmitters: Potential as a new functional brain mapping technique. *International Journal of Imaging Systems and Technology* 21 (2):232–240. doi: 10.1002/ima.20284.

Morris, G. L. and W. M. Mueller. 1999. Long-term treatment with vagus nerve stimulation in patients with refractory epilepsy. The Vagus Nerve Stimulation Study Group E01-E05. *Neurology* 53 (8):1731–1735.

Nitsche, M. A., P. S. Boggio, F. Fregni, and A. Pascual-Leone. 2009. Treatment of depression with transcranial direct current stimulation (tDCS): A review. *Experimental Neurology* 219 (1):14–19. doi: 10.1016/j.expneurol.2009.03.038.

O'Brien, W. D. 2007. Ultrasound—Biophysics mechanisms. *Progress in Biophysics & Molecular Biology* 93 (1–3):212–255. doi: 10.1016/j.pbiomolbio.2006.07.010.

Ostrow, L. W., T. M. Suchyna, and F. Sachs. 2011. Stretch induced endothelin-1 secretion by adult rat astrocytes involves calcium influx via stretch-activated ion channels (SACs). *Biochemical and Biophysical Research Communications* 410 (1):81–86. doi: 10.1016/j.bbrc.2011.05.109.

Plaksin, M., S. Shoham, and E. Kimmel. 2014. Intramembrane cavitation as a predictive bio-piezoelectric mechanism for ultrasonic brain stimulation. *Physical Review X* 4 (1):011004. doi: 10.1103/PhysRevX.4.011004.

Rieke, V. and K. B. Pauly. 2008. MR thermometry. *Journal of Magnetic Resonance Imaging* 27 (2):376–390. doi: 10.1002/jmri.21265.

Rinaldi, P. C., J. P. Jones, F. Reines, and L. R. Price. 1991. Modification by focused ultrasound pulses of electrically evoked responses from an in vitro hippocampal preparation. *Brain Research* 558 (1):36–42. doi: 16/0006-8993(91)90711-4.

Rothwell, J. 2007. Transcranial magnetic stimulation as a method for investigating the plasticity of the brain in Parkinson's disease and dystonia. *Parkinsonism & Related Disorders* 13 (Suppl 3):S417–S420. doi: 10.1016/S1353-8020(08)70040-3.

Schiff, N. D., J. T. Giacino, and J. J. Fins. 2009. Deep brain stimulation, neuroethics, and the minimally conscious state: Moving beyond proof of principle. *Archives of Neurology* 66 (6):697–702. doi: 10.1001/archneurol.2009.79.

Song, J. and K. Hynynen. 2009. A 1372-element large scale hemispherical ultrasound phased array transducer for noninvasive transcranial therapy. *Eighth International Symposium on Therapeutic Ultrasound* Minneapolis, MN, 2009/04/14/.

Song, J., A. Pulkkinen, Y. Huang, and K. Hynynen. 2012. Investigation of standing-wave formation in a human skull for a clinical prototype of a large-aperture, transcranial MR-guided focused ultrasound (MRgFUS) phased array: An experimental and simulation study. *IEEE Transactions on Biomedical Engineering* 59 (2):435–444. doi: 10.1109/TBME.2011.2174057.

Tempany, C. M. C., N. J. McDannold, K. Hynynen, and F. A. Jolesz. 2011. Focused ultrasound surgery in oncology: Overview and principles. *Radiology* 259 (1):39–56. doi: 10.1148/radiol.11100155.

Thomas, J.-L. and M. A. Fink. 1996. Ultrasonic beam focusing through tissue inhomogeneities with a time reversal mirror: Application to transskull therapy. *IEEE Transactions on Ultrasonics, Ferroelectrics and Frequency Control* 43 (6):1122–1129. doi: 10.1109/58.542055.

Treat, L. H., N. McDannold, N. Vykhodtseva, Y. Zhang, K. Tam, and K. Hynynen. 2007. Targeted delivery of doxorubicin to the rat brain at therapeutic levels using MRI-guided focused ultrasound. *International Journal of Cancer* 121 (4):901–907. doi: 10.1002/ijc.22732.

Tufail, Y., A. Matyushov, N. Baldwin, M. L. Tauchmann, J. Georges, A. Yoshihiro, S. I. H. Tillery, and W. J. Tyler. 2010. Transcranial pulsed ultrasound stimulates intact brain circuits. *Neuron* 66 (5):681–694. doi: 16/j.neuron.2010.05.008.

Uhlhaas, P. J. and W. Singer. 2010. Abnormal neural oscillations and synchrony in schizophrenia. *Nature Reviews Neuroscience* 11 (2):100–113. doi: 10.1038/nrn2774.

Vallancien, G., M. Harouni, B. Veillon, A. Mombet, D. Prapotnich, J. M. Brisset, and J. Bougaran. 1992. Focused extracorporeal pyrotherapy: Feasibility study in man. *Journal of Endourology* 6 (2):173–181. doi: 10.1089/end.1992.6.173.

Walter, M., A. Henning, S. Grimm et al. 2009. The relationship between aberrant neuronal activation in the pregenual anterior cingulate, altered glutamatergic metabolism, and anhedonia in major depression. *Archives of General Psychiatry* 66 (5):478–486. doi: 10.1001/archgenpsychiatry.2009.39.

Yang, F.-Y. and S.-C. Horng. 2013. Chemotherapy of glioblastoma by targeted liposomal platinum compounds with focused ultrasound. *Thirty fifth Annual International Conference of the IEEE Engineering in Medicine and Biology Society*. Osaka International Convention Center, Osaka, Japan

Yang, P. S., H. Kim, W. Lee, M. Bohlke, S. Park, T. J. Maher, and S.-S. Yoo. 2012. Transcranial focused ultrasound to the thalamus is associated with reduced extracellular GABA levels in rats. *Neuropsychobiology* 65 (3):153–160.

Yang, R., N. T. Sanghvi, F. J. Rescorla, C. A. Galliani, F. J. Fry, S. L. Griffith, and J. L. Grosfeld. 1992. Extracorporeal liver ablation using sonography-guided high-intensity focused ultrasound. *Investigative Radiology* 27 (10):796–803.

Yoo, S.-S., A. Bystritsky, J.-H. Lee, Y. Zhang, K. Fischer, B.-K. Min, N. J. McDannold, A. Pascual-Leone, and F. A. Jolesz. 2011a. Focused ultrasound modulates region-specific brain activity. *NeuroImage* 56 (3):1267–1275. doi: 16/j.neuroimage.2011.02.058.

Yoo, S.-S., H. Kim, B.-K. Min, E. Franck, and S. Park. 2011b. Transcranial focused ultrasound to the thalamus alters anesthesia time in rats. *Neuroreport* 22 (15):783–787. doi: 10.1097/WNR.0b013e32834b2957.

Younan, Y., T. Deffieux, B. Larrat, M. Fink, M. Tanter, and J.-F. Aubry. 2013. Influence of the pressure field distribution in transcranial ultrasonic neurostimulation. *Medical Physics* 40 (8):082902. doi: 10.1118/1.4812423.

24 Advances in Portable Neuroimaging and Their Effect on Novel Therapies

Eric M. Bailey, Ibrahim Bechwati, Sonal Ambwani,
Matthew Dickman, Joseph Fonte, and Geethika Weliwitigoda

CONTENTS

INTRODUCTION ON PORTABLE NEUROIMAGING

Portable forms of imaging, mainly ultrasound and 2D x-ray, have been around for many decades. But neither of these modalities has been very useful for neuroimaging. Portable ultrasound has played a small neuroimaging role in bedside transcranial Doppler and carotid artery scanning. Also portable 2D x-ray has played a role in intraoperative imaging during spine surgery. The major neuroimaging modalities, namely, CT, MRI, and single-photon emission computed tomography (SPECT)/PET, have not existed in portable form. However, the first decade of the twentieth century has seen portable forms of CT and SPECT imaging systems hit the market. These new systems, while still in their clinical infancy, have opened up new usages for clinical neuroimaging. In addition, they have allowed for new forms of therapy, particularly therapy delivered at the "bedside." In particular, these systems have allowed neuroimaging in hitherto new places, such as the intensive care unit (ICU), emergency department, operating room (OR), and physician office and even in ambulances. This new form of "in situ" imaging is beginning to play a strong role in neurological issues such as stroke, traumatic brain injury, instrumented spine surgery, brain tumor surgery, deep brain stimulation (DBS) surgery, and refractory epilepsy. Perhaps the most exciting new application is intra-ambulance CT scanning, enabling rapid diagnosis and intra-ambulance thrombolysis treatment of ischemic stroke patients. This chapter will attempt to explore these new neuroimaging applications by first giving a background on the design and development of these new portable imaging systems. Then we will give a background on the clinical need for portable imaging of several of these neurological diseases. Finally, we will concentrate on four sections on the state of accomplishments portable CT scanning in the ICU, portable intraoperative CT, intra-ambulance CT for ischemic stroke, and portable SPECT imaging of refractory epilepsy patients. A conclusion will then summarize the portable neuroimaging accomplishments to date and also try and predict where portable neuroimaging may go in the next decade.

MEDICAL NECESSITY FOR PORTABLE NEUROIMAGING

Many millions of people in the world are afflicted with a variety of neurological diseases and the costs are approximated to be greater than $500B annually (Brain Institute). If one looks worldwide, these numbers become almost astronomical. Because the brain is such a complex organ, it is also quite difficult and costly just to diagnose these disorders. As for treatment, many of these disorders such as dementia and amyotrophic lateral sclerosis (ALS) lack any form of treatment. Medicine has made some progress in treatment; however, these treatments require time, urgent diagnoses, and/or precise surgical deliveries. Organ complexity requires neuroimaging to be used for diagnosis, and precision delivery generally requires neuroimaging to guide such therapy accurately. This presents a dilemma because complex neuroimaging, until this previous decade, could only be performed in a fixed radiology suite. The advent of portable neuroimaging has broken the paradigm shift whereas a patient must be brought to a scanner. Now it is possible to bring the scanner to the patient. This is a great concept particularly for patients unable to go near a scanner due to some obstacles such as being in ICU in an unstable condition while tethered to other life support machinery, an example being a patient undergoing neurosurgery. Other examples include any patient undergoing surgery, including neurosurgery where rapid imaging of the neuraxis is needed to guide the surgeon appropriately. Therefore, this section will explore what we know today to be these areas and instances whereby portable imaging has not only made sense but may also be the only choice.

MEDICAL NECESSITY FOR PORTABLE CT IN THE ICU

Although stroke and traumatic brain injury are the third and fourth leading causes of death; they are also the first and second leading cause of disability. A large majority of the deaths occur several days following the event; remarkably 80% of patients survive such events (Wikipedia). The result of this is that most of these patients have some type of ICU stay with an average 3–10-day period. All of these patients regardless of trauma or stroke can suffer further brain damage in the days following the event from edema, mass effect, and hemorrhage to mention a few. A large majority of these patients require intensive monitoring and/or life support machinery such as ventilator. In addition, a majority is in coma condition either due to the acute brain injury or induced by drug. Further ischemic injury can occur from other phenomena related to the acute injury such as cerebral vasospasm. ICU intensivists have few tools to monitor for such events at the bedside, and thus it required transport of complex, often unstable patients to the radiology department for brain imaging. Portable neuroimaging eliminates the need for the dangerous and labor-intensive patient transport.

MEDICAL NECESSITY FOR PORTABLE CT IN THE OR

Neurosurgery has become a very busy service in major hospitals worldwide. Surgeries such as tumor resections, complex instrumented spine surgeries, DBS electrodes, and placement of catheters and shunts are just few of the more common surgeries. Even the most minor neurosurgery carries a high risk of surgical morbidity and mortality. Patients undergo an extensive pre and post neuroimaging workup, often consisting of many modalities. However, there is a lack of available neuroimaging during these surgeries. Without real-time imaging, the surgeon has limited capability to detect important anatomical and physical changes during neurosurgery caused by phenomena such as brain shift, patient movement, tissue resection, and blood and other fluids. It is unfortunate that a large number of surgeries are unsuccessful and require additional procedures due to a lack of real-time imaging. Portable intraoperative CT allows surgeons to image the brain in real time, assess the success of the surgical procedure, and identify and ameliorate complications when they arise allowing surgeons to make good quality assurance decisions prior to surgical completion.

MEDICAL NECESSITY FOR PORTABLE CT IN THE AMBULANCE

Stroke, despite being ranked as the top neurological disorder, has only one acute therapy. That therapy is for the majority of patients that have acute ischemic stroke as opposed to hemorrhagic stroke. Acute cerebral ischemia can benefit from the only approved therapy being thrombolysis with tissue plasminogen activation (tPA). But tPA must be administered within the first few hours after symptoms, and must not be given to patients with any evidence of acute intracranial hemorrhage, otherwise the therapy could cause further brain damage or kill the patient. The quickest and easiest way to diagnose acute ischemic stroke is with a CT scan. However, it takes a long time for a patient to be transported to an ER, undergo neurological workup, and go to a CT scan, to interpret the scan, and then to administer therapy. As a result, many patients who arrive in the emergency department for work-up of stroke are often outside the window for treatment and therefore never receive tPA. Much time can be saved by bringing the necessary items to a patient's point of origin and perform the workup in situ. Hitherto, the limiting factor was the availability of an ambulance-sized CT machine. But now, portable CT can be installed in ambulances turning them into stroke mobiles for mobile diagnosis and treatment. Bringing the team and ability to diagnose acute stroke with portable CT to the patient enables a much larger percentage of acute stroke patients to receive tPA and therefore have their stroke treated limiting the extent of brain infarction.

MEDICAL NECESSITY FOR PORTABLE SPECT IN EPILEPSY MONITORING UNITS

Almost 30% of epilepsy patients are diagnosed with intractable, drugresistant, MRI-negative forms of epilepsy (AHRQ). These very ill patients not only cost a lot of money to continue to treat, but they have a high mortality and morbidity. However, surgical resection surgery is becoming an option for these patients. In fact some hospitals have boasted as high as 70% cure rate on such surgeries (Seton Brain and Spine Institute). But the hard part is diagnosing which tissue to resect prior to surgery. This often involves a costly and complex stay in an epilepsy monitoring unit where many types of modalities such as EEG, both extra cranial and intracranial, and with ictal and interictal SPECT scanning. However, the long transport to SPECT machine in radiology becomes dangerous as most of these patients are off their antiseizure meds. Portable high-resolution SPECT allows for safe imaging of patients within the unit.

DEVELOPMENT OF THE PORTABLE CT AND SPECT

3D imaging technologies such as MRI, CT, and SPECT are quite large and expensive in their physical form. These machines often weigh thousands of pounds. They also require large instantaneous and precise electrical feed, air conditioning, humidity control, etc. They are further restricted by the necessity to shield the machines from the operator and public due to issues such as radiation leakage. In order to develop a portable 3D neuroimaging device, one had to overcome these normal limitations. Otherwise, this type of imaging could only be offered in the basement of hospitals in specific radiology suites thus requiring the transport of patients to the device.

PORTABLE CT DEVELOPMENT

In 2004, a company named NeuroLogica Corp. was formed with a mission to invent a small portable CT scanner for neuroimaging. By using batteries, it overcame the electrical restrictions. By putting the x-ray tube close to the patients head, the size and weight of the overall machine is reduced by the inverse-square law of radiation. This in turn reduced the radiation leakage, but in addition, shielding was added inside the machine covers to bring external exposure to safe levels. The result was an 800 lb. system that was on wheels and can be pushed by hand around a hospital. Its rugged design allowed it to go outside the hospital in the rugged environment of an ambulance. In addition, the scanner contained a subsystem to translate the entire scanner on the ground over the patient, thus eliminating the need to move the patient with a moving table. The patient could remain in their hospital bed and be scanned. Figure 24.1 shows NeuroLogica portable head scanner.

PORTABLE SPECT DEVELOPMENT

Several years later, this same company, NeuroLogica Corp., underwent the challenge to create high-resolution portable neuro-SPECT. The scanner was developed around the CereTom frame, but contained high-resolution SPECT detectors containing "focused" collimators, which revolved around the patient. This translation of the detector "focal spots" allowed for 3D iterative reconstruction. The collimators were fabricated with very small holes allowing for spatial resolution of almost 3 mm in each direction. Normal SPECT is 7–10 mm resolution on a side. Again this design ran on batteries and can be pushed around on wheels allowing for scanning in situ to an OR, ER, ICU, etc. Figure 24.2 shows the InSPira portable SPECT scanner.

PORTABLE CT IN ICU

The standard for intrahospital transport is to provide the same level of care and intervention that are available in the ICU. Transporting a patient to and from the ICU is a common procedure, yet very risky especially for sicker patients. Patient may need to leave the ICU for regular therapeutic treatment or in response to an emergency situation such as a quick decline in the current medical status. Outside the ICU, critically ill patients will be exposed to clinically unsafe environment, for example, hallways and waiting room. The average time a patient spent outside the ICU is 62–95

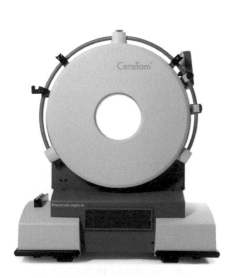

FIGURE 24.1   NeuroLogica CereTom head scanner.

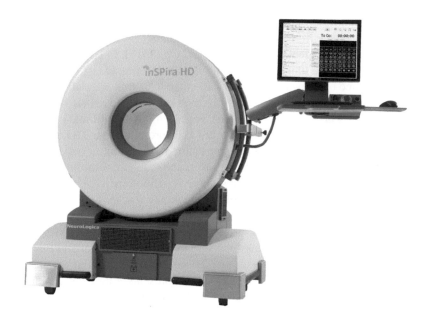

FIGURE 24.2 NeuroLogica mobile SPECT scanner.

minutes where the time range is 20–225 minutes (Stevenson and Hass 2002). The transport time of 157 patients were recorded, the average transport time was 47 minutes with transport time ranging from 20 to 204 minutes (Peace and Maloney-Wilensky, 2011). Intrahospital transport was the subject of several medical studies between 1999 and 2009 (Day, 2010) . These studies showed that some 60%–70% of the patients suffered complications during transports. Adverse events resulting from mishaps during patient transports include airway obstructions, respiratory arrest, hospital-acquired infections, cardiac arrest, bleeding, and finally disability related to neurological events such as increased intracranial pressure (ICP) or spinal cord destabilization. Ventilator-associated pneumonia (VAP) is a leading cause of death from hospital-acquired infections, with estimated mortality rates between 20% and 70%. Intrahospital transport is shown to increase the risk of acquiring VAP, a study by Bercault et al. Nicolas Bercault (2005) shows that 26% of patients who underwent an intrahospital transport have suffered from VAP, while only 10% of nontransported patients have suffered from VAP. A clinical study by Swanson and Mascitelli (2010) showed that the brain oxygen level is decreased in 54% of patients that underwent transport from the neurointensive care unit (NICU). Studies showed that sicker patients are at higher risks during transports; however, sicker patients are the most likely to require transports out of the ICU. Some studies also linked longer stays in the ICU to intrahospital stays (Day, 2010). Masaryk et al. state that 13% morbidity is associated with transporting critically ill patients (Thomas Masaryk, 2008). Intrahospital transportation, especially to and from the ICU, can be very hazardous to critically ill patients. Transporting patients out of the ICU may require disconnection from a ventilator which may aggravate the respiratory function of intubated patients.

Transporting critically ill patient is inherently risky as such ICU nurses and staffs will usually experience a heightened level of stress and anxiety while transporting or planning to transport a critically ill patient out of the ICU. Studies showed that the key to a safe intrahospital transport is better planning and trained staff. Hospitals have implemented guidelines for intrahospital transport (Jonathan Warren, 2004); maintaining and enforcing these guidelines requires additional resources. The Agency for Healthcare and Quality (AHRQ) provides the 5 Ws (why, who, what, when, where). The questions that should be answered before any transport are as follows: *Why* is the transfer needed? *Who* is the patient being transported and *who* will be in charge of caring of the patient? *What* is needed to make sure that the transport is safe? *When* will transport happen? *Where* will the patient be taken to?

Planning a transfer to and from the ICU is very costly considering the number of equipment needed to provide patient critical care. In few cases, a specialized team of highly experienced staff may be called upon to help with the patient transport (Lora and Ott, 2011). In addition, transporting patient out of the ICU will deprive the unit from some of its staff. In some cases, critically ill patients will require to be accompanied by some of the senior staff of the ICU unit. Specialists may on a rare occasion be called to assist in the transport. Some of the nurses may have to remain with the patient during his stay outside the ICU. This will represent a challenge to the ICU staff and management trying to deal with the partial loss of the resources that needed to be present in the ICU at all times (Agrawal, 2010). Monitoring should be maintained during the entire transport time.

The three key elements to mobile scanning are scanner maneuverability, image quality, and radiation safety. The "maneuverability" of the scanner is defined by its size, its power source, and its accessibility. The scanner should be capable of producing *image quality* that is comparable to stationary scanners. "Radiation safety" is defined by the level of radiation released by the scanner during patient scanning. Mobile scanning was introduced earlier in the 1990s; the TomoScan, a mobile scanner that was used earlier, is bulky and hard to maneuver (Butler and Piaggio, 1998). The TomoScan was used as intraoperative CT scanner. However, the size of the scanner was a limiting factor on its mobility and accessibility. The CereTom represents a new generation of mobile CT scanners that embodied all three key elements of mobile imaging. Its size makes it very suitable for mobile scanning. The scanner size is 1.5 m ×1.33 × 0.73 allowing it to be stationed in the ICU for ease of access. The scanner size allows it to be stored in the ICU next to other mobile devices as shown in Figure 24.3. The scanner is battery powered as such the scanner does not need to be plugged during scanning. The batteries can be charged by plugging the scanner to a wall outlet when

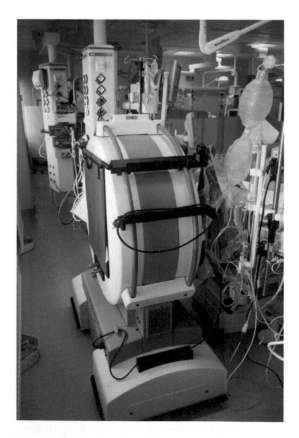

FIGURE 24.3 The CereTom in the ICU unit.

in storage mode. These features make the scanner highly *maneuverable*. During patient scanning the scanner is moved to the patient room. A head board is attached to the end of the patient bed. The patient is then carefully moved, so their head is resting on the head board. The scanner is then positioned over the patient head. Figure 24.4 shows the scanning of critically ill patient in the ICU.

The CereTom image quality was assessed using the American College of Radiology (ACR) accreditation phantom. Rumboldt et al. reviewed the image quality of several portable scanners including the CereTom. The CereTom was evaluated using five key image quality parameters: (1) the CT number accuracy, (2) slice thickness, (3) low-contrast detectability, (4) uniformity, and (5) high-contrast resolution (Rumboldt and Huda, 2009). The study concludes that the CereTom images are clinically acceptable and diagnostically accurate. The CereTom image quality was also assessed using the Catphan 600. The assessment showed that the scanner produced clinically acceptable images (Gingold and Dave, 2008). In both cases the image quality assessments were done using scan protocol with scan doses that are below the recommended ACR dose values. Figures 24.5 and 24.6 show scans of a normal and diseased brain using the CereTom.

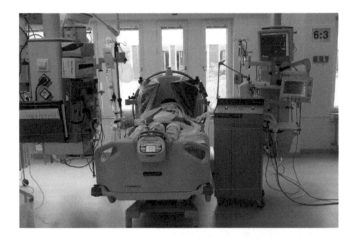

FIGURE 24.4 Patient scanning using the CereTom.

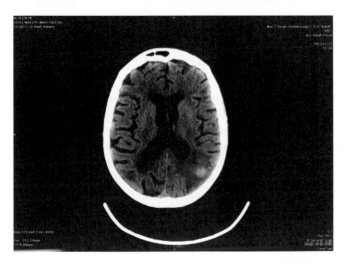

FIGURE 24.5 Scan of a patient with acute intraparenchymal hemorrhage involving the left parieto-occipital lobe.

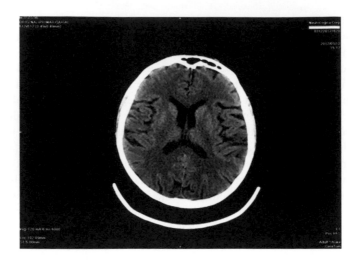

FIGURE 24.6 Normal brain scan.

The increased awareness to the dose delivered by CT scanners to both the patient and the public makes radiation safety the main concern in mobile scanning. The CereTom is designed with enhanced features to increase *radiation safety* to the public around the scanner. The scanner is self-shielded to minimize the scatter radiation. Lead curtains on the front and the back of the scanner reduce the scatter radiation from the scanned patients. The scanner is also equipped with a mobile workstation that can connect wirelessly to the scanner providing the CT technician an ample space away from the scanner. A programmable time delay allows the CT technician to initiate the scan from the scanner and provide enough time to reach a safe distance away from the scanner. Additional features prevent accidental initiation of x-ray. For instance, x-ray is disabled during transfer or storage mode. Access to the scanner is limited; a password is needed to enable the scanner (Gingold and Dave, 2008) .

Since its introduction in 2004, the CereTom is being used in ICUs in the United States and around the world. Its dedication to head scanning makes it very useful in NICU. After traumatic brain injury, a follow-up head CT scan is needed to monitor expanding epidural or subdural hematoma. Follow-up head CT is also needed to monitor cases of acute hydrocephalus. Brain swelling or cerebral edema is another reason for follow-up head CT. These patients often need serial head CT scans to monitor the brain for complications and therefore may require frequent trips to the radiology department jeopardizing patient safety (Day, 2010). The use of the scanner in ICUs reduces the patient's complications due to transport from the ICU to the CT room and back, especially patients attached to respirators and other equipment. For instance, following traumatic brain injuries or tumor removal, follow-up head CT scans are required for monitoring the brain for complications. Figure 24.4 shows the scanner in use. The patient appears to be intubated, yet the CT technician and the ICU staff were able to position the scanner and scan the patient.

A good example for the benefit of using the CereTom in the NICU can be illustrated using the study of Swanson and Mascitelli (2010), which showed that the oxygen level in the brain tissue is reduced after undergoing a transfer from the ICU. A follow-up study by Kaitlin Peace and Maloney-Wilensky (2011) showed that the portable head CT had little effect on ICP; it also showed that the mean level of oxygen in the brain tissue was similar before and after the head CT scan.

A total of 3421 portable head CTs were performed in the period extending from June 2006 to December 2009, and 3278 scans were done in the NICU at the University of New Mexico, Department of Neurosurgery. The scanner was used for both routine and emergency scans. It was found that the head CT scan can be reliably performed using the CereTom (Yonas, 2011). At the JPN Apex Trauma Center in India, a total of 1292 head scans were done over a period of 9 months. The average scan time was 12.6 min with scan time ranging from 7.8 to 47 minutes. Image quality was judged to be excellent. Despite the occasional breakdown of the scanner, the study shows that the mobile CT is very useful in the management of patients with severe traumatic brain injuries, and it can be recommended for high-volume scanning (Deepak Agrawal, 2011).

The economic and the clinical benefits of portable head CT scans were the subject of a study done by Cleveland Clinic. The study was done on a total of 502 head scans performed on ICU patients using the CereTom portable scanner. The study showed that the introduction of a portable CT scanner in the ICU is economically feasible. The scanner provided a full return on investment in 6.9 months; the 5-year expected economic benefit is $2,619,290. The economic benefit is mainly due to the following factors: (1) the reduction of adverse events that can result from intra-hospital transports, (2) the reduction of the number of personnel needed for intrahospital transfer, and (3) increasing the throughput of stationary scanner through the use of portable CT in the ICU (Thomas Masaryk, 2008).

Mobile CT scanners brings CT imaging to the patient's bedside reducing the risk of any adverse clinical events that can result from a mishap during intrahospital transfer to the radiology department. Mobile imaging reduces interruptions to patient's critical care. It also provides doctors and ICU staff with real-time imaging that allow them to adjust the patient treatment management as needed. Furthermore, mobile imaging is proven to be economically feasible.

PORTABLE CT IN THE OPERATING ROOM

Simple forms of imaging, such as 2D x-ray and ultrasound, have been used during surgery for many decades. However complex (3D) forms of imaging, such as MRI and CT, have been used sparsely, mainly due to the limitation of having a special OR built with a large fixed 3D system installed. Even in such rooms the patient is often transported during surgery from an OR table to the imaging table and back. MRI has also had the limitations with limiting the presence of ferromagnetic material, weak magnet strength, and lack of sufficient head coils to accommodate open cranial patients, just to name a few. But the advent of portable CT has allowed the presence of intraoperative CT in almost any OR and for any form of surgery. Still in its infancy, it has been quickly adopted for various forms of neurosurgery. The following paragraphs will highlight three main areas of intraoperative usage, namely, tumor resection, DBS surgery, and complex spine surgery.

Brain Tumor Resection

There are estimated to be approximately 70,000 new cases of primary brain tumors each year in the United States (American Brain Tumor Association, n.d.). This includes tumors in the brain and central nervous system (CNS). It is the second leading cause of death in adolescents under the age of 20. Approximately, one-third of these tumors are malignant. With an average 10-year survival rate, this results in a population of over 700,000 people living in the United States with a brain tumor diagnosis. This does not include secondary metastatic tumors in the brain and CNS that are much more prevalent. Resective surgery is sometimes a great option for a patient, and thus thousands of such surgeries are being done every year. The morbidity and mortality associated with brain tumor resection surgery are high. But one of the great challenges with brain tumor surgery is the ability to visualize and resect the entire tumor during the surgery. Often the surgeon can only rely on visual sight of the tumors with the help of microscopes. Two-dimensional versions of imaging, such as planar x-ray and ultrasound, are entirely inadequate. This results in a large number of surgeries that result in only partial resection. Residual tumor is located by means of postoperative MRI, CT, or PET/CT. This leads to the added expense and danger of repeat surgery.

Portable CT imaging in the OR has become a fast growing tool in the visualization and quality control of brain tumor resection. Anytime during surgery, or at least prior to close, the surgeon can have a brain CT of the patient to visualize remaining tumor. Tumor visualization can often be

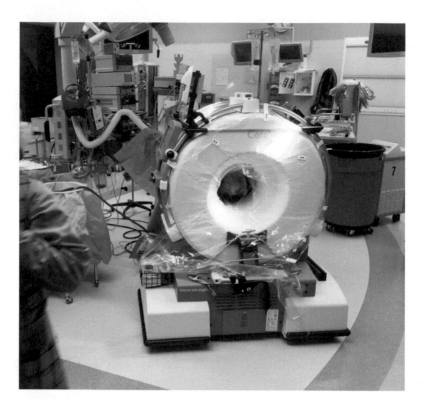

FIGURE 24.7   The scanner in the operating room.

enhanced by injection of a small (50–100 cc) bolus of iodinated contrast. The CT scans can also be coregistered with surgical navigation systems, biopsy needles, robotic surgery devices, and preop MRI images. The result is a very high confidence rate of surgery and low number of repeat surgeries. An added benefit is detecting dangerous complications such as hemorrhage, ischemia, and severe edema (Figures 24.7 and 24.8).

DBS SURGERY

Neurological motor disorders are very prevalent worldwide, and are increasing in prevalence. This includes disorders such as Parkinson's disease (PD), essential tremor, and dystonia. It is estimated that over 1 million people in the United States are affected by PD and the number is growing as the population becomes more aged (Parkinson's Disease Foundation, n.d.). The exact cause of these disorders remains a mystery, but in the 1970s a link between PD and the prevalence of a neurotransmitter called dopamine was discovered. This discovery led to the only pharmacological treatment of PD with a drug called levodopa, which increases the available dopamine and thus minimizes side effects of the disease such as shaking and tremors. In the late 1990s, Medtronic Corporation released a medical device that consisted of an electrode implanted deep in the basal ganglia and controlled by an external programmable electrical stimulator. This low-level electrical stimulation has proven to be effective in the treatment of PD and essential tremor. It is also being used in dystonia albeit that it is not FDA approved for such treatment as yet. In the last decade alone over 80,000 surgeries have been performed (Dr. Andres Lozano, 2010).

Despite its effectiveness, the number of DBS surgeries has been limited for several reasons: The surgery is long and costly; it requires a high level of proficiency and training by the surgeon; and the patient must remain awake for periods of the surgery. All of these limitations are caused by the need to accurately guide and place the electrode using microelectrode recordings, that is, listening to the

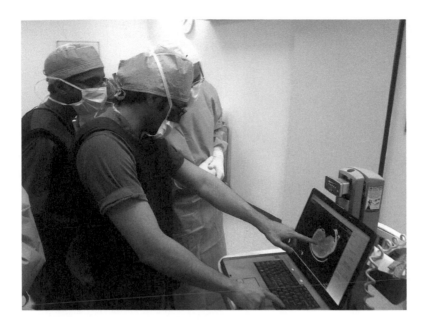

FIGURE 24.8   Intraoperative scan using the CereTom in operating room.

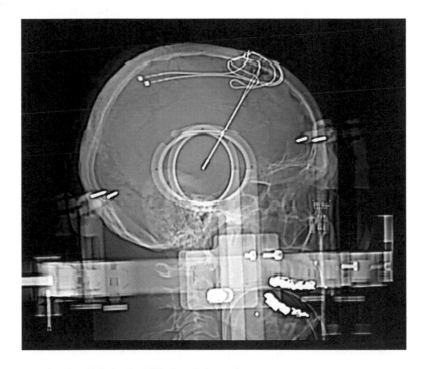

FIGURE 24.9   Brain stimulation using CT-placed electrodes.

electrical activity of the electrode to decide where it is located in tissue. Also the target area is quite small, effectively being less than a few millimeters. But since the advent of the portable CT, some surgeons have opted to "guide" the electrode into location with the assistance of CT imaging. The results so far have shown a much shorter and easier operation while also keeping the patient sedated (Kim Burchiel, 2013) (Figure 24.9).

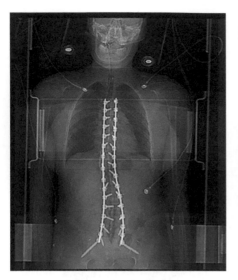

FIGURE 24.10   Spine alignment.

INSTRUMENTED SPINE SURGERY

Back pain is one of the most common medical conditions in the United States. It is reported that approximately four out of five people in the United States will at some time miss work due to back pain (Dartmouth-Hitchcock, 2014). While most cases only last a short while, some patients experience permanent pain and immobility due to conditions such as degenerative disk disease, severe disk herniation, bone spurs, and fractures, just to name a few. In the case of severe mechanical damage, most patients opt for some type of fusion surgery that involves the implantation of metal in various forms such as rods, screws, cages, and plates. The surgeries are often accompanied by some type of bone graft so as to form a more solid and lasting fusion in the months to come. In addition, there may be some removal of bone, laminectomy or laminotomy, to provide more space for the nerve root to exit the spinal column without impingement.

Even though instrumented spine surgery dates back into the 1950s for the cases of severe deformities such as scoliosis, it is not a simple procedure and has not always achieved effectiveness. One major cause of ineffectiveness is the nonoptimal placement of the metal hardware, namely, pedicle screws and interbody cages. Because they have to be drilled and screwed at such oblique angles, with minimal room for angular error, they have been hard to achieve 100% placement accuracy. X-ray C-arms have been the staple for imaging in the last two decades but suffer from only 2D accuracy.

The advent of portable CT has allowed surgeons to now make 3D images at any time during a complex spine procedure so as to maximize screw placement accuracy. Further integration of the CT images with surgical navigation systems and surgical robots has improved the procedure even further. The portable CT can thus account for any possible movement of the patient's spine during the surgery. The portable CT provides the added benefit of doing the post op scan to help visualize any complications such as hemorrhage or misplaced screws or fracture. It is the hope and promise that portable CT will not only make these procedures safer for the patient, but will also contribute to better outcome due to the improved mechanical accuracy of the fusion procedure (Figure 24.10).

PORTABLE CT IN THE AMBULANCE

Ischemic stroke is one of the leading causes of death and disability worldwide. It occurs when an artery delivering blood to the brain becomes occluded causing cerebral hypoxia. About 90% of all strokes are ischemic in nature. Presently, thrombolysis with recombinant tissue plasminogen activator (r-tPA) is

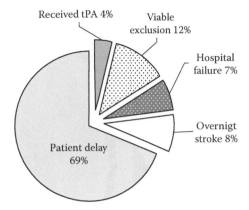

Received tPA 4%
Viable exclusion 12%
Hospital failure 7%
Overnigt stroke 8%
Patient delay 69%

FIGURE 24.11 Pie chart representing the different clinical outcomes from stroke care. (From CASPR, *Neurology*, 64, 654, 2005. With permission.)

the only approved treatment for ischemic stroke. Its chief function is to facilitate blood flow through the occluded region until surgical intervention can alleviate the blockage. However, according to the American Stroke Association, the "benefit of [intravenous r-tPA] therapy is time-dependent and treatment should be initiated as quickly as possible" for a Class I acute stroke (NINDS, 1995; Jauch, 2013; Saver, 2013). In fact, neurons in the order of millions are lost each second during an ischemic event. Ischemic events that are not treated within an hour of symptom onset generally result in either death or long-term disabilities. Even so, only 15%–40% of patients arrive at the hospital early enough to be considered for thrombolysis treatment (Walter, 2012) (Figure 24.11).

Recent efforts have been focused on expediting the treatment process by bringing the personnel and medical equipment directly to the stroke victim. As such, specialized ambulances have been developed with the imaging, diagnostic, and communication capabilities necessary for acute ischemic stroke evaluation and treatment. These mobile stroke units (MSUs) are outfitted with a mobile CT scanner and a point-of-care laboratory along with well-trained staff of neurologists, physicians, and radiology technicians. Typically the emergency response sequence begins when the decision is made to deploy the MSU. Once the emergency response team first engages the patient, an initial evaluation will be performed to further assess the patient's condition. Upon the results of the initial evaluation, the staff can decide whether to administer r-tPA to the patient, which can be carried out inside the MSU while the patient is being transported (Fassbender, 2013). During the transportation phase of the emergency response sequence, the response team also has the ability to transfer real-time patient data from the ambulance to the hospital destination through the MSU's telemedicine capabilities. Thus, the destination of the patient will have all the relevant information needed to further accelerate treatment.

Perhaps the most important component of an MSU is the mobile CT scanner. In fact, even if during the initial evaluation the response team suspects an ischemic stroke event, thrombolysis treatment cannot be performed without the validation that there is no evidence for acute intracranial hemorrhage, a validation that requires the analysis of CT images of the patient's head. With a readily available CT scanner, the emergency response team can promptly perform this validation. In addition, CT images can provide an immediate diagnosis of other brain diseases/conditions that may be affecting the patient, such as trauma, subarachnoid hemorrhage, epidural/subdural hematoma, and tumor, which will impact the treatment decisions during the rest of the response sequence (Figure 24.12).

Installing a CT scanner inside an MSU, however, is not a trivial task. It is a challenging undertaking, both in terms of feasibility and efficacy. The relatively small space within an MSU demands that the CT scanner be smaller than conventional scanners. In addition, the scanner must be mobile in that it must be able to translate in relation to the stationary patient bed during a scan protocol and

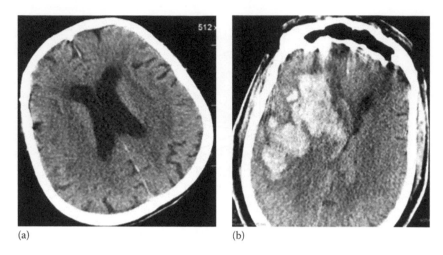

(a)   (b)

FIGURE 24.12 CT images of patient's head taken by MSU developed by System Strobel (a). The image demonstrates an extensive right frontal lobe and basal ganglia acute intraparenchymal hemorrhage with leftward midline shift and effacement of the right lateral ventricle (b). (From Kostopoulos, P.E., *Neurology*, 78(23), 1849, 2012. With permission.)

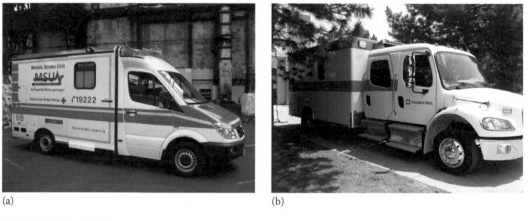

(a)   (b)

FIGURE 24.13 MSU developed by System Strobel (a) and Cleveland Clinic MSU (b).

still produce stable images that can be interpreted with ease by the radiologist. This is especially difficult given the fact that the ambulance may not be situated on the most ideal terrain for such a procedure. Two MSUs currently being employed in Europe, the MSU developed by System Strobel and the Cleveland Clinic Stroke Mobile Unit (Figure 24.13), use the CereTom CT scanner (NeuroLogica Corporation, Danvers, MA) for their brain imaging procedures. Past studies (Rumboldt and Huda, 2009; Peace, 2010) have proven that the CereTom scanner (Figure 24.4) is robust and reliable enough to handle such rigorous conditions in the field while producing images of acceptable image quality. Furthermore, the small dimensions of the CereTom in relation to conventional scanners make it an ideal candidate to be integrated into an MSU.

Another challenge with point-of-care stroke treatment arises when considering the logistics involved with such a specialized emergency response system. For instance, the decision as to whether to dispatch the MSU versus a traditional emergency vehicle such as an ambulance cannot be made lightly. Dispatching the MSU when it is not required will render it inaccessible for other legitimate stroke cases. Additionally, the MSU must also be able to reach the stroke victim as quickly as possible for the advantages of point-of-care evaluation, diagnosis, and treatment to be felt. Even after

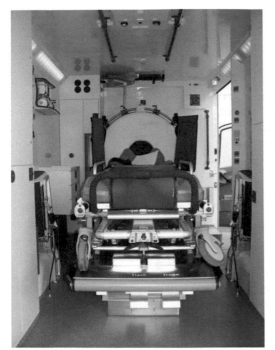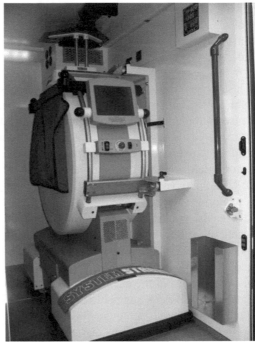

FIGURE 24.14 Images of the CereTom CT scanner inside the MSU developed by System Strobel.

reaching the victim, a protocol must be developed to decide which hospital to transport the patient to and the appropriate data links must be established with that site to ensure proper transfer of patient data. Much of these factors are highly dependent on the environment in which the stroke emergency response system is being implemented. Urban areas, for example, may stipulate a higher number of MSUs than rural areas due to the higher population density. In contrast, in more rural areas where hospitals and treatment clinics are sparse, the strategic selection of the location of the MSU deployment centers may prove more vital than the number of units they hold. Age demographics could also play a factor with the selection of the ideal infrastructure. For instance, areas that have predominantly older inhabitants may have a higher rate of stroke and therefore more of a need for MSU response systems than areas with younger inhabitants. To truly understand the impact of stroke emergency response, it is imperative that studies evaluating these systems take place in different environmental settings to account for the demographic/geographical discrepancies (Figure 24.14).

In a 3-month study conducted in Berlin, Germany, in 2011 using the sroke emergency mobile (STEMO), the feasibility of prehospital stroke care was assessed (Weber, 2013). During this 3 month period, the dispatch center deployed the STEMO based on the results of a stroke identification algorithm for each alarm that was raised. The weeks in which the STEMO was used were randomly selected, with the conventional ambulances being used for the remaining weeks. Upon completion of the study, it was deemed that once the decision was made to deploy the STEMO, there was a 75% probability that it will reach the emergency site within 16 minutes of symptom onset. Out of the 45 patients treated in STEMO for an acute ischemic stroke, 23 patients were administered r-tPA treatment. The average time between the alarm-to-*treatment* was 36 minutes less (~62 minutes) than that from 50 consecutive patients treated in 2010 without the use of STEMO (~98 minutes). A separate study conducted in 2012 investigating the efficacy of the MSU response system compared to the conventional system found similar results (Saver, 2013). The study recorded the alarm-to-*therapy* decision time for several cases of MSU deployment and found that, on average, this time was reduced by 50% when the MSU was deployed (~35 min compared to 76 min). Fifty-seven percent of the patients that were engaged by the MSU had a symptom onset-to-therapy decision time of less than one hour compared to only 4% of patients who received conventional care. Ultimately, both studies concluded that the MSU concept provides a safer, more efficacious alternative than conventional stroke emergency response and could potentially be integrated into a metropolitan EMS system.

With the introduction of mobile units specializing in stroke treatment, the modern healthcare industry has rejected the notion that ambulances are merely conduits between the patient and treatment center, but can also be effectively utilized as treatment vessels. However, like many novel ideas, gaining global traction requires continual testing of the stroke emergency response systems that deploy these ambulances, especially in different environments. Further trials are presently being conducted to discern treatment efficacy as well as whether the healthcare benefits justify the increase in costs that come with the implementation of such systems. Most recently, Russia developed an MSU in anticipation of the 2014 Winter Olympic Games in Sochi. An MSU program is also set to begin in Alberta, Canada as well as in the Health Science Center at Houston Medical School (UTHealth) and the Cleveland Clinic. In addition to the evaluation of treatment efficacy and infrastructural factors, there is a need for more testing of the stroke identification algorithms that are used to deploy MSUs in the first place. Presently, very few studies have gone into a detailed analysis of these algorithms (Krebes, 2012). Since many of the algorithms involve parameters that are based on the observations of the emergency caller and are rather subjective, there is the potential possibility for false positives. A comparative ROC analysis of these algorithms may prove beneficial not only in terms of accurate deployment of the MSU, but also in terms of the infrastructural requirements for a given area. Algorithms that are proven to be reliable may decrease the number of MSUs that are needed to service an area since the chances of false deployment will be low. Ultimately, the algorithms will never perfectly identify every case of stroke. This uncertainty will manifest in a tradeoff; more conservative algorithms will ensure that the majority of stroke victims are cared for by the stroke emergency response system, while algorithms with stricter restrictions may mean less false positives but more false negatives.

Along with ambulances, there have also been talks of integrating the components of an MSU with other modes of transportation such as helicopters and boats. This could be potentially advantageous for the treatment of stroke victims that are in more remote places only accessible by air or water transportation. Of course, incorporating a mobile CT scanner with these vehicles will entail different challenges than those associated with an ambulance, that is, the weight restrictions of the helicopter and the motion of the boat in water. Nevertheless, effective stroke response demands that the stroke victim be engaged with the necessary personnel, equipment, and therapy in as short a time window as possible. As the medical community continues to gain confidence with this idea, it is inevitable that even these challenges will be overcome, offering a future where stroke treatment success depends less on how quickly the patient is administered to a treatment center and more on the inherent condition of the patient.

PORTABLE SPECT IN THE EPILEPSY MONITORING UNIT

Epilepsy is the general term for a group of long-term neurological disorders affecting approximately 2.3 million people in the United States and costs upwards of $15 billion in medical and other indirect expenses (Center for Disease Control and Prevention, 2011). About a third of patients develop refractory epilepsy, which means that medication does not work well, or at all, to control the seizures. Resective surgery has been effective in treating refractory epilepsy, but needs accurate localization of the epileptogenic focus. Here, we describe the merits of a specialized, high-resolution, and portable SPECT scanner to enable this localization.

Characteristic symptoms of epilepsy are debilitating, involuntary seizures, and an overall detrimental effect on the patient's quality of life. The clinical signs and symptoms of epilepsy depend on the location of the electrical discharge and its propagation in the brain. Most seizures arise from the temporal lobes and are usually accompanied by symptoms such as aphasia, swallowing, tooth grinding, and staring spells (Camargo, 2001). Patients often have to live with these symptoms since childhood and endure the stigma associated with suffering from a brain disorder.

To formulate a treatment plan for refractory epilepsy, it is vital to locate the area of the brain that initially "seizes" during an epileptic episode. In focal epilepsy, the area that initially seizes is called the "ictus." Ideally, if the *ictus* is located accurately and surgically removed or isolated then the epileptic circuit will never be triggered. The functionality of the brain is restored with time as the brain rewires its connections, but it is desirable to remove as little as possible for minimal side-effects. In the past, localization of the seizure focus was done with the help of invasive intracranial electroencephalography (EEG) and/or scalp EEG and their correlation with structural imaging modalities such as MR. The introduction of noninvasive, functional imaging modalities such as SPECT has dramatically changed the presurgical epilepsy evaluation process (Lynch, 2011).

SPECT Imaging in Epilepsy

In recent decades, there has been a great deal of study toward optimizing two primary functional imaging techniques, PET and SPECT, for imaging epileptic patients with the goal of detecting focal disturbances. The role of these modalities in epilepsy imaging is not the diagnosis of the disease but the localization of the seizure focus for surgical intervention. PET is primarily used to image glucose metabolism rate, while SPECT images the regional cerebral blood flow (Kim, 2011). PET studies are performed when the patient is in the *interictal* or seizure-free state, while SPECT has the unique capability of imaging blood flow functional changes during a seizure or in the *ictal* state. PET scans with flourine-18 (fludeoxyglucose) show an area of focal activity reduction, while SPECT images show hyperperfusion in the epileptic circuit. Often, the patient also undergoes an *interictal* SPECT scan. When the *ictal* and *interictal* scans are compared, the area containing the *ictus* shows a much higher tracer uptake in the SPECT image, whereas the rest of the brain image shows minimal deviation. In clinical practices, the difference image is super-imposed on the MRI or CT of the brain to gain an anatomical reference for the *ictus*. Brain SPECT scans are generally performed with tracers such as technetium-99m (Tc-99m) hexamethyl-propylene-amine-oxime (HMPAO). It is important to note that SPECT scanners provide the only practical means of imaging brain function during or soon after seizures. They are also less expensive than a PET scanner and use longer-lived, more easily available radioisotopes in comparison. The disadvantage, with respect to PET, has traditionally been the low resolution of SPECT images (<10 mm SPECT versus 2–3 mm PET), but modern mechanical and algorithmic innovations are slowly bridging the resolution gap.

The InSPira HD Portable SPECT Scanner

It is important that the SPECT images used for diagnosing and localizing the ictus are accurate and of high resolution. While accuracy deals with correctly identifying the location of the ictus, resolution pertains to the size of the affected region such that only the necessary part of the brain is resected for minimizing side-effects to the patient upon recovery. The standard gamma camera SPECT has poor resolution and subpar accuracy when it comes to identifying the location of the ictus. Another requirement that often goes unfulfilled is the portability of SPECT scanners, which is important because patients are often tethered to other monitoring devices.

In 2009, the InSPira HD Scanning Collimator SPECT scanner (Figure 24.15) was introduced as a dedicated, portable brain imaging device that uses a novel method of data acquisition combined with an advanced iterative reconstruction algorithm (Batis, 2011). It has the advantages of superior accuracy, resolution, sensitivity, and portability over existing standard gamma camera SPECT devices. It has been developed by NeuroLogica Corp. (a subsidiary of Samsung Electronics Co.,

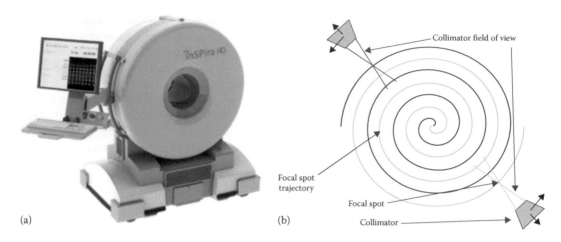

(a)   (b)

FIGURE 24.15   (a) The InSPira portable SPECT system and (b) the helical path inscribed by the focused collimator detectors.

Ltd). Its unique capabilities provide the highest quality SPECT images wherever needed, including the clinic, ICU, OR, and the emergency department. The InSPira HD is capable of imaging for clinical applications such as epilepsy, PD, stroke, and Alzheimer's disease. Its applications also extend to the fields of pediatric and preclinical imaging.

The InSPira uses a 72-detector spiral scanning focused collimator system that comprises two clam-shell halves with 12 detector heads each. Each head, in turn, has 3 PMT detectors. Each detector head has a common focal spot that starts at the isocenter of the field of view (FOV). This is illustrated in Figure 24.15b. The clam shells move in a helical motion as the scan progresses, with each shell rotating and moving outward simultaneously, thereby scanning the entire FOV. The 360-degree coverage enhances system sensitivity, while the moving focus scanning improves the spatial resolution.

InSPira HD is fitted with a battery for uninterrupted power supply that is crucial for areas where a steady electrical supply is unreliable or even unavailable. When used in a hospital setting, the scanner can easily be unplugged without turning it off and transported to the patient for a scan; a feature especially useful when scanning ictal scans where time is of the essence. The interface is simple and intuitive, capable of being used by paramedics, EMTs, neurologists, and technicians. It also has wireless connectivity to remain connected to an image analysis work station when the scanner is in transit.

CLINICAL STUDIES

Dr. M.A. Rossi and his team at Rush University Medical Center (RUMC) carried out an eight patient study with medically refractory epileptic sources (Rossi, 2012). The aim of this study was to compare the resolution and accuracy of InSPira's high-resolution 72-detector focused collimator SPECT technology with a conventional 2-detector fan beam collimator system (Siemens). They also compared Siemens' Subtraction Ictal SPECT Co-registered to MRI (SISCOM) post-processing to InSPira's Relative Ictal SPECT Co-registered to MRI (RISCOM) (Figure 24.16).

InSPira HD's RISCOM post-processing analysis highlighted certain hyperperfused areas not seen in the SISCOM images. For all 8 patients, these new regions of hyperperfusion were concordant with chronic and/or intraoperative electrocorticography (ECoG). This shows that InSPira images showed an improved identification of the extent of the ictal onset zones for resective surgery. All 8 patients showed an Engel's class I or II postsurgical recovery outcome.

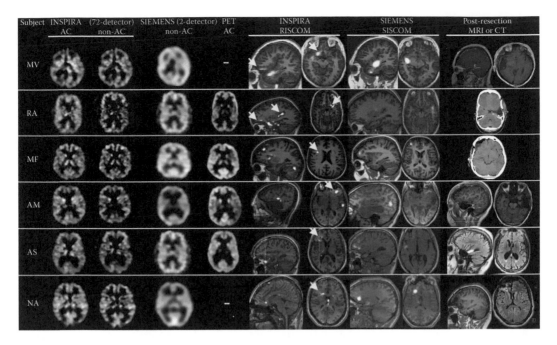

FIGURE 24.16 Comparison of InSPira RISCOM with Siemens' SISCOM. Each row corresponds to a single patient suffering from extratemporal and/or temporal epileptic circuits (Rossi, 2012). The results of attenuation corrected (AC) SPECT images from both InSPira and Siemens' 2-detector SPECT were used in conjunction with the corresponding PET (with AC) and post-resection MRI or CT, as shown in the last column. The yellow arrows indicate additional regions of hyperperfusion at the ictal onset seen only on the InSPira images.

SUMMARY

The application of SPECT imaging in epilepsy is well studied and with the advent of high resolution scanners such as InSPira HD, there has been a demonstrable improvement in localizing the source of this disease and managing patient care. In fact, the introduction of InSPira HD portable scanner as a dedicated brain SPECT imaging device has filled a huge gap in the modern healthcare industry and the treatment of certain brain disorders in particular. The unique data acquisition system allows high sensitivity during data acquisition and high resolution and high contrast during image reconstruction.

It has been shown that the RISCOM processing using InSPira facilitates the identification of the ictal onset in extensive epileptic circuits that were otherwise not seen in a standard gamma camera SPECT or a 2-detector fan beam collimator system. It enables improved localization of the extent of the ictal onset zone for resective surgery.

Novel epilepsy treatment techniques such as responsive cortical stimulation (RMS) can benefit from InSPira's high resolution SPECT imaging in the ictal localization, epileptic circuit mapping and postimplant validation stages (Rossi, 2008) (Engel, 1993). The need for a high-resolution SPECT system is not limited to epilepsy but to other brain disorders and neurodegenerative diseases such as dementia, schizophrenia, and Alzheimer's disease.

CONCLUSION ON PORTABLE NEUROIMAGING

Neuroimaging most often consists of some form of 3D imaging such as MRI or CT. Yet these machines, since their advent in the 1970s and early 1980s have only existed in a large fixed form and placed in the basements of most hospitals. Thus if one had a patient requiring neuroimaging you had to "take the patient to the scanner." In the early 2000s, we have seen the invention of portable forms of 3D neuroimaging that allows one to "take the scanner to the patient". As has been described in this chapter, this allows for a whole host of novel therapies to come about at the patient's location albeit ICU, OR, or even ambulance. Still in its infancy, portable 3D neuroimaging will likely bring about more improvements in neurological treatments in the decades to come.

REFERENCES

Agrawal, S. (August 2010). A portable CT scanner in the pediatric intensive care unit decreeases transfer-associated adverse events and staff disruption. *European Journal of Trauma and Emergency Surgery* 36(4):346–352. doi: 10.1007/s00068-009-9127-8. Epub November 2, 2009.

AHRQ. (n.d.). Epilepsy: Percentage of patients with a diagnosis of intractable epilepsy who were considered for referral for a neurological evaluation of appropriateness for surgical therapy and the consideration. http://www.qualitymeasures.ahrq.gov/content.aspx?id=34262 (accessed January 2011).

American Brain Tumor Association. (n.d.). Brain tumor statistics. http://www.abta.org/about-us/news/brain-tumor-statistics/ (accessed 2014).

Andres Lozano, U. O. (2010). Ask the expert—Deep brain stimulation surgery. http://parkinsonpost.com/2010/06/page/2/.

Batis, J. Z. (May 2011). Performance characteristics of the InSPira HD, a new generation brain-dedicated SPECT camera. *Journal of Nuclear Medicine* 52(1):1934.

Brain Institute. (n.d.). Neuroscience Research at OHSU Brain Institute Gives Hope. http://www.ohsu.edu/xd/health/services/brain/research-training/ (accessed 2014).

Butler, W. E. and C. Piaggio. (1998). A mobile CT scanner with intraoperative and ICU applications. *Neurosurgery* 42(6):1304–1310.

Camargo, E. (2001). Brain SPECT in neurology and psychiatry. *Journal of Nuclear Medicine* 42:611–623.

CASPR. (2005). *Prioritizing interventions to improve rates of thrombolysis for ischemic stroke. Neurology* 64(4):654–659.

Center for Disease Control and Prevention. (2015). At a glance 2015. Targeting Epilepsy. CDC, Atlanta, GA. http://www.cdc.gov/chronicdisease/resources/publications/aag/epilepsy.htm (accessed 2016).

Dartmouth-Hitchcock, S. C. (2014). *Chronic Low Back Pain*. The Dartmouth-Hitchcock Health Difference. http://patients.dartmouth-hitchcock.org/spine/low_back_pain.html (accessed 2014).

Day, D. (August 2010). Keeping patients safe during intrahospital transport. *Critical Care Nurse* 30(4):17–33.

Deepak Agrawal, S. S. (2011). Initial experience with mobile computed tomogram in neurosurgery intensive care unit in a level I trauma center in India. *Neurology India* 59(5):739–742.

Engel, J. (1993). *Surgical Treatment of the Epilepsies*, 2nd edn. Lippincott Williams & Wilkins, New York, NY. ISBN 0-88167-988-7.

Fassbender, K. E. (2013). Streamlining of prehospital stroke management: The Golden Hour. *Lancet Neurology* 12:585–596.

Gingold, E. and J. Dave. (2008). Dose and image quality evaluation of a mobile CT scanner for head and neck imaging. *50th AAPM Annual Meeting*, Houston, TX. Thomas Jefferson University Hospital, Philadelphia, PA.

Jauch, E. E. (2013). Guidelines for the early management of patients with acute ishcemic stroke: A guideline for health-care professionals from the American Heart Association/American Stroke Association. *Stroke* 44(3):870–947.

Jonathan Warren, R. E. (2004). Guidelines for the inter- and intrahospital transport of critically ill. *Critical Care Medicine* 32(1):256–262.

Kim Burchiel, M. O. (2013). New technique for deep brain stimulation surgery proves accurate and safe. https://news.ohsu.edu/2013/06/05/new-technique-for-deep-brain-stimulation-surgery-proves-accurate-and-safe (June 05, 2013).

Kim, S. A. (2011). SPECT imaging of epilepsy: An overview and comparison with F-18 FDG PET. *International Journal of Molecular Imaging 2011*, Article ID 813028.

Kostopoulos, P. E. (2012). Mobile stroke unit for diagnosis-based triage of persons with suspected stroke. *Neurology* 78(23):1849–1852.

Krebes, S. E. (2012). Development and validation of a dispatcher identification algorithm for stroke emergencies. *Stroke* 43(3):776–781.

Lora, K. and L. A. Ott. (June 2011). Intrahospital Transport to the Radiology Department: Risk for adverse events, nursing surveillance, Utilization of a MET and practice implications. *Journal of Radiology Nursing* 30(2):49–52.

Lynch, B. O. (November–December 2011). Correlation of 99mTc-HMPAO SPECT with EEG monitoring: prognostic value for outcome of epilepsy surgery in children. *Brain and Development 17*:409–417.

Nicolas Bercault, M. W.-C. (November 2005). Intrahospital transport of critically ill ventilated patients: A risk factor for ventilator-associated pneumonia—A matched cohort study. *Critical Care Medicine 33*(11):2471–2478.

NINDS. (December 1995). *Tissue Plasminogen Activator for Acute Ischemic Stroke.* The National Institute of Neurological Disorders and Stroke rt-PA Stroke Study Group. *The New England Journal of Medicine 333*(24):1581–1587.

Parkinson's Disease Foundation. (n.d.). Statistics on Parkinson's. http://www.pdf.org/en/parkinson_statistics (accessed 2014).

Peace, K. (2010). The use of portable head CT scanner in the intensive care unit. *American Association of Neuroscience Nurses 24*(2):109–116.

Peace, K. and E. Maloney-Wilensky. (2011). Portable head CT scan and its effect on intercranial pressure, cerebral perfusion pressure and brain oxygen. *Journal of Neurosergury 114*:1479–1484.

Rossi, M. A. (2012). Improved localization of extratemporal ictal onset-associated blood flow changes using a 72-detector scanning focused collimator SPECT system. *American Epilepsy Society*, abstract no. 1.182.

Rossi, M. H. (2008). Subtracted activated SPECT validates depth lead placement in white matter for responsive neurostimulation therapy in refractory partial-onset epilepsy. *Epilepsia 49*(7):380, abstract no. 3.102.

Rumboldt, Z. and W. Huda. (2009). Review of portable CT with assessment of a dedicated head CT scanner. *American Journal of Neuroradiology 30*:1630–1636.

Saver, J. L. (2013). Time to treatment with intravenous tissue plasminogen activator and outcome from acute ischemic stroke. *Journal of the American Medical Association 309*(23):2480–2488.

Seton Brain and Spine Institute. (n.d.). Evaluation and referral for epilepsy surgery. http://www.setonbrainand-spine.com/treatment/epilepsy/epilepsy-surgery.

Stevenson, V. and C. F. Hass. (2002). Intrahospital transport of the adult mechanically ventilated patient. *Respiratory Care Clinics of North America 8*(1):1–35.

Swanson, E. W. and J. Mascitelli. (May 2010). Patient transport and brain oxygen. *Neurosurgery 66*(5):925–932.

Thomas Masaryk, R. K. (March–April 2008). The economics and clinical benefits of portable head/neck CT imaging in the intensive care unit. *Radiology Management 30*(2):50–54.

Walter, S. E. (2012). Diagnosis and treatment of patients with stroke in a mobile stroke unit versus in hospital: A randomised control trial. *Lancet Neurology 11*:397–404.

Weber, J. E. (2013). Prehospital thrombolysis in acute stroke results of the phantom-S pilot study. *Neurology 80*:1–7.

Wikipedia. (n.d.). *Traumatic Brain Injury.* https://en.wikipedia.org/wiki/Traumatic_brain_injury (accessed 2014).

Yonas, A. P. (2011). Portable head computed tomography scanner–technology and applications: Experience with 3421 scans. *Journal of Neuroimaging XX*(X):1–8.

25 The Deep Brain Connectome

Ifije E. Ohiorhenuan, Vance L. Fredrickson, and Mark A. Liker

CONTENTS

INTRODUCTION

The basal ganglia are a set of subcortical nuclei with strong recurrent connections between the cerebral cortex, the thalamus, and the brainstem. Abnormalities in basal ganglia circuitry have been implicated in a diverse set of neurological and psychiatric disorders. These include movement disorders, such as Parkinson's disease, Huntington's disease, dystonia, and essential tremor; behavioral disorders, such as Tourette's syndrome and obsessive–compulsive disorders; and psychiatric disorders such as schizophrenia, depression, and addiction (Ring and Serra-Mestres 2002). Many of these disorders have been treated by surgically targeting the basal ganglia using deep brain stimulation (DBS), pallidotomies, or thalamotomies. Consequently, a thorough understanding of the basal ganglia can provide insight not only into the functioning of the brain but also into the treatment of a number of neurological and psychiatric disorders. Recent advances in human brain mapping techniques offer to advance our understanding significantly by generating detailed connectivity maps of the entire brain—the connectome. The deep brain connectome, the wiring diagram of the basal ganglia, its projections, and adjacent white matter tracts, will provide a number of insights into the diagnosis and surgical treatment of basal ganglia disorders. This chapter will review the reasons for studying the deep brain connectome, the current approaches for analyzing the connectome, and the potential of the deep brain connectome for advancing the surgical treatment of basal ganglia disorders.

BASAL GANGLIA CIRCUITRY

The basal ganglia consist of the caudate nucleus, the putamen, the globus pallidus, and the nucleus accumbens. The caudate and putamen, which together comprise the striatum, receive the majority of afferent input to the basal ganglia, including projections from the cortex, the substantia nigra, and the amygdala. The basal ganglia circuitry is notable for consisting of multiple parallel loops involving connections between cortical association areas, through the basal ganglia to the thalamus and back to the cortex. Initially, five parallel circuits were defined: the "motor circuit," the "oculo-motor circuit," the "dorsolateral prefrontal circuit," the "lateral orbitofrontal circuit," and the "anterior cingulate circuit" (Alexander et al. 1986). Since then additional basal ganglia–thalamocortical circuits have been identified. These circuits are integral to motor learning, sequencing of movements, attention, working memory, and learning. Our understanding of the basal ganglia has relied on not only animal models and imaging methods but also, importantly, clinical studies of patient undergoing surgical procedures, which have, sometimes serendipitously, uncovered critical roles for the basal ganglia in modulating human disease.

SURGICAL TREATMENT OF BASAL GANGLIA DISORDERS

Surgical interventions for basal ganglia disorders dates back to the late 1930s where resections of the caudate head were used to treat Parkinson's disease (Goetz 2011). After an incidental sacrifice of the anterior choroidal artery (which supplies the globus pallidus) led to an improvement in a patient's tremor and rigidity, ligations of the anterior choroidal artery were used to treat patients with Parkinson's disease (Cooper 1961). In the early 1950s, the rise of the stereotactic frame allowed for the precise lesioning of the globus pallidus and thalamus were found to improve the motor symptoms of Parkinson's disease (Goetz 2011). In the late 1980s, the accidental discovery that high-frequency stimulation of the ventral intermediate nucleus of the thalamus decreased tremor in Parkinson's disease patients, led to the advent of DBS (Benabid et al. 1987). Surprisingly, despite being in use for almost 30 years, how DBS modulates brain activity is incompletely understood. Early hypotheses for the mechanism of DBS centered on the activation (Lozano et al. 2002) or inhibition (McIntyre et al. 2004) of activity within the nuclei being stimulated. More recently, however, it has begun to emerge that DBS jams aberrant network activity, disrupting pathological oscillations and facilitating function (McIntyre and Hahn 2010). This finding highlights the importance of understanding the connectivity pattern as well as the dynamics of basal ganglia circuits and help understand and treat disorders of the basal ganglia.

INTRODUCING THE CONNECTOME

The "connectome" was first proposed as the "comprehensive structural description of the network of elements and connections forming the human brain" in 2005 (Sporns et al. 2005). Inspired by the seminal work of mapping the connectivity diagram of the nematode (White et al. 1986), Sporns et al. proposed a similar comprehensive mapping of connections in the human brain. The motivation for this proposal was that, at the time, despite considerable research into neuroanatomy, an understanding of the fine connectivity patterns and laminar projections between segregated cortical areas was lacking (Sporns et al. 2005). The connectome was proposed as a standardized, publicly available database mapping the structure of neuronal units and their connections. The connectome would therefore modernize and quantify the study of neuron anatomy, an area of inquiry dating back to the ancient Egyptians (Atta 1999).

Constructing a true connectome of the human brain is a daunting prospect. It is estimated that the brain contains 10^{11} neurons with an average of 8000 synapses per neuron and 10 million kilometers of wiring (Murre and Sturdy 1995). The experimental techniques to achieve a complete synapse by synapse map of the brain as well as the computational capacity to store this information vastly outstrip our current capabilities. Consequently, any "connectome" is necessarily an approximation. These approximations, however, are complicated by the observation that the brain displays many of the features of a small-world network (Watts and Strogatz 1998) with groups of densely connected neurons spanned by sparse, long-range connections (Bassett and Bullmore 2006).

TOOLS FOR CONNECTOMICS

The tools for constructing a connectome can be broadly separated into two groups: histological preparation and neuroimaging. Histological studies of the nervous system date back to over a hundred years ago with the seminal work of Cajal (DeFelipe 2002). Axonal tracing has been a mainstay in elucidating connectivity patterns in the animal models. Currently, this is performed using two-photon imaging of fluorescent labeled axons combined with automated imaging segmentation and registration (Oh et al. 2014). This approach successfully led to the construction of the Allen Mouse Brain Connectivity Atlas, a mesoscale connectome of the mouse brain (Oh et al. 2014). Similarly, postmortem tract tracing has been used in humans to study intrinsic connections within a brain region (Tardif and Clarke 2001). However, this tends to be limited to fairly short ranges and is of minimal use in understanding connections between distant brain regions (Beach and McGeer 1987). As a result, the most informative studies of connectivity in the human brain have relied on magnetic resonance imaging (MRI) modalities including task-based functional MRI, resting-state functional MRI (rfMRI), and diffusion MRI (dMRI) (Ugurbil et al. 2013).

Functional MRI (fMRI) uses cerebral blood-oxygen-level-dependent changes in an MRI signal to construct a spatial map of the brain. If these signals are recorded while a subject performs a task, this generates a map of brain areas engaged by that task, the tfMRI. rfMRI uses temporal correlations in the spontaneous fluctuations of a fMRI time series to construct a spatial map of functional connectivity (van de Ven et al. 2004). dMRI uses the anisotropy (directional dependence of diffusion) of water molecules within fiber tracts to construct a map of structural connections within the brain (Jbabdi and Johansen-Berg 2011). These methods are complementary and can be used to construct connectivity maps of the brain.

Although widely used and accepted, a critical caveat to the use of fMRI techniques for the construction of connectomes is that, inherently, they rely on indirect measures of connectivity. Unlike the direct visualization of axonal trajectories provided by microscopy, voxels in a diffusion MR image often contain thousands of axons requiring that the tractography algorithm choose a dominant orientation out of a number of possibilities. This can occasionally lead to spurious or missing connections (Jbabdi and Johansen-Berg 2011). These errors are compounded by variations in imaging equipment, scanning protocols, and tractography algorithms leading to difficulty in directly comparing results from different laboratories. The National Institutes of Health (NIH) Human Connectome Project aimed to minimize these confounds and potential sources for error by using a well-thought-out, standardized set of protocols and equipment to generate high fidelity imaging data from a large number of individuals (Ugurbil et al. 2013).

THE HUMAN CONNECTOME PROJECT

Launched in 2012, with an NIH-sponsored budget of approximately $40 million, the Human Connectome Project aims to create a wiring diagram of the human brain using high-resolution scanners and standardized protocols. Currently, the Human Connectome Project is gathering data on 1200 individuals using a combination of MRI modalities from high-resolution 3T and 7T magnetic resonance scanners. Despite being in its early stages, this project has already yielded some interesting insights that point to the potential impact of this dataset on our understanding of the human brain and behavior. For instance, it has been shown that an individual's frontoparietal network both is a fingerprint, uniquely identifying him or her out of a population, and can be used to predict his or her fluid intelligence, a measure of reasoning and problem solving (Finn et al. 2015). Moreover, it has been demonstrated that certain connectivity patterns are associated with positive behavioral traits such as cognition, memory, years of education, and income level, while others are associated with negative behavioral traits such as substance use, rule-breaking, and aggression (Smith et al. 2015). Thus, even in its earliest stages, the human connectome can provide insights with far-reaching implications.

BASAL GANGLIA CONNECTOME

While much progress has been made with respect to global patterns of connectivity, a similar analysis of connectivity of the basal ganglia and deep white matter tracts is still nascent (Lenglet et al. 2012). Due to their small size and proximity to other subcortical structures, imaging of the basal ganglia is challenging and relies on ultrahigh 7T MRI modalities that are not yet widely available (Keuken et al. 2014). Furthermore, the precise visualization of subcortical structures often requires novel MR sequences such as quantitative susceptibility mapping (Keuken et al. 2014). Consequently, the atlases of deep brain nuclei used for surgical planning are constructed from group-averaged MRI studies. Nevertheless, generating a basal ganglia and thalamic connectome for an individual is possible and has led to the identification and reconstruction of canonical basal ganglia pathways (Lenglet et al. 2012).

Diffusion MRI studies of deep brain nuclei have helped elucidate poorly studied pathways in the human brain. For instance, a diffusion tensor imaging study confirmed connections between two relay thalamic nuclei, the mediodorsal and centromedian nuclei, and cortical and subcortical targets in humans, whose existence had only been postulated by anatomical tracing studies in animals (Eckert et al. 2012). Similarly, an unusually dense connectivity pattern was found between the frontal lobes and cingulate gyri with the claustrum, a poorly characterized structure that is highly conserved across various mammalian species (Torgerson et al. 2015). While the detailed connectivity patterns of the basal ganglia provided by the deep brain connectome are helpful for understanding the function these structures, it is critical to extend these findings to the study of disease states to help predict outcomes and advance the treatment of disorders associated with the basal ganglia.

CLINICAL IMPACT OF THE DEEP BRAIN CONNECTOME

Although a full set of connectomes across the human life span (accounting for brain atrophy) or across a range of diseases does not yet exist, diffusion tractography studies in different disease states highlight the potential utility of this type of information. Specifically, there are three areas in which the deep brain connectome can influence treatment of movement, behavioral, and psychiatric disorders. First, imaging could uncover the neuropathology of these disorders and act as a marker of severity. Second, it could generate new targets for DBS. Finally, it could help predict a patient's response to intervention helping personalize a patient's treatment. Here, we examine each of these possibilities in turn.

BIOMARKERS OF DISEASE SEVERITY

Neurological and psychiatric disorders involving the basal ganglia by definition engage a diverse set of cortical and subcortical connections. A number of recent diffusion MRI studies have shown that deficiencies in the quality of basal ganglia and deep white matter connectivity are associated with different disease states. Decreases in fractional anisotropy and increases in mean diffusivity, indicators of microsurgical integrity and axonal membrane density, respectively, have been found in movement disorders (cervical dystonia [Blood et al. 2012], Huntington's disease [Novak et al. 2015], progressive supranuclear palsy [Surova et al. 2015], Parkinsonian syndromes [Cochrane and Ebmeier 2013], essential tremor [Saini et al. 2012], Tourette's syndrome [Chelse and Blackburn 2015]), psychiatric disorders (obsessive–compulsive disorder [Chiu et al. 2011]), and bipolar disorder (Teng et al. 2014). Interestingly, for some of these disorders (progressive supranuclear palsy, Huntington's disease, Tourette's syndrome, and obsessive–compulsive disorder), there is a correlation between the degree of impairment and the extent of white matter injury. Thus, damage to basal ganglia circuitry and its projections are intimately related to the pathophysiology of certain movement and psychiatric disorders and could possibly be used as biomarkers for clinical severity.

IDENTIFICATION OF NOVEL TREATMENT TARGETS

Prior to the advent of DBS, stereotactic thalamotomies and pallidotomies were mainstays of the surgical treatment of Parkinsonian tremor. The use of DBS for movement disorders grew out of the serendipitous discovery that high-frequency stimulation of the ventral intermediate nucleus of the thalamus led to a significant reduction in tremor in patients with Parkinson's disease (Benabid et al. 1987). Since then candidate targets for DBS have been generated by combining observations from clinical studies, animal models of disease, and functional models of the basal ganglia (Benabid and Torres 2012). By providing high-resolution connectivity of the basal ganglia, the deep brain connectome may help facilitate the identification of new targets for surgical intervention. For instance, the medial pulvinar of the thalamus is densely connected with the hippocampus (Barron et al. 2014), and activity in this area correlates with ictal activity in temporal lobe epilepsy patients (Rosenberg et al. 2006) suggesting that stimulation of the medial pulvinar could modulate seizure activity (Barron et al. 2014). In a similar fashion, comparisons of white matter connectivity in normal controls and patients with basal ganglia disorders could provide a principled approach for generating new DBS targets.

PREDICTING EFFICACY OF SURGERY

In addition to acting as a biomarker of disease severity and identifying new treatment targets, connectome identified in post-DBS patients may be helpful for improving targeting as well as tracking responsiveness to treatment. In one illustrative case, the subthalamopontocerebellar and dentatothalamic tracts were mapped out in patients with Parkinson's disease undergoing implantation of subthalamic electrodes (Sweet et al. 2014). The authors found a trend toward better postoperative tremor control in patients where the active contact was closer to the dentatothalamic tract. Although, due to the small sample size, this trend did not reach significance, this study offers an example of how fiber tractography may one day be used to help refine current surgical practices. Similarly, pre- and post-surgical connectomes are starting to shed light on the underpinnings of patient's responsiveness to DBS and suggesting imaging markers for tracking treatment effectiveness. In one case series, DBS-mediated improvement of certain Parkinsonian symptoms was associated with structural and functional changes in connectivity (van Hartevelt et al. 2014). In another, patients with mesial temporal lobe epilepsy who were seizure-free following surgery were found to have a different pre-surgical pattern of connectivity within the thalamocortical network from patients with persistent seizures after surgery (Ji et al. 2015). Moreover, the surgical control of seizures was associated with widespread reorganization of these patients' connectomes. These studies relied on brain connectivity measures from graph theory as well as computational modeling, underscoring the fact that the true power of connectomics can only be realized by combining structural connectivity maps with theoretical models of brain activity.

SUMMARY

The basal ganglia are key neurological structures subserving a number of critical functions. Disruption of the basal ganglia circuits has been implicated in the pathogenesis of a number of motor, behavioral, and psychiatric disorders. Consequently, the detailed wiring map provided by the deep brain connectome is an invaluable resource for understanding and treating these disorders. Investigating how the deep brain connectome is altered by these disease processes and by surgical intervention can help advance the care of patients with basal ganglia disorders. Much of this research is still in its infancy. However, when combined with computational models for whole brain dynamics, the connectome will unlock a wealth of new opportunities for the treatment of neurological diseases.

REFERENCES

Alexander, G. E., M. R. DeLong et al. (1986). Parallel organization of functionally segregated circuits linking basal ganglia and cortex. *Annu Rev Neurosci* **9**: 357–381.

Atta, H. M. (1999). Edwin Smith Surgical Papyrus: The oldest known surgical treatise. *Am Surg* **65**(12): 1190–1192.

Barron, D. S., N. Tandon et al. (2014). Thalamic structural connectivity in medial temporal lobe epilepsy. *Epilepsia* **55**(6): e50–e55.

Bassett, D. S. and E. Bullmore (2006). Small-world brain networks. *Neuroscientist* **12**(6): 512–523.

Beach, T. G. and E. G. McGeer (1987). Tract-tracing with horseradish peroxidase in the postmortem human brain. *Neurosci Lett* **76**(1): 37–41.

Benabid, A. L., P. Pollak et al. (1987). Combined (thalamotomy and stimulation) stereotactic surgery of the VIM thalamic nucleus for bilateral Parkinson disease. *Appl Neurophysiol* **50**(1–6): 344–346.

Benabid, A. L. and N. Torres (2012). New targets for DBS. *Parkinsonism Relat Disord* **18**(Suppl 1): S21–S23.

Blood, A. J., J. K. Kuster et al. (2012). Evidence for altered basal ganglia-brainstem connections in cervical dystonia. *PLoS One* **7**(2): e31654.

Chelse, A. B. and J. S. Blackburn (2015). Structural connectivity in Gilles de la Tourette syndrome. *Pediatr Neurol Briefs* **29**(4): 26.

Chiu, C. H., Y. C. Lo et al. (2011). White matter abnormalities of fronto-striato-thalamic circuitry in obsessive-compulsive disorder: A study using diffusion spectrum imaging tractography. *Psychiatry Res* **192**(3): 176–182.

Cochrane, C. J. and K. P. Ebmeier (2013). Diffusion tensor imaging in parkinsonian syndromes: A systematic review and meta-analysis. *Neurology* **80**(9): 857–864.

Cooper, I. S. (1961). *Parkinsonism: Its Medical and Surgical Therapy*. Springfield, IL: Thomas.

DeFelipe, J. (2002). Sesquicentenary of the birthday of Santiago Ramon y Cajal, the father of modern neuroscience. *Trends Neurosci* **25**(9): 481–484.

Eckert, U., C. D. Metzger et al. (2012). Preferential networks of the mediodorsal nucleus and centromedian-parafascicular complex of the thalamus—A DTI tractography study. *Hum Brain Mapp* **33**(11): 2627–2637.

Finn, E. S., X. Shen et al. (2015). Functional connectome fingerprinting: Identifying individuals using patterns of brain connectivity. *Nat Neurosci* **18**(11): 1664–1671.

Goetz, C. G. (2011). The history of Parkinson's disease: Early clinical descriptions and neurological therapies. *Cold Spring Harb Perspect Med* **1**(1): a008862.

Jbabdi, S. and H. Johansen-Berg (2011). Tractography: Where do we go from here? *Brain Connect* **1**(3): 169–183.

Ji, G. J., Z. Zhang et al. (2015). Connectome reorganization associated with surgical outcome in temporal lobe epilepsy. *Medicine (Baltimore)* **94**(40): e1737.

Keuken, M. C., P. L. Bazin et al. (2014). Quantifying inter-individual anatomical variability in the subcortex using 7 T structural MRI. *Neuroimage* **94**: 40–46.

Lenglet, C., A. Abosch et al. (2012). Comprehensive in vivo mapping of the human basal ganglia and thalamic connectome in individuals using 7T MRI. *PLoS One* **7**(1): e29153.

Lozano, A. M., J. Dostrovsky et al. (2002). Deep brain stimulation for Parkinson's disease: Disrupting the disruption. *Lancet Neurol* **1**(4): 225–231.

McIntyre, C. C. and P. J. Hahn (2010). Network perspectives on the mechanisms of deep brain stimulation. *Neurobiol Dis* **38**(3): 329–337.

McIntyre, C. C., M. Savasta et al. (2004). Uncovering the mechanism(s) of action of deep brain stimulation: Activation, inhibition, or both. *Clin Neurophysiol* **115**(6): 1239–1248.

Murre, J. M. and D. P. Sturdy (1995). The connectivity of the brain: Multi-level quantitative analysis. *Biol Cybern* **73**(6): 529–545.

Novak, M. J., K. K. Seunarine et al. (2015). Basal ganglia-cortical structural connectivity in Huntington's disease. *Hum Brain Mapp* **36**(5): 1728–1740.

Oh, S. W., J. A. Harris et al. (2014). A mesoscale connectome of the mouse brain. *Nature* **508**(7495): 207–214.

Ring, H. A. and J. Serra-Mestres (2002). Neuropsychiatry of the basal ganglia. *J Neurol Neurosurg Psychiatry* **72**(1): 12–21.

Rosenberg, D. S., F. Mauguiere et al. (2006). Involvement of medial pulvinar thalamic nucleus in human temporal lobe seizures. *Epilepsia* **47**(1): 98–107.

Saini, J., B. S. Bagepally et al. (2012). Diffusion tensor imaging: Tract based spatial statistics study in essential tremor. *Parkinsonism Relat Disord* **18**(5): 477–482.

Smith, S. M., T. E. Nichols et al. (2015). A positive-negative mode of population covariation links brain connectivity, demographics and behavior. *Nat Neurosci* **18**(11): 1565–1567.

Sporns, O., G. Tononi et al. (2005). The human connectome: A structural description of the human brain. *PLoS Comput Biol* **1**(4): e42.

Surova, Y., M. Nilsson et al. (2015). Disease-specific structural changes in thalamus and dentatorubrothalamic tract in progressive supranuclear palsy. *Neuroradiology* **57**(11): 1079–1091.

Sweet, J. A., B. L. Walter et al. (2014). Fiber tractography of the axonal pathways linking the basal ganglia and cerebellum in Parkinson disease: Implications for targeting in deep brain stimulation. *J Neurosurg* **120**(4): 988–996.

Tardif, E. and S. Clarke (2001). Intrinsic connectivity of human auditory areas: A tracing study with DiI. *Eur J Neurosci* **13**(5): 1045–1050.

Teng, S., C. F. Lu et al. (2014). Altered resting-state functional connectivity of striatal-thalamic circuit in bipolar disorder. *PLoS One* **9**(5): e96422.

Torgerson, C. M., A. Irimia et al. (2015). The DTI connectivity of the human claustrum. *Hum Brain Mapp* **36**(3): 827–838.

Ugurbil, K., J. Xu et al. (2013). Pushing spatial and temporal resolution for functional and diffusion MRI in the Human Connectome Project. *Neuroimage* **80**: 80–104.

van Hartevelt, T. J., J. Cabral et al. (2014). Neural plasticity in human brain connectivity: The effects of long term deep brain stimulation of the subthalamic nucleus in Parkinson's disease. *PLoS One* **9**(1): e86496.

van de Ven, V. G., E. Formisano et al. (2004). Functional connectivity as revealed by spatial independent component analysis of fMRI measurements during rest. *Hum Brain Mapp* **22**(3): 165–178.

Watts, D. J. and S. H. Strogatz (1998). Collective dynamics of 'small-world' networks. *Nature* **393**(6684): 440–442.

White, J. G., E. Southgate et al. (1986). The structure of the nervous system of the nematode *Caenorhabditis elegans*. *Philos Trans R Soc Lond B Biol Sci* **314**(1165): 1–340.

26 Neurovascular Photonics

*Nimer Adeeb, Babak Kateb, Salman Abbasi Fard,
and Martin M. Mortazavi*

CONTENTS

INTRODUCTION

Photonics is the field of science that deals with light, in terms of generation, emission, transmission, amplification, detection, and signal processing (Tuan, 2003).

Throughout history, light has played a significant role in medicine as a healing tool. Temples were built to worship the therapeutic power of light, mainly sunlight. The healing "god" was often referred to as the "sun god." With the advances in science and technology, the contributions of light in medicine have evolved. A major breakthrough occurred in the seventeenth century with the invention of microscope, which, overtime, has become a central tool in research and treatment. In the late nineteenth century, Wilhelm Roentgen made the main contribution in the discovery of x-rays, which became a fundamental tool in medical diagnostics (McClellan and Dorn, 1999; Tuan, 2003).

Photonics is often used interchangeably with optics. The field of photonics, however, is broader than that of optics. Optics refers to a type of electromagnetic radiation that is visible to the naked eye, "the visible light." On the other hand, photonics refers to the smallest quanta of energy in the entire spectrum of electromagnetic radiation. This spectrum ranges from cosmic rays, gamma rays, and x-rays throughout ultraviolet, visible, infrared, microwave, and radiofrequency energy (Tuan, 2003).

Biomedical photonics is the field of science and technology that uses the entire spectrum of electromagnetic radiation, beyond visible light, for medical applications, including diagnosis, treatment, and disease prevention. These technologies include lasers, fiber optics, electro-optical instruments, microelectromechanical systems, and nanosystems (Tuan, 2003).

LASER

PRINCIPLE

The word laser is an acronym for "light amplification by stimulated emission of radiation." A laser device consists of a lasing (amplification) medium enclosed between two, fully and partially reflecting, mirrors. A source of light is present within the laser tube. When the light energy is applied to the lasing medium, more atoms within the medium attain an excited state. A photon that strikes an atom in the excited state will cause it to fall to a lower energy state and release a photon of the same wavelength, traveling in the same direction and in phase with the first photon "stimulated emission." As light bounces back and forth between the two mirrors, through the amplifying material, the gained amplification becomes large enough so the light escapes through the partially reflecting mirror, producing laser light (Fuller, 1980; Wiersma, 2000). The first successful pulsed-wave laser was designed by Maiman in 1960, using ruby medium to produce a laser pulse. This was replaced by helium–neon gas medium by Javan et al. in 1961.

APPLICATIONS

Laser has become a milestone in the development of biomedical photonics. It has been used as a light source to excite tissues for disease diagnosis and as optical scalpel in interventional surgery for pathological tissue removal. Due to its monochromaticity and intensity, laser forms an ideal light source, which can be coupled with optic fibers and be used in endoscopes and microscopes. Its precision and intensity can be highly controlled by computers, which would reduce side effects and human errors.

Medical applications for laser include its use to treat skin conditions, including wrinkles, tattoos, birthmarks, warts, and tumors. It is also used to treat eye conditions, including refractive errors and glaucoma, and to open clogged blood vessels (Tuan, 2003). Laser has also become a mainstay of treatment in many other surgical fields, including plastic surgery, vascular surgery, otorhinolaryngology, and neurosurgery.

The main basis for laser application in surgery is the conversion of laser light into heat. Laser photon absorption by the tissue stimulates oscillation and rotation of tissue molecules. The subsequent collision between these molecules release kinetic energy to produce heat. The produced heat results in variable biological effects, including the following (Goldman and Rockwell, 1966; Edwards et al., 1983; Devin et al., 2003; Lucroy and Bartels, 2003):

- Hyperthermia (42°C–45°C): Leading to reversible or irreversible tissue damage.
- Coagulation (60°C–65°C): Leading to irreversible tissue damage by coagulative necrosis and protein denaturation.
- Vaporization (100°C): Leading to tissue dehydration and reduces thermal conductivity into the surrounding tissue. The solid tissue becomes vapor and smoke plume.
- Carbonization (>100°C): With continued heating and impaired tissue conduction, the tissue temperature rises rapidly. The tissue burns and becomes charred. This level should be avoided in surgery, as carbon acts as foreign material, leading to inflammation and impaired wound healing. If it occurs, the charred tissue should be carefully removed.

The amount of tissue heating, the area of laser effect, and the type of lesion created depends on several factors:

1. Duration of laser exposure
2. Laser properties
   a. Wavelength (determined by the lasing medium, such as CO_2)
   b. Power output (watts)
   c. Beam density (spot size)
3. Exposed tissue thermal and optical properties
   a. Absorption coefficient (absorbed light is rapidly converted to heat)
   b. Extinction length (the depth that light will penetrate)
   c. Presence of light-absorbing chromophores such as water, hemoglobin, or melanin
4. Degree of tissue vascularity, which affects the tissue cooling after hyperthermia

LASERS IN NEUROSURGERY

The first experimental use of lasers in neurosurgery was on animal models by Earle et al. (1965) and Fine et al. (1965). Pulsed-wave ruby laser was used, and a single high-energy pulse was applied to intact craniums of mice. The mice were immediately killed due to rapid expansion of the intracranial contents and subsequent cerebral herniation. Although herniation could be prevented by wide craniectomy in later studies, the destructive neural effect was intolerable, leading to wedge-shaped lesions and intracranial bleeding (Ryan et al., 2009). In 1966, Rosomoff and Carroll, for the first time, used the pulsed ruby laser on human glioma. Although there was an evidence of tumor necrosis, the thermal effects of the laser were difficult to control.

CO_2 Lasers

Introduction

A major advance in laser application in neurosurgery came with the introduction of continuous-wave lasers. The use of CO_2 as a lasing medium to produce a continuous-wave laser was first described by Patel in 1964. Due to the risk associated with pulsed-wave laser, the continuous-wave CO_2 did not gain immediate popularity. Later, it was extensively studied by Stellar et al. (1970, 1974) who showed that precise incisions could be made on cat brain and spinal cord, with minimal collateral injury. They were also the first to use this technique to gain subtotal resection by vaporizing a glioblastoma.

Thereafter, Ascher and Heppner (Ascher, 1979; Ascher and Heppner, 1984) performed over 650 cases using CO_2 laser, including brain and spinal cord tumors. To increase the precision of the laser beam, they incorporated a pilot light marker to indicate the path and focus of the invisible CO_2 laser beam. The authors remarked that no real contraindication would exist for its use and also suggested that gliosis and painful neuroma formation were reduced when the laser was used (Ascher, 1979; Ascher and Heppner, 1984).

Further improvement included laser coupling with the surgical microscope, instead of the earlier freehand use of the laser machine (Ascher and Heppner, 1984; Devin et al., 2003). In addition, a computer-assisted, stereotactic laser for resection of tumors has also been developed (Kelly, 1989). Several articles have later reported on CO_2 use in treatment of central and peripheral nervous system tumors, with promising outcomes (Beck, 1980; Edwards et al., 1983; Takizawa, 1984; Cozzens and Cerullo, 1985). These positive feedbacks have led to the establishment of the journal *Lasers in Surgery and Medicine* in 1980 and the *First American Congress on Lasers In Neurosurgery*, held in Chicago in 1981. Currently, it is especially used to coagulate and devascularize vascular tumors in order to decrease perioperative bleeding.

Tissue Effect

Upon neural tissue exposure to radiation, three distinct areas develop, depending on the degree of heating. A central crater with a carbonized border (high temperature), a middle zone of desiccated tissue (moderate heating), and an outer zone of edematous tissue (minimal heating) are detected in histological study. However, this damage is often confined to within a millimeter of the beam application. Small blood vessels are readily coagulated by the CO_2 laser, making oozing surfaces easier to handle (Stellar et al., 1970, 1974).

Continuous-wave lasers were found to have high absorption in tissue and blood with rapid conversion of light energy into heat in a small volume of tissues. This allows for accurate cutting using focused beams, without the need to handle or retract the tissue. A wider, defocused beam allows for some hemostatic function, with minimal thermal damage (Fox et al., 1967; Hall et al., 1971).

Limitations

The main disadvantages of CO_2 laser include the inability of the laser beam to traverse any tissues without damage, not even the transparent media of the eye. Therefore, a wide surgical exposure of the deep structures must be done before laser irradiation, to avoid unnecessary damage to the superficial structures (Stellar et al., 1970). Moreover, foreign tissue exposure to laser will result in its death but not removal (Devin et al., 2003).

Due to the high water absorption, CO_2 laser energy does not travel through fluids, but is rather quickly converted to local heat. This limits its use around fluid medium, including cerebrospinal fluid (CSF), and, thus, requires a dry environment. This can be overcome in the future if a dry path to the target can be found, such as by bubbling gas through the delivery fiber or by partial drainage of CSF, to benefit from the other safety advantages (Ryan et al., 2010).

The long wavelength of the laser (10.6 µm) results in loss of its power when transmitted through conventional solid fiber-optic cables. That is when the laser has been integrated with fiber-optic endoscopes or microscopes. Therefore, bulky articulating arms with mirrors were required to transmit sufficient energy to the surgical site in direct line of sight, restricting freedom of movement. Constant refocusing of the CO_2 beam during surgery was also required when coupled to a microscope (Ryan et al., 2010).

Flexible Omnidirectional CO_2 Laser

These previous factors have rendered rigid fiber-optic CO_2 laser technique less favorable among neurosurgeons, despite its many potential benefits in coagulation and tumor resection. Many prominent neurosurgeons have, therefore, preferred the carefully applied bipolar cautery, and microdissection using surgical scalpel for the removal of tumors, especially those adjacent to neural or vascular structures (Rhoton, 2003).

To overcome these drawbacks, especially those of the troublesome ergonomics of the laser system, a flexible hollow-core fiber, lined with an omnidirectional mirror for the delivery of CO_2 laser energy, was developed (Fink et al., 1998; Hart et al., 2002). This photonic bandgap fiber assembly allows for the flexible delivery of CO_2 laser energy in power settings similar to those used with rigid delivery systems. However, it adds the advantage of using a laser delivery apparatus that is on a small scale and does not use a large, inconvenient assembly (Ryan et al., 2010). It allows for bringing the laser energy directly to the surgical site, while working under the microscope or endoscope, with low absorptive losses (Ryan et al., 2009). This system, however, lacks a secondary visible pilot laser, in contrast to earlier CO_2 laser system, where helium–neon marker was usually used (Ascher and Heppner, 1984). Nevertheless, a constant stream of helium gas can be used for aiming, as it produces a slight depression in most tissues and surfaces (Ryan et al., 2010).

In comparing flexible CO_2 laser (OmniGuide Inc., MA, USA) to bipolar cautery (Codman and Shurtleff Inc., MA, USA), the laser, although unable to coagulate large vessels, allows for delivery of thermal energy without the need for direct contact. This may be efficacious in controlling diffuse bleeding from small vessels such as in a tumor bed. The cautery, on the other hand, requires direct application

TABLE 26.1

Comparison between Flexible Co_2 Laser and Bipolar Cautery

	Laser	Bipolar Cautery
Vessel size	More effective in small vessels bleeding	More effective in larger vessels bleeding
Tissue contact	Does not require direct application to deliver energy	Require direct application to deliver energy
Lateral spread	Less lateral spread of energy	More lateral spread of energy
Neuromonitoring	Allows for continuous electrophysiological monitoring without interruptions	Interferes with electrophysiological monitoring

of the forceps on the tissue to deliver energy. This may result in building up of char and coagulum that can stick to and tear tissues, causing bleeding. Moreover, bipolar cautery is associated with more lateral spread of energy that increases with bipolar cautery power and duration. This has a clinical importance when operating near cranial nerves. Changes in electrophysiological recordings might be induced, despite the absence of direct contact with the nerve, indicating some degree of nerve injury (Ryan et al., 2010). Therefore, laser could offer the potential to reduce postoperative neurological morbidity. On the contrary, CO_2 laser does not interfere with the electrophysiological signal and allows for continuous monitoring during dissection within critical structures. This potentially helps in earlier recognition of monitoring changes and avoidance of neurological deficit (Choudhri et al., 2014) (Table 26.1).

In comparing its cutting properties to surgical scalpel, the incision created by the laser can be clearly observed, and the shape of the cut will be largely determined by the power settings of the device. However, care should be taken to avoid continuous application of the laser on one site, which may then penetrate into deeper tissues. On the other hand, the depth of incision is made blindly using the surgical scalpel by its downward mechanical force. The surgeon must control the depth of the cut by feel to avoid deeper injury. No lateral tissue effect is seen with the scalpel, as no energy is applied (Ryan et al., 2010).

CO₂ Laser in Neurovascular Diseases

The ability of CO_2 laser to dissect the periphery of the lesions with high precision, and minimal disruption of the surrounding structures made it a desirable choice for lesions located near eloquent structures. Furthermore, the laser allows for simultaneous coagulation and cutting of the small feeding vessels, reducing the time needed using standard techniques. It also reduces the risk of bleeding from these feeding vessels. Therefore, it has become a promising instrument for vascular lesions.

In earlier studies, rigid CO_2 laser was occasionally used for resection of cavernous malformation (Seifert and Gaab, 1989) and arteriovenous malformation (AVM) (Gil-Salú et al., 2004). Despite the positive outcomes, laser was not favored due to the previously mentioned reasons. With the recent development of the flexible omnidirectional CO_2 laser, more cases were treated using this method. Killory et al. (2010) used it for resection of 11 cavernous malformations in variable locations including brainstem, with successful resection of all cases. Choudhri et al. (2014) reported the largest series with 58 cavernous malformation, including 30 in the brainstem, with total resection in all but one case. It was therefore associated with higher rates of total resection and low postoperative morbidity and mortality compared to other treatment options.

Limitations of laser use in these series include the need for straight corridor to the lesion, the inability of the laser to control bleeding from large vessels, and the high cost of the equipments. Bleeding from large vessels warrants using bipolar coagulation at low power and cutting the vessels with micro-scissors (Choudhri et al., 2014).

Other Laser Techniques

Different laser techniques, using various lasing medium, have also been developed. These include the Argon laser (*wl.* 488–516 nm) (Edwards et al., 1983) and Nd:YAG laser (*wl.* 1060–1340 nm)

(Eggert et al., 1985). These laser types have shorter wavelengths compared to the CO_2 laser. They are well absorbed by hemoglobin, producing effective coagulation. However, they are scattered more widely in tissues, creating a broader zone of heating and tissue effect (Ryan et al., 2009).

Both Argon and Nd:YAG lasers were also applied within endoscopic neurosurgery (Powers, 1992; Vandertop et al., 1998; van Beijnum et al., 2008), but the serious risk of damage to adjacent structures has limited their application.

INDOCYANINE GREEN VIDEO ANGIOGRAPHY IN NEUROSURGERY

INTRODUCTION

Maintaining the integrity of the tissue perfusion is one of the basic principles of surgery and is particularly important in neurosurgery. Disruption of the vascular supply or drainage may lead to devastating neurological complications. The surgeon's assessment of intra-operative anatomy is visually dependent on the microscope, with a sole use of the white-light reflex. Therefore, the extent of vascular injury, or obliteration of intracranial vascular anomaly, depends mainly on postoperative digital subtraction (DS) angiography, which is the gold standard assessment. If any abnormalities were found on this imaging, either a second operation is warranted or it has already caused tissue infarction and reoperation is pointless. Therefore, the need of intra-operative evaluation was raised.

Intra-operative Doppler ultrasound has been traditionally used for evaluation of vascular patency. It is quick and inexpensive and has good correlation with postoperative DS angiography (Bailes et al., 1997; Firsching et al., 2000; Akdemir et al., 2006). However, the interpretation of Doppler ultrasound remains operator dependent and subjective. It might be controversial when evaluating clip-induced stenosis. Moreover, evaluation of small perforating vessels may be difficult and unreliable (Raabe et al., 2003).

Many authors have, therefore, recommended intra-operative DS angiography for more accurate results, at least for difficult cases, such as those involving complex aneurysms. Nevertheless, this technique is invasive, expensive, and time consuming and might need to be repeated if a deformity was detected and corrected. Moreover, complication rate, including stroke and permanent neurological deficit, ranges around 1%–3% (Derdeyn et al., 1995; Alexander et al., 1996).

Therefore, a simpler method, yet clear and effective, was needed for intra-operative evaluation of blood flow. In 1967, Feindel et al. were the first to introduce fluorescence angiography, using fluorescein dye, for intra-operative visualization of cerebral circulation. In 1994, Wrobel et al. reported using intra-operative fluorescein angiography during aneurysm surgery. The process was easy and inexpensive compared to intra-operative DS angiography, with good correlation to postoperative DS angiography. The authors, however, had technical difficulties with low-quality images.

Indocyanine green (ICG) fluorescence angiography has been widely used in ophthalmologic surgery, for the assessment of retinal microcirculation. In 2003, Raabe et al. were the first to use this technique in brain surgery.

PRINCIPLE

ICG is a fluorescent dye that is injected intravenously for angiography or for organ function diagnosis. The absorption and emission range of ICG is within near-infrared region. The absorption ranges between 600 and 900 nm (peak 805 nm), and the florescence emission ranges between 750 and 950 nm (peak 835 nm) (Raabe et al., 2003). ICG is also advantageous over fluorescein, as it has less adverse reactions (Yannuzzi et al., 1986), and the light emission is more intense and easier to detect (Balamurugan et al., 2011). The recommended intravenous dose of ICG is 0.2–0.5 mg/kg, with a maximum daily dose of 5 mg/kg. Upon injection, the ICG binds to intravascular proteins (mainly globulins) within 1–2 s and remains intravascular. With a half-life of 3–4 min, ICG is excreted exclusively by the liver and is not absorbed by the intestine. The time course of the ICG after intravenous injection, similar to DS angiography, is divided into arterial, capillary, and venous phases. When used intra-operatively, a near-infrared laser light source is applied on the area of interest. This will then induce ICG fluorescence, allowing for the visualization of large and small caliber vessels (<0.5 mm). Due to the short half-life, ICG should be injected after the laser is set and the area is illuminated. The images will then be detected and viewed in real time with high quality (Raabe et al., 2003, 2005).

LIMITATIONS

Vessels that are covered by blood clots, bone, dura, or brain tissue, and are not visible to the surgeon (within the surgical field), cannot be observed with ICG angiography. Therefore, lesions in these deep-seated or covered vessels cannot be excluded. Moreover, in thick-walled or thrombosed vessels, fluorescence signals may be poor to be detected properly. False-positive findings may also be seen following repeated injections of ICG within short intervals (Raabe et al., 2003, 2005; Balamurugan et al., 2011)

NEUROVASCULAR SURGERY

The most important indication for intra-operative ICG angiography is for evaluation of treatment success during neurovascular surgery. In aneurysm clipping, it is used to detect complete obliteration of the aneurysm neck, along with preservation of main, branching, and perforating arteries. In AVM surgery, it is used to confirm complete resection and the absence of residues. In extracranial–intracranial (EC–IC) bypass surgery, it is used to confirm patency of the anastomosis site and to avoid early bypass graft failure (Balamurugan et al., 2011). It has also been used recently in other vascular anomalies, including cavernous malformation (Endo et al., 2013) and dural arteriovenous fistulas (DAVF) (Thind et al., 2015).

Aneurysm Clipping

Following aneurysm clipping, the surgeon might be unable to confirm complete obliteration of the neck and patency of the main stem and branching arteries via visual inspection. Postoperative DS angiography is the gold standard assessment. The incidence of residual filing of the aneurysm has been around 2%–8%, and occlusion of the main stem or branching arteries was around 4%–12% (Thornton et al., 2000; Raabe et al., 2003; Balamurugan et al., 2011). This might be associated with high risk of aneurysm regrowth and significant morbidity, respectively. A second operation, which is time and money consuming, might not be effective in reversing such neurological morbidity. Intra-operative Doppler ultrasound was effectively used in detecting incomplete neck obliteration, and parent or branching artery stenosis, guiding repositioning of aneurysm clip in around 25%–30% of cases (Bailes et al., 1997; Akdemir et al., 2006). Using intra-operative DS angiography, the rates of clip repositioning ranged from 3% to 27% for arterial stenosis, and 6% to 26% for residual aneurysm filling, with combined rates of 7% to 34% (Derdeyn et al., 1995; Alexander et al., 1996; Raabe et al., 2003; Balamurugan et al., 2011). Following the introduction of ICG angiography, this technique has evolved rapidly. It has also been integrated into surgical microscope, allowing for rapid evaluation and clip repositioning, which would lower the associated morbidity. The rate of clip repositioning was 4%–9% (Raabe et al., 2005; Washington et al., 2013; Roessler et al., 2014; Sharma et al., 2014). The results of ICG angiography were comparable to those of intra-operative or postoperative DS angiography. More than 90% of ICG angiography findings were confirmed by DS angiography, with 2.3% rate of false-negative results compared to postoperative DS angiography (Raabe et al., 2005). Intra-operative DS angiography may also be associated with false-negative findings compared to postoperative DS angiogram. Payner et al. (1998) reported 7.9% residual aneurysms and a 4.8% rate of distal arterial branch occlusion missed by intra-operative and detected on postoperative DS angiography. These missed findings would be expected, as two different angiographic views are often required to find the neck remnants covered by the clip or branching artery (Balamurugan et al., 2011).

In studies comparing intra-operative ICG and DS angiography, the concordance in findings between these two was 90%–100%, and the discordance was not clinically significant. On the other hand, Washington et al. (2013) found a 14.3% discordance rate requiring clip readjustment. Therein, ICG angiography demonstrated aneurysm obliteration and normal vessel flow, while subsequent DS angiography showed the opposite. This discordance was mostly in the deep-seated aneurysms, which is a known limitation of the ICG angiography. Therefore, in general, ICG angiography may be used as the main intra-operative imaging modality in regular aneurysm surgeries. In complicated, large, or deep-seated aneurysms, an intra-operative supporting study might also be used.

Adverse effects including hypotension, tachycardia, nausea, pruritus, syncope, and skin eruptions were previously reported in 0.05%–0.2% of cases after ICG injection (Hope-Ross et al., 1994). However, following the introduction of ICG angiography, no adverse effects were reported in recent studies (Raabe et al., 2003, 2005; Balamurugan et al., 2011).

AVM Resection

Microsurgery is treatment of choice for resection of AVM and the only available way to immediately eliminate the risk of hemorrhage. The main challenge in the surgery is to accurately identify the AVM nidus and achieve total resection, in order to reduce the risk of recurrence and re-rupture (Hoh et al., 2004). To achieve that, intra-operative DS angiography has been used (Anegawa et al., 1994; Munshi et al., 1999; Gaballah et al., 2014). However, as discussed previously, this technique is expensive and time consuming and requires surgery interruption. Therefore, ICG angiography is a favorable alternative, with comparable outcomes (Killory et al., 2009; Takagi et al., 2012). It has also been more effective in detecting residual nidus (Takagi et al., 2007, 2012). However, it was less effective in detecting the deep-seated lesions.

EC–IC Bypass

Cerebral bypass surgery is being used for treatment of different neurovascular conditions, including aneurysms, moyamoya disease, and cerebral ischemia. Intra-operative assessment of the bypass patency is essential (Quinones-Hinojosa and Lawton, 2005; Kubo et al., 2006). Early bypass occlusion is the most fearful obstacle of the surgery, which might be associated with ischemia and severe neurological deficit. Therefore, both intra-operative and postoperative assessments are needed (Mendelowitsch et al., 2004).

Similar to other vascular conditions, ICG angiography is becoming increasingly preferred over DS angiography for intra-operative assessment, with promising outcomes (Woitzik et al., 2005; Takagi et al., 2008). Moreover, the patency rate of EC–IC bypass, which was below 90% previously, has now become 100% with the usage of ICG angiography (Woitzik et al., 2005).

Cavernous Malformation

Endo et al. (2013) have reported using ICG angiography during surgical resection of cavernous malformation. Cavernous malformation is filled with slow-flowing blood and is therefore angiographically negative (Haque et al., 2008). In this study, when ICG angiography was used, the cavernous malformation appeared as an avascular area surrounded by neural parenchyma with high ICG fluorescence uptake. It was also effective in demonstrating the margins between the cavernous malformation and the associated vascular anomalies, like AVM and deep venous anomalies, and normal superficial vessels. This allowed for safe and maximal resection of the cavernous malformation.

Dural Arteriovenous Fistulas

Surgical intervention along with endovascular treatment remains the treatment of choice for DAVF. Successful treatment of DAVF requires complete obliteration of the blood flow through fistulous connection. To identify this point, intra-operative DS angiography has been used, which is associated with numerous limitations (see previous discussion). Thind et al. (2015) reported using ICG angiography during surgical obliteration of cerebral DAVF. ICG angiography was effective in identifying 96% of the DAVF before obliteration and correctly identifying complete disconnection of DAVF in 91% of the cases. Moreover, in comparing intra-operative ICG and DS angiography, the former was associated with lower false-positive rate (8.7% vs. 10.5%).

Similar outcomes were also reported with surgical obliteration of spinal DAVF (Schuette et al., 2010; Beynon et al., 2012).

Tumors

Intra-operative ICG angiography has also been effectively used for evaluation of the peritumoral vessels during and after tumor resection (Kim et al., 2011). In highly vascularized tumors, hidden residuals can also be identified using this method (Hwang et al., 2010).

REFERENCES

Akdemir H, Oktem IS, Tucer B, Menku A, Basaslan K, Gunaldi O. 2006. Intraoperative microvascular Doppler sonography in aneurysm surgery. *Minimally Invasive Neurosurgery: MIN* 49:312–316.

Alexander TD, Macdonald RL, Weir B, Kowalczuk A. 1996. Intraoperative angiography in cerebral aneurysm surgery: A prospective study of 100 craniotomies. *Neurosurgery* 39:10–17; discussion 17–18.

Anegawa S, Hayashi T, Torigoe R, Harada K, Kihara S. 1994. Intraoperative angiography in the resection of arteriovenous malformations. *Journal of Neurosurgery* 80:73–78.

Ascher PW. 1979. Newest ultrastructural findings after the use of a CO_2-laser on CNS tissue. *Acta Neurochirurgica Supplement (Wien)* 28:572–581.

Ascher PW, Heppner F. 1984. CO_2-laser in neurosurgery. *Neurosurgery Review* 7:123–133.

Bailes JE, Tantuwaya LS, Fukushima T, Schurman GW, Davis D. 1997. Intraoperative microvascular Doppler sonography in aneurysm surgery. *Neurosurgery* 40:965–970; discussion 970–962.

Balamurugan S, Agrawal A, Kato Y, Sano H. 2011. Intra operative indocyanine green video-angiography in cerebrovascular surgery: An overview with review of literature. *Asian Journal of Neurosurgery* 6:88–93.

Beck OJ. 1980. The use of the Nd-YAG and the CO_2 laser in neurosurgery. *Neurosurgery Review* 3:261–266.

Beynon C, Herweh C, Rohde S, Unterberg AW, Sakowitz OW. 2012. Intraoperative indocyanine green angiography for microsurgical treatment of a craniocervical dural arteriovenous fistula. *Clinical Neurology and Neurosurgery* 114:696–698.

Choudhri O, Karamchandani J, Gooderham P, Steinberg GK. 2014. Flexible omnidirectional carbon dioxide laser as an effective tool for resection of brainstem, supratentorial, and intramedullary cavernous malformations. *Neurosurgery* 10(Suppl 1):34–34; discussion 43–35.

Cozzens JW, Cerullo LJ. 1985. Comparison of the effect of the carbon dioxide laser and the bipolar coagulator on the cat brain. *Neurosurgery* 16:449–453.

Derdeyn CP, Moran CJ, Cross DT, Grubb RL, Jr., Dacey RG, Jr. 1995. Intraoperative digital subtraction angiography: A review of 112 consecutive examinations. *AJNR American Journal of Neuroradiology* 16:307–318.

Devin KB, Meic HS, Rose D, Mitchel SB. 2003. Lasers in neurosurgery. In: Tuan V-D, ed. *Biomedical Photonics Handbook*. CRC Press, Baco Raton, FL.

Earle KM, Carpenter S, Roessmann U, Ross MA, Hayes JR, Zeitler E. 1965. Central nervous system effects of laser radiation. *Federation Proceedings* 24 (Suppl 14):129.

Edwards MS, Boggan JE, Fuller TA. 1983. The laser in neurological surgery. *Journal of Neurosurgery* 59:555–566.

Eggert HR, Kiessling M, Kleihues P. 1985. Time course and spatial distribution of neodymium: Yttrium-aluminum-garnet (Nd:YAG) laser-induced lesions in the rat brain. *Neurosurgery* 16:443–448.

Endo T, Aizawa-Kohama M, Nagamatsu K, Murakami K, Takahashi A, Tominaga T. 2013. Use of microscope-integrated near-infrared indocyanine green videoangiography in the surgical treatment of intramedullary cavernous malformations: Report of 8 cases. *Journal of Neurosurgery Spine* 18:443–449.

Feindel W, Yamamoto YL, Hodge CP. 1967. Intracarotid fluorescein angiography: A new method for examination of the epicerebral circulation in man. *Canadian Medical Association Journal* 96:1–7.

Fine S, Klein E, Nowak W, Scott RE, Laor Y, Simpson L, Crissey J, Donoghue J, Derr VE. 1965. Interaction of laser radiation with biologic systems I. Studies on interaction with tissues. *Federation Proceedings* 24 (Suppl 14):35–47.

Fink Y, Winn JN, Fan S, Chen C, Michel J, Joannopoulos JD, Thomas EL. 1998. A dielectric omnidirectional reflector. *Science* 282:1679–1682.

Firsching R, Synowitz HJ, Hanebeck J. 2000. Practicability of intraoperative microvascular Doppler sonography in aneurysm surgery. *Minimally Invasive Neurosurgery: MIN* 43:144–148.

Fox J, Hayes J, Stein M, Green R, Paananen R. 1967. Experimental cranial and vascular studies of the effects of pulsed and continuous wave laser radiation. *Journal of Neurosurgery* 27:126–137.

Fuller TA. 1980. The physics of surgical lasers. *Lasers in Surgery and Medicine* 1:5–14.

Gaballah M, Storm PB, Rabinowitz D, Ichord RN, Hurst RW, Krishnamurthy G, Keller MS, McIntosh A, Cahill AM. 2014. Intraoperative cerebral angiography in arteriovenous malformation resection in children: A single institutional experience. *Journal of Neurosurgery Pediatrics* 13:222–228.

Gil-Salú JL, González-Darder JM, Vera-Román JM. 2004. Intraorbital arteriovenous malformation: Case report. *Skull Base* 14:31–36.

Goldman L, Rockwell R. 1966. Laser action at the cellular level. *JAMA* 198:641–644.

Hall RR, Beach AD, Baker E, Morison PCA. 1971. Incision of tissue by carbon dioxide laser. *Nature* 232:131–132.

Haque R, Kellner CP, Solomon RA. 2008. Cavernous malformations of the brainstem. *Clinical Neurosurgery* 55:88–96.

Hart SD, Maskaly GR, Temelkuran B, Prideaux PH, Joannopoulos JD, Fink Y. 2002. External reflection from omnidirectional dielectric mirror fibers. *Science* 296:510–513.

Hoh BL, Carter BS, Ogilvy CS. 2004. Incidence of residual intracranial AVMs after surgical resection and efficacy of immediate surgical re-exploration. *Acta Neurochirurgica* 146:1–7; discussion 7.

Hope-Ross M, Yannuzzi LA, Gragoudas ES, Guyer DR, Slakter JS, Sorenson JA, Krupsky S, Orlock DA, Puliafito CA. 1994. Adverse reactions due to indocyanine green. *Ophthalmology* 101:529–533.

Hwang SW, Malek AM, Schapiro R, Wu JK. 2010. Intraoperative use of indocyanine green fluorescence videography for resection of a spinal cord hemangioblastoma. *Neurosurgery* 67:ons300–303; discussion ons303.

Javan A, Bennett WR, Herriott DR. 1961. Population inversion and continuous optical maser oscillation in a gas discharge containing a He-Ne mixture. *Physical Review Letters* 6:106–110.

Kelly PJ. 1989. Future perspectives in stereotactic neurosurgery: Stereotactic microsurgical removal of deep brain tumors. *Journal of Neurosurgical Science* 33:149–154.

Killory BD, Chang SW, Wait SD, Spetzler RF. 2010. Use of flexible hollow-core CO_2 laser in microsurgical resection of CNS lesions: Early surgical experience. *Neurosurgery* 66:1187–1192.

Killory BD, Nakaji P, Gonzales LF, Ponce FA, Wait SD, Spetzler RF. 2009. Prospective evaluation of surgical microscope-integrated intraoperative near-infrared indocyanine green angiography during cerebral arteriovenous malformation surgery. *Neurosurgery* 65:456–462; discussion 462.

Kim EH, Cho JM, Chang JH, Kim SH, Lee KS. 2011. Application of intraoperative indocyanine green videoangiography to brain tumor surgery. *Acta Neurochirurgica* 153:1487–1495; discussion 1494–1485.

Kubo Y, Ogasawara K, Tomitsuka N, Otawara Y, Kakino S, Ogawa A. 2006. Revascularization and parent artery occlusion for giant internal carotid artery aneurysms in the intracavernous portion using intraoperative monitoring of cerebral hemodynamics. *Neurosurgery* 58:43–50; discussion 43–50.

Lucroy MD, Bartels KE. 2003. Surgical lasers. In: Slatter D, ed. *Textbook of Small Animal Surgery*, 3rd edn. St. Louis, MO: Elsevier Health Sciences.

Maiman TH. 1960. Stimulated optical radiation in Ruby. *Nature* 187:493–494.

McClellan JEI, Dorn H. 1999. *Science and Technology in World History*. London, U.K.: Johns Hopkins University Press.

Mendelowitsch A, Taussky P, Rem JA, Gratzl O. 2004. Clinical outcome of standard extracranial-intracranial bypass surgery in patients with symptomatic atherosclerotic occlusion of the internal carotid artery. *Acta Neurochirurgica* 146:95–101.

Munshi I, Macdonald RL, Weir BK. 1999. Intraoperative angiography of brain arteriovenous malformations. *Neurosurgery* 45:491–497; discussion 497–499.

Patel C. 1964. Continuous-wave laser action on vibrational-rotational transitions of CO_2. *Physical Review* 136A:1187–1193.

Payner TD, Horner TG, Leipzig TJ, Scott JA, Gilmor RL, DeNardo AJ. 1998. Role of intraoperative angiography in the surgical treatment of cerebral aneurysms. *Journal of Neurosurgery* 88:441–448.

Powers SK. 1992. Fenestration of intraventricular cysts using a flexible, steerable endoscope. *Acta Neurochirurgica Supplement (Wien)* 54:42–46.

Quinones-Hinojosa A, Lawton MT. 2005. In situ bypass in the management of complex intracranial aneurysms: Technique application in 13 patients. *Neurosurgery* 57:140–145; discussion 140–145.

Raabe A, Beck J, Gerlach R, Zimmermann M, Seifert V. 2003. Near-infrared indocyanine green video angiography: A new method for intraoperative assessment of vascular flow. *Neurosurgery* 52:132–139; discussion 139.

Raabe A, Nakaji P, Beck J, Kim LJ, Hsu FP, Kamerman JD, Seifert V, Spetzler RF. 2005. Prospective evaluation of surgical microscope-integrated intraoperative near-infrared indocyanine green videoangiography during aneurysm surgery. *Journal of Neurosurgery* 103:982–989.

Rhoton AL, Jr. 2003. Operative techniques and instrumentation for neurosurgery. *Neurosurgery* 53:907–934; discussion 934.

Roessler K, Krawagna M, Dorfler A, Buchfelder M, Ganslandt O. 2014. Essentials in intraoperative indocyanine green videoangiography assessment for intracranial aneurysm surgery: Conclusions from 295 consecutively clipped aneurysms and review of the literature. *Neurosurgical Focus* 36:E7.

Rosomoff HL, Carroll F. 1966. Reaction of neoplasm and brain to laser. *Archives of Neurology* 14:143–148.

Ryan RW, Spetzler RF, Preul MC. 2009. Aura of technology and the cutting edge: A history of lasers in neurosurgery. *Neurosurgical Focus* 27:E6.

Ryan RW, Wolf T, Spetzler RF, Coons SW, Fink Y, Preul MC. 2010. Application of a flexible CO_2 laser fiber for neurosurgery: Laser-tissue interactions. *Journal of Neurosurgery* 112:434–443.

Schuette AJ, Cawley CM, Barrow DL. 2010. Indocyanine green videoangiography in the management of dural arteriovenous fistulae. *Neurosurgery* 67:658–662; discussion 662.

Seifert V, Gaab MR. 1989. Laser-assisted microsurgical extirpation of a brain stem cavernoma: Case report. *Neurosurgery* 25:986–990.

Sharma M, Ambekar S, Ahmed O, Nixon M, Sharma A, Nanda A, Guthikonda B. 2014. The utility and limitations of intraoperative near-infrared indocyanine green videoangiography in aneurysm surgery. *World Neurosurgery* 82:e607–613.

Stellar S, Polanyi T, Bredemeier H. 1974. Lasers in surgery. In: Wolbarsht M, ed. *Laser Applications in Medicine and Biology*. Springer, New York, pp. 241–293.

Stellar S, Polanyi TG, Bredemeier HC. 1970. Experimental studies with the carbon dioxide laser as a neurosurgical instrument. *Medical and Biological Engineering* 8:549–558.

Takagi Y, Kikuta K, Nozaki K, Sawamura K, Hashimoto N. 2007. Detection of a residual nidus by surgical microscope-integrated intraoperative near-infrared indocyanine green videoangiography in a child with a cerebral arteriovenous malformation. *Journal of Neurosurgery* 107:416–418.

Takagi Y, Sawamura K, Hashimoto N. 2008. Intraoperative near-infrared indocyanine green videoangiography performed with a surgical microscope—Applications in cerebrovascular surgery. *European Neurological Review* 3:66–68

Takagi Y, Sawamura K, Hashimoto N, Miyamoto S. 2012. Evaluation of serial intraoperative surgical microscope-integrated intraoperative near-infrared indocyanine green videoangiography in patients with cerebral arteriovenous malformations. *Neurosurgery* 70:34–42; discussion 42–33.

Takizawa T. 1984. The carbon dioxide laser surgical unit as an instrument for surgery of brain tumours—Its advantages and disadvantages. *Neurosurgery Review* 7:135–144.

Thind H, Hardesty DA, Zabramski JM, Spetzler RF, Nakaji P. 2015. The role of microscope-integrated near-infrared indocyanine green videoangiography in the surgical treatment of intracranial dural arteriovenous fistulas. *Journal of Neurosurgery* 122:876–882.

Thornton J, Bashir Q, Aletich VA, Debrun GM, Ausman JI, Charbel FT. 2000. What percentage of surgically clipped intracranial aneurysms have residual necks? *Neurosurgery* 46:1294–1298; discussion 1298–1300.

Tuan V-D. 2003. Biomedical photonics. In: Tuan V-D, ed. *Biomedical Photonics Handbook*. CRC Press, Baco Raton, FL.

van Beijnum J, Hanlo PW, Fischer K, Majidpour MM, Kortekaas MF, Verdaasdonk RM, Vandertop WP. 2008. Laser-assisted endoscopic third ventriculostomy: Long-term results in a series of 202 patients. *Neurosurgery* 62:437–443; discussion 443–434.

Vandertop WP, Verdaasdonk RM, van Swol CF. 1998. Laser-assisted neuroendoscopy using a neodymium-yttrium aluminum garnet or diode contact laser with pretreated fiber tips. *Journal of Neurosurgery* 88:82–92.

Washington CW, Zipfel GJ, Chicoine MR, Derdeyn CP, Rich KM, Moran CJ, Cross DT, Dacey RG, Jr. 2013. Comparing indocyanine green videoangiography to the gold standard of intraoperative digital subtraction angiography used in aneurysm surgery. *Journal of Neurosurgery* 118:420–427.

Wiersma D. 2000. Laser physics: The smallest random laser. *Nature* 406:132–135.

Woitzik J, Horn P, Vajkoczy P, Schmiedek P. 2005. Intraoperative control of extracranial-intracranial bypass patency by near-infrared indocyanine green videoangiography. *Journal of Neurosurgery* 102:692–698.

Wrobel CI, Meltzer H, Lamond R, Alksne JF. 1994. Intraoperative assessment of aneurysm clip placement by intravenous fluorescein angiography. *Neurosurgery* 35:970–973; discussion 973.

Yannuzzi LA, Rohrer KT, Tindel LJ, Sobel RS, Costanza MA, Shields W, Zang E. 1986. Fluorescein angiography complication survey. *Ophthalmology* 93:611–617.

27 Neurophotonics for Peripheral Nerves

Ashfaq Ahmed, Yuqiang Bai, Jessica C. Ramella-Roman, and Ranu Jung

CONTENTS

INTRODUCTION

In a broad sense, photonics is the science of light (photon) generation, detection, and manipulation through emission, transmission, modulation, switching, amplification, and signal processing. Although photonic applications can cover the entire light spectrum, most lie in the range of visible and near-infrared light. Photonic applications started in earnest in the 1960s with the invention of the laser diode and optical fibers. Neurophotonics, photonics for probing and activating the nervous system, has been used in biomedical applications for quite a while and the use of this modality in the central nervous system stimulation and imaging has been a major field of research. Lately, it has found its way into the peripheral nervous system (PNS) stimulation and imaging. The PNS is the part of the nervous system that consists of the nerves and ganglia outside the brain and spinal cord. Its function is to connect the brain to the limbs and organs successfully, essentially creating a communication bridge between them. Although, for peripheral nerve stimulation, electrical methods have been the gold standard so far, they have some inherent limitations such as reverse recruitment of the larger diameter nerve fibers by the electric field. Optical methods are transcending these limitations.

There has long been interest in using light to influence the behavior of neurons partially and particularly due to its noninvasive nature. Light can trigger action potentials similarly to the ones obtained through electrical stimulation. This book chapter reviews different photonics applications in peripheral nerve stimulation and imaging.

ORGANIZATION OF THE PERIPHERAL NERVOUS SYSTEM

The PNS consists of two major components, autonomic and somatic (Horch and Dhillon, 2004). The autonomic nervous system is mainly concerned with functions not normally under voluntary control. It consists of a preganglionic component (with nerve cells originating in the spinal column or skull) and a postganglionic component (with nerve cells lying wholly outside the axial skeleton). The somatic system is concerned with voluntary control of muscles. It consists of nerve cells with somata located within or near the spinal column or skull. The autonomic nervous system is further subdivided into sympathetic and parasympathetic components on the basis of anatomical, biochemical, and functional features. On the same basis, the somatic nervous system is divided into sensory and motor components.

The sensory system consists of nerve cells with somata located outside the spinal cord in aggregates called dorsal root ganglia (Horch and Dhillon, 2004). These nerve processes are called afferent nerve fibers because they conduct action potentials and, therefore, information from the periphery to the central nervous system. Afferent sensory fibers can be myelinated or unmyelinated, the latter ranging from 2 to 20 μm in diameter, and convey various sensory inputs, mainly mechanical, thermal, and noxious stimuli (Navarro et al., 2005). Myelinated axons are used for tasks (such as control of skeletal muscle contraction or signaling of temporally rapid and brief events) where speed is required or where fine tactile or proprioceptive discriminations are to be made. Unmyelinated fibers are normally associated with control of smooth muscle or signaling diffuse, temporally sluggish events such as pain and temperature. The motor system consists of nerve cells with somata located in the ventral quadrant of the spinal cord (Horch and Dhillon, 2004). These nerve processes are called efferent nerve fibers because they conduct information in terms of action potentials from the central nervous system to the periphery. They can be divided into two types: alpha-motor fibers that innervate the skeletal extrafusal muscle fibers and gamma-motor fibers that innervate the spindle muscle fibers (Navarro et al., 2005).

Nerve fibers, both afferent and efferent, are grouped in fascicles surrounded by connective tissue in the peripheral nerve (Peters and Palay, 1991). The fascicular architecture changes with an increasing number of fascicles of smaller size from the proximal to the distal end of the nerve. These fascicles eventually give origin to branches that innervate distinct targets, either muscular or cutaneous. In addition to bundles of nerve fibers, the peripheral nerves are composed of three supportive sheaths: epineurium, perineurium, and endoneurium. The epineurium is the outermost layer, composed of loose connective tissue and carries blood vessels that supply the nerve. The perineurium surrounds each fascicle in the nerve. It consists of inner layers of flat perineurial cells and an outer layer of collagen fibers organized in longitudinal, circumferential, and oblique bundles. The perineurium is the main contributor to the tensile strength of the nerve, acts as a diffusion barrier, and maintains the endoneurial fluid pressure. The endoneurium is composed of fibroblasts, collagen and reticular fibers, and extracellular matrix, occupying the space between nerve fibers within the fascicle. The endoneurial collagen fibrils are packed around each nerve fiber to form the walls of the endoneurial tubules. Inside these tubules, axons are accompanied by Schwann cells, which either myelinate or just surround the axons (Peters and Palay, 1991).

PHOTONICS FOR NEURAL STIMULATION

A number of optical methods are now available to generate neural stimulation. Early modalities included introducing light sensitive receptors in the neurons by genetically modifying the target neurons and introducing photoactive molecules whose behavior could be changed upon exposure to light. Recently, infrared stimulation has become a viable approach made possible by advancements in laser diode technology.

OPTOGENETICS

With optogenetics light sensitive ion channels like channelrhodopsin-2 (ChR2) and halorhodopsin are switched on and off optically with tight temporal and spatial confinement (Zhang et al., 2007a; Kramer et al., 2009). When the technology was first introduced, it allowed narrow spatial localization, but temporal localization was on the order of seconds (Zemelman et al., 2002; Banghart et al., 2004). Later, the expression of ChR2, a fast-gated blue-light-sensitive (470 nm) cation channel, in oocytes of *Xenopus laevis* and mammalian cells allowed the depolarization of cells in timescales on the order of milliseconds (Nagel et al., 2003).

So far, optogenetics has been used to great effect in the brain (Yizhar et al., 2011). Its application in the PNS has been limited to a few studies (Wang and Zylka, 2009; Llewellyn et al., 2010; Sharp and Fromherz, 2011; Ji et al., 2012; Liske et al., 2013). It has been used to both activate (Llewellyn et al., 2010) and inhibit (Liske et al., 2013) motor neuron axons in anesthetized transgenic mice. Channelrhodopsin-2 has been used to excite neurons by depolarization. Halorhodopsin, which responds to light near 580 nm is used to inhibit excitation of neurons by hyperpolarization (Zhang et al., 2007b; Gradinaru et al., 2010). Optogenetic approaches have also been utilized to further our understanding of neural disorders (Tye and Deisseroth, 2012), neural systems, and encoding (Monesson-Olson et al., 2014). Recently optogenetic approaches have been proposed for the cure of blindness and Parkinson's disease (Gradinaru et al., 2009; Kramer et al., 2009; Carter and de Lecea, 2011).

One advantage of optogenetics over electrical stimulation is the ability to control neuronal subpopulations that innervate different muscles. Controlling the subpopulation of neurons independently would allow researchers to quantify the contribution of each muscle during normal function or after a neurological injury such as stroke (Towne et al., 2013). Furthermore, chronic optogenetic stimulation of specific muscles devoid of any interfering nonspecific activation may provide insight into how surrounding nonstimulated muscles compensate for excess activity in a nearby muscle. Optogenetic activation is also devoid of the stimulus artifact potential that accompanies electrical stimulation. Using ChR2 is preferable due to the bioavailability of its chromophore in mammalian brain that greatly facilitates experiments, particularly those performed *in vivo* (Kramer et al., 2009).

Although ChR2 has improved temporal response compared to previously used opsins, it is still limited to 40 Hz (Gunaydin et al., 2010). Another limitation of optogenetics is that it requires transfection of genes into neuronal cells, which, besides technical challenges, face significant regulatory hurdles for human use (Knopfel and Boyden, 2012; Sahel and Roska, 2013).

PHOTOACTIVE MOLECULES

Another technique to optically activate neurons is based on neurotransmitters that are held in a photosensitive cage and can be liberated upon exposure to light. The first caged neurotransmitter agonists were o-nitrobenzyl derivatives of carbamylcholine, an activator of acetylcholine receptors that were released upon ultraviolet light exposure (Walker et al., 1986; Milburn et al., 1989). Then came caged glutamate that was uncaged by laser from intact brain slice (Dalva and Katz, 1994; Wieboldt et al., 1994).

However, when the technology was in its infancy, there was high level of light scattering and strong absorption in tissue limiting, respectively, the spatial resolution and penetration depth of this technique (Warther et al., 2010). The development of two-photon-responsive glutamate cages like MNI-caged glutamate (4-methoxy 7-nitroindolinyl-caged L-glutamate) allowed both a finer spatial resolution and deeper penetration into tissue (Rial Verde et al., 2008). Moreover, MNI-caged glutamate has a very low rate of spontaneous glutamate liberation in the dark having no apparent effect on normal neuronal function. Caged versions of other neurotransmitters have also been developed.

Another strategy to control neurons is to couple a photoisomerizable molecule (i.e., photoswitch) onto an ordinary ion channel or receptor so that it becomes sensitive to light of different wavelengths (Banghart et al., 2004). The photoswitch is attached in such a way that photoisomerization exerts force on the channel causing it to open. Several chemical photoswitches are available, but azobenzene has emerged as the best one for biological applications (Beharry and Woolley, 2011). In addition to control of neurons, techniques are emerging to read neural activity with light, which include sensing of cell membrane potential, calcium, and neurotransmitter release (Matsuzaki et al., 2001).

These photoactive molecules have emerged as key for determining functional connectivity of neurons due to some advantages they provide (Miesenböck and Kevrekidis, 2005; Warther et al., 2010). For example, caged glutamate-based photostimulation eliminates artifacts and limitations inherent in conventional electrical stimulation methods, including stimulation of axons of passage (Farber and Grinvald, 1983; Callaway and Katz, 1993), and poor temporal and current artifacts of iontophoretic application. Consequently, one of its versions, two-photon-sensitive caged GABA has been used to inhibit neural activity (Rial Verde et al., 2008; Warther et al., 2010). It has also been used to map the functional synaptic connections in the visual cortex (Miesenböck and Kevrekidis, 2005), the cerebral cortex (Matsuzaki et al., 2004), and many other areas (Nikolenko et al., 2007). Photoswitches are synthesized from the ground up, so by exploiting the right chemistry, it might be possible to generate the exact molecule to activate or inhibit the function of an ion channel or receptor. Also, a successful hunt for the right photoswitchable ligands might create a pathway to regulate not only voltage-gated and ligand-gated ion channels but also G protein–coupled receptors, growth factor receptors and other receptor tyrosine kinases, and transporters (Kramer et al., 2009).

Despite the success of these molecules as a tool for fundamental neuroscience studies, they have limited potential for *in vivo* studies due to the potential toxicity of the cage (Kramer et al., 2009). Moreover, in photosensitive cage–based stimulation, the glutamate is gradually depleted by repeated photolysis, so this method is less effective for prolonged stimulation or high repetition rates.

INFRARED LIGHT

Infrared light extends from 700 nm to 1 mm in the electromagnetic spectrum. Only mid- to far-infrared range (>1.4 μm) has been used to stimulate neurons. Infrared neural stimulation (INS) is considered the most clinically translatable method of optical stimulation since it does not require any genetic manipulation or modification of target tissue.

Pulses of mid-infrared light were first observed to elicit responses in mammalian nerves by Wells (Wells et al., 2005b). These researchers exposed the rat's sciatic nerve to laser irradiation. Laser wavelengths between 2 and 10 μm were utilized in this approach. In this study, compound nerve action potentials (CNAP) and compound muscle action potentials (CMAP) with a strong spatial specificity were observed. Since then, the technique has been extended and demonstrated in a number of other experiments.

INS has been extensively used in both the central and peripheral nervous system stimulation (Teudt et al., 2007; Fried et al., 2008; Richter et al., 2008; Duke et al., 2009). In both cases, neurons have responded well to INS and shown strong spatial resolution. INS has been used successfully to stimulate the cochlea of acute and chronic deafened animals in which acoustic thresholds are significantly elevated. The evoked potentials from optical stimulation in acutely deafened animals and chronically deafened animals were not significantly different from optical stimulation in normal animals, which demonstrates the feasibility of optical stimulation in neural systems that have undergone trauma resulting in neural degeneration (Richter et al., 2008). Different investigators have adopted this approach and demonstrated its feasibility in stimulation of the facial nerve (Teudt et al., 2007), the cavernous nerve (Fried et al., 2008), and vagus nerve (Duke et al., 2009).

Infrared stimulation provides the opportunity to confine optical energy to a small volume making it suitable for precise neural stimulation. This improved spatial precision overcomes the limitation of electrical stimulation that is prone to current leakage. Moreover, the lack of a stimulation artifact on the recording electrodes enables a whole new area of investigation (Fiore et al., 1996; Wagenaar and Potter, 2002; Andreasen and Struijk, 2003). Simply put, this approach will allow recording of neural potentials close to the source of stimulation. Additionally, fewer stimuli need to be applied, while in electrical stimulation hundreds of stimuli need to be averaged, resulting in higher throughput of mapping. Another group of applications originates from the fact that optical stimulation, unlike its electrical equivalent, does not require direct contact between the stimulating probe (optical fiber) and the target tissue. This obviates issues associated with properties of impedance, current shunting, and field distortion around the area of contact between the electrode and the tissue in the acute setting (Palanker et al., 2005). In the chronic setting, issues of half-cell potential differences, metal toxicity, and tissue reaction to various implanted electrodes significantly limit the materials and sizes of electrodes used for chronic implants for electrical stimulation. In principle, a fiber optic–based chronic implant may be advantageous due to longer tissue stability (no unstable impedance characteristics) and safer (i.e., inert) interface materials (glass/fiber optic cable vs. metal) (Agnew et al., 1989).

Disadvantages of INS include heating of the tissue to a level that could cause damage (Wells et al., 2007b; Thompson et al., 2012) and a restriction on the maximum depth of stimulation due to absorption of light in the intervening tissue (Thompson et al., 2012).

USE OF INFRARED LIGHT FOR PERIPHERAL NERVE STIMULATION

INS in mammalian PNS was first demonstrated by Wells (Wells et al., 2005b). Response was elicited in rat sciatic nerve by applying pulses of mid-infrared light. An infrared pulsed laser source coupled into a fiber optic was employed to provide a specific spot size for stimulating a nerve. A holmium:YAG laser ($\lambda = 2.12$ μm, $\tau_p = 350$ μs (FWHM)) was used to characterize optical stimulation in the PNS. Measurement of optical stimulus intensity was performed by placing a beam splitter coupled to an energy meter in the laser path to sample and measure 10% of the energy delivered for each pulse. The remaining 90% was transmitted through a focusing lens. The fiber optic was mounted on a three-dimensional micromanipulator and precisely positioned over the nerve at the site of electrical stimulation (Wells et al., 2007c). Figure 27.1 depicts a typical experimental setup for this study.

CHOICE OF PARAMETERS

The size of fiber is chosen based on the relative size of the nerve fascicle it is targeting. A variety of optical fibers have been used with different diameters—600 μm, 400 μm, and 200 μm diameters are typical. Stimulation experiments in the rat sciatic nerve reveal that 400–600 μm diameter fibers produce most efficient excitation while maintaining precision in stimulation. Although optical fiber is the standard method of delivering light to tissue, some other technologies are emerging to deliver light closer to the nerves. Arrayed light source is one of them. Optrode array and LED array fall into this category (Thompson et al., 2014).

The wavelength used in this approach was shown to be optimal for peripheral nerve stimulation through a wavelength optimization study performed using a free-electron laser. Theoretically, the most appropriate wavelengths for stimulation will depend on the tissue geometry. A typical rat sciatic nerve section stimulated in this study was approximately 1.5 mm in diameter, with a 100–200 μm epineural and perineural sheath between the actual axons and the nerve surface. Despite the fact that the number of fascicles per nerve varies greatly across all mammalian species, the typical fascicle thickness tends to be in a range of 200–400 μm (Paxinos, 2014). Thus, to theoretically achieve selective stimulation of individual fascicles within the main nerve, the penetration depth of the laser must be greater than the thickness of the outer protective tissue (200 μm) and in between the thickness of the underlying fascicle (penetration depth of 300–500 μm). Ultraviolet wavelengths ($\lambda = 1$ nm–0.45 μm) are strongly absorbed by tissue constituents such as amino acids, fats, proteins,

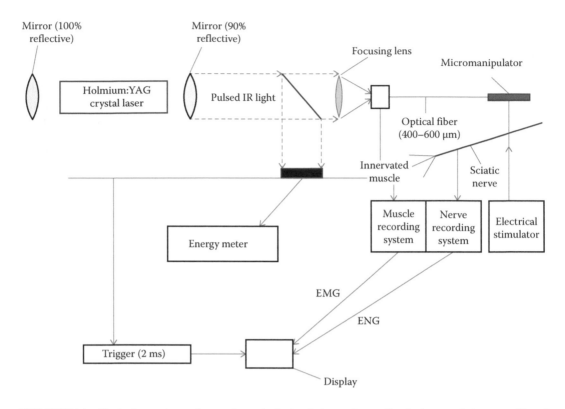

FIGURE 27.1 Typical experimental setup for optical stimulation and recording in the rat sciatic nerve. (Based on Wells, J. et al., *J. Neurosci. Methods*, 163, 326, 2007c.)

and nucleic acids, while in the visible part of the spectrum ($\lambda = 0.40$–0.70 μm), absorption is dominated by (oxy)hemoglobin and melanin, and the near-infrared part of the spectrum (700–1300 nm) represents an area where light is relatively poorly absorbed. Only in the mid- to far-infrared range ($\lambda > 1400$ nm) tissue water dominates absorption and results in shallow penetration (Vogel and Venugopalan, 2003). Thus, the wavelength of the laser should ideally be in mid- to far-infrared range (>1.4 μm).

Considering all these restrictions, the most appropriate wavelengths for stimulation of the sciatic nerve occurs at optical penetration depth of 300–500 μm that corresponds to the optical penetration depth of 2.12 μm. In this scenario, the optical penetration depth matches up with the target geometry in order to stimulate one fascicle within the nerve. It should be clear that this finding also implies that other parts of the nervous system (with different morphology) may require a different wavelength for optimal stimulation such that the optical penetration depth is matched to the morphology of the targeted excitable tissue.

Using the tunable free-electron laser (Edwards and Hutson, 2003), the stimulation threshold (defined as the minimum radiant exposure required for a visible muscle contraction) as well as the ablation/damage threshold (defined as the minimum radiant exposure required for visible cavitation or ejection of material from the nerve) for a number of wavelengths ($\lambda = 2.1$, 3.0, 4.0, 4.5, 5.0, and 6.1 μm) that cover a range of optical penetration depths was determined (Querry and Hale, 1972). Compared to electrical stimulation, the stimulation threshold is lower, so the heat load to the tissue is reduced leading to less functional damage in tissue. Stimulation threshold (0.3–0.4 J/cm^2) is at least two times lower than the damage threshold (0.8–1.0 J/cm^2) (Wells et al., 2005b). In fact, stimulation thresholds reported for stimulation of sensory neurons in the gerbil cochlea were as low as 0.005 J/cm^2 (Izzo et al., 2007). Using histological analysis of exposed tissue, Wells found damage thresholds to be 0.7 J/cm² for a 1% probability of damage and 0.91 J/cm² for a 50% probability (Wells et al., 2007a). Higher pulse rate is likely to cause more damage than lower pulse rate (Chernov et al., 2014). Additionally, higher frequency causes a lowering of damage threshold.

Safety ratio (defined as the ratio of threshold radiant exposure for ablation to that for stimulation) is a better indicator of optimal wavelength (Wells et al., 2010). The Ho:YAG laser was successfully used to determine safety ratio and a number of other parameters. Neural stimulation with this laser resulted in an average stimulation threshold radiant exposure of 0.32 J/cm² and an associated ablation threshold of 2.0 J/cm² (n = 10), yielding a safety ratio of greater than 6 (Wells et al., 2005a).

To determine, if optical stimulation was safe for peripheral nerves, histological analysis was performed (Wells et al., 2007a) on excised rat sciatic nerves, extracted acutely (<1 h after stimulation) or 3–5 days following stimulation. Indications of damage include, but are not limited to, collagen hyalinization, collagen swelling, coagulated collagen, decrease or loss of birefringence image intensity, spindling of cells in perineurium and in nerves (thermal coagulation of cytoskeleton), disruption and vacuolization of myelin sheaths of nerves, disruption of axons, and ablation crater formation.

Nerves from the survival study do not reveal damage to the nerve or surrounding perineurium in 8 of the 10 specimens with damage occurring at radiant exposures above two times the stimulation threshold. These histological findings suggest that nerves can be consistently stimulated using optical means at or near threshold without causing any neural tissue damage.

APPLICATIONS AND ADVANTAGES

The Ho:YAG laser at 2.12 µm is commercially available, can be utilized with fiber optics, and is currently used for a variety of clinical applications (Razvi et al., 1995; Topaz et al., 1995; Kabalin et al., 1998; Fong et al., 1999; Jones et al., 1999). There are few lasers that emit light at 4.0 µm in wavelength, and fiber optic delivery at this wavelength is problematic as regular glass fibers do not transmit beyond 2.5 µm.

Optical stimulation results in artifact free signals even when the recording signal is close to the stimulus. Figure 27.2 illustrates a conceptual comparison between waveforms after electrical and optical stimulation.

Infrared stimulation, due to its wavelength-dependent optical penetration depth, variable spot size, and laser radiant exposure, excels in applications where a precisely controlled and quantifiable volume of action in biological tissue is required (van Hillegersberg, 1997; Vogel and

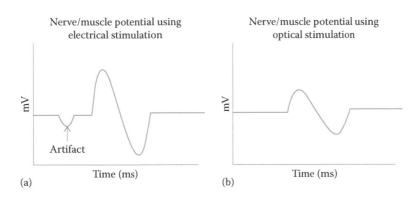

FIGURE 27.2 Conceptual diagram of compound nerve and muscle action potentials versus time based on the recordings from sciatic nerve in rat. (a) CNAP/CMAP recorded using optical stimulation at a wavelength in infrared range; (b) CNAP/CMAP from electrical stimulation. All these responses are achieved after proper amplification and filtering. Note that signals from optical stimulation are artifact free. (Based on Wells, J. et al., *J. Biomed. Opt.*, 10, 064003, 2005a.)

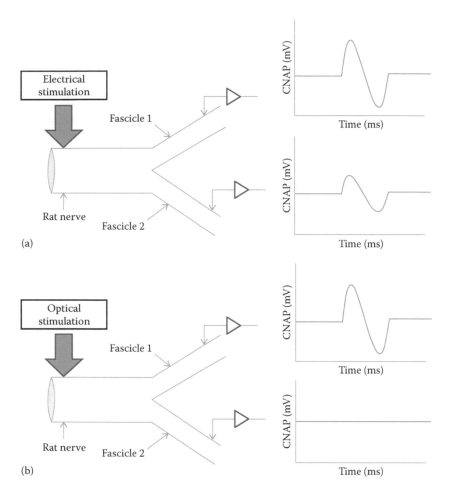

FIGURE 27.3 Conceptual diagram of selective recruitment of isolated nerve fascicles within a large peripheral nerve (sciatic nerve of the rat) using electrical versus optical stimulation techniques. (a) Electrical stimulation results in contraction of both fascicles. The closest one has higher amplitude and the far one has lower. (b) Optical stimulation creates response only where it is applied. The other fascicle shows no contraction.

Venugopalan, 2003). As a consequence of this spatial confinement of energy deposition, optical stimulation is used in cases where closely spaced fascicles (small groups of axons within a nerve) need to be selectively targeted, resulting in selective muscle contraction (Wells et al., 2007c). Figure 27.3 illustrates one such situation and compares it with that of electrical stimulation.

MECHANISM

There are different mechanisms at play in infrared stimulation. One of these is the photochemical effect from laser–tissue interaction. The stimulation thresholds in the infrared part of the spectrum in essence follow the water absorption curve (Wells et al., 2005a) suggesting that no "magical wavelength" has been identified, effectively excluding a single tissue chromophore responsible for any direct photochemical effects. Also one can predict that a photochemical phenomenon is not responsible since infrared photon energy (<0.1 eV) is too low for a direct photochemical effect of laser–tissue interactions, and the laser radiant exposures used are insufficient for any multiphoton effects (Thomsen, 1991).

Another probable mechanism is photomechanical effects (i.e., pressure wave generation) leading to nerve stimulation (Shusterman et al., 2002; Norton, 2003). Results showed that increase in pressure wave through increase in pulse width, almost 3 orders of magnitude (5 μs–5 ms), did not change the stimulation threshold radiant exposure. Moreover, all pulse durations lie well outside the stress confinement zone providing strong evidence that laser-induced pressure waves are not responsible for the optical stimulation mechanism.

The most promising mechanism behind laser stimulation is photothermal interactions between light and nerves. Photothermal interactions include a large group of interaction types resulting from the transformation of absorbed light energy to heat, leading to coagulation or destruction of tissue. Photothermal effects generally appear to be nonspecific and are mediated primarily by absorption of optical energy and secondarily by tissue thermal properties, heat capacity, and diffusivity (Thomsen, 1991; Jacques, 1992).

Infrared camera measurements during laser stimulation of rat sciatic nerves showed a peak temperature rise at the center of the spot of 8.95°C, yielding an average temperature rise of 3.66°C across the Gaussian laser spot. The thermal relaxation time of the rat peripheral nerve was measured to be about 90 ms, which corresponds well to the theoretical value of about 100 ms (given the penetration depth of 0.25 mm).

It was further shown that the nerve temperature increases linearly with laser radiant exposure. Recent literature suggests that thermal changes to mitochondria begin to occur as low as 43°C (protein denaturation begins at tissue temperature close to 57°C). This temperature corresponds to the radiant exposure associated with the onset of thermal damage found in histological analysis of short-term laser nerve stimulation (0.8–1.0 J/cm^2). These results suggest that optical stimulation of peripheral nerves is mediated through a thermal gradient as a result of laser–tissue interaction and that this phenomenon is safe at radiant exposures of at least two times the threshold required for action potential generation (Wells et al., 2010).

USE OF PHOTONICS IN PERIPHERAL NERVE IMAGING

Multiple optical modalities are available for nerve imaging such as two-photon microscopy, fluorescence imaging, and optical coherence tomography (OCT). The primary application of most of them has been so far in the central nervous system. In PNS, their use is limited. Still, some modalities are finding their way into peripheral nerve imaging.

OCT has been used in peripheral nerve imaging. Spectral domain polarization-sensitive OCT (PS-OCT) imaging was used to evaluate uninjured rat sciatic nerves both qualitatively and quantitatively (Islam et al., 2012). OCT and its functional extension, PS-OCT, were used to image sciatic nerve structure with clear delineation of the nerve boundaries to muscle and adipose tissues. A long-known optical effect, bands of Fontana, was also observed. Moving forward, three-dimensional OCT was used for the assessment of the microsurgical anastomoses of vessels and nerves (Boppart et al., 1998). Cross-sectional *in vitro* images of rabbit and human vessels and nerves were obtained in as little as 125 ms at 10 μm resolution by using a solid-state laser as a light source. A surgical microscope was integrated with OCT to perform simultaneous imaging with *en face* visualization. Cross-sectional images were assembled to produce three-dimensional reconstructions of microsurgical specimens. Three-dimensional OCT reconstructions depicted the structure within an arterial anastomosis and helped to identify sites of luminal obstruction. The longitudinal spatial orientation of individual nerve fascicles was tracked in three dimensions to identify changes in position. *In vitro* human arteries and nerves embedded in highly scattering tissue and not visible at microscopy were located and imaged with OCT at eight frames per second (Boppart et al., 1998).

Fluorescence microscopy has also been used for peripheral nerve imaging. Peripheral sensory axons marked with the yellow fluorescent protein in transgenic mice were viewed transcutaneously in superficial nerves (Pan et al., 2003). Degenerating and regenerating axons were followed in live animals with a dissecting microscope and then, after fixation, studied at high resolution by confocal microscopy. Using this approach, differences in regenerative ability after nerve transection, crush injury, and crush injury after a previous "conditioning" lesion were identified. It was also shown that the chemotherapeutic drug vincristine rapidly but transiently blocks regeneration and that the immunosuppressive drug FK506 modestly enhances regeneration. Fibred fluorescence microscopy, an updated technique has been successfully used in imaging axon regeneration–degeneration in peripheral nerves (Vincent et al., 2006). It was used to image an injured the saphenous nerve of a Thy-1 eYFP transgenic mouse. They imaged the fiber bundles forming the main nerve trunk *in vivo*. The arrangement was the same as observed in the fixed nerve, which was visualized by standard fluorescence microscopy. These authors observed an absence of fluorescence immediately after the injury, which is attributed to the loss of the YFP protein through the damaged cell membranes.

Fiber-optic confocal microendoscopy (CME) of yellow fluorescent protein (YFP) expression was used for live imaging in motor neurons to observe and monitor axonal and neuromuscular synaptic phenotypes in mutant mice (Wong et al., 2009). Slow degeneration of axons and motor nerve terminals at neuromuscular junctions was visualized following sciatic nerve injury in WldS mice with slow Wallerian degeneration.

Two-photon microscopy was used as a noninvasive method for imaging single nerve endings within the skin of anesthetized transgenic mice (Yuryev and Khiroug, 2012). Besides excellent signal-to-background ratio and nanometer-scale spatial resolution, this method offers time-lapse "movies" of pathophysiological changes in nerve fine structure over minutes, hours, days or weeks. Structure of keratinocytes and dermal matrix was visualized simultaneously with nerve endings, providing clear landmarks for longitudinal analysis. Feasibility of dissecting individual nerve fiber was demonstrated with infrared laser along with monitoring their degradation and regeneration. In summary, this excision-free optical biopsy technique is ideal for longitudinal microscopic analysis of animal skin and skin innervations *in vivo* and can be applied widely in preclinical models of chronic pain, allergies, skin cancers, and a variety of dermatological disorders.

CHALLENGES AND FUTURE PROSPECTS

The primary challenge with optical stimulation is to make it viable for a large number of medical applications. To date, this approach has been demonstrated using large, cumbersome, and expensive laboratory laser sources (FEL, Ho:YAG). For its regular use in medical applications, a simple, user-friendly, low-cost device must be developed. A research-grade optical stimulator (Lockheed-Martin-Aculight, Capella) based on solid-state laser technology has recently been used commercially to an extent (Wells et al., 2010). This is a 1.85–1.87 μm (resulting in similar absorption as Ho:YAG laser in soft tissue) diode laser with a fiber-coupled output, representing a >95% reduction in size compared to the Ho:YAG laser.

Another limitation of infrared stimulation is that it requires high light intensities and has demonstrated limited localization and depth penetration due to the prevalence of water in all tissues. This is a problem where nonselective, whole nerve stimulation is needed. Emerging technologies exploiting extrinsic absorbers might overcome some of these deficiencies. Moreover, the development of a wider range of small molecule photoswitches promises neurostimulation to advance to a great extent.

This technique is limited to procedures in which the nerve preparation is exposed and clear from debris, such as saline, blood, or tissue, covering the target irradiation zone for stimulation. Furthermore, if the mechanism is thermal, long-term tissue tolerance with optical stimulation may be a potential obstacle for implantable implementation. Chronic safety studies are required to assess the full clinical potential of this technique.

New technologies are on the way to address these issues. Electrical–optical hybrid stimulation is one of them. Subthreshold electrical depolarization has been proposed to reduce the optical energy required for stimulation and the strongest response was observed when the two pulses ended simultaneously (Duke et al., 2009). His work suggests the optical energy requirements could be reduced by a factor of 2 if electrical stimulus is applied at 90% of the electrical threshold and by a factor of 3 if the percentage is 95. In both cases, spatial localization is preserved comparing to INS alone with the added benefit of increasing the ratio of damage threshold to stimulation threshold resulting in a reduction of heat load to the tissue. However, this reduction is dependent upon electrode position relative to the target neurons, and some recent research (Liljemalm et al., 2013; Peterson and Tyler, 2013) suggests that this process may not be applicable to all neural targets.

Adding light absorbing materials to tissue is another way to enhance INS. Photo absorbers along with patterned light were employed near cultured rat cortical cells (Farah et al., 2013). This method allowed to use wavelengths both in visible and near-infrared range. Most importantly, significant reductions in the energy required for stimulation was achieved although at the cost of introducing exogenous substance to the target tissue.

Optoelectric neural stimulation is another novel technique to stimulate neurons using light (Bareket-Keren and Hanein, 2014b). This technique uses photoactive materials that generate an electric field in response to exposure to light. These photoactive materials include, but are not limited to quantum dots (Pappas et al., 2007; Lugo et al., 2012), photoconductive silicon (Suzurikawa et al., 2007), organic polymers (Ghezzi et al., 2011, 2013), and carbon nanotubes (Bareket-Keren and Hanein, 2014a,b). All of them have biocompatibility issues.

In summary, different photonic approaches could be used for peripheral nerve stimulation and imaging. Other than already existing modalities of photonics, novel approaches are emerging to stimulate and image PNS with more sensitivity and specificity. Use of photonics in this field is still in its infancy. Despite its limitations, we can justifiably expect that more concentrated efforts from all fronts of researchers might alleviate many technical hurdles, thereby opening new pathways and possibilities to move the field forward.

REFERENCES

Agnew WF, McCreery DB, Yuen TG, Bullara LA (1989) Histologic and physiologic evaluation of electrically stimulated peripheral nerve: Considerations for the selection of parameters. *Ann Biomed Eng* 17:39–60.

Andreasen LN, Struijk JJ (2003) Artefact reduction with alternative cuff configurations. *IEEE Trans Biomed Eng* 50:1160–1166.

Banghart M, Borges K, Isacoff E, Trauner D, Kramer RH (2004) Light-activated ion channels for remote control of neuronal firing. *Nat Neurosci* 7:1381–1386.

Bareket-Keren L, Hanein Y (2014a) Carbon nanotube-based multi electrode arrays for neuronal interfacing: Progress and prospects. *Frontiers in Neural Circuits* 6:10–25.

Bareket-Keren L, Hanein Y (2014b) Novel interfaces for light directed neuronal stimulation: Advances and challenges. *Int J Nanomed* 9:65–83.

Beharry AA, Woolley GA (2011) Azobenzene photoswitches for biomolecules. *Chem Soc Rev* 40:4422–4437.

Boppart SA, Bouma BE, Pitris C, Tearney GJ, Southern JF, Brezinski ME, Fujimoto JG (1998) Intraoperative assessment of microsurgery with three-dimensional optical coherence tomography. *Radiology* 208:81–86.

Callaway EM, Katz LC (1993) Photostimulation using caged glutamate reveals functional circuitry in living brain slices. *Proc Natl Acad Sci U S A* 90:7661–7665.

Carter ME, de Lecea L (2011) Optogenetic investigation of neural circuits in vivo. *Trends Mol Med* 17:197–206.

Chernov MM, Chen G, Roe AW (2014) Histological assessment of thermal damage in the brain following infrared neural stimulation. *Brain Stimul* 7:476–482.

Dalva MB, Katz LC (1994) Rearrangements of synaptic connections in visual cortex revealed by laser photostimulation. *Science* 265:255–258.

Duke AR, Cayce JM, Malphrus JD, Konrad P, Mahadevan-Jansen A, Jansen ED (2009) Combined optical and electrical stimulation of neural tissue in vivo. *J Biomed Opt* 14:060501–060501-3.

Edwards GS, Hutson MS (2003) Advantage of the Mark-III FEL for biophysical research and biomedical applications. *J Synchr Radiat* 10:354–357.

Farah N, Zoubi A, Matar S, Golan L, Marom A, Butson CR, Brosh I, Shoham S (2013) Holographically patterned activation using photo-absorber induced neural? Thermal stimulation. *J Neural Eng* 10:056004.

Farber IC, Grinvald A (1983) Identification of presynaptic neurons by laser photostimulation. *Science* 222:1025–1027.

Fiore L, Corsini G, Geppetti L (1996) Application of non-linear filters based on the median filter to experimental and simulated multiunit neural recordings. *J Neurosci Methods* 70:177–184.

Fong M, Clarke K, Cron C (1999) Clinical applications for the holmium: YAG laser in disorders of the paediatric airway. *J Otolaryngol Head Neck Surg* 28:337.

Fried NM, Lagoda GA, Scott NJ, Su L, Burnett AL (2008) Optical stimulation of the cavernous nerves in the rat prostate. In: *Proc. SPIE 6842, Photonic Therapeutics and Diagnostics IV, 684213, International Society for Optics and Photonics*, San Jose, CA, January 19, 2008.

Ghezzi D, Antognazza MR, Dal Maschio M, Lanzarini E, Benfenati F, Lanzani G (2011) A hybrid bioorganic interface for neuronal photoactivation. *Nat Commun* 2:166.

Ghezzi D, Antognazza MR, Maccarone R, Bellani S, Lanzarini E, Martino N, Mete M, Pertile G, Bisti S, Lanzani G (2013) A polymer optoelectronic interface restores light sensitivity in blind rat retinas. *Nat Photon* 7:400–406.

Gradinaru V, Mogri M, Thompson KR, Henderson JM, Deisseroth K (2009) Optical deconstruction of parkinsonian neural circuitry. *Science* 324:354–359.

Gradinaru V, Zhang F, Ramakrishnan C, Mattis J, Prakash R, Diester I, Goshen I, Thompson KR, Deisseroth K (2010) Molecular and cellular approaches for diversifying and extending optogenetics. *Cell* 141:154–165.

Gunaydin LA, Yizhar O, Berndt A, Sohal VS, Deisseroth K, Hegemann P (2010) Ultrafast optogenetic control. *Nat Neurosci* 13:387–392.

Horch KW, Dhillon GS (2004) *Neuroprosthetics: Theory and Practice*. World Scientific, Hackensack, NJ.

Islam MS, Oliveira MC, Wang Y, Henry FP, Randolph MA, Park BH, de Boer JF (2012) Extracting structural features of rat sciatic nerve using polarization-sensitive spectral domain optical coherence tomography. *J Biomed Opt* 17:056012-1–056012-9.

Izzo AD, Walsh Jr JT, Jansen ED, Bendett M, Webb J, Ralph H, Richter C (2007) Optical parameter variability in laser nerve stimulation: A study of pulse duration, repetition rate, and wavelength. *IEEE Trans Biomed Eng* 54:1108–1114.

Jacques SL (1992) Laser-tissue interactions. Photochemical, photothermal, and photomechanical. *Surg Clin North Am* 72:531–558.

Ji Z, Ito S, Honjoh T, Ohta H, Ishizuka T, Fukazawa Y, Yawo H (2012) Light-evoked somatosensory perception of transgenic rats that express channelrhodopsin-2 in dorsal root ganglion cells. *PLoS ONE* 7:e32699.

Jones JW, Schmidt SE, Richman BW, Miller CC, Sapire KJ, Burkhoff D, Baldwin JC (1999) Holmium: YAG laser transmyocardial revascularization relieves angina and improves functional status. *Ann Thorac Surg* 67:1596–1601.

Kabalin JN, Gilling PJ, Fraundorfer MR (1998) Application of the holmium: YAG laser for prostatectomy. *J Clin Laser Med Surg* 16:21–27.

Knopfel T, Boyden ES (2012) *Optogenetics: Tools for Controlling and Monitoring Neuronal Activity*. Elsevier, New York.

Kramer RH, Fortin DL, Trauner D (2009) New photochemical tools for controlling neuronal activity. *Curr Opin Neurobiol* 19:544–552.

Liljemalm R, Nyberg T, von Holst H (2013) Heating during infrared neural stimulation. *Lasers Surg Med* 45:469–481.

Liske H, Towne C, Anikeeva P, Zhao S, Feng G, Deisseroth K, Delp S (2013) Optical inhibition of motor nerve and muscle activity in vivo. *Muscle Nerve* 47:916–921.

Llewellyn ME, Thompson KR, Deisseroth K, Delp SL (2010) Orderly recruitment of motor units under optical control in vivo. *Nat Med* 16:1161–1165.

Lugo K, Miao X, Rieke F, Lin LY (2012) Remote switching of cellular activity and cell signaling using light in conjunction with quantum dots. *Biomed Opt Exp* 3:447–454.

Matsuzaki M, Ellis-Davies GC, Nemoto T, Miyashita Y, Iino M, Kasai H (2001) Dendritic spine geometry is critical for AMPA receptor expression in hippocampal CA1 pyramidal neurons. *Nat Neurosci* 4:1086–1092.

Matsuzaki M, Honkura N, Ellis-Davies GC, Kasai H (2004) Structural basis of long-term potentiation in single dendritic spines. *Nature* 429:761–766.

Miesenböck G, Kevrekidis IG (2005) Optical imaging and control of genetically designated neurons in functioning circuits. *Annu Rev Neurosci* 28:533–563.

Milburn T, Matsubara N, Billington AP, Udgaonkar JB, Walker JW, Carpenter BK, Webb WW, Marque J, Denk W (1989) Synthesis, photochemistry, and biological activity of a caged photolabile acetylcholine receptor ligand. *Biochemistry (N Y)* 28:49–55.

Monesson-Olson BD, Browning-Kamins J, Aziz-Bose R, Kreines F, Trapani JG (2014) Optical stimulation of zebrafish hair cells expressing channelrhodopsin-2. *PloS One* 9:e96641.

Nagel G, Szellas T, Huhn W, Kateriya S, Adeishvili N, Berthold P, Ollig D, Hegemann P, Bamberg E (2003) Channelrhodopsin-2, a directly light-gated cation-selective membrane channel. *Proc Natl Acad Sci U S A* 100:13940–13945.

Navarro X, Krueger TB, Lago N, Micera S, Stieglitz T, Dario P (2005) A critical review of interfaces with the peripheral nervous system for the control of neuroprostheses and hybrid bionic systems. *J Peripher Nerv Syst* 10:229–258.

Nikolenko V, Poskanzer KE, Yuste R (2007) Two-photon photostimulation and imaging of neural circuits. *Nat Methods* 4:943–950.

Norton SJ (2003) Can ultrasound be used to stimulate nerve tissue? *Biomed Eng Online* 2:6.

Palanker D, Vankov A, Huie P, Baccus S (2005) Design of a high-resolution optoelectronic retinal prosthesis. *J Neural Eng* 2:S105.

Pan YA, Misgeld T, Lichtman JW, Sanes JR (2003) Effects of neurotoxic and neuroprotective agents on peripheral nerve regeneration assayed by time-lapse imaging in vivo. *J Neurosci* 23:11479–11488.

Pappas TC, Wickramanyake WS, Jan E, Motamedi M, Brodwick M, Kotov NA (2007) Nanoscale engineering of a cellular interface with semiconductor nanoparticle films for photoelectric stimulation of neurons. *Nano Lett* 7:513–519.

Paxinos G (2014) *The Rat Nervous System.* Academic Press, San Diego, CA.

Peters A, Palay SL (1991) *The Fine Structure of the Nervous System: Neurons and Their Supporting Cells.* Oxford, U.K.: Oxford University Press.

Peterson E, Tyler D (2013) Motor neuron activation in peripheral nerves using infrared neural stimulation. *J Neural Eng* 11:016001.

Querry M, Hale G (1972) Optical-constants of water in 200-nm to 0.2-mm wavelength region. *Journal of the Optical Society of America* 62:1381–1381.

Razvi HA, Chun SS, Denstedt JD, Sales JL (1995) Soft-tissue applications of the holmium: YAG laser in urology. *J Endourol* 9:387–390.

Rial Verde E, Zayat L, Etchenique R, Yuste R (2008) Photorelease of GABA with visible light using an inorganic caging group. *Front Neural Circuits* 2:2.

Richter C, Bayon R, Izzo AD, Otting M, Suh E, Goyal S, Hotaling J, Walsh JT (2008) Optical stimulation of auditory neurons: Effects of acute and chronic deafening. *Heart Res* 242:42–51.

Sahel J, Roska B (2013) Gene therapy for blindness. *Annu Rev Neurosci* 36:467–488.

Sharp AA, Fromherz S (2011) Optogenetic regulation of leg movement in midstage chick embryos through peripheral nerve stimulation. *J Neurophysiol* 106:2776–2782.

Shusterman V, Jannetta PJ, Aysin B, Beigel A (2002) Direct mechanical stimulation of brainstem modulates cardiac rhythm and repolarization in humans. *J Electrocardiol* 35:247.

Suzurikawa J, Takahashi H, Kanzaki R, Nakao M, Takayama Y, Jimbo Y (2007) Light-addressable electrode with hydrogenated amorphous silicon and low-conductive passivation layer for stimulation of cultured neurons. *Appl Phys Lett* 90:093901.

Teudt IU, Nevel AE, Izzo AD, Walsh JT, Richter C (2007) Optical stimulation of the facial nerve: A new monitoring technique? *Laryngoscope* 117:1641–1647.

Thompson AC, Stoddart PR, Jansen ED (2014) Optical stimulation of neurons. *Curr Mol Imaging* 3:162.

Thompson AC, Wade SA, Brown WG, Stoddart PR (2012) Modeling of light absorption in tissue during infrared neural stimulation. *J Biomed Opt* 17:0750021–0750026.

Thomsen S (1991) Pathologic analysis of photothermal and photomechanical effects of laser–tissue interactions. *Photochem Photobiol* 53:825–835.

Topazo, Rozenbaum EA, Luxenberg MG, Schumacher A (1995) Laser-assisted coronary angioplasty in patients with severely depressed left ventricular function: Quantitative coronary angiography and clinical results. *J Interv Cardiol* 8:661–670.

Towne C, Montgomery KL, Iyer SM, Deisseroth K, Delp SL (2013) Optogenetic control of targeted peripheral axons in freely moving animals. *PLoS ONE* 8:e72691.

Tye KM, Deisseroth K (2012) Optogenetic investigation of neural circuits underlying brain disease in animal models. *Nat Rev Neurosci* 13:251–266.

van Hillegersberg R (1997) Fundamentals of laser surgery. *Eur J Surg* 163:3–12.

Vincent P, Maskos U, Charvet I, Bourgeais L, Stoppini L, Loresche N, Changeux JP, Lambert R, Meda P, Paupardin-Tritsch D (2006) Live imaging of neural structure and function by fibred fluorescence microscopy. *EMBO Rep* 7:1154–1161.

Vogel A, Venugopalan V (2003) Mechanisms of pulsed laser ablation of biological tissues. *Chem Rev* 103:577–644.

Wagenaar DA, Potter SM (2002) Real-time multi-channel stimulus artifact suppression by local curve fitting. *J Neurosci Methods* 120:113–120.

Walker JW, McCray JA, Hess GP (1986) Photolabile protecting groups for an acetylcholine receptor ligand Synthesis and photochemistry of a new class of o-nitrobenzyl derivatives and their effects on receptor function. *Biochemistry (N Y)* 25:1799–1805.

Wang H, Zylka MJ (2009) Mrgprd-expressing polymodal nociceptive neurons innervate most known classes of substantia gelatinosa neurons. *J Neurosci* 29:13202–13209.

Warther D, Gug S, Specht A, Bolze F, Nicoud J, Mourot A, Goeldner M (2010) Two-photon uncaging: New prospects in neuroscience and cellular biology. *Bioorg Med Chem* 18:7753–7758.

Wells J, Kao C, Jansen ED, Konrad P, Mahadevan-Jansen A (2005a) Application of infrared light for in vivo neural stimulation. *J Biomed Opt* 10:064003–064003-12.

Wells J, Kao C, Konrad P, Milner T, Kim J, Mahadevan-Jansen A, Jansen ED (2007b) Biophysical mechanisms of transient optical stimulation of peripheral nerve. *Biophys J* 93:2567–2580.

Wells J, Kao C, Mariappan K, Albea J, Jansen ED, Konrad P, Mahadevan-Jansen A (2005b) Optical stimulation of neural tissue in vivo. *Opt Lett* 30:504–506.

Wells J, Konrad P, Kao C, Jansen ED, Mahadevan-Jansen A (2007c) Pulsed laser versus electrical energy for peripheral nerve stimulation. *J Neurosci Methods* 163:326–337.

Wells JD, Cayce JM, Mahadevan-Jansen A, Konrad PE, Jansen ED (2010). Infrared nerve stimulation: A novel therapeutic laser modality. In: Welch AJ, van Gemert MJC (eds.), *Optical-Thermal Response of Laser-Irradiated Tissue*, pp. 915–939. Springer, New York.

Wells JD, Thomsen S, Whitaker P, Jansen ED, Kao CC, Konrad PE, Mahadevan-Jansen A (2007a) Optically mediated nerve stimulation: Identification of injury thresholds. *Lasers Surg Med* 39:513–526.

Wieboldt R, Gee KR, Niu L, Ramesh D, Carpenter BK, Hess GP (1994) Photolabile precursors of glutamate: Synthesis, photochemical properties, and activation of glutamate receptors on a microsecond time scale. *Proc Natl Acad Sci U S A* 91:8752–8756.

Wong F, Fan L, Wells S, Hartley R, Mackenzie FE, Oyebode O, Brown R, Thomson D, Coleman MP, Blanco G (2009) Axonal and neuromuscular synaptic phenotypes in Wld S, SOD1 G93A and ostes mutant mice identified by fiber-optic confocal microendoscopy. *Mol Cellul Neurosci* 42:296–307.

Yizhar O, Fenno LE, Davidson TJ, Mogri M, Deisseroth K (2011) Optogenetics in neural systems. *Neuron* 71:9–34.

Yuryev M, Khiroug L (2012) Dynamic longitudinal investigation of individual nerve endings in the skin of anesthetized mice using in vivo two-photon microscopy. *J Biomed Opt* 17:046007–046007.

Zemelman BV, Lee GA, Ng M, Miesenböck G (2002) Selective photostimulation of genetically chARGed neurons. *Neuron* 33:15–22.

Zhang F, Aravanis AM, Adamantidis A, de Lecea L, Deisseroth K (2007a) Circuit-breakers: Optical technologies for probing neural signals and systems. *Nat Rev Neurosci* 8:577–581.

Zhang F, Wang L, Brauner M, Liewald JF, Kay K, Watzke N, Wood PG, Bamberg E, Nagel G, Gottschalk A (2007b) Multimodal fast optical interrogation of neural circuitry. *Nature* 446:633–639.

28 EEG Biomarkers in Depression

*Nevzat Tarhan, Fatma Zehra Keskin Krzan,
Merve Çebi, Şehadet Ekmen, Türker Ergüzel,
Serhat Özekes, Baris Ünsalver, and Cumhur Taş*

CONTENTS

INTRODUCTION

Depression is a multifactorial mental health disorder that is influenced by various social and clinical factors, including childhood parental relations, trauma, gender, age, social status, and life events, together with biological factors, such as genetic vulnerability and impairments in biological stress regulation. To date, researchers and clinicians have extensively utilized symptom-based classification systems such as DSM and ICD to diagnose patients suffering from mental health disorder. Importantly, the reliability of symptom-based classifications has always been a debate, and thus, there is a strong necessity for identifying biological criteria for diagnosing our patients. This will not only resolve the problems in the reliability of clinicians' diagnosis but also may help to predict treatment response and long-term prognosis of our patients. One possible tool that could have potential as a biomarker in depression and other mental health disorders is quantitative electroencephalogram (qEEG). Since the 1980s, following the studies in frontal alpha band asymmetry, many researchers attempted to use qEEG to diagnose and follow up patients with depression. Although ever-growing data underlines the benefits of qEEG in depression, the specificity of qEEG in depression has always been questionable.

The aim of this chapter is to review the state of the art regarding the potential use of qEEG–based biomarkers in depression. Within this perspective, we first reviewed the early studies using EEG absolute and relative power and then presented the findings on recent EEG markers such as cordance and connectivity in depression. As a future suggestion, combining artificial intelligence (AI) methods with qEEG data appears to have a crucial importance as a biomarker in depression. Therefore, the last part of this chapter includes a section reviewing the recent state of the art in regards to AI methods to diagnose and identify predictors of treatment in depression.

ABSOLUTE AND RELATIVE POWER IN DEPRESSION

Power spectral analysis is one of the most common measurement methods of quantitative EEG topography in which absolute and relative power of specified frequency bands are obtained (Pivik et al. 1993). Absolute power is the actual amount of power of a band at a given electrode that has no combination with other frequency bands and can be evaluated in microvolts (Hunter et al. 2007; Knott et al. 2001). In contrast, relative power is the percentage of power contained in a frequency band that is relative to the total EEG spectrum. The literature presents numerous findings for each frequency band. However, alpha power has been studied as one of the most remarkable indicators of depression or increased risk for depression (Henriques and Davidson 1990; Putnam and McSweeney 2008). The relative alpha power is measured by taking a percentage of the alpha within the combination of the total activity of alpha, beta, delta, and theta bands (Hunter et al. 2007; Knott et al. 2001; Leuchter et al. 1994a). Mathematically, the relative spectrum is defined as

$$Pr(f) = \frac{Pa(f)}{\left(\sum Pa(f_i)\right)}$$

where Pr(f) refers to the relative EEG waves at frequency f, Pa(f) represents the absolute EEG waves at the same frequency, and the $\sum$ is the total of EEG power of the entire spectrum or a portion of it (Kropotov 2010).

The relationship between EEG power measurements and metabolism or perfusion in the brain may vary across frequency bands (Buchsbaum et al. 1986; Hunter et al. 2007; Nagata 1988). As a note, in comparison with other qEEG measures, the absolute power and relative power appear to have a limited correlation with cerebral local perfusion (Hughes and John 1999; Niedermeyer and Lopes da Silva 2004; Tas et al. 2015). Nagata (1988) presented that power ratios between low- and high-frequency bands have been the most accurate estimates to detect changes in EEG activity as well as metabolic changes.

The absolute and relative power differences have been found to be stable over time and demonstrated as a reflection of oscillatory patterns of brain function by means of different approaches (Fingelkurts et al. 2006). Particularly, absolute and relative power alterations have been demonstrated in responder and nonresponder to specific drugs in depression. Ulrich et al. (1984) found that among other EEG parameters in which patients were separated, there is a decrease in absolute alpha power and increases in other slower frequencies over 4 weeks of drug treatment (see also Schaffer et al. 1983). Filipović et al. (1998) investigated whether there are specific spectral EEG characteristics that distinguish depressed from nondepressed patients. They found that depressed patients differed from the spectral EEG characteristic of nondepressed patients presenting less absolute and relative power in the alpha1 (7.5–10 Hz) and relatively more relative power in the alpha 2 (10.513 Hz) band. In another study, Iosifescu et al. (2004) showed the predictive capability of the frontal theta band relative power in individuals with major depressive disorder (MDD). More recently, Grin-Yatsenko et al. (2009) found an increased alpha and theta power in patients in early stages of depression. Jaworska et al. (2012a) also demonstrated that depressed individuals displayed an increased frontal and parietal alpha power and left midfrontal hypoactivity. In line with this, one of the previous studies on depression and its comorbidity with internet addiction was conducted recording resting and eyes-closed qEEG and computing absolute and relative power. Accordingly, for all the brain regions, increased relative theta power and decreased relative alpha power were reported in the comorbid group (Lee et al. 2014).

Moreover, a large body of literature in depression has been dedicated to the computation of the EEG power differences between the left and right hemisphere, particularly in the alpha power that is referred to as the frontal alpha asymmetry (Alhaj et al. 2011; Henriques and Davidson 1990; Olbrich and Arns 2013). This "amplitude asymmetry" is an alternative method that describes the interlead power variation and fundamentally measures the variation in power among two EEG leads and the outcome can be presented as a mathematical difference or ratio. This measure when used for homologous left and right recording sites would provide a measure of the lateralized differences among the two sites (Myslobodsky et al. 1991). For instance, Bruder et al. (2008) found increased alpha power and alpha asymmetry in SSRI medication responders. Lastly, in an earlier study Schaffer et al. (1983) demonstrated that there is greater brain activity in the right frontal regions in depressed subjects as compared with healthy controls supporting the contribution of frontal EEG asymmetry in the development of depression.

The detected increased pretreatment alpha activity might be a biomarker of the association between low serotonergic activity and low arousal due to the important role of the biological mechanism over the increased alpha power and alpha asymmetry as it is seen in SSRI responders (Bruder et al. 2008; Tenke and Kayser 2005). Recent studies suggest that alterations in absolute and relative powers might be useful in predicting antidepressant response and hence may provide a tool for better treatment approaches (Bares et al. 2008; Hunter et al. 2007). However, several studies have failed to replicate that increased alpha power in depressed patients. In contrary, they reported a decreased alpha power at the right frontal sites relative to the left side (Chang et al. 2012; Henriques and Davidson 1991; Schaffer et al. 1983). There are also several studies that demonstrated decreased relative alpha activity in patients with MDD compared to other patient groups (Olbrich and Arns 2013; Pozzi et al. 1995; Knott and Lapierre 1987).

As for alpha asymmetry, compared with other different types of analysis techniques, it has a number of limitations. A potential problem with the measure of the amplitude asymmetry is that the interpretation of recordings between the two sites is likely to be affected by structural or functional asymmetry as the shape of the head is generally not always symmetric (Myslobodsky et al. 1991). Particularly, differences in cranial and bone thickness (brain parenchymal asymmetries) influence the frontal alpha asymmetry. Moreover, frontal alpha asymmetry has been related to negative feelings independent from depression in a number of experimental studies. Nevertheless, several studies demonstrated that the frontal alpha asymmetry is moderately stable over time in MDD (Debener et al. 2000; see also Tomarken et al. 1992).

Taken as a whole, despite the value of early studies, which ultimately improved our insights for understanding depression, the inconsistencies across studies and findings that are not solely specific to depression suggest that this absolute and relative power may not be reliably used as a clinical biomarker in depression.

EEG CORDANCE IN DEPRESSION

Currently, cordance has been started to take place in the literature as another qEEG parameter and is developed by Leuchter et al. in 1994 for measuring brain activity indirectly (Leuchter et al. 1994a). In brief, cordance is calculated by combining z-transformed scores of absolute and relative power of EEG spectra and found to be sensitive in the detection of cortical differentiation (Leuchter et al. 1994b, 1999). The mathematical model of cordance measure includes three steps: First, EEG power values are calculated by means of reattributional electrode montage, which yields a reattributed power that is averaged from electrode pairs sharing a common electrode. As compared to other reference approaches, the reattributional electrode montage has been found to have stronger associations with qEEG and positron emission tomography measures. Second, the absolute (A) and relative (R) power values for each frequency band (f) in each electrode site (s) are normalized via z-transformation statistics and Anorm(s, f) and Rnorm(s, f) are yielded in order to measure deviation from the mean values for each power value. Finally, a cordance value (Z) is created by summing the z-scores for absolute and relative powers for each electrode site and each frequency band where Z(s, f) = Anorm(s, f) + Rnorm(s, f) (Cook et al. 1998; Leuchter et al. 1999).

Regional cerebral perfusion has been associated with concordant activity, which is represented by positive cordance values, whereas negative cordance values have been found to reflect discordant activity, which is in turn thought to be an indicator of low cerebral metabolism (Leuchter et al. 1999). Furthermore, as compared to either absolute power or relative power alone, cordance has been proven to have stronger correlations with cerebral metabolism (Leuchter et al. 1999) and found to be less affected by covariate factors such as age and gender (Baskaran et al. 2012; Morgan et al. 2005).

Recently, the relationship between prefrontal theta rhythm (4–8 Hz) and anterior cingulate cortex has been reported by ample studies (Asada et al. 1999; Ishii et al. 1999, Pizzagalli et al. 2003). Previous neuroimaging studies reported that prefrontal theta activity is associated with cerebral metabolism in rostral anterior cingulate cortex (Mulert et al. 2007; Pizzagalli et al. 2003). Researchers demonstrated that prefrontal theta cordance can be utilized as an indicator of rostral part of anterior cingulate cortex activity, which is strongly involved in affective processes (Bares et al. 2012). In parallel to this, several neuroimaging studies replicated the role of anterior cingulate cortex in depression (Cook et al. 2014; Drevets 1998; Mayberg et al. 1997). Accordingly, depression has been associated with hypoactivity in dorsal anterior cingulate region (Ebert and Ebmeier 1996) and rostral anterior cingulate activity has been linked to the degree of clinical response to antidepressants (Chen et al. 2007; Hunter et al. 2013; Mayberg et al. 1997; Mulert et al. 2007).

Related EEG literature suggests that cordance has the potential to be used as a biomarker for predicting and treating depression using antidepressants. In 1999, Cook et al. provided the first evidence for the pretreatment predictive value of cordance in depression. In their study, patients who received antidepressant and placebo treatment for 8 weeks were divided into two groups defined as concordant and discordant using the number of electrodes exhibiting discordance. As mentioned earlier, cordance calculation refers to the two separate measurements: First is a categorical index for each electrode and the localization of the underlying brain region that depends on the type of link between absolute and relative power; second is a representation of global brain state, depending on the sections of electrodes showing concordance and discordance (Cook et al. 1999). Results showed that patients who had concordant activity in theta frequency band prior to fluoxetine treatment were more responsive to antidepressant treatment as compared with patients with discordant activity. These results are consistent with other studies, which have shown that frontal activity in theta band is linked to the function of the anterior cingulate cortex that is involved in the pathophysiology of depression (Bares et al. 2007). Furthermore, response to place was not predicted by dividing the place group into concordant and discordant. Leuchter et al. (2002), however, later associated increased theta cordance with responsiveness to placebo treatment. Several studies have also revealed that decreased prefrontal theta cordance compared to baseline was successful in discriminating antidepressant treatment responders from nonresponders (Bares et al. 2008, 2010; Hunter et al. 2006). Hunter et al. (2010) investigated patterns of change in symptom severity in major depression. They revealed that compared with nonresponders, responders showed decreased midline and right frontal theta cordance, indicating the predictive value of theta cordance measures. Regarding comparative studies that have used cordance for diagnostic purposes, prefrontal theta cordance was found to be significantly higher among depressed patients as compared with control subjects (Broadway et al. 2012). In another comparative EEG study, our group reported a decreased right parietal theta cordance in patients with bipolar depression as compared with unipolar depression (Tas et al. 2015). In addition to these studies, investigating the prediction of depression and treatment response to antidepressants, another study found that prefrontal theta cordance can be also a predictor for 6-month antidepressant response to deep brain stimulation of subcallosal cingulate cortex in patients with treatment resistance depression (Broadway et al. 2012). Recently, Bares et al. (2015) compared antidepressant medication treatment with 1 Hz repetitive transcranial magnetic stimulation (rTMS) in patients with major depression. They reported that prefrontal theta cordance is not only a biomarker for antidepressant medication treatment, but also rTMS treatment responsiveness in depression. Contrary to the general findings focusing on theta cordance, Arns et al. (2012) indicated that decreased prefrontal delta and beta cordance in nonresponders to rTMS treatment as compared to responders. Finally, using AI methods, researchers also provided evidence for the use of prefrontal theta cordance as a biomarker in rTMS response in major depression (Ergüzel et al. 2015a). As a note, all of these studies were conducted during awake state and our knowledge regarding the concordant and discordant activities during sleep is very limited. A very recent study investigated EEG patterns of patients with depression during REM phase of sleep at the first and fourth week following the antidepressant treatment. Adamczyk et al. (2015) found that responders showed significant higher prefrontal theta cordance compared to nonresponders group after the first week of medication treatment. Their findings showed that prefrontal theta cordance during REM sleep may also have predictive value for responsiveness to antidepressant treatment in depressed patients (Adamczyk et al. 2015).

Despite the aforementioned potential use of EEG cordance as a biomarker in depression there also some limitations that are worth mentioning. Most of the cordance studies were conducted in relatively low sample sizes and larger samples are required to draw firm conclusions. Besides, mathematically combining absolute and relative power using nonlinear approaches may be more meaningful. In fact, antidepressant treatment response (ATR) index was developed to fill this gap (Leuchter et al. 2009). ATR is a measure that combines absolute and relative power nonlinearly and performs better in the early course of antidepressant treatment in major depressive disorder compared with cordance measures (Kuo and Tsai 2010, see also Olbrich and Arns 2013). It has been demonstrated that nonlinear measures as global and local scaling exponents and quantification of phase space dynamics might be used to differ time series of human physiology and performance between healthy groups and individuals with pathology (Arns et al. 2014). Besides, unlike cordance measures, ATR index was found to be the only significant predictor of remission in depression (Leuchter et al. 2009). However, it is important to underline that ATR is based on typical montage and measured with a complex algorithm. For this reason, the biological meaning of high or low ATR scores is not entirely clear. Besides, frontal electrodes might be problematic in calculating ATR index and cordance values due to well-known artifacts (e.g., eye and facial muscles) (Pariante et al. 2009).

Taken as a whole, ATR index have yielded positive results but these results needs to be replicated in future studies. Meta-analyses that include unpublished results in ATR and cordance studies in depression may also be crucial to estimate the real value using cordance and ATR index in depression. Finally, future researches combining multiple neuroimaging tools are required to understand the contradictory findings about different patterns of cordance changes.

EEG CONNECTIVITY IN DEPRESSION

The working brain can be understood not only through the location and signal magnitude of activation sites but also through the connectivity of neural activity between distant regions (Olbrich and Arns, 2013). The coupling of local and distant neurons is a product of the synchronized brain that summates into the rhythmic patterns of the EEE signal (Buzsáki and Draguhn 2004). Among the various types of EEG measurements, connectivity is a measure of temporal correlations between different cortical brain regions (Olbrich and Arns 2013; Wang et al. 2013), providing quantitative analyses of relevant network activity through phase synchronization (nonlinear) and amplitude (linear) properties of a time series of data (Leuchter et al. 2012; Olbrich and Arns 2013; Pereda et al. 2005).

Coherence and phase synchronization are indicators of the brain connectivity within a time series corresponding to different cortical locations (Srinivasan et al. 1998). Phase synchronization is a process in which two or more cyclic signals oscillate together (Fell and Axmacher 2011; Lachaux et al. 1999; Mezeiova and Paluš 2012; Mormann et al. 2000; Rosenblum et al. 1996; Winterhalder et al al 2006). Phase synchronization ranges from 0, indicating no synchronization, to 1, indicating complete synchronization (Li et al. 2015). On the other hand, coherence is a linear association between identical frequency band–related neuronal synchrony, indicating that oscillatory signals of two points or nodes have a fixed oscillatory phase relationship

(Leuchter et al. 2012; Rosenblum et al. 1996; Srinivasan et al. 1998; Thatcher et al. 2008; Weiss and Mueller 2003). Mathematically, coherence is defined as follows:

$$Cxy(f) = \frac{\left|\left(Pxy(f)\right)\right|^2}{Pxx(f)Pyy(f)}$$

where

Cxy(f) is the normalized cross-power spectrum

Pxx and Pyy are the power spectral densities of two signals (x and y)

Pxy refers to the cross-power spectral density (Leuchter et al. 2012; Mezeiova and Paluš 2012)

Following frequency bands were computed for coherence; δ: 1–3 Hz, θ: 4–8 Hz, α: 8–10 Hz, σ: 11–17 Hz, β: 15–29 Hz, and γ: 28–42 Hz (Mezeiova and Paluš 2012). Coherence values range from 0 (no interaction between neuronal populations) to 1 (complete synchronicity) (Leuchter et al. 2012).

Neuroimaging studies have demonstrated that abnormalities in connectivity may help us to understand underlying pathological dysfunction of depression and to determine the optimal treatment options (Kito et al. 2014; Price and Drevets 2010; Rosenblum et al. 1996). Numerous studies have revealed that patients with depression showed altered connectivity patterns when compared with healthy control group (Chorlian et al. 2009; Fingelkurts et al. 2007; Fox et al. 2012; Greicius et al. 2007; Li et al. 2015; Lieber 1988; Olbrich and Arns 2013; Pizzagalli 2011). In particular, alterations within the beta frequency band have been associated with abnormally increased EEG coherence, which can be considered a measure of functional connectivity (Li et al. 2015; Miskovic and Schmidt 2010; Olbrich and Arns 2013). Fingelkurts et al. also showed an increased synchronization in the alpha and theta bands in individuals with MDD (Fingelkurts et al. 2007). In the mentioned study, positively correlating with the severity of depression were short-range anterior, posterior, and left hemisphere functional connections for the alpha frequency band and short-range anterior connections for the theta frequency band (Fingelkurts et al. 2007). Leutcher et al. also demonstrated an increased topographic EEG coherence in alpha and theta bands, particularly between frontal brain regions in MDD (Leuchter et al. 2012). Regarding other frequency bands, Li et al. found abnormally increased connectivity, especially in gamma band in patients with depression compared with the global brain network of healthy controls (Li et al. 2015). Differences in connectivity were also present between treatment responders and nonresponders for depression. Lee et al. (2011) reported that compared with nonresponders, responders to antidepressants showed functional connectivity in the form of increased phase synchronization in the alpha band in frontoparietal regions (the subgenual prefrontal cortex, the left dorsolateral cortex, the left medial prefrontal cortex). Finally, in a very recent study, Tas et al. (2015) found greater coherence in central-temporal theta bands and parietal-temporal alpha and theta bands in patients with bipolar depression as compared with unipolar depressed individuals. These results indicate that EEG connectivity may also help clinicians to diagnose different forms of depression.

These findings have shown that EEG connectivity measurements have potential value as biomarkers in clinical settings. However, there are several limitations in studying connectivity between brain regions. A major issue is that coupling EEG signals do not necessarily imply corresponding activity between surface EEG and brain functional activity in underlying structures (Gómez-Herrero et al. 2008; Leuchter et al. 2012). The reason for this is that scalp EEG potentials do not reflect averaged postsynaptic activity from localized cortical regions beneath a given electrode; instead they reflect activity recorded at surface sites overlying the various cortical regions (Gómez-Herrero et al. 2008; Leuchter et al. 2012; Malmivuo and Plonsey 1995). Therefore, the connectivity of brain regions is inferred from the superposition of all active neural sources located throughout the brain (Gómez-Herrero et al. 2008; Leuchter et al. 2012; Malmivuo and Plonsey 1995). Analysis techniques such as blind source separation have been successfully used to overcome this issue (see Gómez-Herrero et al. 2008). Another important concern is that the results on EEG connectivity within specific frequency bands (e.g., alpha) are inconsistent and scattered (Fingelkurts et al. 2007, see also Fingelkurts et al. 2004). Many studies have reported increased EEG connectivity in depression (Fingelkurts et al. 2007; Leuchter et al. 2012), while many other studies have found decreased EEG coherence in depression (Knott et al. 2001; Lee et al. 2011; Sun et al. 2008). These findings are principally found in the alpha band. Other frequency bands were also significantly correlated with the severity of depressive symptoms, for example, anterior short-range functional connections within the theta band (Fingelkurts et al. 2007; Li et al. 2015). Li et al. (2015) demonstrated that the global coherence in gamma bands was significantly higher in patients with depression than those in healthy controls. Fingelkurts et al. (2007) also found the impaired functional connectivity in both alpha and theta frequency band in major depression. Additionally, several studies showed that increased beta activity in depressed patients (Knott et al. 2001; Lieber 1988).

Taken as a whole, the findings discussed here suggest that in order to eliminate the inconsistencies across studies, it is crucial to standardize recording condition, preprocessing, and data analysis in order to use EEG connectivity as a reliable biomarker. Further research will also be necessary to investigate different neuroimaging modalities especially those utilizing current density maps or estimated sources of scalp EEG.

EEG AND ARTIFICIAL INTELLIGENCE IN DEPRESSION

The ultimate goal for EEG studies in depression is to enlighten the neurophysiological networks and identify potential biomarkers for clinical use. However, many of the studies underlined throughout this chapter have utilized conventional statistical methods that are based on group comparisons using the mean and standard deviations assuming the parametric distribution of patients with depression. However, it is clear that a disorder that is classified by symptoms could not be homogeneous. Therefore, alternative statistical techniques, which are robust to heterogeneous samples and include individual-based evaluations, are needed. At the intersection of medicine and computer science, using AI, information technologists can deploy intelligent algorithms to help improving human health, disease prevention, and healthcare service delivery. By having the capacity to learn and to help discovering new phenomena, AI can assist physicians in diagnosing and their patients and provide new solutions to help making better clinical decisions.

Today, there is significant interest on AI and its branch, machine learning, in the analysis of EEG datasets for decision-making processes like depression diagnosis and treatment response prediction. In general, these studies follow several steps: (1) acquiring brain signals, (2) signal preprocessing, (3) feature extraction, and (4) classification steps. Once the data is collected, selected inputs (features) are subjected to the mathematical algorithms. Within this approach, feature extraction is an important step that directly affects the success of classification step. Feature extraction extracts the embedded information in EEG signal and describes it by values that are named as features. As Bashashati et al. (2007) mentioned, some of the feature extraction methods are band powers (BP), cross-correlation between EEG BP, frequency representation, time–frequency representation, Hjorth parameters, parametric modeling, inverse model, peak picking, and slow cortical potentials calculation. As the final step, classification uses the feature set as input and outputs the mental state of the user. As a note, the most commonly used classification algorithms are artificial neural networks, support vector machines, decision trees, k-nearest neighbors, and naive Bayes.

AI approaches for medical decision-making processes are valuable when both high classification accuracy and less feature requirements are satisfied. Machine learning methods successfully meet the first goal with its adaptive engine, while nature inspired algorithms are focusing on the feature selection (FS) process in order to eliminate less informative and less discriminant features. Besides engineering applications of ANN and FS algorithms, medical informatics is another emerging field using similar methods for medical data processing. Classification of psychiatric disorders is one of the major focuses of medical informatics using AI approaches. Being one of the most debilitating psychiatric diseases, bipolar disorder is often misdiagnosed as unipolar disorder, leading to suboptimal treatment and poor outcomes. Thus, dichotomizing the depressed subjects at earlier stages of illness could help to facilitate efficient and specific treatment process. Recent studies wrapping classification and FS methods to reveal more informative feature subset to discriminate patients suffering from depression underline the promising future of AI approach in neuroscience. Besides, Erguzel et al. (2015a) used pretreatment qEEG cordance as a biomarker to predict rTMS response of patients with depression. In another study, they combined particle swarm optimization (PSO) to improve the classification performance of treatment response in depression (Erguzel et al. 2015b).

Due to its computational shortcomings of existing statistical methods, researchers appreciated metaheuristic algorithms based on simulations to solve complex optimization problems. A prevalent feature of metaheuristic algorithms is that they combine the algorithm's principles, probability, and randomness to imitate natural phenomena similar to the biological evolutionary process or swarm intelligence. In this context, genetic algorithm, differential evolution, PSO, ant colony optimization, firefly algorithm, physical annealing process (e.g., simulated annealing is widely used nature-inspired FS algorithms. The improved and modified versions of those methods mentioned earlier have also been recently used to overcome the drawbacks of the standard applications of the methods (Rizk-Allah et al. 2013). Promising performances of those methods are appreciated by scientists and extensively applied to many optimization problems. We believe that AI approaches could have potential to resolve many diagnostic drawbacks of psychiatric disorders.

CONCLUSIONS

Depression is affecting more than 350 million people in the world and is a significant contributor to the global burden of disease. In this chapter, we presented the state of art of using EEG as a biomarker in depression. As a conclusion, it is clear that we are not very close to the place where EEG biomarkers are accepted as a routine clinical application for depression. However, the current data undisputedly suggests that qEEG is a noninvasive neuroimaging technique that might be used in combination with the conventional symptom-based interviews in depression.

REFERENCES

Adamczyk, M., Gazea, M., Wollweber, B. et al. Cordance derived from REM sleep EEG as a biomarker for treatment response in depression—A naturalistic study after antidepressant medication. *Journal of Psychiatric Research* 63 (2015): 97–104.

Alhaj, H., Wisniewski, G., and McAllister-Williams, R. H. The use of the EEG in measuring therapeutic drug action: Focus on depression and antidepressants. *Journal of Psychopharmacology* 25(9) (2011): 1175–1191.

Arns, M., Cerquerac, A., Gutiérrezc, R. M., Hasselmane, F., and Freund, J. A. Non-linear EEG analyses predict non-response to rTMS treatment in major depressive disorder. *Clinical Neurophysiology* 125(7) (2014): 1392–1399.

Arns, M., Drinkenburg, W. H., Fitzgerald, P. B., and Kenemans, J. L. Neurophysiological predictors of non-response to rTMS in depression. *Brain Stimulation* 5(4) (2012): 569–576.

Asada, H., Fukuda, Y., Tsunoda, S., Yamaguchi, M., and Tonoike, M. Frontal midline theta rhythms reflect alternative activation of prefrontal cortex and anterior cingulate cortex in humans. *Neuroscience Letters* 274(1) (1999): 29–32.

Bares, M., Brunovsky, M., Kopecek, M. et al. Changes in QEEG prefrontal cordance as a predictor of response to antidepressants in patients with treatment resistant depressive disorder: A pilot study. *Journal of Psychiatric Research* 41(3–4) (2007): 319–325.

Bares, M., Brunovskyi, M., Miloslav, K. et al. Early reduction in prefrontal theta QEEG cordance value predicts response to venlafaxine treatment in patients with resistant depressive disorder. *European Psychiatry* 23(5) (2008): 350–355. Accessed March 9, 2008. doi:10.1016/j.eurpsy.2008.03.001.

Bares, M., Brunovsky, M., Novak, T. et al. The change of prefrontal QEEG theta cordance as a predictor of response to bupropion treatment in patients who had failed to respond to previous antidepressant treatments. *European Neuropsychopharmacology* 20 (2010): 459–466.

Bares, M., Brunovsky, M., Novak, T. et al. QEEG theta cordance in the prediction of treatment outcome to prefrontal repetitive transcranial magnetic stimulation or venlafaxine ER in patients with Major Depressive Disorder. *Clinical EEG and Neuroscience* 46(2) (2015): 73–80.

Bares, M., Novak, T., Brunovsky, M. et al. The change of QEEG prefrontal cordance as a response predictor to antidepressive intervention in bipolar depression. A pilot study. *Journal of Psychiatric Research* 46 (2012): 219–225.

Bashashati, A., Fatourechi, M., Ward, R. K., and Birch, G. E. A survey of signal processing algorithms in brain-computer interfaces based on electrical brain signals. *Journal of Neural Engineering* 4(2) (2007): 32–57.

Baskaran, A., Milev, R., and McIntrye, R. S. The neurobiology of the EEG biomarker as a predictor of treatment response in depression. *Neuropharmacology* 63 (2012): 507–513.

Broadway, J. M., Holtzheimer, P. E., and Hilimire, M. R. Frontal theta cordance predicts 6-month antidepressant response to subcallosal cingulate deep brain stimulation for treatment-resistant depression: A pilot study. *Neuropsychopharmacology* 37(7) (2012): 1764–1772.

Bruder, G. E., Sedoruk, J. P., Stewart, J. W., McGrath, P. J., Quitkin, F. M., and Tenke, C. E. Electro-encephalographic alpha measures predict therapeutic response to a selective serotonin reuptake inhibitor antidepressant: Pre- and post-treatment findings. *Biological Psychiatry* 63 (2008): 1171–1177.

Buchsbaum, M. S., Hazlett, E., Nancy, S. et al. Geometric and scaling issues in topographic electroencephalography. In *Topographic Mapping of Brain Electrical Activity*, ed. Duffy, F.H., pp. 325–371. Stoneham, MA: Butterworth Publishers, 1986.

Buzsáki, G. and Draguhn, A. Neuronal oscillations in cortical networks. *Science* 25(304) (2004): 1926–1929.

Chang, J. S., Yoo, C., Yi, S. H. et al. An integrative assessment of the psychophysiologic alterations in young women with recurrent major depressive disorder. *Psychosomatic Medicine* 74 (2012): 495–500.

Chen, C. H., Ridler, K., Suckling, J. et al. Brain imaging correlates of depressive symptom severity and predictors of symptom improvement after antidepressant treatment. *Biological Psychiatry* 62 (2007): 407–414.

Chorlian, D. B., Rangaswamy, M., and Porjesz, B. EEG coherence: Topography and frequency structure. *Experimental Brain Research* 198(1) (2009): 59–83. Accessed July 22, 2009. doi:10.1007/s00221-009-1936-9.

Cook, I. A., Hunter, A. M., Korb, A. S., and Leuchter, A. F. Do prefrontal midline electrodes provide unique neurophysiologic information in major depressive disorder? *Journal of Psychiatric Research* 53 (2014): 69–75.

Cook, I. A., Leuchter, A. F., Uijtdehaage, S. H. J., Abrams, M., Anderson-Hanley, C., and Rosenberg-Thompson, S. Neurophysiologic predictors of treatment response to fluoxetine in major depression. *Psychiatry Research* 85 (1999): 263–273.

Cook, I. A., O'Hara, R., Uijtdehaage, S. H., Mandelkern, M., and Leuchter, A. F. Assessing the accuracy of topographic EEG mapping for determining local brain function. *Electroencephalography and Clinical Neurophysiology* 107 (1998): 408–414.

Debener, S., Beauducel, A., Nessler, D., Brocke, B., Heilemann, H., and Kayser, J. Is resting anterior EEG alpha asymmetry a trait marker for depression? Findings for healthy adults and clinically depressed patients. *Neuropsychobiology* 41 (2000): 31–37.

Drevets, W. C. Functional neuroimaging studies of depression: The anatomy of melancholia. *Annual Reviews of Medicine* 49 (1998): 341–361.

Ebert, D. and Ebmcier, K. P. The role of the cingulate gyrus in depression: From functional anatomy to neurochemistry. *Biological Psychiatry* 39 (1996): 1044–1050.

Erguzel, T. T., Gökben, H. S., and Tarhan, N. Artificial intelligence approach to classify unipolar and bipolar depressive disorders. *Neural Computing & Applications* (2015b). Accessed June 5, 2015. doi:10.1007/s00521-015-1959-z.

Erguzel, T. T., Ozekes, S., Gultekin, S., Tarhan, N., Gökben, H. S., and Bayram, A. Neural network based response prediction of rTMS in major depressive disorder using QEEG cordance. *Psychiatry Investigation* 12(1) (2015a): 61–65.

Fell, J. and Axmacher, N. The role of phase synchronization in memory processes. *Nature Reviews Neuroscience* 12 (2011): 105–118. doi:10.1038/nrn2979.

Filipovići, S. R., Covickovič-Sternići, N., Stojanović-Sveteli, M., Lecići, D., and Kostići, V. S. Depression in Parkinson's disease: An EEG frequency analysis study. *Parkinsonism Related Disorders* 4(4) (1998): 171–178.

Fingelkurts, A. A., Fingelkurts, A. A., Ermolaev, V. A., and Kaplan, A. Y. Stability, reliability and consistency of the compositions of brain oscillations. *International Journal of Psychophysiology* 59(2) (2006): 116–126.

Fingelkurts, A. A., Fingelkurts, A. A., Kivisaari, R. et al. Enhancement of GABA-related signalling is associated with increase of functional connectivity in human cortex. *Human Brain Mapping* 22 (2004): 27–39.

Fingelkurts, A. A., Fingelkurts, A. A. Rytsälä, H., Suominen, K., Isometsä, E., and Kähkönen, S.. Impaired functional connectivity at EEG alpha and theta frequency bands in major depression. *Human Brain Mapping* 28 (2007): 247–261.

Fox, M. D., Buckner, R. L., White, M. P., Greicius, M. D., and Pascual-Leone, A. Efficacy of transcranial magnetic stimulation targets for depression is related to intrinsic functional connectivity with the subgenualcingulate. *Biological Psychiatry* 72 (2012): 595–603.

Gómez-Herrero, G., Atienza, M., Egiazarian, K., and Cantero, J. L. Measuring directional coupling between EEG sources. *Neuroimage* 43 (2008): 497–508. doi:10.1016/j.neuroimage.2008.07.032.

Greicius, M. D., Flores, B. H., Menon, V. et al. Resting-state functional connectivity in major depression: Abnormally increased contributions from subgenual cingulate cortex and thalamus. *Biological Psychiatry* 62 (2007): 429–437.

Grin-Yatsenko, V. A., Baas, I., Ponomarev, V. A., and Kropotov, J. D. EEG power spectra at early stages of depressive disorders. *Journal of Clinical Neurophysiology* 2 (2009): 401–406.

Henriques, J. B. and Davidson, R. J. Regional brain electrical asymmetries discriminate between previously depressed and healthy control subjects. *Journal of Abnormal Psychology* 99 (1990): 22–31.

Henriques, J. B. and Davidson, R. J. Left frontal hypoactivation in depression. *Journal of Abnormal Psychology* 100 (1991): 535–545.

Hughes, J. R. and John, E. R. Conventional and quantitative electroencephalography in psychiatry. *Journal of Neuropsychiatry and Clinical Neuroscience* 11 (1999): 190–208.

Hunter, A. M., Cook, I. A., and Leuchter, A. F. The promise of the quantitative electroencephalogram as a predictor of antidepressant treatment outcomes in major depressive disorder. *Psychiatric Clinics of North America* 30(1) (2007): 105–124.

Hunter, A. M., Korb, A. S., Cook, I. A., and Leuchter, A. F. Rostral anterior cingulate activity in major depressive disorder: State or trait marker of responsiveness to medication? *Journal of Neuropsychiatry and Clinical Neuroscience* 25 (2013): 126–133.

Hunter, A. M., Leuchter, A. F., Morgan, M. L., and Cook, I. A. Changes in brain function (quantitative EEG cordance) during placebo lead-in and treatment outcomes in clinical trials for major depression. *American Journal of Psychiatry* 163 (2006): 1426–1432.

Hunter, A. M., Muthén, B. O., Cook, I. A., and Leuchter, A. F. Antidepressant response trajectories and quantitative electroencephalography (QEEG) biomarkers in major depressive disorder. *Journal of Psychiatric Research* 44(2) (2010): 90–98. doi:10.1016/j.jpsychires.2009.06.006.

Iosifescu, D., Greenwald, S. D., Devlin, P. et al. Frontal EEG predicts clinical response to SSRI treatment in MDD [abstract #170]. *Poster Presented at the 44th Annual Meeting of the New Clinical Drug Evaluation Unit*, Phoenix, AZ, June 1–4, 2004.

Ishii, R., Shinosaki, K., Ukai, S. et al. Medial prefrontal cortex generates frontal midline theta rhythm. *Neuroreport* 10 (1999): 675–679.

Jaworska, N., Blier, P., Fusee, W., and Knott, V. α Power, α asymmetry and anterior cingulate cortex activity in depressed males and females. *Journal of Psychiatric Research* 46(1) (2012a): 1483–1491.

Kito, S., Pascual-Marqui, R. D., Hasegawa, T., and Koga, Y. High-frequency left prefrontal transcranial magnetic stimulation modulates resting EEG functional connectivity for gamma band between the left dorsolateral prefrontal cortex and precuneus in depression. *Brain Stimulation* 7(1) (2014): 145–146. doi: 10.1016/j.brs.2013.09.006.

Knott, V. J., Colleen, M., Sidney, K., and Kens, E. EEG power, frequency, asymmetry and coherence in male depression. *Psychiatry Research* 106 (2001): 123–140.

Knott, V. J. and Lapierre, Y. D. Computerized EEG correlates of depression and antidepressant treatment. *Progress in Neuro Psychopharmacology & Biological Psychiatry* 11 (1987): 213–221.

Kropotov, J. *Quantitative EEG, Event-Related Potentials and Neurotherapy*. San Diego, CA: Academic Press, 2010.

Kuo, C.-C. and Tsai, J.-F. Cordance or antidepressant treatment response (ATR) index? *Psychiatry Research* 180(1) (2010): 60.

Lachaux, J. P., Rodriguez, E., Martinerie, J., and Varela, F. J. Measuring phase synchrony in brain signals. *Human Brain Mapping* 8(4) (1999): 194–208.

Lee, J., Hwang, J. Y., Park, S. M. et al. Differential resting-state EEG patterns associated with comorbid depression in Internet addiction. *Progress in Neuro-Psychopharmacology and Biological Psychiatry* 50 (2014): 21–26.

Lee, T. W., Yu, Y. W., Chen, M.-C., and Chen, T.-J. Cortical mechanisms of the symptomatology in major depressive disorder: A resting EEG study. *Journal of Affective Disorders* 131 (2011): 243–250 .

Leuchter, A. F., Cook, I. A., Lufkin, R. B. et al. Cordance: A new method for assessment of cerebral perfusion and metabolism using quantitative electroencephalography. *Neuroimage* 1 (1994a): 208–219.

Leuchter, A. F., Cook, I. A., Mena, I. et al. Assessing cerebral perfusion using quantitative EEG cordance. *Psychiatry Research: Neuroimaging* 558 (1994b): 141–152.

Leuchter, A. F., Cook, I. A., Hunter, A. M., Cai, C., and Horvath, S. Resting-state quantitative electroencephalography reveals increased neurophysiologic connectivity in depression. *PLoS One* 7(2) (2012): e32508. doi:10.1371/journal.pone.0032508.

Leuchter, A. F., Cook, I. A., Marangell, L. B. et al. Comparative effectiveness of biomarkers and clinical indicators for predicting outcomes of SSRI treatment in Major Depressive Disorder: Results of the BRITE-MD study. *Psychiatry Research* 169(2) (2009):124–131.

Leuchter, A. F., Cook, I. A., Witte, E. A., Morgan, M., and Abrams, M. Changes in brain function of depressed subjects during treatment with Placebo. *American Journal of Psychiatry* 159 (2002): 122–129.

Leuchter, A. F., Uijtdehaage, S. H., Cook, I. A., O'Hara, R., and Mandelkern, M. Relationship between brain electrical activity and cortical perfusion in normal subjects. *Psychiatry Research* 90 (1999): 125–140.

Li, Y., Dan, C., Wei, L., Yingying, T., and Jijun, W. Abnormal functional connectivity of EEG gamma band in patients with depression. *Clinical Neurophysiology* pii: S1388-2457(15)00019-X. (2015). doi:10.1016/j.clinph.2014.12.026.

Lieber, A. L. Diagnosis and subtyping of depressive disorders by quantitative electroencephalography: II. Interhemispheric measures are abnormal in major depressives and frequency analysis may discriminate certain subtypes. *Hillside Journal of Clinical Psychiatry* 10 (1988): 84–89.

Malmivuo, J. and Plonsey, R. *Bioelectromagnetism: Principles and Applications of Bioloectric and Biomagnetic Fields.* New York: Oxford University Press, 1995. Avaliable online: http://www.bem.fi/book/.

Mayberg, H. S., Brannan, S. K., Mahurin, R. K. et al. Cingulate function in depression: A potential predictor of treatment response. *NeuroReport* 8 (1997): 1057–1061.

Mezeiova, K. and Paluš, M. Comparison of coherence and phase synchronization of the human sleep electroencephalogram. *Clinical Neurophysiology* 123(9) (2012): 1821–1830.

Miskovic, V. and Schmidt, L. A. Cross-regional cortical synchronization during affective image viewing. *Brain Research* 1362 (2010): 102–111.

Morgan, M. L., Witte, E. A., Cook, I. A., Leuchter, A. F., Abrams, M., and Siegman, B. Influence of age, gender, health status, and depression on quantitative EEG. *Neurophysiology* 52 (2005): 71–76.

Mormann, F., Lehnertz, K., David, P., and Elger, C. E. Mean phase coherence as a measure for phase synchronization and its application to the EEG of epilepsy patients. *Physica D* 144(3–4) (2000): 358–369.

Mulert, C., Juckel, G., Brunnmeier, M. et al. Rostral anterior cingulate cortex activity in the theta band predicts response to antidepressive medication. *Clinical EEG and Neuroscience* 38 (2007): 78–81.

Myslobodsky, M. S., Coppola, R., and Weinberger, D. R. EEG laterality in the era of structural brain imaging. *Brain Topography* 3 (1991): 381–390.

Nagata, K. Topographic EEG in brain ischemia: Correlation with blood flow and metabolism. *Brain Topogrophy* 1 (1988): 97–106.

Niedermeyer, E. and F. H. Lopes da Silva. *Electroencephalography: Basic Principles, Clinical Applications, and Related Fields.* Baltimore, MD: Lippincont Williams & Wilkins, 2004.

Olbrich, S. and Arns, S. M. EEG biomarkers in major depressive disorder: Discriminative power and prediction of treatment response. *International Review of Psychiatry* 25(5) (2013): 604–618. doi:10.3109/09540261.2013.816269.

Pariante, C. M., Nesse, R. M., Nutt, D., and Wolpert, L. *Understanding Depression: A Translational Approach.* New York: Oxford University Press. Inc., 2009, Published Online: November 4.

Pereda, E., Quiroga, R. Q., Bhattacharya, J. Causal Influence: Nonlinear multivariate analysis of neurophysical signals. *Proogress in Neurobiology* 77(1–2) (2005): 1–37. doi:10.1016/j.pneurobio.2005.10.003.PMID 16289760.

Pivik, R. T., Broughton, R. J., Coppola, R. et al. Guidelines for the recording and quantitative analysis of electroencephalographic activity in research context. *Psychophysiology* 30 (1993): 547–558.

Pizzagalli, D. A. Fronto-cingulate dysfunction in depression: Toward biomarkers of treatment response. *Neuropsychopharmacology* 36 (2011): 183–206.

Pizzagalli, D. A., Oakes, T. R., and Davidson, R. J. Coupling of theta activity and glucose metabolism in the human rostral anterior cingulate cortex: An EEG/PET study of normal and depressed subjects. *Psychophysiology* 40 (2003): 939–949.

Pozzi, D., Golimstock, A., Petracchi, M., García, H., and Starkstein, S. Quantified electroencephalographic changes in depressed patients with and without dementia. *Biological Psychiatry* 38 (1995): 677–683.

Price, J. L. and Drevets, W. C. Neurocircuitry of mood disorders. *Neuropsychopharmacology* 35(1) (2010): 192–216. doi:10.1038/npp.2009.104.

Putnam, K. M. and McSweeney, L. B. Depressive symptoms and baseline prefrontal EEG alpha activity: A study utilizing ecological momentary assessment. *Biological Psychology* 77(2) (2008): 237–240.

Rizk-Allah, R. M., Zaki, E. M., and El-Sawy, A. A. Hybridizing ant colony optimization with firefly algorithm for unconstrained optimization problems. *Applied Mathematics and Computation* 224 (2013): 473–483.

Rosenblum, M. G., Pikovsky, A. S., and Kurths, J. Phase synchronization of chaotic oscillators. *Physical Review Letters* 76(11) (1996): 1804–1807.

Schaffer, C. E., Davidson, R. J., and Saron, C. Frontal and parietal electroencephalogram asymmetry in depressed and nondepressed subjects. *Biological Psychiatry* 18 (1983): 753–762.

Srinivasan, R., Nunez, P. L., and Silberstein, R. B. Spatial filtering and neocortical dynamics: Estimates of EEG coherence. *IEEE Transactions on Biomedical Engineering* 45 (1998): 814–825.

Sun, Y., Li, Y., Zhu, Y., Chen, X. and Tong, S. Electroencephalographic differences between depressed and control subjects: An aspect of interdependence analysis. *Brain Research Bulletin* 76 (2008): 559–564.

Tas, C., Cebi, M., Tan, O., Hızlı-Sayar, G., Tarhan, N., and Brown, E. C. EEG power, cordance and coherence differences between unipolar and bipolar depression. *Journal of Affective Disorders* 172 (2015): 184–190.

Tenke, C. E. and Kayser, J. Reference-free quantification of EEG spectra: Combining current source density (CDS) and frequency principal components analysis (fPCA). *Clinical Neurophysiology* 116 (2005): 2826–2846.

Thatcher, R. W., North, D. M., and Biver, C. J. Development of cortical connections as measured by EEG coherence and phase delays. *Human Brain Mapping* 29(12) (2008): 1400–1415.

Tomarken, A. J., Davidson, R. J., Wheeler, R. E., and Kenney, L. Psychometric properties of resting anterior EEG asymmetry: Temporal stability and internal consistency. *Psychophysiology* 29 (1992): 576–592.

Ulrich, G., Renfordt, E., Zeller, G., and Frick, K. Interrelation between changes in the EEG and psychopathology under pharmacotherapy for endogenous depression. A contribution to the predictor question. *Pharmacopsychiatry* 17(6) (1984): 178–183.

Wang, L., Li, K., Zhang, Q.-E. et al. Interhemispheric functional connectivity and its relation-ships with clinical characteristics in major depressive disorder: A resting state FMRI study. *PLoS One* 8 (2013): e60191.

Weiss, S. and Mueller, H. M. The contribution of EEG coherence to the investigation of language. *Brain and Language* 85 (2003): 325–343.

Winterhalder, M., Björn, S., Kurths, J., Andreas, S.-B., and Timmer, J. Sensitivity and specificity of coherence and phase synchronization analysis. *Physics Letters A* 356 (2006): 26–34.

29 fMRI as a Biomarker in Neuroimaging

Asimina Lazaridou, Babak Kateb, and A. Aria Tzika

CONTENTS

INTRODUCTION

This chapter explores functional magnetic resonance imaging (fMRI) modality as an applicable tool for diagnosis and clinical research in humans. We begin this chapter by establishing strong foundation in the physics and biology of fMRI. We present fMRI applications and functional connectivity (FC) in different disciplines. In addition, we review various studies being conducted using fMRI as a biomarker and present findings on stroke rehabilitation. In recent years, there has been increasing interest in studying connectivity using the fMRI method. Understanding brain function and mapping areas using fMRI is a powerful tool. Studies on healthy volunteers usually requires a different approach while in clinical studies complex problems usually arise because of patients' noncompliance (e.g., dementia, ADHD, brain tumors). Therefore, the application of fMRI in the clinical setting is a different challenge, depicted in the study designs as well as in the analysis of neuroimaging data algorithms. More recent clinical applications include mapping of recovery from stroke or traumatic brain injury and developmental changes or changes during a medical condition.

MRI

Magnetic resonance imaging (MRI) is a method to explore the structure and function of the brain by measuring changes in the magnetic properties of specific molecules. The most important reason for the vast use of MRI is a routine medical procedure for diagnostic and scientific purposes. Nowadays, the field of statistics has played an integral role in the advancement of functional neuroimaging research. Moreover, MRI is noninvasive and carries few health risks. MRI was a significant improvement over other available neuroimaging techniques. For example, computed tomography scanning uses x-rays and positron emission tomography scanning requires injecting the subject with a drug containing a radioactive label.

MRI can detect various medical diseases in the brain such as tumors, bleeding, swelling, structural and developmental abnormalities, infections, problems with the blood vessels, or inflammatory disorders. In addition, it can detect structural damage in the brain caused by an injury or a stroke. MRI of the brain can be useful in evaluating problems such as persistent headaches, dizziness, or seizures, and it can help to detect certain chronic diseases related to the nervous system, such as multiple sclerosis. In some cases, MRI can provide clear images of the brain that can't be acquired through an x-ray, CAT scan, or ultrasound, making it exceptionally useful for diagnosing issues with the pituitary gland and brain stem. fMRI is a functional neuroimaging procedure using MRI technology that measures brain activity by detecting associated changes in blood flow, and it is explained in more detail in the following text.

FUNCTIONAL MRI

fMRI has enjoyed a great increase in its use as a tool for cognitive neuroscience research. Since its invention in the early 1990s to the end of 2014, more than 14,000 articles have been published that mention fMRI in the abstract or title (according to PubMed), and this number is now growing by roughly 40–50 scientific papers every week. fMRI has offered new avenues and directions to research topics that needed rigorous scientific investigation. In the last couple of years, fMRI has been a useful tool to study the nature of consciousness (Lloyd et al. 2011), the impact of mindfulness on brain function (Holzel et al. 2011), the neural basis of moral judgments, and the neural networks relative to rehabilitation. However, researchers that collect fMRI data face many challenges when it comes to the point of analyzing their data. An fMRI experiment could create massive amounts of complicated data. The statistical methods that are used need to be far beyond analysis of variance and regression. What is more, a lot of statistical methods such as Granger causality, Gaussian random field theory, false discovery rate, coherence analysis, and independent component analysis that are used in fMRI analysis are rarely taught in traditional statistics courses.

Most of the fMRI experiments measure the blood oxygen level–dependent (BOLD) signal. The BOLD signal is a measure of the ratio of oxygenated to deoxygenated hemoglobin.

More specifically, the cylindrical tube of an MRI scanner has a very powerful electro magnet. A typical research scanner has a field strength of 3 T (tesla) about 50,000 times greater than the Earth's field. The magnetic field inside the scanner affects the magnetic nuclei of atoms. Atomic nuclei are usually randomly oriented but under the influence of a magnetic field of the nuclei and become aligned with the direction of the field. The stronger the field, the greater the alignment is. When pointing in the same direction, the tiny magnetic signals from individual nuclei add up resulting in a signal that is large enough to measure. In fMRI, it is the magnetic signal from hydrogen nuclei in water that is detected. The most important part of the MRI is that the signal from hydrogen nuclei varies in strength depending on the environment. This provides a means of distinguishing white matter, gray matter, and cerebral spinal fluid in structural images of the brain.

Oxygen is delivered to neurons by hemoglobin in capillary red blood cells. When neuronal activity increases, there is an increased demand for oxygen, and the local response is an increase in blood flow to regions of increased neural activity.

Hemoglobin is diamagnetic when oxygenated but paramagnetic when deoxygenated. This difference in magnetic properties leads to subtle differences in the MR signal of blood depending on the degree of oxygenation. Since blood oxygenation varies according to the levels of neural activity, these differences can be used to detect activity in different brain regions. This form of MRI is known as blood oxygenation level–dependent (BOLD) imaging.

Someone might think that blood oxygenation will usually decrease with activation, but the reality is a little more complicated. There is a momentary decrease in blood oxygenation immediately after neural activity increases, known as the initial dip (decrease) in the hemodynamic response. This is followed by a period where the blood flow increases, not to a level where oxygen demand is met, but overcompensating for the increased need. This means the blood oxygenation actually increases following neural activation. The blood flow increases after around 6 s and then returns back to baseline, often followed by a "post-stimulus" undershoot.

Researchers look at the activity on a scan in voxels—or "volume pixels", the smallest distinguishable box-shaped part of a three-dimensional image. The activity in a voxel is defined as how closely the time course of the signal from that voxel matches the expected time course. Voxels whose signal corresponds closer are given a high activation score, voxels showing no correlation have a low score, and voxels showing deactivation are given a negative score. These can then be translated into activation brain maps.

For the past 10 years, fMRI based on imaging of the endogenous BOLD contrast (Kwong et al. 1992) has been suggested to be a very important tool for exploring brain function during complex cognitive processes such as memory consolidation in humans (Stern et al. 1996; Gabrieli et al. 1997; Brewer et al. 1998). For example, these fMRI studies, together with several other related studies, have confirmed the central role of the medial temporal lobe structures, in other words, the hippocampus and neighboring parahippocampal cortices, in encoding new events into long-term memory (Eichenbaum 2000; Scoville and Milner 2000). Typically, fMRI experiments compare the BOLD signal during one cognitive condition (e.g., encoding new information) to a control task (e.g., viewing learned information) or to a passive baseline condition (e.g., fixed cross). This can be done using "block design" paradigms, in which cognitively similar kinds of stimuli are grouped together in blocks usually lasting 20–40 s, or using "event-related" paradigms, in which single stimuli from different cognitive conditions are randomly presented. In addition to fMRI studies during task performance, there has recently been considerable interest in exploring neuronal activity during awake resting state, also called the "default mode" activity of the human brain (Raichle and Mintun 2006; Horovitz et al. 2009).

Clinical fMRI research into the neurologic or psychiatric brain diseases has become established more recently (Rombouts et al. 2000; Sperling et al. 2003). FC, explained in the following text, has helped researchers understand the functionality of different brain regions as a network.

FUNCTIONAL CONNECTIVITY (FC)

Imaging neuroscience has firmly established functional segregation as a principle of brain organization in humans. The integration of segregated areas has proven more difficult to assess. FC has been defined as the temporal correlation between spatially remote neurophysiologic events. There are various statistical methods for examining FC ranging from descriptive approaches to inferential statistical procedures. Functional integration is the study of connected processes. Methods for functional integration can be broadly divided into FC (finding statistical patterns) and effective connectivity (model how regions interacts).

FC reveals correlations in activation among spatially distinct brain regions, either in a resting state or when processing external stimuli. FC has been extensively evaluated with several functional neuroimaging methods, particularly fMRI. Yet these relationships have been quantified using very different measures and the extent to which they index the same constructs is unclear. FC measures are usually categorized into two groups: whole time-series and trial-based approaches. We evaluate these measures via simulations with different patterns of FC and provide recommendations for their use.

More specifically, in the analysis of neuroimaging, time-series FC is defined as the temporal correlations between spatially remote neurophysiological events (Friston et al. 1993). This definition provides a simple characterization of functional relationships between different brain regions. The alternative is to refer explicitly to effective connectivity (i.e., the influence one neuronal system exerts over another). FC is simply a term being used about the observed correlations between different brain regions. It does not suggest any direct understanding into how these correlations are mediated. For example, at the level of multiunit microelectrode recordings, correlations can result from stimulus-locked transients, evoked by a common afferent input, or reflect stimulus-induced oscillations, phasic coupling of neural assemblies, mediated by synaptic connections (Gerstein and Kirkland 2001). To examine the integration within a distributed system, defined by FC, one turns to effective connectivity. Effective connectivity is closer to the intuitive notion of a connection and can be defined as the influence on neural system exerts over another, at either a synaptic (cf. synaptic efficacy) or cortical level. In electrophysiology, there is a close relationship between effective connectivity and synaptic efficacy.

However, brain activation or activation of networks does not always depict or reveal the same information. It is usually suggested that cognitive capacities reflect the "local processing of inputs" or the "output" of a region, represented in the patterns of action potentials, with their characteristic frequency and timing. In general, brain structures can be visualized as information processing entities, with an input, a local-processing capacity, and an output. Yet, although such a scheme may depict the function of subcortical nuclei, its application in different areas of cortex is not direct. In fact the traditional cortical input–elaboration–output scheme, commonly presented as an instantiation of the tripartite perception–cognition–action model, is probably a misleading oversimplification (Braitenberg 1998). Research indicates that the subcortical input to cortex is weak, the feedback is massive, the local connectivity reveals strong excitatory and inhibitory recurrence, and the output presents changes in the balance between excitation and inhibition, other than simple feedforward integration of subcortical inputs (Douglas and Martin 2004).

As early as 2002, Koch and his colleagues (2002) directly compared functional and anatomical connectivity (the probability that a tract can be traced between two points using diffusion tensor imaging [DTI]) by combining fMRI and DTI techniques. They found FC tended to be high when structural connectivity was high between voxels, and pairs of regions situated around the central sulcus indicated a dependence of the two connectivity measures on each other. Following studies confirmed this functional–structural nonperfect coupling (FC reflects structural connectivity, but could exist where there is little or no structural connectivity) either at the level of several brain regions (Greicius et al. 2007; van den Heuvel et al. 2008), at the level of whole-brain parcellations (Honey et al. 2007; Hagmann et al. 2008), or at voxel-wise level. DTI is the most popular method that can detect changes in anatomical connectivity in vivo (Le Bihan 2003). In normal white matter, water molecules move relatively freely in a direction parallel to fiber tracts and opposite to the restricted movement across the tracts. This process is known as diffusion anisotropy. Based on diffusion anisotropy, the white matter fiber tracts can be reconstructed using diffusion tensor tractography (Mori et al. 1999). Anatomical connectivity can be assessed by either visualizing fiber tracts or evaluating diffusion aspects. Fractional anisotropy (FA) and mean diffusivity (MD) are the most commonly used DTI measures: MD reflects the average diffusion amplitude, and FA represents the degree of directionality of microstructures (Qin et al. 2012). Therefore, fMRI is usually being used along with DTI to detect both anatomical and FC in clinical studies.

FC AND STROKE REHABILITATION

Stroke is the third leading cause of death and the leading cause of long-term disability in the United States (1). Approximately 4 million Americans live with the negative consequences of stroke (2,3). In addition, the lives of caregivers including spouses, children, and friends are personally affected because of this significant disease. Current investigations have focused on stroke rehabilitation and brain plasticity as a mechanism in recovery (4).

Therefore, plasticity following stroke remains a crucial issue for stroke survivors and there is invariably some degree of functional recovery (5). In other words, when neurons are damaged by stroke, other neurons take over for them. This adaptive behavior assists in the reorganization of the brain and recovery of lost skills. Brain plasticity is therefore the reason why intensive therapy is such a critical component of stroke rehabilitation (6–8).

Plasticity after stroke has traditionally been studied by observing changes only in the spatial distribution and laterality of focal brain activation during affected limb movement (9). However, neural reorganization is multifaceted, and our understanding may be enhanced by examining dynamics of activity within large-scale networks involved in the sensorimotor control of the limbs. In stroke

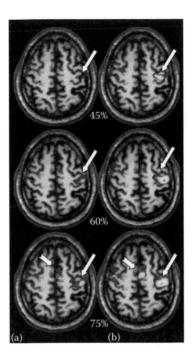

FIGURE 29.1 fMRI conducted with MR_CHIROD shows functional cortical plasticity in a chronic stroke patient. (a) Patient performance halfway through training. (b) Patient performance after full 8-week training. Patient squeezed the MR_CHIROD at 45%, 60%, and 75% maximum grip force. Activation threshold $P < 0.05$ corrected; activation maps are superimposed on the patient's T1-weighted anatomical images. Somato-motor cortex (SMC) activation, longer arrow; SMA activation, shorter arrow). Red color (t-score = 10, $P < 0.0001$); Blue color (t-score = 4.8, $P = 0.05$).

rehabilitation, functional imaging studies of the motor system have described task-related brain activation in recovered patients over and above control subjects in contralesional sensorimotor and premotor cortex, ipsilesional cerebellum, bilateral supplementary motor area (SMA), and parietal cortex (Figure 29.1).

Our recent work on stroke rehabilitation reveals that functional cortical plasticity is feasible beyond 6 months. We recruited stroke patients that underwent training at home with exercise gel balls (Cando gel hand exercise balls; www.bpp2.com/physical_therapy_products/2932.html) underwent fMRI using our second-prototype MR-compatible hand-induced robotic device (MR_CHIROD) (Figure 29.2).

The design and testing of the hand device have been described previously in published research (Khanicheh et al. 2008; Mintzopoulos et al. 2009; Astrakas et al. 2012). The hand device consists of three main subsystems: (1) an electrorheological fluid (ERF)-resistive element, (2) handles, and (3) two sensors, one works as an optical encoder to measure patient-induced motion and the other works as a force sensor. Unlike previously described devices (Siekierka et al. 2007; Tsekos et al. 2007), MR_CHIROD is the first ERF-based device that has been demonstrated to function in conjunction with fMRI for brain mapping in chronic stroke patients (Khanicheh et al. 2008; Mintzopoulos et al. 2008). Of note, MR_CHIROD is capable of limiting and controlling a number of factors that affect its function, making it particularly useful for home-based training given the low level of expert clinical support in the home environment, which can be accompanied by low extrinsic motivation. MR_CHIROD can be reengineered to improve the cost-to-benefit ratio and therapy effectiveness by providing autonomous and recordable training programs with extrinsic motivation through virtual reality technology.

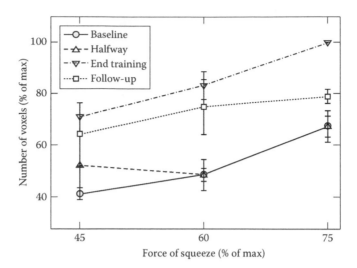

FIGURE 29.2 Number of activated voxels in the contralateral SMC as a function of squeezing force in chronic stroke patients. Online mapping was performed at four time points: baseline (solid line); halfway through training (lower dashed line); at the end of the training (upper dashed line), and 4 weeks after training (do Figure 29.3a–c DCM Model and resultstted line). Note the persistence of increased cortical activation observed during and after the training period.

Training consisted of squeezing one exercise ball from a set of 6 balls at approximately 75% of maximum strength with the paretic hand for 1 h/day, 3 day/week during the 8-week training period. The appropriate hand exercise ball was selected after measuring maximum hand-grip strength with a dynamometer. MRI exams were performed at baseline (before training started); 4 weeks later, halfway through the exercise period; another 4 weeks later, at the end of the training period; and again 4 weeks after completing the training period. Results are shown for a representative patient (63-year-old, right-handed male with subcortical middle cerebral artery [MCA] stroke, 4 years post stroke). The number of activated voxels had increased overall and as a function of effort level at completion of the 8-week training period. Figure 29.1 summarizes results from 5 patients squeezing at three performance levels and over four time points (baseline, halfway through training, end of training, and at follow-up 4 weeks after completion of training). There were a higher number of activated voxels upon the completion of training than at baseline or halfway through training for all three submaximal performance levels. For example, squeezing at 60% effort at the completion of training resulted in $83.25\% \pm 5.45\%$ activated voxels, compared with $48.74\% \pm 2.53\%$ at baseline ($P < 0.0001$). Significant behavioral gains were also found at the end of the treatment. For example, mean arm motor Fugl-Meyer score at the end of treatment increased from 42 ± 7 at baseline to 55 ± 6 after treatment ($P < 0.05$). Likewise, mean action research arm test score increased from 37 ± 15 at baseline to 40 ± 14 after treatment ($P < 0.05$). Somato-motor cortex (SMC) activation with 60% effort squeezing 4 weeks after training completion remained higher ($74.94\% \pm 10.71\%$) than at baseline ($P < 0.05$). A similar trend was observed with 75% effort, though the comparison to baseline data did not reach statistical significance. These results suggest that the increased SMC activation persists at a reduced degree 4 weeks after training. Importantly, these data demonstrate functional cortical plasticity in chronic stroke. They also confirm a previous report by Fasoli et al. (2004) in which chronic stroke patients subjected to goal-directed robotic therapy showed significantly improved motor abilities assessed by traditional motor evaluation; these improvements were sustained 4 months after discharge. It is thus hypothesized that the functional cortical plasticity we observed here is accompanied by recovery of motor performance.

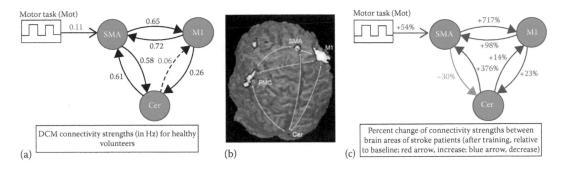

FIGURE 29.3 (a–c) DCM model and results.

FC ALTERATIONS IN MOTOR-RELATED AREAS SUGGEST REORGANIZATION OF MOTOR SYSTEMS IN STROKE

Our additional findings suggest that training can alter effective connectivity and promote functional recovery in chronic stroke patients (Mintzopoulos et al. 2009). The DCM model was constructed using brain regions that were activated in all subjects (Figure 29.3a) and comprised three regions: M1, SMA, and cerebellum (Cer). Volumes of interest were defined in these regions using a sphere centered at the maximum activation from the second-level analysis and with a radius of two voxels. The cognitive input encoding the motor task (Mot) was represented by a step function. The fMRI analysis revealed activations in M1, SMA, premotor cortex, and Cer in both stroke patients and controls (Figure 29.3b).

Connectivity strengths of healthy subjects are shown in Figure 29.3a and percent changes in connectivity strengths after training relative to baseline are shown in Figure 29.3c. The DCM analysis produced the following four noteworthy results. (1) In healthy subjects performing a simple motor task, there is minimum effective connectivity from Cer to M1 (Figure 29.3a). This result has been reproduced by an independent electromyographic study (Gross et al. 2002), validating our analysis. (2) Training significantly increased coupling between M1 and SMA, suggesting an induction of SMA recruitment (Figure 29.3c). This possibility has been suggested by earlier fMRI studies in healthy volunteers (Cramer et al. 2002). (3) Finally, SMA-Cer coupling and Cer-M1 coupling were induced by training (Figure 29.3c). Cer hyperactivity has been documented in Parkinson disease patients, where it was suggested to represent a compensatory mechanism for defective basal ganglia (Yu et al. 2007). Here, it most likely reflects efforts by stroke patients to improve motor control and function. Enhancement of SMA activity (i.e., by high-frequency transcranial current stimulation) has been suggested as a potential means for ameliorating M1 dysfunction after stroke (Hummel et al. 2005). Our findings suggest that rehabilitative exercise training can induce a similar effect.

DIFFUSION TENSOR IMAGING AND MAGNETIC RESONANCE TRACTOGRAPHY USING MR_CHIROD IN CHRONIC STROKE

DTI fiber tract reconstruction was performed with the DTI Studio software package. Deterministic tractography was performed using the FACT algorithm (Lazar et al. 2003). All tracts were visualized and subsequently visually inspected for directionality and location. Regions of interest were designated in the ascending fibers of the pons to visualize the corticospinal tract (CST). The purpose of this analysis was to probe alterations in diffusion-based tractography and consequently to demonstrate changes in structural plasticity in addition to the functional changes we observed in the brains of chronic stroke patients as a result of hand training. We have reconstructed the CST tract choosing

TABLE 29.1

Comparison Using *t*-Test for the Corticospinal Tract Fibers of the Affected Hemisphere before and after 2 Months of Training

Affected Fibers	Average Number ± SD	Average Length ± SD (mm)	*P* Value
Before training	46 ± 8.1	43.6 ± 3.6	<0.001
After training	96.8 ± 7.1	71.4 ± 4.5	<0.001

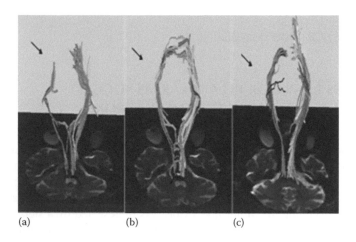

(a)   (b)   (c)

FIGURE 29.4 Reconstructed CST tracts from the same stroke patient, before training (a), 4 weeks (b), and 8 weeks (c) after training one month after training. Clearly CST reconstructed fibers increased dramatically with training on the right side (arrow).

as seeding areas the pons, the posterior limb of the internal capsule, and the motor cortex. We calculated the number of fibers and average tract length, and the results are presented in Table 29.1.

In Figure 29.4, DTI images are presented from one representative patient who suffered a single left-sided ischemic subcortical MCA stroke ≥6 months prior and did not have spasticity or joint stiffness. The patient was subjected to MR evaluation at baseline (before training) and after 4 and 8 weeks of training. All studies were performed on a Siemens Tim Trio (3T). Training was done in the patient's home and consisted of squeezing a gel exercise ball with the paretic hand at approximately 75% of maximum strength for 1 h/day, 3 day/week.

These results show alterations in fiber density over the patient's hand training period, which may be evidence for neuroplasticity. New CST fiber tracts (arrows) projecting closer to the motor cortex appeared through training. New DTI fibers may be indicative of structural neuroplasticity. In summary, our preliminary results reveal structural neuroplasticity in rehabilitation of chronic stroke. These results are in agreement with reports of structural DTI fiber tract alterations in stroke and are consistent with the possibility that functional neuroplasticity in chronic stroke patients may be concomitant with connectivity alterations (Mintzopoulos et al. 2008, 2009; Schaechter et al. 2008).

ANATOMIC MRI CHANGES

Our preliminary data analysis using volumetric techniques showed decreases in cortical thickness, volume, and neural density extending far beyond the stroke infarct and included most of the sensorimotor regions of the stroke and intact hemispheres. Movement of residual tissue toward the infarct was observed promoting the idea that extensive time-dependent morphological changes that occur in residual tissue must be considered when evaluating plasticity-related cortical changes associated with

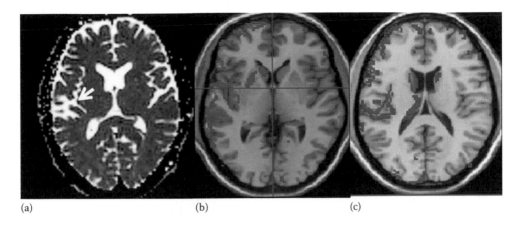

(a)   (b)   (c)

FIGURE 29.5 Apparent diffusion coefficient (ADC) maps (a) and VBM results (b, c) of a 55 year old patient with a left temporal stroke (white arrow). VBM results are overlaid on a template image and red color encodes areas of atrophy.

poststroke recovery of function. As an example, we present a typical case with a stroke at the left temporal lobe showing intense signal at the apparent diffusion coefficient (ADC) map (Figure 29.5b, arrow).

Voxel-based morphometry conducted with SPM8 calculated deviations of brain volume of this patient from 11 age- and sex-matched controls and showed cortical atrophy mainly in the affected hemisphere and noticeably even beyond the stroke region (middle and left images). The results of a recently published study on rats are consistent with our findings, and the idea that extensive time-dependent anatomical changes that occur in residual tissue must be considered when evaluating plasticity-related cortical changes associated with poststroke recovery of function (Karl et al. 2010). Indeed, our preliminary data showed a significant ($P < 0.05$) increase in the cortical thickness in the ventral postcentral gyrus areas of patients after training relative to the cortical thickness before training.

CONCLUSION

In conclusion, we suggest that assessing changes in connectivity by means of fMRI and MR_CHIROD might be used in the future to demonstrate the neural network plasticity that underlies functional recovery in chronic stroke patients. Our findings suggest that rehabilitative exercise training might induce FC alterations after training in both stroke and healthy subjects. Thus, we purport that fMRI as a molecular imaging biomarker of functional reorganization of motor systems in stroke is a clinically relevant molecular medicine approach, and it may allow caregivers to select the most appropriate rehabilitation approach for each patient and to fine-tune this approach based on brain maps obtained before and after a short trial of therapy. This is a new concept of personalized molecular medicine combining motor fMRI with a novel MR_CHIROD training in chronic stroke that can be applied in other motor pathologies.

Overall, fMRI is currently the mainstay of neuroimaging in cognitive neuroscience. fMRI is a technique for measuring and mapping brain activity that is noninvasive and safe. It is being used successfully in many studies to better understand how the healthy brain works, and in a growing number of studies, it is being applied to understand how that normal function is disrupted in disease. Future studies should target mainly on the combination and effectiveness of different modalities along with fMRI such as EEG, MEG, and PET. Therefore, despite its disadvantages, fMRI is currently one of the most important tools for gaining insights into brain function and formulating interesting and eventually testable hypotheses, even though the plausibility of these hypotheses critically depends on used magnetic resonance technology, experimental protocol, statistical analysis, and effective modeling.

REFERENCES

Astrakas, L. G., Naqvi, S. H., Kateb, B., and Tzika, A. A. (2012). Functional MRI using robotic MRI compatible devices for monitoring rehabilitation from chronic stroke in the molecular medicine era (Review). *Int J Mol Med* **29**(6): 963–973.

Braitenberg, V. (1998). Selection, the impersonal engineer. *Artif Life* **4**(4): 309–310.

Brewer, J. B., Zhao, Z., Desmond, J. E., Glover, G. H., and Gabrieli, J. D. (1998). Making memories: Brain activity that predicts how well visual experience will be remembered. *Science* **281**(5380): 1185–1187.

Cramer, S. C., Weisskoff, R. M., Schaechter, J. D., Nelles, G., Foley, M., Finklestein, S. P., and Rosen, B. R. (2002). Motor cortex activation is related to force of squeezing. *Hum Brain Mapp* **16**(4): 197–205.

Douglas, R. J. and Martin, K. A. (2004). Neuronal circuits of the neocortex. *Annu Rev Neurosci* **27**: 419–451.

Eichenbaum, H. (2000). A cortical-hippocampal system for declarative memory. *Nat Rev Neurosci* **1**(1): 41–50.

Fasoli, S. E., Krebs, H. I., Stein, J., Frontera, W. R., Hughes, R., and Hogan, N. (2004). Robotic therapy for chronic motor impairments after stroke: Follow-up results. *Arch Phys Med Rehabil* **85**(7): 1106–1111.

Friston, K. J., Frith, C. D., Liddle, P. F., and Frackowiak, R. S. (1993). Functional connectivity: The principal-component analysis of large (PET) data sets. *J Cereb Blood Flow Metab* **13**(1): 5–14.

Gabrieli, J. D., Brewer, J. B., Desmond, J. E., and Glover, G. H. (1997). Separate neural bases of two fundamental memory processes in the human medial temporal lobe. *Science* **276**(5310): 264–266.

Gerstein, G. L. and Kirkland, K. L. (2001). Neural assemblies: Technical issues, analysis, and modeling. *Neural Netw* **14**(6–7): 589–598.

Greicius, M. D., Flores, B. H., Menon, V., Glover, G. H., Solvason, H. B., Kenna, H., Reiss, A. L., and Schatzberg, A. F. (2007). Resting-state functional connectivity in major depression: Abnormally increased contributions from subgenual cingulate cortex and thalamus. *Biol Psychiatry* **62**(5): 429–437.

Gross, J., Timmermann, L., Kujala, J., Dirks, M., Schmitz, F., Salmelin, R., and Schnitzler, A. (2002). The neural basis of intermittent motor control in humans. *Proc Natl Acad Sci U S A* **99**(4): 2299–2302.

Hagmann, P., Cammoun, L., Gigandet, X., Meuli, R., Honey, C. J., Wedeen, V. J., and Sporns, O. (2008). Mapping the structural core of human cerebral cortex. *PLoS Biol* **6**(7): e159.

Holzel, B. K., Carmody, J., Vangel, M., Congleton, C., Yerramsetti, S. M., Gard, T., and Lazar, S. W. (2011). Mindfulness practice leads to increases in regional brain gray matter density. *Psychiatry Res* **191**(1): 36–43.

Honey, C. J., Kotter, R., Breakspear, M., and Sporns, O. (2007). Network structure of cerebral cortex shapes functional connectivity on multiple time scales. *Proc Natl Acad Sci U S A* **104**(24): 10240–10245.

Horovitz, S. G., Braun, A. R., Carr, W. S., Picchioni, D., Balkin, T. J., Fukunaga, M., and Duyn, J. H. (2009). Decoupling of the brain's default mode network during deep sleep. *Proc Natl Acad Sci U S A* **106**(27): 11376–11381.

Hummel, F., Celnik, P., Giraux, P., Floel, A., Wu, W. H., Gerloff, C., and Cohen, L. G. (2005). Effects of non-invasive cortical stimulation on skilled motor function in chronic stroke. *Brain* **128**(Pt 3): 490–499.

Karl, J. M., Alaverdashvili, M., Cross, A. R., and Whishaw, I. Q. (2010). Thinning, movement, and volume loss of residual cortical tissue occurs after stroke in the adult rat as identified by histological and magnetic resonance imaging analysis. *Neuroscience* **170**(1): 123–137.

Khanicheh, A., Mintzopoulos, D., Weinberg, B., Tzika, A. A., and Mavroidis, C. (2008). MR_CHIROD v.2: Magnetic resonance compatible smart hand rehabilitation device for brain imaging. *IEEE Trans Neural Syst Rehabil Eng* **16**(1): 91–98.

Koch, M. A., Norris, D. G., and Hund-Georgiadis, M. (2002). An investigation of functional and anatomical connectivity using magnetic resonance imaging. *Neuroimage* **16**(1): 241–250.

Kwong, K. K., Belliveau, J. W., Chesler, D. A. et al. (1992). Dynamic magnetic resonance imaging of human brain activity during primary sensory stimulation. *Proc Natl Acad Sci U S A* **89**(12): 5675–5679.

Lazar, M., Weinstein, D. M., Tsuruda, J. S. et al. (2003). White matter tractography using diffusion tensor deflection. *Hum Brain Mapp* **18**(4): 306–321.

Le Bihan, D. (2003). Looking into the functional architecture of the brain with diffusion MRI. *Nat Rev Neurosci* **4**(6): 469–480.

Lloyd, D. M., McKenzie, K. J., Brown, R. J., and Poliakoff, E. (2011). Neural correlates of an illusory touch experience investigated with fMRI. *Neuropsychologia* **49**(12): 3430–3438.

Mintzopoulos, D., Astrakas, L. G., Khanicheh, A., Konstas, A. A., Singhal, A., Moskowitz, M. A., Rosen, B. R., and Tzika, A. A. (2009). Connectivity alterations assessed by combining fMRI and MR-compatible hand robots in chronic stroke. *NeuroImage* **47**(Suppl 2): T90–T97.

Mintzopoulos, D., Khanicheh, A., Konstas, A. A., Astrakas, L. G., Singhal, A. B., Moskowitz, M. A., Rosen, B. R., and Tzika, A. A. (2008). Functional MRI of rehabilitation in chronic stroke patients using novel MR-compatible hand robots. *Open Neuroimag J* **2**: 94–101.

Mori, S., Crain, B. J., Chacko, V. P., and van Zijl, P. C. (1999). Three-dimensional tracking of axonal projections in the brain by magnetic resonance imaging. *Ann Neurol* **45**(2): 265–269.

Qin, W., Zhang, M., Piao, Y., Guo, D., Zhu, Z., Tian, X., Li, K., and Yu, C. (2012). Wallerian degeneration in central nervous system: Dynamic associations between diffusion indices and their underlying pathology. *PLoS One* **7**(7): e41441.

Raichle, M. E. and Mintun, M. A. (2006). Brain work and brain imaging. *Annu Rev Neurosci* **29**: 449–476.

Rombouts, S. A., Barkhof, F., Veltman, D. J., Machielsen, W. C., Witter, M. P., Bierlaagh, M. A., Lazeron, R. H., Valk, J., and Scheltens, P. (2000). Functional MR imaging in Alzheimer's disease during memory encoding. *AJNR Am J Neuroradiol* **21**(10): 1869–1875.

Schaechter, J. D., Perdue, K. L., and Wang, R. (2008). Structural damage to the corticospinal tract correlates with bilateral sensorimotor cortex reorganization in stroke patients. *Neuroimage* **39**(3): 1370–1382.

Scoville, W. B. and Milner, B. (2000). Loss of recent memory after bilateral hippocampal lesions. 1957. *J Neuropsychiatry Clin Neurosci* **12**(1): 103–113.

Siekierka, E. M., Eng, K., Bassetti, C. et al. (2007). New technologies and concepts for rehabilitation in the acute phase of stroke: A collaborative matrix. *Neurodegener Dis* **4**(1): 57–69.

Sperling, R. A., Bates, J. F., Chua, E. F., Cocchiarella, A. J., Rentz, D. M., Rosen, B. R., Schacter, D. L., and Albert, M. S. (2003). fMRI studies of associative encoding in young and elderly controls and mild Alzheimer's disease. *J Neurol Neurosurg Psychiatry* **74**(1): 44–50.

Stern, C. E., Corkin, S., Gonzalez, R. G., Guimaraes, A. R., Baker, J. R., Jennings, P. J., Carr, C. A., Sugiura, R. M., Vedantham, V., and Rosen, B. R. (1996). The hippocampal formation participates in novel picture encoding: Evidence from functional magnetic resonance imaging. *Proc Natl Acad Sci U S A* **93**(16): 8660–8665.

Tsekos, N., Khanicheh, A., Christoforou, E., and Mavroidis, C. (2007). Magnetic resonance-compatible robotic and mechatronics systems for image-guided interventions and rehabilitation: A review study. *Annu Rev Biomed Eng* **9**: 351–387.

van den Heuvel, M., Mandl, R., Luigjes, J., and Hulshoff Pol, H. (2008). Microstructural organization of the cingulum tract and the level of default mode functional connectivity. *J Neurosci* **28**(43): 10844–10851.

Yu, H., Sternad, D., Corcos, D. M., and Vaillancourt, D. E. (2007). Role of hyperactive cerebellum and motor cortex in Parkinson's disease. *Neuroimage* **35**(1): 222–233.

30 Single-Neuron Recordings from Depth Electrodes with Embedded Microwires in Epilepsy Patients
Techniques and Examples

Ueli Rutishauser and Adam N. Mamelak

CONTENTS

INTRODUCTION

Temporal lobe epilepsy (TLE) is one of the most common forms of epilepsy in humans (Ojemann, 1997; Engel, 2001). However, TLE is often difficult to control with antiepileptic drugs and as a result, about 50% of TLE patients (Ojemann, 1997) require other treatments. Often, surgical removal of the epileptogenic parts of the medial temporal lobe (MTL) is an option. After surgical treatment, 70%–90% of patients become free of disabling seizures (Engel, 2001). However, planning for such resective surgery requires extensive preparation to evaluate the location and extent of the resection as well as an evaluation of possible loss of function resulting from the resection. Accurate localization can be challenging and time consuming. This is particularly true for hippocampal sclerosis, which is one of the most common pathologies that results in TLE. The neuronal loss in the hippocampus is often hard to detect using conventional structural MRI, and noninvasive techniques, chiefly electroencephalography (EEG) and magnetoencephalography, have limited ability to monitor hippocampal activity. If noninvasive methods fail to clearly localize the seizure onset zone, invasive recording techniques are necessary (Spencer et al., 2007). These include subdural grids of electrodes placed on the surface of the cortex as well as depth electrodes, which are implanted semichronically for up to several weeks for seizure monitoring. Depth electrodes are implanted to record intracranial EEG signals directly from areas suspected to generate seizures, such as the hippocampus. Apart from their clinical utility, depth electrodes provide a rare and unique opportunity to study the human brain. These clinical situations provide one of the very few possibilities to record invasively from the human brain with techniques, otherwise only feasible with animal models. In this chapter, we provide an overview of the techniques by which such recordings can be performed to study single-neuron activity in awake humans for research purposes.

DEPTH ELECTRODES WITH EMBEDDED MICROWIRES
TO RECORD SINGLE NEURONS

Two types of signals acquired from epilepsy patients with implanted depth electrodes can be used for research purposes. First, the signals originating from the low-impedance clinical contacts can be utilized to record low-frequency (<100 Hz) intracranial EEG. Due to their low impedance (<1 kΩ) and large size, these electrodes do not allow the recording of local, small, and fast extracellular currents such as those evoked by action potentials ("spikes"). Second, microwires embedded in the depth electrode (the so-called hybrid depth electrode, AD-Tech Medical Instrument Corp, Racine, WI) can be used to record local field potentials and single-neuron activity. The embedded microwires are made of isolated platinum/iridium and are 40 μm in diameter (Fried et al., 1999). We used electrodes with 8–9 embedded microwires. Their tip is exposed by cutting the wires to appropriate length (5 mm typical) during surgery (Misra et al., 2014). Studies have repeatedly shown that insertion of microwires poses no additional risk to patients (Carmichael et al., 2008; Hefft et al., 2013). Due to their much smaller exposed tip, microwires have high impedance (100–500 kΩ at 1 kHz; see Rutishauser et al. [2006b] for measurements of implanted electrodes). This allows the measurement of single-unit activity with high reliability.

Electrodes are implanted bilaterally at likely epileptogenic sites using a lateral approach (Figure 30.1a). Typical implantation sites are the hippocampus, amygdala, and anterior cingulate cortex. The implantation site was determined based on coregistered CT and structural MRI. Implantation was guided by a stereotactic frame fixed to the head of the patient (Spencer et al., 2007). Implantation location was confirmed postoperatively using structural MRI (Figure 30.1a through c). Monitoring can take up to several weeks of continuous observation and recording. Since activity has to be recorded continuously (while waiting for a seizure to occur), there are large periods of time where brain function is normal but intracranial signals can be recorded. This gives scientists the unique opportunity to directly observe the electrical activity of the awake human brain during the behavior (Mamelak, 2014). The data reported in this chapter are all recorded during these periods of time.

DATA PROCESSING: SPIKE SORTING AND DETECTION

This section summarizes the procedures for an experiment to proceed from the acquisition of continuously sampled data (Figure 30.2a, top) from microwires (Figure 30.1d) all the way to points of time a putative single neuron fired a spike (Figure 30.2b). See Rutishauser et al. (2014b) for a more detailed description.

Microwires record the extracellular voltage at a particular point in space. This signal, a voltage as a function of time, is the linear superposition of a great number of current sources generated by electrically active components of the brain (Buzsaki et al., 2012), including synaptic events and action potentials propagating down an axon or backpropagating inside dendrites. In general, it is extremely difficult to decompose the extracellular signal into the single events that give rise to it. An important aspect of the extracellular signals that can sometimes lead to a reasonable interpretation is the occurrence of action potentials in the near vicinity of the microwire tip. In the rat hippocampus CA1 region, it has been estimated that electrodes can distinguish extracellular spikes from neuronal processes located as far away as 140 μm from the tip of the electrode (Buzsaki, 2004). The peak amplitude depends on the physical size of the neurons under study, making it most likely that a majority of extracellular recordings involve pyramidal cells (Henze et al., 2000). There are no such quantitative estimates for human neurons. However, since the peak amplitude of extracellular spikes decreases rapidly as a function of distance and cell sizes are roughly comparable in rodents and humans, it is reasonable to assume that the basic properties of extracellular recordings summarized in Buzsaki et al. (2012) are also applicable in human neurophysiology. Thus, if action potential sources occur sufficiently close to the microwire

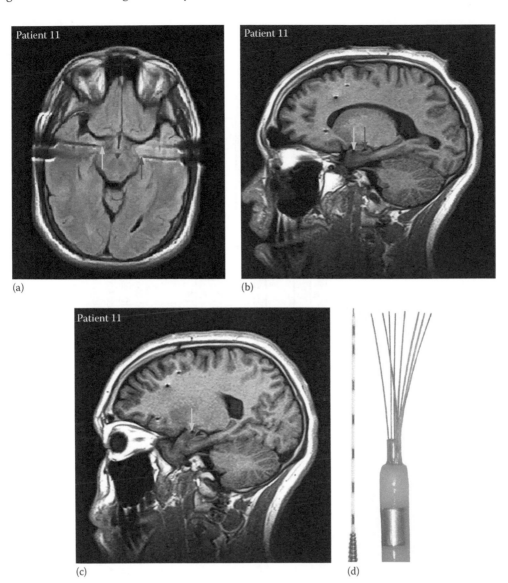

FIGURE 30.1 (a–c) Post-implantation structural MRI (1.5T T1 FLAIR, 0.75 × 0.75 mm in-plane resolution) used to localize electrodes. Arrows indicate an electrode tip in the hippocampus (red) and amygdala (green). (Modified from Rutishauser, U. et al., *Nature*, 464, 903, 2010.) (d) Examples of a depth electrode with macro contacts used for clinical recordings (left) and microwires protruding at the tip (right). The electrode shown is a Behnke–Fried depth electrode (Ad-Tech Medical Instrument Corp., Racine, WI) made with polyurethane (outer diameter 1.3 mm) with seven Pt/Ir contacts each 1.5 mm in length. The microwire bundle (Ad-Tech Medical Instrument Corp., Racine, WI) consists of eight polyimide insulated Pt/Ir (80/20%) microwires inserted through the lumen of depth electrode. Each microwire is 40 μm in diameter. (Reproduced from Staba, R.J. et al., Subchronic in vivo human microelectrode recording, in *Single Neuron Studies of the Human Brain*, Fried, I., Rutishauser, U., Cerf, M., and Kreiman, G., eds., MIT Press, Boston, MA, 2014. With permission.)

and the sources are sparse in space and time, the extracellular signal can distinguish the shapes of the single waveforms sufficiently well to allow a clustering process that groups waveforms of sufficient similarity into clusters that originate from *putative* single neurons.

For the reasons outlined earlier, the variety of extracellular waveforms encountered is large. This variability can be used to distinguish between different neurons, a process commonly referred to as spike

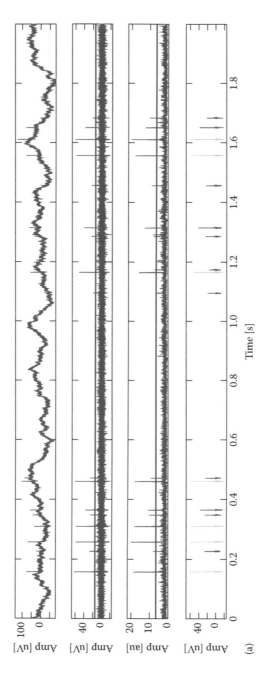

FIGURE 30.2 Example of a microwire recording and its processing (spike detection and sorting). Shown are recordings obtained from a single wire implanted in the right anterior cingulate cortex of human epileptic patient. The detection and sorting shown was done automatically using OSort (Rutishauser et al., 2006a). (a) From top to bottom: raw data (2 Hz high pass), band-pass filtered (300–3 kHz; red line is 4× estimated standard deviation; see text), energy signal used for spike detection (line shows 5 × SD as used for spike detection in this example), and detected and sorted spikes (color indicates cluster identity, as computed by OSort). *(Continued)*

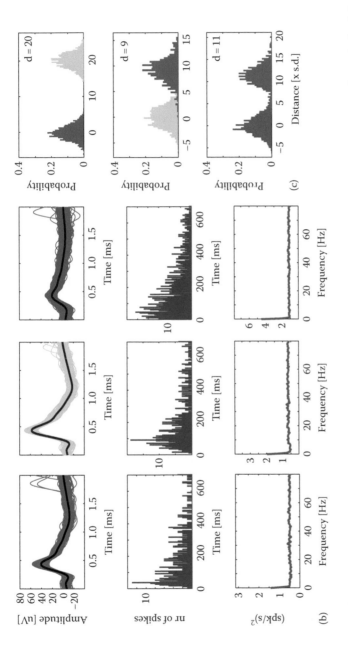

FIGURE 30.2 (Continued) Example of a microwire recording and its processing (spike detection and sorting). Shown are recordings obtained from a single wire implanted in the right anterior cingulate cortex of human epileptic patient. The detection and sorting shown was done automatically using OSort (Rutishauser et al., 2006a). (b) Metrics of three of the identified clusters: raw waveforms (top), histogram of interspike intervals (middle), and power spectrum of the spike train (bottom). All clusters were well separated, with the percentage of interspike intervals (ISIs) <3 ms equal to 0%, 0%, and 0.19% respectively. (c) Pairwise projection test for all possible combinations shows that the clusters are well separated. The distance d is indicated. Amp, amplitude; Nr, number; Spk/s, spikes per second. (Reproduced from Rutishauser, U. et al., Subchronic in vivo human microelecterode recording, in *Single Neuron Studies of the Human Brain*, Fried, I., Rutishauser, U., Cerf, M., and Kreiman, G., eds., MIT Press, Boston, MA, 2014b, Figure 1. With permission.)

sorting (Lewicki, 1998; Pouzat et al., 2002; Quiroga et al., 2004; Rutishauser et al., 2006a; Gibson et al., 2012). The first analysis tasks after the continuously sampled raw signals have been acquired (Figure 30.2a) are to (1) detect the spikes and (2) to identify which putative neuron the spikes originated from.

The goal of spike sorting is to assign each detected waveform to a putative single neuron that generated this spike (Figure 30.2). The number of possible unique neurons that could be present in a recording is unknown and also has to be estimated—that is, spike sorting is an unsupervised clustering problem where the number of the clusters is unknown. Spike sorting is a complex process, the details of which have been described extensively in general (Lewicki, 1998; Pouzat et al., 2002; Quiroga et al., 2004; Rutishauser et al., 2006a; Gibson et al., 2012) and specifically for human recordings (Rutishauser et al., 2014b). A number of ready-to-use software packages are available commercially or as open source. Examples of software packages and their mode of operation that have been used for human recordings are MClust, KlustaKwik (Harris et al., 2000), OSort (Rutishauser et al., 2006a), and Wave_clus (Quiroga et al., 2004). For the data reported in this chapter, we used the spike detection and sorting methods implemented in OSort, which is frequently used to sort human single-neuron data and is available as open source (Rutishauser et al., 2006a, 2013a).

EXAMPLE STUDY: DESIGN

Studying the human brain at the level of single neurons during normal behavior offers unique opportunities to gain new insights into the neural mechanisms of cognition. To illustrate this approach, we next review a series of experiments investigating the mechanisms of memory formation that we performed. See Rutishauser et al. (2014a) for a more detailed description.

Mammalian brains evolved many specialized learning systems that operate in parallel and jointly support behavior (Knowlton and Squire, 1993; Schacter and Tulving, 1994; Poldrack et al., 2001; Squire, 2004). One such system provides declarative memory, which enables the rapid transformation of experiences (episodes) into long-term memories (Tulving, 2002). These memories can later be accessed (i.e., declared) either by free recall or by familiarity. In humans, the ability to rapidly form new memories even after a single-trial learning depends on the MTL, in particular the hippocampus and its surrounding structures (Squire and Alvarez, 1995; Squire et al., 2004). The capacity to store new information after a single exposure of the declarative memory system is large. For example, studies show that human subjects can easily encode 10,000 novel stimuli after a single exposure lasting only seconds each (Standing et al., 1970; Standing, 1973). The neural mechanisms that support this capacity are poorly understood. Episodic memory has been suggested to be largely unique to humans, with some features already present in primates (Hampton, 2001; Fortin et al., 2004). Human single-unit recordings are thus uniquely positioned to contribute toward our understanding of the MTL-dependent neural mechanisms mentioned earlier and beyond what can be learned from animal models.

The experiments summarized in this chapter are based on a common experimental paradigm (Figure 30.3). We presented complex novel visual stimuli (photographs) for a short period of time (2–4 s) on the screen of a notebook computer situated at the patient's bedside. Patients were asked to carefully look at the stimuli for a later memory test. After viewing the to-be-learned stimuli, subjects performed an attention-demanding nonmemory task to prevent active rehearsal. Following the delay, subjects were shown another sequence of images. These were either identical to the images seen during learning (familiar images) or novel (not seen before). After each image, patients were asked to indicate whether they had seen the image before or not (Figure 30.3).

EXAMPLE STUDY: RESULTS ON NOVELTY RESPONSES AND SINGLE-TRIAL LEARNING IN HUMAN MTL

Neurons that respond differently when subjects are repeatedly presented with an identical stimulus are good candidates to explore the neural mechanisms of rapid learning.

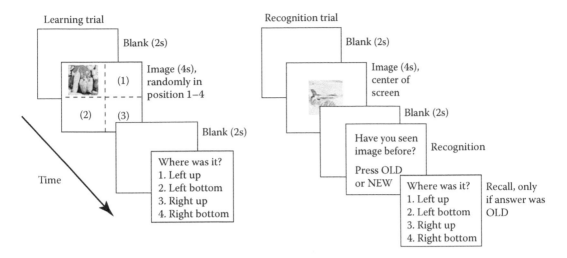

FIGURE 30.3 Structure of the memory task. (Modified from Rutishauser, U. et al., Subchronic in vivo human microelecterode recording, in *Single Neuron Studies of the Human Brain*, Fried, I., Rutishauser, U., Cerf, M., and Kreiman, G., eds., MIT Press, Boston, MA, 2014a. With permission.)

We identified a subpopulation of single neurons in the hippocampus and the amygdala that showed striking differences in their spiking response when comparing novel with familiar stimuli (Rutishauser et al., 2006b, 2008). There were two subgroups of cells, one which increased firing (relative to baseline) for novel stimuli only and the other which increased their firing rate only when familiar stimuli were presented. In total, approximately 20% of recorded cells showed this behavior (40 out of 244 from 6 subjects in one study). Figure 30.4 shows an example. The difference could be seen as soon as 10 min after presenting the novel stimulus (the shortest period tested) and can be seen at least 24 h later (the longest period tested). The tuning of the cells was not stimulus specific as novelty- and familiarity-sensitive neurons responded to stimuli of different categories (such as animals, houses, trees, landscapes, toys, fruits). That is, the same cell would indicate the novelty of a stimulus regardless of which category the stimulus was from. This is in contrast to the subpopulation of cells that show highly selective responses to particular familiar stimuli such as a particular person or place. These cells are visually selective (to visual categories, individuals or specific objects), but nevertheless also show changes in firing rate as a function of repeated presentation even when subjects are only passively viewing stimuli without an explicit memory task and shorter time intervals between repetitions (Pedreira et al., 2010). Interestingly, similar novelty-/familiarity-sensitive responses in the monkey MTL that have been described were also not stimulus selective (Rolls et al., 1982, 1993; Riches et al., 1991; Fahy et al., 1993; Xiang and Brown, 1998). Viskontas et al. (2006) performed a similar task to ours and found novelty- and familiarity-sensitive cells in the parahippocampal gyrus, entorhinal cortex, hippocampus, and amygdala. These cells were similarly abstract and untuned. Furthermore, they (Viskontas et al., 2006) also found a subset of cells that signaled novelty selectively only for a particular category (scenes or faces). Together with our results, it seems that there are at least two distinct classes of novelty-sensitive neurons in the human MTL: abstract and visually tuned. The first class of untuned general novelty detectors could serve to signal the significance of stimuli during the acquisition of new memories (Lisman and Otmakhova, 2001). It has been suggested that such neurons trigger dopaminergic release through projections to the ventral tegmental area (Lisman and Grace, 2005), which in turn strengthens plasticity in the hippocampus.

Episodic memories allow us to remember not only whether we have seen something before but also where and when. To further investigate the underlying mechanisms, we asked patients to remember for a sequence of stimuli: both which pictures they had seen and also where they were presented on the screen. Stimuli appeared randomly in one of four quadrants on the screen. Later, during retrieval, stimuli where shown at the center of the screen, and patients were asked to indicate the location the

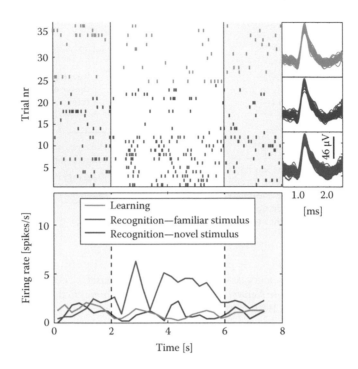

FIGURE 30.4 Example of a neuron in the hippocampus that increases its firing rate to familiar but not to novel stimuli (a "familiarity detector"). The same images were presented twice, initially during learning (green trials) and later again during retrieval (red trials). Also during retrieval other novel stimuli (blue trials) were shown. Trials were randomly intermixed but are shown sorted for display purposes only. The waveforms associated with the trials of the three different categories are shown on the right, confirming that the same neuron was present in both parts of the experiment. The average firing rate histogram at the bottom summarizes the firing rate difference between novel (green) and familiar (red) trials. Notice that the visual input for both trials was the same. The only difference was that stimuli had been seen before (red trials). This cell is thus a memory cell. (Modified from Rutishauser, U. et al., *Neuron*, 49, 805, 2006b; Rutishauser, U. et al., Subchronic in vivo human microelecterode recording, in *Single Neuron Studies of the Human Brain*, Fried, I., Rutishauser, U., Cerf, M., and Kreiman, G., eds., MIT Press, Boston, MA, 2014a. With permission.)

stimulus was shown in when it was first presented if they indicated that a stimulus was familiar. We then compared trials where both recognition and recollections succeeded with trials where only recognition succeeded. We refer to the first and second category as R+ and R−, respectively (Figure 30.5). The same criteria to select a population of novelty and familiarity signaling neurons as described in the previous section were used in this analysis (114 out of the 412 well-separated units from a total of 17 sessions from 8 patients distinguished novel from familiar stimuli; see Rutishauser et al. [2008]). The response was significantly different between R+ and R− trials for novelty as well as familiarity signaling neurons. The differences, however, were of different polarity for the two classes of neurons. Neurons that increased their firing rate for familiar stimuli had a higher firing rate for R+ trials compared to R− trials. In contrast, neurons that increased their firing rate for novel stimuli had a lower firing rate for R+ trials compared to R− trials (Figure 30.5). This indicates that both classes of neurons fire proportionally to the strength of the memory, but with opposite slopes. This leads to the following strength-of-memory gradient hypothesis: the stronger the memory, the stronger (weaker) the firing response for familiarity (novelty) neurons, respectively (Figure 30.5a). This gradient would start with errors (stimuli forgotten, i.e., a very weak memory), continue with stimuli that were only recognized but had no recollection (R−), and end with stimuli that were both recognized and recollected (R+).

We used a single-trial measure of the response of each neuron to quantify this gradient, described as follows.

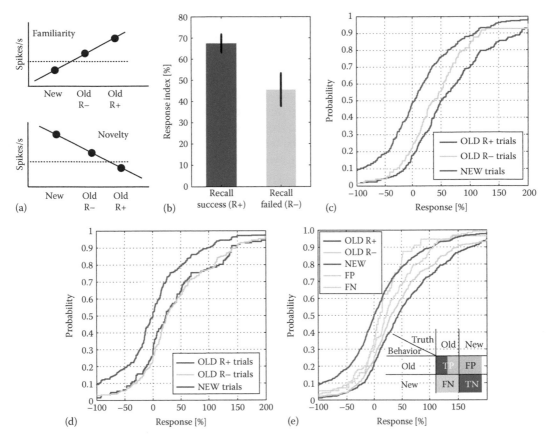

FIGURE 30.5 (a) Continuous strength of memory representation hypothesis illustrated for neurons that increase their firing rate in response to the presentation of familiar (top) and novel (bottom) stimuli. (b) Average response index for trials with successful (R+) and failed (R−) recall. The difference was significantly different p < 0.01. Error bars are ±s.e. over n = 386 and 123 trials, respectively. (c) The same data as shown in (b) but replotted as a cumulative distribution. Notice the two shifts to the right of the entire distribution: the first due to familiarity (to R−) and the other by recollection (to R+). (d) For patients that could not recollect, there was no significant difference between R+ and R− trials (p = 0.53, ks-test). (e) Activity during errors reflects true memory rather than behavior. Response indices were significantly different between forgotten stimuli and novel stimuli p < 0.001 as well as between falsely remembered stimuli and novel stimuli p < 0.01 (both ks-test). (Modified from Rutishauser, U. et al., *Proc. Natl. Acad. Sci. USA*, 105, 329, 2008; Rutishauser, U. et al., Subchronic in vivo human microelecterode recording, in *Single Neuron Studies of the Human Brain*, Fried, I., Rutishauser, U., Cerf, M., and Kreiman, G., eds., MIT Press, Boston, MA, 2014a. With permission.)

We defined a response index as $R_{i,j} = \dfrac{N(i) - \text{mean(NEW)}}{\text{mean(baseline)}} C_j \times 100\%$, where N(i) is the number of spikes after stimulus onset in trial i, j is the neuron index, and C_j is equal to −1 for neurons that increase their rate for novel stimuli and equal to 1 for neurons that increase their rate for familiar stimuli. The response index quantifies how different the response to a familiar image was relative to baseline and the response to novel stimuli. It is, by definition, negative for neurons that increase their firing rate in response to novel stimuli. To allow pooling of both classes, the response index was multiplied by the constant C_j as indicated.

Some patients with poor memory could not recollect the spatial location above chance level, whereas others performed significantly above chance level (77% ± 6% vs 35% ± 4%; chance level of 25%). Both groups had good recognition memory (see Figure 30.1 in [Rutishauser et al., 2008] for details). Comparing R+ and R− trials in both groups separately, we found that R+ trials had a significantly stronger response index compared to R− trials (67% vs 45%; see Figure 30.5b and c for statistics). The 22% difference in firing rate was due to successful spatial recollection. Patients with at-chance recollection performance still had R+ as well as R− trials either due to some remaining memory or due to chance. In contrast to the above-chance patients, however, there was no difference between R+ and R− trials in this patient group (response index of 45% vs 46%, Figure 30.5d). The at-chance recollection patient group's ability to recognize stimuli was statistically not different from the above-chance recollection group. Thus, the inability to recollect in the at-chance patients was likely not due to a general capacity reduction in memory but rather a selective deficit to recollect. We thus conclude that (1) novelty-sensitive MTL neurons code for both familiarity- and recollection-based memories and (2) this difference was abolished in patients who were not able to recollect, indicating that their MTL was not able to support spatial recall.

Trials where patients made errors also followed the same gradient (Figure 30.5e) for both false negatives (forgotten) and false positives (falsely remembering). In both cases, the response index was significantly larger compared to the response to novel stimuli (14% and 19%, respectively). At the same time, the response was significantly weaker in both cases relative to familiar stimuli that were correctly recognized (Figure 30.5e). This pattern of activity during behavioral errors is consistent with the idea that the neurons represent the strength of a memory continuously.

Were the neuronal responses reported earlier influenced by tissue damage due to epilepsy? All our subjects have been diagnosed with epilepsy of unknown localization, requiring bilateral implantation of electrodes in all candidate locations. After localization, some electrodes were typically determined to be in nonepileptic tissue, which allows us to directly compare memory-related responses in tissue with and without epilepsy. We found that novelty-sensitive neurons recorded from nonepileptic tissue had a stronger and more reliable response. Nevertheless, we found that neurons from the "to be resected" tissue (presumed epileptic) still exhibited a response that distinguishes new from old stimuli, but this response was weaker and did not differentiate between recollected and not recollected stimuli. The absence of a significant difference in response to R+ and R− trials in epileptic tissue indicates that MTL areas that are in the seizure origin are damaged and have only reduced capacity to support memory function. This finding also indicates that it might be possible to use indices of novelty responses for diagnostic purposes. Of particular interest is the possibility to use this approach to assess whether presumed "healthy" tissue is functional enough to support memory after resection. No studies to date have been conducted to assess whether this is the case, that is, whether single-neuron response strength to novel stimuli predicts memory capacity or other neuropsychological measures.

CONCLUSION

The ability to perform single-neuron recordings in awake human patients capable of performing cognitive tasks is a powerful research paradigm that has contributed valuable and otherwise unobtainable insights into many areas of neuroscience, including memory, learning, visual recognition, perception, motor control, decision-making, and emotions (Halgren et al., 1978; Heit et al., 1988; Fried et al., 1997; Kreiman et al., 2002; Sederberg et al., 2003; Quiroga et al., 2005; Kreiman, 2007; Mormann et al., 2008, 2011; Rutishauser et al., 2011, 2013b; Mukamel and Fried, 2012). For a researcher, human single-unit recordings are a unique opportunity to generate fundamental insights not obtainable with noninvasive methods. For a clinician, single-unit recordings represent a unique opportunity to contribute to cutting-edge research on basic cognitive mechanisms as well as to advance understanding of seizure generation and mechanisms of epilepsy. For a surgical epilepsy program as a whole, single-unit studies are an opportunity to utilize the large amount of time spent by patients in the unit waiting for seizures productively and to contribute invaluable new knowledge on how the human nervous system works. In this chapter, we have reviewed previous studies on memory that have provided new mechanistic understanding of the physiology of memory. This and similar work on the basic physiology of memory with human patients can make fundamental contributions toward understanding at the level of neuronal circuits, which will facilitate development of better learning methods as well as treatments for the learning problems that so often accompany diseases of the brain, including epilepsy.

REFERENCES

Buzsaki G (2004) Large-scale recording of neuronal ensembles. *Nature Neuroscience* 7:446–451.

Buzsaki G, Anastassiou CA, Koch C (2012) The origin of extracellular fields and currents—EEG, ECoG, LFP and spikes. *Nature Reviews Neuroscience* 13:407–420.

Carmichael DW, Thornton JS, Rodionov R, Thornton R, McEvoy A, Allen PJ, Lemieux L (2008) Safety of localizing epilepsy monitoring intracranial electroencephalograph electrodes using MRI: Radiofrequency-induced heating. *Journal of Magnetic Resonance Imaging* 28:1233–1244.

Engel J, Jr. (2001) Mesial temporal lobe epilepsy: What have we learned? *Neuroscientist* 7:340–352.

Fahy FL, Riches IP, Brown MW (1993) Neuronal activity related to visual recognition memory: Long-term memory and the encoding of recency and familiarity information in the primate anterior and medial inferior temporal and rhinal cortex. *Experimental Brain Reserch* 96:457–472.

Fortin NJ, Wright SP, Eichenbaum H (2004) Recollection-like memory retrieval in rats is dependent on the hippocampus. *Nature* 431:188–191.

Fried I, MacDonald KA, Wilson CL (1997) Single neuron activity in human hippocampus and amygdala during recognition of faces and objects. *Neuron* 18:753–765.

Fried I, Wilson CL, Maidment NT, Engel J, Behnke E, Fields TA, MacDonald KA, Morrow JW, Ackerson L (1999) Cerebral microdialysis combined with single-neuron and electroencephalographic recording in neurosurgical patients—Technical note. *Journal of Neurosurgery* 91:697–705.

Gibson S, Judy JW, Markovi D (2012) Spike sorting the first step in decoding the brain. *IEEE Signal Proceedings Magazine* 29:124–143.

Halgren E, Babb TL, Crandall PH (1978) Activity of human hippocampal formation and amygdala neurons during memory testing. *Electroencephalography and Clinical Neurophysiology* 45:585–601.

Hampton RR (2001) Rhesus monkeys know when they remember. *Proceedings of the National Academic Sciences USA* 98:5359–5362.

Harris KD, Henze DA, Csicsvari J, Hirase H, Buzsaki G (2000) Accuracy of tetrode spike separation as determined by simultaneous intracellular and extracellular measurements. *Journal of Neurophysiology* 84:401–414.

Hefft S, Brandt A, Zwick S, von Elverfeldt D, Mader I, Cordeiro J, Trippel M, Blumberg J, Schulze-Bonhage A (2013) Safety of hybrid electrodes for single-neuron recordings in humans. *Neurosurgery* 73:78–85; discussion 85.

Heit G, Smith ME, Halgren E (1988) Neural encoding of individual words and faces by the human hippocampus and amygdala. *Nature* 333:773–775.

Henze DA, Borhegyi Z, Csicsvari J, Mamiya A, Harris KD, Buzsaki G (2000) Intracellular features predicted by extracellular recordings in the hippocampus in vivo. *Journal of Neurophysiology* 84:390–400.

Knowlton BJ, Squire LR (1993) The learning of categories: Parallel brain systems for item memory and category knowledge. *Science* 262:1747–1749.

Kreiman G (2007) Single unit approaches to human vision and memory. *Current Opinion in Neurobiology* 17:471–475.

Kreiman G, Fried I, Koch C (2002) Single-neuron correlates of subjective vision in the human medial temporal lobe. *Proceedings of the National Academy of Sciences of the United States of America* 99:8378–8383.

Lewicki MS (1998) A review of methods for spike sorting: The detection and classification of neural action potentials. *Network* 9:R53–R78.

Lisman JE, Grace AA (2005) The hippocampal-VTA loop: Controlling the entry of information into long-term memory. *Neuron* 46:703–713.

Lisman JE, Otmakhova NA (2001) Storage, recall, and novelty detection of sequences by the hippocampus: Elaborating on the SOCRATIC model to account for normal and aberrant effects of dopamine. *Hippocampus* 11:551–568.

Mamelak A (2014) Ethical and practical considerations for human microelectrode recording studies. In: *Single Neuron Studies of the Human Brain* (Fried I, Rutishauser U, Cerf M, Kreiman G, eds). Boston, MA: MIT Press.

Misra A, Burke JF, Ramayya AG, Jacobs J, Sperling MR, Moxon KA, Kahana MJ, Evans JJ, Sharan AD (2014) Methods for implantation of micro-wire bundles and optimization of single/multi-unit recordings from human mesial temporal lobe. *Journal of Neural Engineering* 11:026013.

Mormann F, Dubois J, Kornblith S et al. (2011) A category-specific response to animals in the right human amygdala. *Nature Neuroscience* 14:1247–1249.

Mormann F, Kornblith S, Quiroga RQ, Kraskov A, Cerf M, Fried I, Koch C (2008) Latency and selectivity of single neurons indicate hierarchical processing in the human medial temporal lobe. *Journal of Neuroscience* 28:8865–8872.

Mukamel R, Fried I (2012) Human intracranial recordings and cognitive neuroscience. *Annual Review of Psychology* 63 63:511–537.

Ojemann GA (1997) Treatment of temporal lobe epilepsy. *Annual Review Medicine* 48:317–328.

Pedreira C, Mormann F, Kraskov A, Cerf M, Fried I, Koch C, Quiroga RQ (2010) Responses of human medial temporal lobe neurons are modulated by stimulus repetition. *Journal of Neurophysiology* 103:97–107.

Poldrack RA, Clark J, Pare-Blagoev EJ, Shohamy D, Creso Moyano J, Myers C, Gluck MA (2001) Interactive memory systems in the human brain. *Nature* 414:546–550.

Pouzat C, Mazor O, Laurent G (2002) Using noise signature to optimize spike-sorting and to assess neuronal classification quality. *Journal of Neuroscience Methods* 122:43–57.

Quiroga RQ, Nadasdy Z, Ben-Shaul Y (2004) Unsupervised spike detection and sorting with wavelets and superparamagnetic clustering. *Neural Computation* 16:1661–1687.

Quiroga RQ, Reddy L, Kreiman G, Koch C, Fried I (2005) Invariant visual representation by single neurons in the human brain. *Nature* 435:1102–1107.

Riches IP, Wilson FA, Brown MW (1991) The effects of visual stimulation and memory on neurons of the hippocampal formation and the neighboring parahippocampal gyrus and inferior temporal cortex of the primate. *Journal of Neuroscience* 11:1763–1779.

Rolls ET, Cahusac PM, Feigenbaum JD, Miyashita Y (1993) Responses of single neurons in the hippocampus of the macaque related to recognition memory. *Experimental Brain Research* 93:299–306.

Rolls ET, Perrett DI, Caan AW, Wilson FA (1982) Neuronal responses related to visual recognition. *Brain* 105 (Pt 4):611–646.

Rutishauser U, Cerf M, Kreiman G (2014b) Data analysis techniques for human microwire recordings: Spike detection and sorting, decoding, relation between neurons and local field potential. In: *Single Neuron Studies of the Human Brain* (Fried I, Rutishauser U, Cerf M, Kreiman G, eds). Boston, MA: MIT Press.

Rutishauser U, Kotowicz A, Laurent G (2013a) A method for closed-loop presentation of sensory stimuli conditional on the internal brain-state of awake animals. *Journal of Neuroscience Methods* 215:139–155.

Rutishauser U, Mamelak AN, Schuman EM (2006b) Single-trial learning of novel stimuli by individual neurons of the human hippocampus-amygdala complex. *Neuron* 49:805–813.

Rutishauser U, Ross IB, Mamelak AN, Schuman EM (2010) Human memory strength is predicted by theta-frequency phase-locking of single neurons. *Nature* 464:903–907.

Rutishauser U, Schuman EM, Mamelak A (2014a) Single neuron correlates of declarative memory formation and retrieval in the human medial temporal lobe. In: *Single Neuron Studies of the Human Brain* (Fried I, Rutishauser U, Cerf M, Kreiman G, eds). Boston, MA: MIT Press.

Rutishauser U, Schuman EM, Mamelak AN (2006a) Online detection and sorting of extracellularly recorded action potentials in human medial temporal lobe recordings, in vivo. *Journal of Neuroscience Methods* 154:204–224.

Rutishauser U, Schuman EM, Mamelak AN (2008) Activity of human hippocampal and amygdala neurons during retrieval of declarative memories. *Proceedings of the National Academy of Sciences of the United States of America* 105:329–334.

Rutishauser U, Tudusciuc O, Neumann D, Mamelak AN, Heller AC, Ross IB, Philpott L, Sutherling WW, Adolphs R (2011) Single-unit responses selective for whole faces in the human amygdala. *Current Biology* 21:1654–1660.

Rutishauser U, Tudusciuc O, Wang S, Mamelak Adam N, Ross Ian B, Adolphs R (2013b) Single-neuron correlates of atypical face processing in autism. *Neuron* 80:887–899.

Schacter DL, Tulving E (1994) *Memory Systems*. Cambridge, MA: MIT Press.

Sederberg PB, Kahana MJ, Howard MW, Donner EJ, Madsen JR (2003) Theta and gamma oscillations during encoding predict subsequent recall. *Journal of Neuroscience* 23:10809–10814.

Spencer SS, Sperling MR, Shewmon DA, Kahane P (2007) Intracranial electrodes. In: *Epilepsy: A Comprehensive Textbook.* (Engel J, Jr., Pedley TA, eds.), pp. 1791–1815. Philadelphia, PA: Lippincott Williams & Williams.

Squire LR (2004) Memory systems of the brain: A brief history and current perspective. *Neurobiology of Learning and Memory* 82:171–177.

Squire LR, Alvarez P (1995) Retrograde amnesia and memory consolidation: A neurobiological perspective. *Current Opinion in Neurobiology* 5:169–177.

Squire LR, Stark CE, Clark RE (2004) The medial temporal lobe. *Annual Review of Neuroscience* 27:279–306.

Staba RJ, Fields T, Behnke E, Wilson CL (2014) Subchronic in vivo human microelectrode recording In: *Single Neuron Studies of the Human*. (Fried I, Rutishauser U, Cerf M, Kreiman G, eds.). Boston, MA: MIT Press.

Standing L (1973) Learning 10,000 pictures. *The Quarterly Journal of Experimental Psychology* 25:207–222.

Standing L, Conezio J, Haber RN (1970) Perception and memory for pictures—Single-trial learning of 2500 visual stimuli. *Psychonomic Science* 19:73–74.

Tulving E (2002) Episodic memory: From mind to brain. *Annual Review of Psychology* 53:1–25.

Viskontas IV, Knowlton BJ, Steinmetz PN, Fried I (2006) Differences in mnemonic processing by neurons in the human hippocampus and parahippocampal regions. *Journal of Cognitive Neuroscience* 18:1654–1662.

Xiang JZ, Brown MW (1998) Differential neuronal encoding of novelty, familiarity and recency in regions of the anterior temporal lobe. *Neuropharmacology* 37:657–676.

Index

T - #0461 - 071024 - C588 - 254/178/26 - PB - 9780367657932 - Gloss Lamination